Springer Series in Solid and Structural Mechanics

Volume 2

Series Editors

Michel Frémond, Rome, Italy
Franco Maceri, Rome, Italy

For further volumes:
http://www.springer.com/series/10616

Einar N. Strømmen

Structural Dynamics

Springer

Einar N. Strømmen
Department of Structural Engineering
Norwegian University of Science
and Technology
Trondheim
Norway

ISSN 2195-3511 ISSN 2195-352X (electronic)
ISBN 978-3-319-37888-6 ISBN 978-3-319-01802-7 (eBook)
DOI 10.1007/978-3-319-01802-7
Springer Cham Heidelberg New York Dordrecht London

Softcover reprint of the hardcover 1st edition 2014

Printed on acid-free paper

Springer is part of Springer Science+Business Media (www.springer.com)

In loving memory of
Alf & Signe

Preface

This text book is intended for studies in the theory of structural dynamics, with focus on civil engineering structures that may be described by line-like beam or beam-column type of systems, or by a system of rectangular plates. Throughout this book the mathematical presentation contains a classical analytical description as well as a description in a discrete finite element format, covering the mathematical development from basic assumptions to the final equations ready for practical dynamic response predictions. Solutions are presented in time domain as well as in frequency domain. It has been my intention to start off at a basic level and step by step bring the reader up to a level where the necessary safety considerations to wind or horizontal ground motion induced dynamic design problems can be performed, i.e. to a level where dynamic displacements and corresponding cross sectional forces can actually be calculated. However, this is not a text book in wind or earthquake engineering, and hence, relevant load descriptions are only included in so far as it has been necessary for the performance of illustrative examples. For more comprehensive descriptions of wind and earthquake induced dynamic load and load effects the reader should consult the literature, e.g. refs. [15] and [16]. Less attention has been given to other load cases, e.g. to any kind of shock or impact loading. Also, a comprehensive description of structural damping properties are beyond the scope of this book, but again, for the sake of completeness, a chapter covering the most important theories behind structural damping has been included. The special theory of the tuned mass damper has been given a comprehensive treatment, as this is a theory not fully covered elsewhere. For the same reason a chapter on the problem of moving loads on beams has been included.

The reading of this book will require some knowledge of structural mechanics, i.e. the basic theory of elasticity. Also, readers unfamiliar with the theory of stochastic processes and time domain simulations should commence their studies by reading Appendices A and B, or another suitable text book.

The drawings have been prepared by Anne Gaarden. Thanks to her and all others who have contributed to the writing of this book.

Trondheim, September 2012 *Einar N. Strømmen*

Notation

Matrices and vectors:

Matrices are in general bold upper case Latin or Greek letters, e.g. $\mathbf{K}$ or $\mathbf{\Phi}$.

Vectors are in general bold lower case Latin or Greek letters, e.g. $\mathbf{q}$ or $\boldsymbol{\varphi}$.

$diag[\cdot]$ is a diagonal matrix whose content is written within the bracket.

$det(\cdot)$ is the determinant of the matrix within the bracket.

$tr(\cdot)$ is the trace of a matrix.

Imaginary quantities:

i is the imaginary unit (i.e. $i = \sqrt{-1}$).

$\mathrm{Re}(\cdot)$ is the real part of the variable within the brackets.

$\mathrm{Im}(\cdot)$ is the imaginary part of the variable within the brackets.

Superscripts and bars above symbols:

Super-script T indicates the transposed of a vector or a matrix.

Super-script * indicates the complex conjugate of a quantity.

Dots above symbols (e.g. $\dot{\mathbf{r}}$, $\ddot{\mathbf{r}}$) indicates time derivatives, i.e. d/dt, d^2/dt^2.

Prime on a variable (e.g. C_L' or ϕ') indicates its derivative with respect to a relevant variable, e.g. $\phi' = d\phi/dx$. Two primes is then the second derivative (e.g. $\phi'' = d^2\phi/dx^2$) and so on.

Bar ($-$) above a variable (e.g. $\bar{r}$) indicates its time invariant average value.

Tilde (~) above a symbol (e.g. $\tilde{M}_n$) indicates a modal quantity.

Hat ($\wedge$) above a symbol (e.g. $\hat{H}_\eta$) indicates a normalised quantity.

The use of indices and superscript:

Index x, y or z refers to the corresponding structural axis.

i and j are general indices on variables.

n and m are mode shape or element numbers.

p and k are in general used as node numbers.

Abbreviations:

CC and SC are short for the centre of cross-sectional neutral axis and the shear centre.

tot is short for total.

max, min are short for maximum and minimum.

$\int_L$ or $\int_A$ means integration over the entire length or the area of the system.

Latin letters:

A	Area, cross sectional area
A_j	Coefficient associated with variable j
$A_1^* - A_6^*$	Aerodynamic derivatives associated with the motion in torsion
$\mathbf{A}, \mathbf{A}_m, \mathbf{A}_n$	Connectivity matrix (associated with element m or n)
a	Coefficient, Fourier coefficient, amplitude
$\mathbf{a}_j$	Fourier coefficient vector associated with variable j
B	Cross sectional width
b	Coefficient, band-width parameter
b_c	Distance between cable planes is a suspension bridge
$\mathbf{b}_q$	Buffeting dynamic load coefficient matrix at cross sectional level
C , $\mathbf{C}$	Damping coefficient or matrix containing damping coefficient
C_{ae} , $\mathbf{C}_{ae}$	Aerodynamic damping, aerodynamic damping matrix
c , c_0	Coefficient,damping coefficient at cross sectional level
$\mathbf{c}_0$	Damping matrix at a cross sectional level
$\mathbf{c}, \mathbf{c}_{ae}$	Damping matrix at element level, aerodynamic damping matrix
$Co, \mathbf{Co}$	Co-spectral density, co-spectral density matrix
$\mathbf{Cov}_j$	Covariance matrix associated with variable j
D , d	Cross sectional depth, Coefficient
$\mathbf{d}$, d_k	Element displacement vector, element end displacement component
E	Modulus of elasticity

e , e_c	Exponential number (≈ 2.718281828), Cable sag
$\mathbf{F}$, F	Element force vector, force
f , f_n	Frequency [Hz], eigen frequency associated with mode n
$f(\cdot)$	Function of variable within brackets
G	Modulus of elasticity in shear
$g(\cdot)$, g	Function of variable within brackets, gravity constant
$H(t)$, $\bar{H}$	Horizontal cable force component
$H_1^* - H_6^*$	Aerodynamic derivatives associated with the across-wind motion
H_n , $\mathbf{H}_r$	Frequency response function, frequency response matrix
$\tilde{H}_\eta$, $\tilde{\mathbf{H}}_\eta$	Modal frequency response functions, matrix containing $\tilde{H}_{\eta_n}$
h_c , h_m	Length of suspension bridge hangers, hanger length at mid span
h_r	Vertical distance between shear centre and hanger attachment
h_0	Height (above girder) of suspension bridge tower
I_t , I_w	St Venant torsion and warping constants
I_j	Turbulence intensity of flow components $j = u, v$ or w
I_y , I_z	Moment of inertia with respect to bending about$y$ or z axis
$\mathbf{I}$	Identity matrix
i	The imaginary unit (i.e. $i = \sqrt{-1}$)
J , $\mathbf{J}$	Joint acceptance function, joint acceptance matrix
j	Index variable
K , $\mathbf{K}$	Stiffness, stiffness matrix
K_{ae} , $\mathbf{K}_{ae}$	Aerodynamic stiffness, aero dynamic stiffness matrix
k	Index variable, node or sample number
k_p	Peak factor
$\mathbf{k}$, $\mathbf{k}_{ae}$	Stiffnessmatrix at element level, aerodynamic stiffness matrix
L	Lagrange function
L	Length (of structural system)
${}^mL_n$	Integral length scales ($m = y$, z or θ, $n = u, v$ or w)
ℓ_e	Effective length
M_g , $\mathbf{M}_g$	Concentrated mass at position x_M , mass matrix containing M_g
M_m	Bending moment ($m=x, y, z$)
m	Index variable
m , $\mathbf{M}$	Mass, mass matrix

$\tilde{m}_n$	Modally equivalent and evenly distributed mass
$\mathbf{m}_0$, $\mathbf{m}$	Mass matrix at a cross sectional level, Mass matrix at element level
N	Number, number of elements in a system
N_r	Number of degrees of freedom in a system
N_x, N_y	Normal force (in x or y directions)
n	Index variable
$\mathbf{n}_n$	Matrix containing time invariant element end forces
P, P_F, P_q	Work performed by external forces acting on the system
$P_1^* - P_6^*$	Aerodynamic derivatives associated with the along-wind motion
p	Index variable, node or sample number
Q_j	External load vector component in directions $j = x, y$ or z
q, $\mathbf{q}$	Pressure, distributed load or load vector at cross sectional level
R, $\mathbf{R}$	External load, reaction force, external load vector at system level
$\tilde{R}, \tilde{\mathbf{R}}$	Modal load, Modal load vector
r, $\mathbf{r}$	Cross sectional displacement or rotation, displacement vector
r_{el}, $\mathbf{r}_{el}$	Element cross sectional displacement, displacement vector
r_p	Polar radius
St	Strouhal number
S, $\mathbf{S}$	Auto or cross spectral density, cross-spectral density matrix
$\mathbf{S}_j$	Cross spectral density matrix associated with variable j
s	General coordinate ($s = x, y$ or z)
T, T_M, T_m	Motion energy of the system body masses
t, T	Time, total length of time window
U, U_M, U_m	Strain energy stored in the material fibres of the system
U	Instantaneous wind velocity in the main flow direction
u	Fluctuating along-wind horizontal velocity component
V	Volume
V, V_R	Mean wind velocity, resonance mean wind velocity
V_y, V_z	Shear forces
v	Fluctuating across wind horizontal velocity component
W_{ext}, W_{int}	External, internal work
w	Fluctuating across wind vertical velocity component
X, Y, Z	Cartesian structural global axis

x, y, z	Cartesian structural element cross sectional main neutral axis (with origo in the shear centre, x in span-wise direction and z vertical)
x_r	Chosen span-wise position for response calculation

Greek letters:

α	Coefficient
β	Phase angle, coefficient
$\boldsymbol{\beta}$	Matrix, matrix containing mode shape derivatives
$\gamma, \gamma_z, \gamma_\theta$	Shear strain, shear strain associated with shear forceor torsion
δ	Incremental displacement operator
∂	Derivative operator
$\varepsilon, \boldsymbol{\varepsilon}, \varepsilon_j$	Strain, strain vector, strain component ($j = x, y$ or z)
ζ or $\boldsymbol{\zeta}$	Damping ratio or damping ratio matrix
$\eta, \boldsymbol{\eta}$	Generalised coordinate, vector containing N_{mod} η components
θ	Index indicating cross sectional rotation or load (about shear centre)
κ	Coefficient
ν	Poisson's ratio, coefficient
λ	Coefficient, wave length
μ	Coefficient, friction coefficient
Π	Total energy
ϑ	Coefficient
ρ, ρ_j	Densityof air, density of component associated with j
σ, σ^2	Standard deviation, variance
σ_x, τ	Normal stress, Shear stress
$\phi_{y_n}, \phi_{z_n}, \phi_{\theta_n}$	Continuous mode shape components in y, z and θ directions
$\varphi(x, y)$	Plate mode shape functions
$\boldsymbol{\Phi}$	$3 \cdot N_{\text{mod}}$ by N_{mod} matrix containing all mode shapes $\boldsymbol{\varphi}_n$
$\boldsymbol{\Phi}_r$	3 by N_{mod} matrix containing the content of $\boldsymbol{\Phi}$ at $x = x_r$
$\boldsymbol{\varphi}_n$	Mode shape number n
ψ	Chosen approximate mode shape function, angle
$\boldsymbol{\psi}, \boldsymbol{\psi}_n$	Chosen approximate mode shape matrix, discrete mode shape
$\widehat{\boldsymbol{\psi}}, \widehat{\widehat{\boldsymbol{\psi}}}$	Contains first and second order derivatives of $\boldsymbol{\psi}$

Ω	Coefficient
ω	Circular frequency (rad/s)
ω_n	Eigenfrequency associated with mode shape n
$\omega_n(V)$	Resonance frequency assoc. with mode n at mean wind velocity V

Symbols with both Latin and Greek letters:

$\Delta f, \Delta\omega$	Frequency segment
Δt	Time step
Δs	Spatial separation ($s = x$, y or z)

Contents

Chapter 1
Basic Theory

1.1 Introduction

This text book focuses on the prediction of dynamic response of slender line-like civil engineering structures. It is a general assumption that structural behaviour is linear elastic and that any non-linear part of the relationship between load and structural displacements may be disregarded. It is taken for granted that the load direction throughout the entire span of the structure is perpendicular to the axis in the direction of its span.

It is assumed that the mean value (static part) of any load is constant such that structural response can be predicted as the sum of a mean value and a fluctuating part, as illustrated in Fig. 1.1.a. As shown in Fig. 1.1 and 1.2 a line-like beam or

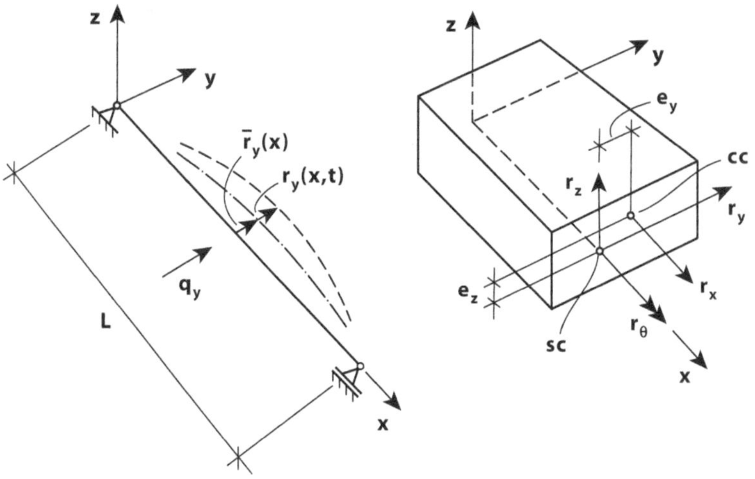

a) Axis definition *b) Displacement components*

Fig. 1.1 Structural axes and displacement components

E.N. Strømmen, *Structural Dynamics*, Springer Series in Solid and Structural Mechanics 2, DOI: 10.1007/978-3-319-01802-7_1, © Springer International Publishing Switzerland 2014

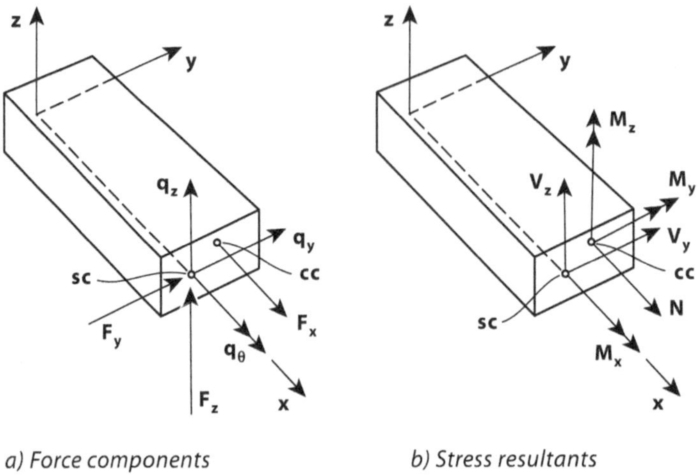

Fig. 1.2 Basic axis and vector definitions

beam-column type of structural element is described in a Cartesian coordinate system $[x, y, z]$, with its origin at the shear centre of the cross section, x is in the span direction and with y and z parallel to the main neutral structural axis CC (i.e. the neutral axis with respect to cross sectional bending according to Hook's law and Navier's hypothesis), which will coincide with the mass centre if material density and modulus of elasticity do not change over the area of the cross section). Response displacements r_y, r_z, r_θ and load components F_y, F_z, q_y, q_z and q_θ are referred to the shear centre (SC), while response displacement r_x and load component F_x are referred to the origin of main neutral axis . Similarly, the cross sectional stress resultants V_y, V_z and M_x are referred to the shear centre, while bending moment and axial stress resultants M_y, M_z and N are referred to the origin of main neutral axis. The basic units are as follows:

- displacement: meter (m)
- time: second (s)
- mass: kilogram (kg)
- force: Newton ($N = kg \cdot m/s^2$, [1, 2]).

1.2 d'Alambert's Principle of Instantaneous Equilibrium

In statics the equilibrium condition of a system subject to a set of constant concentrated forces is given by $\sum F_i = 0$, where F_i (with unit N) are all the relevant forces in the direction of $i = x, y$ or z. It comes from the requirement that a system in static equilibrium must be at rest or in a situation of constant speed, i.e. that its acceleration in any direction $\ddot{r}_i$ (with unit m/s^2) is equal to zero. Newton's second law will then require that $\sum F_i = M \cdot \ddot{r}_i = 0$, where M is the mass of the body. However, in dynamics any equilibrium consideration will have to include the motion of the system. This is done by adopting the principle of d'Alambert (first published by Lagrange [3]) that equilibrium for a system in motion can be established by considering an instantaneous situation where the system is frozen at an arbitrary position in space and time, and that the acceleration of the system can be interpreted as an inertia force in accordance with Newton's second law, i.e. as a resistance against being accelerated.

Discrete Systems

Below, examples of discrete systems are illustrated in Figs. 1.3, 1.5-1.9 and 1.11. For such systems the relevant equilibrium requirements are most conveniently established in a vector-matrix description. Let a system of a simple mass M and a linear elastic spring with stiffness K be suspended in a vertical position as illustrated in Fig. 1.3. To the left the system is shown at its unloaded position. Let the system then be subject to gravity Mg (where g is the gravity acceleration constant) and a constant time invariant force $\overline{F}$. In this position the system is at rest in its static position and it has been displaced a distance $\overline{r}$. As shown in Fig. 1.3.b the equilibrium requirement is then that $K \cdot \overline{r} = M \cdot g + \overline{F}$, from which $\overline{r}$ may be calculated if all other quantities are known. Let the system then be subject to an additional dynamic force $F(t)$, which is accompanied by a corresponding dynamic displacement $r(t)$. The equilibrium condition is then that the external forces $M \cdot g + \overline{F} + F(t)$ must be equal to the sum of the elastic spring force $K \cdot r_{tot}$ and a resistance inertia force $M \cdot \ddot{r}_{tot}$ in accordance with Newton's second law and the principle of d'Alambert, i.e. that

$$M \cdot \ddot{r}_{tot} + K \cdot r_{tot} = M \cdot g + \overline{F} + F(t) \tag{1.1}$$

Introducing that $r_{tot} = \overline{r} + r(t)$ then

$$M \cdot \ddot{r} + K \cdot (\overline{r} + r) = M \cdot g + \overline{F} + F(t) \tag{1.2}$$

Since the static equilibrium condition is that

$$K \cdot \overline{r} = M \cdot g + \overline{F} \tag{1.3}$$

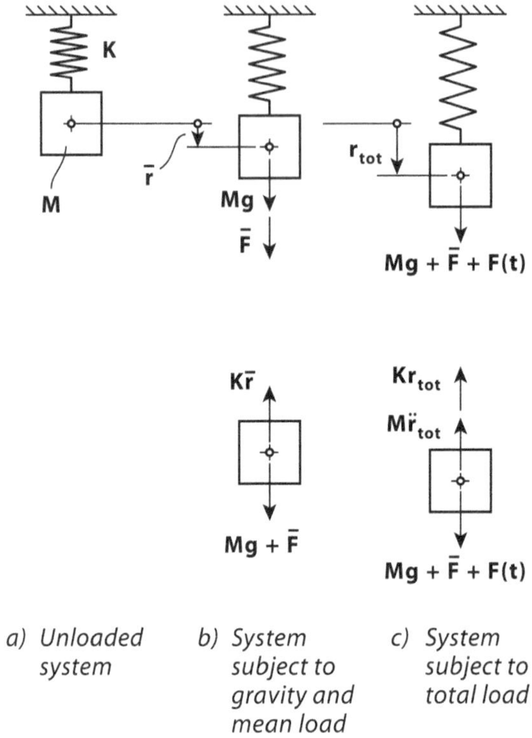

Fig. 1.3 Simple spring-mass system

it is seen that Eq. 1.2 may be reduced accordingly into a purely dynamic equilibrium condition

$$M \cdot \ddot{r} + K \cdot r = F(t) \tag{1.4}$$

Thus, it may be concluded that the equilibrium condition for such a linear elastic system may be split into two, a static time invariant condition and a dynamic equilibrium condition where only dynamic loads are included and where the forces due to the instantaneous acceleration of the system itself is represented by a set of inertia forces acting in the opposite direction of the motion. Hence, by splitting the load (concentrated or evenly distributed) into a mean time invariant part and a fluctuating part

$$\mathbf{F}_{tot} = \bar{\mathbf{F}} + \mathbf{F}(t) = \begin{bmatrix} \bar{F}_x \\ \bar{F}_y \\ \bar{F}_z \end{bmatrix} + \begin{bmatrix} F_x(t) \\ F_y(t) \\ F_z(t) \end{bmatrix} \text{ or } \mathbf{q}_{tot} = \bar{\mathbf{q}} + \mathbf{q} = \begin{bmatrix} \bar{q}_y(x) \\ \bar{q}_z(x) \\ \bar{q}_\theta(x) \end{bmatrix} + \begin{bmatrix} q_y(x,t) \\ q_z(x,t) \\ q_\theta(x,t) \end{bmatrix} \tag{1.5}$$

then the mean and fluctuating parts of the response displacements as well as the corresponding cross sectional stress resultants

$$\bar{\mathbf{r}} + \mathbf{r} = \begin{bmatrix} \bar{r}_y(x) \\ \bar{r}_z(x) \\ \bar{r}_\theta(x) \end{bmatrix} + \begin{bmatrix} r_y(x,t) \\ r_z(x,t) \\ r_\theta(x,t) \end{bmatrix} \quad \text{and} \quad \begin{bmatrix} \bar{V}_y(x) \\ \bar{V}_z(x) \\ \bar{M}_x(x) \\ \bar{M}_y(x) \\ \bar{M}_z(x) \\ \bar{N}(x) \end{bmatrix} + \begin{bmatrix} V_y(x,t) \\ V_z(x,t) \\ M_x(x,t) \\ M_y(x,t) \\ M_z(x,t) \\ N(x,t) \end{bmatrix} \tag{1.6}$$

may be obtained by separately satisfying the relevant static and dynamic equilibrium requirements of the system.

Let us first assume that $F(t)=0$, but that the system in Fig. 1.3 has been set into an oscillating motion by imposing an initial displacement $r(0)$ and $\dot{r}(0)$. The solution to Eq. 1.4 is then given by

$$r(t) = b \cdot \sin(\omega_n t) + c \cdot \cos(\omega_n t) \tag{1.7}$$

where b and c are coefficients which may be determined from the position and velocity conditions at $t=0$, i.e. that $r(0)=c$ and $\dot{r}(0)=\omega_n b$, and where the frequency of motion ω_n may be obtained by introducing Eq. 1.7 into Eq. 1.4, from which it is obtained that

$$\left(K - \omega_n^2 M\right) \cdot r(t) = 0 \tag{1.8}$$

A non-trivial solution $r(t) \neq 0$ can then only be obtained if $K - \omega_n^2 M = 0$, and thus, the frequency of a free unloaded and oscillatory motion is given by $\omega_n = \sqrt{K/M}$. The motion is harmonic because it contains only a single and stationary frequency. This is what we call the eigenfrequency of the system. It has the unit rad/s. [In some cases it may be convenient to convert it into $f_n = \omega_n/(2\pi)$ Hz (1/s), while yet another option is to introduce the period of the motion $T_n = 1/f_n$.] Furthermore, since for two arbitrary angles α_1 and α_2

$$\sin\alpha_1 \cdot \sin\alpha_2 + \cos\alpha_1 \cdot \cos\alpha_2 = \cos(\alpha_1 - \alpha_2) \tag{1.9}$$

then the solution in Eq. 1.7 may readily be converted into

$$r(t) = a \cdot \cos(\omega t - \beta) \text{ where } \begin{cases} a = \sqrt{b^2 + c^2} = \sqrt{[r(0)]^2 + [\dot{r}(0)/\omega_n]^2} \\ \tan\beta = b/c = \dot{r}(0)/[\omega_n \cdot r(0)] \end{cases} \tag{1.10}$$

Or, and most often more conveniently, $r(t)$ may be expressed in a complex format by defining $a = c - i \cdot b = |a| e^{-i\beta} = (a^* a)^{1/2} e^{-i\beta}$ where i is the complex unit ($i = \sqrt{-1}$) and $\tan\beta = \mathrm{Im}(a)/\mathrm{Re}(a) = b/c$. Thus, it is seen that

$$\begin{aligned} r(t) &= \mathrm{Re}(a \cdot e^{i\omega t}) = \mathrm{Re}[(c - i \cdot b) \cdot e^{i\omega t}] \\ &= \mathrm{Re}(|a| e^{-i\beta} e^{i\omega t}) = |a| \cdot \mathrm{Re}[e^{i(\omega t - \beta)}] = |a| \cdot \cos(\omega t - \beta) \end{aligned} \tag{1.11}$$

and, as shown above, $c = r(0)$ and $b = \dot{r}(0)/\omega_n$. A plot of $r(t)$ is illustrated in Fig. 1.4.

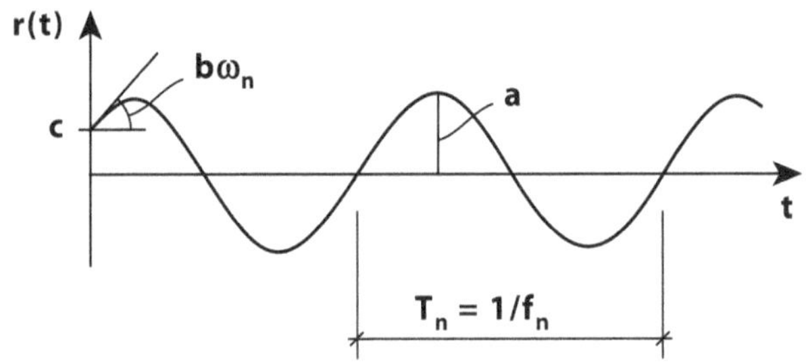

Fig. 1.4 Simple unloaded and undamped motion of single degree of freedom system

Example 1.1

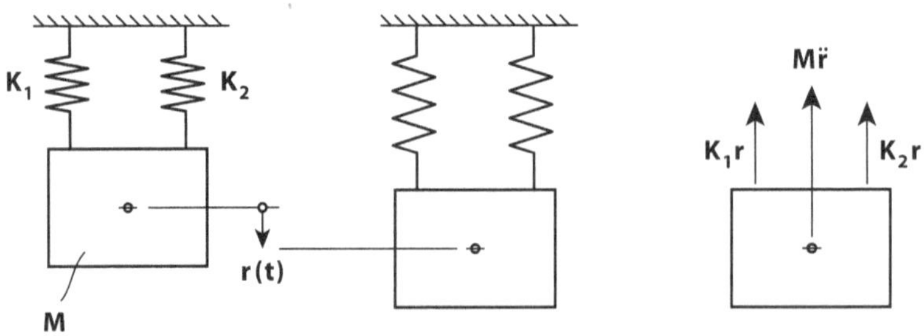

Fig. 1.5 Single mass with two parallel springs

A single mass with two parallel springs is shown on the left hand side in Fig. 1.5 above. Next, it has been given an arbitrary harmonic displacement $r(t)=\mathrm{Re}\left[ae^{i\omega t}\right]$, *and to the far right is shown the corresponding free body diagram of forces acting on the mass. Thus, equilibrium will require*

$$M\ddot{r}+(K_1+K_2)r=0 \quad \textit{and thus, introducing } r(t)=\mathrm{Re}\left[ae^{i\omega t}\right], \textit{ then}$$

$$K_1+K_2-\omega^2 M=0 \qquad \textit{rendering} \qquad \omega_n=\sqrt{(K_1+K_2)/M}$$

from which it may be concluded that stiffness contributions in parallel are additive.

Example 1.2

A single mass with two springs in sequence is shown on the left hand side of Fig. 1.6. Next, it has been given an arbitrary harmonic displacement $r_2(t)=\mathrm{Re}\left[a_2e^{i\omega t}\right]$. *During this motion the connection between the two springs has undergone a corresponding harmonic displacement* $r_1(t)=\mathrm{Re}\left[a_1e^{i\omega t}\right]$. *The resisting force in the upper spring is* K_1r_1, *while the resisting force in the lower spring is* $F_k=K_2(r_2-r_1)$. *The force throughout the sequence of springs must be unchanged, and thus*

$$K_1r_1=K_2(r_2-r_1) \quad \Rightarrow \quad r_1=\frac{K_2}{K_1+K_2}r_2 \quad \Rightarrow \quad F_k=\frac{K_1K_2}{K_1+K_2}r_2$$

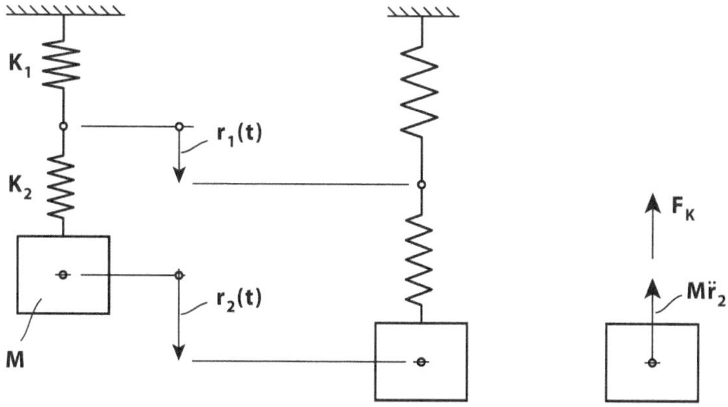

Fig. 1.6 Single mass with two springs in sequence

Equilibrium of the mass (see right hand side of Fig. 1.6) will then require

$$M\ddot{r}_2 + \frac{K_1 K_2}{K_1 + K_2} r_2 = 0 \qquad \textit{which with} \qquad r_2(t) = \mathrm{Re}\left[a_2 e^{i\omega t}\right]$$

$$\Rightarrow \qquad \frac{K_1 K_2}{K_1 + K_2} - \omega^2 M = 0 \qquad \textit{and thus} \qquad \omega_n = \sqrt{K_{tot}/M}$$

$$\textit{where } K_{tot} = \left(\frac{1}{K_1} + \frac{1}{K_2}\right)^{-1}$$

It may be concluded that stiffness contributions in sequel are inversely additive.

Example 1.3

A single mass with a spring on either side is shown on the left hand side of Fig. 1.7 below. The springs have been pre-stretched by a constant (time invariant) normal force $\bar{N}$ *such that prior to any displacement the system is in a state of equilibrium. It is taken for granted that the system displacements are never larger than that which will cause the springs to slacken. Next, it has been given an arbitrary harmonic displacement* $r(t) = \mathrm{Re}\left[ae^{i\omega t}\right]$, *and to the far right is shown the corresponding free body diagram of forces acting on the mass. Thus, equilibrium will require*

$$M\ddot{r} + \left(\bar{N} + K_1 r\right) - \left(\bar{N} - K_2 r\right) = 0 \qquad \textit{which with} \qquad r(t) = \mathrm{Re}\left[ae^{i\omega t}\right]$$

$$\Rightarrow \qquad K_1 + K_2 - \omega^2 M = 0 \qquad \textit{and thus} \qquad \omega_n = \sqrt{(K_1 + K_2)/M}$$

from which it may be concluded that stiffness contributions are additive.

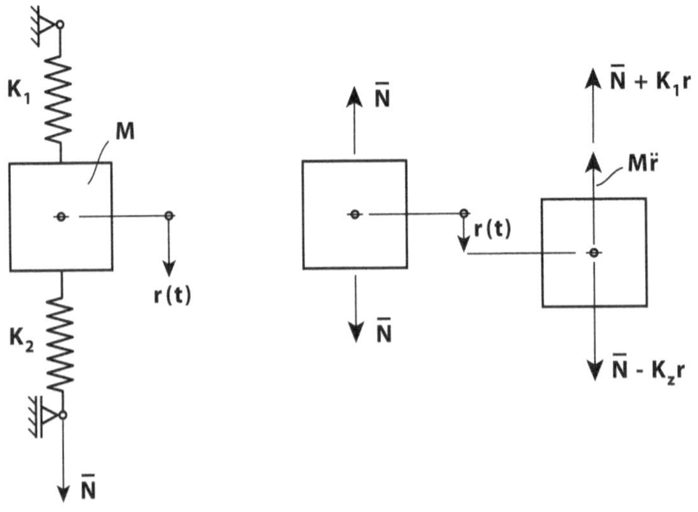

Fig. 1.7 Single mass with springs on either side

Example 1.4

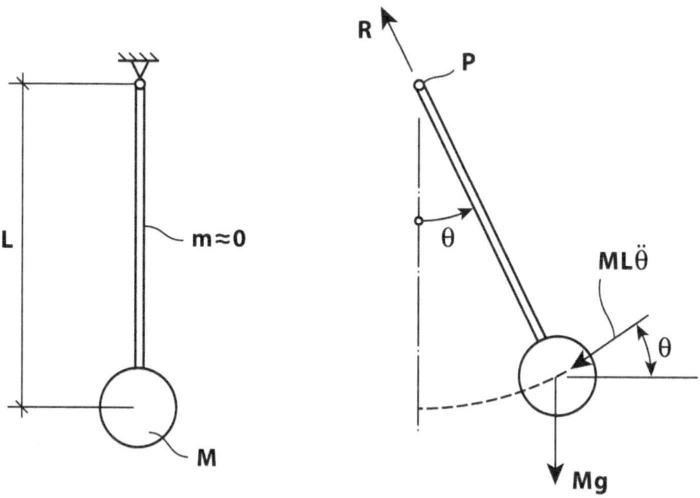

Fig. 1.8 Small displacement pendulum

The classic case of a simple pendulum is shown in Fig. 1.8 above. For simplicity, the mass and bending stiffness of the rod are assumed negligible. At an arbitrary rotation $\theta(t)=\mathrm{Re}\left[a_\theta e^{i\omega t}\right]$ *a free body diagram of the system is shown to the right. In this situation the mass* M *is subject to the gravity force* Mg *and tangential acceleration*

$$\frac{d}{dt^2}(\theta L) \text{ and corresponding restoring inertia force } M\cdot\frac{d}{dt^2}(\theta L)=ML\ddot{\theta}$$

Instantaneous moment equilibrium about the rotation point p *will then require*

$$\left(ML\ddot{\theta}\right)\cdot L+Mg\cdot L\sin\theta=0 \qquad \Rightarrow \qquad \ddot{\theta}+(g/L)\cdot\sin\theta=0$$

which cannot be analytically solved unless we assume θ *small such that* $\sin\theta\approx\theta$, *in which case*

$$\ddot{\theta}+(g/L)\cdot\theta=0 \qquad \text{and thus} \qquad \theta=\mathrm{Re}\left[a_\theta e^{i\omega t}\right]$$

$$\text{rendering} \quad g/L-\omega^2=0 \quad \text{from which} \quad \omega_n=\sqrt{g/L} \quad \text{is obtained.}$$

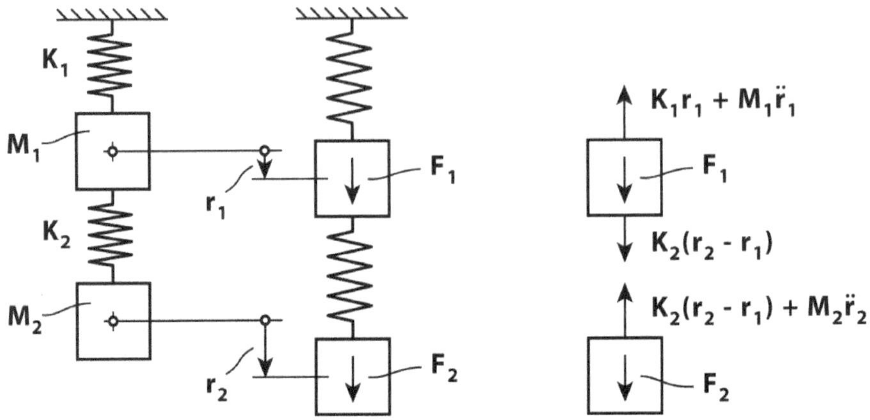

Fig. 1.9 Spring mass system with two degrees of freedom

The system in Fig. 1.3, as well as all the examples above, contains only one unknown displacement component. We say they have one degree of freedom. A system with two degrees of freedom is illustrated in Fig. 1.9. I.e.,

- *the number of degrees of freedom in a system is equal to the number of unknown displacement components that are necessary in order to enable a complete depiction of the position of the system at all times.*

The system in Fig. 1.9 has two degrees of freedom, r_1 and r_2. The equilibrium requirements (see right hand side free body diagram of M_1 and M_2 in Fig. 1.9 above) are then given by

$$\left.\begin{aligned} K_1 \cdot r_1 + M_1 \cdot \ddot{r}_1 - F_1 - K_2 \cdot (r_2 - r_1) = 0 \\ K_2 \cdot (r_2 - r_1) + M_2 \cdot \ddot{r}_2 - F_2 = 0 \end{aligned}\right\} \tag{1.12}$$

This may more conveniently be written in a matrix-vector format

$$\begin{bmatrix} M_1 & 0 \\ 0 & M_2 \end{bmatrix} \begin{bmatrix} \ddot{r}_1 \\ \ddot{r}_2 \end{bmatrix} + \begin{bmatrix} K_1 + K_2 & -K_2 \\ -K_2 & K_2 \end{bmatrix} \begin{bmatrix} r_1 \\ r_2 \end{bmatrix} = \begin{bmatrix} F_1 \\ F_2 \end{bmatrix} \tag{1.13}$$

which, by defining

$$\mathbf{r} = \begin{bmatrix} r_1 \\ r_2 \end{bmatrix} \quad \mathbf{M} = \begin{bmatrix} M_1 & 0 \\ 0 & M_2 \end{bmatrix} \quad \mathbf{K} = \begin{bmatrix} K_1 + K_2 & -K_2 \\ -K_2 & K_2 \end{bmatrix} \quad \text{and} \quad \mathbf{F} = \begin{bmatrix} F_1 \\ F_2 \end{bmatrix} \tag{1.14}$$

may be reduced into a most compact format

$$\mathbf{M}\cdot\ddot{\mathbf{r}}+\mathbf{K}\cdot\mathbf{r}=\mathbf{F} \tag{1.15}$$

If $\mathbf{F}=\mathbf{0}$ then the solution is a harmonic motion which may be described by

$$\mathbf{r}=\operatorname{Re}\left(\boldsymbol{\varphi}\cdot e^{i\omega t}\right) \quad \text{where} \quad \boldsymbol{\varphi}=\begin{bmatrix}a_1 & a_2\end{bmatrix}^T \tag{1.16}$$

By introducing this into Eq. 1.15 then the following requirement is obtained

$$\left(\mathbf{K}-\omega^2\cdot\mathbf{M}\right)\cdot\boldsymbol{\varphi}=\mathbf{0} \tag{1.17}$$

Thus, $\mathbf{r}\neq\mathbf{0}$ can only be obtained if $\det\left[\left(\mathbf{K}-\omega^2\cdot\mathbf{M}\right)\right]=0$, rendering

$$\left(K_1+K_2-\omega^2M_1\right)\cdot\left(K_2-\omega^2M_2\right)-K_2^2=0 \tag{1.18}$$

$$\Rightarrow \omega^4-\left[\left(K_1+K_2\right)/M_1+K_2/M_2\right]\omega^2+\left(K_1/M_1\right)\left(K_2/M_2\right)=0 \tag{1.19}$$

which has the following roots

$$\left.\begin{aligned}\omega_1^2&=\frac{1}{2}\left(\frac{K_1+K_2}{M_1}+\frac{K_2}{M_2}\right)-\sqrt{\frac{1}{4}\left(\frac{K_1+K_2}{M_1}+\frac{K_2}{M_2}\right)^2-\frac{K_1}{M_1}\frac{K_2}{M_2}}\\ \omega_2^2&=\frac{1}{2}\left(\frac{K_1+K_2}{M_1}+\frac{K_2}{M_2}\right)+\sqrt{\frac{1}{4}\left(\frac{K_1+K_2}{M_1}+\frac{K_2}{M_2}\right)^2-\frac{K_1}{M_1}\frac{K_2}{M_2}}\end{aligned}\right\} \tag{1.20}$$

Eq. 1.17 is an eigenvalue problem whose solution is given by ω_1 and ω_2. They are the *eigenfrequencies* of the system. The number of eigenfrequencies will always be the same as the number of degrees of freedom in the system. They are usually presented in ascending order because in almost all practical cases it is a few of the lowest that are of primary interest. For each eigenfrequency there is a corresponding eigenvector. Introducing ω_1 back into Eq. 1.17 we obtain

$$\boldsymbol{\varphi}_1=a_1\begin{bmatrix}1\\ \left(K_1+K_2-\omega_1^2M_1\right)/K_2\end{bmatrix} \tag{1.21}$$

If we introduce ω_2 back into Eq. 1.16 we obtain

$$\boldsymbol{\varphi}_2 = a_1 \begin{bmatrix} 1 \\ \left(K_1 + K_2 - \omega_2^2 M_1\right) / K_2 \end{bmatrix} \tag{1.22}$$

It is seen from Eq. 1.17 that $\boldsymbol{\varphi}_1$ and $\boldsymbol{\varphi}_2$ may be arbitrarily scaled (e.g. by setting $a_1 = 1$). I.e., they do not represent the actual displacements of the system, only its shape. We call them the *mode shapes* of the system. It is only if we have a forcing action on the system that we can quantify a corresponding displacement response. Let for instance $K_1 = K_2 = 2 \cdot 10^7\ Nm$ and $M_1 = M_2 = 10^6\ kg$. Then $\omega_1 = 2.76$ rad/s and corresponding mode shape $\boldsymbol{\varphi}_1 = a_1 \left[1 \quad 1.618\right]^T$, while $\omega_2 = 7.24$ rad/s and its corresponding mode shape $\boldsymbol{\varphi}_2 = a_1 \left[1 \quad -0.618\right]^T$. The motion represented by ω_1 and $\boldsymbol{\varphi}_1$ is shown in the upper diagram in Fig. 1.10, and similarly, the motion represented by ω_2 and $\boldsymbol{\varphi}_2$ is shown in the lower diagram in Fig. 1.10. In both cases $r_1(t=0) = 0.5$ and $\dot{r}_1(t=0) = 0.5$

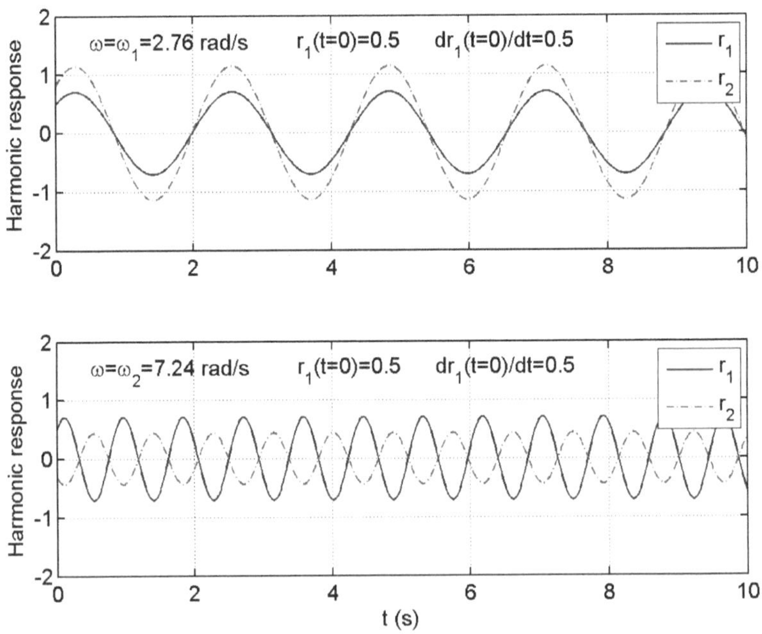

Fig. 1.10 Harmonic motion of two degree of freedom system at eigenfrequencies ω_1 and ω_2

Example 1.5

To illustrate the effects of rotational inertia, a beam on flexible supports K_1 *and* K_2 *is shown in Fig. 1.11. For simplicity, it is assumed infinitely rigid, i.e. its bending stiffness is large. The free body diagram at arbitrary displacements* $r_1(t) = \mathrm{Re}\left[a_1 e^{i\omega t}\right]$ *and* $r_2(t) = \mathrm{Re}\left[a_2 e^{i\omega t}\right]$ *is illustrated on the right hand side of Fig. 1.11. In this case it is necessary to demand vertical as well as moment equilibrium. First it is seen that*

the beam displacement is given by $r(x,t) = r_1 + (r_2 - r_1)x/L$

while the support forces $R_1 = K_1 r_1$ *and* $R_2 = K_2 r_2$

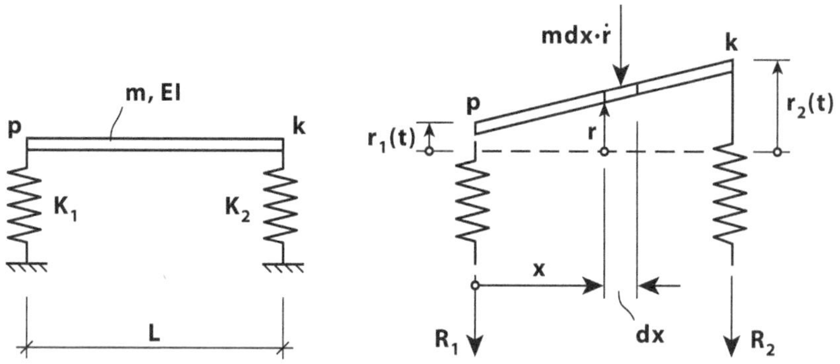

Fig. 1.11 Rigid beam on flexible supports

Thus, vertical equilibrium will require

$$R_1 + R_2 + \int_0^L m\ddot{r}dx = K_1 r_1 + K_2 r_2 + m\int_0^L \left[\ddot{r}_1 + (\ddot{r}_2 - \ddot{r}_1)\frac{x}{L}\right]dx$$

$$= K_1 r_1 + K_2 r_2 + (\ddot{r}_1 + \ddot{r}_2)\frac{mL}{2} = 0$$

while moment equilibrium about the beam end p *will require*

$$R_2 L + \int_0^L m\ddot{r}xdx = K_2 r_2 L + m\int_0^L \left[\ddot{r}_1 + (\ddot{r}_2 - \ddot{r}_1)\frac{x}{L}\right]xdx = K_2 r_2 L + \left(\frac{\ddot{r}_1}{6} + \frac{\ddot{r}_2}{3}\right)mL^2 = 0$$

Introducing $r_1(t) = \mathrm{Re}\left(a_1 e^{i\omega t}\right)$ *and* $r_2(t) = \mathrm{Re}\left(a_2 e^{i\omega t}\right)$ *then these equations turn into*

$$K_1 a_1 + K_2 a_2 - \omega^2 (a_1 + a_2)\frac{mL}{2} = 0 \quad and \quad K_2 a_2 - \omega^2\left(\frac{a_1}{6} + \frac{a_2}{3}\right)mL = 0$$

which may more conveniently be written

$$\begin{bmatrix} \left(K_1 - \omega^2 \frac{mL}{2}\right) & \left(K_2 - \omega^2 \frac{mL}{2}\right) \\ -\omega^2 \frac{mL}{6} & \left(K_2 - \omega^2 \frac{mL}{3}\right) \end{bmatrix} \begin{bmatrix} a_1 \\ a_2 \end{bmatrix} = \begin{bmatrix} 0 \\ 0 \end{bmatrix}$$

A non-trivial solution can only be obtained if the determinant to the coefficient matrix is zero, i.e. that

$$\det \begin{bmatrix} \left(K_1 - \omega^2 \frac{mL}{2}\right) & \left(K_2 - \omega^2 \frac{mL}{2}\right) \\ -\omega^2 \frac{mL}{6} & \left(K_2 - \omega^2 \frac{mL}{3}\right) \end{bmatrix} = 0$$

$$\Rightarrow \qquad \omega^4 - 4\left(\frac{K_1}{mL} + \frac{K_2}{mL}\right)\omega^2 + 12\frac{K_1}{mL}\frac{K_2}{mL} = 0$$

and thus, the following eigenfrequencies (in ascending order) are obtained

$$\omega_n = \sqrt{2\left[\frac{K_1}{mL} + \frac{K_2}{mL} \mp \sqrt{\left(\frac{K_1}{mL}\right)^2 - \frac{K_1}{mL}\frac{K_2}{mL} + \left(\frac{K_2}{mL}\right)^2}\right]}$$

If, for instance $K_1 = K_2 = K$ *, then* $\omega_n = \sqrt{2\left(\frac{2K}{ml} \mp \frac{K}{mL}\right)} \Rightarrow \begin{cases} \omega_1 = \sqrt{2K/(mL)} \\ \omega_2 = \sqrt{6K/(mL)} \end{cases}$

Introducing $\omega = \omega_1 = \sqrt{2K/(mL)}$ *into the second row of the matrix-vector relationship above*

$$-\omega_1^2 \frac{mL}{6} a_1 + \left(K - \omega_1^2 \frac{mL}{3}\right) a_2 = 0 \qquad \Rightarrow \qquad a_1 = a_2$$

implying that the motion is purely translational (in vertical direction). Introducing $\omega = \omega_2 = \sqrt{6K/(mL)}$ *into the second row of the matrix-vector relationship above*

$$-\omega_2^2 \frac{mL}{6} a_1 + \left(K - \omega_2^2 \frac{mL}{3}\right) a_2 = 0 \qquad \Rightarrow \qquad a_1 = -a_2$$

implying that the motion is purely rotational (about the mid-span of the beam).

Continuous Systems

Below, a continuous line-like system is illustrated in Fig 1.12. For such a system the relevant equilibrium requirements are at this stage most conveniently established in the form of one or several differential equations.

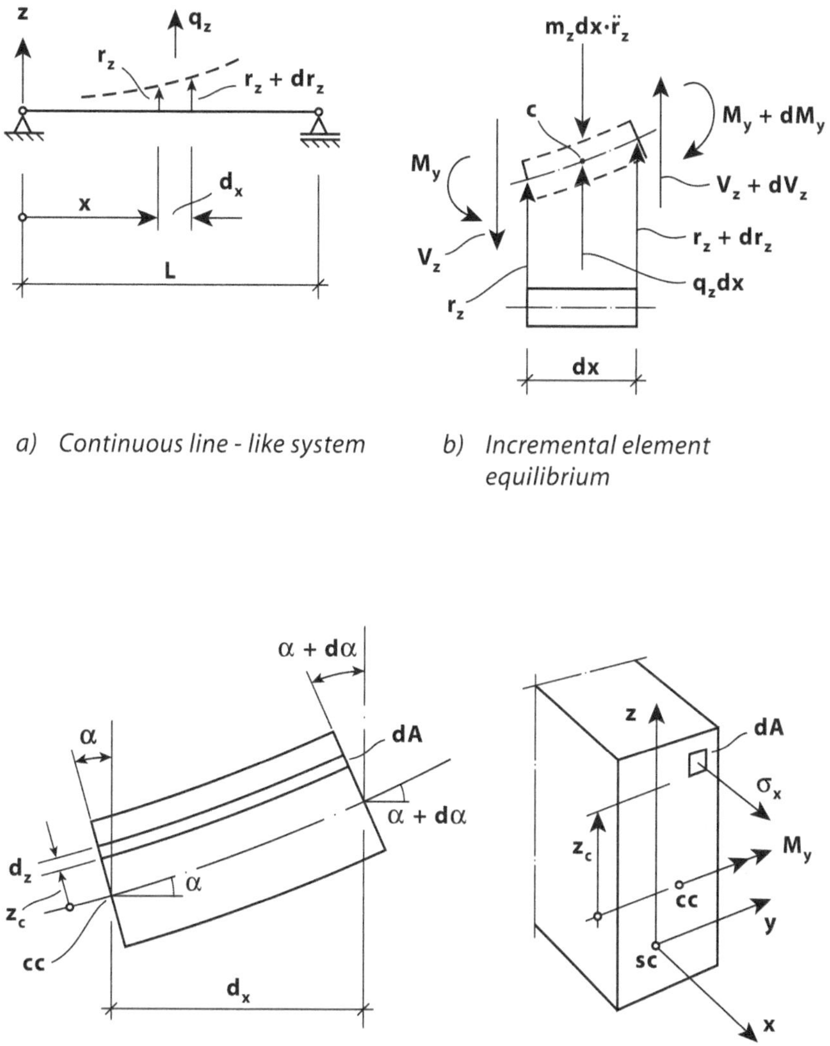

a) *Continuous line - like system* b) *Incremental element equilibrium*

c) *Navier's hypothesis* d) *Stress component*

Fig. 1.12 Line-like continuous beam subject to distributed dynamic load

The situation of a simple beam subject to a distributed dynamic load $q_z(x,t)$ (i.e. with unit N/m) is illustrated in Fig. 1.12.a. It is taken for granted that the system is symmetric about the z-axis such that the response motion is exclusively taking place in the direction of z. Since the system itself is continuous, so is the displacement $r_z(x,t)$, and therefore, it will be necessary to determine the dynamic displacement response at all positions x in order to enable a complete description of its motion at all times. Thus, in this case it is meaningless to introduce the concept of degrees of freedom in the system. Rather, as mentioned above, we resort to calculus to obtain a solution. As shown in Fig. 1.12.b, an incremental element dx will require moment equilibrium (about the mid-point c)

$$dM_y - V_z dx = 0 \tag{1.23}$$

as well as force equilibrium in the z direction

$$q_z \cdot dx - m_z \cdot dx \cdot \ddot{r}_z + dV_z = 0 \tag{1.24}$$

where m_z is the mass per unit length of the beam associated with motion in the z direction. From Eq. 1.23 we obtain $V_z = M'_y$, while from Eq. 1.24 we obtain $V'_z = -q_z + m_z \cdot \ddot{r}_z$. Thus

$$M''_y = -q_z + m_z \cdot \ddot{r}_z \tag{1.25}$$

Since the dynamic motion exclusively takes place in the direction of z, the cross section of the beam is subject to pure bending about the y axis (i.e. $M_y \neq 0$ and $V_z \neq 0$ while all other cross sectional stress resultants are equal to zero). The cross sectional neutral axis is defined by the axis through which there is zero strain. Adopting Navier's hypothesis [4,5] that a cross section that is perpendicular to the system neutral axis prior to bending will remain perpendicular to the neutral axis after bending. This will render a strain distribution (see Fig. 1.12)

$$\varepsilon_x = \frac{\alpha \cdot z_c - (\alpha + d\alpha) \cdot z_c}{dx} = -\frac{d\alpha}{dx} \cdot z_c \tag{1.26}$$

where z_c is the distance from the neutral axis (CC) to an arbitrary cross sectional element dA. Let us assume linear elasticity and take it for granted that displacements are small such that $\alpha \approx r'_z$. Then $\varepsilon_x = -r''_z \cdot z_c$, and thus:

$$\sigma_x = E \cdot \varepsilon_x \quad \Rightarrow \quad \begin{cases} N = \int_A \sigma_x dA = -r''_z E \int_A z_c dA = 0 \\ M_y = \int_A z_c \cdot \sigma_x dA = -r''_z E \int_A z_c^2 dA = -r''_z E I_y \end{cases} \tag{1.27}$$

The requirement that $N = 0$ is used to determine the position of the neutral axis (i.e. that $\int_A z_c dA = 0$), while $I_y = \int_A z_c^2 dA$ is defined as the second area moment of the cross section. Thus

$$\frac{d^2}{dx^2}\left(-r_z'' EI_y\right) = -q_z + m_z \cdot \ddot{r}_z \tag{1.28}$$

which, provided EI_y is constant along the span, may be simplified into

$$m_z \cdot \ddot{r}_z + EI_y \cdot r_z'''' = q_z \tag{1.29}$$

This is the dynamic equilibrium condition for a perfectly elastic and continuous line-like beam type of system with constant properties (m_z and EI_y) along its span, and whose load and cross sectional properties are such that the motion will only take place in the direction of z. Let us assume that the beam in Fig. 1.12 is unloaded, i.e. $q_z = 0$. The general solution to Eq. 1.29 is then given by a harmonic motion

$$r_z(x,t) = \mathrm{Re}\left[\phi_z(x) \cdot e^{i\omega t}\right] \Rightarrow EI_y \phi_y'''' - \omega^2 m_z \phi_y = 0 \tag{1.30}$$

where $\phi_z(x)$ is a shape function whose fourth derivative must be shapewise congruent to itself, i.e.

$$\phi_z(x) = a_1 \sin\left(\lambda \frac{x}{L}\right) + a_2 \cos\left(\lambda \frac{x}{L}\right) + a_3 \sinh\left(\lambda \frac{x}{L}\right) + a_4 \cosh\left(\lambda \frac{x}{L}\right) \tag{1.31}$$

and where λ is a non-dimensional wave length dependent of the system boundary conditions.

Example 1.6

Let us for simplicity assume that the beam in Fig. 1.12 is simply supported. This implies that $r_z(x=0,t) = r_z(x=L,t) = 0$ and that the cross sectional bending moments $M_y(x=0,t) = M_y(x=L,t) = 0$, which will require (see the expression of M_y in Eq. 1.27) that $r_z''(x=0,t) = r_z''(x=L,t) = 0$. Thus,

$$\left.\begin{array}{l}\phi_z(x=0)=0\\ \phi_z''(x=0)=0\end{array}\right\} \Rightarrow \left.\begin{array}{l}a_2+a_4=0\\ a_4-a_2=0\end{array}\right\} \Rightarrow a_1=a_2=0$$

and

$$\left.\begin{array}{l}\phi_z(x=L)=0\\ \phi_z''(x=L)=0\end{array}\right\} \Rightarrow \begin{bmatrix}\sin(\lambda) & \sinh(\lambda)\\ -\sin(\lambda) & \sinh(\lambda)\end{bmatrix}\begin{bmatrix}a_1\\ a_3\end{bmatrix}=\begin{bmatrix}0\\ 0\end{bmatrix} \Rightarrow \sin(\lambda)\cdot\sinh(\lambda)=0$$

It is readily seen that $\sinh(\lambda)=0$ *will require* $\lambda=0$, *which is a non-relevant solution rendering* $r_z(x,t)=0$, *and thus, we must demand* a_3 *and* $\sin(\lambda)$ *equal to zero, which implies that* $\lambda=n\pi$ *where* $n=1,2,3,\ldots$. *Thus*

$$r_z(x,t)=\sin\left(n\pi\frac{x}{L}\right)\cdot\mathrm{Re}\left[a_1\cdot e^{i\omega t}\right]$$

Introducing this into Eq. 1.30 then the following is obtained:

$$EI_y(n\pi/L)^4-\omega^2 m_z=0$$

Thus, the eigenfrequency and corresponding eigenmode of the system are given by

$$\omega_{z_n}=(n\pi/L)^2\sqrt{EI_y/m_z} \quad \text{and} \quad \phi_z(x)=\sin(n\pi x/L)$$

The three first mode shapes are illustrated in Fig. 1.13.

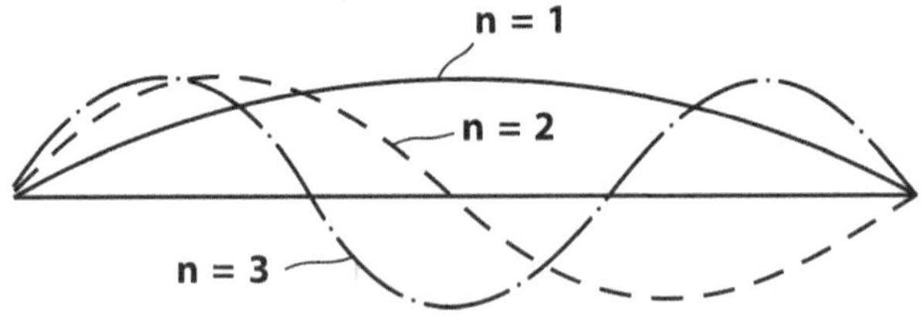

Fig. 1.13 Mode shapes of simply supported beam

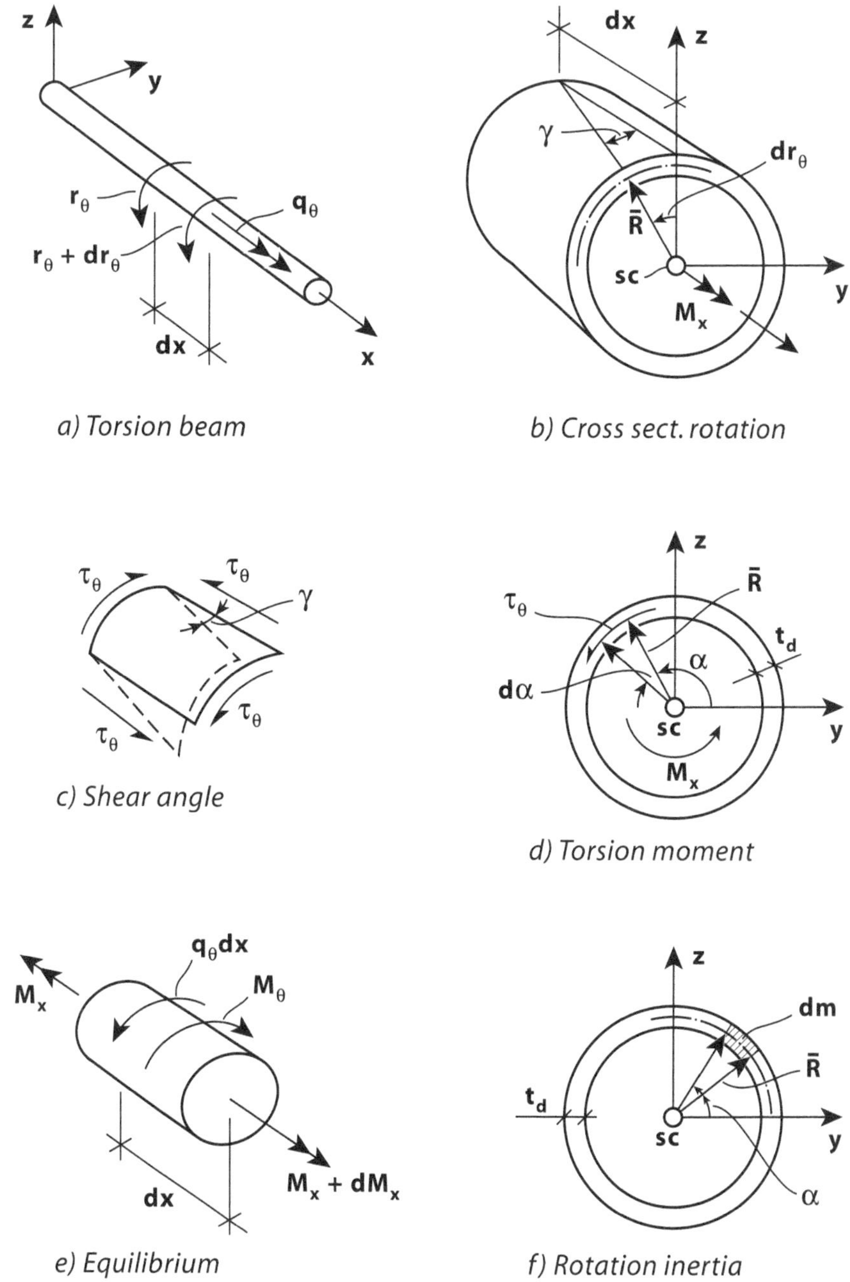

Fig. 1.14 Line-like continuous thin walled beam subject to torsion

A system containing the problem of torsion is illustrated in Fig. 1.14.a. For simplicity a beam with a thin walled tube type of cross section is chosen $\left(t_d \ll \bar{R}\right)$. The beam is assumed to have constant physical properties along its span. It is subject to a distributed torsion load $q_\theta\left(x,t\right)$ (i.e. with unit Nm/m). As shown in Fig. 1.14.b the connection between shear angle γ and incremental change of cross sectional rotation dr_θ is given by $\gamma \cdot dx = \bar{R} \cdot dr_\theta$, i.e. $\gamma = \bar{R} r_\theta'$. Assuming that τ_θ is constant across the tube thickness t_d and independent of α (see Fig. 1.14.d), then the torsion moment

$$M_x \approx \int_A \tau_\theta \bar{R} dA = \int_0^{2\pi} \tau_\theta \bar{R}^2 t_d d\alpha = \left(2\pi \bar{R}^2 t_d\right) \cdot \tau_\theta \tag{1.32}$$

Introducing $\tau_\theta = G \cdot \gamma = G\bar{R} r_\theta'$, then

$$\left.\begin{aligned} M_x &= G I_t r_\theta' \\ \tau_\theta &= M_x \bar{R} / I_t \end{aligned}\right\} \quad \text{where} \quad I_t = 2\pi \bar{R}^3 t_d \tag{1.33}$$

The dynamic inertia (per unit length, see Fig. 1.14.e and f) is given by

$$M_\theta = \int_M dm \cdot \frac{d^2}{dt^2}\left(\bar{R} r_\theta\right) \cdot \bar{R} = \ddot{r}_\theta \left(\int_A \rho_m \bar{R}^2 dA \right) dx = \ddot{r}_\theta m_\theta dx \tag{1.34}$$

where ρ_m is the material density and m_θ is the cross section rotational mass (with unit kgm^2/m). (In this case the cross section is a tube and then $m_\theta = 2\pi \bar{R}^3 t_d \rho_m$.) The equilibrium requirement for an incremental element dx is then given by (see Fig. 1.14.e)

$$dM_x + q_\theta dx - M_\theta = 0 \quad \Rightarrow \quad m_\theta \ddot{r}_\theta - G I_t r_\theta'' = q_\theta \tag{1.35}$$

Let us assume that the beam in Fig. 1.14 is unloaded, i.e. $q_\theta = 0$. The general solution to Eq. 1.35 is then given by a harmonic motion

$$r_\theta\left(x,t\right) = \mathrm{Re}\left[\phi_\theta\left(x\right) \cdot e^{i\omega t}\right] \tag{1.36}$$

Example 1.7

Let us also assume that the torsion beam in Fig. 1.14 is simply supported and that it has fork bearings at its ends, such that the torsion boundary conditions are $r_\theta(x=0,t)=r_\theta(x=L,t)=0$. *The general solution is then given by*

$$r_\theta(x,t)=\sin(\lambda x/L)\cdot \mathrm{Re}\left[a\cdot e^{i\omega t}\right]$$

where λ *is the non-dimensional wave length of the mode shape. From the boundary condition* $r_\theta(x=L,t)=0$ *it is seen that* $\sin(\lambda)=0$, *which implies that* $\lambda=n\pi$. *Thus*

$$r_\theta(x,t)=\sin(n\pi x/L)\cdot \mathrm{Re}\left[a\cdot e^{i\omega t}\right]$$

Introducing this into Eq. 1.35 will then render

$$(n\pi/L)^2 GI_t-\omega^2 m_\theta=0$$

Thus, the eigenfrequency and corresponding eigenmode of the system are given by

$$\omega_{\theta_n}=\frac{n\pi}{L}\sqrt{\frac{GI_t}{m_\theta}} \qquad \textit{and} \qquad \phi_\theta(x)=\sin(n\pi x/L)$$

See further elaboration below.

In the special case of a thin walled tube it was shown that $m_\theta=2\pi\bar{R}^3 t_d\rho_m$ and $I_t=2\pi\bar{R}^3 t_d$. It should be noted that in a more general case (see Fig. 1.15)

$$\left.\begin{aligned} &M_\theta=\int_M dm\cdot\frac{d^2}{dt^2}(r\cdot r_\theta)\cdot\bar{R}=\ddot{r}_\theta\left(\int_A \rho_m r^2 dA\right)dx=\ddot{r}_\theta m_\theta dx\\ &\Rightarrow m_\theta=\int_A \rho_m r^2 dA\end{aligned}\right\} \qquad (1.37)$$

while for a closed box type of cross section (see Fig. 1.15)

$$\left.\begin{aligned} M_x &= \int_A \tau_\theta a dA = \int_A G a r'_\theta a dA = G\left(\int_A a^2 dA\right) r'_\theta = G\left[\int_A \left(\frac{2A_m}{ds}\right)^2 tds\right] r'_\theta \\ &\Rightarrow I_t = 4A_m^2 \Big/ \oint_s ds/t \end{aligned}\right\} \tag{1.38}$$

It should also be noted that in general the torsion stiffness will contain contributions from warping. As illustrated in Fig. 1.16, the phenomenon is caused by shear forces V that occur in flanges at a distance a from the shear centre, rendering a torsion moment contribution

$$M_x = \sum_j V_j \cdot a_j \tag{1.39}$$

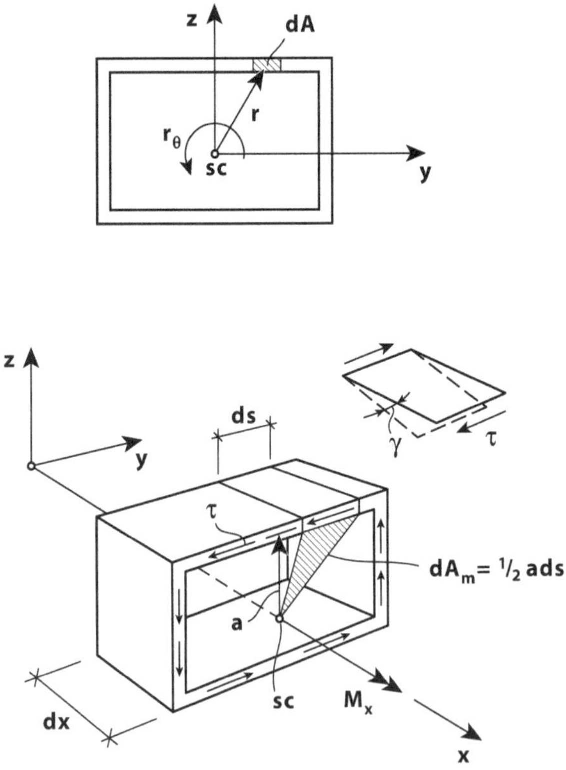

Fig. 1.15 Rotation inertia of a box type of cross section

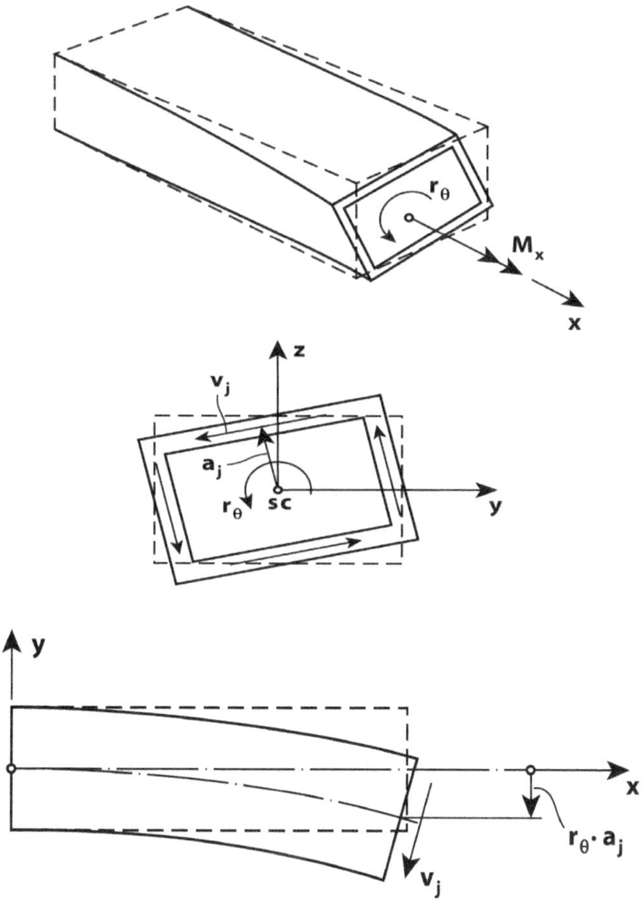

Fig. 1.16 Torsion due to cross sectional warping

For a simple beam element the connection between the shear force and the bending moment is shown in Eq. 1.23, while the connection between the bending moment and the second derivative of the displacement is shown in Eq. 1.27. Thus, for an arbitrary flange shown in Fig. 1.16.c

$$V_j = \frac{d}{dx}\left[-EI_j \frac{d^2}{dx^2}\left(r_\theta a_j\right)\right] = -EI_j a_j r_\theta''' \tag{1.40}$$

Thus, the contribution from warping is given by

$$M_x = \sum_j V_j \cdot a_j = \sum_j \left(EI_j a_j r_\theta'''\right) \cdot a_j = E \cdot \left(\sum_j I_j a_j^2\right) \cdot r_\theta''' \tag{1.41}$$

which, by defining the warping constant $I_w = \sum_j I_j a_j^2$, renders $M_x = EI_w r_\theta'''$.

Thus

$$M_{x_{tot}} = -GI_t r_\theta' + EI_w r_\theta''' \tag{1.42}$$

Returning to the equilibrium requirement in Eq. 1.35 (i.e. that $dM_x + q_\theta dx - M_\theta = 0$) will then render the following more general differential equation for torsion

$$m_\theta \ddot{r}_\theta - GI_t r_\theta'' + EI_w r_\theta'''' = q_\theta \tag{1.43}$$

The calculation of I_t and I_w may be found in the literature of structural mechanics. Setting $q_\theta = 0$ and solving Eq. 1.43 rather than the simpler version in Eq. 1.35 will then provide a more accurate eigenfrequency

$$\omega_{\theta_n} = \frac{n\pi}{L}\sqrt{\frac{GI_t}{m_\theta}\left[1+\left(\frac{n\pi}{L}\right)^2\frac{EI_w}{GI_t}\right]} \tag{1.44}$$

than that which has been developed in Example 1.7.

1.3 The Principle of Energy Conservation

As previously mentioned it is in the following taken for granted that material behaviour with respect to axial and shear strain is linear elastic, as indicated in Fig. 1.17.a. In addition to this, it is a general requirement that any force, concentrated or distributed, is conservative, i.e. during a motion from position A to another position B through any path s, the size and direction of the force will remain unchanged. See Fig. 1.17.b.

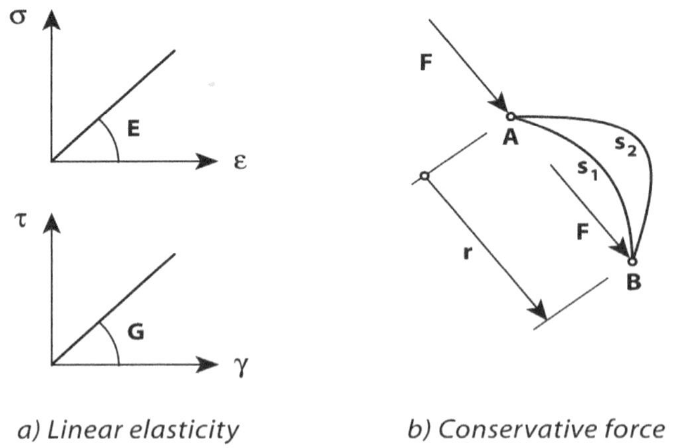

Fig. 1.17 Linear elasticity and conservative forces

Energy in Line-Like Type of Systems

In general the principle of d'Alambert is adopted such that at all times and positions the system is in a condition of physical equilibrium, having included inertia body forces according to Newton's second law. It is taken for granted that the system is in a permanent condition of thermal equilibrium, i.e. that thermal contributions to the energy balance may be disregarded. Thus, the instantaneous energy variation in the system at any time and position is defined by

- *the energy P held by external forces acting on the system,*
- *the motion (kinetic) energy T of the system body masses, and*
- *the strain energy U stored in the material fibres of the system.*

The relevant energy considerations in a simple mass-spring system are illustrated in Fig. 1.18. As can be seen a force $F(t)$ moving a positive displacement r has lost its ability to perform the work $F \cdot r$, i.e. it has reduced its energy level by $F \cdot r$. The only restriction to $F(t)$ is that it is independent of the path of motion (see Fig. 1.17.b above). Thus, if a force and the direction of motion coincide, then the force is losing ability to perform work, $P = -F \cdot r$, while if the force and direction of motion are opposite to each other (the force is 'lifted') then the force has gained energy i.e. it has gained ability to perform work $P = F \cdot r$. In vector description a concentrated force vector $\mathbf{F}$ moving a displacement $\mathbf{r}$ is gaining the energy

$$P_F(\mathbf{r}) = \int_0^{\mathbf{r}} -\mathbf{F}^T d\mathbf{r} = -\mathbf{F}^T \mathbf{r} \tag{1.45}$$

while for a distributed force vector $\mathbf{q}$ the gain of energy is given by

$$P_q(\mathbf{r}) = \int_L \int_0^{\mathbf{r}} -\mathbf{q}^T(t,x) \cdot dx \cdot d\mathbf{r} = -\int_L \mathbf{q}^T(t,x) \cdot \mathbf{r} \cdot dx \tag{1.46}$$

If a body with mass M, unconstrained and small enough to be mathematically treated as a single particle is moving with an acceleration $\ddot{\mathbf{r}}$, then its motion energy is identical to the work that has been performed in order to obtain this acceleration. Since the force (inertia) resistance to an acceleration is $\mathbf{M}\ddot{\mathbf{r}}$, then the work performed from rest to position $\mathbf{r}$ is given by

$$\int_0^{\mathbf{r}} (M\ \ddot{\mathbf{r}})^T d\mathbf{r} = \int_0^t M\ \ddot{\mathbf{r}}^T \dot{\mathbf{r}} \cdot dt = \int_0^t \frac{d}{dt}\left(\frac{1}{2} M\ \dot{\mathbf{r}}^T \dot{\mathbf{r}}\right) dt = \frac{1}{2} M\ \dot{\mathbf{r}}^T \dot{\mathbf{r}} \tag{1.47}$$

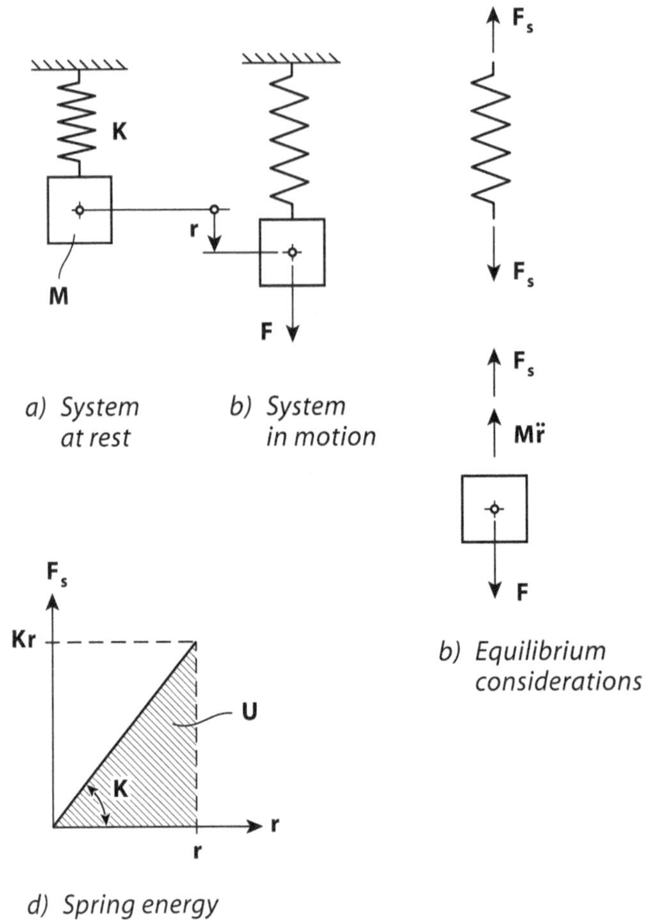

Fig. 1.18 Energy considerations in a simple mass-spring system

Thus, at any time t and velocity condition $\dot{\mathbf{r}}$ of the system, it has gained a motion energy defined by

$$T_M(\dot{\mathbf{r}}) = \frac{1}{2} M \; \dot{\mathbf{r}}^T \dot{\mathbf{r}} \tag{1.48}$$

This is called the kinetic energy of the mass M. Similarly, if a continuous line-like system with distributed mass $m(x)$ is in a plane motion e.g. in the z direction then its kinetic energy is given by

$$T_m(\dot{r}_z) = \frac{1}{2} \int_L m(x) \cdot \left[\dot{r}_z(x,t) \right]^2 dx \tag{1.49}$$

Strain energy is the energy stored in the system because material has been stretched or compressed. A simple illustration is shown in Fig. 1.18, where a spring with stiffness K has been stretched a displacement r, rendering a spring energy $K \cdot r^2/2$ (see Fig. 1.18.d). By adopting linear elasticity

$$\boldsymbol{\sigma} = \mathbf{E}\boldsymbol{\varepsilon} \tag{1.50}$$

where $\boldsymbol{\sigma}(\mathbf{r}) = \begin{bmatrix} \sigma_x \\ \tau_{yz} \end{bmatrix}$ $\quad \boldsymbol{\varepsilon}(\mathbf{r}) = \begin{bmatrix} \varepsilon_x \\ \gamma_{yz} \end{bmatrix}$ $\quad$ and $\quad \mathbf{E} = \begin{bmatrix} E & 0 \\ 0 & G \end{bmatrix}$ (1.50)

then the strain energy for an incremental material element $dAdx$ is defined by

$$\begin{aligned} dU_M(\mathbf{r}) &= \int_0^{\boldsymbol{\varepsilon}} \boldsymbol{\sigma}^T \cdot d\boldsymbol{\varepsilon} = \int_0^{\boldsymbol{\varepsilon}} (\mathbf{E} \cdot \boldsymbol{\varepsilon})^T \cdot d\boldsymbol{\varepsilon} = \int_0^{\boldsymbol{\varepsilon}} \boldsymbol{\varepsilon}^T \cdot \mathbf{E} \cdot d\boldsymbol{\varepsilon} \\ &= E \int_0^{\varepsilon_x} \varepsilon_x d\varepsilon_x + G \int_0^{\gamma_{yz}} \gamma_{yz} d\gamma_{yz} = E\frac{\varepsilon_x^2}{2} + G\frac{\gamma_{yz}^2}{2} = \frac{\sigma_x^2}{2E} + \frac{\tau_{yz}^2}{2G} \end{aligned} \tag{1.51}$$

Thus, the total strain energy in the system is given by

$$U_M(\mathbf{r}) = \int_L \left[\int_A \left(\int_0^{\boldsymbol{\varepsilon}} \boldsymbol{\sigma}^T \cdot d\boldsymbol{\varepsilon} \right) dA \right] dx = \frac{1}{2} \int_L \left[\int_A \left(\frac{\sigma_x^2}{E} + \frac{\tau_{yz}^2}{G} \right) dA \right] dx \tag{1.52}$$

Let us for instance assume that σ_x is caused by pure bending about the y axis (i.e. by M_y), and that τ_{yz} is caused by torsion (i.e. by M_x) on a tube type of cross section. For such a continuous system it has previously been shown (see Eq. 1.27) that

$$\left. \begin{array}{l} \sigma_x = E \cdot \varepsilon_x \\ \varepsilon_x = -r_z'' \cdot z_c \end{array} \right\} \Rightarrow \int_A \frac{\sigma_x^2}{E} dA = \int_A \frac{\left(-E r_z'' z_c\right)^2}{E} dA = E\left(\int_A z_c^2 dA \right) \left(r_z''\right)^2 = EI_y \left(r_z''\right)^2 \tag{1.53}$$

and similarly (see Eq. 1.33) that

$$\left. \begin{array}{l} \tau_{yz} = \tau_\theta = G\gamma \\ \gamma = \bar{R} r_\theta' \end{array} \right\} \Rightarrow \int_A \frac{\tau_\theta^2}{G} dA = \int_A \frac{\left(G\bar{R} r_\theta'\right)^2}{G} dA = G\left(\int_A \bar{R}^2 dA \right) \left(r_\theta'\right)^2 = GI_t \left(r_\theta'\right)^2 \tag{1.54}$$

Thus, the strain energy in a continuous system subject to mono-axial bending and torsion is given by

$$U_m = \frac{1}{2} \int_L \left[EI_y \left(r_z''\right)^2 + GI_t \left(r_\theta'\right)^2 \right] dx \tag{1.55}$$

It should be noted that Eq. 1.55 is applicable for any continuous line-like beam subject to mono-axial bending and St. Venant torsion, i.e. it is not restricted to a beam with a tube type of cross section as long as I_t is correctly calculated for the cross section in question. For the case of bi-axial bending it is necessary to add $EI_z\left(r_y''\right)^2$ into the integration of U_m.

Basic Ideas

The basic idea behind all energy methods is that under the conditions mentioned above (physical and thermal equilibrium) energy can neither be created nor destroyed; it can only be transformed from one state to another. This idea manifests itself in three alternative ways. First, the *Rayleigh-Ritz method* [6] is based on d'Alambert's perception of inertia. The observation is that from the point of view of an observer outside of a system the total energy at any time

$$\Pi = T + U + P \tag{1.56}$$

is constant, and therefore the variation δ of Π with respect to its variables (t , $\mathbf{r}$ and $\dot{\mathbf{r}}$) is equal to zero, i.e. that

$$\delta\Pi = 0 \tag{1.57}$$

Secondly, there is the principle of *Hamilton & Euler/Lagrange* [7]. This is not based on d'Alambert's perception of inertia, i.e. the observer is not standing still considering the energy account at a particular time t . Rather, it is based on the balance of energy transfer between T and $(U+P)$. The observer is himself sitting on the system and his observation is that the energy in the system is changing between exclusively kinetic (where $\mathbf{r}=0$) and exclusively the sum of strain and load energies (where $\dot{\mathbf{r}}=0$). The transfer of energy between these two extremes may be described by the Lagrange function

$$L(\mathbf{r},\dot{\mathbf{r}}) = T(\dot{\mathbf{r}}) - \left[U(\mathbf{r}) + P(\mathbf{r})\right] \tag{1.58}$$

Hamilton's assumption is that in the time interval between t_1 and t_2 the energy transfer (i.e. the difference between that which comes in and that which goes out as you travel with the system in time)

$$f(\mathbf{r},\dot{\mathbf{r}}) = \int_{t_1}^{t_2} L(\mathbf{r},\dot{\mathbf{r}})\,dt \tag{1.59}$$

will occur along a functional ridge (there is no waste of energy in any direction) such that the variation of $f(\mathbf{r},\dot{\mathbf{r}})$ with respect to its variables will always be zero, i.e. that

$$\delta f(\mathbf{r},\dot{\mathbf{r}})=0 \tag{1.60}$$

Finally, and most importantly, there is the *principle of virtual work* which is usually attributed to d'Alambert and Lagrange. It is like the Rayleigh-Ritz method based on d'Alambert's perception of inertia. The observation is that from the point of view of an observer outside of a system, a free body in physical equilibrium will not change its energy level for any incremental change $\delta\mathbf{r}$ of its position. The only restriction imposed on $\delta\mathbf{r}$ is that it is time invariant. Otherwise, $\delta\mathbf{r}$ is arbitrary or virtual. Hence, it has been labelled the principle of virtual work.

Example 1.8

Let us for simplicity consider the system in Fig. 1.18. Imposing a time invariant and non-zero virtual displacement δr to the mass M of the rigid body diagram on the right hand side of the illustration, then $dT=(M\ddot{r})\delta r$ (because δr is time invariant), $dU=F_s\delta r$ and $dP=-F\delta r$. Since $F_s=Kr$ then the change of energy level is given by

$$d\Pi=(M\ddot{r}+Kr-F)\delta r=0$$

rendering the system equilibrium condition: $M\ddot{r}+Kr=F$.

While the methods of Rayleigh-Ritz (Chapter 1.4) and Hamilton-Euler-Lagrange (Chapter 1.5) are usually only used in cases of calculating the eigenfrequency of special systems, the principle of virtual work is widely used throughout structural mechanics for development of relevant equilibrium requirements of general systems, as well as the development of the finite element method. The principle of virtual work is presented in chapter 1.6 below.

1.4 The Rayleigh-Ritz Method

The Rayleigh-Ritz method is usually used to determine approximate values of eigenfrequencies. It may be used in a continuous as well as a discrete system. For the simple single degree of freedom system illustrated in Fig. 1.18, in motion at an arbitrary position r, the mass has at this position gained a kinetic energy of $T=M\dot{r}^2/2$, while at the same time the spring has been stretched such that it has obtained an elastic energy of $U=Kr^2/2$ (as illustrated in Fig. 1.18.d). The force F at the arbitrary positive position r has gained an energy of $P=-Fr$

(negative because F and r are in the same direction, and thus, it has lost the ability to perform that work). In this position the system is in equilibrium and the total energy is given by

$$\Pi(r,\dot{r}) = U + T + P = \frac{1}{2}Kr^2 + \frac{1}{2}M\dot{r}^2 - Fr \tag{1.61}$$

Obviously, the variation of Π can be performed on any of its variables t, r and $\dot{r}$. Let us in this case perform the variation on t. Thus

$$\begin{aligned} \delta\Pi &= \frac{d}{dt}\left(\frac{1}{2}Kr^2 + \frac{1}{2}M\dot{r}^2 - Fr\right)dt \\ &= \left(Kr\dot{r} + M\dot{r}\ddot{r} - F\dot{r}\right)dt = \dot{r}\left(M\ddot{r} + Kr - F\right)dt = 0 \end{aligned} \tag{1.62}$$

which can only be fulfilled at all times if

$$\Rightarrow \quad M\ddot{r} + Kr = F \tag{1.63}$$

again rendering the equilibrium condition of the system Since we are primarily interested in using the method to calculate eigenfrequencies and associated eigenmodes we shall in the following take it for granted that the system is unloaded and undamped. Let

$$\mathbf{r} = \begin{bmatrix} r_1 & r_2 & \cdots & r_i & \cdots & r_N \end{bmatrix}^T \tag{1.64}$$

be the displacement vector that is required for a sufficient description of the motion of the system. Let

$$\mathbf{M} = \begin{bmatrix} \ddots & & \cdot^{\cdot^{\cdot}} \\ & M_{ij} & \\ \cdot^{\cdot^{\cdot}} & & \ddots \end{bmatrix} \quad \text{and} \quad \mathbf{K} = \begin{bmatrix} \ddots & & \cdot^{\cdot^{\cdot}} \\ & K_{ij} & \\ \cdot^{\cdot^{\cdot}} & & \ddots \end{bmatrix} \quad \left.\begin{matrix} i = 1,2,...,N \\ j = 1,2,...,N \end{matrix}\right\} \tag{1.65}$$

be the corresponding mass and stiffness matrices. The strain and kinetic energies are then given by

$$U = \frac{1}{2}\mathbf{r}^T\mathbf{K}\mathbf{r} \quad \text{and} \quad T = \frac{1}{2}\dot{\mathbf{r}}^T\mathbf{M}\dot{\mathbf{r}} \tag{1.66}$$

and thus

$$\Pi(\mathbf{r},\dot{\mathbf{r}}) = \frac{1}{2}\left(\mathbf{r}^T\mathbf{K}\mathbf{r} + \dot{\mathbf{r}}^T\mathbf{M}\dot{\mathbf{r}}\right) \tag{1.67}$$

For such an unloaded and undamped system the motion will be harmonic, i.e. the motion is given by the product of a mode shape $\boldsymbol{\varphi} = \begin{bmatrix} \phi_1 & \phi_2 & \cdots & \phi_i & \cdots & \phi_N \end{bmatrix}^T$ and a harmonic function, which here will be described in a complex format, i.e.

$$\mathbf{r} = \mathrm{Re}\left[\boldsymbol{\varphi} \cdot e^{i\omega t}\right] \quad \text{and} \quad \dot{\mathbf{r}} = \mathrm{Re}\left[i\omega\boldsymbol{\varphi} \cdot e^{i\omega t}\right] \tag{1.68}$$

Thus

$$\Pi(\boldsymbol{\varphi}) = \mathrm{Re}\left[\frac{e^{i2\omega t}}{2}\left(\boldsymbol{\varphi}^T\mathbf{K}\boldsymbol{\varphi} - \omega^2\boldsymbol{\varphi}^T\mathbf{M}\boldsymbol{\varphi}\right)\right] = \mathrm{Re}\left[\frac{e^{i2\omega t}}{2}\boldsymbol{\varphi}^T\left(\mathbf{K} - \omega^2\mathbf{M}\right)\boldsymbol{\varphi}\right] \tag{1.69}$$

Since there is no other energy in the system it is seen that any variation

$$\delta\Pi(\boldsymbol{\varphi}) = 0 \quad \text{is obtained if} \quad \left(\mathbf{K} - \omega^2 \cdot \mathbf{M}\right) \cdot \boldsymbol{\varphi} = \mathbf{0} \tag{1.70}$$

which implies that for a system in harmonic motion in absence of external forces the sum of kinetic and strain energy are always zero. It is an eigenvalue problem identical to that which has previously been obtained by using the equilibrium requirements, see Eq. 1.17. However, in many cases in structural engineering a good insight into the particular mass and stiffness distribution in a system may permit an adequate guess of what the mode shapes will look like. Thus, an approximate solution may readily be obtained. Let for instance

$$\boldsymbol{\varphi} \approx \boldsymbol{\psi}^T\mathbf{a} \tag{1.71}$$

where $\mathbf{a} = \begin{bmatrix} a_1 & a_2 & \cdots & a_i & \cdots & a_N \end{bmatrix}^T$ is a vector of N unknown coefficients and where $\boldsymbol{\psi} = \begin{bmatrix} \boldsymbol{\psi}_1 & \boldsymbol{\psi}_2 & \cdots & \boldsymbol{\psi}_i & \cdots & \boldsymbol{\psi}_N \end{bmatrix}^T$ in a discrete format is a matrix of N conveniently chosen and fully known vectors $\boldsymbol{\psi}_i$, or in a continuous format $\boldsymbol{\psi} = \begin{bmatrix} \psi_1 & \psi_2 & \cdots & \psi_i & \cdots & \psi_N \end{bmatrix}^T$ is a vector containing fully known functions ψ_i (e.g. polynomials, or a combination of harmonic sinus, cosine or hyperbolic type of functions), whichever is most convenient. The only restriction we shall impose on the content of $\boldsymbol{\psi}$ is that none of the vectors $\boldsymbol{\psi}_i$ or functions ψ_i violates the physical boundary conditions of the system. Then

$$\begin{aligned}\Pi(\mathbf{a}) &= \mathrm{Re}\left[\frac{e^{i2\omega t}}{2}\left(\boldsymbol{\varphi}^T\mathbf{K}\boldsymbol{\varphi} - \omega^2\boldsymbol{\varphi}^T\mathbf{M}\boldsymbol{\varphi}\right)\right] \\ &= \mathrm{Re}\left\{\frac{e^{i2\omega t}}{2}\left[\left(\boldsymbol{\psi}^T\mathbf{a}\right)^T\mathbf{K}\left(\boldsymbol{\psi}^T\mathbf{a}\right) - \omega^2\left(\boldsymbol{\psi}^T\mathbf{a}\right)^T\mathbf{M}\left(\boldsymbol{\psi}^T\mathbf{a}\right)\right]^T\right\} \\ &= \mathrm{Re}\left[\frac{e^{i2\omega t}}{2}\left(\mathbf{a}^T\boldsymbol{\psi}\mathbf{K}\boldsymbol{\psi}^T\mathbf{a} - \omega^2\mathbf{a}^T\boldsymbol{\psi}\mathbf{M}\boldsymbol{\psi}^T\mathbf{a}\right)\right]\end{aligned} \tag{1.72}$$

Imposing such an approximation to the displacement function is a restriction which is equivalent to adding artificial stiffness to the system, in which case $\Pi(\mathbf{a}) \geq \Pi(\boldsymbol{\varphi})$. Obviously, the closer $\boldsymbol{\psi}$ is to the exact solution $\boldsymbol{\varphi}$ the closer $\Pi(\mathbf{a})$ is to $\Pi(\boldsymbol{\varphi})$, but $\Pi(\mathbf{a})$ cannot become smaller than $\Pi(\boldsymbol{\varphi})$. Thus, a best fit solution will be obtained if $\Pi(\mathbf{a})$ is minimised, i.e. if

$$\frac{\partial \Pi(\mathbf{a})}{\partial a_i} = 0 \quad \Rightarrow \quad \frac{\partial}{\partial a_i}\left(\mathbf{a}^T\boldsymbol{\psi}\mathbf{K}\boldsymbol{\psi}^T\mathbf{a} - \omega^2\mathbf{a}^T\boldsymbol{\psi}\mathbf{M}\boldsymbol{\psi}^T\mathbf{a}\right) = 0 \tag{1.73}$$

This will then render N equation for the determination of the content of the unknown $\mathbf{a}$ vector. It is advantageous that the ψ_i functions are as close to orthogonal and representing the actual mode shapes as possible.

Example 1.9

Let us for simplicity apply the method to the mass spring system shown in Fig. 1.9, whose solution is previously obtained from equilibrium considerations in Eqs. 1.12 – 1.22. Let us further simplify the mathematics by setting $K_1 = K_2 = K$ and $M_1 = M_2 = M$. Referring to Eq. 1.14, the mass and stiffness matrices of the system are then given by

$$\mathbf{M} = M\begin{bmatrix} 1 & 0 \\ 0 & 1 \end{bmatrix} \quad \textit{and} \quad \mathbf{K} = K\begin{bmatrix} 2 & -1 \\ -1 & 1 \end{bmatrix}$$

Let us assume $\boldsymbol{\psi}_1 = \begin{bmatrix} 1 & 1 \end{bmatrix}^T$ *and* $\boldsymbol{\psi}_2 = \begin{bmatrix} 1 & -1 \end{bmatrix}^T$, *i.e. that* $\boldsymbol{\psi} = \begin{bmatrix} 1 & 1 \\ 1 & -1 \end{bmatrix}$

Then $\mathbf{a}^T \boldsymbol{\psi} \mathbf{K} \boldsymbol{\psi}^T \mathbf{a} = K\left(a_1^2 + 2a_1a_2 + 5a_2^2\right)$ *and* $\mathbf{a}^T \boldsymbol{\psi} \mathbf{M} \boldsymbol{\psi}^T = \omega^2 M\left(2a_1^2 + 2a_2^2\right)$, *and thus:*

$$\Rightarrow \begin{cases} \dfrac{\partial}{\partial a_1}\left(\mathbf{a}^T \boldsymbol{\psi} \mathbf{K} \boldsymbol{\psi}^T \mathbf{a} - \omega^2 \mathbf{a}^T \boldsymbol{\psi} \mathbf{M} \boldsymbol{\psi}^T \mathbf{a}\right) = K\left(2a_1 + 2a_2\right) - \omega^2 M\left(4a_1\right) = 0 \\ \dfrac{\partial}{\partial a_2}\left(\mathbf{a}^T \boldsymbol{\psi} \mathbf{K} \boldsymbol{\psi}^T \mathbf{a} - \omega^2 \mathbf{a}^T \boldsymbol{\psi} \mathbf{M} \boldsymbol{\psi}^T \mathbf{a}\right) = K\left(2a_1 + 10a_2\right) - \omega^2 M\left(4a_2\right) = 0 \end{cases}$$

which may be rewritten into the following matrix format

$$\left(\begin{bmatrix} 1 & 1 \\ 1 & 5 \end{bmatrix} - \omega^2 \frac{M}{K}\begin{bmatrix} 2 & 0 \\ 0 & 2 \end{bmatrix}\right)\begin{bmatrix} a_1 \\ a_2 \end{bmatrix} = \begin{bmatrix} \left(1 - 2\hat{\omega}^2\right) & 1 \\ 1 & \left(5 - 2\hat{\omega}^2\right) \end{bmatrix}\begin{bmatrix} a_1 \\ a_2 \end{bmatrix} = \begin{bmatrix} 0 \\ 0 \end{bmatrix}$$

where $\hat{\omega} = \omega / \sqrt{K/M}$. *This requirement can only be fulfilled if the determinant to the coefficient matrix is zero, i.e. if* $\left(1 - 2\hat{\omega}^2\right)\left(5 - 2\hat{\omega}^2\right) - 1 = 0$.

The solution is given by $\hat{\omega}^2 = \frac{3}{2} \pm \frac{\sqrt{5}}{2}$ *from which, in ascending order, the following eigenvalues are obtained*

$$\left.\begin{aligned} \omega_1 &= \sqrt{\frac{1}{2}\left(3 - \sqrt{5}\right)\frac{K}{M}} \approx 0.618\sqrt{K/M} \\ \omega_2 &= \sqrt{\frac{1}{2}\left(3 + \sqrt{5}\right)\frac{K}{M}} \approx 1.618\sqrt{K/M} \end{aligned}\right\}$$

The corresponding eigenmodes may be obtained by consecutively introducing either of those back into the eigenvalue equation above. I. e., if $\omega = \omega_1$ *then*

$$\left[1 - 2\left(\omega_1 / \sqrt{K/M}\right)^2\right] a_1 + a_2 = 0 \quad \Rightarrow \quad a_2 \approx -0.236 a_1$$

and thus $$\boldsymbol{\varphi}_1 \approx \boldsymbol{\psi}^T \mathbf{a} = \begin{bmatrix} 1 & 1 \\ 1 & -1 \end{bmatrix}\begin{bmatrix} a_1 \\ -0.236 a_1 \end{bmatrix} = a_1 \begin{bmatrix} 0.764 \\ 1.236 \end{bmatrix}$$

If $\omega = \omega_2$ *then* $$\left[1 - 2\left(\omega_2 / \sqrt{K/M}\right)^2\right] a_1 + a_2 = 0 \quad \Rightarrow \quad a_2 \approx 4.236 a_1$$

and thus $$\boldsymbol{\varphi}_2 \approx \boldsymbol{\psi}^T \mathbf{a} = \begin{bmatrix} 1 & 1 \\ 1 & -1 \end{bmatrix} \begin{bmatrix} a_1 \\ 4.236a_1 \end{bmatrix} = a_1 \begin{bmatrix} 5.236 \\ -3.236 \end{bmatrix}$$

The method may equally effectively be applied to continuous systems. Let us for instance consider an undamped and unloaded line-like beam in plane motion e.g. in the z direction. Then (see Eqs. 1.49 and 1.55)

$$\Pi = T_m + U_m = \frac{1}{2} \int_L m_z \left(\dot{r}_z\right)^2 dx + \frac{1}{2} \int_L EI_y \left(r_z''\right)^2 dx \tag{1.74}$$

It has previously been shown that in this case the motion is given by $r_z(x,t) = \mathrm{Re}\left[\phi_z(x) \cdot e^{i\omega t}\right]$. Introducing this into Eq. 1.74 the following is obtained

$$\Pi = \mathrm{Re}\left\{ \frac{e^{2i\omega t}}{2} \left[\int_L EI_y \left(\phi_z''\right)^2 dx - \omega^2 \int_L m_z \phi_z^2 dx \right] \right\} \tag{1.75}$$

$\delta\Pi = 0$ at any value of t and any variation of ϕ_z will then require

$$\int_L EI_y \left(\phi_z''\right)^2 dx - \omega^2 \int_L m_z \phi_z^2 dx = 0 \tag{1.76}$$

from which the eigenfrequency

$$\omega_z = \left[\int_L EI_y \left(\phi_z''\right)^2 dx \Big/ \int_L m_z \phi_z^2 dx \right]^{1/2} \tag{1.77}$$

is obtained. This is called the Rayleigh quotient (in its most simple format).

Example 1.10

From Eq. 1.77 an approximate value of the eigenfrequency may be obtained by assuming a mode shape $\psi_z(x)$ as close to the exact solution as possible.

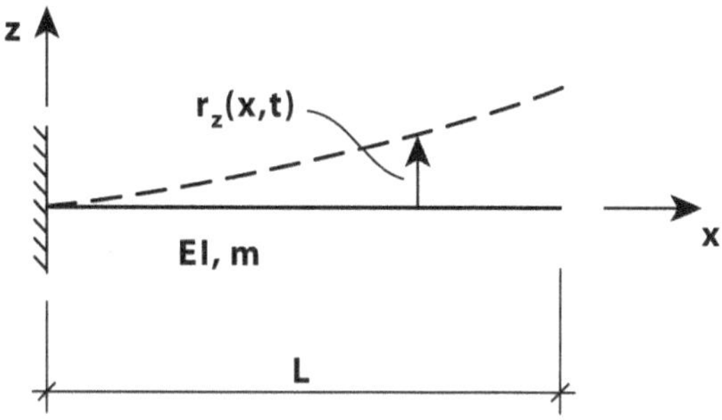

Fig. 1.19 Cantilevered beam

The case of a cantilevered beam with constant cross section properties m and EI_y is illustrated in Fig. 1.19.

Let us assume $\phi_z(x) \approx \psi_z(x)$ where $\psi_z(x) = 1 - \cos\left(\frac{\pi x}{2L}\right)$. Thus

$$\omega_z = \left[\frac{EI_y}{m} \cdot \frac{\int_0^L (\psi_z'')^2 dx}{\int_0^L \psi_z^2 dx}\right]^{\frac{1}{2}} = \left(\frac{\pi}{2}\right)^2 \left\{\frac{EI_y}{m_z L^4} \frac{\int_0^L \left[\cos\left(\frac{\pi x}{2L}\right)\right]^2 dx}{\int_0^L \left[1 - \cos\left(\frac{\pi x}{2L}\right)\right]^2 dx}\right\}^{\frac{1}{2}}$$

$$\Rightarrow \omega_z = \left(\frac{\pi}{2}\right)^2 \left(\frac{EI_y}{m_z L^4} \cdot \frac{\frac{\pi}{4}}{\frac{3\pi}{4} - 2}\right)^{\frac{1}{2}} \approx 3.67\sqrt{\frac{EI_y}{m_z L^4}}$$

A more exact solution is $\omega_z = 3.52\sqrt{EI_y / \left(m_z L^4\right)}$.

1.5 The Principle of Hamilton and Euler-Lagrange

While Rayleigh-Ritz method is based on variation on the total energy $\Pi = T + U + P$ in a system at a particular time and position, the method of Hamilton and Euler-Lagrange is based on variation of the Lagrange function of instantaneous energy balance (i.e. the difference between that which comes in and that which goes out) as you travel with the system in time:

$$L(\mathbf{r},\dot{\mathbf{r}})=T(\dot{\mathbf{r}})-\left[U(\mathbf{r})+P(t,\mathbf{r})\right] \tag{1.78}$$

Assuming for simplicity that the displacement is a continuous function $r(x,t)$ rather than a discrete vector, it is seen that the integral of $L(r,\dot{r})$ over an arbitrary time interval from t_1 to t_2 :

$$f(r,\dot{r})=\int_{t_1}^{t_2} L(r,\dot{r})dt \tag{1.79}$$

is only dependent on the path along which it has changed during the motion from t_1 to t_2. The basic principle of Hamilton [40] is then based on the assumption that the change of energy balance from t_1 to t_2 will occur along a path (a functional ridge) where the variation of $f(r,\dot{r})$ with respect to r and $\dot{r}$ is zero, rendering

$$\delta f(r,\dot{r})=\delta\int_{t_1}^{t_2} L(r,\dot{r})dt=\int_{t_1}^{t_2}\left[\frac{\partial L(r,\dot{r})}{\partial r}+\frac{\partial L(r,\dot{r})}{\partial \dot{r}}\right]dt=0 \tag{1.80}$$

Let $\varepsilon(t)$ be a small arbitrary perturbation, independent of the path itself, but such that $\varepsilon=0$ at t_1 and t_2. Then a first order expansion

$$\delta\int_{t_1}^{t_2} L(r+\varepsilon,\dot{r}+\dot{\varepsilon})dt=\int_{t_1}^{t_2}\left[\varepsilon\frac{\partial L(r,\dot{r})}{\partial r}+\dot{\varepsilon}\frac{\partial L(r,\dot{r})}{\partial \dot{r}}\right]dt \tag{1.81}$$

and integrating the second term by parts, then

$$\begin{aligned}\delta\int_{t_1}^{t_2} L(r+\varepsilon,\dot{r}+\dot{\varepsilon})dt&=\left[\varepsilon\frac{\partial L(r,\dot{r})}{\partial \dot{r}}\right]_{t_1}^{t_2}+\int_{t_1}^{t_2}\left\{\varepsilon\frac{\partial L(r,\dot{r})}{\partial r}-\varepsilon\frac{d}{dt}\left[\frac{\partial L(r,\dot{r})}{\partial \dot{r}}\right]\right\}dt\\&=\int_{t_1}^{t_2}\varepsilon\left\{\frac{\partial L(r,\dot{r})}{\partial r}-\frac{d}{dt}\left[\frac{\partial L(r,\dot{r})}{\partial \dot{r}}\right]\right\}dt\end{aligned} \tag{1.82}$$

For all $\varepsilon\neq 0$ this can only become zero if

$$\frac{\partial L(r,\dot{r})}{\partial r}-\frac{d}{dt}\left[\frac{\partial L(r,\dot{r})}{\partial \dot{r}}\right]=0 \tag{1.83}$$

This is known as the Euler equation.

Example 1.11

Let us for instance consider an undamped single degree of freedom system with stiffness K and mass M, subject to a force F (see Fig. 1.3). The Lagrange function is then given by:

$$L = \frac{1}{2}M\dot{r}^2 - \left(\frac{1}{2}Kr^2 - Fr\right)$$

Introducing this into Euler's equation will then render

$$-(Kr - F) - \frac{d}{dt}(M\dot{r}) = F - Kr - M\ddot{r} = 0 \qquad \text{and thus} \qquad M\ddot{r} + Kr = F$$

which is identical to the equilibrium condition that was developed in Eq. 1.4, and which was also derived from variation of Π (with respect to t) in Example 1.8.

Euler's equation (Eq. 1.83) may be further expanded if an approximate series solution is adopted

$$r(x,t) = \sum_{i=1}^{N} r_i(x,t) \tag{1.84}$$

where the variables of $r_i(x,t)$ are split into a set of shape functions $\psi_i(x)$ and corresponding time domain functions $\eta_i(t)$ such that

$$r_i(x,t) = \psi_i(x) \cdot \eta_i(t) \tag{1.85}$$

It is taken for granted that $r_i(x,t)$ are independent functions and then Hamilton's principle can only be fulfilled if Euler's equation is satisfied independently for all $r_i(x,t)$. Thus

$$\frac{\partial L}{\partial r_i} - \frac{d}{dt}\left(\frac{\partial L}{\partial \dot{r}_i}\right) = 0 \tag{1.86}$$

Furthermore, it is observed that

$$\partial r_i = \frac{dr_i}{d\eta_i}\partial\eta_i = \psi_i \partial\eta_i \qquad \text{and} \qquad \partial\dot{r}_i = \frac{d\dot{r}_i}{d\dot{\eta}_i}\partial\dot{\eta}_i = \psi_i \partial\dot{\eta}_i \tag{1.87}$$

Thus, Eq. 1.86 may be replaced by

$$\frac{\partial L}{\partial \eta_i} - \frac{d}{dt}\left(\frac{\partial L}{\partial \dot{\eta}_i}\right) = 0 \tag{1.88}$$

Introducing Eq. 1.58 into Eq. 1.88 (acknowledging that $\partial U/\partial \dot{\eta}$ and $\partial P/\partial \dot{\eta}$ are both zero)) will then render

$$\frac{d}{dt}\left(\frac{\partial T}{\partial \dot{\eta}_i}\right) - \frac{\partial T}{\partial \eta_i} + \frac{\partial U}{\partial \eta_i} + \frac{\partial P}{\partial \eta_i} = 0 \tag{1.89}$$

Obviously, there will altogether by N such equations. They are called Lagrange's equations. It is an obvious requirement to the $\psi_i(x)$ functions that they satisfy the physical boundary conditions of the system, but it is not a requirement that they are orthogonal.

Example 1.12

Let us for instance consider the undamped and unloaded cantilevered beam shown in Fig. 1.19 (Example 1.10). Its mass m and bending stiffness EI are assumed constants along its entire span L. The following series solution is adopted

$$r_z(x,t) = \psi_1(x)\cdot\eta_1(t) + \psi_2(x)\cdot\eta_2(t) \qquad \text{where} \qquad \left.\begin{aligned}\psi_1 &= x^2\\ \psi_2 &= x^3\end{aligned}\right\}$$

The two functions $\psi_1 = x^2$ and $\psi_2 = x^3$ will both satisfy the physical boundary conditions $\psi(x=0) = \psi'(x=0) = 0$. Then

$$\left.\begin{aligned} T &= \int_0^L \frac{1}{2} m_z \dot{r}_z^2 dx = \frac{m_z}{2}\int_0^L \left(x^2\dot{\eta}_1 + x^3\dot{\eta}_2\right)^2 dx = \frac{m_z L^5}{2}\left(\frac{\dot{\eta}_1^2}{5} + \frac{L\dot{\eta}_1\eta_2}{3} + \frac{L^2\dot{\eta}_2^2}{7}\right)\\ U &= \int_0^L \frac{1}{2} EI_y \left(r_z''\right)^2 dx = \frac{EI_y}{2}\int_0^L \left(2\eta_1 + 6x\eta_2\right)^2 dx = 2EI_y L\left(\eta_1^2 + 3L\eta_1\eta_2 + 3\eta_2^2\right)\end{aligned}\right\}$$

while $P = 0$. Introducing this into Eq. 1.89, with i consecutively equal to 1 and 2, then the following is obtained:

$$\left.\begin{aligned}&\frac{m_z L^5}{2}\left(\frac{2}{5}\ddot{\eta}_1+\frac{L}{3}\ddot{\eta}_2\right)+2EI_yL(2\eta_1+3L\eta_2)=0\\&\frac{m_z L^5}{2}\left(\frac{L}{3}\ddot{\eta}_1+\frac{2L^2}{7}\ddot{\eta}_2\right)+2EI_yL\left(3L\eta_1+6L^2\eta_2\right)=0\end{aligned}\right\}$$

which may more conveniently be written

$$2EI_yL\begin{bmatrix}2 & 3L\\3L & 6L^2\end{bmatrix}\begin{bmatrix}\eta_1\\\eta_2\end{bmatrix}+\frac{m_zL^5}{2}\begin{bmatrix}2/5 & L/3\\L/3 & 2L^2/7\end{bmatrix}\begin{bmatrix}\ddot{\eta}_1\\\ddot{\eta}_2\end{bmatrix}=\begin{bmatrix}0\\0\end{bmatrix}$$

Assuming a simple harmonic motion $\begin{bmatrix}\eta_1\\\eta_2\end{bmatrix}=\mathrm{Re}\{\begin{bmatrix}a_1\\a_2\end{bmatrix}e^{i\omega t}\}$ *then the following is obtained*

$$\left(2EI_yL\begin{bmatrix}2 & 3L\\3L & 6L^2\end{bmatrix}-\omega^2\frac{m_zL^5}{2}\begin{bmatrix}2/5 & L/3\\L/3 & 2L^2/7\end{bmatrix}\right)\begin{bmatrix}a_1\\a_2\end{bmatrix}=\mathbf{0}$$

This is a classical eigenvalue problem similar to that which has previously been seen for two degrees of freedom systems. A non-trivial solution can only be obtained if $\mathbf{a}=\begin{bmatrix}a_1 & a_2\end{bmatrix}^T\neq\mathbf{0}$, *and thus, we must have that*

$$\det\left(2EI_yL\begin{bmatrix}2 & 3L\\3L & 6L^2\end{bmatrix}-\omega^2\frac{m_zL^5}{2}\begin{bmatrix}2/5 & L/3\\L/3 & 2L^2/7\end{bmatrix}\right)=0$$

rendering the following quadratic equation

$$\left(\omega^2\frac{m_zL^4}{EI_y}\right)^2-1224\left(\omega^2\frac{m_zL^4}{EI_y}\right)+15120=0$$

whose solution is given by $\Rightarrow\quad\left(\omega^2\frac{m_zL^4}{EI_y}\right)=612\pm\sqrt{359424}\approx\begin{cases}1211.5\\12.5\end{cases}$

Thus, its solution is in ascending order rendering the following eigenfrequencies of the system

$$\omega_{z1}=3.54\sqrt{\frac{EI_y}{m_zL^4}}\qquad and\qquad \omega_{z2}=34.81\sqrt{\frac{EI_y}{m_zL^4}}$$

More exact values are $\omega_{z_1} = 3.52\sqrt{EI/mL^4}$ *and* $\omega_{z_2} = 22.03\sqrt{EI/mL^4}$. *Obviously, the accuracy of the solution is entirely dependent of how well the approximation* $r(x,t) = \psi_1(x)\cdot\eta_1(t) + \psi_2(x)\cdot\eta_2(t)$ *is able to describe the exact shape of the dynamic displacement of the system. Thus, it may be concluded that the choice of* $\psi_1 = x^2$ *and* $\psi_2 = x^3$ *are well suited to describe the first mode shape, but that they are incapable of describing anything in the vicinity of the second mode shape.*

1.6 The Principle of Virtual Work

The principle of virtual work is usually attributed to d'Alambert, but it was first presented in a variation format by Lagrange [8]. It contains d'Alambert's perception of instantaneous inertia. The basic assumption is that for an observer at rest outside of a system that is in a physical condition of equilibrium the total energy level will not change for any incremental change $\delta\mathbf{r}$ of the position of the system. The only restriction imposed on $\delta\mathbf{r}$ is that it is time invariant. Otherwise, $\delta\mathbf{r}$ is arbitrary or virtual. Hence, it has been labelled the principle of virtual work. The upper left hand side illustration on Fig. 1.20 shows the free-body-diagram of a line-like continuous type of system. For simplicity, its displacement is in the z direction alone and with corresponding bending about the y axis. At arbitrary time t the system has an instantaneous position $r_z(x,t)$ and corresponding support reaction forces R_n $, n = 1, 2, \ldots, N_R$. It is subject to external loads $F_z(t)$ and $q_z(x,t)$. The lower left hand side illustration shows the same system, but now with an additional time invariant and arbitrary (virtual) displacement $\delta r_z(x)$.

In our case we shall assume that $\delta r_z(x)$ complies with the geometric boundary conditions of the system, i.e. that δr_z, $\delta r_z'$, $\delta r_z''$ or $\delta r_z'''$ are zero at support positions wherever this may be required of the boundaries of the system itself. [This is not a general requirement to the application of the method, but in our case it is merely a convenient choice as it implies that no energy changes will take place by the support forces R_n during the virtual motion $\delta r_z(x)$.] While external forces have performed work and thereby had their energy level reduced, the inertia forces have been lifted to a higher position in the force field of the system and thereby had their energy level increased. Thus, during the motion from $r_z(x,t)$ to $r_z(x,t) + \delta r_z(x)$ the energy level of external force $F_z(t)$ is changed by

$$\delta P_F = -\int_{r_z(x_F,t)}^{r_z(x_F,t)+\delta r_z(x_F)} F_z(t)\,dr = -\delta r_z(x_F)\cdot F_z(t) \tag{1.90}$$

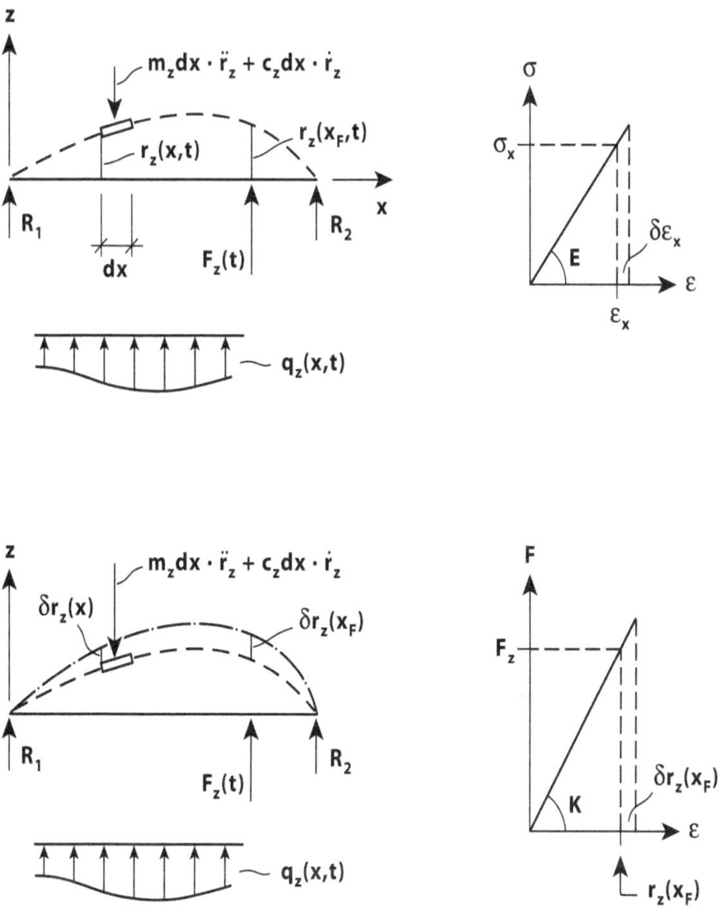

Fig. 1.20 The principle of virtual work applied to a line-like continuous system

Similarly, the change of energy level for the external force $q_z(x,t)$ is given by

$$\delta P_q = -\int_L \left\{ \int_{r_z(x,t)}^{r_z(x,t)+\delta r_z(x)} \left[q_z(x,t)\,dx \right] dr \right\} = -\int_L \delta r_z(x) \cdot q_z(x,t)\,dx \tag{1.91}$$

The sum of energy changes to the distributed mass inertia force $m_z(x)dx \cdot \ddot{r}_z(x,t)$ and possible concentrated mass contribution $M \cdot \ddot{r}_z(x_M,t)$ are given by

$$\delta P_m = \int_L \left\{ \int_{r_z(x,t)}^{r_z(x,t)+\delta r_z(x)} \left[m_z(x)\ddot{r}_z(x,t)dx \right] dr \right\} = \int_L \delta r_z(x) m_z(x) \ddot{r}_z(x,t) dx \quad (1.92)$$

and

$$\delta P_M = \int_{r_z(x_M,t)}^{r_z(x_M,t)+\delta r_z(x_M)} \left[M \cdot \ddot{r}_z(x_M,t) \right] dr = \delta r_z(x_M) \cdot \left[M \cdot \ddot{r}_z(x_M,t) \right] \quad (1.93)$$

In addition to this we shall now include a resistance force commonly attributed to internal damping in the system. The origin and effects of structural damping is discussed in Chapter 9. Here, we shall conveniently assume a viscous type of damping which will generate an internal cross sectional force component $c_z(x)dx \cdot \dot{r}_z(x,t)$ (see Fig. 1.21) whose energy change is given by

$$\delta P_c = \int_L \left\{ \int_{r_z(x,t)}^{r_z(x,t)+\delta r_z(x)} \left[c_z(x)dx \cdot \dot{r}_z(x,t) \right] dr \right\} = \int_L \delta r_z(x) \cdot c_z(x) \cdot \dot{r}_z(x,t) dx \quad (1.94)$$

and where c_z is a cross sectional viscous damping coefficient associated with motion in the vertical direction.

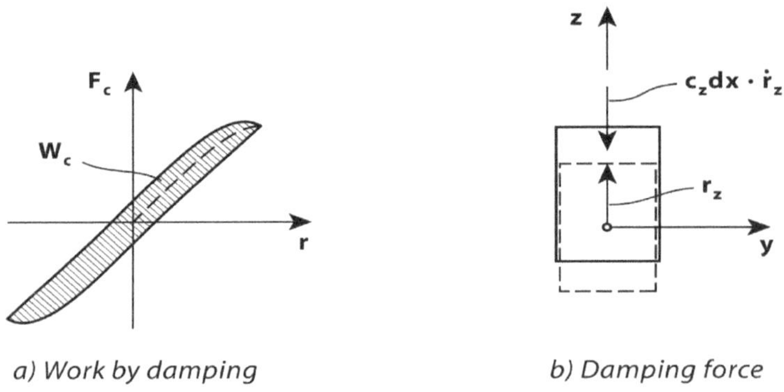

Fig. 1.21 Resistant viscous damping force due to vertical motion

During the motion from $r_z(x,t)$ to $r_z(x,t)+\delta r_z(x)$ the strain in the system has changed from $\varepsilon_x(x,t)$ to $\varepsilon_x(x,t)+\delta\varepsilon_x(x)$ as shown on the upper right hand side illustration in Fig. 1.20. The corresponding change of strain energy is given by

$$\delta U = \int_L \int_A \left[\int_{\varepsilon_x(x,t)}^{\varepsilon_x(x,t)+\delta\varepsilon_x(x)} \left\{ \left[\sigma_x(x,t) dA \right] \left(d\varepsilon_x dx \right) \right\} \right] = \int_L \int_A \delta\varepsilon_x(x) \sigma_x(x,t) dA dx \quad (1.95)$$

Since the virtual motion δr_z is time invariant then the change of kinetic energy in the motion from $r_z(x,t)$ to $r_z(x,t)+\delta r_z(x)$ is zero, and thus, the total change of energy is given by

$$\delta\Pi = \delta P_F + \delta P_q + \delta P_m + \delta P_M + \delta P_c + \delta U \quad (1.96)$$

The basic idea is that during the motion from $r_z(x,t)$ to $r_z(x,t)+\delta r_z(x)$ the total energy in the system cannot have changed, i.e. that

$$\begin{aligned} \delta\Pi = & -\delta r_z(x_F) F_z(t) - \int_L \delta r_z(x) q_z(x,t) dx + \int_L \delta r_z(x) m_z(x) \ddot{r}_z(x,t) dx \\ & + \delta r_z(x_M) M \ddot{r}_z(x_M,t) + \int_L \delta r_z(x) c_z(x) \dot{r}_z(x,t) dx \\ & + \int_L \int_A \delta\varepsilon_x(x) \sigma_x(x,t) dA dx = 0 \end{aligned} \quad (1.97)$$

Since we have so far restricted ourselves to in-plane motion and corresponding normal stress and strain components (see Eq. 1.27)

$$\sigma_x = E \cdot \varepsilon_x = E \cdot \left(-r_z'' \cdot z_c \right) \qquad \text{and} \qquad \delta\varepsilon_x = -\delta r_z'' \cdot z_c \quad (1.98)$$

then

$$\begin{aligned} \delta U & = \int_L \int_A \delta\varepsilon_x(x) \sigma_x(x,t) dA dx = \int_L \int_A \left[-\delta r_z''(x) z_c \right] \cdot \left[-E r_z''(x,t) z_c \right] dA dx \\ & = \int_L E \left(\int_A z_c^2 dA \right) \delta r_z''(x) r_z''(x,t) dx = \int_L EI_y \delta r_z''(x) r_z''(x,t) dx \end{aligned} \quad (1.99)$$

and thus, $\delta\Pi = 0$ will require

$$\begin{aligned} & \delta r_z(x_F) \cdot F_z(t) + \int_L \delta r_z(x) \cdot q_z(x,t) dx - \delta r_z(x_M) \cdot \left[M \cdot \ddot{r}_z(x_M,t) \right] \\ & - \int_L \delta r_z(x) \cdot \left[m(x) \cdot \ddot{r}_z(x,t) \right] dx - \int_L \delta r_z(x) \cdot \left[c_z(x) \cdot \dot{r}_z(x,t) \right] dx \\ & = \int_L EI_y \cdot \delta r_z''(x) \cdot r_z''(x,t) dx \end{aligned} \quad (1.100)$$

Example 1.13

Let us for instance consider the case of an unloaded and undamped case of an arbitrary continuous system where the response is harmonic, i.e. that

$$\left.\begin{array}{l} F_z(t)=0 \\ q_z(x,t)=0 \end{array}\right\} \qquad c_z=0 \qquad \text{and} \qquad r_z(x,t)=\operatorname{Re}\left[\phi_z(x)\cdot e^{i\omega t}\right]$$

where $\phi_z(x)$ *is the mode shape of the system. Let us choose the virtual displacement*

$$\delta r_z(x)=\phi_z(x)\cdot\delta a$$

where δa *is an incremental amplitude variation. Introducing this into Eq. 1.100 will then render*

$$-\int_L\left[\phi_z(x)\delta a\right]\cdot\left[m_z(x)\cdot(i\omega)^2\phi_z(x)e^{i\omega t}\right]dx=\int_L EI_y\cdot\left[\phi_z''(x)\delta a\right]\cdot\left[\phi_z''(x)e^{i\omega t}\right]dx$$

$$\text{I.e.:}\quad\Rightarrow\qquad \omega^2\int_L\left[\phi_z(x)\right]^2\cdot m_z(x)dx=\int_L EI_y\cdot\left[\phi_z''(x)\right]^2dx$$

Thus, the eigenfrequency of the system is given by

$$\omega_z=\left\{\int_L EI_y\cdot\left[\phi_z''(x)\right]^2dx\Big/\int_L\left[\phi_z(x)\right]^2\cdot m_z(x)dx\right\}^{\frac{1}{2}}$$

which is the Rayleigh quotient, identical to that which was obtained by the use of Rayleigh-Ritz method, see Eq. 1.77.

While we above have been restricted to consider a system with in-plane loads and motion in the z direction (with corresponding normal stress and strain components, σ_x and ε_x), a more general case will comprise displacements

$$\mathbf{r}(x,t)=\begin{bmatrix} r_x & r_y & r_z & r_\theta \end{bmatrix}^T \tag{1.101}$$

and corresponding stresses

$$\boldsymbol{\sigma}_{tot}(x,y,z,t)=\bar{\boldsymbol{\sigma}}(x,y,z)+\boldsymbol{\sigma}(x,y,z,t) \tag{1.102}$$

where $\bar{\boldsymbol{\sigma}}(x,y,z)=\begin{bmatrix}\bar{\sigma}_x & \bar{\tau}_\theta\end{bmatrix}^T$ is the mean (static) part of the total stress vector on an arbitrary cross section along the span, and where $\boldsymbol{\sigma}(x,y,z,t)=\begin{bmatrix}\sigma_x & \tau_\theta\end{bmatrix}^T$ is the corresponding fluctuating (dynamic) part. (The reason why $\bar{\boldsymbol{\sigma}}$ need to be included is that the mean time invariant cross sectional forces will affect the stiffness properties of the system, as will become more evident later. If desirable, initial structural displacements may also be included.) Furthermore, the system may be subject to arbitrary external concentrated and distributed forces

$$\mathbf{F}_i(x_F,t)=\begin{bmatrix}F_y & F_z & F_\theta\end{bmatrix}_i^T \quad \text{and} \quad \mathbf{q}_i(x,t)=\begin{bmatrix}q_y & q_z & q_\theta\end{bmatrix}_i^T \tag{1.103}$$

and it may contain concentrated masses $M_{g_j}(x_M)$. Similarly, the virtual displacement and corresponding virtual strain may contain components

$$\delta\mathbf{r}(x)=\begin{bmatrix}\delta r_x & \delta r_y & \delta r_z & \delta r_\theta\end{bmatrix}^T \quad \text{and} \quad \delta\boldsymbol{\varepsilon}(x,y,z,t)=\begin{bmatrix}\delta\varepsilon_x & \delta\gamma\end{bmatrix}^T \tag{1.104}$$

Thus, in a more general format Eq. 1.100 may be written

$$\begin{gathered}\sum_{i=1}^{N_F}\left\{\left[\delta\mathbf{r}(x_F)\right]^T\mathbf{F}(t)\right\}_i+\sum_{i=1}^{N_q}\left\{\int_L\left[\delta\mathbf{r}(x)\right]^T\mathbf{q}(x,t)\,dx\right\}_i-\sum_{j=1}^{N_M}\left\{\left[\delta\mathbf{r}(x_M)\right]^T\mathbf{M}_g\ddot{\mathbf{r}}(x_M)\right\}_j \\ -\int_L\left[\delta\mathbf{r}(x)\right]^T\mathbf{m}_g\ddot{\mathbf{r}}(x,t)\,dx-\int_L\left[\delta\mathbf{r}(x)\right]^T\mathbf{c}_0\dot{\mathbf{r}}(x)\,dx=\int_L\int_A\delta\boldsymbol{\varepsilon}^T\boldsymbol{\sigma}_{tot}\,dA\,dx\end{gathered} \tag{1.105}$$

where $\mathbf{M}_{g_j}=diag\left\{M_{g_y} \quad M_{g_z} \quad M_{g_\theta}\right\}_j$ is a concentrated mass matrix at an arbitrary position x_M and identified by index j, and where

$$\mathbf{m}_g=\begin{bmatrix}m_x & 0 & 0 & 0\\ 0 & m_y & 0 & -m_y e_z\\ 0 & 0 & m_z & m_z e_y\\ 0 & -m_y e_z & m_z e_y & m_\theta\end{bmatrix} \tag{1.106}$$

is the cross sectional mass matrix (see Figs. 1.1 and 1.2). The damping matrix $\mathbf{c}_0$ is defined by

$$\mathbf{c}_0=diag\begin{bmatrix}c_x & c_y & c_z & c_\theta\end{bmatrix} \tag{1.107}$$

where c_x, c_y, c_z and c_θ are cross sectional viscous damping coefficients associated with motion velocities in x, y, z, θ directions. As can be seen, the right hand side of Eq. 1.105 contains the total change of strain energy during the virtual displacements $\delta\mathbf{r}(x)$:

$$\int_L\int_A \delta\boldsymbol{\varepsilon}^T \boldsymbol{\sigma}_{tot} dAdx = \int_L\int_A \delta\boldsymbol{\varepsilon}^T \left(\bar{\boldsymbol{\sigma}} + \boldsymbol{\sigma}\right) dAdx = \int_L\int_A \delta\boldsymbol{\varepsilon}^T \bar{\boldsymbol{\sigma}} dAdx + \int_L\int_A \delta\boldsymbol{\varepsilon}^T \boldsymbol{\sigma} dAdx \tag{1.108}$$

Let us first focus on the second term, i.e. on the dynamic stress contributions to the change of strain energy. We have previously adopted elastic material behaviour and Navier's hypothesis. In addition to bi-axial bending, we shall here also include strain due to axial elongation (see Fig. 1.22.a) and general shear strain due to torsion (see Fig. 1.22.b), but for simplicity, shear strain due to shear forces is assumed negligible.

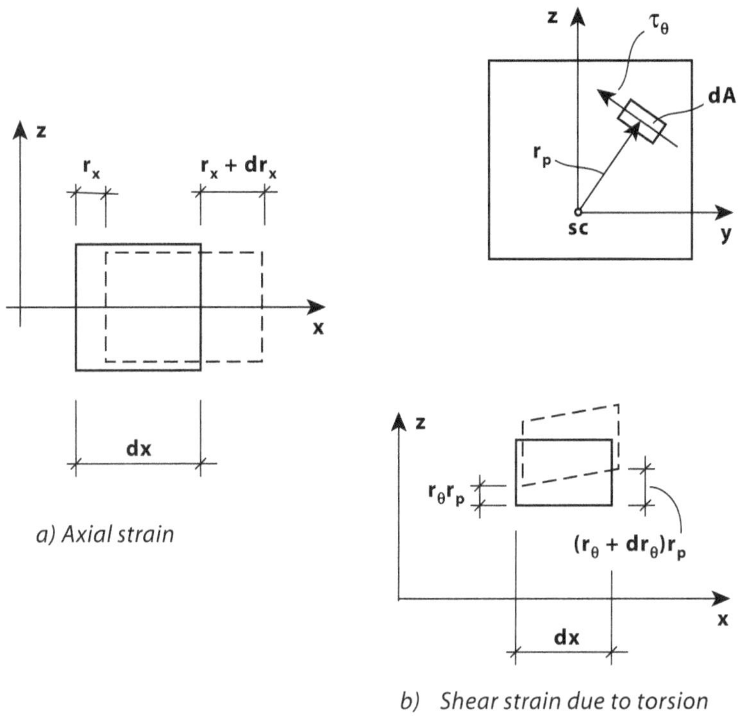

Fig. 1.22 Strain due to axial elongation and shear strain due to torsion

Thus

$$\left.\begin{aligned} \sigma_x &= E \cdot \varepsilon_x \\ \varepsilon_x &= r_x' + r_y'' \cdot y_c - r_z'' \cdot z_c \end{aligned}\right\} \quad \text{and} \quad \left.\begin{aligned} \tau_{yz} &= \tau_\theta = G \cdot \gamma \\ \gamma &= r_p \cdot r_\theta' \end{aligned}\right\} \tag{1.109}$$

where $y_c = y - e_y$ and $z_c = z - e_z$ (see Figs. 1.1 and 1.2). Then

$$\int_L \int_A \delta \boldsymbol{\varepsilon}^T \boldsymbol{\sigma} dA dx = \int_L \int_A \begin{bmatrix} \delta \varepsilon_x \\ \delta \gamma \end{bmatrix}^T \begin{bmatrix} \sigma_x \\ \tau_\theta \end{bmatrix} dA dx$$
$$= \int_L \int_A \begin{bmatrix} \delta\left(r_x' + r_y'' y_c - r_z'' z_c\right) \\ \delta\left(r_p r_\theta'\right) \end{bmatrix}^T \begin{bmatrix} E \cdot \left(r_x' + r_y'' y_c - r_z'' z_c\right) \\ G \cdot \left(r_p r_\theta'\right) \end{bmatrix} dA dx \tag{1.110}$$

Performing the vector multiplication and cross sectional integration we obtain

$$\int_L \int_A \delta \boldsymbol{\varepsilon}^T \cdot \boldsymbol{\sigma} dA dx =$$
$$\int_L \left[\delta r_x' E \left(r_x' \int_A dA + r_y'' \int_A y_c dA - r_z'' \int_A z_c dA \right) + \delta r_y'' E \left(r_x' \int_A y_c dA + r_y'' \int_A y_c^2 dA - r_z'' \int_A y_c z_c dA \right) \right.$$
$$\left. - \delta r_z'' E \left(r_x' \int_A z_c dA + r_y'' \int_A y_c z_c dA - r_z'' \int_A z_c^2 dA \right) + \delta r_\theta' G r_\theta' \int_A r_p^2 dA \right] dx$$
$$\tag{1.111}$$

The cross sectional neutral axis centre with respect to elastic bending is defined by

$$\int_A \begin{bmatrix} y_c \\ z_c \\ y_c z_c \end{bmatrix} dA = \mathbf{0} \quad \text{and} \quad \int_A \begin{bmatrix} 1 \\ y_c^2 \\ z_c^2 \\ r_p^2 \end{bmatrix} dA = \begin{bmatrix} A \\ I_z \\ I_y \\ I_t \end{bmatrix} \tag{1.112}$$

and thus

$$\int_L \int_A \delta \boldsymbol{\varepsilon}^T \boldsymbol{\sigma} dA dx = \int_L \left[\delta r_x' E A r_x' + \delta r_y'' E I_z r_y'' + \delta r_z'' E I_y r_z'' + \delta r_\theta' G I_t r_\theta' \right] dx \tag{1.113}$$

Let us then turn to the first part of the right hand side of Eq. 1.108, i.e. to the contribution of the mean (static) part of the total stress vector to the change of strain energy during the virtual displacement $\delta \mathbf{r}(x) = \begin{bmatrix} \delta r_x & \delta r_y & \delta r_z & \delta r_\theta \end{bmatrix}^T$.

It is in the following focused on the contributions from cross sectional stress resultants axial load $\bar{N}$ and bending moments $\bar{M}_y$ and $\bar{M}_z$. (I.e., any shear force and torsion moment contributions are disregarded.) The effects of $\bar{N}$ is illustrated in Figs. 1.23.a and b, and in Fig. 1.24. As shown in Fig. 1.24 an

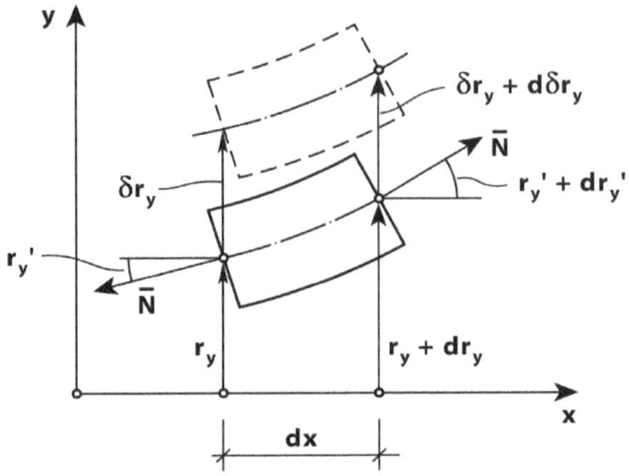

a) $\bar{N}$ *contribution in a* δr_y *motion*

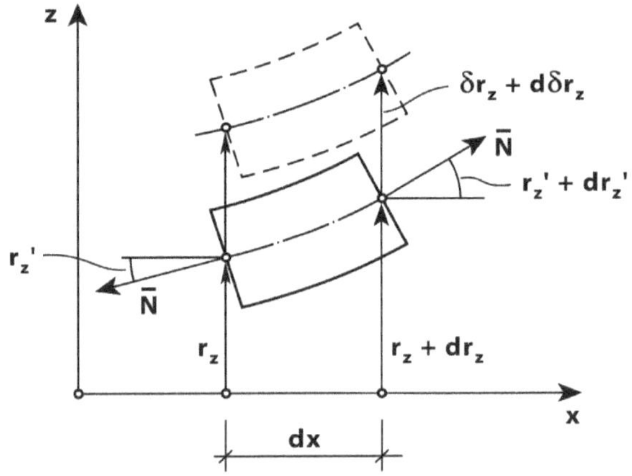

b) $\bar{N}$ *contribution in a* δr_z *motion*

Fig. 1.23 The effects of a time invariant axial force $\bar{N}$

additional shear stress contribution $\bar{\tau}_\theta$ will occur due to the normal stress component $\bar{N}/A$. It may readily be obtained from simple moment equilibrium of an infinitesimal element $dA \cdot dx$, i.e. from the condition that $(\bar{\tau}_\theta \cdot dA) \cdot dx = \left[(\bar{N}/A)dA\right] \cdot (r_{sc} dr_\theta)$ where r_{sc} $(= r_p)$ is the radial distance between the infinitesimal element and the shear centre.

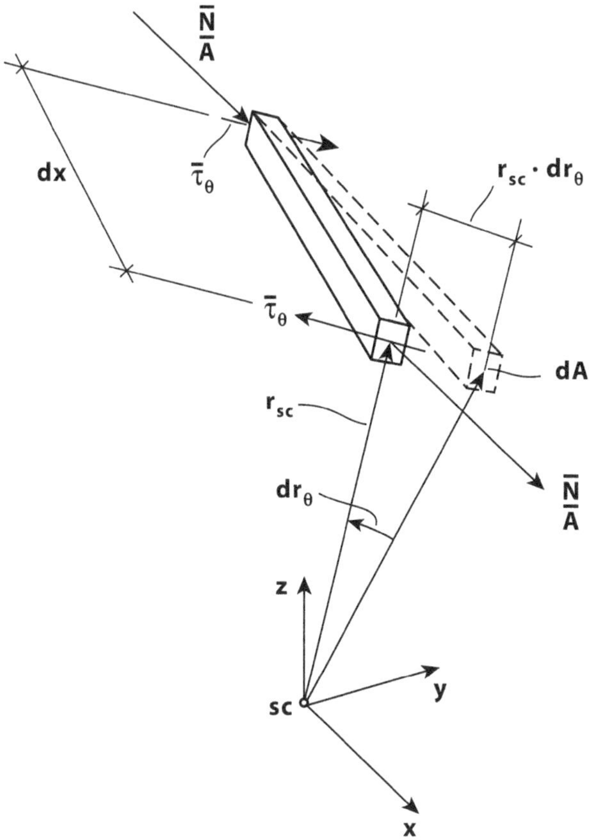

Fig. 1.24 The effect of $\overline{N}$ on torsion shear stresses

Omitting all higher order terms, it is seen that the total change of energy contribution from $\overline{N}$ is then given by

$$\begin{gathered}\int_L \left[-\delta r_y \overline{N} r_y' + \left(\delta r_y + d\delta r_y\right)\overline{N}\left(r_y' + dr_y'\right)\right] + \int_L \left[-\delta r_z \overline{N} r_z' + \left(\delta r_z + d\delta r_z\right)\overline{N}\left(r_z' + dr_z'\right)\right] \\ + \int_L \int_A \left(r_{sc}\delta r_\theta' dx\right)\left(\frac{\overline{N}}{A} r_{sc} dA r_\theta'\right) \approx \int_L \left(\delta r_y' \overline{N} r_y' + \delta r_z' \overline{N} r_z' + \delta r_\theta' \overline{N} e_0^2 r_\theta'\right) dx\end{gathered} \tag{1.114}$$

where $e_0^2 = \dfrac{1}{A}\displaystyle\int_A r_{sc}^2 dA = \dfrac{1}{A}\int_A \left[\left(y_c + e_y\right)^2 + \left(z_c + e_z\right)^2\right] dA = e_p^2 + e_y^2 + e_z^2$ (1.115)

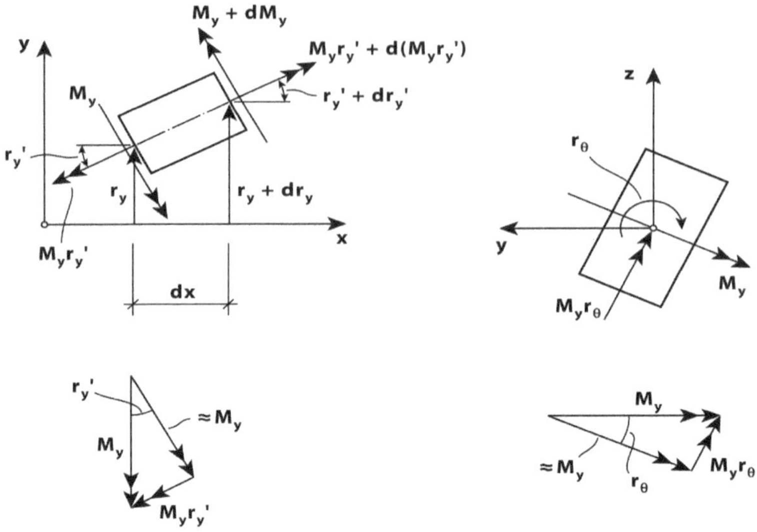

a) Bending moment M_y

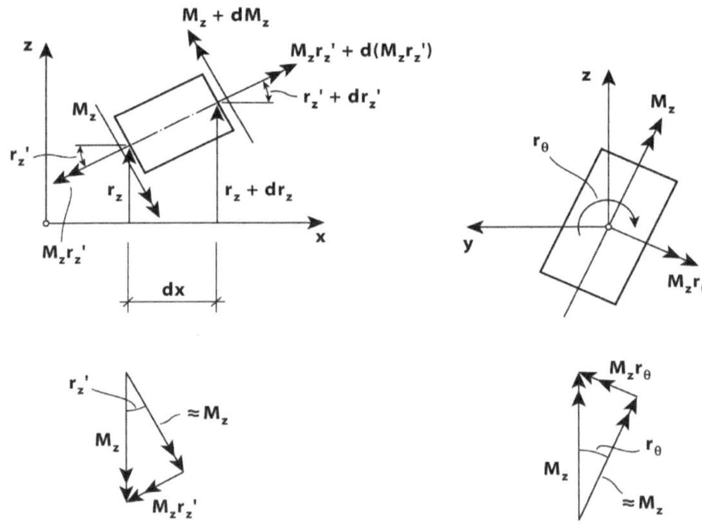

b) Bending moment M_z

Fig. 1.25 The effects of time invariant bending moments $\bar{M}_y$ and $\bar{M}_z$

and $$e_p^2 = \frac{1}{A}\int_A \left(y_c^2 + z_c^2 \right) dA = \frac{1}{A}\left(I_y + I_z \right) \tag{1.116}$$

The effects of biaxial bending moments $\bar{M}_y$ and $\bar{M}_z$ are illustrated in Figs. 1.25.a and b. Omitting all higher order terms, it is seen that the total change of energy contribution from $\bar{M}_y$ and $\bar{M}_z$ are given by

$$\begin{aligned}
&\int_L \Big[\delta r_y' \bar{M}_y r_\theta - \left(\delta r_y' + d\delta r_y' \right)\left(\bar{M}_y + d\bar{M}_y \right)\left(r_\theta + dr_\theta \right) \\
&\quad + \delta r_z' \bar{M}_z r_\theta - \left(\delta r_z' + d\delta r_z' \right)\left(\bar{M}_z + d\bar{M}_z \right)\left(r_\theta + dr_\theta \right) \\
&\quad + \delta r_\theta \bar{M}_y r_y' - \left(\delta r_\theta + d\delta r_\theta \right)\left(\bar{M}_y + d\bar{M}_y \right)\left(r_y' + dr_y' \right) \\
&\quad + \delta r_\theta \bar{M}_z r_z' - \left(\delta r_\theta + d\delta r_\theta \right)\left(\bar{M}_z + d\bar{M}_z \right)\left(r_z' + dr_z' \right) \Big] \\
&\approx - \int_L \left(\delta r_y' \bar{M}_y r_\theta' + \delta r_z' \bar{M}_z r_\theta' + \delta r_\theta' \bar{M}_y r_y' + \delta r_\theta' \bar{M}_z r_z' \right) dx
\end{aligned} \tag{1.117}$$

By the joining of Eqs. 1.113, 1.114 and 1.117, then

$$\begin{aligned}
&\int_L \int_A \delta \boldsymbol{\varepsilon}^T \boldsymbol{\sigma}_{tot} dA dx = \int_L \left[\delta r_x' EA r_x' + \delta r_y'' EI_z r_y'' + \delta r_z'' EI_y r_z'' + \delta r_\theta' GI_t r_\theta' \right] dx + \\
&\int_L \left(\delta r_y' \bar{N} r_y' + \delta r_z' \bar{N} r_z' + \delta r_\theta' \bar{N} e_0^2 r_\theta' - \delta r_y' \bar{M}_y r_\theta' - \delta r_z' \bar{M}_z r_\theta' - \delta r_\theta' \bar{M}_y r_y' - \delta r_\theta' \bar{M}_z r_z' \right) dx
\end{aligned} \tag{1.118}$$

Thus, in a general format the principle of virtual displacements for continuous line-like structures is given by

$$\begin{aligned}
&\sum_{i=1}^{N_F} \left\{ \left[\delta \mathbf{r}(x_F) \right]^T \cdot \mathbf{F} \right\}_i + \sum_{i=1}^{N_q} \left\{ \int_L \delta \mathbf{r}^T \cdot \mathbf{q} dx \right\}_i - \sum_{j=1}^{N_M} \left\{ \left[\delta \mathbf{r}(x_M) \right]^T \cdot \mathbf{M}_g \ddot{\mathbf{r}}(x_M) \right\}_j \\
&\qquad - \int_L \delta \mathbf{r}^T \cdot \mathbf{c} \cdot \dot{\mathbf{r}} dx - \int_L \delta \mathbf{r}^T \cdot \mathbf{m}_g \cdot \ddot{\mathbf{r}} dx = \\
&\int_L \Big[\delta r_x' EA r_x' + \delta r_y' \left(\bar{N} r_y' - \bar{M}_y r_\theta' \right) + \delta r_z' \left(\bar{N} r_z' - \bar{M}_z r_\theta' \right) + \delta r_y'' EI_z r_y'' \\
&\qquad + \delta r_z'' EI_y r_z'' + \delta r_\theta' \left(GI_t r_\theta' + \bar{N} e_0^2 r_\theta' - \bar{M}_y r_y' - \bar{M}_z r_z' \right) \Big] dx
\end{aligned} \tag{1.119}$$

Although by first glance this may seem rather complex, it is in fact a remarkably powerful toolbox.

Example 1.14 The Beam Column

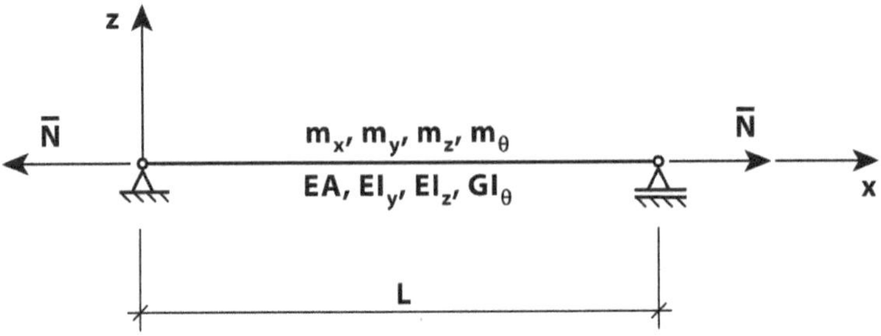

Fig. 1.26 Simply supported beam subject to a time invariant axial force $\bar{N}$

Let us for instance consider a simply supported beam with no external dynamic load $\mathbf{F}$ *or* $\mathbf{q}$ *as illustrated in Fig. 1.26. It is assumed that* $e_y = e_z = 0$ *(i.e. that the shear centre coincides with the centroid) and that its end supports have fork bearings, i.e. that* $r_\theta(x=0) = r_\theta(x=L) = 0$. *The beam is subject to a time invariant axial force* $\bar{N}$. *Otherwise, all time invariant moments* $\bar{M}_x$, $\bar{M}_y$ *and* $\bar{M}_z$ *as well as possible concentrated masses* $\mathbf{M}_g$ *are equal to zero. Let us also disregard any axial displacement and assume* $\mathbf{r}(x,t) = \begin{bmatrix} 0 & r_y & r_z & r_\theta \end{bmatrix}^T$. *Thus,* $\delta\mathbf{r}(x) = \begin{bmatrix} 0 & \delta r_y & \delta r_z & \delta r_\theta \end{bmatrix}^T$.

I.e., our investigation is limited to the search of its lowest undamped eigenfrequencies. Then Eq. 1.119 reduces to

$$-\int_L \begin{bmatrix} \delta r_y \\ \delta r_z \\ \delta r_\theta \end{bmatrix}^T \begin{bmatrix} m_y & 0 & 0 \\ 0 & m_z & 0 \\ 0 & 0 & m_\theta \end{bmatrix} \begin{bmatrix} \ddot{r}_y \\ \ddot{r}_z \\ \ddot{r}_\theta \end{bmatrix} dx =$$

$$\int_L \left[\bar{N}\left(\delta r_y' r_y' + \delta r_z' r_z'\right) + \delta r_y'' EI_z r_y'' + \delta r_z'' EI_y r_z'' + \delta r_\theta' \left(GI_{t\theta} + \bar{N} e_0^2 \right) r' \right] dx$$

Based on the results in Chapter 1.2 (see Examples 1.6 and 1.7) it seems reasonable to assume that the mode shapes associated with these eigenfrequencies are simple sinus functions, i.e. that a harmonic motion is given by

$$\begin{bmatrix} r_y \\ r_z \\ r_\theta \end{bmatrix} = \mathrm{Re}\left\{ \begin{bmatrix} a_y \\ a_z \\ a_\theta \end{bmatrix} \sin\left(\frac{n\pi x}{L}\right) e^{i\omega t} \right\} \text{ and thus we choose } \begin{bmatrix} \delta r_y \\ \delta r_z \\ \delta r_\theta \end{bmatrix} = \begin{bmatrix} \delta a_y \\ \delta a_z \\ \delta a_\theta \end{bmatrix} \sin\left(\frac{n\pi x}{L}\right)$$

Introducing this into the equation above, then the following is obtained

$$\delta a_y \left(\frac{n\pi}{L}\right)^4 EI_z a_y + \delta a_z \left(\frac{n\pi}{L}\right)^4 EI_y a_z + \delta a_\theta \left(\frac{n\pi}{L}\right)^2 GI_t a_\theta + \delta a_y \left(\frac{n\pi}{L}\right)^2 \bar{N} a_y +$$
$$\delta a_z \left(\frac{n\pi}{L}\right)^2 \bar{N} a_z + \delta a_\theta \left(\frac{n\pi}{L}\right)^2 \bar{N} e_0^2 a_\theta - \omega^2 \left(\delta a_y m_y a_y + \delta a_z m_z a_z + \delta a_\theta m_\theta a_\theta\right) = 0$$

This may more conveniently be written

$$\begin{bmatrix} \delta a_y \\ \delta a_z \\ \delta a_\theta \end{bmatrix} \left\{ \left(\frac{n\pi}{L}\right)^2 \begin{bmatrix} \left(\frac{n\pi}{L}\right)^2 EI_z + \bar{N} & 0 & 0 \\ 0 & \left(\frac{n\pi}{L}\right)^2 EI_y + \bar{N} & 0 \\ 0 & 0 & GI_t + \bar{N} e_0^2 \end{bmatrix} \right.$$
$$\left. -\omega^2 \begin{bmatrix} m_y & 0 & 0 \\ 0 & m_z & 0 \\ 0 & 0 & m_\theta \end{bmatrix} \right\} \begin{bmatrix} a_y \\ a_z \\ a_\theta \end{bmatrix} = 0$$

from which it is seen that the pre-multiplication by $\begin{bmatrix} \delta a_y & \delta a_z & \delta a_\theta \end{bmatrix}^T$ *is obsolete, and thus the following eigenvalue problem is obtained*

$$\left\{ \begin{bmatrix} \left(\frac{n\pi}{L}\right)^2 EI_z + \bar{N} & 0 & 0 \\ 0 & \left(\frac{n\pi}{L}\right)^2 EI_y + \bar{N} & 0 \\ 0 & 0 & GI_t + \bar{N} e_0^2 \end{bmatrix} - \hat{\omega}^2 \begin{bmatrix} m_y & 0 & 0 \\ 0 & m_z & 0 \\ 0 & 0 & m_\theta \end{bmatrix} \right\} \begin{bmatrix} a_y \\ a_z \\ a_\theta \end{bmatrix} = 0$$

where $\hat{\omega} = \omega L/(n\pi)$. *From this a non-trivial solution can only be obtained if the determinant to the coefficient matrix is zero, rendering*

$$\left[\left(\frac{n\pi}{L}\right)^2 EI_z + \bar{N} - \hat{\omega}^2 m_y\right]\left[\left(\frac{n\pi}{L}\right)^2 EI_y + \bar{N} - \hat{\omega}^2 m_z\right]\left(GI_t + \bar{N}e_0^2 - \hat{\omega}^2 m_\theta\right) = 0$$

Introducing the Euler buckling forces

$$N_{E_y} = \frac{(n\pi)^2 EI_y}{L^2}, \qquad N_{E_z} = \frac{(n\pi)^2 EI_z}{L^2} \qquad \text{and} \qquad N_{E_x} = \frac{GI_t}{e_0^2}$$

associated with bending about the y *and* z *axes and with pure torsion, then the solution above is reduced to*

$$\left[\left(\frac{n\pi}{L}\right)^2 EI_z\left(1+\frac{\bar{N}}{N_{E_z}}\right) - \hat{\omega}^2 m_y\right]\cdot\left[\left(\frac{n\pi}{L}\right)^2 EI_y\left(1+\frac{\bar{N}}{N_{E_y}}\right) - \hat{\omega}^2 m_z\right]$$
$$\cdot\left[GI_t\left(1+\frac{\bar{N}}{N_{E_x}}\right) - \hat{\omega}^2 m_\theta\right] = 0$$

As can be seen, the solution will contain three independent eigenfrequencies $\omega = \hat{\omega}(n\pi/L)$, *one associated with motion in the* y *direction*

$$\hat{\omega}_y^2 = \left(\frac{n\pi}{L}\right)^2 \frac{EI_z}{m_y}\left(1+\frac{\bar{N}}{N_{E_z}}\right) \quad \Rightarrow \quad \omega_y = \left(\frac{n\pi}{L}\right)^2 \sqrt{\frac{EI_z}{m_y}\left(1+\frac{\bar{N}}{N_{E_z}}\right)}$$

one associated with motion in the z *direction*

$$\hat{\omega}_z^2 = \left(\frac{n\pi}{L}\right)^2 \frac{EI_y}{m_z}\left(1+\frac{\bar{N}}{N_{E_y}}\right) \quad \Rightarrow \quad \omega_z = \left(\frac{n\pi}{L}\right)^2 \sqrt{\frac{EI_y}{m_z}\left(1+\frac{\bar{N}}{N_{E_y}}\right)}$$

and one associated with pure torsion

$$\hat{\omega}_\theta^2 = \frac{GI_t}{m_\theta}\left(1+\frac{\bar{N}}{N_{E_x}}\right) \quad \Rightarrow \quad \omega_\theta = \frac{n\pi}{L}\sqrt{\frac{GI_t}{m_\theta}\left(1+\frac{\bar{N}}{N_{E_x}}\right)}$$

This solution is identical to that which has been obtained in Chapter 1.2 (see Examples 1.6 and 1.7), except the additional effect of the time invariant axial force $\bar{N}$. *If* $\bar{N}$ *is positive, then it has a positive contribution to the sideway or torsion stiffness of the system due to element stretching. However, if* $\bar{N}$ *is negative, i.e. compression, then its contribution to system stiffness is negative due to sideway or torsion buckling.*

Let us also consider the special case of a taught string, i.e. the case that bending and torsion stiffness $(\pi/L)^2 EI_y$, $(\pi/L)^2 EI_z$ *and* GI_t/e_0^2 *are all insignificant as compared to* $\bar{N}$, *then the solution above reduces to*

$$\left\{\left(\frac{n\pi}{L}\right)^2 \bar{N} - \omega^2 m_y\right\}\left\{\left(\frac{n\pi}{L}\right)^2 \bar{N} - \omega^2 m_z\right\}\left\{\left(\frac{n\pi}{L}\right)^2 \bar{N} e_0^2 - \omega^2 m_\theta\right\} = 0$$

from which the following eigenfrequencies are obtained

$$\omega_y = \frac{n\pi}{L}\sqrt{\frac{\bar{N}}{m_y}} \qquad \omega_z = \frac{n\pi}{L}\sqrt{\frac{\bar{N}}{m_z}} \qquad \text{and} \qquad \omega_\theta = \frac{n\pi}{L}\sqrt{\frac{\bar{N} e_0^2}{m_\theta}}$$

1.7 Galerkin's Method

In Galerkin's method [9] the equilibrium requirement in the form of a differential equation (or an interconnected group of differential equations) of an unloaded and undamped dynamic system in harmonic motion

$$f(r, \ddot{r}) = 0 \tag{1.120}$$

is converted into a numerical eigenvalue problem

$$\mathbf{A} \cdot \mathbf{a} = \mathbf{0} \tag{1.121}$$

with unknown coefficients $\mathbf{a} = \begin{bmatrix} a_1 & \cdots & a_i & \cdots & a_{N_\psi} \end{bmatrix}^T$, by imposing an approximate solution comprising a linear combination of unknown coefficients a_i and a corresponding set of known functions $\psi_i(x)$, such that

$$r \approx \text{Re} \sum_{i=1}^{N_\psi} a_i \cdot \psi_i(x) \cdot e^{i\omega t} \tag{1.122}$$

and then applying the method of weighted residuals in its functional space (i.e. its length L, its surface A or its volume V). Thus, the approximate solution

$$f\left(\sum_{i=1}^{N_\psi} a_i \psi_i e^{i\omega t}\right) = 0 \tag{1.123}$$

is successively weighed with the same functions $\psi_j(x)$, $j = 1,2,......,N_\psi$, and integrated over L (or A), rendering the set of numeric equations in Eq. 1.121, which fully written is then given by

$$\begin{bmatrix} A_{11} & \cdots & A_{1j} & \cdots & A_{1N_\psi} \\ \vdots & \ddots & \vdots & \iddots & \vdots \\ A_{i1} & \cdots & A_{ij} & \cdots & A_{iN_\psi} \\ \vdots & \iddots & \vdots & \ddots & \vdots \\ A_{N_\omega 1} & \cdots & A_{N_\psi j} & \cdots & A_{N_\psi N_\psi} \end{bmatrix} \begin{bmatrix} a_1 \\ \vdots \\ a_i \\ \vdots \\ a_{N_\psi} \end{bmatrix} = \mathbf{0} \tag{1.124}$$

where

$$A_{ij} = \int_L \psi_i \cdot f\left(\psi_j, \omega\right) dx \qquad \text{or} \qquad A_{ij} = \int_A \psi_i \cdot f\left(\psi_j, \omega\right) dA \tag{1.125}$$

This is a general method which may offer an approximate solution to often complex systems. It is a requirement to the accuracy of the solution that $\psi_i(x)$ fulfils (more or less) the geometric boundary conditions of the system. It is advantageous that they are as close to orthogonal as possible.

To illustrate the use of Galerkin's method let us consider an undamped and unloaded continuous beam, whose motion is restricted to displacements in the z direction. Its differential equation has been developed in Eq. 1.29, i.e.:

$$f\left(r_z, \ddot{r}_z\right) = m_z \cdot \ddot{r}_z + EI_y \cdot r_z'''' = 0 \tag{1.126}$$

Let us for simplicity assume that its mass m_z and bending stiffness EI_y are constants along its entire span L. The following harmonic solution is adopted

$$r_z(x,t) = \mathrm{Re}\left\{\left[a_1 \cdot \psi_1(x) + a_2 \cdot \psi_2(x)\right] \cdot e^{i\omega t}\right\} \tag{1.127}$$

Introducing this into Eq. 1.126 then the following is obtained:

$$f\left(a_i, \psi_i\right) = EI_y\left(a_1 \cdot \psi_1'''' + a_2 \cdot \psi_2''''\right) - \omega^2 m_z\left(a_1 \cdot \psi_1 + a_2 \cdot \psi_2\right) = 0 \tag{1.128}$$

Pre-multiplying Eq. 1.128 by ψ_1 and integrating over the entire length L renders

$$EI_y \int_L \left(a_1 \cdot \psi_1 \cdot \psi_1'''' + a_2 \cdot \psi_1 \cdot \psi_2'''' \right) dx - \omega^2 m_z \int_L \left(a_1 \cdot \psi_1^2 + a_2 \cdot \psi_1 \cdot \psi_2 \right) dx = 0 \quad (1.129)$$

Similarly, pre-multiplying Eq. 1.128 by ψ_2 and integrating over the length L:

$$EI_y \int_L \left(a_1 \cdot \psi_2 \cdot \psi_1'''' + a_2 \cdot \psi_2 \cdot \psi_2'''' \right) dx - \omega^2 m_z \int_L \left(a_1 \cdot \psi_2 \cdot \psi_1 + a_2 \cdot \psi_2^2 \right) dx = 0 \quad (1.130)$$

These two equations may then more conveniently be written

$$\int_L \left\{ EI_y \begin{bmatrix} \psi_1 \psi_1'''' & \psi_1 \psi_2'''' \\ \psi_2 \psi_1'''' & \psi_2 \psi_2'''' \end{bmatrix} - \omega^2 m_z \begin{bmatrix} \psi_1^2 & \psi_1 \psi_2 \\ \psi_2 \psi_1 & \psi_2^2 \end{bmatrix} \right\} dx \cdot \begin{bmatrix} a_1 \\ a_2 \end{bmatrix} = \begin{bmatrix} 0 \\ 0 \end{bmatrix} \quad (1.131)$$

It is usually more convenient to express ψ_1 and ψ_2 by the non-dimensional coordinate $\hat{x} = x/L$. Introducing that $d\psi/dx = (d\psi/d\hat{x})(d\hat{x}/dx) = (d\psi/d\hat{x}) L^{-1}$ and defining

$$\mathbf{A} = \int_0^1 \begin{bmatrix} \psi_1 \psi_1'''' & \psi_1 \psi_2'''' \\ \psi_2 \psi_1'''' & \psi_2 \psi_2'''' \end{bmatrix} d\hat{x} \qquad \text{and} \qquad \mathbf{B} = \int_0^1 \begin{bmatrix} \psi_1^2 & \psi_1 \psi_2 \\ \psi_2 \psi_1 & \psi_2^2 \end{bmatrix} dx \quad (1.132)$$

then the following eigenvalue problem is obtained

$$(\mathbf{A} - \lambda \mathbf{B}) \mathbf{a} = \mathbf{0} \quad (1.133)$$

where

$$\lambda = \omega^2 m_z L^4 / EI_y \qquad \text{and} \qquad \mathbf{a} = \begin{bmatrix} a_1 & a_2 \end{bmatrix}^T \quad (1.134)$$

Example 1.15

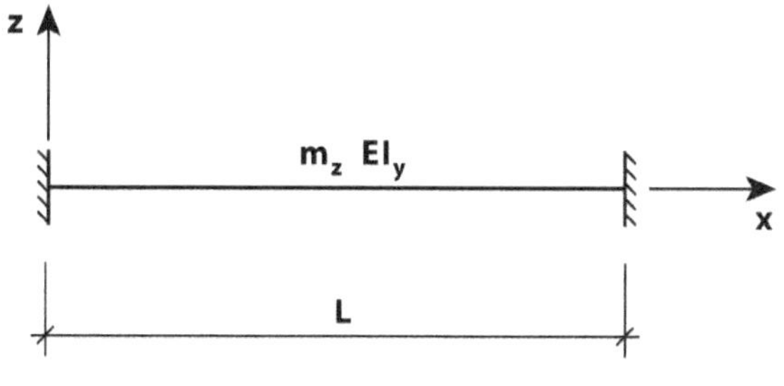

Fig. 1.27 Continuous beam with fixed end supports

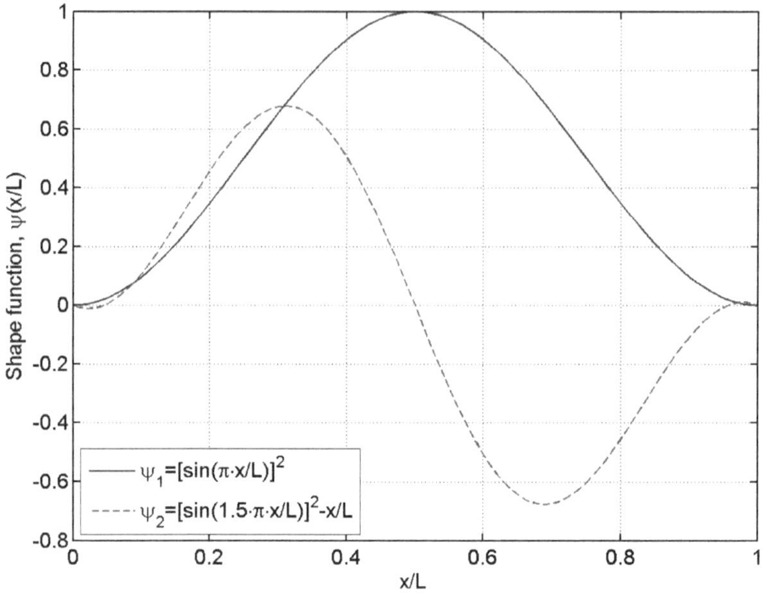

Fig. 1.28 Shape functions ψ_1 and ψ_2

Let us for instance consider the case of a continuous beam with fixed end supports as shown in Fig. 1.27 [i.e. $r_z(x=0)=r_z(x=L)=0$ *and* $r_z'(x=0)=r_z'(x=L)=0$*], and assume the following shape functions (see Fig. 1.28)*

$$\psi_1(\hat{x})=\left[\sin(\pi\hat{x})\right]^2 \quad \textit{and} \quad \psi_2(\hat{x})=\left[\sin(1.5\pi\hat{x})\right]^2-\hat{x}$$

whose fourth derivatives are given by

$$\psi_1''''(\hat{x})=-8\pi^4\cos(2\pi\hat{x}) \quad \textit{and} \quad \psi_2''''(\hat{x})=-\frac{1}{2}(3\pi)^4\cos(3\pi\hat{x})$$

Introducing this into Eq. 1.132 and performing the integration will then render

$$\mathbf{A}\approx\begin{bmatrix}195 & 0\\ 0 & 897\end{bmatrix} \quad \textit{and} \quad \mathbf{B}\approx\begin{bmatrix}0.375 & 0\\ 0 & 0.1858\end{bmatrix}$$

It is seen that due to the orthogonality properties between the chosen shape functions **A** *and* **B** *are diagonal. Thus, from Eq. 1.133 the following eigenvalues are obtained*

$$\text{Thus} \qquad \lambda_1 = 520 \quad \Rightarrow \quad \omega_{z_1} = 22.8\sqrt{\frac{EI_y}{m_z L^4}} \qquad \text{and} \qquad \mathbf{a}_1 = \begin{bmatrix} 1 & 0 \end{bmatrix}^T$$

$$\text{and} \qquad \lambda_2 = 4828 \quad \Rightarrow \quad \omega_{z_2} = 69.5\sqrt{\frac{EI_y}{m_z L^4}} \qquad \text{and} \qquad \mathbf{a}_2 = \begin{bmatrix} 0 & 1 \end{bmatrix}^T$$

More exact values are

$$\omega_{z_1} = 22.4\sqrt{\frac{EI_y}{m_z L^4}} \qquad \text{and} \qquad \omega_{z_2} = 61.7\sqrt{\frac{EI_y}{m_z L^4}} \tag{1.191}$$

I.e., the error is only about 2% for ω_{z_1} *and about 12% for* ω_{z_2}. *Obviously, the accuracy of the method will improve with the ability of the shape functions to represent the correct mode shapes of the system.*

Chapter 2
One and Two Degree of Freedom Systems

2.1 Introduction

While we in Chapter 1 focused on the basic mathematical methods of determining the structural properties of free vibrations, we shall now turn to the more realistic cases of including the effects of damping and the possibility of an external fluctuating force. However, in this chapter we limit ourselves to only consider the cases of single or two degrees of freedom systems, subject to a single harmonic force. Such a load case is virtually absent in the field of structural dynamics. Nonetheless, the case of a single harmonic force on simple systems is an illustrative overture to the ensuing chapters in this book. It is taken for granted that forces are rectilinear.

The addition of damping stems from the observation that any structural system which is initially given a displacement or impact and then left to oscillate by itself will more or less slowly lowering the size of its motion and finally return to a condition of rest. That which causes this diminishing effect of oscillations is an internal force attributed to what we call damping. The concept of damping was first presented in Chapter 1, see Fig. 1.21. Damping in general is further discussed in Chapter 9. Here we shall only mention that for a full scale structure its cause is complex and often difficult to identify. It will in general include contributions from friction in joints and supports, material nonlinearities and submerged flow resistance (e.g. in air or water). In structural mechanics, damping has usually been represented by an internal force conveniently assumed proportional to the velocity of the system. Such a force effect is what we call viscous damping. In general, this is adhered to throughout this book, except a single case of pure friction damping included in Chapter 9.

2.2 Unloaded Single Degree of Freedom System

The system of a single mass, spring and viscous damper is shown in Fig. 2.1.a. The corresponding free body diagram in accordance with Newton's second law and the principle of d'Alambert is shown in Fig. 2.1.b. Hence, equilibrium comprise the contributions from

E.N. Strømmen, *Structural Dynamics*, Springer Series in Solid and Structural Mechanics 2,
DOI: 10.1007/978-3-319-01802-7_2, © Springer International Publishing Switzerland 2014

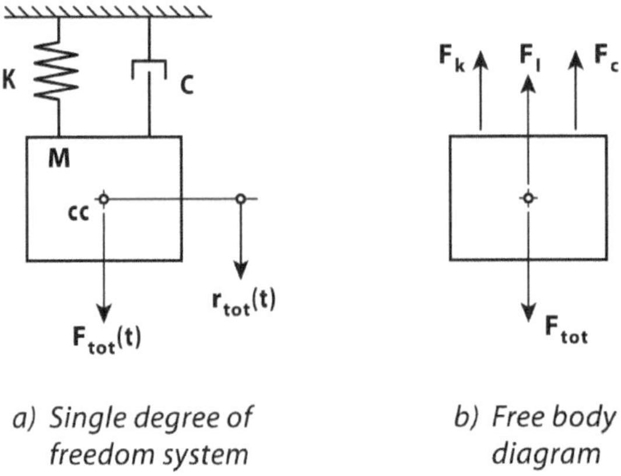

Fig. 2.1 Single mass, spring and viscous damper system

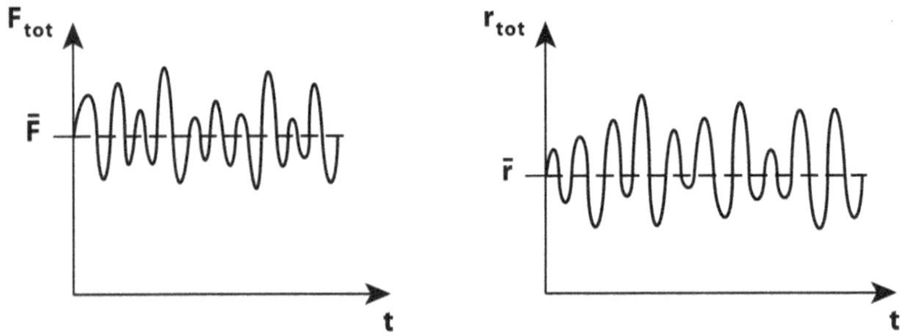

Fig. 2.2 Static and dynamic load and response

- an external force $F_{tot}(t)$ (with unit N),
- a spring force proportional to the body displacement $F_K = K \cdot r_{tot}(t)$, where K (with unit N/m) is the elastic spring constant,
- a damping force proportional to the body velocity $F_C = C \cdot \dot{r}_{tot}(t)$, where C (with unit Ns/m) is the viscous damping constant, and
- an inertia force proportional to and in the opposite direction of the body acceleration $F_I = M \cdot \ddot{r}_{tot}(t)$, where M (with unit kg) is the mass.

It is seen that equilibrium $F_I + F_C + F_K = F_{tot}$ will then require the solution of the following equation

$$M \cdot \ddot{r}_{tot}(t) + C \cdot \dot{r}_{tot}(t) + K \cdot r_{tot}(t) = F_{tot}(t) \tag{2.1}$$

As illustrated in Fig. 2.2, it is taken for granted that the fluctuating load is stationary during the entire time window considered relevant for the calculations performed throughout this book, i.e. the external force will comprise a time invariant (static) part $\bar{F}$ and a fluctuating (dynamic) part $F(t)$. It was shown in Chapter 1.2 that the displacement will then also comprise the sum of a time invariant part $\bar{r}$ and a fluctuating part $r(t)$. Introducing this into Eq. 2.1 will then render

$$M \cdot \ddot{r}(t) + C \cdot \dot{r}(t) + K \cdot \left[\bar{r} + r(t)\right] = \bar{F} + F(t) \tag{2.2}$$

and thus, it is seen that response calculations may be split into the solution of a time invariant equilibrium requirement

$$\bar{F} = K \cdot \bar{r} \tag{2.3}$$

and the solution of a purely dynamic equilibrium requirement

$$M \cdot \ddot{r}(t) + C \cdot \dot{r}(t) + K \cdot r(t) = F(t) \tag{2.4}$$

Thus, the principle of superposition between time invariant and dynamic load effects is applicable. The static load cases will in the following not be pursued.

Let us first consider the unloaded case $F(t) = 0$. The general solution of a freely oscillating but damped system is given by

$$r(t) = \text{Re}\left(a \cdot e^{\alpha t}\right) \tag{2.5}$$

where a is the amplitude and α is an unknown constant, which, after introduction of Eq. 2.5 into Eq. 2.4, may be determined from the condition

$$M\alpha^2 + C\alpha + K = 0 \tag{2.6}$$

Dividing by M and introducing the undamped eigenfrequency $\omega_n = \sqrt{K/M}$ and the damping ratio

$$\zeta_n = C/(2M\omega_n) \tag{2.7}$$

then

$$(\alpha/\omega_n)^2 + 2\zeta_n(\alpha/\omega_n) + 1 = 0 \tag{2.8}$$

Thus, the following solution is obtained

$$\alpha/\omega_n = -\zeta_n \pm \sqrt{\zeta_n^2 - 1} \tag{2.9}$$

It is seen that there are two values of α that will satisfy Eq. 2.6. The solution to $r(t)$ will then comprise the sum of these alternatives,

$$r(t) = r_1(\alpha_1, t) + r_2(\alpha_2, t) \tag{2.10}$$

and it contains two alternative types of motion, depending on $\zeta_n \geq 1$ or $\zeta_n < 1$.

Case 1: $\zeta_n > 1$

Let us first consider the case that $\zeta_n > 1$. Then the expression under the root sign in Eq. 2.9 is positive, and thus,

$$\alpha_1 = \omega_n\left(-\zeta_n + \sqrt{\zeta_n^2 - 1}\right) \qquad \text{and} \qquad \alpha_2 = \omega_n\left(-\zeta_n - \sqrt{\zeta_n^2 - 1}\right) \tag{2.11}$$

The corresponding solution is given by

$$r(t) = r_1 + r_2 = a_1 e^{\alpha_1 t} + a_2 e^{\alpha_2 t} = \left(a_1 e^{\omega_n t\sqrt{\zeta_n^2 - 1}} + a_2 e^{-\omega_n t\sqrt{\zeta_n^2 - 1}} \right) e^{-\zeta_n \omega_d t} \tag{2.12}$$

Introducing the initial conditions that

$$\left.\begin{aligned} r(t=0) &= a_1 + a_2 = r_0 \\ \dot{r}(t=0) &= a_1\omega_n\left(-\zeta_n + \sqrt{\zeta_n^2 - 1}\right) + a_2\omega_n\left(-\zeta_n - \sqrt{\zeta_n^2 - 1}\right) = \dot{r}_0 \end{aligned}\right\} \tag{2.13}$$

and solving with respect to a_1 and a_2, then the following solution is obtained

$$\begin{aligned} r(t) = \Bigg\{ & \left[r_0\left(\zeta_n + \sqrt{\zeta_n^2 - 1}\right) + \frac{\dot{r}_0}{\omega_n} \right] e^{\omega_n t\sqrt{\zeta_n^2 - 1}} \Big/ 2\sqrt{\zeta_n^2 - 1} \\ & - \left[r_0\left(\zeta_n - \sqrt{\zeta_n^2 - 1}\right) + \frac{\dot{r}_0}{\omega_n} \right] e^{-\omega_n t\sqrt{\zeta_n^2 - 1}} \Big/ 2\sqrt{\zeta_n^2 - 1} \Bigg\} \cdot e^{-\zeta_n \omega_d t} \end{aligned} \tag{2.14}$$

This solution is illustrated in Fig. 2.3.a, and as can be seen, there are no oscillations of the system, the response is more or less rapidly dropping towards a condition where the system is at rest in its original position. Hence, we call it the *over-damped solution.*

Case 2: $\zeta_n = 1$

Let us then consider the case that $\zeta_n = 1$. Then $\alpha_1 = \alpha_2 = -\omega_n$, and hence an obvious solution is that $r_1 = a_1 e^{-\omega_n t}$. Since r_2 must contain the same boundary conditions as r_1 it is a reasonable assumption that $r_2 = f(t) \cdot r_1$, where $f(t)$ is an unknown function, whose properties must be such that the solution satisfies the equilibrium condition given in Eq. 2.4 [with $F(t)=0$], i.e. that

$$M \cdot \left(f \cdot \ddot{r}_1 + 2\dot{f} \cdot \dot{r}_1 + \ddot{f} \cdot r_1\right) + C \cdot \left(f \cdot \dot{r}_1 + \dot{f} \cdot r_1\right) + K \cdot f \cdot r = 0 \tag{2.15}$$

This may be rewritten into

$$f \cdot (M\ddot{r}_1 + C\dot{r}_1 + Kr_1) + \dot{f} \cdot (2M\dot{r}_1 + Cr_1) + \ddot{f} \cdot Mr_1 = 0 \tag{2.16}$$

It is seen that the first bracket contains r_1 inserted into Eq. 2.4. It must therefore be equal to zero. Since $C = 2M\,\omega_n \zeta_n$ (see Eq. 2.7) and $\zeta_n = 1$, then the second bracket is also zero because

$$\left(2M\dot{r}_1 + Cr_1\right) = -2M\,\omega_n a_1 e^{-\omega_n t} + 2M\,\omega_n a_1 e^{-\omega_n t} = 0 \tag{2.17}$$

and thus, we must demand that $\ddot{f} = 0$, which implies that $r_2(t) = a_2 t e^{-\omega_n t}$. Hence

$$r(t) = (a_1 + a_2 t)e^{-\omega_n t} \tag{2.18}$$

Introducing that $r(t=0) = r_0$ and $\dot{r}(t=0) = \dot{r}_0$, then the following is obtained

$$r(t) = \left[r_0 + (\dot{r}_0 + \omega_n r_0)t\right]e^{-\omega_n t} \tag{2.19}$$

This solution is illustrated in Fig. 2.3.a, and as can be seen, there are still no oscillations of the system, but this is the transition between that which is shown in Fig. 2.3.a and that which is shown in Fig. 2.3.b, where oscillations actually occur. Hence, we call it the *critically damped solution*, and $C_{cr} = 2M\,\omega_n$ is called the *critical damping coefficient.*

Case 3: $\zeta_n < 1$

Finally, let us consider the case that $\zeta_n < 1$. Then the expression under the root in Eq. 2.9 is negative, and thus, the solution will contain the complex roots

$$\alpha_1 = \omega_n\left(-\zeta_n + i\sqrt{1-\zeta_n^2}\right) \qquad \text{and} \qquad \alpha_2 = \omega_n\left(-\zeta_n - i\sqrt{1-\zeta_n^2}\right) \tag{2.20}$$

Dividing throughout the equilibrium condition in Eq. 2.4 by M and introducing that $C/M = 2\omega_n \zeta_n$ (see Eq. 2.7) and $K/M = \omega_n^2$ then

$$\ddot{r} + 2\omega_n \zeta_n \dot{r} + \omega_n^2 r = 0 \tag{2.21}$$

Again, it is assumed that

$$r(t) = r_1(t) + r_2(t) \qquad \text{where} \qquad \begin{cases} r_1 = a_1 e^{\alpha t} \sin \gamma t \\ r_2 = a_2 e^{\alpha t} \cos \gamma t \end{cases} \tag{2.22}$$

Introducing r_1 into Eq. 2.21 will then require that

$$\left(\alpha^2 - \gamma^2 + 2\omega_n \zeta_n \alpha + \omega_n^2\right) \sin \gamma t + 2\gamma\left(\alpha + \omega_n \zeta_n\right) \cos \gamma t = 0 \tag{2.23}$$

for all values of t, which can only be fulfilled if

$$\alpha = -\omega_n \zeta_n \qquad \text{and} \qquad \gamma = \omega_n \sqrt{1 - \zeta_n^2} \tag{2.24}$$

Similarly, introducing r_2 into Eq. 2.21 will then require that

$$\left(\alpha^2 - \gamma^2 + 2\omega_n \zeta_n \alpha + \omega_n^2\right) \cos \gamma t - 2\gamma\left(\alpha + \omega_n \zeta_n\right) \sin \gamma t = 0 \tag{2.25}$$

for all values of t, which will be satisfied for the same solution that was obtained in Eq. 2.24. Thus

$$r(t) = \left[a_1 \sin\left(\omega_n t \sqrt{1 - \zeta_n^2}\right) + a_2 \cos\left(\omega_n t \sqrt{1 - \zeta_n^2}\right)\right] \cdot e^{-\omega_n \zeta_n t} \tag{2.26}$$

Introducing that $r(t = 0) = r_0$ and $\dot{r}(t = 0) = \dot{r}_0$, then the following is obtained

$$r(t) = \left[\frac{r_0 \zeta_n + \dot{r}_0 / \omega_n}{\sqrt{1 - \zeta_n^2}} \sin\left(\omega_n t \sqrt{1 - \zeta_n^2}\right) + r_0 \cos\left(\omega_n t \sqrt{1 - \zeta_n^2}\right)\right] \cdot e^{-\omega_n \zeta_n t} \tag{2.27}$$

Using the trigonometric property that $\cos(x_1 - x_2) = \sin x_1 \cdot \sin x_2 + \cos x_1 \cdot \cos x_2$ then this solution may more conveniently be written

$$r(t) = a \cdot e^{-\omega_n \zeta_n t} \cdot \cos(\omega_d t - \beta_n) \tag{2.28}$$

where $\omega_d = \omega_n \sqrt{1 - \zeta_n^2}$, and where

$$\left. \begin{aligned} a &= \sqrt{r_0^2 + \left(\dot{r}_0 / \omega_d + r_0 \zeta_n \Big/ \sqrt{1 - \zeta_n^2}\right)^2} \\ \tan \beta_n &= \dot{r}_0 / (r_0 \omega_d) + \zeta_n \Big/ \sqrt{1 - \zeta_n^2} \end{aligned} \right\} \tag{2.29}$$

This solution (setting $\dot{r}_0 T_n / r_0 = 1$ where $T_n = 2\pi/\omega_n$) is illustrated in Fig. 2.3.b, and as can be seen, the system is oscillating with a frequency of ω_d , which we call the damped eigenfrequency, and it decays exponentially with a rate determined by ζ_n. We call this the *under-damped solution*. The special case that $r_0 = 0$ implies that $a = \dot{r}_0/\omega_d$ and $\beta_n = \pi/2$, rendering

$$r(t) = \frac{\dot{r}_0}{\omega_d} e^{-\omega_n \zeta_n t} \sin(\omega_d t) \tag{2.30}$$

which is a useful result in connection with impact loading. It is in the following consistently taken for granted that $\zeta_n << 1$.

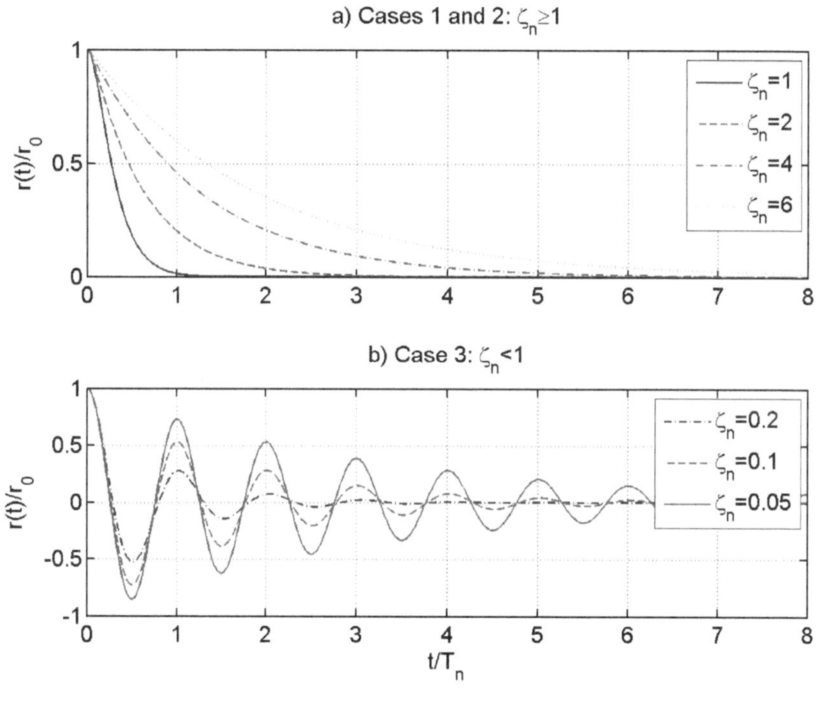

Fig. 2.3 Free oscillations of single degree of freedom system

2.3 Single Degree of Freedom System with Harmonic Load

Let us consider the case of a harmonic sinusoidal load with amplitude F_0 and an arbitrary frequency ω, i.e. the case that $F(t) = F_0 \cdot \sin \omega t$. The solution $r(t)$ to the equilibrium condition in Eq. 2.4

$$M \cdot \ddot{r}(t) + C \cdot \dot{r}(t) + K \cdot r(t) = F(t) \tag{2.31}$$

may then be split into the sum of a solution $r_h(t)$ to the homogeneous equation

$$M \cdot \ddot{r}_h(t) + C \cdot \dot{r}_h(t) + K \cdot r_h(t) = 0 \tag{2.32}$$

and a particular solution $r_p(t)$ to

$$M \cdot \ddot{r}_p(t) + C \cdot \dot{r}_p(t) + K \cdot r_p(t) = F(t) \tag{2.33}$$

As mentioned above, it is taken for granted that $\zeta_n \ll 1$, and as shown in Chapter 2.2 above, the homogeneous solution is then given in Eq. 2.28, i.e.

$$r_h(t) = a_h \cdot e^{-\omega_n \zeta_n t} \cdot \cos(\omega_d t - \beta_h) \tag{2.34}$$

where $\omega_d = \omega_n \sqrt{1 - \zeta_n^2}$, and where a_h and β_h will be determined from the conditions that $r(t = 0) = 0$ and $\dot{r}(t = 0) = 0$. A particular solution to Eq. 2.33 is given by

$$r_p(t) = b_p \sin \omega t + c_p \cos \omega t \tag{2.35}$$

Introducing this into Eq. 2.33 will then imply that

$$\left(K - \omega^2 M\right)\left(b_p \sin \omega t + c_p \cos \omega t\right) + C\omega\left(b_p \cos \omega t - c_p \sin \omega t\right) = F_0 \sin \omega t \tag{2.36}$$

which may also written

$$\left[\left(K - \omega^2 M\right) b_p - C\omega c_p - F_0\right] \sin \omega t + \left[C\omega b_p + \left(K - \omega^2 M\right) c_p\right] \cos \omega t = 0 \tag{2.37}$$

Since this requirement can only be achieved at all times if both terms are simultaneously equal to zero, it is seen that

$$\begin{bmatrix} \left(K - \omega^2 M\right) & -C\omega \\ C\omega & \left(K - \omega^2 M\right) \end{bmatrix} \cdot \begin{bmatrix} b_p \\ c_p \end{bmatrix} = \begin{bmatrix} F_0 \\ 0 \end{bmatrix} \tag{2.38}$$

rendering

$$\begin{bmatrix} b_p \\ c_p \end{bmatrix} = \frac{1}{\left(K-\omega^2 M\right)^2+(C\omega)^2} \begin{bmatrix} \left(K-\omega^2 M\right) & C\omega \\ -C\omega & \left(K-\omega^2 M\right) \end{bmatrix} \cdot \begin{bmatrix} F_0 \\ 0 \end{bmatrix}$$
$$= \frac{F_0}{\left(K-\omega^2 M\right)^2+(C\omega)^2} \begin{bmatrix} \left(K-\omega^2 M\right) \\ -C\omega \end{bmatrix} \tag{2.39}$$

Introducing $\omega_n^2 = K/M$ and $\zeta_n = C/(2\omega_n M) = \omega_n \cdot C/(2K)$, see Eq. 2.7, then

$$\begin{bmatrix} b_p \\ c_p \end{bmatrix} = \frac{F_0}{K} \cdot \frac{1}{\left[1-(\omega/\omega_n)^2\right]^2 + (2\zeta_n\,\omega/\omega_n)^2} \cdot \begin{bmatrix} 1-(\omega/\omega_n)^2 \\ -2\zeta_n\,\omega/\omega_n \end{bmatrix} \tag{2.40}$$

Thus:
$$r_p(t) = \frac{F_0}{K} \cdot \frac{\left(1-(\omega/\omega_n)^2\right)\sin\omega t - (2\zeta_n\,\omega/\omega_n)\cos\omega t}{\left[1-(\omega/\omega_n)^2\right]^2 + (2\zeta_n\,\omega/\omega_n)^2} \tag{2.41}$$

Since $a \cdot \sin(\alpha-\beta) = a\cdot\cos\beta\cdot\sin\alpha - a\cdot\sin\beta\cdot\cos\alpha = a_1\sin\alpha - a_2\cos\alpha$ where $\sqrt{a_1^2+a_2^2} = \sqrt{(a\cos\beta)^2+(a\sin\beta)^2} = a$, it is seen that

$$r_p(t) = a_p \cdot \sin\left(\omega t - \beta_p\right) \tag{2.42}$$

where

$$a_p = \frac{F_0}{K}\sqrt{\frac{\left[1-(\omega/\omega_n)^2\right]^2 + (2\zeta_n\,\omega/\omega_n)^2}{\left\{\left[1-(\omega/\omega_n)^2\right]^2 + (2\zeta_n\,\omega/\omega_n)^2\right\}^2}} = \frac{F_0/K}{\sqrt{\left[1-(\omega/\omega_n)^2\right]^2 + (2\zeta_n\,\omega/\omega_n)^2}} \tag{2.43}$$

and
$$\tan\beta_p = \frac{2\zeta_n\,\omega/\omega_n}{1-(\omega/\omega_n)^2} \tag{2.44}$$

Thus, the total solution $r(t) = r_h(t) + r_p(t)$ is given by

$$r(t) = a_h e^{-\omega_n\zeta_n t}\cos\left(\omega_d t - \beta_h\right) + \frac{F_0}{K}\frac{\sin\left(\omega t - \beta_p\right)}{\sqrt{\left[1-(\omega/\omega_n)^2\right]^2 + (2\zeta_n\,\omega/\omega_n)^2}} \tag{2.45}$$

As mentioned above, it is assumed that the system is starting off at rest where

$$r(t=0)=a_h\cos(\beta_h)-\frac{(F_0/K)\cdot\sin(\beta_p)}{\sqrt{\left[1-(\omega/\omega_n)^2\right]^2+(2\zeta_n\,\omega/\omega_n)^2}}=0 \tag{2.46}$$

and

$$\dot{r}(t=0)=a_h\left[\omega_d\sin(\beta_h)-\zeta\omega_n\cos(\beta_h)\right]+\frac{(F_0/K)\cdot\omega\cdot\cos(\beta_p)}{\sqrt{\left[1-\left(\frac{\omega}{\omega_n}\right)^2\right]^2+\left(2\zeta_n\frac{\omega}{\omega_n}\right)^2}}=0 \tag{2.47}$$

from which it is obtained that

$$a_h=\frac{F_0}{K}\cdot\frac{\sin(\beta_p)/\cos(\beta_h)}{\sqrt{\left[1-(\omega/\omega_n)^2\right]^2+(2\zeta_n\,\omega/\omega_n)^2}} \tag{2.48}$$

$$\tan(\beta_h)=\left[-1+2\zeta^2+(\omega/\omega_n)^2\right]\Big/\left(2\zeta\sqrt{1-\zeta^2}\right) \tag{2.49}$$

Thus:
$$r(t)=\frac{F_0}{K}\cdot\frac{\dfrac{\sin(\beta_p)}{\cos(\beta_h)}\cdot e^{-\omega_n\zeta_n t}\cdot\cos(\omega_d t-\beta_h)+\sin(\omega t-\beta_p)}{\sqrt{\left[1-(\omega/\omega_n)^2\right]^2+(2\zeta_n\,\omega/\omega_n)^2}} \tag{2.50}$$

It is seen that the response may be split into a transient part

$$r_{trans}(t)=\frac{F_0}{K}\cdot\frac{\sin(\beta_p)}{\cos(\beta_h)}\cdot\frac{e^{-\omega_n\zeta_n t}\cdot\cos(\omega_d t-\beta_h)}{\sqrt{\left[1-(\omega/\omega_n)^2\right]^2+(2\zeta_n\,\omega/\omega_n)^2}} \tag{2.51}$$

and a steady state part

$$r_{st}(t)=\frac{F_0}{K}\cdot\frac{\sin(\omega t-\beta_p)}{\sqrt{\left[1-(\omega/\omega_n)^2\right]^2+(2\zeta_n\,\omega/\omega_n)^2}} \tag{2.52}$$

Due to the decaying exponential function in $r_{trans}(t)$ it is a transient contribution to the response because it only lasts for a limited period of time, i.e.

$\lim_{t\to\infty}\left[r_{trans}(t)\right]=0$ (see also the plot in Fig. 2.4), while $r_{st}(t)$ is a steady state contribution because, after a certain period of time, it will constitute the entire response on its own and then remain unchanged with increasing time. The larger ζ_n the quicker $r_{trans}(t)$ dies out.

Example 2.1

Introducing $\omega_n = 0.3$ rad/s, $\omega = \omega_n$ *and* $\zeta_n = 0.02$ *then the reduced response* $r(t)/r_0$ *(where* $r_0 = F_0/K$ *) according to Eq. 2.50 is shown in Fig. 2.4 below.*

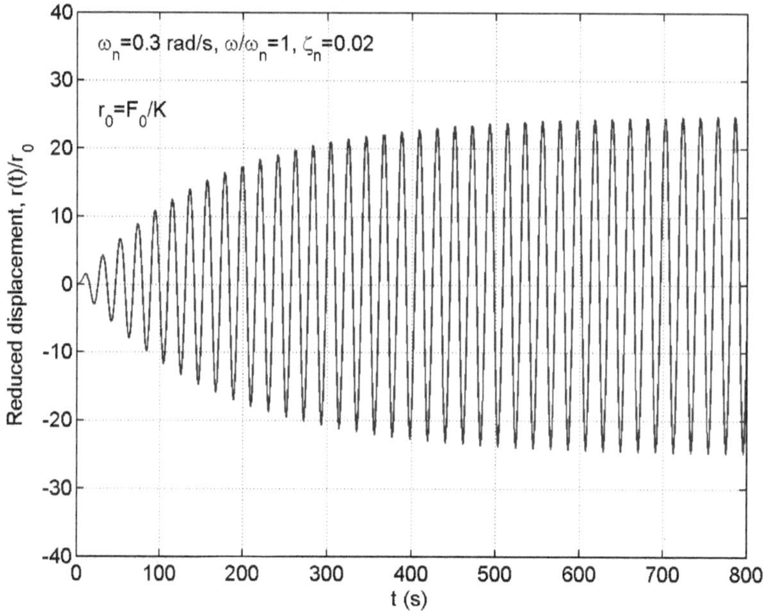

Fig. 2.4 Forced response of simple single degree of freedom system, $F = F_0 \sin(\omega t)$

Elaboration 2.1: The Phenomenon of Beating

If damping is very low and the forcing frequency ω is very close to the eigenfrequency ω_n, then the particular phenomenon of "beating" may occur. Let for simplicity $\zeta_n \approx 0$. Let also $\omega/\omega_n \approx 1$, but $\omega \neq \omega_n$. Then $\beta_p \approx 0$ (see Eq. 2.44) and $\omega_d \approx \omega_n$. Thus

$$r(t) = a_h \cdot \cos(\omega_n t - \beta_h) + \frac{F_0}{K} \cdot \frac{\sin(\omega t)}{1 - (\omega/\omega_n)^2}$$

Imposing the requirements that $r(t=0)=0$ and $\dot{r}(t=0)=0$ will then render

$$\beta_h = \frac{\pi}{2} \quad \text{and} \quad a_h = -\frac{F_0}{K} \cdot \frac{1}{1-(\omega/\omega_n)^2} \cdot \frac{\omega}{\omega_n} \approx -\frac{F_0}{K} \cdot \frac{1}{1-(\omega/\omega_n)^2}$$

Thus, the following approximate solution is obtained

$$r(t) \approx \frac{F_0}{K} \cdot \frac{\sin(\omega t) - \sin(\omega_n t)}{1-(\omega/\omega_n)^2} \approx \frac{2 \cdot F_0}{K} \cdot \frac{\sin(\omega_b t/2) \cdot \cos(\bar{\omega} t)}{1-(\omega/\omega_n)^2}$$

where $\bar{\omega} = (\omega + \omega_n)/2$ is the average frequency and $\omega_b = \omega - \omega_n$ is the frequency of the "beat". The special case of $\omega_n = 0.3\ rad/s$ and $\omega/\omega_n = 0.9$ is illustrated in Fig. 2.5 below.

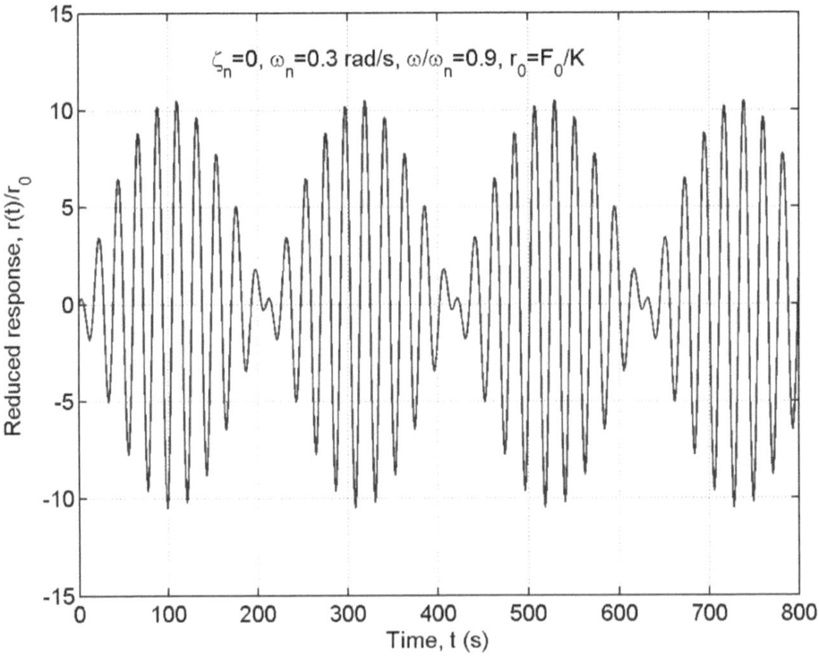

Fig. 2.5 Typical "beat" response

2.4 The Steady State Response in a Complex Format

In some cases of impact or shock load effects it is the transient part of the response which is of primary interest for structural safety considerations. But, in most cases the load will render a more or less stationary type of response, in which case it is the steady state part which is of main concern. Let $F(t)$ be a harmonic function which in a complex format may be expressed by

$$F(t) = \operatorname{Re}\left(F_0 \cdot e^{i\omega t}\right) = F_0 \cdot \operatorname{Re}\left(\cos \omega t + i \cdot \sin \omega t\right) = F_0 \cdot \cos \omega t \tag{2.53}$$

where it has been taken for granted that the amplitude F_0 is real. [It should be acknowledged that while we in Chapter 2.3 above adopted a forcing function $F(t) = F_0 \cdot \sin \omega t$, rendering a $r(t) = a(\omega) \cdot \sin(\omega t - \beta)$ type of steady state response, it is of no consequence that we now adopt $F(t) = F_0 \cdot \cos \omega t$, as this will simply render a $r(t) = a(\omega) \cdot \cos(\omega t - \beta)$ type of response.] The corresponding steady state response function may then be expressed by

$$r(t) = \operatorname{Re}\left[A(\omega) \cdot e^{i\omega t}\right] \tag{2.54}$$

where $A(\omega)$ is a complex amplitude $A(\omega) = b + ic$. Thus

$$r(t) = \operatorname{Re}\left[(b + ic)(\cos \omega t + i \sin \omega t)\right] = b \cdot \cos \omega t - c \cdot \sin \omega t \tag{2.55}$$

Introducing

$$\left.\begin{aligned} & a = \left|A(\omega)\right| = \sqrt{A^* A} = \sqrt{b^2 + c^2} \\ & \left.\begin{aligned} a \cdot \cos \beta_n &= b \\ a \cdot \sin \beta_n &= -c \end{aligned}\right\} \Rightarrow \tan \beta_n = \frac{-c}{b} \end{aligned}\right\} \tag{2.56}$$

then

$$\begin{aligned} r(t) &= b \cdot \cos \omega t + (-c) \cdot \sin \omega t \\ &= a \cdot \left(\cos \omega t \cdot \cos \beta_n + \sin \omega t \cdot \sin \beta_n\right) = a \cdot \cos\left(\omega t - \beta_n\right) \end{aligned} \tag{2.57}$$

Thus, it is seen that

$$r(t) = \mathrm{Re}\left[A(\omega) \cdot e^{i\omega t}\right] = |A(\omega)| \cdot \cos(\omega t - \beta_n) \tag{2.58}$$

In general, the complex amplitude $A(\omega)$ and the phase angle β_n are determined by fulfilment of the equilibrium condition. Thus, introducing Eq. 2.54 into

$$M \cdot \ddot{r}(t) + C \cdot \dot{r}(t) + K \cdot r(t) = F(t) \tag{2.59}$$

then the following is obtained

$$\left(-M\omega^2 + Ci\omega + K\right) A(\omega) e^{i\omega t} = F_0 e^{i\omega t} \tag{2.60}$$

Dividing throughout the equation by K and introducing that $\omega_n^2 = K/M$ and $\zeta_n = C/(2\omega_n M) = \omega_n \cdot C/(2K)$, then

$$A(\omega) = \frac{F_0}{K} \cdot \left[1 - (\omega/\omega_n)^2 + 2i\zeta_n \omega/\omega_n\right]^{-1} \tag{2.61}$$

which, by multiplication and division by the complex conjugate of the denominator, may be expressed by

$$A(\omega) = \frac{F_0/K}{\left[1 - (\omega/\omega_n)^2\right]^2 + (2\zeta_n \omega/\omega_n)^2} \cdot \left\{\left[1 - (\omega/\omega_n)^2\right] - i \cdot 2\zeta_n \omega/\omega_n\right\} \tag{2.62}$$

As shown in Eqs. 2.56 and 2.57 above, the response $r(t) = |A(\omega)| \cdot \cos(\omega t - \beta_n)$ is then defined by its amplitude

$$|A(\omega)| = \sqrt{A^* A} = \sqrt{b^2 + c^2} = \frac{F_0}{K}\left\{\left[1 - (\omega/\omega_n)^2\right]^2 + (2\zeta_n \omega/\omega_n)^2\right\}^{-\frac{1}{2}} \tag{2.63}$$

and phase angle

$$\tan \beta_n = \frac{-c}{b} = \frac{2\zeta_n \omega/\omega_n}{1 - (\omega/\omega_n)^2} \tag{2.64}$$

$F_0/K = r_0$ is the static displacement of the system if load amplitude F_0 had acted on its own. Defining the non-dimensional frequency response function (complex and associated with eigenfrequency ω_n) by

$$\hat{H}_n(\omega) = \left[1 - (\omega/\omega_n)^2 + 2i\,\zeta_n\omega/\omega_n\right]^{-1} \tag{2.65}$$

it is seen that

$$r(t) = (F_0/K) \cdot \left|\hat{H}(\omega)\right| \cdot \cos(\omega t - \beta_n) \tag{2.66}$$

I.e., the steady state dynamic response of a simple single degree of freedom system subject to a harmonic load $F(t) = F_0 \cdot \cos\omega t$ is also a cosine, but delayed by a phase β_n (determined by Eq. 2.64), and an amplitude which is equal to the static effect of F_0 magnified by the absolute value of the frequency response function $\left|\hat{H}(\omega)\right|$. A plot of $\left|\hat{H}(\omega)\right|$ and β_n at $\zeta_n = 0.02$ are shown in Fig. 2.6.

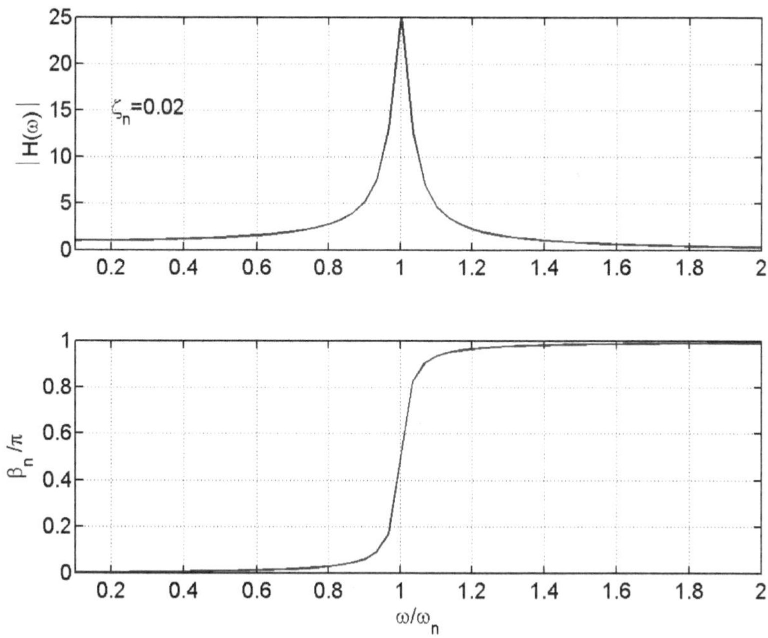

Fig. 2.6 Plots of $\left|\hat{H}(\omega)\right|$ and β_n (at $\zeta_n = 0.02$)

Example 2.2 Single Storey Shear Frame

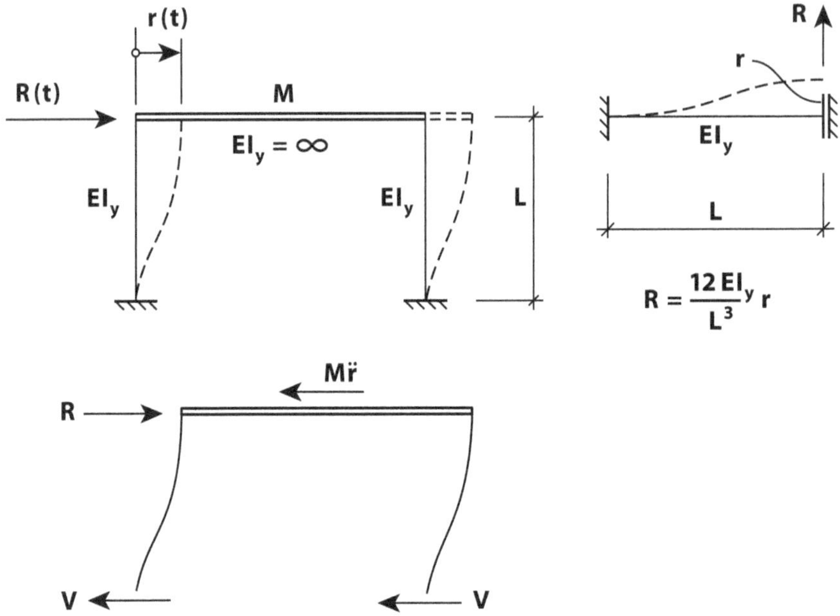

Fig. 2.7 Single storey shear frame

A single storey shear frame is shown in Fig. 2.7 above. It is called a shear frame because the bending stiffness of the beam is assumed infinitely large, and, as illustrated below, its displacement pattern will create a condition of large shear forces (as well as bending) in the columns. The connection between the shear force and the sideway column displacement is shown in the upper right hand side of Fig. 2.7. For simplicity, the distributed mass of the columns (m_c *) is assumed negligible. At an arbitrary displacement* $r(t)=\operatorname{Re}\left[ae^{i\omega t}\right]$ *a free body diagram of the system is shown in the lower illustration of Fig. 2.7. Since* $V(t)=\left(12EI_y/L^3\right)\cdot r(t)$ *horizontal equilibrium is expressed by* $M\ddot{r}+Kr=0$, *where* M *is the total mass of the beam, and* $K=2\cdot\left(12EI_y/L^3\right)=24EI_y/L^3$. *Introducing* $r=\operatorname{Re}\left[ae^{i\omega t}\right]$ *and thus*

$$24EI_y/L^3-\omega^2M=0 \qquad \Rightarrow \omega_n=\sqrt{24EI_y/\left(ML^3\right)}$$

Let us assume $L = 2.4\ m$, $M = 960\ kg$, $EI_y = 3.5 \cdot 10^4\ N/m$ *and* $\zeta_n = 0.01$ *then* $\omega_n = 7.96\ rad/s$. *Let us also assume* $F(t) = F_0 \cos(\omega t)$ *where* $F_0 = 2 \cdot 10^3\ N$.

Then $r(t) = \dfrac{0.033 \cdot \cos(\omega t - \beta_n)}{\sqrt{\left[1 - (\omega/\omega_n)^2\right]^2 + (2\zeta_n \omega/\omega_n)^2}}$ *where* $\tan\beta_n = \dfrac{0.02\,\omega/\omega_n}{1 - (\omega/\omega_n)^2}$

It is seen that if $\omega/\omega_n \ll 1$ *then* $\beta_n \approx 0$ *and* $r(t) \approx F_0/K = 0.033\ m$. *This is what we call a quasi-static type of response. It is the stiffness which is decisive. If* $\omega/\omega_n = 1$ *then* $\beta_n = \pi/2$ *and* $r(t) = (F_0/K) \cdot (1/2\zeta_n) \cdot \sin(\omega_n t)$. *We call this response resonant. It is the damping ratio that is decisive. If* $\omega/\omega_n \gg 1$ *then* $\beta_n \approx \pi$, *while the response* $r(t) \approx (F_0/K)\cos(\omega t - \beta_n)/(\omega/\omega_n)^2 = \left(F_0/\omega^2 M\right)\cos(\omega t - \beta_n)$, *i.e. it is the mass that is decisive.*

2.5 Response to a General Periodic Load

A simple single degree of freedom system subject to a fluctuating load $F(t)$ with period T_F is illustrated in Fig. 2.8 below. It is taken for granted that its mean value is zero, i.e. that

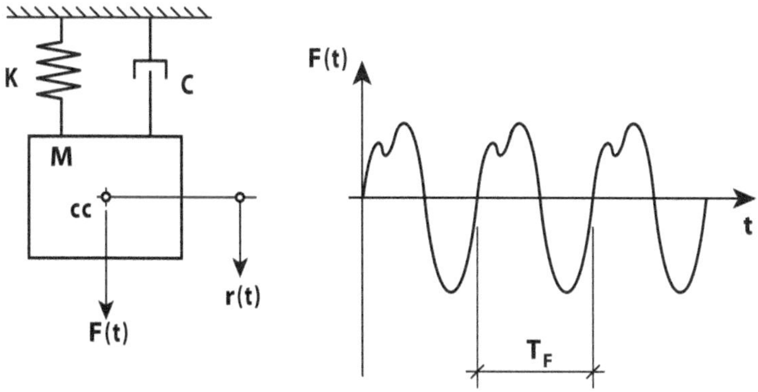

Fig. 2.8 Single degree of freedom system subject to periodic load

$$\bar{F} = \frac{1}{T_F} \int_0^{T_F} F(t)\,dt = 0 \tag{2.67}$$

Let a Fourier transform of the load (in simple harmonic cosine and sinus functions) be given by

$$F(t) = \sum_{j=1}^{N_j} \left(B_j \cos \omega_j t + C_j \sin \omega_j t \right) \tag{2.68}$$

where

$$\begin{bmatrix} B_j \\ C_j \end{bmatrix} = \frac{2}{T_F} \int_0^{T_F} F(t) \cdot \begin{bmatrix} \cos \omega_j t \\ \sin \omega_j t \end{bmatrix} dt \tag{2.69}$$

and where $\omega_j = j \cdot 2\pi / T_F = j \cdot \omega_F$ (and where the obvious formal requirement that $N_j \to \infty$ is omitted for the sake of simplicity). By defining

$$F_j = \sqrt{B_j^2 + C_j^2} \quad \text{and} \quad \begin{cases} C_j = F_j \sin \beta_{F_j} \\ B_j = F_j \cos \beta_{F_j} \end{cases} \quad \Rightarrow \quad \tan \beta_{F_j} = \frac{C_j}{B_j} \tag{2.70}$$

then

$$\begin{aligned} F(t) &= \sum_{j=1}^{N_j} \left(B_j \cos \omega_j t + C_j \sin \omega_j t \right) = \sum_{j=1}^{N_j} \left(F_j \cos \beta_{F_j} \cos \omega_j t + F_j \sin \beta_{F_j} \sin \omega_j t \right) \\ &= \sum_{j=1}^{N_j} F_j \cos\left(\omega_j t - \beta_{F_j} \right) = \operatorname{Re} \sum_{j=1}^{N_j} F_j \cdot e^{i\left(\omega_j t - \beta_{F_j}\right)} \end{aligned} \tag{2.71}$$

We have seen in Chapter 2.4 above that the steady state response of a single degree of freedom system to a harmonic load $F(t) = \operatorname{Re}\left(F_0 \cdot e^{i\omega t} \right)$ was given by

$$r(t) = \frac{F_0}{K} \left| \hat{H}_n(\omega) \right| \cos(\omega t - \beta_n) = \operatorname{Re}\left[\frac{F_0}{K} \hat{H}_n(\omega) e^{i\omega t} \right] = \operatorname{Re}\left[a(\omega) e^{i\omega t} \right] \tag{2.72}$$

Therefore, for the more general case that

$$F(t) = \operatorname{Re} \sum_{j=1}^{N_j} F_j \cdot e^{i\left(\omega_j t - \beta_{F_j}\right)} \tag{2.73}$$

the corresponding response will be given by

$$r(t) = \mathrm{Re}\sum_{j=1}^{N_j} a_j(\omega)\cdot e^{i\left(\omega_j t-\beta_{F_j}\right)} = \sum_{j=1}^{N_j}\left|a_j(\omega)\right|\cos\left(\omega_j t-\beta_{F_j}-\beta_{n_j}\right) \tag{2.74}$$

where $a_j(\omega)=b_j+ic_j$, i.e. $\left|a_j(\omega)\right|=\sqrt{b_j^2+c_j^2}$ and $\tan\beta_{n_j}=c_j/b_j$. Taking the time domain Fourier transform throughout the dynamic equilibrium condition

$$M\cdot\ddot{r}(t)+C\cdot\dot{r}(t)+K\cdot r(t)=F(t) \tag{2.75}$$

will then require that for every ω_j setting

$$\mathrm{Re}\sum_{j=1}^{N_j}\left\{\left[-M\omega_j^2+Ci\omega_j+K\right]a_j-F_j\right\}e^{i\left(\omega_j t-\beta_{F_j}\right)}=0 \tag{2.76}$$

and thus, from an equilibrium point of view we must have that

$$a_j=F_j\left[-M\omega_j^2+Ci\omega_j+K\right]^{-1} \tag{2.77}$$

Introducing that $\omega_n^2=K/M$ and $\zeta_n=C/(2\omega_n M)=\omega_n\cdot C/(2K)$, and defining the non-dimensional frequency response function

$$\hat{H}_n(\omega_j)=\left[1-\left(\omega_j/\omega_n\right)^2+2i\zeta_n\omega_j/\omega_n\right]^{-1} \tag{2.78}$$

then the following is obtained

$$a_j=\left(F_j/K\right)\cdot\hat{H}(\omega_j) \tag{2.79}$$

and thus

$$r(t)=\mathrm{Re}\sum_{j=1}^{N_j}\frac{F_j}{K}\hat{H}(\omega_j)e^{i\left(\omega_j t-\beta_{F_j}\right)}=\sum_{j=1}^{N_j}\frac{F_j}{K}\left|\hat{H}(\omega_j)\right|\cos\left(\omega_j t-\beta_{F_j}-\beta_{n_j}\right) \tag{2.80}$$

$$\left|\hat{H}\left(\omega_j\right)\right| = \left\{\left[1-\left(\frac{\omega_j}{\omega_n}\right)^2\right]^2 + \left(2\zeta_n\frac{\omega_j}{\omega_n}\right)^2\right\}^{-\frac{1}{2}}$$

where (2.81)

$$\tan\beta_{n_j} = \frac{2\zeta_n\,\omega_j/\omega_n}{1-\left(\omega_j/\omega_n\right)^2}$$

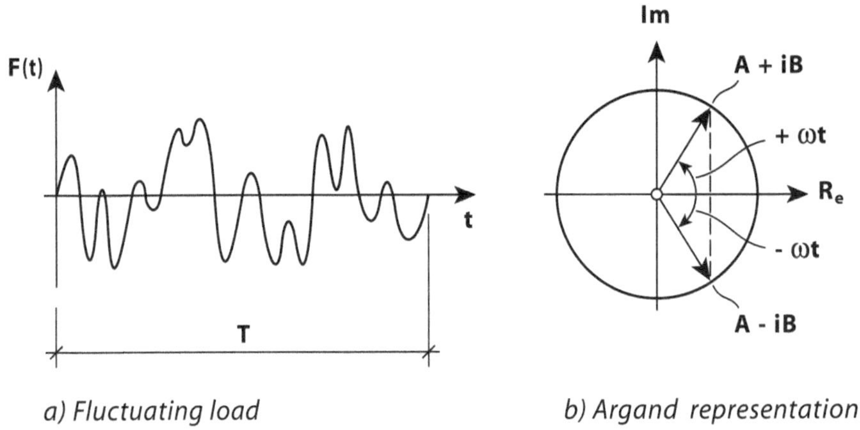

a) Fluctuating load *b) Argand representation*

Fig. 2.9 Single degree of freedom system subject to fluctuating load

The development above may readily be made applicable to a more general fluctuating load (see Fig. 2.9.a) by letting $T_F = T$ become large, formally $T \to \infty$, and transformed into a continuous format where $N_j \to \infty$, $\omega = j\cdot 2\pi/T$ and $2\pi/T \to d\omega$). Introducing the Euler equations

$$\cos\omega_j t = \frac{1}{2}\left(e^{i\omega_j t} + e^{-i\omega_j t}\right) \quad \text{and} \quad \sin\omega_j t = -i\frac{1}{2}\left(e^{i\omega_j t} - e^{-i\omega_j t}\right) \tag{2.82}$$

into Eq. 2.68 renders

$$\begin{aligned} F(t) &= \frac{1}{2}\sum_{j=1}^{\infty}\left[B_j\left(e^{i\omega_j t} + e^{-i\omega_j t}\right) - iC_j\left(e^{i\omega_j t} - e^{-i\omega_j t}\right)\right] \\ &= \frac{1}{2}\sum_{j=1}^{\infty}\left[\left(B_j - iC_j\right)e^{i\omega_j t} + \left(B_j + iC_j\right)e^{-i\omega_j t}\right] \end{aligned} \tag{2.83}$$

which, by defining $$D_j = \left(B_j - iC_j\right)/2 \tag{2.84}$$

may be expanded into positive and fictitious negative ω domain, such that imaginary parts in the positive ω domain will consistently cancel out the same imaginary parts in the negative ω domain (see Fig. 2.9.b), and then

$$F(t) = \sum_{-\infty}^{\infty} D_j \cdot e^{i\omega_j t} = \sum_{-\infty}^{\infty} \left(\frac{T}{2\pi} D_j\right) \cdot e^{i\omega_j t} \left(2\pi/T\right) \tag{2.85}$$

Defining
$$\begin{cases} G\left(\omega_j\right) = \dfrac{T}{2\pi} D_j = \dfrac{T}{2\pi}\dfrac{1}{2}\left(B_j - iC_j\right) \\ = \dfrac{T}{2\pi}\dfrac{1}{2}\dfrac{2}{T}\displaystyle\int_0^T F(t)\left(\cos\omega_j t - i\sin\omega_j t\right)dt = \dfrac{1}{2\pi}\displaystyle\int_0^T F(t) e^{-i\omega_j t} dt \end{cases} \tag{2.86}$$

and letting T and $j \to \infty$, $\omega = j \cdot 2\pi/T$ and $2\pi/T \to d\omega$, then the following is obtained:

$$\left.\begin{aligned} F(t) &= \int_{-\infty}^{\infty} G(\omega) \cdot e^{i\omega t} d\omega \\ G(\omega) &= \frac{1}{2\pi} \int_0^{\infty} F(t) \cdot e^{-i\omega t} dt \end{aligned}\right\} \tag{2.87}$$

[It should be noted that if $F(t)$ has the unit N, then the Fourier function $G(\omega)$ will have the unit Ns/rad.] Similarly, in a complex format, the dynamic response may be expressed by

$$r(t) = \int_{-\infty}^{\infty} g(\omega) \cdot e^{i\omega t} d\omega \tag{2.88}$$

where the Fourier function $g(\omega)$ may be determined by demanding fulfilment of the equilibrium requirement in Eq. 2.75 at every ω setting (which is equivalent to taking the Fourier transform throughout the equation and demanding fulfilment in frequency domain). Thus

$$\left[-M\omega^2 + Ci\omega + K\right] g(\omega) = G(\omega) \tag{2.89}$$

from which the following is obtained

$$r(t)=\int_{-\infty}^{\infty}\frac{G(\omega)}{K}\cdot\hat{H}_n(\omega)\cdot e^{i\omega t}d\omega \tag{2.90}$$

where
$$\left.\begin{aligned}\hat{H}_n(\omega)&=\left[1-(\omega/\omega_n)^2+2i\zeta_n\omega/\omega_n\right]^{-1}\\ G(\omega)&=\frac{1}{2\pi}\int_0^{\infty}F(t)\cdot e^{-i\omega t}dt\end{aligned}\right\} \tag{2.91}$$

and $\omega_n^2=K/M$ and $\zeta_n=C/(2\omega_n M)$. This is what constitutes the basis for the frequency domain dynamic response calculation of structures subject to stochastic load. It is seen that the necessity of operating on the real values vanish with the introduction of a double sided ω domain, as all imaginary quantities cancel out with the fictitious double sided integration (as shown in Fig. 2.8).

2.6 Systems with Two Degrees of Freedom

A simple system with two degrees of freedom is shown in Fig. 2.10.a. The necessary equilibrium considerations for each of the two bodies are illustrated in Fig. 2.10.b. Thus

$$\left.\begin{aligned}M_1\ddot{r}_1+C_1\dot{r}_1+K_1r_1-C_2(\dot{r}_2-\dot{r}_1)-K_2(r_2-r_1)-F_1&=0\\ M_2\ddot{r}_2+C_2(\dot{r}_2-\dot{r}_1)+K_2(r_2-r_1)-F_2&=0\end{aligned}\right\} \tag{2.92}$$

which may more conveniently be written

$$\mathbf{M}\cdot\ddot{\mathbf{r}}+\mathbf{C}\cdot\dot{\mathbf{r}}+\mathbf{K}\cdot\mathbf{r}=\mathbf{F} \tag{2.93}$$

where the displacement and load vectors are defined by $\mathbf{r}=[r_1\quad r_2]^T$ and $\mathbf{F}=[F_1\quad F_2]^T$, and where the mass, damping and stiffness matrices are given by

$$M=\begin{bmatrix}M_1 & 0\\ 0 & M_2\end{bmatrix}\quad \mathbf{C}=\begin{bmatrix}(C_1+C_2) & -C_2\\ -C_2 & C_2\end{bmatrix}\quad \mathbf{K}=\begin{bmatrix}(K_1+K_2) & -K_2\\ -K_2 & K_2\end{bmatrix} \tag{2.94}$$

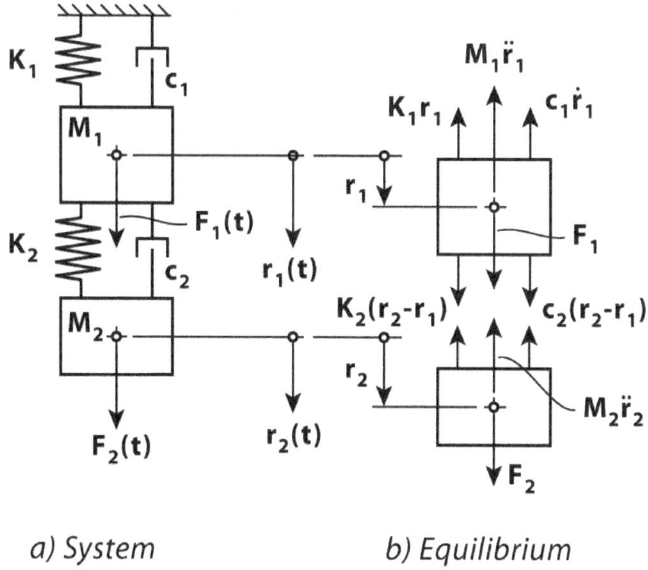

Fig. 2.10 Simple system with two degrees of freedom

The situation of undamped free oscillations (where $\mathbf{F}=0$ and $\mathbf{C}=0$) has previously been solved in Chapter 1.2, rendering the eigenfrequencies ω_1 and ω_2 (see Eq. 1.20), as well as the corresponding eigenmodes $\boldsymbol{\varphi}_1=\begin{bmatrix}\phi_{11} & \phi_{21}\end{bmatrix}^T$ and $\boldsymbol{\varphi}_2=\begin{bmatrix}\phi_{12} & \phi_{22}\end{bmatrix}^T$ given in Eqs. 1.21 – 1.22.

Elaboration 2.2: The Dynamic Absorber

Let us for simplicity assume that the damping is insignificant (i.e. that $C_1=C_2\approx 0$) and that the system is subject to a single harmonic load $F_1=F_{1_0}\cdot\text{Re}\left(e^{i\omega t}\right)$ and $F_2=0$. Thus, the equilibrium requirement (Eq. 2.93) becomes

$$\begin{bmatrix} M_1 & 0 \\ 0 & M_2 \end{bmatrix}\begin{bmatrix} \ddot{r}_1 \\ \ddot{r}_2 \end{bmatrix}+\begin{bmatrix} (K_1+K_2) & -K_2 \\ -K_2 & K_2 \end{bmatrix}\begin{bmatrix} r_1 \\ r_2 \end{bmatrix}=\begin{bmatrix} F_{1_0} \\ 0 \end{bmatrix}\text{Re}\left(e^{i\omega t}\right)$$

We have previously seen that the response to such a harmonic excitation is given by

$$\mathbf{r} = \mathrm{Re}\left(\begin{bmatrix} r_{1_0} \\ r_{2_0} \end{bmatrix} e^{i\omega t}\right)$$

Introduced into the equilibrium equation this will render

$$\begin{bmatrix} \left(K_1 + K_2 - \omega^2 M_1\right) & -K_2 \\ -K_2 & \left(K_2 - \omega^2 M_2\right) \end{bmatrix} \begin{bmatrix} r_{1_0} \\ r_{2_0} \end{bmatrix} = \begin{bmatrix} F_{1_0} \\ 0 \end{bmatrix}$$

and thus

$$\begin{bmatrix} r_{1_0} \\ r_{2_0} \end{bmatrix} = \frac{F_{1_0}}{\left(K_1 + K_2 - \omega^2 M_1\right)\left(K_2 - \omega^2 M_2\right) - K_2^2} \begin{bmatrix} \left(K_2 - \omega^2 M_2\right) \\ K_2 \end{bmatrix}$$

Introducing $\omega_{1_0} = \sqrt{K_1/M_1}$ and $\omega_{2_0} = \sqrt{K_2/M_2}$, then the following is obtained:

$$\begin{bmatrix} r_{1_0} \\ r_{2_0} \end{bmatrix} = \frac{F_{1_0}}{K_1} \cdot \frac{1}{\left[1 + K_2/K_1 - \left(\omega/\omega_{1_0}\right)^2\right]\left[1 - \left(\omega/\omega_{2_0}\right)^2\right] - K_2/K_1} \begin{bmatrix} \left(1 - \left(\omega/\omega_{2_0}\right)^2\right) \\ 1 \end{bmatrix}$$

which is the basic idea behind the dynamic absorber.

To illustrate its effect let us for simplicity choose $\omega_{1_0} \approx \omega_{2_0}$ (i.e. that $K_1/M_1 \approx K_2/M_2$) and $M_2/M_1 \approx K_2/K_1 = \mu$. Then

$$\begin{bmatrix} r_{1_0} \\ r_{2_0} \end{bmatrix} = \frac{F_{1_0}}{K_1} \cdot \frac{1}{\left[1 + \mu - \left(\omega/\omega_{1_0}\right)^2\right]\left[1 - \left(\omega/\omega_{1_0}\right)^2\right] - \mu} \begin{bmatrix} \left(1 - \left(\omega/\omega_{1_0}\right)^2\right) \\ 1 \end{bmatrix}$$

The frequency response function of the main system, r_{1_0}, is illustrated in Fig. 2.11 below. As can be seen, if the load frequency ω is close to ω_{1_0}, which is the eigenfrequency of the body subject to the load, then r_{1_0} becomes unduly large if $\mu = 0$, but, if for instance $\mu = 0.01$ then r_{1_0} is reduced to virtually nothing. This is known as the dynamic absorber effect. It is the basic idea behind the more useful concept of the tuned mass damper, which is given an extensive presentation in Chapter 9.4.

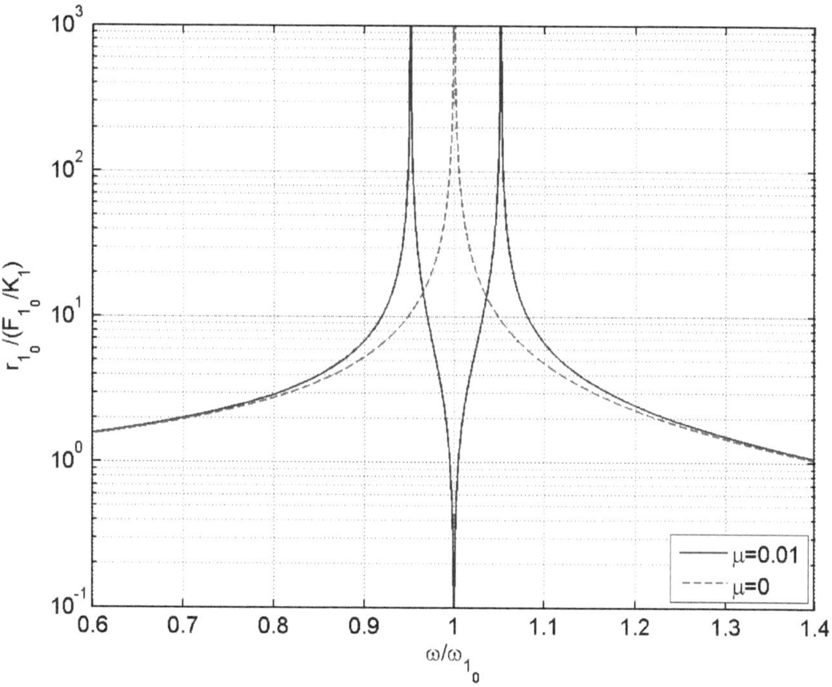

Fig. 2.11 The dynamic absorber, $\omega_{1_0} \approx \omega_{2_0}$

Example 2.3 Two Storey Shear Frame

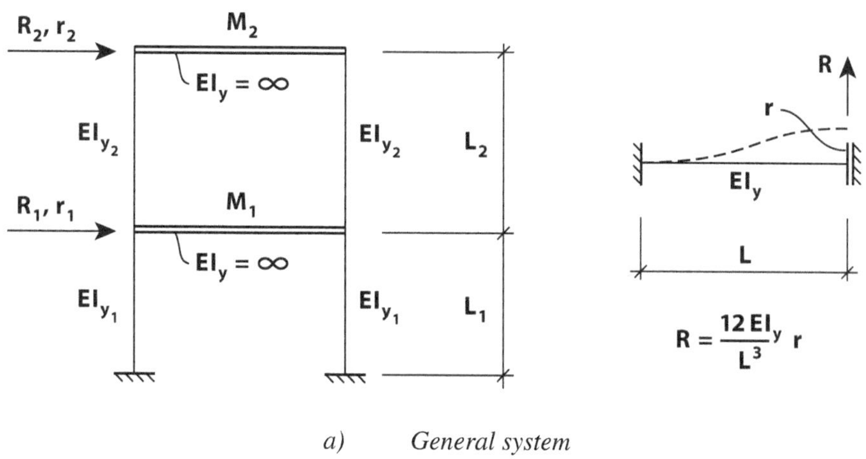

Fig. 2.12 Two storey shear frame

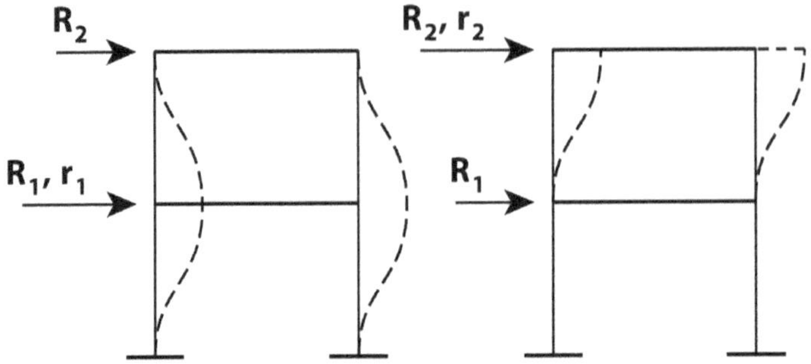

b) Establishing content to the equilibrium equations

Fig. 2.12 *(continued)*

For an effective development of the equilibrium conditions of systems with more than one degree of freedom it is convenient to introduce a more robust approach. We know that in a matrix format the equilibrium requirements to the two storey shear frame in Fig. 2.12 above may be written

$$\begin{bmatrix} M_{11} & M_{12} \\ M_{21} & M_{22} \end{bmatrix} \begin{bmatrix} \ddot{r}_1 \\ \ddot{r}_2 \end{bmatrix} + \begin{bmatrix} C_{11} & C_{12} \\ C_{21} & C_{22} \end{bmatrix} \begin{bmatrix} \dot{r}_1 \\ \dot{r}_2 \end{bmatrix} + \begin{bmatrix} K_{11} & K_{12} \\ K_{21} & K_{22} \end{bmatrix} \begin{bmatrix} r_1 \\ r_2 \end{bmatrix} = \begin{bmatrix} R_1 \\ R_2 \end{bmatrix}$$

Demanding equilibrium in the fictitious setting that $r_1 \neq 0$ *and* $r_2 = 0$ *(see Fig. 2.12.b), then*

$$R_1 = M_{11}\ddot{r}_1 + C_{11}\dot{r}_1 + K_{11}r_1$$
$$R_2 = M_{21}\ddot{r}_1 + C_{21}\dot{r}_1 + K_{21}r_1$$

rendering the first column in the matrices in the equilibrium condition above. Similarly, if we demand equilibrium in the fictitious setting that $r_1 = 0$ *and* $r_2 \neq 0$, *then*

$$R_1 = M_{12}\ddot{r}_2 + C_{12}\dot{r}_2 + K_{12}r_2$$
$$R_2 = M_{22}\ddot{r}_2 + C_{22}\dot{r}_2 + K_{22}r_2$$

rendering the second column in the matrices in the equilibrium condition above. Let us in the following for simplicity assume that damping is exclusively associated with the motion of masses M_1 *and* M_2, *and define*

$$K_1 = 24EI_y / L_1^3 \qquad \text{and} \qquad K_2 = 24EI_y / L_2^3$$

then

1) $r_1 \neq 0$ *and* $r_2 = 0$ *(see Fig. 2.21.b)* $\Rightarrow \begin{cases} R_1 = M_1\ddot{r}_1 + C_1\dot{r}_1 + (K_1 + K_2) r_1 \\ R_2 = -K_2 r_1 \end{cases}$

2 $r_1 = 0$ *and* $r_2 \neq 0$ $\Rightarrow \begin{cases} R_1 = -K_2 r_2 \\ R_2 = M_2\ddot{r}_2 + C_2\dot{r}_2 + K_2 r_2 \end{cases}$

Thus, the equilibrium condition of the two storey frame in Fig. 2.12 is given by

$$\begin{bmatrix} M_1 & 0 \\ 0 & M_2 \end{bmatrix}\begin{bmatrix} \ddot{r}_1 \\ \ddot{r}_2 \end{bmatrix} + \begin{bmatrix} C_1 & 0 \\ 0 & C_2 \end{bmatrix}\begin{bmatrix} \dot{r}_1 \\ \dot{r}_2 \end{bmatrix} + \begin{bmatrix} (K_1 + K_2) & -K_2 \\ -K_2 & K_2 \end{bmatrix}\begin{bmatrix} r_1 \\ r_2 \end{bmatrix} = \begin{bmatrix} R_1 \\ R_2 \end{bmatrix}$$

which may be written in the following more compact way

$$\mathbf{M}\ddot{\mathbf{r}} + \mathbf{C}\dot{\mathbf{r}} + \mathbf{K}\mathbf{r} = \mathbf{R}$$

where $\mathbf{r} = [r_1 \quad r_2]^T$, $\mathbf{R} = [R_1 \quad R_2]^T$ *and*

$$\mathbf{M} = \begin{bmatrix} M_1 & 0 \\ 0 & M_2 \end{bmatrix} \qquad \mathbf{C} = \begin{bmatrix} C_1 & 0 \\ 0 & C_2 \end{bmatrix} \qquad \mathbf{K} = \begin{bmatrix} (K_1 + K_2) & -K_2 \\ -K_2 & K_2 \end{bmatrix}$$

The undamped and unloaded eigenvalues and corresponding eigenmodes may as usual be obtained by setting $\mathbf{C}$ *and* $\mathbf{R}$ *both equal to zero, and impose the harmonic motion*

$$\mathbf{r}(t) = \begin{bmatrix} r_1 \\ r_2 \end{bmatrix} = \mathrm{Re}\left(\mathbf{a}e^{i\omega t}\right) \qquad \textit{where} \qquad \mathbf{a} = [a_1 \quad a_2]^T$$

Thus, the following eigenvalue problem is obtained

$$\left(\mathbf{K} - \omega^2\mathbf{M}\right)\mathbf{a} = \mathbf{0}$$

which, fully written is $\begin{bmatrix} \left(K_1 + K_2 - \omega^2 M_1\right) & -K_2 \\ -K_2 & \left(K_2 - \omega^2 M_2\right) \end{bmatrix}\begin{bmatrix} a_1 \\ a_2 \end{bmatrix} = \begin{bmatrix} 0 \\ 0 \end{bmatrix}$

Its solution may be obtained by setting the determinant to the coefficient matrix equal to zero, rendering the following fourth order polynomial solution

$$\omega^4 - \left(\frac{K_1 + K_2}{M_1} + \frac{K_2}{M_2}\right)\omega^2 + \frac{K_1}{M_1}\frac{K_2}{M_2} = 0$$

whose roots are given by

$$\omega = \left\{ \frac{1}{2}\left(\frac{K_1 + K_2}{M_1} + \frac{K_2}{M_2}\right) \pm \left[\frac{1}{4}\left(\frac{K_1 + K_2}{M_1} + \frac{K_2}{M_2}\right)^2 - \frac{K_1}{M_1}\frac{K_2}{M_2} \right]^{1/2} \right\}^{1/2}$$

Let for instance $K_1 = 2K$, $K_2 = K$, $M_1 = 2M$ *and* $M_2 = M$, *then the solution to the eigenvalue problem*

$$\begin{bmatrix} \left(3K - \omega^2 2M\right) & -K \\ -K & \left(K - \omega^2 M\right) \end{bmatrix} \begin{bmatrix} a_1 \\ a_2 \end{bmatrix} = \begin{bmatrix} 0 \\ 0 \end{bmatrix}$$

is given by $\dfrac{\omega^2 M}{K} = \dfrac{5}{4} \pm \dfrac{3}{4}$ *and thus* $\begin{cases} \omega_1 = \sqrt{K/(2M)} \\ \omega_2 = \sqrt{2K/M} \end{cases}$

From the second row of the eigenvalue problem it is seen that

$$-a_1 + \left(1 - \frac{\omega^2 M}{K}\right)a_2 = 0$$

It is seen from the eigenvalue problem that one is free to scale the content of the eigenmodes $\boldsymbol{\varphi} = \begin{bmatrix} a_1 & a_2 \end{bmatrix}^T$. *Eigenmodes associated with eigenvalues* ω_1 *and* ω_2 *are then obtained by successively introducing* ω_1 *and* ω_2 *into the equation above. Thus:*

1) $\omega = \omega_1 \Rightarrow a_2/a_1 = 2$ *and thus* $\boldsymbol{\varphi} = \begin{bmatrix} 1 & 2 \end{bmatrix}^T$

2) $\omega = \omega_2 \Rightarrow a_2/a_1 = -1$ *and thus* $\boldsymbol{\varphi} = \begin{bmatrix} 1 & -1 \end{bmatrix}^T$

The numerical eigenvalue problem is more thoroughly presented in Chapter 4.4.

Chapter 3
Eigenvalue Calculations of Continuous Systems

3.1 Eigenvalue Calculations of Simple Beams

A simple beam is defined as a single span beam whose cross section is symmetric about the y- as well as the z- axis, i.e. its shear centre and its centre of pure linear bending coincides with its mass centre ($e_y = e_z = 0$). It is homogenous and line-like in the sense that along the entire span it contains only one type of material, and the cross section is small as compared to the length of the beam such that it may mathematically be modelled as a single line through its shear centre. Furthermore, it is assumed that time invariant mean cross sectional forces ($\bar{N}$, $\bar{M}_y$ and $\bar{M}_z$) are zero, and that the motion of the system is restricted such that it is only r_z that is unequal to zero. The more general case of beams with non-symmetric cross section where r_y, r_z and r_θ are simultaneously unequal to zero is covered in Chapter 3.2, while the effects of time invariant forces are included in the cases covered in Chapter 3.3. The most typical four cases of boundary conditions for simple beams are illustrated in Fig. 3.1.c. As shown in Chapter 1.1 (see Eqs. 1.23 – 1.36), an exact solution to the problem of undamped and unloaded free oscillations

$$r_z(x,t) = \mathrm{Re}\left[\phi_z(x) \cdot e^{i\omega t}\right] \tag{3.1}$$

of a continuous system can be obtained by solving the differential equation of dynamic equilibrium

$$m_z \ddot{r}_z + EI_y r_z'''' = 0 \qquad \Rightarrow \qquad EI_y \phi_z'''' - \omega^2 m_z \phi_z = 0 \tag{3.2}$$

which can only be obtained for all values of x if the fourth derivative of ϕ_z is shapewise congruent to itself. Thus, the general solution (conveniently expressed in the non-dimensional coordinate $\hat{x} = x/L$) is given by

$$\phi_z(x) = a_1 \sin(\lambda \hat{x}) + a_2 \cos(\lambda \hat{x}) + a_3 \sinh(\lambda \hat{x}) + a_4 \cosh(\lambda \hat{x}) \tag{3.3}$$

E.N. Strømmen, *Structural Dynamics*, Springer Series in Solid and Structural Mechanics 2,
DOI: 10.1007/978-3-319-01802-7_3, © Springer International Publishing Switzerland 2014

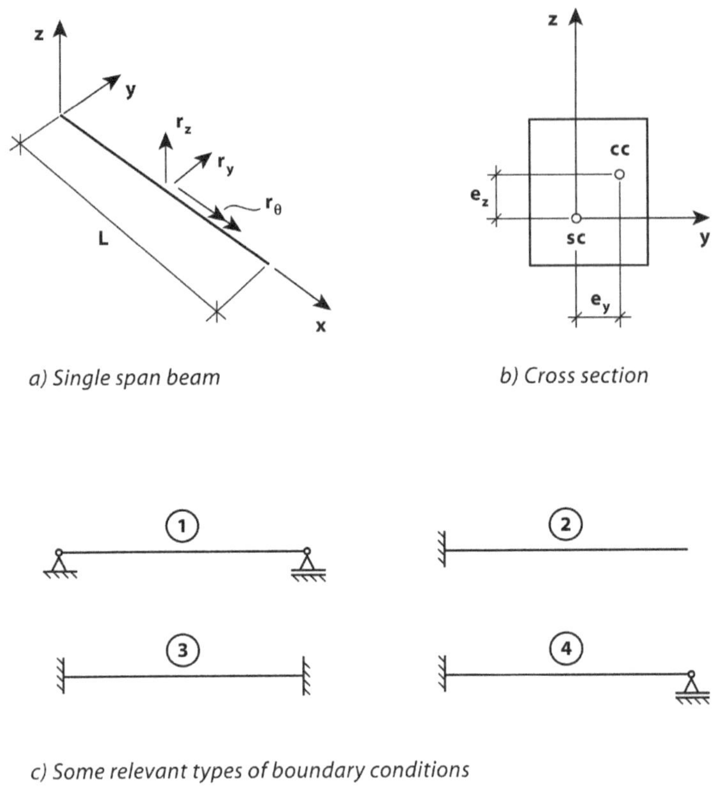

Fig. 3.1 Typical continuous system and some relevant boundary conditions

where the coefficients a_j, $j = 1, 2, 3$ or 4, and the non-dimensional wave length λ are all determined from the relevant boundary conditions at the beam element ends, i.e. at $\hat{x} = 0$ and $\hat{x} = 1$. Apart from boundary requirements to $\phi_z(x)$ itself, the boundary conditions may entail requirements to the slope

$$\phi_z'(x) = (\lambda/L)\left[a_1 \cos(\lambda\hat{x}) - a_2 \sin(\lambda\hat{x}) + a_3 \cosh(\lambda\hat{x}) + a_4 \sinh(\lambda\hat{x})\right] \tag{3.4}$$

to the bending moment

$$\begin{aligned} M_y(x,t) = -EI_y r_z'' = -EI_y \operatorname{Re}\left[\phi_z''(x) e^{i\omega t}\right] = \\ -EI_y \operatorname{Re}\left\{(\lambda/L)^2 \left[-a_1 \sin(\lambda\hat{x}) - a_2 \cos(\lambda\hat{x}) + a_3 \sinh(\lambda\hat{x}) + a_4 \cosh(\lambda\hat{x})\right] e^{i\omega t}\right\} \end{aligned} \tag{3.5}$$

or to the shear force

$$V_z(x,t) = M'_y(x,t) = \\ -EI_y \operatorname{Re}\left\{(\lambda/L)^3\left[-a_1\cos(\lambda\hat{x}) + a_2\sin(\lambda\hat{x}) + a_3\cosh(\lambda\hat{x}) + a_4\sinh(\lambda\hat{x})\right]e^{i\omega t}\right\} \tag{3.6}$$

In general, there will be four boundary conditions, two at either ends of the beam, rendering a relative connection between the coefficients a_j, $j = 1, 2, 3$ or 4 and a transcendental equation $f(\lambda) = 0$. Thus, since it is only the relative connection between the coefficients that can be obtained it is only the shape of ϕ_z that can be determined, i.e. one of the coefficients is arbitrary and may conveniently be chosen equal to unity. Furthermore, the transcendental equation obtained from the four boundary conditions will have an infinite number of solutions $\lambda = \lambda_n$, n = 1, 2, 3,…, each representing a possible satisfaction of the relevant boundary conditions. Thus, there will be an infinite number of possible mode shapes $\phi_{z_n}(\lambda_n\hat{x})$, which, introduced back into the differential equation of dynamic equilibrium (Eq. 3.2), will render

$$\left[EI_y(\lambda_n/L)^4 - \omega^2 m_z\right] \cdot \phi_{z_n}(\lambda_n\hat{x}) = 0 \tag{3.7}$$

Thus, a corresponding set of eigenfrequencies are obtained

$$\omega_{z_n} = \lambda_n^2 \sqrt{EI_y/(m_z L^4)} \tag{3.8}$$

As mentioned above (see also Eq. 3.7), the size of ϕ_{z_n} is arbitrary, i.e. they are merely shapes that may be scaled up or down at will. In the following a_1 is conveniently set equal to unity. From a structural safety point of view it is in general only a few of the lowest values of ω_{z_n} and corresponding set of mode shapes ϕ_{z_n} that are of interest. The reason for this is that higher eigenvalues are likely to lie beyond the frequency band of possible load excitation.

Beam Type 1, Simple Supports at Either Ends (See Fig. 3.1.c)

For beam type 1 in Fig. 3.1.c the boundary conditions require

$$\left.\begin{array}{l}\left.\begin{array}{l}\phi_z(x=0)=0\\ M_y(x=0)=0\end{array}\right\}\Rightarrow\left.\begin{array}{r}a_2+a_4=0\\ -a_2+a_4=0\end{array}\right\}\Rightarrow a_2=a_4=0\\ \left.\begin{array}{l}\phi_z(x=L)=0\\ M_y(x=L)=0\end{array}\right\}\Rightarrow\left.\begin{array}{r}a_1\sin\lambda+a_3\sinh\lambda=0\\ -a_1\sin\lambda+a_3\sinh\lambda=0\end{array}\right\}\Rightarrow\left\{\begin{array}{l}a_3=0\\ \sin\lambda\cdot\sinh\lambda=0\end{array}\right.\end{array}\right\} \tag{3.9}$$

The non-trivial solution to the transcendental equation $f(\lambda)=\sin\lambda\cdot\sinh\lambda=0$ is that $\lambda=\lambda_n=n\pi$, $n=1,2,3,\ldots$, rendering n mode shape functions

$$\phi_{z_n}(x)=\sin(\lambda_n\hat{x}) \tag{3.10}$$

Introducing Eq. 3.10 back into Eq. 3.1 and 3.2 will then render Eq. 3.8, from which the eigenfrequencies may be obtained. The four first mode shapes and corresponding λ-values are shown in Fig. 3.2 below.

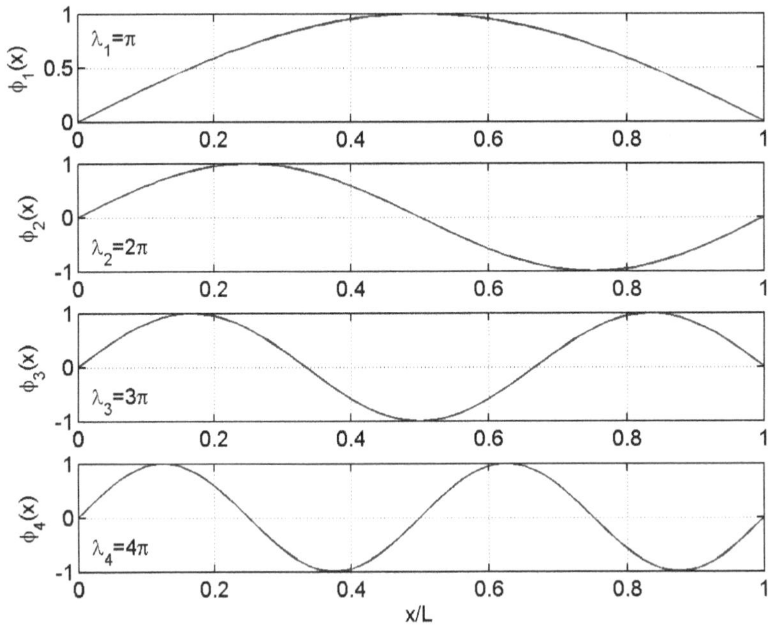

Fig. 3.2 Beam type 1, beam with simple supports at either ends

Beam Type 2, Fixed Support on One Side, Free on the Other (See Fig. 3.1.c)

For beam type 2 in Fig. 3.1.c the boundary conditions require

$$\left.\begin{array}{l}\left.\begin{array}{l}\phi_z(x=0)=0\\ \phi_z'(x=0)=0\end{array}\right\}\Rightarrow \left.\begin{array}{l}a_2+a_4=0\\ a_1+a_3=0\end{array}\right\}\Rightarrow\begin{cases}a_4=-a_2\\ a_3=-a_1\end{cases}\\ \left.\begin{array}{l}M_y(x=L)=0\\ V_z(x=L)=0\end{array}\right\}\Rightarrow\begin{cases}a_1(\sin\lambda+\sinh\lambda)+a_2(\cos\lambda+\cosh\lambda)=0\\ a_1(\cos\lambda+\cosh\lambda)-a_2(\sin\lambda-\sinh\lambda)=0\end{cases}\end{array}\right\} \tag{3.11}$$

$$\Rightarrow\begin{cases}\dfrac{a_2}{a_1}=-\dfrac{(\sin\lambda-\sinh\lambda)}{(\cos\lambda+\cosh\lambda)}\\ \cos\lambda\cdot\cos\mathrm{h}\lambda+1{=}0\end{cases} \tag{3.12}$$

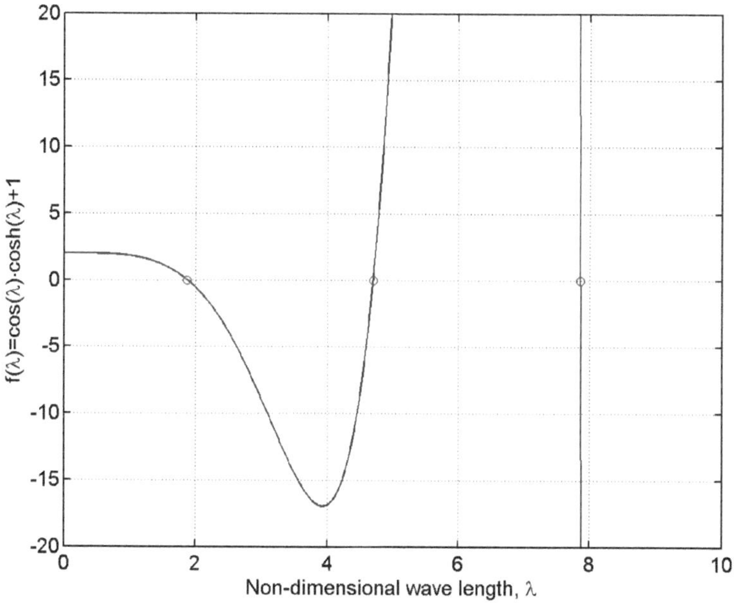

Fig. 3.3 The three first zero crossing points of $f(\lambda)=\cos\lambda\cdot\cosh\lambda+1$

In Fig. 3.3 are shown the three first zero crossings of the transcendental equation $f(\lambda)=\cos\lambda\cdot\cosh\lambda+1$. In general, there are an infinite number of such crossings. The non-trivial solutions $\lambda=\lambda_n$, $n=1,2,3,\ldots$ to the transcendental equation $f(\lambda)=\cos\lambda\cdot\cosh\lambda+1=0$ will then render the relevant mode shapes

$$\phi_{z_n} = \sin(\lambda_n \hat{x}) - \sinh(\lambda_n \hat{x}) + \frac{\sin\lambda_n + \sinh\lambda_n}{\cos\lambda_n + \cosh\lambda_n}\left[\cos(\lambda_n \hat{x}) - \cosh(\lambda_n \hat{x})\right] \quad (3.13)$$

The corresponding set of eigenfrequencies are given in Eq. 3.8. The four first mode shapes and corresponding λ-values are shown in Fig. 3.4.

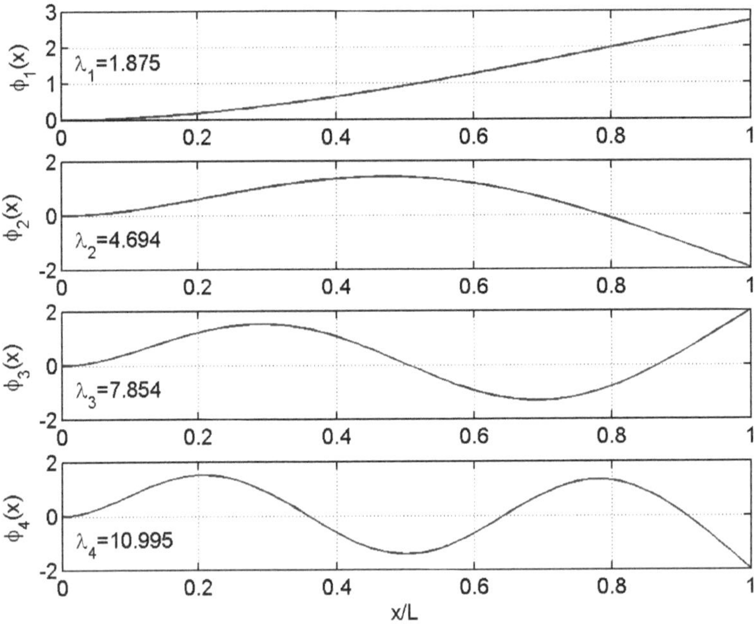

Fig. 3.4 Beam type 2, fixed support on one side and free on the other

Beam Type 3, Fixed Supports at Either Ends (See Fig. 3.1.c)

For beam type 3 in Fig. 3.1.c the boundary conditions require

$$\left.\begin{aligned}
&\left.\begin{aligned}\phi_z(x=0)=0\\ \phi_z'(x=0)=0\end{aligned}\right\} \Rightarrow \left.\begin{aligned}a_2+a_4=0\\ a_1+a_3=0\end{aligned}\right\} \Rightarrow \left\{\begin{aligned}a_4&=-a_2\\ a_3&=-a_1\end{aligned}\right.\\
&\left.\begin{aligned}\phi_z(x=L)=0\\ \phi_z'(x=L)=0\end{aligned}\right\} \Rightarrow \left\{\begin{aligned}a_1(\sin\lambda-\sinh\lambda)+a_2(\cos\lambda-\cosh\lambda)=0\\ a_1(\cos\lambda-\cosh\lambda)-a_2(\sin\lambda+\sinh\lambda)=0\end{aligned}\right.
\end{aligned}\right\} \quad (3.14)$$

$$\Rightarrow \begin{cases} \dfrac{a_2}{a_1} = \dfrac{(\cos\lambda - \cosh\lambda)}{(\sin\lambda + \sinh\lambda)} \\ \cos\lambda \cdot \cosh\lambda - 1 = 0 \end{cases} \tag{3.15}$$

The relevant mode shapes

$$\phi_{z_n} = \sin(\lambda_n \hat{x}) - \sinh(\lambda_n \hat{x}) + \frac{(\cos\lambda_n - \cosh\lambda_n)}{(\sin\lambda_n + \sinh\lambda_n)}\left[\cos(\lambda_n \hat{x}) - \cosh(\lambda_n \hat{x})\right] \tag{3.16}$$

may then be obtained from the non-trivial solutions $\lambda = \lambda_n$, $n = 1, 2, 3, \ldots$ to the transcendental equation $f(\lambda) = \cos\lambda \cdot \cosh\lambda - 1 = 0$, while the corresponding set of eigenfrequencies given in Eq. 3.8. The four first mode shapes and corresponding λ-values are shown in Fig. 3.5.

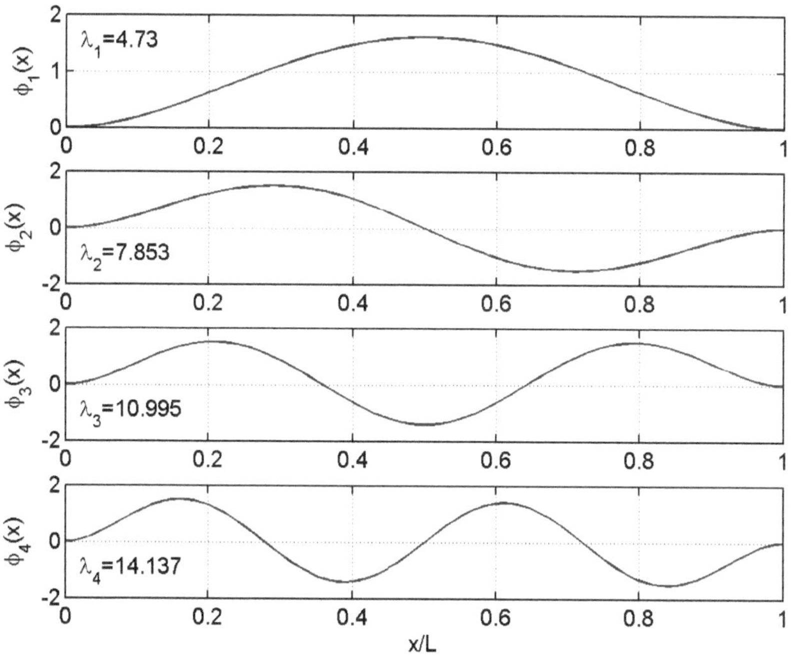

Fig. 3.5 Beam type 3, fixed supports at either ends

Beam Type 4, Fixed Support at One End, Simple at the Other (Fig. 3.1.c):

For beam type 4 in Fig. 3.1.c the boundary conditions require

$$\left.\begin{array}{l}\left.\begin{array}{l}\phi_z(x=0)=0\\ \phi_z'(x=0)=0\end{array}\right\}\Rightarrow \left.\begin{array}{l}a_2+a_4=0\\ a_1+a_3=0\end{array}\right\}\Rightarrow\left\{\begin{array}{l}a_4=-a_2\\ a_3=-a_1\end{array}\right.\\ \left.\begin{array}{l}\phi_z(x=L)=0\\ M_y(x=L)=0\end{array}\right\}\Rightarrow\left\{\begin{array}{l}a_1(\sin\lambda-\sinh\lambda)+a_2(\cos\lambda-\cosh\lambda)=0\\ a_1(\sin\lambda+\sinh\lambda)+a_2(\cos\lambda+\cosh\lambda)=0\end{array}\right.\end{array}\right\} \tag{3.17}$$

$$\Rightarrow\left\{\begin{array}{l}\dfrac{a_2}{a_1}=-\dfrac{(\sin\lambda-\sinh\lambda)}{(\cos\lambda+\cosh\lambda)}\\ \cos\lambda\cdot\sin\mathrm{h}\lambda-\sin\lambda\cdot\cosh\lambda=0\end{array}\right. \tag{3.18}$$

The non-trivial solutions $\lambda=\lambda_n$, $n=1,2,3,\ldots$ to the transcendental equation $f(\lambda)=\cos\lambda\cdot\sin\mathrm{h}\lambda-\sin\lambda\cdot\cosh\lambda=0$ will then render the relevant mode shapes

$$\phi_{z_n}=\sin(\lambda_n\hat{x})-\sinh(\lambda_n\hat{x})-\frac{(\sin\lambda_n-\sinh\lambda_n)}{(\cos\lambda_n-\cosh\lambda_n)}\left[\cos(\lambda_n\hat{x})-\cosh(\lambda_n\hat{x})\right] \tag{3.19}$$

and the corresponding set of eigenfrequencies are given in Eq. 3.8. The four first mode shapes and corresponding λ-values are shown in Fig. 3.6.

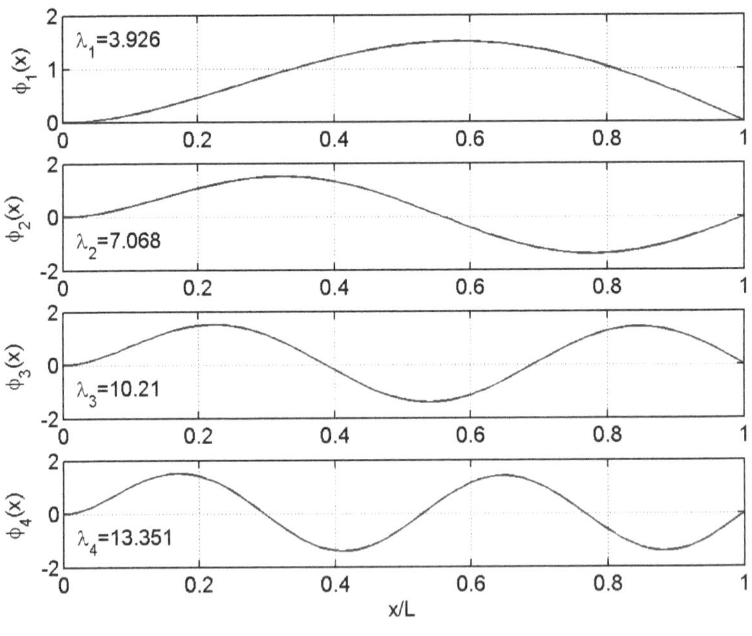

Fig. 3.6 Beam type 4, fixed support at one side and simple at the other

3.2 Beams with Non-symmetric Cross Section

In the previous chapter we dealt with simple unloaded and undamped single span beams whose cross section was assumed to be symmetric about the y as well as the z axis, such that $e_y = e_z = 0$. It was also taken for granted that all time invariant cross sectional forces (e.g. $\bar{N}$, $\bar{M}_y$ and $\bar{M}_z$) were zero. In that case there was no coupling between response components r_y, r_z and r_θ, and therefore, any eigenvalue or response calculation could be handled separately for each of these components, and we chose to focus on r_z. In this chapter we shall adopt the same restrictions, except that we shall now consider the possibility of a non-symmetric cross section, i.e. the situation that e_y and e_z are unequal to zero. This is a more complex case than that which was dealt with in Chapter 3.1 above. Rather than making an attempt to solve the differential equation, it is more convenient to turn to the principle of virtual work, which was presented in Chapter 1.6, rendering the general expression in Eq. 1.119. Under the assumptions given above, Eq. 1.119 is reduced into

$$-\int_L \delta\mathbf{r}^T \mathbf{m}_g \ddot{\mathbf{r}} dx = \int_L \left(\delta r_x' EA r_x' + \delta r_y'' EI_z r_y'' + \delta r_z'' EI_y r_z'' + \delta r_\theta' EA r_\theta' \right) dx$$
$$= \int_L \begin{bmatrix} \delta r_x' \\ \delta r_y'' \\ \delta r_z'' \\ \delta r_\theta' \end{bmatrix}^T \cdot \begin{bmatrix} EA & 0 & 0 & 0 \\ 0 & EI_z & 0 & 0 \\ 0 & 0 & EI_y & 0 \\ 0 & 0 & 0 & GI_t \end{bmatrix} \cdot \begin{bmatrix} r_x' \\ r_y'' \\ r_z'' \\ r_\theta'' \end{bmatrix} dx \tag{3.20}$$

where $\mathbf{r}(x,t) = \begin{bmatrix} r_x & r_y & r_z & r_\theta \end{bmatrix}^T$ and $\delta\mathbf{r}(x) = \begin{bmatrix} \delta r_x & \delta r_y & \delta r_z & \delta r_\theta \end{bmatrix}^T$ and where

$$\mathbf{m}_g = \begin{bmatrix} m_x & 0 & 0 & 0 \\ 0 & m_y & 0 & -m_y e_z \\ 0 & 0 & m_z & m_z e_y \\ 0 & -m_y e_z & m_z e_y & m_\theta \end{bmatrix} \tag{3.21}$$

Since we are here only considering an unloaded and undamped system, the solution is a purely harmonic motion, which may be written on the format

$$\mathbf{r}(x,t) = \mathrm{Re}\left(\begin{bmatrix} a_x\psi_x(x) \\ a_y\psi_y(x) \\ a_z\psi_z(x) \\ a_\theta\psi_\theta(x) \end{bmatrix} \cdot e^{i\omega t}\right) = \mathrm{Re}\left(\boldsymbol{\Psi} \cdot \mathbf{a} \cdot e^{i\omega t}\right) \tag{3.22}$$

where $\boldsymbol{\Psi} = diag\begin{bmatrix} \psi_x & \psi_y & \phi_z & \psi_\theta \end{bmatrix}$ is a four by four matrix containing all the relevant mode shape functions and where $\mathbf{a} = \begin{bmatrix} a_x & a_y & a_z & a_\theta \end{bmatrix}^T$ is a four by one vector containing the corresponding amplitudes of motion. The choice of ψ_x, ψ_y, ψ_z and ψ_θ may be based on the results obtained in Chapter 3.1 above or on chosen approximations fulfilling the boundary conditions. The virtual displacement is arbitrary, and therefore, we conveniently choose $\delta\mathbf{r} = \boldsymbol{\Psi} \cdot \delta\mathbf{a}$ where $\delta\mathbf{a} = \begin{bmatrix} \delta a_x & \delta a_y & \delta a_z & \delta a_\theta \end{bmatrix}^T$, and thus, Eq. 3.20 becomes

$$-\int_L (\boldsymbol{\psi}\delta\mathbf{a})^T \mathbf{m}_g (i\omega)^2 \boldsymbol{\psi}\mathbf{a}e^{i\omega t} dx$$
$$= \int_L \delta\mathbf{a}^T \begin{bmatrix} EA\psi_x'^2 & 0 & 0 & 0 \\ 0 & EI_z\psi_y''^2 & 0 & 0 \\ 0 & 0 & EI_y\psi_z''^2 & 0 \\ 0 & 0 & 0 & GI_t\psi_\theta'^2 \end{bmatrix} \mathbf{a}e^{i\omega t} dx \tag{3.23}$$

rendering

$$\delta\mathbf{a}^T \omega^2 \int_L \boldsymbol{\psi}^T \mathbf{m}_g \boldsymbol{\psi} dx \mathbf{a} = \delta\mathbf{a}^T \int_L \begin{bmatrix} EA\psi_x'^2 & 0 & 0 & 0 \\ 0 & EI_z\psi_y''^2 & 0 & 0 \\ 0 & 0 & EI_y\psi_z''^2 & 0 \\ 0 & 0 & 0 & GI_t\psi_\theta'^2 \end{bmatrix} dx\mathbf{a} \tag{3.24}$$

which, by defining

$$\tilde{\mathbf{M}} = \int_L \boldsymbol{\psi}^T \mathbf{m}_g \boldsymbol{\psi} dx = \int_L \begin{bmatrix} m_x\psi_x^2 & 0 & 0 & 0 \\ 0 & m_y\psi_y^2 & 0 & -m_y e_z \psi_y \psi_\theta \\ 0 & 0 & m_z\psi_z^2 & m_z e_y \psi_z \psi_\theta \\ 0 & -m_y e_z \psi_y \psi_\theta & m_z e_y \psi_z \psi_\theta & m_\theta \psi_\theta^2 \end{bmatrix} dx \tag{3.25}$$

and

$$\tilde{\mathbf{K}} = \int_L \begin{bmatrix} EA\psi_x'^2 & 0 & 0 & 0 \\ 0 & EI_z\psi_y''^2 & 0 & 0 \\ 0 & 0 & EI_y\psi_z''^2 & 0 \\ 0 & 0 & 0 & GI_t\psi_\theta'^2 \end{bmatrix} dx \tag{3.26}$$

and acknowledging that the pre-multiplication by $\delta\mathbf{a}^T$ is obsolete, then the following eigenvalue problem is obtained

$$\left(\tilde{\mathbf{K}} - \omega^2\tilde{\mathbf{M}}\right) \cdot \mathbf{a} = \mathbf{0} \tag{3.27}$$

It is readily seen that the r_x component is independent of the other three r_y, r_z and r_θ components. Thus, it is given a separate treatment below.

Case 1: Along Span Wave Propagation

Along span structural oscillations $r_x(x,t)$ in a line like beam is what we more generally associate with wave propagation. As mentioned above, the r_x displacements may be handled separately from the other components, and thus, extracting the first row and column from Eq. 3.27, the following is obtained:

$$\left(\int_L EA\psi_x'^2 dx - \omega^2 \int_L m_x\psi_x^2 dx\right) a_x = 0 \tag{3.28}$$

and thus, the eigenfrequency associated with along span wave propagation is given by

$$\omega_x = \left(\int_L EA\psi_x'^2 dx \Big/ \int_L m_x\psi_x^2 dx\right)^{\frac{1}{2}} \tag{3.29}$$

The mode shape function covering most relevant cases may generally be expressed by

$$\psi_x = \sin(\lambda x/L) + b \cdot \cos(\lambda x/L) \tag{3.30}$$

where the coefficient b and the wave length λ may be found from the boundary conditions. For beam types 1, 2 and 4 in Fig. 3.1, the boundary conditions $r_x(x=0,t)=0$ and $N(x=L,t)=EAr_x'(x=L,t)=0$ will require $\psi_x(x=0)=\psi_x'(x=L)=0$. The first boundary condition $\psi_x(x=0)=0$ will require $b=0$, The solution is then $\psi_x=\sin(\lambda x/L)$. The wave length λ may be found from the second boundary condition $\psi_x'(x=L)=(\lambda/L)\cos\lambda=0$, rendering the non-trivial solution $\lambda=n\pi-\pi/2$, $n=1,2,3,\ldots$. Thus, since $\sin(2\lambda)=\sin\left[\pi(2n-1)\right]=0$, the following is obtained

$$\omega_{x_n}=\sqrt{\frac{EA(\lambda/L)^2\int_0^L\cos^2(\lambda x/L)dx}{m_x\int_0^L\sin^2(\lambda x/L)dx}}=\lambda\sqrt{\frac{EA}{m_xL^2}}=\pi\left(n-\frac{1}{2}\right)\sqrt{\frac{EA}{m_xL^2}} \tag{3.31}$$

For beam type 3 in Fig. 3.1 the boundary conditions $r_x(x=0,t)=0$ and $r_x(x=L,t)=0$ will require $\psi_x(x=0)=\psi_x(x=L)=0$. From the first boundary condition $b=0$. The second will require $\sin(\lambda)=0$, and thus, $\lambda=n\pi$. The solution is then $\psi_x=\sin(n\pi x/L)$, rendering

$$\omega_{x_n}=\sqrt{EA\left(\frac{n\pi}{L}\right)^2\int_0^L\cos^2\left(\frac{n\pi x}{L}\right)dx\Big/\left[m_x\int_0^L\sin^2\left(\frac{n\pi x}{L}\right)dx\right]}=n\pi\sqrt{\frac{EA}{m_xL^2}} \tag{3.32}$$

Case 2: Bi-axial Bending and Torsion

After the extraction of the r_x component, the mass and stiffness matrices in Eqs. 3.25 and 3.26 are reduced into

$$\tilde{\mathbf{M}}=\int_L\begin{bmatrix} m_y\psi_y^2 & 0 & -m_ye_z\psi_y\psi_\theta \\ 0 & m_z\psi_z^2 & m_ze_y\psi_z\psi_\theta \\ -m_ye_z\psi_y\psi_\theta & m_ze_y\psi_z\psi_\theta & m_\theta\psi_\theta^2 \end{bmatrix}dx \tag{3.33}$$

and
$$\tilde{\mathbf{K}} = \int_L \begin{bmatrix} EI_z \psi_y''^2 & 0 & 0 \\ 0 & EI_y \psi_z''^2 & 0 \\ 0 & 0 & GI_t \psi_\theta'^2 \end{bmatrix} dx \tag{3.34}$$

and accordingly, the eigenvalue problem in Eq. 3.27 may be written

$$\int_L \begin{bmatrix} \left(EI_z \psi_y''^2 - \omega^2 m_y \psi_y^2\right) & 0 & \omega^2 m_y e_z \psi_y \psi_\theta \\ 0 & \left(EI_y \psi_z''^2 - \omega^2 m_z \psi_z^2\right) & -\omega^2 m_z e_y \psi_z \psi_\theta \\ \omega^2 m_y e_z \psi_y \psi_\theta & -\omega^2 m_z e_y \psi_z \psi_\theta & \left(GI_t \psi_\theta'^2 - \omega^2 m_\theta \psi_\theta^2\right) \end{bmatrix} dx \cdot \mathbf{a} = \mathbf{0} \tag{3.35}$$

As can be seen, if the cross section is symmetric about both axis y and z (i.e. $e_y = e_z = 0$), then there are three independent solutions, one associated with pure motion in the y direction

$$\omega_y = \sqrt{\int_L EI_z \psi_y''^2 dx \Big/ \int_L m_y \psi_y^2 dx} \tag{3.36}$$

one associated with pure motion in the z direction

$$\omega_z = \sqrt{\int_L EI_y \psi_z''^2 dx \Big/ \int_L m_z \psi_z^2 dx} \tag{3.37}$$

and one associated with pure cross sectional twisting (torsion)

$$\omega_\theta = \sqrt{\int_L GI_t \psi_\theta'^2 dx \Big/ \int_L m_\theta \psi_\theta^2 dx} \tag{3.38}$$

The same will occur if for instance m_y and m_z are constants along the span of the beam and ψ_y and ψ_θ as well as ψ_z and ψ_θ are orthogonal. Otherwise, i.e. in the case of no cross sectional symmetry about neither y nor z axes nor any mode shape orthogonality, then the motion will contain some coupling between motion in y and z directions and cross sectional twisting (torsion). By pre-multiplication of Eq. 3.35 by

$$\left\{ \int_L \begin{bmatrix} m_y \psi_y^2 & 0 & 0 \\ 0 & m_z \psi_z^2 & 0 \\ 0 & 0 & m_\theta \psi_\theta^2 \end{bmatrix} dx \right\}^{-1} \tag{3.39}$$

it may be written in the following more convenient format

$$\begin{bmatrix} \left(\omega_y^2 - \omega^2\right) & 0 & \omega^2 e_z \tilde{M}_{y\theta} / \tilde{M}_y \\ 0 & \left(\omega_z^2 - \omega^2\right) & -\omega^2 e_y \tilde{M}_{z\theta} / \tilde{M}_z \\ \omega^2 e_z \tilde{M}_{y\theta} / \tilde{M}_\theta & -\omega^2 e_y \tilde{M}_{z\theta} / \tilde{M}_\theta & \left(\omega_\theta^2 - \omega^2\right) \end{bmatrix} \cdot \mathbf{a} = \mathbf{0} \tag{3.40}$$

$$\text{where} \begin{cases} \tilde{M}_y = \int_L m_y \psi_y^2 dx \qquad \tilde{M}_z = \int_L m_z \psi_z^2 dx \quad \text{and} \quad \tilde{M}_\theta = \int_L m_\theta \psi_\theta^2 dx \\ \tilde{M}_{y\theta} = \int_L m_y \psi_y \psi_\theta dx \quad \text{and} \quad \tilde{M}_{z\theta} = \int_L m_z \psi_z \psi_\theta dx \end{cases} \tag{3.41}$$

As usual, a non-trivial solution requires $\mathbf{a} \neq \mathbf{0}$, and therefore the eigenvalues may be obtained by setting the determinant to the coefficient matrix in Eq. 3.40 equal to zero, rendering

$$\begin{aligned} &\left[1 - \frac{\left(e_z \tilde{M}_{y\theta}\right)^2}{\tilde{M}_y \tilde{M}_\theta} - \frac{\left(e_y \tilde{M}_{z\theta}\right)^2}{\tilde{M}_z \tilde{M}_\theta} \right] \omega^6 - \\ &\left\{ \omega_y^2 \left[1 - \frac{\left(e_y \tilde{M}_{z\theta}\right)^2}{\tilde{M}_z \tilde{M}_\theta} \right] + \omega_z^2 \left[1 - \frac{\left(e_z \tilde{M}_{y\theta}\right)^2}{\tilde{M}_y \tilde{M}_\theta} \right] + \omega_\theta^2 \right\} \omega^4 + \\ &\left[\left(\omega_y \omega_z\right)^2 + \left(\omega_y \omega_\theta\right)^2 + \left(\omega_z \omega_\theta\right)^2 \right] \omega^2 - \left(\omega_y \omega_z \omega_\theta\right)^2 = 0 \end{aligned} \tag{3.42}$$

As can be seen, there are three possible eigenvalues which are determined by the zero crossings of a third order polynomial, each representing an eigenvalue and a corresponding coupled motion. In general, only a numerical solution can be obtained. However, analytical solutions may be obtained for the special cases that either e_y or e_z are equal to zero. The solution to the case that $e_y = 0$ is dealt with in Case 3 below.

Case 3: Mono-axial Bending and Torsion

Let for instance $e_y = 0$. Then Eq. 3.40 is further reduced into

$$\begin{bmatrix} \left(\omega_y^2 - \omega^2\right) & 0 & \omega^2 e_z \tilde{M}_{y\theta} / \tilde{M}_y \\ 0 & \left(\omega_z^2 - \omega^2\right) & 0 \\ \omega^2 e_z \tilde{M}_{y\theta} / \tilde{M}_\theta & 0 & \left(\omega_\theta^2 - \omega^2\right) \end{bmatrix} \cdot \mathbf{a} = \mathbf{0} \tag{3.43}$$

from which at is readily seen that the first eigenvalue $\omega_1 = \omega_z$, representing pure motion in the z direction, while the other two, representing a coupled horizontal and torsion motion, may be determined from

$$\left(\omega_y^2 - \omega^2\right)\left(\omega_\theta^2 - \omega^2\right) - \omega^4 \left(e_z \tilde{M}_{y\theta}\right)^2 / \left(\tilde{M}_y \tilde{M}_\theta\right) = 0 \tag{3.44}$$

$$\Rightarrow \quad \left[1 - \left(e_z \tilde{M}_{y\theta}\right)^2 / \left(\tilde{M}_y \tilde{M}_\theta\right)\right] \omega^4 - \left(\omega_y^2 + \omega_\theta^2\right)\omega^2 + \omega_y^2 \omega_\theta^2 = 0 \tag{3.45}$$

If $\alpha = \left(e_z \tilde{M}_{y\theta}\right)^2 / \left(\tilde{M}_y \tilde{M}_\theta\right) = 1$, then there is only one positive root

$$\omega_2 = \omega_\theta \Big/ \sqrt{1 + \left(\omega_\theta / \omega_y\right)^2} \tag{3.46}$$

If $\alpha = \left(e_z \tilde{M}_{y\theta}\right)^2 / \left(\tilde{M}_y \tilde{M}_\theta\right) > 1$, there is still only one positive root

$$\omega_2 = \omega_\theta \sqrt{\frac{-\left[1 + \left(\omega_y / \omega_\theta\right)^2\right] + \sqrt{\left[1 + \left(\omega_y / \omega_\theta\right)^2\right]^2 + 4(\alpha - 1)\left(\omega_y / \omega_\theta\right)^2}}{2(\alpha - 1)}} \tag{3.47}$$

The solution to these two cases (i.e. that $\alpha \geq 1$) is shown in Fig. 3.7 below.

If $\alpha = \left(e_z \tilde{M}_{y\theta}\right)^2 / \left(\tilde{M}_y \tilde{M}_\theta\right) < 1$, then there are two positive roots

$$\omega_{2,3} = \omega_\theta \sqrt{\left\{1 + \left(\omega_y / \omega_\theta\right)^2 \mp \sqrt{\left[1 - \left(\omega_y / \omega_\theta\right)^2\right]^2 + 4\alpha\left(\omega_y / \omega_\theta\right)^2}\right\} \Big/ \left[2(1 - \alpha)\right]} \tag{3.48}$$

The solution is shown in Fig. 3.8 below.

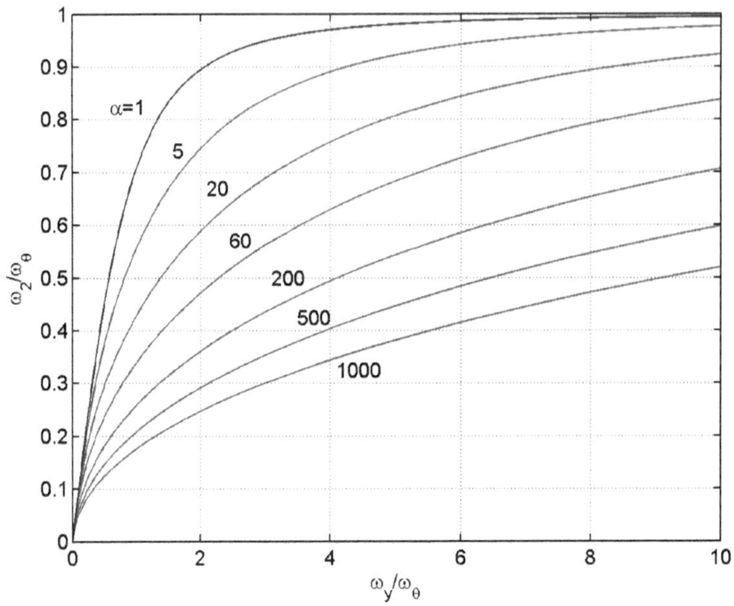

Fig. 3.7 Non-sym. cross section, ω_2/ω_θ for $\alpha = \left(e_z \tilde{M}_{y\theta}\right)^2 / \left(\tilde{M}_y \tilde{M}_\theta\right) \geq 1$

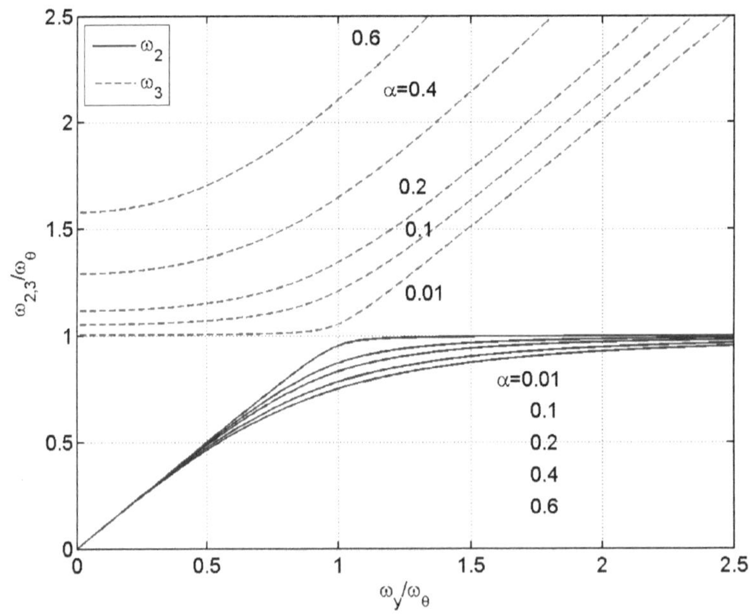

Fig. 3.8 Non-sym. cross section, ω_2/ω_θ , ω_3/ω_θ , $\alpha = \left(e_z \tilde{M}_{y\theta}\right)^2 / \left(\tilde{M}_y \tilde{M}_\theta\right) < 1$

Example 3.1

Let us for instance consider a simply supported beam (type 1 in Fig. 3.1.c), and still assume that $e_y = 0$*. Then* $\psi_y = \psi_z = \psi_\theta = \sin(n\pi x/L)$*. Let us also for simplicity assume constant cross section properties along the entire span of the beam.*

Acknowledging that $\int_0^L \sin^2(n\pi x/L)dx = \int_0^L \cos^2(n\pi x/L)dx = L/2$ *then:*

$$
\tilde{M}_y = m_y \int_0^L \phi_y^2 dx = m_y \frac{L}{2} \qquad \tilde{M}_\theta = m_\theta \int_0^L \phi_\theta^2 dx = m_\theta \frac{L}{2} \qquad \tilde{M}_{y\theta} = m_y \int_0^L \phi_y \phi_\theta dx = m_y \frac{L}{2}
$$

and

$$
\begin{cases}
\omega_y^2 = EI_z \int_0^L \phi_y''^2 dx \Big/ m_y \int_0^L \phi_y^2 dx = \left(\frac{n\pi}{L}\right)^4 \frac{EI_z}{m_y} \\
\omega_z^2 = EI_y \int_0^L \phi_z''^2 dx \Big/ m_z \int_0^L \phi_z^2 dx = \left(\frac{n\pi}{L}\right)^4 \frac{EI_y}{m_z} \\
\omega_\theta^2 = GI_t \int_0^L \phi_\theta'^2 dx \Big/ m_\theta \int_0^L \phi_\theta^2 dx = \left(\frac{n\pi}{L}\right)^2 \frac{GI_t}{m_\theta}
\end{cases}
$$

and thus $\Rightarrow$

$$
\begin{cases}
\omega_y = \left(\int_L EI_z \phi_y''^2 dx \Big/ \int_L m_y \phi_y^2 dx \right)^{\frac{1}{2}} = (n\pi)^2 \sqrt{EI_z / m_y L^4} \\
\omega_y = \left(\int_L EI_y \phi_z''^2 dx \Big/ \int_L m_z \phi_z^2 dx \right)^{\frac{1}{2}} = (n\pi)^2 \sqrt{EI_y / m_z L^4} \\
\omega_\theta = \left(\int_L GI_t \phi_\theta'^2 dx \Big/ \int_L m_\theta \phi_\theta^2 dx \right)^{\frac{1}{2}} = n\pi \sqrt{GI_t / m_\theta L^2}
\end{cases}
$$

As shown above, in the special case that $\alpha = e_z^2 m_y / m_\theta = 1$*, then there is only one positive root:* $\omega_2 = \omega_\theta \Big/ \sqrt{1 + (\omega_\theta/\omega_y)^2}$*.*

If $\alpha = e_z^2 m_y / m_\theta > 1$*, then there is still only one positive root*

$$
\omega_2 = \omega_\theta \sqrt{\left\{ -\left[1 + (\omega_y/\omega_\theta)^2\right] + \sqrt{\left[1 + (\omega_y/\omega_\theta)^2\right]^2 + 4(\alpha - 1)(\omega_y/\omega_\theta)^2} \right\} \Big/ \left[2(\alpha - 1)\right]}
$$

However, if $\alpha = e_z^2 m_y / m_\theta < 1$*, then there are two positive roots*

$$
\omega_{2,3} = \omega_\theta \sqrt{\left\{ 1 + (\omega_y/\omega_\theta)^2 \mp \sqrt{\left[1 - (\omega_y/\omega_\theta)^2\right]^2 + 4\alpha(\omega_y/\omega_\theta)^2} \right\} \Big/ \left[2(1 - \alpha)\right]}
$$

Introducing either of these eigenvalue solutions ω_j *back into Eq. 3.43*

$$\begin{bmatrix} \left(\omega_y^2-\omega_j^2\right) & 0 & \omega_j^2 e_z \tilde{M}_{y\theta}/\tilde{M}_y \\ 0 & \left(\omega_z^2-\omega_j^2\right) & 0 \\ \omega_j^2 e_z \tilde{M}_{y\theta}/\tilde{M}_\theta & 0 & \left(\omega_\theta^2-\omega_j^2\right) \end{bmatrix} \cdot \begin{bmatrix} a_y \\ a_z \\ a_\theta \end{bmatrix} = \begin{bmatrix} 0 \\ 0 \\ 0 \end{bmatrix}$$

it is seen that a_y *and* a_z *are arbitrary and may conveniently be chosen equal to unity, in which case (from the first row)* $a_\theta = \left[1-\left(\omega_y/\omega_j\right)^2\right] / \left(e_z m_y / m_\theta\right)$

Thus, the unloaded undamped harmonic motion for such a system is given by

1) either purely vertical $r_z = \text{Re}\left[a \cdot \sin\left(n\pi x/L\right) \cdot e^{i\omega_1 t}\right]$ *where* $\omega_1 = \omega_z$

2) or a combined sideway and torsion

$$\begin{bmatrix} r_y \\ r_\theta \end{bmatrix}_j = \text{Re}\left\{ a \begin{bmatrix} 1 \\ \left\{1-\left(\omega_y/\omega_j\right)^2\right\} / \left(e_z m_y / m_\theta\right) \end{bmatrix} \sin\left(\frac{n\pi x}{L}\right) e^{i\omega_j t} \right\} \text{ where } \omega_j = \begin{cases} \omega_2 \\ \text{or } \omega_3 \end{cases}$$

(i.e. for combined sideway and torsion there are two alternative eigenfrequencies and corresponding mode shapes).

3.3 The Beam Column

In Chapter 3.1 we dealt with a simple unloaded and undamped single span beam whose cross section was assumed to be symmetric about the y as well as the z axis, such that $e_y = e_z = 0$. It was also taken for granted that all time invariant cross sectional forces were zero. In this chapter we shall adopt the same restrictions, except that we shall now consider the possible presence of time invariant cross sectional forces $\bar{N}$, $\bar{M}_y$ and $\bar{M}_z$. Again, this is a more complex case than that which was dealt with in Chapter 3.1, and rather than making an attempt to solve the differential equation, it is, like we did in Chapter 3.2, more convenient to turn to the principle of virtual work, which was presented in Chapter 1.6, rendering the general expression in Eq. 1.119. The r_x will not be included, simply because there is no interaction between r_x and cross sectional forces $\bar{N}$, $\bar{M}_y$ and $\bar{M}_z$, and thus, r_x may be handled separately as shown in Chapter 3.2. (Though, it should be noted that stresses due to wave propagation will be augmented by the presence of initial stresses due to $\bar{N}$, $\bar{M}_y$ and $\bar{M}_z$.) Under the assumptions given above Eq. 1.119 is reduced to

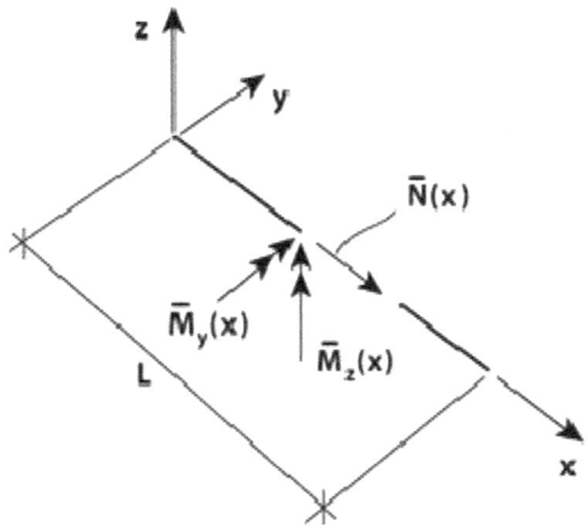

Fig. 3.9 The beam column

$$-\int_L \delta\mathbf{r}^T \mathbf{m}_g \ddot{\mathbf{r}} dx = \int_L \Big[\delta r_y' \left(\bar{N} r_y' - \bar{M}_y r_\theta' \right) + \delta r_z' \left(\bar{N} r_z' - \bar{M}_z r_\theta' \right) + \delta r_y'' E I_z r_y'' \\ \delta r_z'' E I_y r'' + \delta r_\theta' \left(G I_t r_\theta' + \bar{N} e_0^2 r_\theta' - \bar{M}_y r_y' - \bar{M}_z r_z' \right) \Big] dx \tag{3.49}$$

where $\mathbf{r}(x,t) = \begin{bmatrix} r_y & r_z & r_\theta \end{bmatrix}^T$ and $\delta\mathbf{r}(x) = \begin{bmatrix} \delta r_y & \delta r_z & \delta r_\theta \end{bmatrix}^T$, and where $\mathbf{m}_g = diag\begin{bmatrix} m_y & m_z & m_\theta \end{bmatrix}$ and $e_0^2 = e_p^2 = \left(I_y + I_z \right)/A$ (see Eq. 1.115). We are here only considering the situation of an unloaded and undamped system, and thus, the solution is purely harmonic, i.e.

$$\mathbf{r}(x,t) = \mathrm{Re}\left(\begin{bmatrix} a_y \psi_y(x) \\ a_z \psi_z(x) \\ a_\theta \psi_\theta(x) \end{bmatrix} \cdot e^{i\omega t} \right) = \mathrm{Re}\left(\mathbf{\Psi} \cdot \mathbf{a} \cdot e^{i\omega t} \right) \tag{3.50}$$

where $\mathbf{\Psi} = diag\begin{bmatrix} \psi_y & \psi_z & \psi_\theta \end{bmatrix}$ is a three by three diagonal matrix containing all the relevant mode shape functions and where $\mathbf{a} = \begin{bmatrix} a_y & a_z & a_\theta \end{bmatrix}^T$ is a three by one vector containing the corresponding amplitudes of motion. The virtual displacement is arbitrary, and therefore, we conveniently choose $\delta\mathbf{r} = \mathbf{\Psi} \cdot \delta\mathbf{a}$ where $\delta\mathbf{a} = \begin{bmatrix} \delta a_y & \delta a_z & \delta a_\theta \end{bmatrix}^T$, and thus, Eq. 3.49 becomes

$$\delta\mathbf{a}^T \omega^2 \int_L \mathbf{\Psi}^T \mathbf{m}_g \mathbf{\Psi} dx \mathbf{a} e^{i\omega t} = \delta\mathbf{a}^T \left\{ \int_L \begin{bmatrix} EI_z \psi_y''^2 & 0 & 0 \\ 0 & EI_y \psi_z''^2 & 0 \\ 0 & 0 & GI_t \psi_\theta'^2 \end{bmatrix} dx \right.$$
$$\left. + \int_L \bar{N} \begin{bmatrix} \psi_y'^2 & 0 & 0 \\ 0 & \psi_z'^2 & 0 \\ 0 & 0 & e_0^2 \psi_\theta'^2 \end{bmatrix} - \begin{bmatrix} 0 & 0 & \bar{M}_y \psi_y' \psi_\theta' \\ 0 & 0 & \bar{M}_z \psi_z' \psi_\theta' \\ \bar{M}_y \psi_y' \psi_\theta' & \bar{M}_z \psi_z' \psi_\theta' & 0 \end{bmatrix} dx \right\} \mathbf{a} e^{i\omega t} \tag{3.51}$$

which, by defining

$$\tilde{\mathbf{M}} = \int_L \mathbf{\Psi}^T \mathbf{m}_g \mathbf{\Psi} dx = diag\begin{bmatrix} \int_L m_y \psi_y^2 dx & \int_L m_z \psi_z^2 dx & \int_L m_\theta \psi_\theta^2 dx \end{bmatrix} \tag{3.52}$$

$$\tilde{\mathbf{K}}_0 = diag\begin{bmatrix} \int_L EI_z \psi_y''^2 dx & \int_L EI_y \psi_z''^2 dx & \int_L GI_t \psi_\theta'^2 dx \end{bmatrix} \tag{3.53}$$

and

$$\tilde{\mathbf{K}}_G = \int_L \begin{bmatrix} \bar{N}\psi_y'^2 & 0 & -\bar{M}_y \psi_y' \psi_\theta' \\ 0 & \bar{N}\psi_z'^2 & -\bar{M}_z \psi_z' \psi_\theta' \\ -\bar{M}_y \psi_y' \psi_\theta' & -\bar{M}_z \psi_z' \psi_\theta' & \bar{N} e_0^2 \psi_\theta'^2 \end{bmatrix} dx \tag{3.54}$$

renders the eigenvalue problem

$$\left(\tilde{\mathbf{K}} - \omega^2 \tilde{\mathbf{M}}\right) \cdot \mathbf{a} = \mathbf{0} \tag{3.55}$$

where

$$\tilde{\mathbf{K}} = \tilde{\mathbf{K}}_0 + \tilde{\mathbf{K}}_G \tag{3.56}$$

Example 3.2

Let us for simplicity consider the simply supported beam (type 1 in Fig. 3.1.c), i.e. that $\psi_y = \psi_z = \psi_\theta = \sin(n\pi x/L)$ *and that* EI_y, EI_z, GI_t, $\bar{N}$, $\bar{M}_y$ *and* $\bar{M}_z$ *are all constants along the span of the beam. Acknowledging that*

$$\left.\begin{aligned}
&\int_L \psi_j^2 dx = \int_0^L \sin^2(n\pi x/L)\,dx = L/2 \\
&\int_L \psi_j'^2 dx = \int_L \psi_j'\psi_k' dx = \int_0^L \cos^2(n\pi x/L)\,dx = (n\pi/L)^2\, L/2 \\
&\int_L \psi_j''^2 dx = \int_0^L \sin^2(n\pi x/L)\,dx = (n\pi/L)^4\, L/2
\end{aligned}\right\}\;\left.\begin{matrix} j \\ k \end{matrix}\right\} = y, z \text{ or } \theta$$

then the mass and stiffness matrices are given by $\tilde{\mathbf{M}} = \frac{L}{2}\begin{bmatrix} m_y & 0 & 0 \\ 0 & m_z & 0 \\ 0 & 0 & m_\theta \end{bmatrix}$ *and*

$$\tilde{\mathbf{K}} = \frac{L}{2}\begin{bmatrix} \left(\frac{n\pi}{L}\right)^4 EI_z + \left(\frac{n\pi}{L}\right)^2 \bar{N} & 0 & -\left(\frac{n\pi}{L}\right)^2 \bar{M}_y \\ 0 & \left(\frac{n\pi}{L}\right)^4 EI_y + \left(\frac{n\pi}{L}\right)^2 \bar{N} & -\left(\frac{n\pi}{L}\right)^2 \bar{M}_z \\ -\left(\frac{n\pi}{L}\right)^2 \bar{M}_y & -\left(\frac{n\pi}{L}\right)^2 \bar{M}_z & \left(\frac{n\pi}{L}\right)^2 GI_t + e_0^2\left(\frac{n\pi}{L}\right)^2 \bar{N} \end{bmatrix}$$

from which the following eigenvalue problem (see Eq. 3.55) is obtained

$$\left\{\begin{bmatrix} \left(\frac{n\pi}{L}\right)^4 EI_z\left(1+\frac{\bar{N}}{N_{E_z}}\right) & 0 & -\left(\frac{n\pi}{L}\right)^2 \bar{M}_y \\ 0 & \left(\frac{n\pi}{L}\right)^4 EI_y\left(1+\frac{\bar{N}}{N_{E_y}}\right) & -\left(\frac{n\pi}{L}\right)^2 \bar{M}_z \\ -\left(\frac{n\pi}{L}\right)^2 \bar{M}_y & -\left(\frac{n\pi}{L}\right)^2 \bar{M}_z & \left(\frac{n\pi}{L}\right)^2 GI_t\left(1+\frac{\bar{N}}{N_{E\theta}}\right) \end{bmatrix} - \omega^2\begin{bmatrix} m_y & 0 & 0 \\ 0 & m_z & 0 \\ 0 & 0 & m_\theta \end{bmatrix}\right\}\cdot \mathbf{a} = \mathbf{0}$$

where $N_{E_y} = (n\pi/L)^2 EI_y$ *and* $N_{E_z} = (n\pi/L)^2 EI_z$ *are the Euler buckling load with respect to bending about the* y *and* z *axes, and where*

$N_{E\theta} = GI_t / e_0^2$ *is the Euler buckling load with respect to torsion. Defining as usual*

$$\omega_y = (n\pi/L)^2 \sqrt{EI_z/m_y} \quad \omega_z = (n\pi/L)^2 \sqrt{EI_y/m_z} \quad \omega_\theta = (n\pi/L)\sqrt{GI_t/m_\theta}$$

and pre-multiplication by $\left\{diag\left[(n\pi/L)^4 EI_z \quad (n\pi/L)^4 EI_y \quad (n\pi/L)^2 GI_t\right]\right\}^{-1}$ *then the eigenvalue problem is given by*

$$\begin{bmatrix} \left\{1+\dfrac{\bar{N}}{N_{E_z}}-\left(\dfrac{\omega}{\omega_y}\right)^2\right\} & 0 & -\dfrac{\bar{M}_y}{(n\pi/L)^2 EI_z} \\ 0 & \left\{1+\dfrac{\bar{N}}{N_{E_y}}-\left(\dfrac{\omega}{\omega_z}\right)^2\right\} & -\dfrac{\bar{M}_z}{(n\pi/L)^2 EI_y} \\ -\dfrac{\bar{M}_y}{GI_t} & -\dfrac{\bar{M}_z}{GI_t} & \left\{1+\dfrac{\bar{N}}{N_{E\theta}}-\left(\dfrac{\omega}{\omega_\theta}\right)^2\right\} \end{bmatrix} \cdot \mathbf{a} = \mathbf{0}$$

If $\bar{M}_y = \bar{M}_z = 0$ *then there are three independent eigenvalues:*

$$\omega_1 = \omega_y\sqrt{1+\bar{N}/N_{E_z}} \qquad \omega_2 = \omega_z\sqrt{1+\bar{N}/N_{E_y}} \qquad \omega_3 = \omega_\theta\sqrt{1+\bar{N}/N_{E\theta}}$$

Obviously, for the special situation that $\bar{N} = 0$ *we are back at the simple bending theory presented in Chapter 3.1 above, where* $\omega_1 = \omega_y$, $\omega_2 = \omega_z$ *and* $\omega_3 = \omega_\theta$ *(not necessarily given in ascending order). It is also seen that if* $\bar{N}$ *is equal to either of* $-N_{E_y}$, $-N_{E_z}$ *or* $-N_{E\theta}$ *then* ω_1, ω_2 *or* ω_3 *is equal to zero.*

If $\bar{M}_z = 0$ *while* $\bar{M}_y \neq 0$ *then the relevant eigenvalue problem becomes*

$$\begin{bmatrix} 1+\bar{N}/N_{E_z}-(\omega/\omega_y)^2 & 0 & -\bar{M}_y/\left\{(n\pi/L)^2 EI_z\right\} \\ 0 & 1+\bar{N}/N_{E_y}-(\omega/\omega_z)^2 & 0 \\ -\bar{M}_y/GI_t & 0 & 1+\bar{N}/N_{E\theta}-(\omega/\omega_\theta)^2 \end{bmatrix} \mathbf{a} = \mathbf{0}$$

As can be seen, in this case $\omega_1 = \omega_z\sqrt{1+\bar{N}/N_{E_z}}$ *while* ω_2 *and* ω_3 *are the solution to the second order equation*

$$\left[1+\frac{\bar{N}}{N_{E_z}}-\left(\frac{\omega}{\omega_y}\right)^2\right]\cdot\left[1+\frac{\bar{N}}{N_{E\theta}}-\left(\frac{\omega}{\omega_\theta}\right)^2\right]-\frac{\bar{M}_y^2}{(n\pi/L)^2 EI_z GI_t}=0$$

which may more conveniently be written

$$\frac{\omega^4}{\omega_y^2\omega_\theta^2}-\left(1+\frac{\bar{N}}{N_{E_z}}\right)\frac{\omega^2}{\omega_\theta^2}-\left(1+\frac{\bar{N}}{N_{E\theta}}\right)\frac{\omega^2}{\omega_y^2}+\left(1+\frac{\bar{N}}{N_{E_z}}\right)\left(1+\frac{\bar{N}}{N_{E\theta}}\right)-\left(\frac{\bar{M}_y}{M_{E_z}}\right)^2=0$$

where $M_{E_z}=\frac{n\pi}{L}\sqrt{EI_z GI_t}$ *is the lateral torsion buckling load for the beam (causing bending about the* z *axis and cross sectional twisting). Thus, the eigenfrequencies are defined by the two roots (in ascending order)*

$$\omega_{2,3}=\frac{\omega_\theta}{\sqrt{2}}\left\{1+\frac{\bar{N}}{N_{E\theta}}+\left(1+\frac{\bar{N}}{N_{E_z}}\right)\frac{\omega_y^2}{\omega_\theta^2}\mp\sqrt{\left[1+\frac{\bar{N}}{N_{E\theta}}-\left(1+\frac{\bar{N}}{N_{E_z}}\right)\frac{\omega_y^2}{\omega_\theta^2}\right]^2+\left(2\frac{\omega_y}{\omega_\theta}\frac{\bar{M}_y}{M_{E_z}}\right)^2}\right\}^{\frac{1}{2}}$$

which, if $\bar{N}=0$, *simplifies into* $\omega_{2,3}=\frac{\omega_\theta}{\sqrt{2}}\sqrt{1+\frac{\omega_y^2}{\omega_\theta^2}\mp\sqrt{\left[1-\frac{\omega_y^2}{\omega_\theta^2}\right]^2+\left(2\frac{\omega_y}{\omega_\theta}\frac{\bar{M}_y}{M_{E_z}}\right)^2}}$

Plots of ω_2 *and* ω_3 *are shown in Figs. 3.10 and 3.11.*

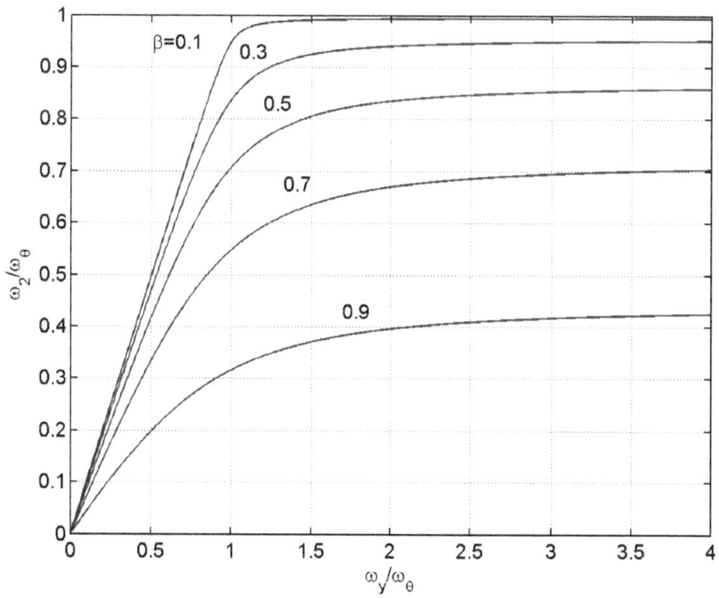

Fig. 3.10 ω_2/ω_θ in the presence of bending moment $\bar{M}_y$, $\beta=\bar{M}_y/M_{E_z}$, $\bar{N}=0$

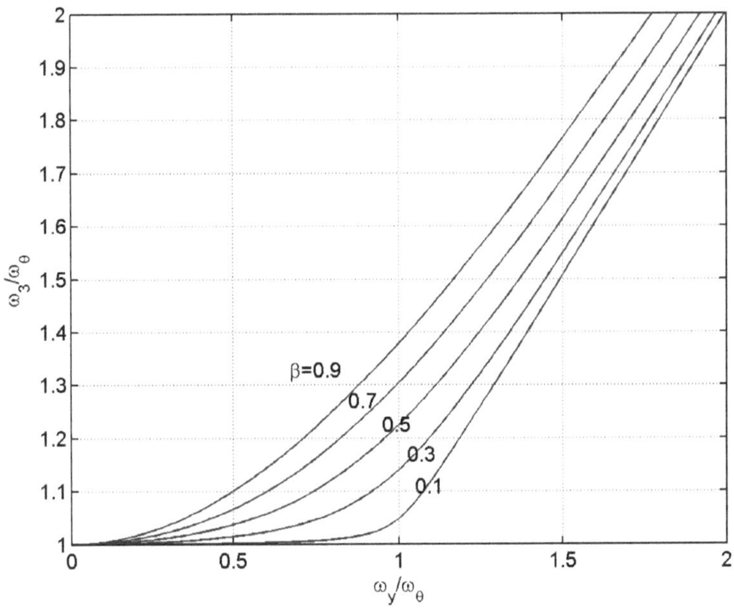

Fig. 3.11 ω_3/ω_θ in the presence of bending moment $\bar{M}_y$, $\beta=\bar{M}_y/M_{E_z}$, $\bar{N}=0$

It is seen that if $\bar{M}_y/M_{E_z}=1$, *then the total torsion stiffness is zero, and thus, there is only one solution* $\omega_2=0$.*)*

3.4 The Shallow Cable Theory

While the stiffness properties of beams or beam-columns mainly relies on cross sectional bending and torsion (EI_y, EI_z and GI_t), a cable relies almost entirely on its axial elastic stiffness property EA and the presence of an axial force N. I.e., it is in the following assumed that we are dealing with a cable whose bending and torsion properties EI_y, EI_z and GI_t are negligible. Thus, a cable will only offer stiffness against displacement r_x, r_y and r_z, while the problem of torsion is irrelevant.

It is taken for granted that the cable catenary (derived from the Latin word for chain, commonly used for the geometry of an idealised cable as described above and suspended in the gravity field) is fairly shallow, i.e. that the sag is less that about a tenth of its suspended length, such that the theory first presented by Irvine & Caughey [10] may be adopted. Since the theory below is primarily included as a prelude to the theory of suspension bridges in Chapter 3.5, it is focused on the situation that the cable supports are at identical levels, see Fig. 3.12. It is convenient to choose coordinate axes as shown in Fig. 3.12.a.

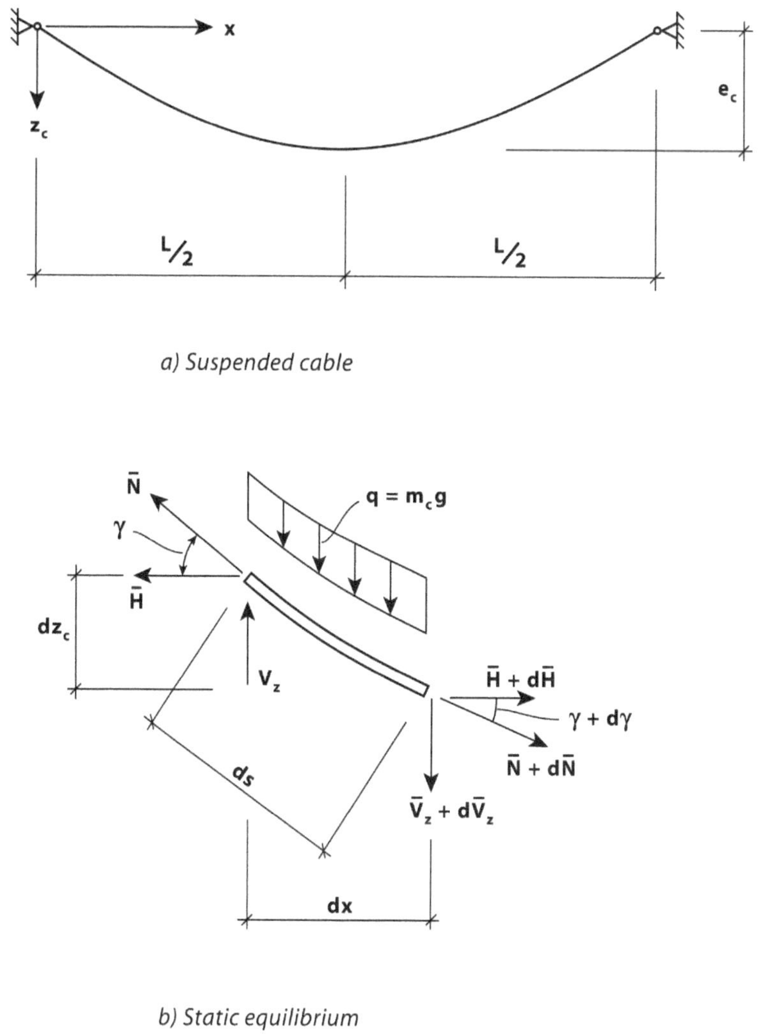

Fig. 3.12 The shallow cable in the gravity field

The Catenary

Let us first consider the situation of a shallow cable in the gravity field (see Fig. 3.12.a) with the aim of determining its static geometry, i.e. the cable position z_c at an arbitrary coordinate x. It is then at rest and all forces are time invariants. For a finite element ds at this position subject to the vertical gravity force $q \cdot ds$ (see Fig. 3.12.b) the force equilibrium requirements are given by

$$\left.\begin{aligned} d\bar{H} &= 0 \\ d\bar{V}_z + q \cdot ds &= 0 \end{aligned}\right\} \tag{3.57}$$

while moment equilibrium about its midpoint is given by

$$\bar{H}\frac{dz_c}{2} - \bar{V}_z\frac{dx}{2} + \left(\bar{H} + d\bar{H}\right)\frac{dz_c}{2} - \left(\bar{V}_z + d\bar{V}_z\right)\frac{dx}{2} \approx \bar{H}dz_c - \bar{V}_z dx = 0 \tag{3.58}$$

It is seen from Eq. 3.57 that $\bar{H}$ is constant along the entire cable span, and from Eqs. 3.58 that

$$\bar{V}_z = \bar{H}\frac{dz_c}{dx} \tag{3.59}$$

Thus (Eq. 3.57)

$$\bar{H} \cdot \frac{d^2 z_c}{dx^2} + q \cdot \frac{ds}{dx} = 0 \tag{3.60}$$

Defining $\alpha = -\bar{H}/q$ (z_c is here defined positive downwards) and acknowledging that $ds/dx = \sqrt{1 + \left(dz_c/dx\right)^2}$, it is seen that

$$z_c'' = (1/\alpha)\sqrt{1 + z_c'^2} \tag{3.61}$$

Introducing $z_c' = \dfrac{dz_c}{dx} = \tan\gamma$ (see Fig. 3.12.b)

$$\Rightarrow \quad z_c'' = \frac{d}{dx}(\tan\gamma) = \frac{1}{\cos^2\gamma}\frac{d\gamma}{dx} = \frac{1}{\alpha}\sqrt{1 + \tan^2\gamma} = \frac{1}{\alpha}\frac{1}{\cos\gamma} \tag{3.62}$$

from which it is obtained that

$$\frac{dx}{\alpha} = \frac{\cos\gamma}{\cos^2\gamma}d\gamma = \frac{\cos\gamma}{1 - \sin^2\gamma}d\gamma \tag{3.63}$$

The integral of the left hand side of this equation is

$$\int \frac{dx}{\alpha} = \frac{x}{\alpha} + C_1 \tag{3.64}$$

where C_1 is an unknown integration constant, while the integral of the right hand side may be obtained by substituting $\tau = \sin\gamma$

$$\int \frac{\cos\gamma}{1-\sin^2\gamma} d\gamma = \int \frac{1}{1-\tau^2} d\tau = \text{Arctanh}(\tau) \tag{3.65}$$

i.e. $\frac{x}{\alpha} + C_1 = Arc\tanh(\tau)$, and thus $\tau = \tanh\left(\frac{x}{\alpha} + C_1\right)$. Furthermore, since

$$\left.\begin{aligned} \cos^2\gamma = 1-\sin^2\gamma = 1-\tau^2 = 1-\tanh^2(x/\alpha + C_1) = 1/\cosh^2(x/\alpha + C_1) \\ \Rightarrow \cos\gamma = 1/\cosh(x/\alpha + C_1) \end{aligned}\right\} \tag{3.66}$$

$$\text{and} \quad z_c' = \tan\gamma = \frac{\sin\gamma}{\cos\gamma} = \frac{\tau}{\cos\gamma} = \frac{\tanh\left(\frac{x}{\alpha} + C_1\right)}{\left[\cosh\left(\frac{x}{\alpha} + C_1\right)\right]^{-1}} = \sinh\left(\frac{x}{\alpha} + C_1\right) \tag{3.67}$$

then the following is obtained

$$z_c = \int \sinh\left(\frac{x}{\alpha} + C_1\right) dx = \alpha\cosh\left(\frac{x}{\alpha} + C_1\right) + C_2 = \frac{-\bar{H}}{q}\cosh\left(\frac{-qx}{\bar{H}} + C_1\right) + C_2 \tag{3.68}$$

Thus, since the cosh function is symmetric, then the cable geometry is given by

$$z_c = -\frac{\bar{H}}{q} \cdot \cosh\left(\frac{q}{\bar{H}} \cdot x - C_1\right) + C_2 \tag{3.69}$$

where C_1 and C_2 are integration constant to be determined from the relevant geometric boundary conditions. If, as we in the following shall take for granted, the boundary conditions are defined by $z_c(x=0)=0$ and $z_c(x=L)=0$, then

$$\left.\begin{aligned} -\frac{\bar{H}}{q}\cosh(-C_1) + C_2 = 0 \\ -\frac{\bar{H}}{q}\cosh\left(\frac{qL}{\bar{H}} - C_1\right) + C_2 = 0 \end{aligned}\right\} \tag{3.70}$$

which is satisfied if $C_1 = \dfrac{qL}{2\bar{H}}$ and $C_2 = \dfrac{\bar{H}}{q}\cosh\left(\dfrac{qL}{2\bar{H}}\right)$. Thus, the cable geometry is given by

$$z_c = \frac{\bar{H}}{q}\left[\cosh\left(\frac{qL}{2\bar{H}}\right) - \cosh\left(\frac{qx}{\bar{H}} - \frac{qL}{2\bar{H}}\right)\right] \tag{3.71}$$

It is seen that
$$z_c' = -\sinh\left(\frac{qx}{\bar{H}} - \frac{qL}{2\bar{H}}\right) = \sinh\left(\frac{qL}{2\bar{H}} - \frac{qx}{\bar{H}}\right) \tag{3.72}$$

is zero at mid span, i.e. at $x = L/2$. The cable sag is given by

$$e_c = z_c\left(x = L/2\right) = \frac{\bar{H}}{q}\left[\cosh\left(\frac{qL}{2\bar{H}}\right) - 1\right] \tag{3.73}$$

An approximate solution to z_c can be obtained by using the series expansion $\cosh(\eta) = 1 + \eta^2/2! + \eta^4/4! + \cdots$, where it for small arguments of η will suffice to include only the two first terms. Thus

$$z_c \approx \frac{\bar{H}}{q}\left\{1 + \frac{1}{2}\left(\frac{qL}{2\bar{H}}\right)^2 - \left[1 + \frac{1}{2}\left(\frac{qx}{\bar{H}} - \frac{qL}{2\bar{H}}\right)^2\right]\right\} = \frac{qL^2}{2\bar{H}}\frac{x}{L}\left(1 - \frac{x}{L}\right) \tag{3.74}$$

which is the well-known parabola solution. It is worth noting that the corresponding approximate expression to the sag is given by

$$e_c \approx qL^2/\left(8\bar{H}\right) \tag{3.75}$$

and that

$$z_c' \approx \frac{qL}{2\bar{H}}\left(1 - 2\frac{x}{L}\right) \qquad \Rightarrow \qquad \begin{cases} z_c'(x=0) = qL/\left(2\bar{H}\right) \\ z_c'(x=L) = -qL/\left(2\bar{H}\right) \end{cases} \tag{3.76}$$

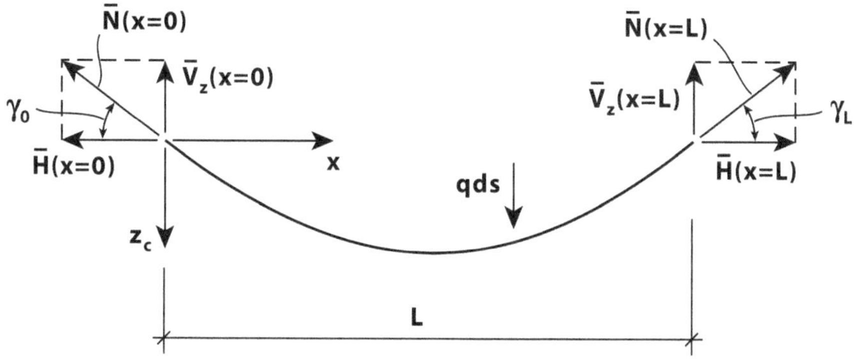

Fig. 3.13 Overall equilibrium

The Tensile Force

The tensile force is given by (see Eqs. 3.58 and 3.72)

$$\begin{aligned}\bar{N} &= \sqrt{\bar{H}^2+\bar{V}_z^2} = \bar{H}\sqrt{1+\left(\frac{\bar{V}_z}{\bar{H}}\right)^2} = \bar{H}\sqrt{1+\left(\frac{dz_c}{dx}\right)^2}\\ &= \bar{H}\sqrt{1+\sinh^2\left(\frac{qx}{\bar{H}}-\frac{qL}{2\bar{H}}\right)} = \bar{H}\cdot\cosh\left(\frac{qx}{\bar{H}}-\frac{qL}{2\bar{H}}\right)\end{aligned} \tag{3.77}$$

rendering

$$\left.\begin{aligned}&\bar{N}(x=0)=\bar{N}(x=L)=\bar{H}\cosh\left(\frac{qL}{2\bar{H}}\right)\\ &\bar{N}(x=L/2)=\bar{H}\end{aligned}\right\} \tag{3.78}$$

The Cable Length

Since $\sinh\eta=\eta+\eta^3/3!+\eta^5/5!+\cdots$ it is seen from Fig. 3.13 that

$$\begin{aligned}\bar{V}_z(x=0)=\bar{V}_z(x=L) &= \sqrt{\left[\bar{H}\cosh\left(\frac{qL}{2\bar{H}}\right)\right]^2-\bar{H}^2} = \bar{H}\sinh\left(\frac{qL}{2\bar{H}}\right)\\ &\approx \bar{H}\left[\left(\frac{qL}{2\bar{H}}\right)+\frac{1}{3!}\left(\frac{qL}{2\bar{H}}\right)^3\right] = \frac{qL}{2}\left[1+\frac{1}{24}\left(\frac{qL}{\bar{H}}\right)^2\right]\end{aligned} \tag{3.79}$$

and that overall equilibrium will require

$$\bar{V}_z(x=0)+\bar{V}_z(x=L)=qL\left[1+\frac{1}{24}\left(\frac{qL}{\bar{H}}\right)^2\right]=\int_{\ell} q\cdot ds=q\cdot\ell \tag{3.80}$$

where ℓ is the length of the cable. From this it may be deduced that

$$\ell=L\left[1+\frac{1}{24}\left(\frac{qL}{\bar{H}}\right)^2\right] \tag{3.81}$$

That this is correct may be shown by integration of ds, i.e.

$$\begin{aligned}\ell&=\int_{\ell} ds=\int_0^L dx\sqrt{1+\left(\frac{dz_c}{dx}\right)^2}=\int_0^L\left[\sqrt{1+\sinh^2\left(\frac{qL}{2\bar{H}}-\frac{qx}{\bar{H}}\right)}\right]dx\\&=\int_0^L\cosh\left(\frac{qL}{2\bar{H}}-\frac{qx}{\bar{H}}\right)dx=\left[\frac{\bar{H}}{q}\sinh\left(\frac{qL}{2\bar{H}}-\frac{qx}{\bar{H}}\right)\right]_0^L=\frac{2\bar{H}}{q}\sinh\left(\frac{qL}{2\bar{H}}\right)\end{aligned} \tag{3.82}$$

and thus

$$\ell\approx\frac{2\bar{H}}{q}\left[\frac{qL}{2\bar{H}}+\frac{1}{3!}\left(\frac{qL}{2\bar{H}}\right)^3\right]=L\cdot\left[1+\frac{1}{24}\left(\frac{qL}{\bar{H}}\right)^2\right] \tag{3.83}$$

Since the sag $e_c\approx qL^2/(8\bar{H})$ this may alternatively be written

$$\ell\approx L\cdot\left[1+\frac{8}{3}\left(\frac{e_c}{L}\right)^2\right] \tag{3.84}$$

The Cable Elongation

Assuming that the cable stiffness is constant along its span, then the cable elongation due to the gravity field is given by

$$\begin{aligned}\Delta\ell&=\int_L \varepsilon\cdot ds=\int_L\frac{\sigma}{E}ds=\int_L\frac{\bar{N}}{EA}ds=\int_0^L\frac{\bar{H}\sqrt{1+(dz_c/dx)^2}}{EA}\sqrt{1+(dz_c/dx)^2}\,dx\\&=\frac{\bar{H}}{EA}\int_0^L\left[1+\left(\frac{dz_c}{dx}\right)^2\right]dx=\frac{\bar{H}}{EA}\int_0^L\cosh^2\left(\frac{qx}{\bar{H}}-\frac{qL}{2\bar{H}}\right)dx\end{aligned} \tag{3.85}$$

Substituting $\beta = qx/\bar{H} - qL/2\bar{H}$ then the following is obtained

$$\begin{aligned}\Delta\ell &= \frac{\bar{H}^2}{qEA}\int_{-qL/2\bar{H}}^{qL/2\bar{H}} \cosh^2\beta \, d\beta = \frac{\bar{H}^2}{qEA}\left[\frac{\sinh 2\beta}{4}+\frac{\beta}{2}\right]_{-qL/2\bar{H}}^{qL/2\bar{H}} \\ &= \frac{\bar{H}L}{2EA}\left[\frac{\bar{H}}{qL}\sinh\left(\frac{qL}{\bar{H}}\right)+1\right]\end{aligned} \tag{3.86}$$

which may be expanded into

$$\Delta\ell \approx \frac{\bar{H}L}{2EA}\left\{\frac{\bar{H}}{qL}\left[\frac{qL}{\bar{H}}+\frac{1}{3!}\left(\frac{qL}{\bar{H}}\right)^3\right]+1\right\} = \frac{\bar{H}L}{EA}\left[1+\frac{1}{12}\left(\frac{qL}{\bar{H}}\right)^2\right] \tag{3.87}$$

and alternatively expressed by the cable sag (see Eq. 3.75)

$$\Delta\ell \approx \frac{\bar{H}L}{EA}\left[1+\frac{16}{3}\left(\frac{e_c}{L}\right)^2\right] \tag{3.88}$$

The Differential Equations of Dynamic Equilibrium

We are now ready to establish the relevant equilibrium requirements for an infinitesimal element ds of the shallow cable in a harmonic type of dynamic motion, see Fig. 3.14. Basically, the problem is highly non-linear. What is shown below is a small displacement linearized theory. It is taken for granted that the cross section axial force comprise a time invariant part $\bar{N}(x)$ from the gravity field plus an additional contribution $N(x,t)$ from the cable motion itself, and thus, the same applies to its horizontal and vertical components H and V_z (see Fig. 3.15), i.e.

$$N_{tot} = \bar{N}(x) + N(x,t) \qquad \Rightarrow \qquad \begin{cases} H_{tot} = \bar{H}(x) + H(x,t) \\ V_{z_{tot}} = \bar{V}_z(x) + V_z(x,t) \end{cases} \tag{3.89}$$

while the out of plane component V_y is only caused by the dynamic motion.

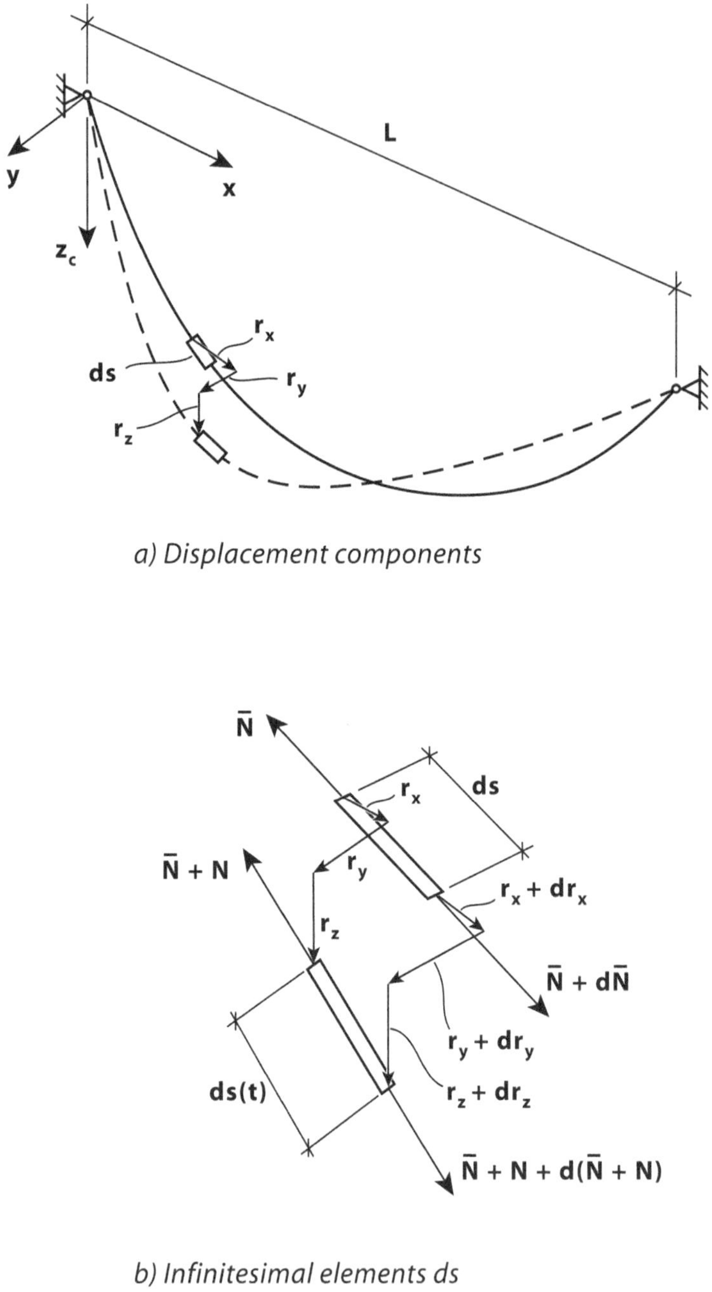

Fig. 3.14 Cable displacements and internal forces

Thus
$$V_{y_{tot}} = V_y(x,t) \tag{3.90}$$

We shall assume shallowness, and for the initial cable geometry adopt the approximate catenary curve given in Eq. 3.74 (see also Eq. 3.75), i.e.

$$z_c \approx \frac{qL^2}{2\bar{H}} \frac{x}{L}\left(1-\frac{x}{L}\right) = 4e_c \frac{x}{L}\left(1-\frac{x}{L}\right) \tag{3.91}$$

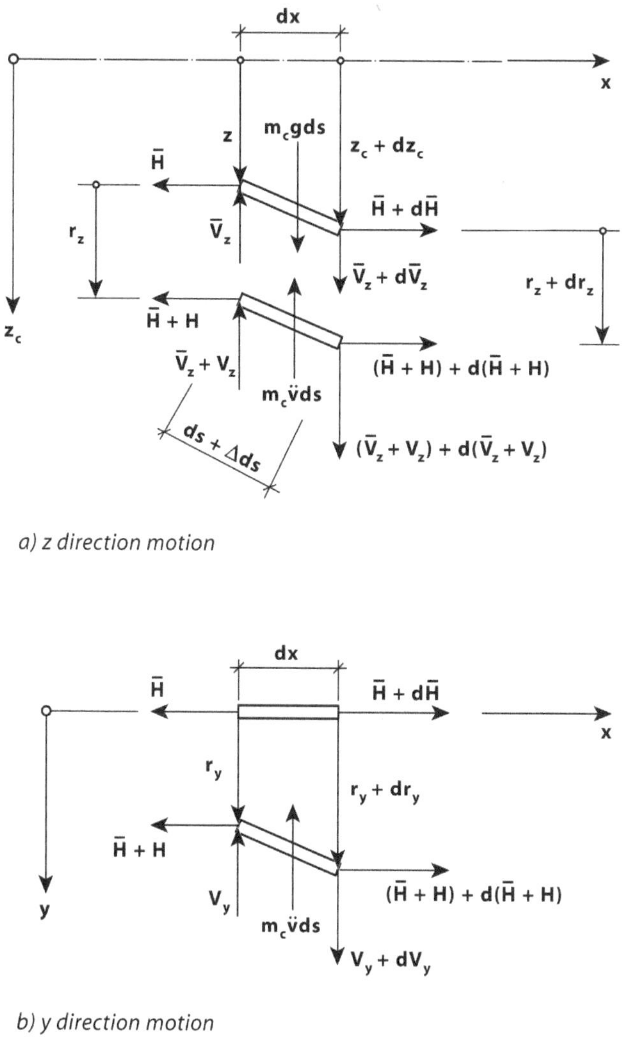

Fig. 3.15 Element equilibrium

In addition to shallowness we shall in the following adopt the simplification that the cable motion in the x direction r_x is negligible. This is a minor sacrifice bearing in mind that our main application of the theory is intended for the use in connection with suspension bridges. It has the advantageous consequence that the equilibrium requirement in the x direction (see Fig. 3.15)

$$\frac{d}{dx}\left(\bar{H}+H\right)=0 \tag{3.92}$$

implies that $\bar{H}$ and $H\left(t\right)$ are independent of x. The corresponding equilibrium requirements in y and z directions are then given by (see Fig. 3.15)

$$\left.\begin{aligned} dV_y - m_c\ddot{r}_y \cdot ds = 0 \\ d\left(\bar{V}_z + V_z\right) + m_c g \cdot ds - m_c\ddot{r}_z \cdot ds = 0 \end{aligned}\right\} \tag{3.93}$$

where m_c is the cable mass per unit length (here assumed constant along the span of the cable). Similarly, moment equilibrium taken about axes through the element midpoint and parallel to the y and z directions, are given by

$$\left.\begin{aligned} \left(\bar{H}+H\right)\cdot d\left(z_c+r_z\right)-\left(\bar{V}_z+V_z\right)\cdot dx=0 \\ \left(\bar{H}+H\right)\cdot dr_y - V_y \cdot dx = 0 \end{aligned}\right\} \tag{3.94}$$

from which the following is obtained

$$\left.\begin{aligned} V_y &= \left(\bar{H}+H\right)\frac{dr_y}{dx} \\ \bar{V}_z + V_z &= \left(\bar{H}+H\right)\frac{d}{dx}\left(z_c+r_z\right) \end{aligned}\right\} \tag{3.95}$$

and thus

$$\left.\begin{aligned} \frac{d}{ds}\left[\left(\bar{H}+H\right)\frac{dr_y}{dx}\right] = m_c\ddot{r}_y \\ \frac{d}{ds}\left[\left(\bar{H}+H\right)\frac{d}{dx}\left(z_c+r_z\right)\right] + m_c g = m_c\ddot{r}_z \end{aligned}\right\} \tag{3.96}$$

Since $\bar{H}$ and $H\left(t\right)$ are independent of x, we see that Eq. 3.96 may be replaced by

$$\left.\begin{aligned}&(\bar{H}+H)\frac{d^2 r_y}{dx^2}\frac{dx}{ds}=m_c\ddot{r}_y\\&(\bar{H}+H)\frac{d^2}{dx^2}(z_c+r_z)\frac{dx}{ds}+m_c g=m_c\ddot{r}_z\end{aligned}\right\}\tag{3.97}$$

from which it is seen that the static case (H, r_y, r_z are all zero) is defined by

$$\bar{H}z_c''+m_c g=0 \quad\Rightarrow\quad \bar{H}z_c''=-m_c g \quad \text{and} \quad Hz_c''=-\frac{m_c g}{\bar{H}}H \tag{3.98}$$

Taking it for granted that dynamic displacements are small, i.e. that

$$Hr_y''<<\bar{H}r_y'' \quad \text{and} \quad Hr_z''<<\bar{H}r_z'' \tag{3.99}$$

and that shallowness justifies $ds\approx dx$, then Eq. 3.97 becomes

$$\left.\begin{aligned}&r_y''-(m_c/\bar{H})\cdot\ddot{r}_y=0\\&r_z''-(m_c/\bar{H})\cdot\ddot{r}_z=(m_c g/\bar{H})\cdot(H/\bar{H})\end{aligned}\right\}\tag{3.100}$$

This is the differential equation for unloaded and undamped motion of a shallow cable. What remains is to find an expression for $H(t)$, i.e. to find the increase of cable tension during a small dynamic vertical motion $r_z(x,t)$ [as $r_y(x,t)$ will not involve any cable stretching, only a sideway skipping rope displacement]. The cable elongation $\Delta s(t)$ may be found by combining $ds^2=dx^2+dz_c^2$ and

$$(ds+\Delta ds)^2=dx^2+(dz_c+dr_z)^2 \tag{3.101}$$

from which the following is obtained:

$$\begin{aligned}\frac{\Delta ds}{ds}&=\frac{\sqrt{dx^2+(dz_c+dr_z)^2}-ds}{ds}=\sqrt{\left(\frac{dx}{ds}\right)^2+\left(\frac{dz_c}{ds}+\frac{dr_z}{ds}\right)^2}-1\\&=\sqrt{1+2\frac{dz_c}{ds}\cdot\frac{dr_z}{ds}+\left(\frac{dr_z}{ds}\right)^2}-1\approx\left\{1+\frac{1}{2}\left[2\frac{dz_c}{ds}\cdot\frac{dr_z}{ds}+\left(\frac{dr_z}{ds}\right)^2\right]\right\}-1\\&=\frac{dz_c}{ds}\cdot\frac{dr_z}{ds}+\frac{1}{2}\left(\frac{dr_z}{ds}\right)^2\approx\frac{dz_c}{ds}\cdot\frac{dr_z}{ds}\end{aligned}\tag{3.102}$$

Since $\dfrac{H}{N}=\dfrac{dx}{ds}$ and $\varepsilon=\dfrac{\Delta ds}{ds}=\dfrac{\sigma}{E}=\dfrac{N}{EA}$ then $\dfrac{\Delta ds}{ds}=\dfrac{H}{EA}\dfrac{ds}{dx}$ and thus

$$\frac{H}{EA}\frac{ds}{dx}\approx\frac{dz_c}{ds}\frac{dr_z}{ds} \tag{3.103}$$

which may also be written

$$\frac{H}{EA}\left(\frac{ds}{dx}\right)^3=\frac{dz_c}{dx}\frac{dr_z}{dx}=z_c'\cdot r_z' \tag{3.104}$$

Recalling that $r_z(x=0)=r_z(x=L)=0$, and that (see Eqs. 3.74 and 3.75)

$$\left.\begin{aligned} z_c' &\approx \frac{qL}{2\bar{H}}\left(1-2\frac{x}{L}\right)=4\frac{e_c}{L}\left(1-2\frac{x}{L}\right)=\frac{8e_c}{L^2}\mathrm{X} \\ z_c'' &= -8e_c/L^2 \end{aligned}\right\} \tag{3.105}$$

where $e_c\approx m_c g L^2/(8\bar{H})$ and $\mathrm{X}=L/2-x$, then spanwise integration of the right hand side of Eq. 3.104 renders

$$\int_L z_c' r_z' dx=\left[z_c' r_z\right]_0^L-\int_0^L z_c'' r_z dx=\frac{m_c g}{\bar{H}}\int_o^L r_z dx=\frac{8e_c}{L^2}\int_o^L r_z dx \tag{3.106}$$

while corresponding integration of the left hand side renders

$$\begin{aligned} \int_L \frac{H}{EA}\left(\frac{ds}{dx}\right)^3 dx &= \frac{H}{EA}\int_L\left(\frac{ds}{dx}\right)^3 dx=\frac{H}{EA}\int_0^L\left[1+\left(\frac{dz_c}{dx}\right)^2\right]^{\frac{3}{2}} dx \\ &\approx \frac{H}{EA}\int_0^L\left[1+\frac{3}{2}\left(z_c'\right)^2\right]dx=\frac{H}{EA}\int_{-L/2}^{L/2}\left[1+\frac{3}{2}\left(\frac{8e_c}{L^2}\mathrm{X}\right)^2\right]d\mathrm{X} \\ &= \frac{H}{EA}\left[\mathrm{X}+32\left(\frac{e_c}{L^2}\right)^2\mathrm{X}^3\right]_{-L/2}^{L/2}=\frac{H}{EA}L\left[1+8\left(\frac{e_c}{L}\right)^2\right] \end{aligned} \tag{3.107}$$

Thus

$$\frac{H}{EA}\cdot L\left[1+8\left(\frac{e_c}{L}\right)^2\right]\approx\frac{8e_c}{L^2}\int_L r_z dx \quad \Rightarrow H(t)=\frac{EA}{\ell_e}\frac{8e_c}{L^2}\int_L r_z(x,t)dx \tag{3.108}$$

where $\ell_e=L\left[1+8\left(e_c/L\right)^2\right]$.

Defining
$$\lambda^2 = \left(\frac{8e_c}{L}\right)^2 \frac{EA}{\bar{H}} \frac{L}{\ell_e} \tag{3.109}$$

and recalling that $e_c \approx m_c g L^2 / \left(8\bar{H}\right)$ then Eq. 3.100 may be replaced by

$$\left.\begin{aligned} & r_y'' - \left(m_c/\bar{H}\right) \cdot \ddot{r}_y = 0 \\ & r_z'' - \left(m_c/\bar{H}\right) \cdot \ddot{r}_z = \lambda^2 \left(1/L^3\right) \int_L r_z dx \end{aligned}\right\} \tag{3.110}$$

These are the differential equations for unloaded and undamped dynamic motion in y and z directions of the shallow cable (as presented by Irvine & Caughey [10]). As can be seen, there is no coupling between r_y and r_z motion, and therefore, they may conveniently be handled separately. In general, the solutions to such second order differential equations are given by

$$\left.\begin{aligned} r_y(x,t) &= \mathrm{Re}\left[\phi_y(x) \cdot e^{i\omega t}\right] \\ r_z(x,t) &= \mathrm{Re}\left[\phi_z(x) \cdot e^{i\omega t}\right] \end{aligned}\right\} \tag{3.111}$$

where ϕ_y and ϕ_z represent the mode shapes of the motion. The solutions are dealt with below.

Horizontal Motion

Let us first consider the case of out of plane horizontal motion, i.e. the situation that $r_y(x,t) \neq 0$ and $r_z(x,t) = 0$. Introducing $r_y(x,t) = \mathrm{Re}\left[\phi_y(x) e^{i\omega t}\right]$ into the differential equation $r_y'' - \left(m_c/\bar{H}\right) \ddot{r}_y = 0$, then

$$\phi_y'' + \omega^2 \left(m_c/\bar{H}\right) \phi_y = 0 \tag{3.112}$$

which can only be satisfied for all values of x if the second derivative of ϕ_y is congruent to ϕ_y itself, and simultaneously satisfy the boundary conditions $\phi_y(x=0) = \phi_y(x=L) = 0$, rendering $\phi_{y_n} = a_{y_n} \sin(n\pi x/L)$, $n = 1,2,3,\ldots$. Thus, the following is obtained:

$$-\left(n\pi/L\right)^2 + \omega^2 m_c/\bar{H} = 0 \tag{3.113}$$

from which the eigenfrequencies

$$\omega_{y_n} = n\pi\sqrt{\bar{H}/\left(m_c L^2\right)} \tag{3.114}$$

are obtained. The corresponding two first eigenmodes are shown in Fig. 3.16.

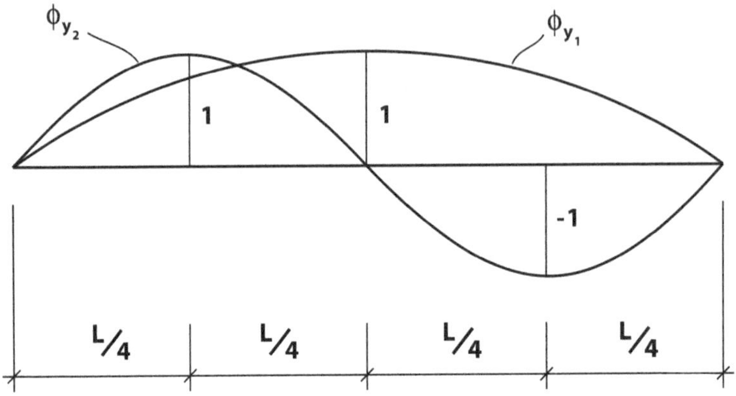

Fig. 3.16 Mode shapes ϕ_{y_n} associated with out of plane horizontal motion

Vertical Anti-symmetric Motion

Let us then consider the case of purely vertical motion, i.e. the situation that $r_y(x,t)=0$ and $r_z(x,t)\neq 0$, and let us also assume that the motion is anti-symmetric with respect to the midpoint of the span, e.g. as indicated in Fig. 3.17. Then the integral $\int_L r_z dx$ is equal to zero (i.e. no cable stretching), and thus

$$r_z'' - \left(m_c/\bar{H}\right)\cdot \ddot{r}_z = 0 \tag{3.115}$$

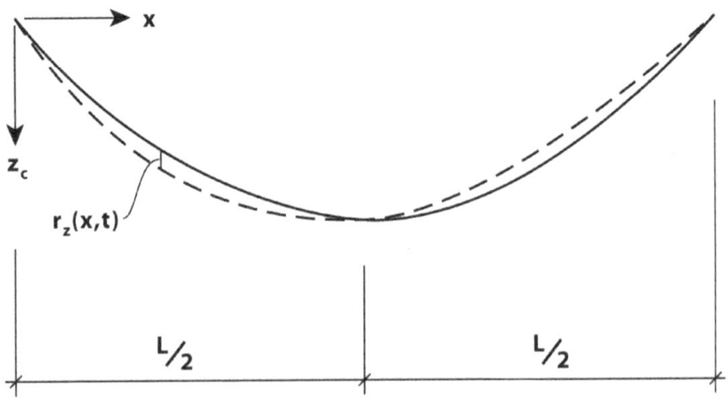

Fig. 3.17 Vertical in plane anti-symmetric motion

Introducing $r_z(x,t) = \mathrm{Re}\left[\phi_z(x) \cdot e^{i\omega t}\right]$ then $\phi_z'' + \omega^2 \frac{m_c}{\bar{H}} \phi_z = 0$ (3.116)

whose anti-symmetric solution is $\phi_{z_n} = a_{z_n} \sin(2n\pi x/L)$, $n = 1, 2, 3, \ldots$, and thus

$$-(2n\pi/L)^2 + \omega^2 m_c / \bar{H} = 0 \tag{3.117}$$

from which the eigenfrequencies

$$\omega_{z_n} = 2n\pi\sqrt{\bar{H}/(m_c L^2)} \tag{3.118}$$

are obtained. The corresponding two first eigenmodes are shown in Fig. 3.18.

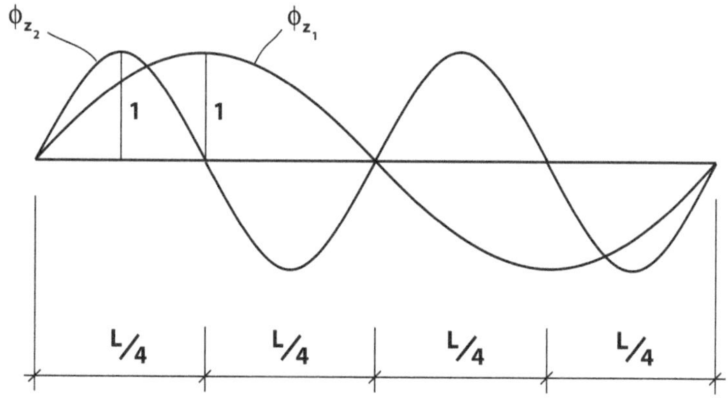

Fig. 3.18 Mode shapes ϕ_{z_n} assoc. with vertical anti-sym. in plane motion

Vertical Symmetric Motion

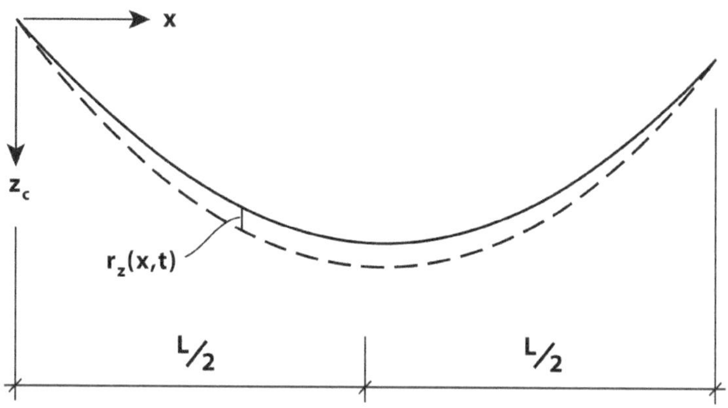

Fig. 3.19 Vertical symmetric motion

Finally, let us consider the case of purely vertical motion, i.e. the situation that $r_y(x,t)=0$ and $r_z(x,t)\neq 0$, and also assume that the motion is symmetric with respect to the midpoint of the span, e.g. as indicated in Fig. 3.19 above. Then the integral $\int_L r_z dx$ is not zero (i.e. cable stretching will occur), and thus

$$r_z'' - \left(m_c/\bar{H}\right)\cdot\ddot{r}_z = \lambda^2\left(1/L^3\right)\int_L r_z dx \tag{3.119}$$

Introducing $r_z(x,t)=\text{Re}\left[\phi_z(x)\cdot e^{i\omega t}\right]$ then

$$\phi_z'' + \beta^2\phi_z = \lambda^2\left(1/L^3\right)\int_L \phi_z dx \tag{3.120}$$

where $\beta^2 = \omega^2 m_c/\bar{H}$. The solution satisfying the boundary conditions as well as Eq. 3.120 for all values of x is given by

$$\phi_z = a_z\left\{1-\cos\left[\beta(x-L/2)\right]/\cos(\beta L/2)\right\} \tag{3.121}$$

Thus
$$\int_L \phi_z dx = a_z\int_0^L\left\{1-\frac{\cos\left[\beta(x-L/2)\right]}{\cos(\beta L/2)}\right\}dx \tag{3.122}$$

which may more easily be solved by substituting $\mathrm{X}=x-L/2$, rendering

$$\begin{aligned}\int_L \phi_z dx &= a_z\int_{-L/2}^{L/2}\left\{1-\frac{\cos[\beta \mathrm{X}]}{\cos(\beta L/2)}\right\}d\mathrm{X}\\ &= a_z\left[\mathrm{X}-\frac{\sin[\beta\mathrm{X}]}{\beta\cdot\cos(\beta L/2)}\right]_{-L/2}^{L/2} = a_z\left[L-\frac{2}{\beta}\tan\left(\frac{\beta L}{2}\right)\right]\end{aligned} \tag{3.123}$$

Introducing Eqs. 3.121 and 3.123 into Eq. 3.120

$$\frac{\beta^2\cos\left[\beta(x-L/2)\right]}{\cos(\beta L/2)} + \beta^2\left\{1-\frac{\cos\left[\beta(x-L/2)\right]}{\cos(\beta L/2)}\right\} = \frac{\lambda^2}{L^3}\left[L-\frac{2}{\beta}\tan\left(\frac{\beta L}{2}\right)\right] \tag{3.124}$$

will then render the following transcendental equation

$$\tan\left(\frac{\beta L}{2}\right) = \frac{\beta L}{2} - \left(\frac{2}{\lambda}\right)^2\left(\frac{\beta L}{2}\right)^3 \tag{3.125}$$

Since $\beta^2 = \omega^2 m_c/\bar{H}$, then any solution $\beta=\beta_n$ to this equation represents an eigenfrequency

$$\omega_n = \beta_n \sqrt{\bar{H}/m_c} \tag{3.126}$$

and a corresponding eigenmode

$$\phi_{z_n} = a_{z_n} \left\{ 1 - \frac{\cos\left[\beta_n (x - L/2)\right]}{\cos(\beta_n L/2)} \right\} \tag{3.127}$$

where a_{z_n} is arbitrary and may conveniently be set at unity. The solution heavily depends on the stiffness parameter λ (defined in Eq. 3.109) as shown in Fig. 3.20, where either side of Eq. 3.125 have been plotted for various values of λ. The solution, i.e. $\beta_n L$ versus λ/π, is plotted in Fig. 3.21, covering the three first eigenmodes of the system. The effect of the stiffness parameter to the modeshapes is illustrated in Fig. 3.22, where the first eigenmode is shown for various values of the stiffness parameter. As can be seen, increasing the stiffness parameter will significantly change the shape of the eigenmode. Finally, the three lowest eigenmodes at $\lambda/\pi = 5$ are shown in Fig. 3.23. They are all symmetric with respect to the mid span.

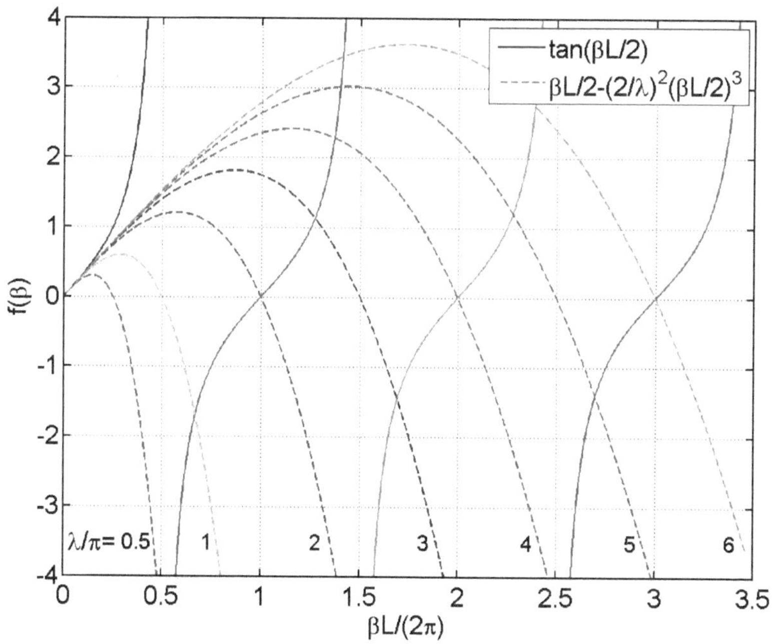

Fig. 3.20 Plots of either side of transcendental Eq. 3.125

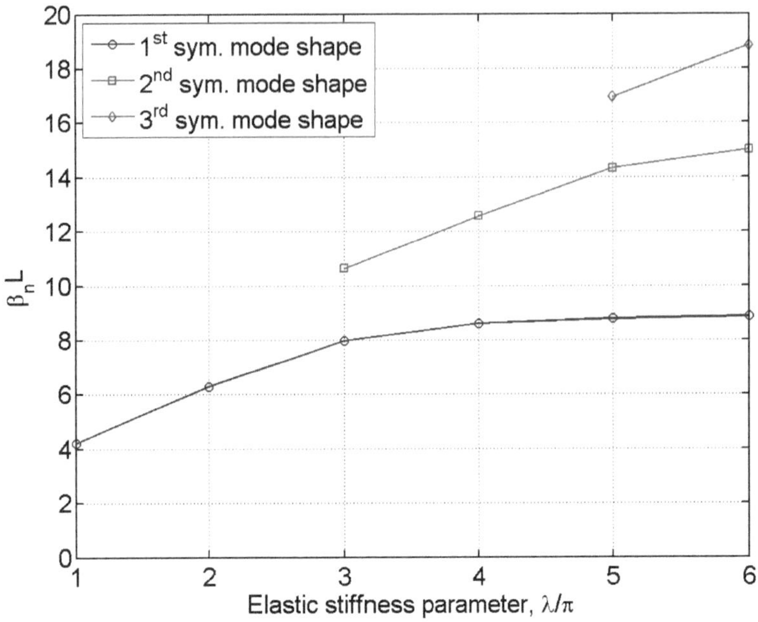

Fig. 3.21 Reduced eigenfrequency $\beta_n L$ vs. stiffness parameter λ/π

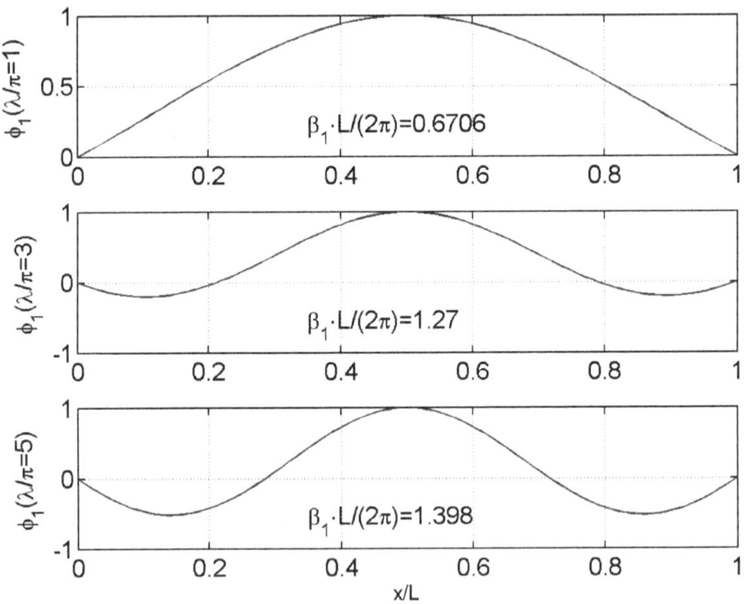

Fig. 3.22 The first symmetric modeshape at increasing stiffness parameters $\lambda/\pi = 1$, 3 and 5

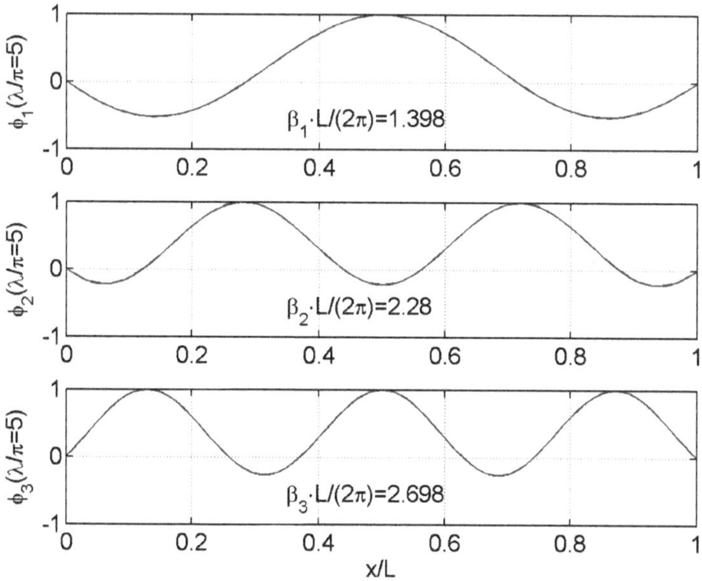

Fig. 3.23 The three lowest symmetric modeshapes at $\lambda/\pi = 5$

If $\lambda \to \infty$, which, in the limit, is identical to the case that the cable is inextensible, then the transcendental expression in Eq. 3.125 becomes

$$\tan(\beta L/2) = \beta L/2 \tag{3.128}$$

rendering the following approximate solution

$$\left.\begin{array}{llll} 1^{st}\ \text{mode} & : \dfrac{\beta_1 L}{2} \approx 1.43 \cdot \pi & \Rightarrow & \omega_1 \approx 8.98\sqrt{\dfrac{\bar{H}}{m_c L^2}} \\ 2^{nd}\ \text{mode} & : \dfrac{\beta_2 L}{2} \approx 2.46 \cdot \pi & \Rightarrow & \omega_2 \approx 15.45\sqrt{\dfrac{\bar{H}}{m_c L^2}} \\ \text{Higher}, n \geq 3 & : \dfrac{\beta_n L}{2} \approx \left(n + \dfrac{1}{2}\right) \cdot \pi & \Rightarrow & \omega_n \approx (2n+1)\pi\sqrt{\dfrac{\bar{H}}{m_c L^2}} \end{array}\right\} \tag{3.129}$$

If $e_c/L \to 0$, which, in the limit, is identical to the case that the cable becomes what in structural mechanics is called a taut string, then $\lambda \to 0$ and the transcendental expression in Eq. 3.125 will require that

$$\tan\left(\beta L/2\right) \to -\infty \tag{3.130}$$

rendering the solution that

$$\beta_n L/2 = \left(n - 1/2\right)\pi \qquad \Rightarrow \qquad \omega_n = \left(2n-1\right)\pi\sqrt{\frac{\bar{H}}{m_c L^2}} \tag{3.131}$$

3.5 The Single Span Suspension Bridge

Having battled through the shallow cable theory it is only a short step to include the problem of eigenvalue calculations for a single span suspension bridge, as, if its cable planes are identical and vertical, it will simply behave as a combination of two cables and a beam. The overall geometry of such a bridge is illustrated in Fig. 3.24. For the sake of simplicity, symmetry is taken for granted and the shear centre of the main beam is assumed to coincide with its centroid.

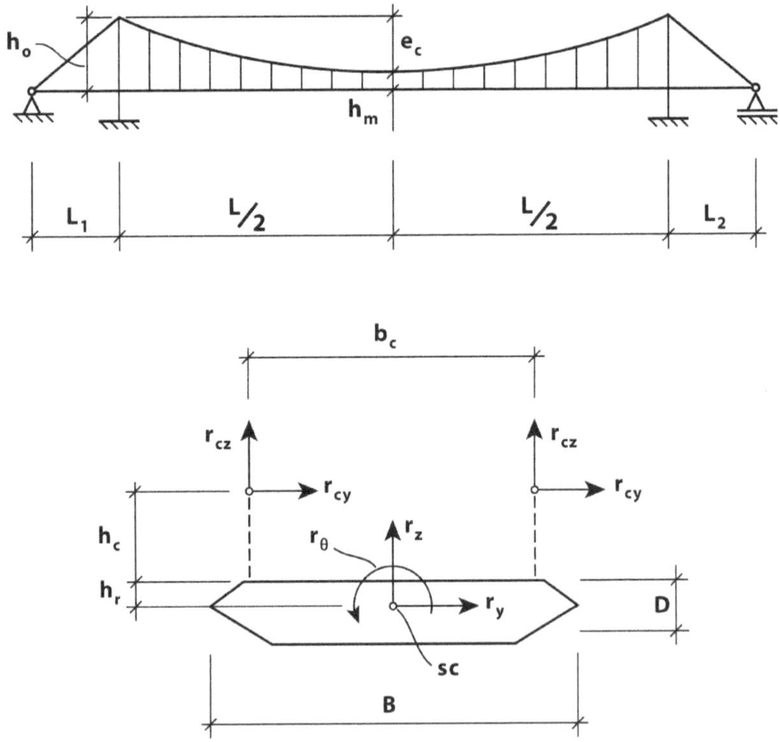

Fig. 3.24 The single span suspension bridge

The flexibility of hangers as well as backstays and towers are for simplicity ignored, usually considered minor discrepancies in the calculation of the eigenvalues of the system. Under these circumstances the main girder and the two cable planes move in perfect harmony, and consequently, the motion may be split into three independent components, one out of plane horizontal, one in plane vertical and one in pure main girder torsion. Below, it has been distinguished between distributed girder mass m_y in motion in the y direction and its distributed mass m_z in motion in the z direction, as it has been taken for granted that the latter contains half the self-weight of the suspension hangers (while the other half has been included in the self-weight of the two cables). The relevant system properties have been defined in Fig. 3.25. The solution strategy is based on Galerkin's method (see Chapter 1.7) and the assumption that the displacement components may be approximately represented by a harmonic series expansion, as first applied by Sigbjørnsson & Hjorth-Hansen [11].

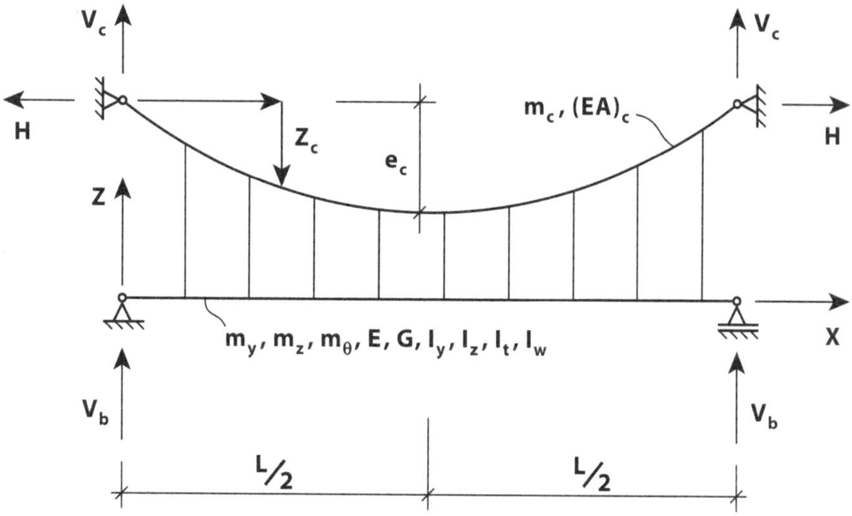

Fig. 3.25 Idealised system properties

Since the stiffness properties of the system heavily depend on the axial force in the cables, it is necessary to start with considering the static (time invariant or mean) situation. As shown in Fig. 3.26, it is assumed that the construction procedure is such that all the weight of the main beam is transferred via the hangers directly into the two cables, and to earth via the backstays and the towers. This is the most usual way of suspension bridge construction, and it has the consequence that the main girder shear forces at its connection to the two towers are negligible, i.e. that $V_b(x=0)=V_b(x=L)\approx 0$. (Even at a different construction procedure this is a reasonable approximation, at least for long span suspension bridges.)

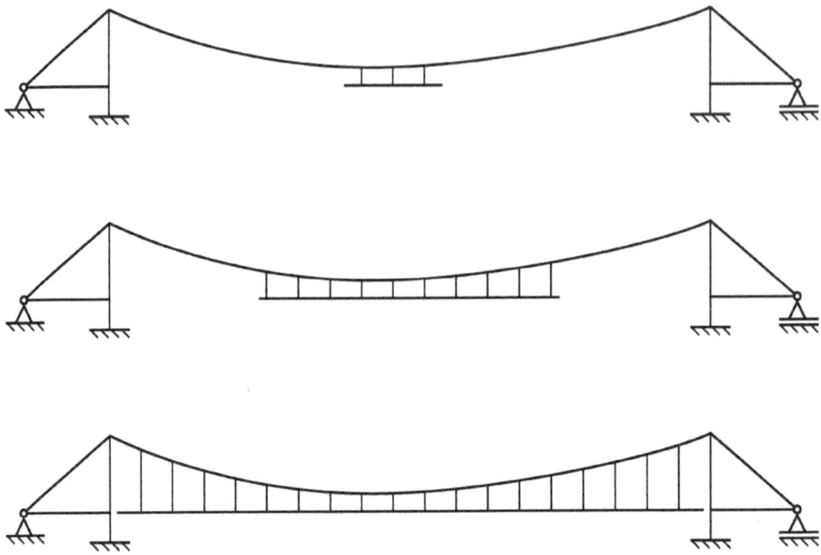

Fig. 3.26 The construction procedure

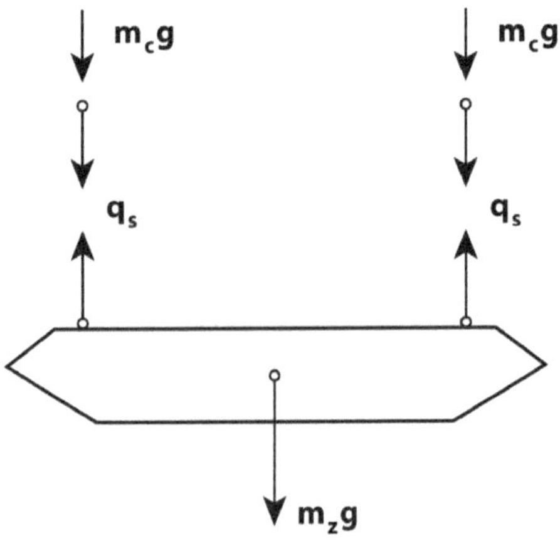

Fig. 3.27 The hanger force distributed per unit length $q_s \approx m_z g/2$

Two aspects regarding the time invariant equilibrium condition are worth noting before we proceed with the problem of dynamics. First, as illustrated in Fig. 3.27, the distributed hanger force (per unit length) is $q_s \approx m_z g/2$. Secondly, as shown in Fig. 3.28, the time invariant vertical cable equilibrium will require

$$d\bar{V} = \left(m_c ds + \frac{1}{2} m_z dx \right) g \tag{3.132}$$

Thus, the time invariant horizontal cable force component $\bar{H}$ may be obtained from the moment equilibrium requirement of half the cable span with respect to its top point (see Fig. 3.29)

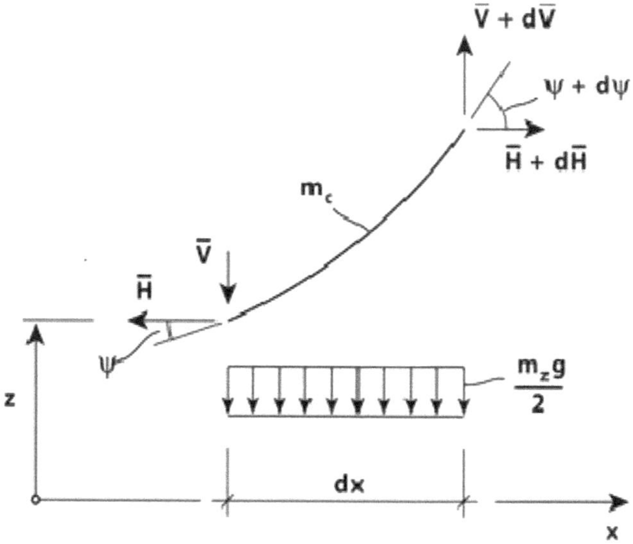

Fig. 3.28 Time invariant vertical cable equilibrium

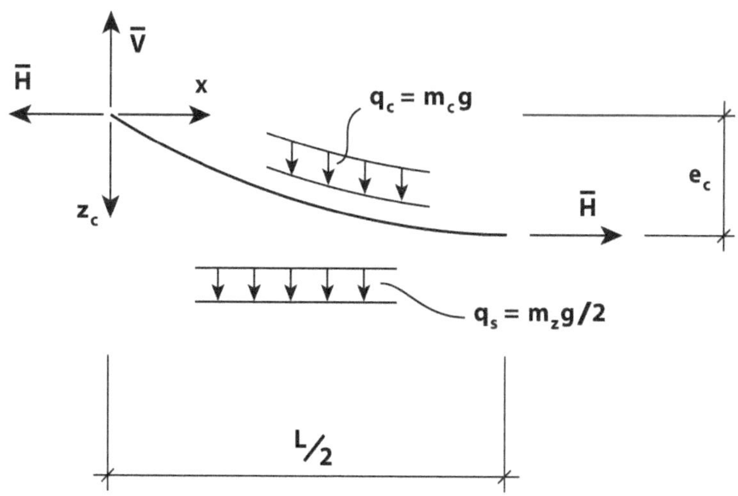

Fig. 3.29 The time invariant horizontal cable force component, $\bar{H}$

$$\bar{H} \cdot e_c = \int_0^{L/2} \left(\frac{m_z g}{2} dx + m_c g ds \right) x = g \int_0^{L/2} \left(\frac{m_z}{2} + m_c \frac{ds}{dx} \right) x dx$$
$$= g \int_0^{L/2} \left\{ \frac{m_z}{2} + m_c \left[1 + \left(\frac{dz_c}{dx} \right)^2 \right]^{\frac{1}{2}} \right\} x dx \approx g \int_0^{L/2} \left[\frac{m_z}{2} + m_c \left(1 + \frac{1}{2} z_c'^2 \right) \right] x dx \tag{3.133}$$

Introducing (see Eqs. 3.74 and 3.75) $z_c \approx 4 e_c \dfrac{x}{L}\left(1 - \dfrac{x}{L}\right)$ then

$$\frac{\bar{H} e_c}{g} = \frac{m_z}{2} \int_0^{L/2} x dx + m_c \int_0^{L/2} \left[1 + 8 \left(\frac{e_c}{L} \right)^2 \left(1 - 2 \frac{x}{L} \right)^2 \right] x dx$$
$$= \frac{m_z L^2}{16} + \frac{m_c L^2}{8} \left[1 + \frac{4}{3} \left(\frac{e_c}{L} \right)^2 \right] \tag{3.134}$$

rendering

$$\bar{H} = \frac{m_z g L^2}{16 e_c} \left\{ 1 + \frac{2 m_c}{m_z} \left[1 + \frac{4}{3} \left(\frac{e_c}{L} \right)^2 \right] \right\} \tag{3.135}$$

Dynamic Motion in the Horizontal y Direction

Let us first consider the case of dynamic motion in the out of plane horizontal direction, as illustrated in Fig. 3.30. Then the equilibrium requirement for an infinitesimal element dx of the main girder is given by (see Fig. 3.30.a and b and Eqs. 1.23 – 1.29)

$$m_y \ddot{r}_y + c_y \dot{r}_y + EI_z \frac{d^4 r_y}{dx^4} + 2 q_s \frac{r_y - r_{cy}}{h_c} = q_y \tag{3.136}$$

while the corresponding equilibrium requirement for each of the cables is given by (see Fig. 3.29.a and c)

$$m_c \ddot{r}_{cy} + c_{cy} \dot{r}_y - \bar{H} \frac{d^2 r_{cy}}{dx^2} - q_s \frac{r_y - r_{cy}}{h_c} = q_{cy} \tag{3.137}$$

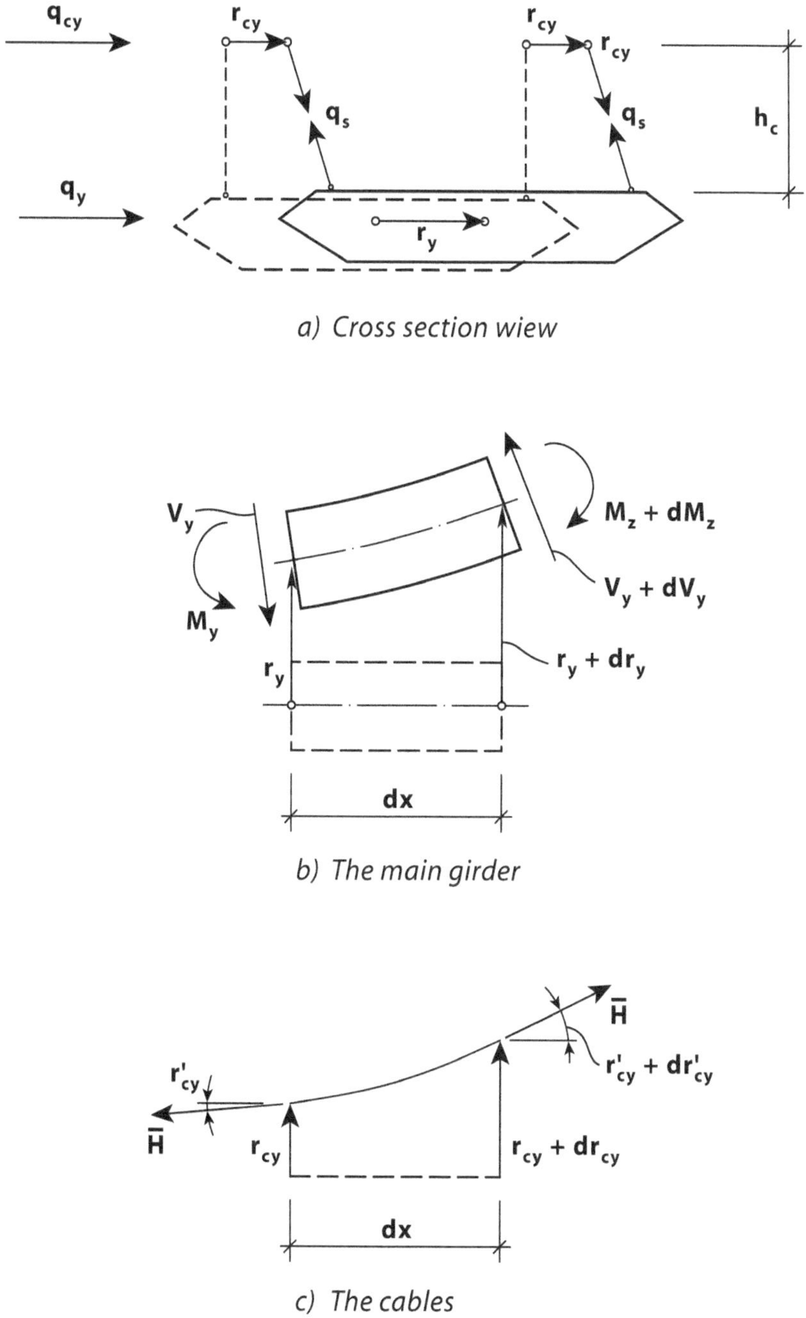

Fig. 3.30 Dynamic motion in horizontal direction

where q_y and q_{cy} are the dynamic loads per unit length on the girder and on each of the cables, r_y and r_{cy} are the displacements, c_y and c_{cy} are the damping coefficients, $q_s = m_z g/2$ is the hanger force (per unit length) and h_c is the hanger length. Since we adopt Eq. 3.91, i.e. that $z_c \approx 4e_c (x/L)(1-x/L)$, then

$$h_c = h_m + e_c - z_c = h_m + e_c (1-2x/L)^2 \tag{3.138}$$

where h_m is defined in Fig. 3.24. Let us consider the unloaded and undamped case (i.e. that c_y and c_{cy} as well as q_y and q_{cy} are all zero), and assume the following harmonic sinus Fourier series solution (taking it for granted that the main girder is simply supported at the towers)

$$\left.\begin{array}{l} r_y(x,t) = \phi_y(x)\cdot e^{i\omega t} \\ r_{cy}(x,t) = \phi_{cy}(x)\cdot e^{i\omega t} \end{array}\right\} \quad \text{where} \quad \left.\begin{array}{l} \phi_y(x) = \sum\limits_{n=1}^{N} a_{y_n} \sin(n\pi x/L) \\ \phi_{cy}(x) = \sum\limits_{n=1}^{N} a_{cy_n} \sin(n\pi x/L) \end{array}\right\} \tag{3.139}$$

Introducing this into Eq. 3.136

$$m_y (i\omega)^2 \phi_y + EI_z \phi_y'''' + m_z g \frac{\phi_y - \phi_{cy}}{h_c} =$$

$$\sum_{n=1}^{N} \left\{ \left[EI_z \left(\frac{n\pi}{L}\right)^4 a_{y_n} + \frac{m_z g \left(a_{y_n} - a_{cy_n}\right)}{h_m + e_c \left(1 - 2\frac{x}{L}\right)^2} - m_y \omega^2 a_{y_n} \right] \sin\left(\frac{n\pi}{L}\right) \right\} = 0 \tag{3.140}$$

we adopt a Galerkin approach by consecutive pre-multiplication of Eq. 3.140 with $(2/L)\sin(p\pi x/L)$, $p = 1,2,3,\ldots$, and integrating over the span renders

$$\sum_{n=1}^{N} \left[\left(\alpha_{pn} + \gamma_{pn}\right) a_{y_n} - \gamma_{pn} a_{cy_n} - \omega^2 \tilde{m}_{y_n} a_{y_n} \right] = 0 \quad p = 1,2,3,\ldots,N \tag{3.141}$$

$$\alpha_{pn} = EI_z\left(\frac{n\pi}{L}\right)^4 \frac{2}{L}\int_0^L \sin\left(\frac{p\pi x}{L}\right)\cdot\sin\left(\frac{n\pi x}{L}\right)dx = \begin{cases} EI_z\left(\frac{n\pi}{L}\right)^4 & \text{for p} = n \\ 0 & \text{for p} \neq \text{n} \end{cases}$$

$$\gamma_{pn} = \frac{m_z g}{e_c}\frac{2}{L}\int_0^L \frac{\sin(p\pi x/L)\cdot\sin(n\pi x/L)}{1 + h_m/e_c - 4(x/L)(1-x/L)}dx \quad \begin{cases} \left.\begin{array}{l}\text{if p even}\\ \&\ n \text{ odd}\end{array}\right\} \Rightarrow \gamma_{pn} = 0 \\ \left.\begin{array}{l}\text{if p odd}\\ \&\ n \text{ even}\end{array}\right\} \Rightarrow \gamma_{pn} = 0 \\ \textit{else} \qquad \Rightarrow \gamma_{pn} \neq 0 \end{cases} \tag{3.142}$$

$$\tilde{m}_{y_n} = \tilde{m}_{y_p} = m_y \frac{2}{L}\int_0^L \sin(p\pi x/L)\cdot\sin(n\pi x/L)dx = \begin{cases} m_y & \text{for p} = n \\ 0 & \text{for p} \neq \text{n} \end{cases}$$

Similarly for the cable equation, i.e. introducing Eq. 3.139 into Eq. 3.137

$$2m_c(i\omega)^2\phi_{cy} - 2\bar{H}\phi_{cy}'' - m_z g\frac{\phi_y - \phi_{cy}}{h_c} =$$

$$\sum_{n=1}^{N}\left\{\left[2\bar{H}\left(\frac{n\pi}{L}\right)^2 a_{cy_n} - \frac{m_z g\left(a_{y_n} - a_{cy_n}\right)}{h_m + e_c\left(1 - 2\frac{x}{L}\right)^2} - 2m_c\omega^2 a_{cy_n}\right]\sin\left(\frac{n\pi}{L}\right)\right\} = 0 \tag{3.143}$$

and then we consecutively pre-multiply Eq. 3.143 with $(2/L)\sin(p\pi x/L)$, $p = 1,2,3,\ldots,N$, [i.e. first pre-multiplying by $(2/L)\sin(1\pi x/L)$, then by $(2/L)\sin(2\pi x/L)$, and so on] and integrate over the entire span, then

$$\sum_{n=1}^{N}\left[-\gamma_{pn}a_{y_n} + \left(\beta_{pn} + \gamma_{pn}\right)a_{cy_n} - \omega^2\tilde{m}_{c_n}a_{cy_n}\right] = 0 \quad , \quad p = 1,2,3,\ldots,N \tag{3.144}$$

where

$$\left.\begin{aligned} \beta_{pn} &= 2\bar{H}\left(\frac{n\pi}{L}\right)^2 \frac{2}{L}\int_0^L \sin\left(\frac{p\pi x}{L}\right)\cdot \sin\left(\frac{n\pi x}{L}\right)dx = \begin{cases} 2\bar{H}\left(\frac{n\pi}{L}\right)^2 & \text{for p} = n \\ 0 & \text{for } \text{p} \neq n \end{cases} \\ \tilde{m}_{c_n} = \tilde{m}_{c_p} &= 2m_c \frac{2}{L}\int_0^L \sin\left(\frac{p\pi x}{L}\right)\cdot \sin\left(\frac{n\pi x}{L}\right)dx = \begin{cases} 2m_c \ \text{for p} = n \\ 0 \qquad\quad \text{for } \text{p} \neq n \end{cases} \end{aligned}\right\} \tag{3.145}$$

The sequence index $p = 1, 2, 3, \ldots, N$ and the series summations in Eqs. 3.141 and 3.144 may conveniently be replaced by

$$\left\{\begin{bmatrix} \alpha_{11}+\gamma_{11} & -\gamma_{11} & \cdots & \gamma_{1p} & -\gamma_{1p} & \cdots & \gamma_{1N} & -\gamma_{1N} \\ -\gamma_{11} & \beta_{11}+\gamma_{11} & \cdots & -\gamma_{1p} & \gamma_{1p} & \cdots & -\gamma_{1N} & \gamma_{1N} \\ \vdots & \vdots & \ddots & \vdots & \vdots & \cdot^{\cdot^{\cdot}} & \vdots & \vdots \\ \gamma_{p1} & -\gamma_{p1} & \cdots & \alpha_{pp}+\gamma_{pp} & -\gamma_{pp} & \cdots & \gamma_{pN} & -\gamma_{pN} \\ -\gamma_{p1} & \gamma_{p1} & \cdots & -\gamma_{pp} & \beta_{pp}+\gamma_{pp} & \cdots & -\gamma_{pN} & \gamma_{pN} \\ \vdots & \vdots & \cdot^{\cdot^{\cdot}} & \vdots & \vdots & \ddots & \vdots & \vdots \\ \gamma_{N1} & -\gamma_{N1} & \cdots & \gamma_{Np} & -\gamma_{Np} & \cdots & \alpha_{NN}+\gamma_{NN} & -\gamma_{NN} \\ -\gamma_{N1} & \gamma_{N1} & \cdots & -\gamma_{Np} & \gamma_{Np} & \cdots & -\gamma_{NN} & \beta_{NN}+\gamma_{NN} \end{bmatrix}\right.$$

$$\left. -\omega^2 \begin{bmatrix} \tilde{m}_{y_1} & & & & & & & \\ & \tilde{m}_{c_1} & & & & & \mathbf{0} & \\ & & \ddots & & & & & \\ & & & \tilde{m}_{y_p} & & & & \\ & & & & \tilde{m}_{c_p} & & & \\ & \mathbf{0} & & & & \ddots & & \\ & & & & & & \tilde{m}_{y_N} & \\ & & & & & & & \tilde{m}_{c_N} \end{bmatrix}\right\} \cdot \begin{bmatrix} a_{y_1} \\ a_{cy_1} \\ \vdots \\ a_{y_p} \\ a_{cy_p} \\ \vdots \\ a_{y_N} \\ a_{cy_N} \end{bmatrix} = 0 \tag{3.146}$$

Thus, defining the displacement amplitude vector

$$\mathbf{a}_y = \begin{bmatrix} \mathbf{a}_{y_1} & \cdots & \mathbf{a}_{y_p} & \cdots & \mathbf{a}_{y_N} \end{bmatrix}^T \tag{3.147}$$

where $\mathbf{a}_{y_p} = \begin{bmatrix} a_{y_p} & a_{cy_p} \end{bmatrix}^T$, and where the stiffness matrix

$$\mathbf{K}_y = \begin{bmatrix} \mathbf{\Omega}_{11} & \cdots & \mathbf{Y}_{1p} & \cdots & \mathbf{Y}_{1N} \\ \vdots & \ddots & & \mathbf{Y}_{pn} & \vdots \\ \mathbf{Y}_{p1} & & \mathbf{\Omega}_{pp} & & \mathbf{Y}_{pN} \\ \vdots & \mathbf{Y}_{np} & & \ddots & \vdots \\ \mathbf{Y}_{N1} & \cdots & \mathbf{Y}_{Np} & \cdots & \mathbf{\Omega}_{NN} \end{bmatrix} \tag{3.148}$$

where $\mathbf{\Omega}_{pp} = \begin{bmatrix} \alpha_{pp} + \gamma_{pp} & -\gamma_{pp} \\ -\gamma_{pp} & \beta_{pp} + \gamma_{pp} \end{bmatrix}$ and $\mathbf{Y}_{pn} = \gamma_{pn} \begin{bmatrix} 1 & -1 \\ -1 & 1 \end{bmatrix}$, and the mass matrix

$$\mathbf{M}_y = diag\begin{bmatrix} \tilde{\mathbf{m}}_{y_1} & \cdots & \tilde{\mathbf{m}}_{y_p} & \cdots & \tilde{\mathbf{m}}_{y_N} \end{bmatrix} \tag{3.149}$$

where $\tilde{\mathbf{m}}_{y_p} = \begin{bmatrix} \tilde{m}_{y_p} & 0 \\ 0 & \tilde{m}_{c_p} \end{bmatrix}$, then Eq. 3.146 may be written

$$\left(\mathbf{K}_y - \omega^2 \mathbf{M}_y\right)\mathbf{a}_y = \mathbf{0} \tag{3.150}$$

which is a classical eigenvalue problem similar to those encountered before.

Elaboration 3.1

An approximate solution may be obtained for the first two eigenfrequencies by setting $N = 2$. Then

$$\alpha_{pn} = EI_z \left(\frac{n\pi}{L}\right)^4 2\int_0^1 \sin(p\pi\hat{x}) \cdot \sin(n\pi\hat{x})\, d\hat{x} \Rightarrow \begin{cases} \alpha_{11} = EI_z \left(\dfrac{\pi}{L}\right)^4 \\ \alpha_{22} = EI_z \left(\dfrac{2\pi}{L}\right)^4 \end{cases}$$

$$\gamma_{pn} = \frac{m_z g}{e_c} 2\int_0^1 \frac{\sin(p\pi\hat{x})\sin(n\pi\hat{x})}{1+\frac{h_m}{e_c}-4\hat{x}(1-\hat{x})} d\hat{x} \Rightarrow \begin{cases} \gamma_{11} = \frac{m_z g}{e_c} 2\int_0^1 \frac{\sin(\pi\hat{x})\sin(\pi\hat{x})}{1+\frac{h_m}{e_c}-4\hat{x}(1-\hat{x})} d\hat{x} \\ \gamma_{22} = \frac{m_z g}{e_c} 2\int_0^1 \frac{\sin(2\pi\hat{x})\sin(2\pi\hat{x})}{1+\frac{h_m}{e_c}-4\hat{x}(1-\hat{x})} d\hat{x} \\ \gamma_{12} = \gamma_{21} = 0 \end{cases}$$

$$\tilde{m}_{y_p} = m_y 2\int_0^1 \sin(p\pi\hat{x})\cdot\sin(n\pi\hat{x})\,dx \Rightarrow \begin{cases} \tilde{m}_{y_1} = m_y \\ \tilde{m}_{y_2} = m_y \end{cases}$$

$$\beta_{pn} = 2\bar{H}\left(\frac{n\pi}{L}\right)^2 2\int_0^1 \sin(p\pi\hat{x})\cdot\sin(n\pi\hat{x})\,d\hat{x} \Rightarrow \begin{cases} \beta_{11} = 2\bar{H}\,(\pi/L)^2 \\ \beta_{22} = 2\bar{H}\,(2\pi/L)^2 \end{cases}$$

$$\tilde{m}_{c_p} = 2m_c \cdot 2\int_0^1 \sin(p\pi\hat{x})\cdot\sin(n\pi\hat{x})\,d\hat{x} = \begin{cases} m_{c_1} = 2m_c \\ m_{c_2} = 2m_c \end{cases}$$

Thus, the eigenvalue problem in Eq. 3.150 is given by $\begin{bmatrix} \mathbf{A}_{11} & \mathbf{0} \\ \mathbf{0} & \mathbf{A}_{22} \end{bmatrix} \begin{bmatrix} a_{y_1} \\ a_{cy_1} \\ a_{y_2} \\ a_{cy_2} \end{bmatrix} = \mathbf{0}$

where
$$\begin{cases} \mathbf{A}_{11} = \begin{bmatrix} \alpha_{11} + \gamma_{11} - \omega^2 \tilde{m}_{y_1} & -\gamma_{11} \\ -\gamma_{11} & \beta_{11} + \gamma_{11} - \omega^2 \tilde{m}_{c_1} \end{bmatrix} \\ \mathbf{A}_{22} = \begin{bmatrix} \alpha_{22} + \gamma_{22} - \omega^2 \tilde{m}_{y_2} & -\gamma_{22} \\ -\gamma_{22} & \beta_{22} + \gamma_{22} - \omega^2 \tilde{m}_{c_2} \end{bmatrix} \end{cases}$$

Its solution is

$$\omega_{y1} = \sqrt{\frac{\alpha_{11}+\gamma_{11}}{2m_y} + \frac{\beta_{11}+\gamma_{11}}{4m_c} - \sqrt{\left(\frac{\alpha_{11}+\gamma_{11}}{2m_y} - \frac{\beta_{11}+\gamma_{11}}{4m_c}\right)^2 + \frac{\gamma_{11}^2}{2m_y m_c}}}$$

$$\omega_{y2} = \sqrt{\frac{\alpha_{22}+\gamma_{22}}{2m_y} + \frac{\beta_{22}+\gamma_{22}}{4m_c} - \sqrt{\left(\frac{\alpha_{22}+\gamma_{22}}{2m_y} - \frac{\beta_{22}+\gamma_{22}}{4m_c}\right)^2 + \frac{\gamma_{22}^2}{2m_y m_c}}}$$

Table 3.1 Numeric example

L (m)	EI_z (Nm^2)	$\bar{H}$ (N)	g (m/s^2)	$m_y = m_z$ (kg/m)	m_c (kg/m)	e_c (m)	h_m (m)
1250	$3.5 \cdot 10^{12}$	$1.12 \cdot 10^8$	9.81	9000	1900	120	2.75

Using the values given in Table 3.1 above, renders $\omega_{y1} = 0.34$ and $\omega_{y2} = 0.68$ rad/s.

Example 3.3

By increasing the number of Fourier components to $N = 4$ a more comprehensive solution may be obtained. Again, adopting the numerical values given in Table 3.1 above, then the solution to the eigenvalue problem in Eq. 3.150 renders the four first eigenmodes and corresponding eigenfrequencies associated with suspension bridge horizontal motion as shown in Fig. 3.31 below.

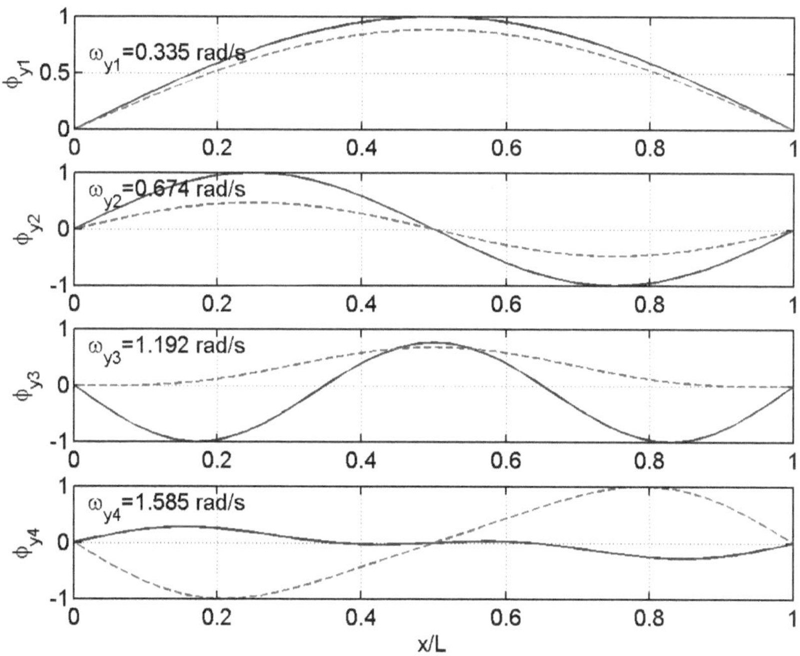

Fig. 3.31 Lowest four eigenvalue solutions for suspension bridge in horizontal motion, $N = 4$; fully drawn lines: main girder, broken lines: the cables

Dynamic in Plane Motion in the Vertical z Direction

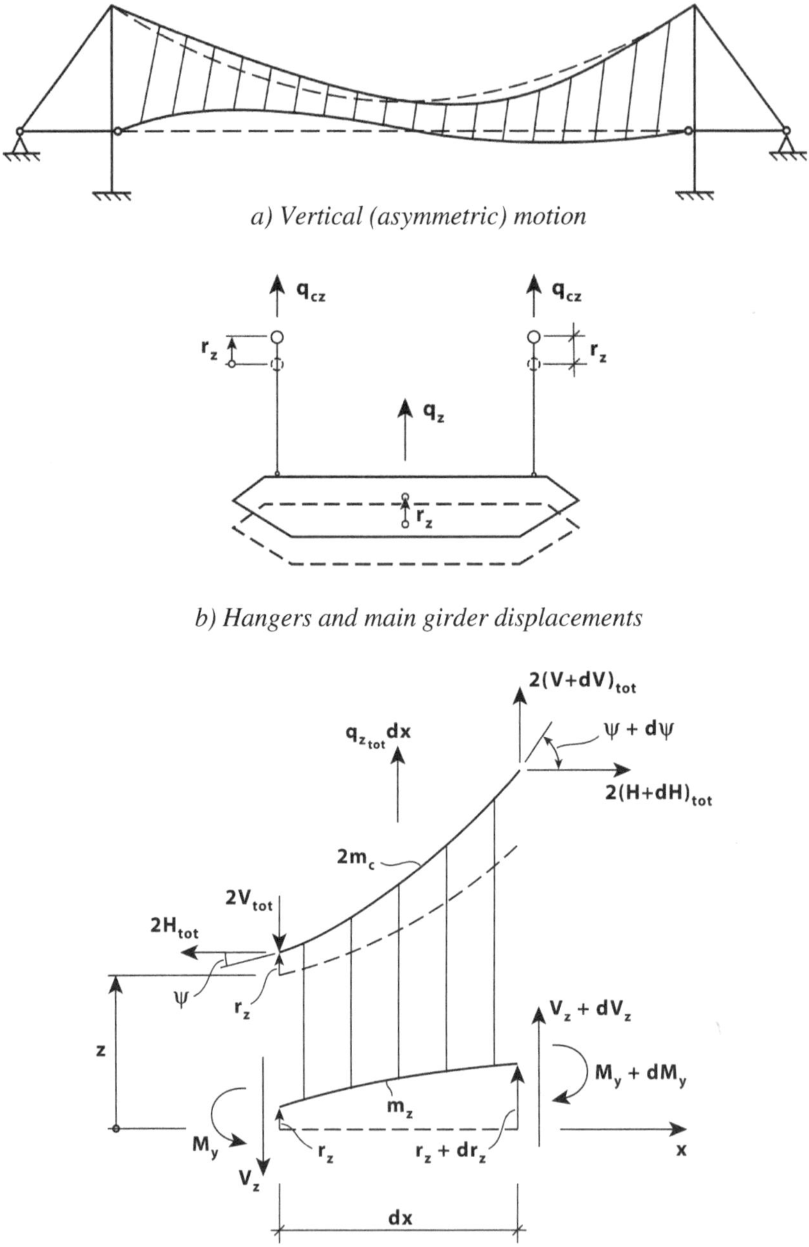

a) Vertical (asymmetric) motion

b) Hangers and main girder displacements

c) Infinitesimal bridge element dx

Fig. 3.32 Dynamic in plane motion in vertical direction

Let us then consider the case of dynamic motion in the vertical direction, as illustrated in Fig. 3.32. Since we neglect the flexibility of the hangers, then the two cables will act in perfect harmony (as if they are one) and in harmony with the main girder as illustrated in Fig. 3.32.b, i.e. $r_{c_z} = r_z$. Therefore, it is not necessary to distinguish between the motion of the girder and the motion of the cables, as they are identical. Thus, defining $H_{tot} = \bar{H} + H(t)$, then the equilibrium requirement for an infinitesimal bridge element dx with respect to forces in the x direction is given by (see Fig. 3.32.c)

$$-2H_{tot} + 2\left(H_{tot} + dH_{tot}\right) = 0 \quad \Rightarrow \quad dH_{tot} = 0 \tag{3.151}$$

Since we know from before that $d\bar{H} = 0$, then also $dH(t) = 0$. The corresponding equilibrium requirement with respect to forces in the z direction is given by (see Fig. 3.32c)

$$\left(2m_c ds + m_z dx\right)\ddot{r}_z + \left(2m_c ds + m_z dx\right)g + c_z dx\dot{r}_z - 2dV_{tot} - dV_z = q_{z_{tot}} dx \tag{3.152}$$

where $q_{z_{tot}} = q_z + 2q_{c_z}$ is the dynamic loads per unit length on the girder and on the two cables, r_z is the vertical displacement, c_z is the damping coefficient, and $V_{tot} = \bar{V} + V(t)$ is the total vertical force component in each of the cables and V_z is the vertical shear force in the girder. Thus, the differential equation associated with vertical motion is given by

$$\left(2m_c \frac{ds}{dx} + m_z\right)\ddot{r}_z + c_z\dot{r}_z - 2V_{tot}' - V_z' + \left(2m_c \frac{ds}{dx} + m_z\right)g = q_{z_{tot}} \tag{3.153}$$

As can be seen from Fig. 3.32.c

$$\begin{aligned} &\frac{V_{tot}}{H_{tot}} = \frac{\bar{V} + V}{\bar{H} + H} = \tan\psi = \frac{d}{dx}\left(z + r_z\right) = z' + r_z' \\ &\Rightarrow \quad \bar{V} + V = \bar{H}z' + Hz' + \left(\bar{H} + H\right)r_z' \end{aligned} \tag{3.154}$$

from which it is seen that

$$\left.\begin{aligned} &\bar{V} = \bar{H}z' \\ &V(t) = Hz' + \left(\bar{H} + H\right)r_z' \end{aligned}\right\} \tag{3.155}$$

Let us introduce that (see Eqs. 1.23 and 1.27)

$$V_z = M'_y = \frac{d}{dx}\left(-EI_y r''_z\right) = -EI_y r'''_z \tag{3.156}$$

and the time invariant equilibrium requirement $2d\bar{V} = 2m_c g ds + m_z g dx$ (see Eq. 3.132), and only consider the unloaded and undamped case, then Eq. 3.153 becomes

$$\left(2m_c \frac{ds}{dx} + m_z\right)\ddot{r}_z - 2\left[Hz'' + \left(\bar{H} + H\right) r''_z\right] + EI_y r''''_z = 0 \tag{3.157}$$

Within the shallow cable theory it is a usual approximation that $H(t) << \bar{H}$, and that $ds/dx = \left[1 + (dz/dx)^2\right]^{1/2} \approx 1$. Furthermore, since (see Eq. 3.91)

$$z = e_c + h_m - z_c \approx e_c + h_m - 4e_c \frac{x}{L}\left(1 - \frac{x}{L}\right) \quad \Rightarrow \quad z'' \approx \frac{8e_c}{L^2} \tag{3.158}$$

and (see Eq. 3.108, noting that in Chapter 3.4 we focused on cable vibrations alone, and then z_c and correspondingly also r_z was defined positive downwards, while we here focus on the entire bridge and therefore more conveniently, we define z and r_z positive upwards)

$$H(t) = \frac{(EA)_c}{\ell_e}\frac{8e_c}{L^2}\int_0^L (-r_z)\,dx \quad \text{where} \quad \ell_e = L\left[1 + 8\left(\frac{e_c}{L}\right)^2\right] \tag{3.159}$$

and where $(EA)_c$ is the product of elastic modulus and cross sectional area assumed identical for each of the cables, then Eq. 3.157 is reduced into

$$(2m_c + m_z)\ddot{r}_z + \frac{128e_c^2}{L^4}\frac{(EA)_c}{\ell_e}\int_0^L r_z dx - 2\bar{H}r''_z + EI_y r''''_z = 0 \tag{3.160}$$

Again, taking it for granted that the main girder is simply supported at the towers, and adopting a harmonic Fourier series solution

$$r_z(x,t) = \phi_z(x)\cdot e^{i\omega t} \quad \text{where} \quad \phi_z(x) = \sum_{n=1}^{N} a_{z_n} \sin(n\pi x/L) \tag{3.161}$$

then Eq. 3.160 becomes

$$\sum_{n=1}^{N}\left\{\frac{128e_c^2}{L^4}\frac{(EA)_c}{\ell_e}a_{z_n}\int_0^L \sin\left(\frac{n\pi x}{L}\right)dx+2\bar{H}\left(\frac{n\pi}{L}\right)^2 a_{z_n}\sin\left(\frac{n\pi x}{L}\right)\right.$$
$$\left.+EI_y\left(\frac{n\pi}{L}\right)^4 a_{z_n}\sin\left(\frac{n\pi x}{L}\right)-\omega^2(2m_c+m_z)a_{z_n}\sin\left(\frac{n\pi x}{L}\right)\right\}e^{i\omega t}=0 \tag{3.162}$$

A Galerkin type of approach by consecutive pre-multiplication of Eq. 3.162 with $(2/L)\sin(p\pi x/L)$, $p=1,2,3,\ldots$, and integrating over the entire span renders

$$\sum_{n=1}^{N}\left[(\kappa_{pn}+\lambda_{pn})a_{z_n}+\mu_{pn}a_{z_n}-\omega^2\tilde{m}_{z_n}a_{z_n}\right]=0, \quad p=1,2,3,\ldots,N \tag{3.163}$$

where

$$\left.\begin{aligned}
\kappa_{pn}&=EI_y\left(\frac{n\pi}{L}\right)^4\frac{2}{L}\int_0^L\sin\left(\frac{p\pi x}{L}\right)\sin\left(\frac{n\pi x}{L}\right)dx=\begin{cases}EI_y(n\pi/L)^4 & \text{for } p=n\\ 0 & \text{for } p\neq n\end{cases}\\
\lambda_{pn}&=2\bar{H}\left(\frac{n\pi}{L}\right)^2\frac{2}{L}\int_0^L\sin\left(\frac{p\pi x}{L}\right)\sin\left(\frac{n\pi x}{L}\right)dx=\begin{cases}2\bar{H}(n\pi/L)^2 & \text{for } p=n\\ 0 & \text{for } p\neq n\end{cases}
\end{aligned}\right\} \tag{3.164}$$

while, recalling that $\int_0^L\sin\left(\frac{j\pi x}{L}\right)dx=\begin{cases}2L/j\pi & \text{for } j=1,3,5,\ldots\\ 0 & \text{for } j=2,4,6,\ldots\end{cases}$

$$\left.\begin{aligned}
\mu_{pn}&=\frac{128e_c^2}{L^4}\frac{(EA)_c}{\ell_e}\frac{2}{L}\int_0^L\sin\left(\frac{p\pi x}{L}\right)dx\int_0^L\sin\left(\frac{n\pi x}{L}\right)dx\\
&=\begin{cases}\left(\dfrac{32e_c}{\pi L}\right)^2\dfrac{(EA)_c}{L\ell_e}\dfrac{1}{pn} & \text{for } \left.\begin{matrix}p\\ \& n\end{matrix}\right\}=1,3,5,\ldots\\[2ex] 0 & \text{for } \left.\begin{matrix}p\\ \text{or } n\end{matrix}\right\}=2,4,6,\ldots\end{cases}
\end{aligned}\right\} \tag{3.165}$$

and where

$$\tilde{m}_{z_p}=(2m_c+m_z)\frac{2}{L}\int_0^L\sin\left(\frac{p\pi x}{L}\right)\sin\left(\frac{n\pi x}{L}\right)dx=\begin{cases}(2m_c+m_z) & \text{for } p=n\\ 0 \ \text{for } p\neq n\end{cases} \tag{3.166}$$

The $p = 1,2,3,\ldots,N$ sequence and series summations in Eq. 3.162 may conveniently be written in the form of a classical eigenvalue problem:

$$\left\{\begin{bmatrix} \kappa_{11}+\lambda_{11} & & & & \\ & \ddots & & \mathbf{0} & \\ & & \kappa_{pp}+\lambda_{pp} & & \\ & \mathbf{0} & & \ddots & \\ & & & & \kappa_{NN}+\lambda_{NN} \end{bmatrix} + \begin{bmatrix} \mu_{11} & \cdots & \mu_{1n} & \cdots & \mu_{1N} \\ \vdots & \ddots & & \iddots & \vdots \\ \mu_{p1} & & \mu_{pn} & & \mu_{pN} \\ \vdots & \iddots & & \ddots & \vdots \\ \mu_{N1} & \cdots & \mu_{Nn} & \cdots & \mu_{NN} \end{bmatrix} - \omega^2 \begin{bmatrix} \tilde{m}_{z_1} & & & & \\ & \ddots & & \mathbf{0} & \\ & & \tilde{m}_{z_p} & & \\ & \mathbf{0} & & \ddots & \\ & & & & \tilde{m}_{z_N} \end{bmatrix}\right\} \begin{bmatrix} a_{z_1} \\ \vdots \\ a_{z_p} \\ \vdots \\ a_{z_N} \end{bmatrix} = 0 \tag{3.167}$$

Thus, defining the displacement amplitude vector

$$\mathbf{a}_z = \begin{bmatrix} a_{z_1} & \cdots & a_{z_p} & \cdots & a_{z_N} \end{bmatrix}^T \tag{3.168}$$

the stiffness matrix

$$\mathbf{K}_z = \begin{bmatrix} \kappa_{11}+\lambda_{11} & & & & \\ & \ddots & & \mathbf{0} & \\ & & \kappa_{pp}+\lambda_{pp} & & \\ & \mathbf{0} & & \ddots & \\ & & & & \kappa_{NN}+\lambda_{NN} \end{bmatrix} + \begin{bmatrix} \mu_{11} & \cdots & \mu_{1n} & \cdots & \mu_{1N} \\ \vdots & \ddots & & \iddots & \vdots \\ \mu_{p1} & & \mu_{pn} & & \mu_{pN} \\ \vdots & \iddots & & \ddots & \vdots \\ \mu_{N1} & \cdots & \mu_{Nn} & \cdots & \mu_{NN} \end{bmatrix} \tag{3.169}$$

and the mass matrix

$$\mathbf{M}_z = (2m_c + m_z)\mathbf{I} \tag{3.170}$$

where $\mathbf{I}$ is the identity matrix, then Eq. 3.167 is given by

$$\left(\mathbf{K}_z - \omega^2 \mathbf{M}_z\right)\mathbf{a}_z = \mathbf{0} \tag{3.171}$$

As shown in Fig. 3.33 below, the solution for the first four eigenvalues converges at $N \geq 4$.

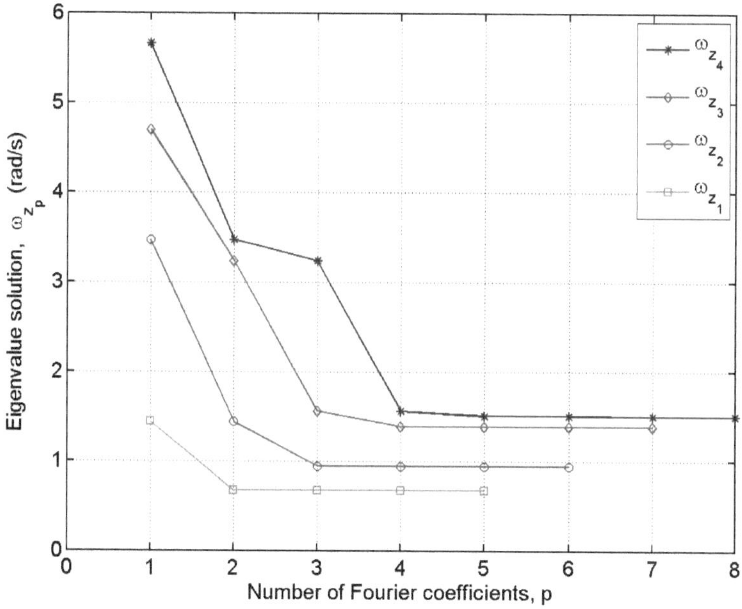

Fig. 3.33 Convergence of eigenvalue solution

Example 3.4

An approximate solution for the first four eigenfrequencies may be obtained by setting $N = 4$. Then Eq. 3.170 becomes

$$\left\{ \left(\frac{\pi}{L}\right)^4 EI_y \begin{bmatrix} 1^4 & & & \\ & 2^4 & \mathbf{0} & \\ & \mathbf{0} & 3^4 & \\ & & & 4^4 \end{bmatrix} + 2\left(\frac{\pi}{L}\right)^2 \bar{H} \begin{bmatrix} 1^2 & & & \\ & 2^2 & \mathbf{0} & \\ & \mathbf{0} & 3^2 & \\ & & & 4^2 \end{bmatrix} + \right.$$

$$\left. \left(\frac{32e_c}{\pi L}\right)^2 \frac{(EA)_c}{L\ell_e} \begin{bmatrix} 1 & 0 & \frac{1}{3} & 0 \\ 0 & 0 & 0 & 0 \\ \frac{1}{3} & 0 & \frac{1}{9} & 0 \\ 0 & 0 & 0 & 0 \end{bmatrix} - \omega^2 (2m_c + m) \begin{bmatrix} 1 & & & \\ & 1 & \mathbf{0} & \\ & \mathbf{0} & 1 & \\ & & & 1 \end{bmatrix} \right\} \begin{bmatrix} a_{z1} \\ a_{z2} \\ a_{z3} \\ a_{z4} \end{bmatrix} = \mathbf{0}$$

Defining

$$c = \frac{2}{\pi^2}\frac{\bar{H}L^2}{EI_y}, \quad d = \left(\frac{32}{\pi^3}\right)^2 \left(\frac{e_c}{L}\right)^2 \frac{L}{\ell_e}\frac{(EA)_c L^2}{EI_y} \text{ and } \hat{\omega} = \frac{\omega}{(\pi/L)^2\sqrt{EI_y/(2m_c+m)}}$$

then

$$\begin{bmatrix} 1+c+d-\hat{\omega}^2 & 0 & d/3 & 0 \\ 0 & 16+4c-\hat{\omega}^2 & 0 & 0 \\ d/3 & 0 & 81+9c+d/9-\hat{\omega}^2 & 0 \\ 0 & 0 & 0 & 256+16c-\hat{\omega}^2 \end{bmatrix} \begin{bmatrix} a_{z1} \\ a_{z2} \\ a_{z3} \\ a_{z4} \end{bmatrix} = \mathbf{0}$$

and thus, the solution to the eigenvalue problem is defined by the zero crossings of the polynomial

$$\left(16+4c-\hat{\omega}^2\right)\left(256+16c-\hat{\omega}^2\right)\left[\left(1+c+d-\hat{\omega}^2\right)\left(81+9c+\frac{d}{9}-\hat{\omega}^2\right)-\left(\frac{d}{3}\right)^2\right]=0$$

It has four real roots, whose indices are referred to below in ascending order. The first, which is associated with an anti-symmetric mode shape, is defined by

$$16+4c-\hat{\omega}^2=0 \quad \Rightarrow \quad \hat{\omega}_1 = 2\sqrt{4+c} \quad \Rightarrow \quad \omega_{z1} = 2\pi\sqrt{\frac{4\pi^2 EI_y + 2\bar{H}L^2}{(2m_c+m)L^4}}$$

The third, also associated with an anti-symmetric mode shape, is defined by

$$256+16c-\hat{\omega}^2=0 \quad \Rightarrow \quad \hat{\omega}_3 = 4\sqrt{16+c} \quad \Rightarrow \quad \omega_{z3} = 4\pi\sqrt{\frac{16\pi^2 EI_y + 2\bar{H}L^2}{(2m_c+m)L^4}}$$

The second and the fourth, both associated with symmetric mode shapes, are defined by the roots of

$$\left(1+c+d-\hat{\omega}^2\right)\left(81+9c+\frac{d}{9}-\hat{\omega}^2\right)-\left(\frac{d}{3}\right)^2=0$$

rendering $\omega_{2,4} = \pi\sqrt{\frac{\pi^2 EI_y}{(2m_c+m)L^4}\left[41+5c+\frac{5d}{9}\mp\sqrt{\left(40+4c-\frac{4d}{9}\right)^2+\left(\frac{d}{3}\right)^2}\right]}$

Example 3.4

Setting the number of Fourier components at $N = 12$ a sufficiently accurate solution may be obtained. Adopting the numerical values given in Table 3.2 below, then the solution to the eigenvalue problem in Eq. 3.171 renders the four first eigenmodes and corresponding eigenfrequencies associated with suspension bridge vertical motion as shown in Fig. 3.34 below.

Table 3.2 Numeric example

L (m)	EI_y (Nm^2)	$\bar{H}$ (N)	m_z (kg/m)	m_c (kg/m)	e_c (m)	A_c (m^2)	E_c (N/m^2)
1250	$0.2 \cdot 10^{12}$	$1.12 \cdot 10^{8}$	9000	1900	120	0.22	$0.2 \cdot 10^{12}$

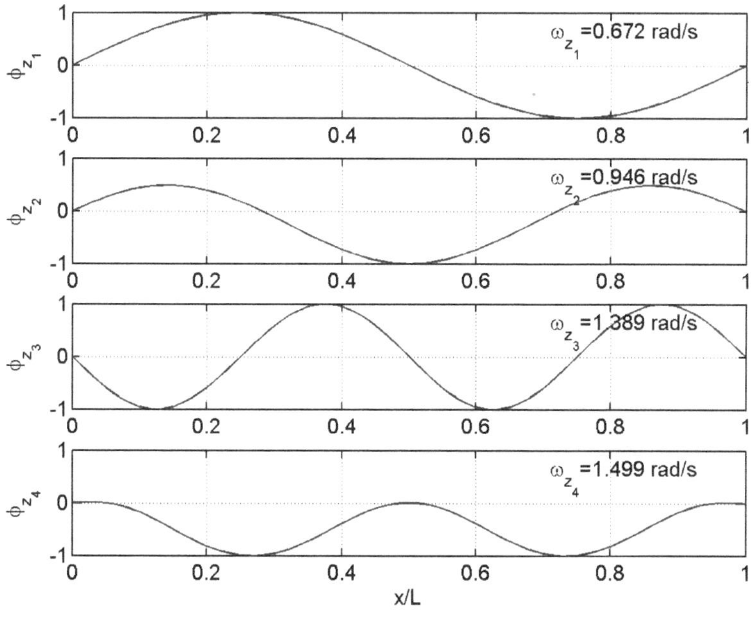

Fig. 3.34 Lowest four eigenvalue solutions for suspension bridge in vertical motion, $N = 12$

Dynamic Motion in Pure Torsion

Finally, we shall consider the case of dynamic motion in pure torsion. Since we assume that the dynamic stretching of hangers is insignificant, then the cables will act in perfect harmony as illustrated in Fig. 3.35.a, i.e. $r_{c_z} = \pm r_\theta b_c/2$, where b_c is the distance between cables. A sideway view is illustrated in Fig. 3.35.b.

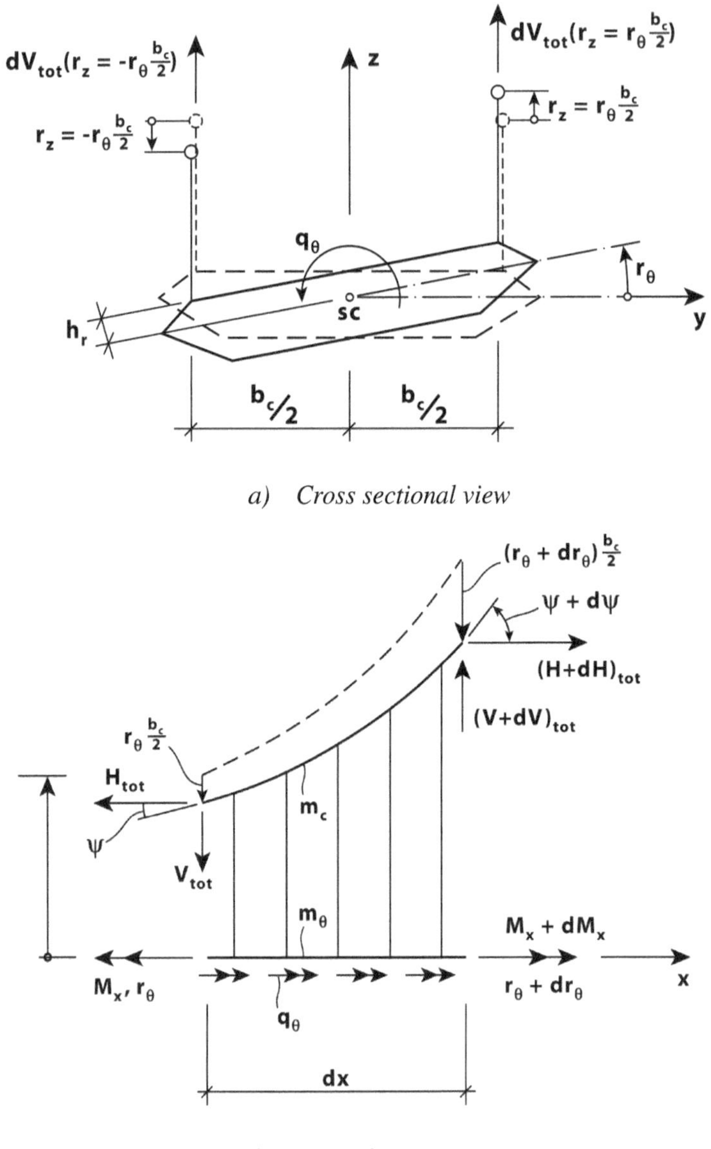

a) Cross sectional view

b) Sideways view

Fig. 3.35 Dynamic motion in torsion

As before, $H_{tot} = \bar{H} + H(t)$, $V_{tot} = \bar{V} + V(t)$, and, as there are no forces acting on the main girder in x and z directions

$$dH_{tot} = 0 \quad \Rightarrow \quad d\bar{H} = 0 \quad \& \quad dH = 0 \tag{3.172}$$

When the system is at rest then (recalling that the hanger force per unit length $q_s \approx m_z g/2$)

$$d\bar{V} - q_s dx - m_c g ds = 0 \quad \Rightarrow \quad \bar{V}' = \frac{m_z g}{2} + m_c g \frac{ds}{dx} \approx \frac{m_z g}{2} + m_c g \tag{3.173}$$

Furthermore, recalling from Eqs. 3.155 that $V = Hz' + \left(\bar{H} + H\right) r_z'$ and from Eq. 3.159 that $H(t) = \left[(EA)_c / \ell_e\right]\left(8e_c/L^2\right)\int_0^L (-r_z)\,dx$, it is seen from Fig. 3.35 that the cable at $y = -b_c/2$ is stretched a vertical displacement $r_z = -(b_c/2)\,r_\theta$, in which case (assuming $H(t) << \bar{H}$)

$$V\left(y = -\frac{b_c}{2}, t\right) \approx Hz' + \bar{H}r_z' = |H|\,z' - \bar{H}\frac{b_c}{2}r_\theta' \tag{3.174}$$

where $|H| = \left[((EA)_c)/(L\ell_e)\right](8e_c/L)(b_c/2)\int_0^L r_\theta dx$. Similarly, at $y = +b_c/2$ the cable is relaxed a vertical displacement $r_z = +(b_c/2)\,r_\theta$, in which case (still assuming $H(t) << \bar{H}$)

$$V\left(y = +\frac{b_c}{2}, t\right) \approx Hz' + \bar{H}r_z' = -|H|\,z' + \bar{H}\frac{b_c}{2}r_\theta' \tag{3.175}$$

Torsion moment equilibrium for the main girder about its shear centre is fulfilled by

$$\begin{aligned} &q_\theta dx + dM_x - c_\theta dx\dot{r}_\theta - \left[m_\theta dx + 2m_c\left(\frac{b_c}{2}\right)^2 ds\right]\ddot{r}_\theta \\ &+dV_{tot}\left(y = \frac{b_c}{2}\right)\cdot\left(\frac{b_c}{2} - r_\theta h_r\right) - dV_{tot}\left(y = -\frac{b_c}{2}\right)\cdot\left(\frac{b_c}{2} + r_\theta h_r\right) = 0 \end{aligned} \tag{3.176}$$

where m_θ is the mass moment of inertia for the main girder, c_θ is its cross sectional damping coefficient (if such an effect is necessary to include) and h_r is the vertical distance from the shear centre to the point of suspender attachment (see Fig. 3.35.a). Thus, assuming $ds/dx \approx 1$, then the differential equation for motion in pure torsion becomes

$$\left[m_\theta + 2m_c\left(\frac{b_c}{2}\right)^2\right]\ddot{r}_\theta + c_\theta \dot{r}_\theta - \left[V'_{tot}\left(y=\frac{b_c}{2}\right) - V'_{tot}\left(y=-\frac{b_c}{2}\right)\right]\frac{b_c}{2}$$
$$+\left[V'_{tot}\left(y=\frac{b_c}{2}\right) + V'_{tot}\left(y=-\frac{b_c}{2}\right)\right]h_r r_\theta - M'_x = q_\theta \tag{3.177}$$

where
$$\begin{cases} V'_{tot}\left(y=\frac{b_c}{2}\right) = \bar{V}' + V'\left(y=\frac{b_c}{2}\right) = m_c g + \frac{m_z g}{2} - |H| z'' + \bar{H}\frac{b_c}{2} r''_\theta \\ V'_{tot}\left(y=-\frac{b_c}{2}\right) = \bar{V}' + V'\left(y=-\frac{b_c}{2}\right) = m_c g + \frac{m_z g}{2} + |H| z'' - \bar{H}\frac{b_c}{2} r''_\theta \end{cases}$$

Thus

$$m_{\theta_{tot}} \ddot{r}_\theta + c_\theta \dot{r}_\theta + |H| b_c z'' - \bar{H}\frac{b_c^2}{2} r''_\theta + (2m_c + m_z) g h_r r_\theta - M'_x = q_\theta \tag{3.178}$$

where $m_{\theta_{tot}} = m_\theta + m_c b_c^2/2$ and $|H| = \left[(EA)_c/(L\ell_e)\right](8e_c/L)(b_c/2)\int_0^L r_\theta dx$.

Introducing $z'' = 8e_c/L^2$ and (see Eq. 1.42) $M'_x = GI_t r''_\theta - EI_w r''''_\theta$, then the following is obtained:

$$m_{\theta_{tot}} \ddot{r}_\theta + c_\theta \dot{r}_\theta + 32\left(\frac{e_c b_c}{L^2}\right)^2 \frac{(EA)_c}{\ell_e}\int_0^L r_\theta dx - \bar{H}\frac{b_c^2}{2} r''_\theta$$
$$+(2m_c + m_z) g h_r r_\theta - GI_t r''_\theta + EI_w r''''_\theta = q_\theta \tag{3.179}$$

In the following we shall only consider the unloaded and undamped situation, i.e. that q_θ and c_θ are zero. Again, taking it for granted that the main girder has a fork type of simple support at the towers, and adopting a harmonic Fourier series solution

$$r_\theta(x,t) = \phi_\theta(x)\cdot e^{i\omega t} \quad \text{where} \quad \phi_\theta(x) = \sum_{n=1}^{N} a_{\theta_n} \sin(n\pi x/L) \tag{3.180}$$

then Eq. 3.179 becomes

$$\sum_{n=1}^{N}\left\{-\omega^2 m_{\theta_{tot}}\sin\left(\frac{n\pi x}{L}\right)+32\left(\frac{e_c b_c}{L^2}\right)\frac{(EA)_c}{\ell_e}\int_0^L\sin\left(\frac{n\pi x}{L}\right)dx+\frac{\bar{H}b_c^2}{2}\left(\frac{n\pi}{L}\right)^2\sin\left(\frac{n\pi x}{L}\right)\right.$$
$$\left.+(2m_c+m_z)gh_r\sin\left(\frac{n\pi x}{L}\right)+\left[GI_t+\left(\frac{n\pi}{L}\right)^2 EI_w\right]\left(\frac{n\pi}{L}\right)^2\sin\left(\frac{n\pi x}{L}\right)\right\}a_{\theta_n}e^{i\omega t}=0 \tag{3.181}$$

A Galerkin type of approach by consecutive pre-multiplication of Eq. 3.181 with $(2/L)\sin(p\pi x/L)$, $p=1,2,3,\ldots$, and integrating over the entire span renders

$$\sum_{n=1}^{N}\left[\left(\Omega_{pn}+\vartheta_{pn}+\nu_{pn}\right)a_{\theta_n}+\chi_{pn}a_{\theta_n}-\omega^2\tilde{m}_{\theta_{tot_n}}a_{\theta_n}\right]=0,\ p=1,2,..,N \tag{3.182}$$

where

$$\begin{aligned}\Omega_{pn}&=\left[GI_t+(n\pi/L)^2 EI_w\right](n\pi/L)^2\frac{2}{L}\int_0^L\sin(p\pi x/L)\cdot\sin(n\pi x/L)dx\\&=\begin{cases}(n\pi/L)^2\left[GI_t+(n\pi/L)^2 EI_w\right] & \text{for p}=n\\ 0 & \text{for p}\neq\text{n}\end{cases}\end{aligned} \tag{3.183}$$

$$\vartheta_{pn}=\frac{\bar{H}b_c^2}{2}\left(\frac{n\pi}{L}\right)^2\frac{2}{L}\int_0^L\sin\left(\frac{p\pi x}{L}\right)\sin\left(\frac{n\pi x}{L}\right)dx=\begin{cases}\bar{H}\dfrac{b_c^2}{2}\left(\dfrac{n\pi}{L}\right)^2 & \text{for p}=n\\ 0 & \text{for p}\neq\text{n}\end{cases} \tag{3.184}$$

$$\nu_{pn}=(2m_c+m_z)gh_r\frac{2}{L}\int_0^L\sin\left(\frac{p\pi x}{L}\right)\sin\left(\frac{n\pi x}{L}\right)dx=\begin{cases}(2m_c+m_z)gh_r & \text{for p}=n\\ 0 & \text{for p}\neq\text{n}\end{cases} \tag{3.185}$$

$$\tilde{m}_{\theta_{tot_p}}=m_{\theta_{tot}}\frac{2}{L}\int_0^L\sin(p\pi x/L)\cdot\sin(n\pi x/L)dx=\begin{cases}m_{\theta_{tot}} & \text{for p}=n\\ 0 & \text{for p}\neq\text{n}\end{cases} \tag{3.186}$$

while, recalling that $\int_0^L\sin\left(\frac{j\pi x}{L}\right)dx=\begin{cases}2L/j\pi & \text{for } j=1,3,5,\ldots\\ 0 & \text{for } j=2,4,6,\ldots\end{cases}$

$$\chi_{pn} = 32\left(\frac{e_c b_c}{L^2}\right)^2 \frac{(EA)_c}{\ell_e} \frac{2}{L} \int_0^L \sin\left(\frac{p\pi x}{L}\right) dx \cdot \int_0^L \sin\left(\frac{n\pi x}{L}\right) dx$$

$$= \begin{cases} \left(\frac{16 e_c b_c}{\pi L}\right)^2 \frac{(EA)_c}{L\ell_e} \frac{1}{pn} & \text{for } \left.\begin{matrix} p \\ \& \, n \end{matrix}\right\} = 1,3,5,\ldots \\ 0 & \text{for } \left.\begin{matrix} p \\ \text{or } n \end{matrix}\right\} = 2,4,6,\ldots \end{cases} \tag{3.187}$$

The $p = 1,2,3,\ldots,N$ sequence and series summations in Eq. 3.182 may conveniently be replaced by

$$\left\{\begin{bmatrix} \Omega_{11}+\vartheta_{11}+\nu_{11} & & & & \\ & \ddots & & \mathbf{0} & \\ & & \Omega_{pp}+\vartheta_{pp}+\nu_{pp} & & \\ & \mathbf{0} & & \ddots & \\ & & & & \Omega_{NN}+\vartheta_{NN}+\nu_{NN} \end{bmatrix} + \right. \tag{3.188}$$

$$\left.\begin{bmatrix} \chi_{11} & \cdots & \chi_{1n} & \cdots & \chi_{1N} \\ \vdots & \ddots & & \ddots & \vdots \\ \chi_{p1} & & \chi_{pn} & & \chi_{pN} \\ \vdots & \ddots & & \ddots & \vdots \\ \chi_{N1} & \cdots & \chi_{Nn} & \cdots & \chi_{NN} \end{bmatrix} - \omega^2 \begin{bmatrix} \tilde{m}_{\theta_{tot_1}} & & & & \\ & \ddots & & \mathbf{0} & \\ & & \tilde{m}_{\theta_{tot_p}} & & \\ & \mathbf{0} & & \ddots & \\ & & & & \tilde{m}_{\theta_{tot_N}} \end{bmatrix}\right\} \begin{bmatrix} a_{z_1} \\ \vdots \\ a_{z_p} \\ \vdots \\ a_{z_N} \end{bmatrix} = 0$$

Thus, defining the torsion amplitude vector

$$\mathbf{a}_\theta = \begin{bmatrix} a_{\theta_1} & \cdots & a_{\theta_p} & \cdots & a_{\theta_N} \end{bmatrix}^T \tag{3.189}$$

the mass matrix

$$\mathbf{M}_\theta = m_{\theta_{tot}} \mathbf{I} \tag{3.190}$$

where $m_{\theta_{tot}} = m_\theta + m_c b_c^2 / 2$ and $\mathbf{I}$ is the identity matrix, the stiffness matrix

$$\mathbf{K}_\theta = \begin{bmatrix} \Omega_{11}+\vartheta_{11}+\nu_{11} & & & & \\ & \ddots & & \mathbf{0} & \\ & & \Omega_{pp}+\vartheta_{pp}+\nu_{pp} & & \\ & \mathbf{0} & & \ddots & \\ & & & & \Omega_{NN}+\vartheta_{NN}+\nu_{NN} \end{bmatrix} + \begin{bmatrix} \chi_{11} & \cdots & \chi_{1p} & \cdots & \chi_{1N} \\ \vdots & \ddots & & \unicode{x22F0} & \vdots \\ \chi_{p1} & & \chi_{pp} & & \chi_{pN} \\ \vdots & \unicode{x22F0} & & \ddots & \vdots \\ \chi_{N1} & \cdots & \chi_{Np} & \cdots & \chi_{NN} \end{bmatrix} \tag{3.191}$$

then Eq. 3.188 is given by

$$\left(\mathbf{K}_\theta - \omega^2 \mathbf{M}_\theta\right)\mathbf{a}_\theta = \mathbf{0} \tag{3.192}$$

which, again is a classical eigenvalue problem similar to those encountered before. The convergence for the first four eigenvalue solutions will occur already at $N > 4$.

Elaboration 3.3

An approximate solution for the first four eigenfrequencies may be obtained by setting $N = 4$. Then Eq. 3.192 becomes

$$\begin{bmatrix} \begin{pmatrix}\Omega_{11}+\vartheta_{11}+\nu_{11} \\ +\chi_{11}-\omega^2 m_{\theta_{tot}}\end{pmatrix} & 0 & \chi_{13} & 0 \\ 0 & \begin{pmatrix}\Omega_{22}+\vartheta_{22}+\nu_{22} \\ -\omega^2 m_{\theta_{tot}}\end{pmatrix} & 0 & 0 \\ \chi_{13} & 0 & \begin{pmatrix}\Omega_{33}+\vartheta_{33}+\nu_{33} \\ +\chi_{33}-\omega^2 m_{\theta_{tot}}\end{pmatrix} & 0 \\ 0 & 0 & 0 & \begin{pmatrix}\Omega_{44}+\vartheta_{44}+\nu_{44} \\ -\omega^2 m_{\theta_{tot}}\end{pmatrix} \end{bmatrix} \mathbf{a}_\theta = \mathbf{0}$$

where $m_{\theta_{tot}} = m_\theta + m_c b_c^2/2$. It is seen that the solutions for the two anti-symmetric modes associated with ω_2 and ω_4 may directly be obtained from

$$\left(\Omega_{22}+\vartheta_{22}+\nu_{22}-\omega^2 m_{\theta_{tot}}\right)=0 \quad \text{and} \quad \left(\Omega_{44}+\vartheta_{44}+\nu_{44}-\omega^2 m_{\theta_{tot}}\right)=0$$

Thus,

$$\omega_2=\sqrt{\frac{\Omega_{22}+\vartheta_{22}+\nu_{22}}{m_{\theta_{tot}}}}=\sqrt{\left(\frac{2\pi}{L}\right)^2\frac{GI_t+(2\pi/L)^2 EI_w+\bar{H}\frac{b_c^2}{2}}{m_{\theta_{tot}}}+\frac{(2m_c+m_z)gh_r}{m_{\theta_{tot}}}}$$

$$\omega_4=\sqrt{\frac{\Omega_{44}+\vartheta_{44}+\nu_{44}}{m_{\theta_{tot}}}}=\sqrt{\left(\frac{4\pi}{L}\right)^2\frac{GI_t+(4\pi/L)^2 EI_w+\bar{H}\frac{b_c^2}{2}}{m_{\theta_{tot}}}+\frac{(2m_c+m_z)gh_r}{m_{\theta_{tot}}}}$$

After elimination of rows and columns two and four, the solutions for the two symmetric eigenmodes associated with ω_1 and ω_3 are defined by the requirement that the determinant to the remaining matrix is zero, i.e. that

$$\left(\Omega_{11}+\vartheta_{11}+\nu_{11}+\chi_{11}-\omega^2 m_{\theta_{tot}}\right)\cdot\left(\Omega_{33}+\vartheta_{33}+\nu_{33}+\chi_{33}-\omega^2 m_{\theta_{tot}}\right)-\chi_{13}^2=0$$

Defining
$$\begin{cases} k_1=10\left(\frac{\pi}{L}\right)^2 GI_t+82\left(\frac{\pi}{L}\right)^4 EI_w+10\left(\frac{\pi}{L}\right)^2\bar{H}\frac{b_c^2}{2} \\ \quad +(2m_c+m_z)gh_r+\frac{10}{9}\left(\frac{16e_c b_c}{\pi L}\right)^2\frac{(EA)_c}{L\ell_e} \end{cases}$$

and $k_2=8\left(\frac{\pi}{L}\right)^2 GI_t+80\left(\frac{\pi}{L}\right)^4 EI_w+8\left(\frac{\pi}{L}\right)^2\bar{H}\frac{b_c^2}{2}-\frac{8}{9}\left(\frac{16e_c b_c}{\pi L}\right)^2\frac{(EA)_c}{L\ell_e}$

and $k_3=\frac{2}{3}\left(\frac{16e_c b_c}{\pi L}\right)^2\frac{(EA)_c}{L\ell_e}$

then the solutions for ω_1 and ω_3 are given by: $\omega_{1,3}=\sqrt{\frac{k_1\mp\sqrt{k_2^2+k_3^2}}{2m_{\theta_{tot}}}}$.

Example 3.5

Increasing the number of Fourier components to $N=8$*, and adopting the numerical values given in Tables 3.3 and 3.4 below, then the solution to the eigenvalue problem in Eq. 3.192 renders the four first eigenmodes and corresponding eigenfrequencies associated with suspension bridge torsion motion as shown in Fig. 3.36 below.*

Table 3.3 Numeric example

L (m)	$\bar{H}$ (N)	m_θ (kgm^2/m)	m_c (kg/m)	e_c (m)	A_c (m^2)	E_c (N/m^2)
1250	$1.12\cdot 1$(	250000	1900	120	0.22	$0.2\cdot 10^{12}$

m_z (kg/m)	GI_t (Nm^2)	EI_w (Nm^4)	b_c (m)	h_r (m)	g (m/s^2)
9000	$0.2\cdot 10^{12}$	$0.9\cdot 10^{12}$	15	1.8	9.81

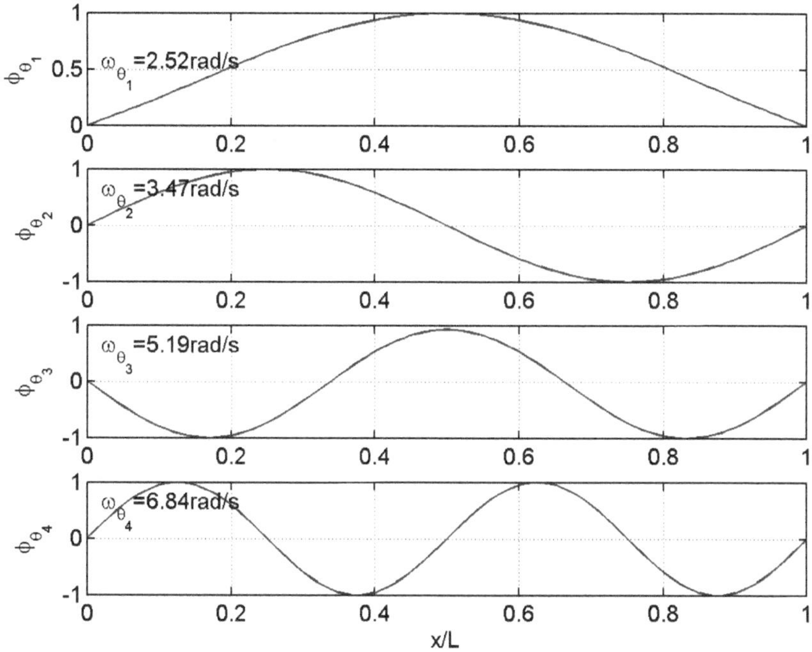

Fig. 3.36 Lowest four eigenvalue solutions for suspension bridge in torsion, $N = 8$

Chapter 4
The Finite Element Method in Dynamics

4.1 Introduction

As linearity has been taken for granted, a formal finite element approach to the problem of structural dynamics will comply with the computational methods usually applied elsewhere in theory of elasticity. However, in dynamics it is necessary to add the effects of mass in motion as well as internal damping, which, for a line like system, will affect the shape of motion into a combination of harmonic and hyperbolic functions, see Chapter 3, and thus also the overall physical properties of the system. Nonetheless, it is in the following assumed that the shape of motion may with sufficient accuracy be described by polynomials. Thus, from a computer programming point of view, all the well-known stiffness properties as well as numeric iteration procedures from other fields of structural mechanics will be applicable. In dynamics it is often the load that needs special attention, e.g. in the form of a frequency domain stochastic description or in a time domain simulation of wind, earthquake ground accelerations or other types of environmental loads. For the special case of wind induced dynamic response calculations it should be noted that due to the fairly short correlation lengths and sharply dropping coherence properties of the wind field there will be demanding requirements for the choice of largest element length. The same applies to the choice of time stepping increment in a time domain solution. For the case of earthquake ground acceleration there may be other challenges, e.g. shock type of excitation effects or the beneficial effects of plasticisation at specially designed joints or other non-linear effects from structural motion. Having adopted a system of six degrees of freedom in each node p (see Fig. 4.1), there is a global load vector

$$\mathbf{R}_{tot} = \begin{bmatrix} \mathbf{R}_1 & \cdots & \mathbf{R}_p & \cdots & \mathbf{R}_{N_p} \end{bmatrix}_{tot}^T \tag{4.1}$$

E.N. Strømmen, *Structural Dynamics*, Springer Series in Solid and Structural Mechanics 2,
DOI: 10.1007/978-3-319-01802-7_4, © Springer International Publishing Switzerland 2014

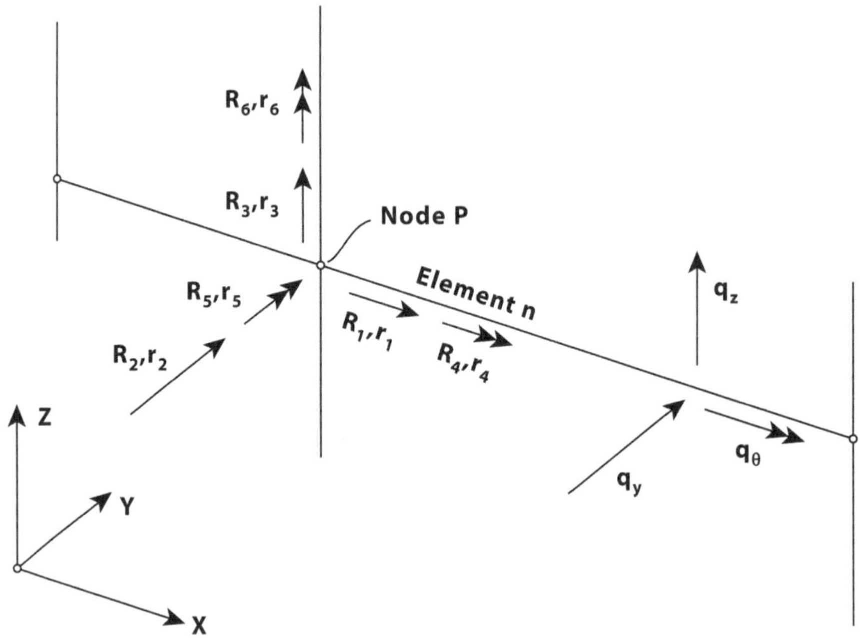

Fig. 4.1 Global degrees of freedom and external loads

where $\mathbf{R}_{Ptot} = \begin{bmatrix} R_1 & R_2 & R_3 & R_4 & R_5 & R_6 \end{bmatrix}_{Ptot}^T$ and a corresponding displacement vector

$$\mathbf{r}_{tot} = \begin{bmatrix} \mathbf{r}_1 & \cdots & \mathbf{r}_p & \cdots & \mathbf{r}_{N_p} \end{bmatrix}_{tot}^T \tag{4.2}$$

where $\mathbf{r}_{Ptot} = \begin{bmatrix} r_1 & r_2 & r_3 & r_4 & r_5 & r_6 \end{bmatrix}_{Ptot}^T$ and N_p is the number of nodes in the system. Thus, the total number of degrees of freedom will in general be $6N_p$. It is taken for granted that forces as well as displacements at global level comprise a time-invariant mean value (the static part) and a fluctuating (dynamic) part, i.e.

$$\left.\begin{aligned} \mathbf{R}_{tot} &= \bar{\mathbf{R}} + \mathbf{R}(t) \\ \mathbf{r}_{tot} &= \bar{\mathbf{r}} + \mathbf{r}(t) \end{aligned}\right\} \tag{4.3}$$

Similarly, it is taken for granted that this also applies at element level, see Eqs. 4.5 – 4.7. In general, the external nodal force vector may contain contributions from forces defined at global level as well as forces defined at element level. I.e., $\mathbf{R}_p$ at node p may contain contributions from globally defined concentrated forces $\mathbf{R}_i$, $i = 1, 2, \ldots, 6$, as well as contributions (see Fig. 4.2)

$$\mathbf{R}_n = \begin{bmatrix} \mathbf{R}_1 \\ \mathbf{R}_2 \end{bmatrix}_n \qquad \begin{cases} \mathbf{R}_{1_n} = \begin{bmatrix} R_1 & R_2 & R_3 & R_4 & R_5 & R_6 \end{bmatrix}^T \\ \mathbf{R}_{2_n} = \begin{bmatrix} R_7 & R_8 & R_9 & R_{10} & R_{11} & R_{12} \end{bmatrix}^T \end{cases} \tag{4.4}$$

defined at element level by distributed loads $\mathbf{q}_n = \begin{bmatrix} q_x & q_y & q_z & q_\theta \end{bmatrix}^T$.

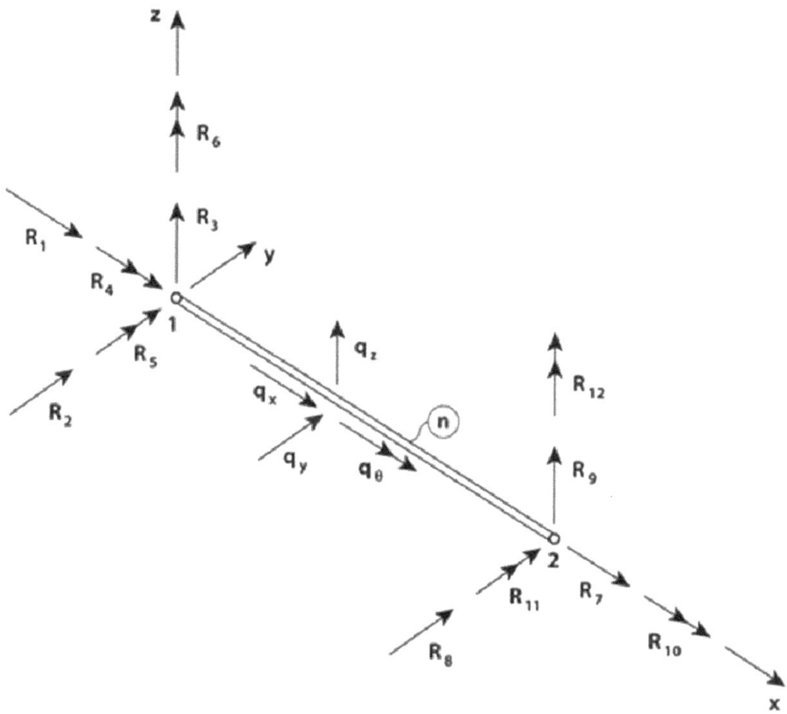

Fig. 4.2 External load $\mathbf{q}_n$ and load vector $\mathbf{R}_i + \mathbf{R}_n$ at element level

For simplicity, it is in the following taken for granted that the structural system is two-dimensional in global X and Z axis, and that Y is perpendicular to the system, as shown in Fig. 4.1. The effects of a time invariant mean axial force $\overline{N}_n$, assumed constant in each element n, together with primary bending moments $\bar{M}_y(x)$ and $\bar{M}_z(x)$ are included to account for the possible stiffness changes associated with axial or lateral torsion buckling. The main focus is on the establishment of the necessary theory, and to the completion of eigenvalue solution strategies. Dynamic excitation will be covered by the introduction of ground motion acceleration in Chapter 7, wind loading in Chapter 8 and spanwise moving loads in Chapter 11.

4.2 The Analysis at Element Level

A free body diagram of an arbitrary beam (line-like) type element n, with local axis x, y and z is illustrated in Fig. 4.3. At position x along its span the cross sectional displacements and rotation (torsion) are defined by

$$\mathbf{r}_{el_{tot}}(x,t) = \begin{bmatrix} r_x & r_y & r_z & r_\theta \end{bmatrix}_{el_{tot}}^T = \bar{\mathbf{r}}_{el}(x) + \mathbf{r}_{el}(x,t) \tag{4.5}$$

where index el indicates quantities within the span of the element. At ends 1 and 2 the element have nodal stress resultants

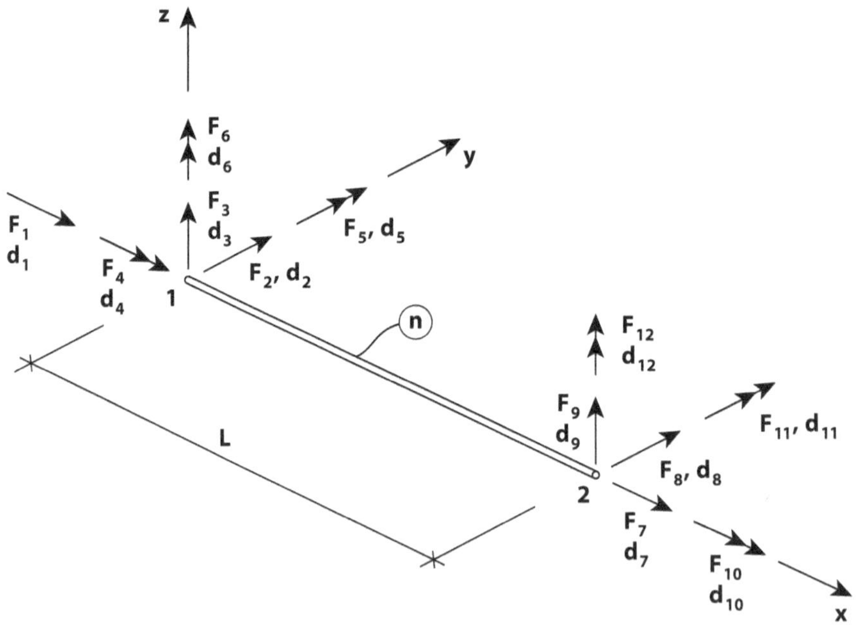

Fig. 4.3 Twelve degrees of freedom element

$$\mathbf{F}_{tot} = \begin{bmatrix} \mathbf{F}_1 \\ \mathbf{F}_2 \end{bmatrix}_{tot} = \bar{\mathbf{F}} + \mathbf{F}(t) \text{ where } \begin{cases} \mathbf{F}_{1_{tot}} = \begin{bmatrix} F_1 & F_2 & F_3 & F_4 & F_5 & F_6 \end{bmatrix}_{tot}^T \\ \mathbf{F}_{2_{tot}} = \begin{bmatrix} F_7 & F_8 & F_9 & F_{10} & F_{11} & F_{12} \end{bmatrix}_{tot}^T \end{cases} \tag{4.6}$$

and corresponding local displacements

$$\mathbf{d}_{tot} = \begin{bmatrix} \mathbf{d}_1 \\ \mathbf{d}_2 \end{bmatrix}_{tot} = \bar{\mathbf{d}} + \mathbf{d}(t) \text{ where } \begin{cases} \mathbf{d}_{1_{tot}} = \begin{bmatrix} d_1 & d_2 & d_3 & d_4 & d_5 & d_6 \end{bmatrix}_{tot}^T \\ \mathbf{d}_{2_{tot}} = \begin{bmatrix} d_7 & d_8 & d_9 & d_{10} & d_{11} & d_{12} \end{bmatrix}_{tot}^T \end{cases} \tag{4.7}$$

It is assumed that the cross sectional displacement vector $\mathbf{r}_{el_{tot}}(x,t)$ may with sufficient accuracy be described by the product of a shape function matrix

$$\mathbf{\Psi}(x) = \begin{bmatrix} \psi_1 & 0 & 0 & 0 & 0 & 0 & \psi_7 & 0 & 0 & 0 & 0 & 0 \\ 0 & \psi_2 & 0 & 0 & 0 & \psi_6 & 0 & \psi_8 & 0 & 0 & 0 & \psi_{12} \\ 0 & 0 & \psi_3 & 0 & \psi_5 & 0 & 0 & 0 & \psi_9 & 0 & \psi_{11} & 0 \\ 0 & 0 & 0 & \psi_4 & 0 & 0 & 0 & 0 & 0 & \psi_{10} & 0 & 0 \end{bmatrix} \tag{4.8}$$

and the nodal displacement vector $\mathbf{d}_{tot}(t)$, i.e.

$$\mathbf{r}_{el_{tot}}(x,t) = \mathbf{\Psi}(x) \cdot \mathbf{d}_{tot}(t) \tag{4.9}$$

where the twelve shape function $\psi_i(x), i = 1-12$, are given in Fig. 4.4. As mentioned in the introduction to this chapter, these are identical to the shape functions commonly used elsewhere in structural mechanics. Since they are polynomial, it should be noted that they will represent an accurate solution to the time-invariant (static) part of the response but not to the dynamic part, as they will not fully satisfy the spanwise differential equation of motion (which, as we have seen before, will require a solution containing a combination of harmonic and hyperbolic functions). However, this will usually not render unduly erroneous results as long as the element length is kept sufficiently short.

Applying the principle of virtual work (see Chapter 1.6) at an arbitrary position of external and internal equilibrium defined by $\mathbf{r}_{el_{tot}}(x,t)$, the system is subject to an incremental virtual displacement (see Fig. 4.5)

$$\delta\mathbf{r}_{el}(x) = \begin{bmatrix} \delta r_x & \delta r_y & \delta r_z & \delta r_\theta \end{bmatrix}^T \tag{4.10}$$

compatible with

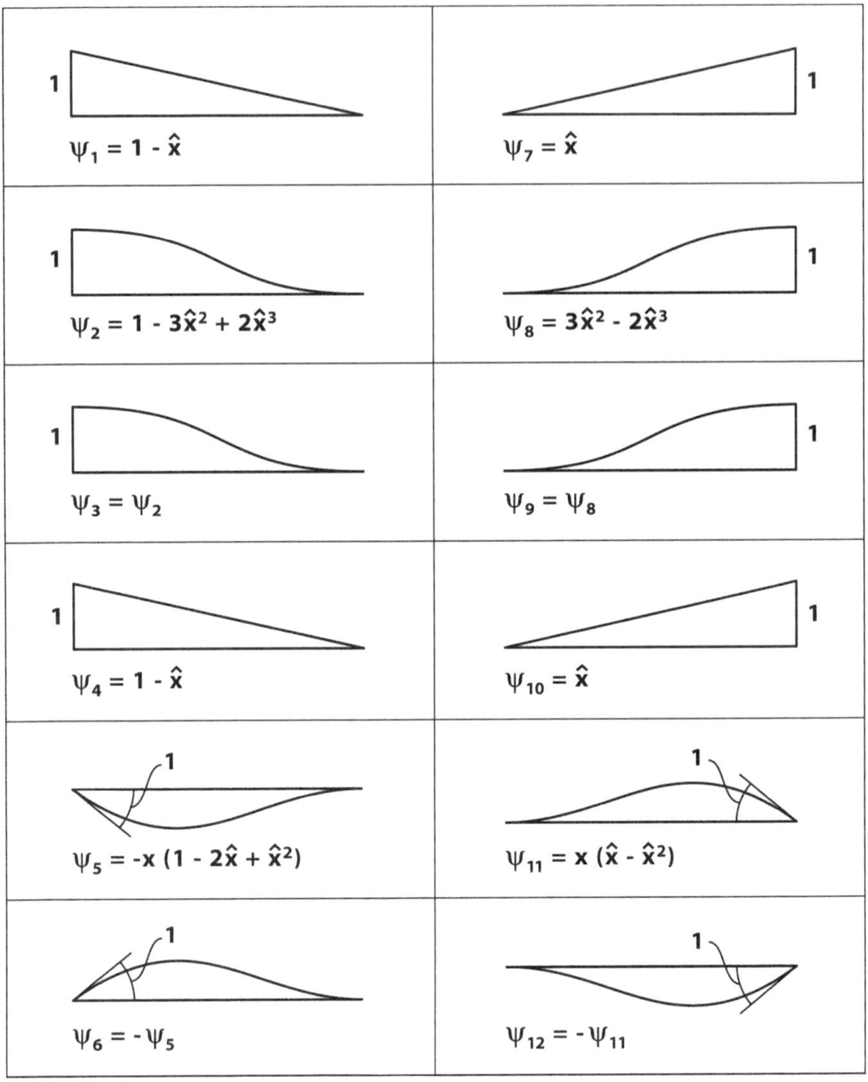

Fig. 4.4 Shape functions ψ_i, $i = 1 - 12$

$$\delta \mathbf{d} = \begin{bmatrix} \delta \mathbf{d}_1 \\ \delta \mathbf{d}_2 \end{bmatrix} \text{ where } \begin{cases} \delta \mathbf{d}_1 = \begin{bmatrix} \delta d_1 & \delta d_2 & \delta d_3 & \delta d_4 & \delta d_5 & \delta d_6 \end{bmatrix}^T \\ \delta \mathbf{d}_2 = \begin{bmatrix} \delta d_7 & \delta d_8 & \delta d_9 & \delta d_{10} & \delta d_{11} & \delta d_{12} \end{bmatrix}^T \end{cases} \tag{4.11}$$

such that

$$\delta \mathbf{r}_{el}(x) = \mathbf{\Psi}(x) \cdot \delta \mathbf{d} \tag{4.12}$$

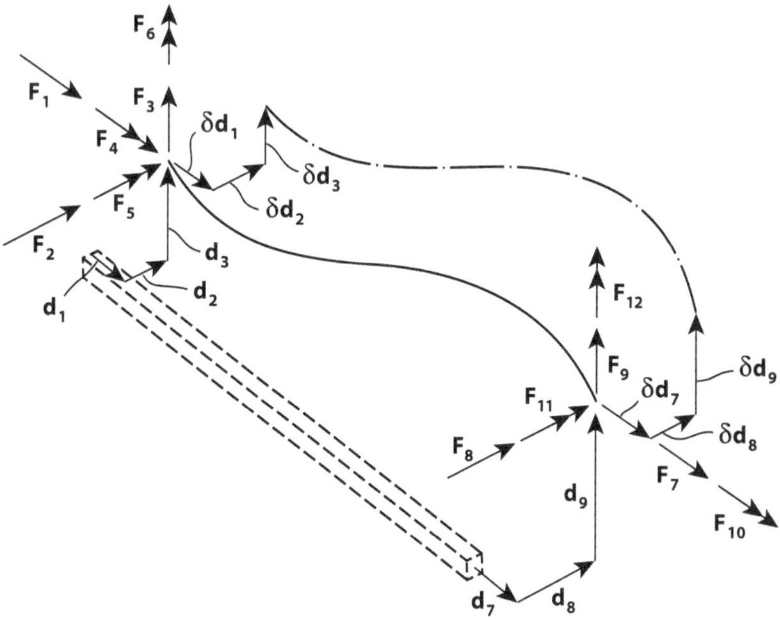

Fig. 4.5 Element and element stress resultants subject to virtual displacement

According to Eq. 1.119 the following work balance applies

$$\begin{gathered}\delta\mathbf{d}^T\mathbf{F}_{tot} - \int_0^L \delta\mathbf{r}_{el}^T\left(\mathbf{c}_v\dot{\mathbf{r}}_{el}\right)dx - \int_L \delta\mathbf{r}_{el}^T\left(\mathbf{m}_g\ddot{\mathbf{r}}_{el}\right)dx = \\ \int_L\left(\delta r_x' EA r_x' + \delta r_y'' EI_z r_y'' + \delta r_z'' EI_y r_z'' + \delta r_\theta' GI_t r_\theta'\right)dx + \\ \int_L\left[\delta r_y'\left(\bar{N}_n r_y' - \bar{M}_y r_\theta'\right) + \delta r_z'\left(\bar{N}_n r_z' - \bar{M}_z r_\theta'\right) + \delta r_\theta'\left(\bar{N}_n e_0^2 r_\theta' - \bar{M}_y r_y' - \bar{M}_z r_z'\right)\right]dx\end{gathered} \tag{4.13}$$

where
$$\mathbf{m}_g = \begin{bmatrix} m_x & 0 & 0 & 0 \\ 0 & m_y & 0 & -m_y e_z \\ 0 & 0 & m_z & m_z e_y \\ 0 & -m_y e_z & m_z e_y & m_\theta \end{bmatrix} \tag{4.14}$$

contains the distributed mass properties of the element, and

$$\mathbf{c}_v = diag\begin{bmatrix} c_x & c_y & c_z & c_\theta \end{bmatrix} \tag{4.15}$$

contains the distributed viscous damping properties of the element. Let us now define the two shape derivative matrices (where primes indicate derivation with respect to x)

$$\hat{\mathbf{\Psi}}(x)=\begin{bmatrix}0&0&0&0&0&0&0&0&0&0&0&0\\0&\psi_2'&0&0&0&\psi_6'&0&\psi_8'&0&0&0&\psi_{12}'\\0&0&\psi_3'&0&\psi_5'&0&0&0&\psi_9'&0&\psi_{11}'&0\\0&0&0&\psi_4'&0&0&0&0&0&\psi_{10}'&0&0\end{bmatrix} \quad (4.16)$$

$$\hat{\hat{\mathbf{\Psi}}}(x)=\begin{bmatrix}\psi_1'&0&0&0&0&0&\psi_7'&0&0&0&0&0\\0&\psi_2''&0&0&0&\psi_6''&0&\psi_8''&0&0&0&\psi_{12}''\\0&0&\psi_3''&0&\psi_5''&0&0&0&\psi_9''&0&\psi_{11}''&0\\0&0&0&\psi_4'&0&0&0&0&0&\psi_{10}'&0&0\end{bmatrix} \quad (4.17)$$

such that

$$\begin{bmatrix}0 & r_y' & r_z' & r_\theta'\end{bmatrix}^T=\hat{\mathbf{\Psi}}\mathbf{d}_{tot} \quad \text{and} \quad \begin{bmatrix}0 & \delta r_y' & \delta r_z' & \delta r_\theta'\end{bmatrix}^T=\hat{\mathbf{\Psi}}\delta\mathbf{d} \quad (4.18)$$

$$\begin{bmatrix}r_x' & r_y'' & r_z'' & r_\theta'\end{bmatrix}^T=\hat{\hat{\mathbf{\Psi}}}\mathbf{d}_{tot} \quad \text{and} \quad \begin{bmatrix}\delta r_x' & \delta r_y'' & \delta r_z'' & \delta r_\theta'\end{bmatrix}^T=\hat{\hat{\mathbf{\Psi}}}\delta\mathbf{d} \quad (4.19)$$

Introducing Eqs. 4.12, 4.18 and 4.19 into Eq. 4.13, then

$$\begin{gathered}\delta\mathbf{d}^T\mathbf{F}_{tot}-\int_L(\mathbf{\Psi}\delta\mathbf{d})^T\mathbf{c}_v\left(\mathbf{\Psi}\dot{\mathbf{d}}_{tot}\right)dx-\int_L(\mathbf{\Psi}\delta\mathbf{d})^T\mathbf{m}_g\left(\mathbf{\Psi}\ddot{\mathbf{d}}_{tot}\right)dx=\\ \int_L\left(\hat{\hat{\mathbf{\Psi}}}\delta\mathbf{d}\right)^T\mathbf{k}_n\left(\hat{\hat{\mathbf{\Psi}}}\mathbf{d}_{tot}\right)dx+\int_L\left(\hat{\mathbf{\Psi}}\delta\mathbf{d}\right)^T\mathbf{n}_n\left(\hat{\mathbf{\Psi}}\mathbf{d}_{tot}\right)dx\end{gathered} \quad (4.20)$$

where

$$\mathbf{k}_n=\begin{bmatrix}EA&0&0&0\\0&EI_z&0&0\\0&0&EI_y&0\\0&0&0&GI_t\end{bmatrix} \quad \text{and} \quad \mathbf{n}_n=\begin{bmatrix}0&0&0&0\\0&\bar{N}_n&0&-\bar{M}_{y_n}\\0&0&\bar{N}_n&-\bar{M}_{z_n}\\0&-\bar{M}_{y_n}&-\bar{M}_{z_n}&\bar{N}_n e_0^2\end{bmatrix} \quad (4.21)$$

This may be further developed into

$$\delta\mathbf{d}^T \int_L \left(\boldsymbol{\Psi}^T \mathbf{m}_g \boldsymbol{\Psi}\right) dx \cdot \ddot{\mathbf{d}}_{tot} + \delta\mathbf{d}^T \int_L \left(\boldsymbol{\Psi}^T \mathbf{c}_v \boldsymbol{\Psi}\right) dx \cdot \dot{\mathbf{d}}_{tot}$$
$$+\delta\mathbf{d}^T \left[\int_L \left(\hat{\hat{\boldsymbol{\Psi}}}^T \mathbf{k}_n \hat{\hat{\boldsymbol{\Psi}}}\right) dx + \int_L \left(\hat{\boldsymbol{\Psi}}^T \mathbf{n}_n \hat{\boldsymbol{\Psi}} \mathbf{d}_{tot}\right) dx \right] \cdot \mathbf{d}_{tot} = \delta\mathbf{d}^T \mathbf{F}_{tot} \tag{4.22}$$

It is seen that the pre-multiplication by $\delta\mathbf{d}^T$ is obsolete and may be discarded. Introducing

$$\mathbf{d}_{tot} = \bar{\mathbf{d}} + \mathbf{d}(t) \quad \text{and} \quad \mathbf{F}_{tot} = \bar{\mathbf{F}} + \mathbf{F}(t) \tag{4.23}$$

and defining

$$\begin{bmatrix} \mathbf{m} \\ \mathbf{k}_0 \\ \mathbf{k}_G \\ \mathbf{c} \end{bmatrix} = \int_L \begin{bmatrix} \boldsymbol{\Psi}^T \mathbf{m}_g \boldsymbol{\Psi} \\ \hat{\hat{\boldsymbol{\Psi}}}^T \mathbf{k}_n \hat{\hat{\boldsymbol{\Psi}}} \\ \hat{\boldsymbol{\Psi}}^T \mathbf{n}_n \hat{\boldsymbol{\Psi}} \\ \boldsymbol{\Psi}^T \mathbf{c}_v \boldsymbol{\Psi} \end{bmatrix} dx \tag{4.24}$$

then the following equilibrium condition at element level is obtained

$$\mathbf{m}\ddot{\mathbf{d}} + \mathbf{c}\dot{\mathbf{d}} + \left(\mathbf{k}_0 + \mathbf{k}_G\right)\left(\bar{\mathbf{d}} + \mathbf{d}\right) = \bar{\mathbf{F}} + \mathbf{F} \tag{4.25}$$

In the case of a time invariant static solution then $\mathbf{d}(t)$ and $\mathbf{F}(t)$ are zero, and thus

$$\mathbf{k}\bar{\mathbf{d}} = \bar{\mathbf{F}} \tag{4.26}$$

where $\mathbf{k} = \mathbf{k}_0 + \mathbf{k}_G$. I.e. the static and the dynamic equilibrium requirements may be handled in separate operation (the superposition principle applies). Thus, the dynamic equilibrium condition is given by

$$\mathbf{m}\ddot{\mathbf{d}} + \mathbf{c}\dot{\mathbf{d}} + \mathbf{k}\mathbf{d} = \mathbf{F} \tag{4.27}$$

Elaboration 4.1: Twelve Degrees of Freedom Beam Element

The element stiffness, damping and mass matrices are defined in Eq. 4.24. By introducing the shape functions in Eq. 4.8 and its derivatives in Eqs. 4.16 and 4.17, then the element mass, damping and stiffness matrices may readily be obtained by integration over the element length. [It should be noted that the development of damping properties at element level is not necessarily a rational choice. Alternatively, damping properties may be introduced at a structural global level (i.e. associated directly with the global degrees of freedom), e.g. in the form of Rayleigh damping or simply a diagonal type of modal damping matrix, see Chapter 9.] Thus, the element mass matrix is given by:

$$\mathbf{m} = \begin{bmatrix} \mathbf{m}_{11} & \mathbf{m}_{12} \\ \mathbf{m}_{21} & \mathbf{m}_{22} \end{bmatrix} \qquad \text{where} \qquad \mathbf{m}_{21} = \mathbf{m}_{12}^T$$

and

$$\mathbf{m}_{11} = \int_0^L \begin{bmatrix} \psi_1^2 m_x & 0 & 0 & 0 & 0 & 0 \\ & \psi_2^2 m_y & 0 & -\psi_2\psi_4 m_y e_z & 0 & \psi_2\psi_6 m_y \\ & & \psi_3^2 m_z & \psi_3\psi_4 m_z e_y & \psi_3\psi_5 m_z & 0 \\ & & & \psi_4^2 m_\theta & \psi_4\psi_5 m_z e_y & -\psi_4\psi_6 m_y e_z \\ & sym. & & & \psi_5^2 m_z & 0 \\ & & & & & \psi_6^2 m_y \end{bmatrix} dx$$

$$\mathbf{m}_{12} = \int_0^L \begin{bmatrix} \psi_1\psi_7 m_x & 0 & 0 & 0 & 0 & 0 \\ 0 & \psi_2\psi_8 m_y & 0 & -\psi_2\psi_{10} m_y e_z & 0 & \psi_2\psi_{12} m_y \\ 0 & 0 & \psi_3\psi_9 m_z & \psi_3\psi_{10} m_z e_y & \psi_3\psi_{11} m_z & 0 \\ 0 & -\psi_4\psi_8 m_y e_z & \psi_4\psi_9 m_z e_y & \psi_4\psi_{10} m_\theta & \psi_4\psi_{11} m_z e_y & -\psi_4\psi_{12} m_y e_z \\ 0 & 0 & \psi_5\psi_9 m_z & \psi_5\psi_{10} m_z e_y & \psi_5\psi_{11} m_z & 0 \\ 0 & \psi_6\psi_8 m_y & 0 & -\psi_6\psi_{10} m_y e_z & 0 & \psi_6\psi_{12} m_y \end{bmatrix} dx$$

$$\mathbf{m}_{22} = \int_0^L \begin{bmatrix} \psi_7^2 m_x & 0 & 0 & 0 & 0 & 0 \\ & \psi_8^2 m_y & 0 & -\psi_8\psi_{10} m_y e_z & 0 & \psi_8\psi_{12} m_y \\ & & \psi_9^2 m_z & \psi_9\psi_{10} m_z e_y & \psi_9\psi_{11} m_z & 0 \\ & & & \psi_{10}^2 m_\theta & \psi_{10}\psi_{11} m_z e_y & -\psi_{10}\psi_{12} m_y e_z \\ & sym. & & & \psi_{11}^2 m_z & 0 \\ & & & & & \psi_{12}^2 m_y \end{bmatrix} dx$$

The element damping matrix is given by: $\mathbf{c} = \begin{bmatrix} \mathbf{c}_{11} & \mathbf{c}_{12} \\ \mathbf{c}_{21} & \mathbf{c}_{22} \end{bmatrix}$ where $\mathbf{c}_{21} = \mathbf{c}_{12}^T$

and

$$\mathbf{c}_{11} = \int_0^L \begin{bmatrix} c_x\Psi_1^2 & 0 & 0 & 0 & 0 & 0 \\ & c_y\Psi_2^2 & 0 & 0 & 0 & c_y\Psi_2\Psi_6 \\ & & c_z\Psi_3^2 & 0 & c_z\Psi_3\Psi_5 & 0 \\ & & & c_\theta\Psi_4^2 & 0 & 0 \\ & sym. & & & c_z\Psi_5^2 & 0 \\ & & & & & c_y\Psi_6^2 \end{bmatrix} dx$$

$$\mathbf{c}_{12} = \int_0^L \begin{bmatrix} c_x\Psi_1\Psi_7 & 0 & 0 & 0 & 0 & 0 \\ 0 & c_y\Psi_2\Psi_8 & 0 & 0 & 0 & c_y\Psi_2\Psi_{12} \\ 0 & 0 & c_z\Psi_3\Psi_9 & 0 & c_z\Psi_3\Psi_{11} & 0 \\ 0 & 0 & 0 & c_\theta\Psi_4\Psi_{10} & 0 & 0 \\ 0 & 0 & c_z\Psi_5\Psi_9 & 0 & c_z\Psi_5\Psi_{11} & 0 \\ 0 & c_y\Psi_6\Psi_8 & 0 & 0 & 0 & c_y\Psi_6\Psi_{12} \end{bmatrix} dx$$

$$\mathbf{c}_{22} = \int_0^L \begin{bmatrix} c_x\Psi_7^2 & 0 & 0 & 0 & 0 & 0 \\ & c_y\Psi_8^2 & 0 & 0 & 0 & c_y\Psi_8\Psi_{12} \\ & & c_z\Psi_9^2 & 0 & c_z\Psi_9\Psi_{11} & 0 \\ & & & c_\theta\Psi_{10}^2 & 0 & 0 \\ & sym. & & & c_z\Psi_{11}^2 & 0 \\ & & & & & c_y\Psi_{12}^2 \end{bmatrix} dx$$

The element stiffness matrix associated with purely material properties is given by:

$$\mathbf{k} = \begin{bmatrix} \mathbf{k}_{11} & \mathbf{k}_{12} \\ \mathbf{k}_{21} & \mathbf{k}_{22} \end{bmatrix} \qquad \text{where} \qquad \mathbf{k}_{21} = \mathbf{k}_{12}^T \qquad \text{and}$$

$$\mathbf{k}_{11} = \int_0^L \begin{bmatrix} EA\psi_1'^2 & 0 & 0 & 0 & 0 & 0 \\ & EI_z\psi_2''^2 & 0 & 0 & 0 & EI_z\psi_2''\psi_6'' \\ & & EI_y\psi_3''^2 & 0 & EI_y\psi_3''\psi_5'' & 0 \\ & & & GI_t\psi_4'^2 & 0 & 0 \\ & sym. & & & EI_y\psi_5''^2 & 0 \\ & & & & & EI_z\psi_6''^2 \end{bmatrix} dx$$

$$\mathbf{k}_{12} = \int_0^L$$

$$\begin{bmatrix} EA\psi_1'\psi_7' & 0 & 0 & 0 & 0 & 0 \\ 0 & EI_z\psi_2''\psi_8'' & 0 & 0 & 0 & EI_z\psi_2''\psi_{12}'' \\ 0 & 0 & EI_y\psi_3''\psi_9'' & 0 & EI_y\psi_3''\psi_{11}'' & 0 \\ 0 & 0 & 0 & GI_t\psi_4'\psi_{10}' & 0 & 0 \\ 0 & 0 & EI_y\psi_5''\psi_9'' & 0 & EI_y\psi_5''\psi_{11}'' & 0 \\ 0 & EI_z\psi_6''\psi_8'' & 0 & 0 & 0 & EI_z\psi_6''\psi_{12}'' \end{bmatrix} dx$$

$$\mathbf{k}_{22} = \int_0^L \begin{bmatrix} EA\psi_7'^2 & 0 & 0 & 0 & 0 & 0 \\ & EI_z\psi_8''^2 & 0 & 0 & 0 & EI_z\psi_8''\psi_{12}'' \\ & & EI_y\psi_9''^2 & 0 & EI_y\psi_9''\psi_{11}'' & 0 \\ & & & GI_t\psi_{10}'^2 & 0 & 0 \\ & sym. & & & EI_y\psi_{11}''^2 & 0 \\ & & & & & EI_z\psi_{12}''^2 \end{bmatrix} dx$$

The element stiffness matrix associated with axial and lateral torsion buckling effects is given by:

$$\mathbf{k}_G = \begin{bmatrix} \mathbf{k}_{G11} & \mathbf{k}_{G12} \\ \mathbf{k}_{G21} & \mathbf{k}_{G22} \end{bmatrix} \qquad \text{where} \qquad \mathbf{k}_{G21} = \mathbf{k}_{G12}^T$$

and

$$\mathbf{k}_{G11}=\int_0^L\begin{bmatrix}0 & 0 & 0 & 0 & 0 & 0\\ & \bar{N}_n\psi_2'^2 & 0 & -\bar{M}_y\psi_2'\psi_4' & 0 & \bar{N}_n\psi_2'\psi_6'\\ & & \bar{N}_n\psi_3'^2 & -\bar{M}_z\psi_3'\psi_4' & \bar{N}_n\psi_3'\psi_5' & 0\\ & & & \bar{N}_n e_0^2\psi_4'^2 & -\bar{M}_z\psi_4'\psi_5' & -\bar{M}_y\psi_4'\psi_6'\\ & sym. & & & \bar{N}_n\psi_5'^2 & 0\\ & & & & & \bar{N}_n\psi_6'^2\end{bmatrix}dx$$

$$\mathbf{k}_{G12}=\int_0^L$$

$$\begin{bmatrix}0 & 0 & 0 & 0 & 0 & 0\\ 0 & \bar{N}_n\psi_2'\psi_8' & 0 & -\bar{M}_y\psi_2'\psi_{10}' & 0 & \bar{N}_n\psi_2'\psi_{12}'\\ 0 & 0 & \bar{N}_n\psi_3'\psi_9' & -\bar{M}_z\psi_3'\psi_{10}' & \bar{N}_n\psi_3'\psi_{11}' & 0\\ 0 & -\bar{M}_y\psi_4'\psi_8' & -\bar{M}_z\psi_4'\psi_9' & \bar{N}_n e_0^2\psi_4'\psi_{10}' & -\bar{M}_z\psi_4'\psi_{11}' & -\bar{M}_y\psi_4'\psi_{12}'\\ 0 & 0 & \bar{N}_n\psi_5'\psi_9' & -\bar{M}_z\psi_5'\psi_{10}' & \bar{N}_n\psi_5'\psi_{11}' & 0\\ 0 & \bar{N}_n\psi_6'\psi_8' & 0 & -\bar{M}_y\psi_6'\psi_{10}' & 0 & \bar{N}_n\psi_6'\psi_{12}'\end{bmatrix}dx$$

$$\mathbf{k}_{G22}=\int_0^L\begin{bmatrix}0 & 0 & 0 & 0 & 0 & 0\\ & \bar{N}_n\psi_8'^2 & 0 & -\bar{M}_y\psi_8'\psi_{10}' & 0 & \bar{N}_n\psi_8'\psi_{12}'\\ & & \bar{N}_n\psi_9'^2 & -\bar{M}_z\psi_9'\psi_{10}' & \bar{N}_n\psi_9'\psi_{11}' & 0\\ & & & \bar{N}_n e_0^2\psi_{10}'^2 & -\bar{M}_z\psi_{10}'\psi_{11}' & -\bar{M}_y\psi_{10}'\psi_{12}'\\ & sym. & & & \bar{N}_n\psi_{11}'^2 & 0\\ & & & & & \bar{N}_n\psi_{12}'^2\end{bmatrix}dx$$

Ready-made integrations are given in Appendix C.1.

Elaboration 4.2: Six Degrees of Freedom Beam Element

For a purely in-plane bending type of problem the six degrees of freedom element shown in Fig. 4.6 will suffice, in which case the following applies:

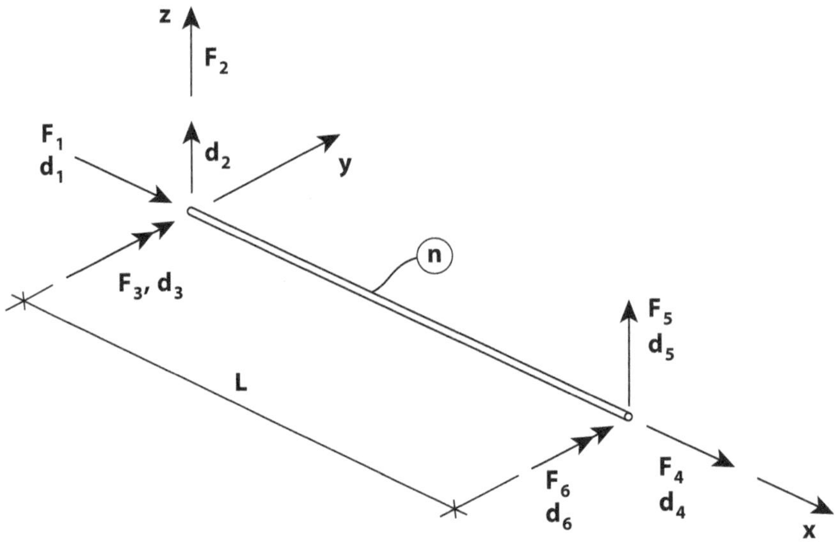

Fig. 4.6 Six degrees of freedom element

$$\mathbf{m} = \int_0^L \begin{bmatrix} \psi_1^2 m_x & 0 & 0 & \psi_1\psi_7 m_x & 0 & 0 \\ & \psi_3^2 m_z & \psi_3\psi_5 m_z & 0 & \psi_3\psi_9 m_z & \psi_3\psi_{11} m_z \\ & & \psi_5^2 m_z & 0 & \psi_5\psi_9 m_z & \psi_5\psi_{11} m_z \\ & & & \psi_7^2 m_x & 0 & 0 \\ & sym. & & & \psi_9^2 m_z & \psi_9\psi_{11} m_z \\ & & & & & \psi_{11}^2 m_y \end{bmatrix} dx$$

$$\mathbf{c} = \int_0^L \begin{bmatrix} \psi_1^2 c_x & 0 & 0 & \psi_1\psi_7 c_x & 0 & 0 \\ & \psi_3^2 c_z & \psi_3\psi_5 c_z & 0 & \psi_3\psi_9 c_z & \psi_3\psi_{11} c_z \\ & & \psi_5^2 c_z & 0 & \psi_5\psi_9 c_z & \psi_5\psi_{11} c_z \\ & & & \psi_7^2 c_x & 0 & 0 \\ & sym. & & & \psi_9^2 c_z & \psi_9\psi_{11} c_z \\ & & & & & \psi_{11}^2 c_y \end{bmatrix} dx$$

$$\mathbf{k} = \int_0^L \begin{bmatrix} EA\psi_1'^2 & 0 & 0 & EA\psi_1'\psi_7' & 0 & 0 \\ & EI_y\psi_3''^2 & EI_y\psi_3''\psi_5'' & 0 & EI_y\psi_3''\psi_9'' & EI_y\psi_3''\psi_{11}'' \\ & & EI_y\psi_5''^2 & 0 & EI_y\psi_5''\psi_9'' & EI_y\psi_5''\psi_{11}'' \\ & & & EA\psi_7'^2 & 0 & 0 \\ & sym. & & & EI_y\psi_9''^2 & EI_y\psi_9''\psi_{11}'' \\ & & & & & EI_y\psi_{11}''^2 \end{bmatrix} dx$$

$$\mathbf{k}_G = \int_0^L \begin{bmatrix} 0 & 0 & 0 & 0 & 0 & 0 \\ & \bar{N}_n\psi_3'^2 & \bar{N}_n\psi_3'\psi_5' & 0 & \bar{N}_n\psi_3'\psi_9' & \bar{N}_n\psi_3'\psi_{11}' \\ & & \bar{N}_n\psi_5'^2 & 0 & \bar{N}_n\psi_5'\psi_9' & \bar{N}_n\psi_5'\psi_{11}' \\ & & & 0 & 0 & 0 \\ & sym. & & & \bar{N}_n\psi_9'^2 & \bar{N}_n\psi_9'\psi_{11}' \\ & & & & & \bar{N}_n\psi_{11}'^2 \end{bmatrix} dx$$

Ready-made integrals are given in Appendix C.2.

Elaboration 4.3: Distributed Forces in the Element Span

Consistent load effects of distributed forces, e.g. $\mathbf{q}(x,t) = \begin{bmatrix} q_x & q_y & q_z & q_\theta \end{bmatrix}^T$ in the span of an element n (see Fig. 4.2) may alternatively be included in the virtual work balance in Eq. 4.13, in which case

$$\int_L \delta\mathbf{r}_{el}^T \mathbf{q} dx = \int_L (\boldsymbol{\psi}\delta\mathbf{d})^T \mathbf{q} dx = \delta\mathbf{d}^T \int_L \boldsymbol{\psi}^T \mathbf{q} dx = \delta\mathbf{d}^T \mathbf{R}_n$$

where $\mathbf{R}_n = \int_L \boldsymbol{\psi}^T \mathbf{q} dx$ should be added to the left hand side of Eq. 4.13. It may readily be shown that

$$\mathbf{R}_n = \begin{bmatrix} \mathbf{R}_1 \\ \mathbf{R}_2 \end{bmatrix}_n \quad \text{where} \quad \begin{cases} \mathbf{R}_{1_n} \approx \dfrac{L}{2}\begin{bmatrix} q_x & q_y & q_z & q_\theta & -q_z L/6 & q_y L/6 \end{bmatrix}^T \\ \mathbf{R}_{2_n} \approx \dfrac{L}{2}\begin{bmatrix} q_x & q_y & q_z & q_\theta & q_z L/6 & -q_y L/6 \end{bmatrix}^T \end{cases}$$

The transformation of $\mathbf{R}_n$ into global load components follow the same transformation as $\mathbf{d}_n$, i.e. the contribution to global loads from distributed forces on element number n is given by $\mathbf{A}_n\mathbf{R}_n$, where $\mathbf{A}_n$ is the connectivity matrix defined in Eq. 4.28 below.

4.3 The Global Analysis

Before proceeding it is necessary to define the six by N_p connectivity matrix $\mathbf{A}_n$ describing the relationship between element degrees of freedom $\mathbf{d}_{n_{tot}}$ and global degrees of freedom $\mathbf{r}_{tot}$, defined such that:

$$\mathbf{d}_{n_{tot}} = \mathbf{A}_n \mathbf{r}_{tot} \Rightarrow \quad \mathbf{d}_{n_{tot}} = \bar{\mathbf{d}}_n + \mathbf{d}_n = \mathbf{A}_n \mathbf{r}_{tot} = \mathbf{A}_n \bar{\mathbf{r}} + \mathbf{A}_n \mathbf{r}_n \tag{4.28}$$

Let us apply to the discrete system (the system as reduced to a non-continuous collection of nodes) a set of virtual displacements $\delta\mathbf{r}$ consistent with $\mathbf{r}_{tot}$

$$\delta\mathbf{d}_n = \mathbf{A}_n \cdot \delta\mathbf{r} \tag{4.29}$$

Since the virtual work exerted by the external forces at global level must be equal to the sum of the virtual work of the internal stress resultants at element nodes, then

$$\delta\mathbf{r}^T \cdot \mathbf{R}_{tot} = \sum_{n=1}^{N} \delta\mathbf{d}_n^T \cdot \mathbf{F}_{n_{tot}} \tag{4.30}$$

where N is the total number of elements in the system. Introducing $\delta\mathbf{d}_n^T = \left(\mathbf{A}_n \cdot \delta\mathbf{r}\right)^T = \delta\mathbf{r}^T \cdot \mathbf{A}_n^T$, then

$$\delta\mathbf{r}^T \cdot \mathbf{R}_{tot} = \delta\mathbf{r}^T \cdot \sum_{n=1}^{N} \mathbf{A}_n^T \cdot \mathbf{F}_{n_{tot}} \tag{4.31}$$

Again, pre-multiplication by $\delta\mathbf{r}^T$ is obsolete, and thus, it is seen that the equilibrium condition at a global structural level is given by

$$\mathbf{R}_{tot} = \sum_{n=1}^{N} \mathbf{A}_n^T \cdot \mathbf{F}_{n_{tot}} \tag{4.32}$$

Recalling from Eq. 4.25 that

$$\mathbf{F}_{n_{tot}} = \mathbf{m}\ddot{\mathbf{d}}_n + \mathbf{c}\dot{\mathbf{d}}_n + (\mathbf{k}+\mathbf{k}_G)\mathbf{d}_{n_{tot}} \tag{4.33}$$

and introducing Eq. 4.28

$$\mathbf{F}_{n_{tot}} = \mathbf{m}\mathbf{A}_n\ddot{\mathbf{r}} + \mathbf{c}\mathbf{A}_n\dot{\mathbf{r}} + (\mathbf{k}+\mathbf{k}_G)\mathbf{A}_n\mathbf{r}_{tot} \tag{4.34}$$

it is seen that Eq. 4.32 becomes

$$\mathbf{R}_{tot} = \mathbf{R}_{p_{tot}} + \sum_{n=1}^{N}\mathbf{A}_n^T\mathbf{R}_{n_{tot}} = \sum_{n=1}^{N}\left\{\mathbf{A}_n^T\mathbf{m}\mathbf{A}_n\ddot{\mathbf{r}} + \mathbf{A}_n^T\mathbf{c}\mathbf{A}_n\dot{\mathbf{r}} + \mathbf{A}_n^T(\mathbf{k}+\mathbf{k}_G)\mathbf{A}_n\mathbf{r}_{tot}\right\} \tag{4.35}$$

where $\mathbf{R}_{p_{tot}}$ are external load contributions added directly in at node p, while $\mathbf{R}_{n_{tot}}$ are load contributions associated with distributed loads on element n (see Fig. 4.2 and Elaboration 4.3 above). Thus, defining the global structural properties

$$\begin{bmatrix}\mathbf{M}\\ \mathbf{C}\\ \mathbf{K}_0\\ \mathbf{K}_G\end{bmatrix} = \sum_{n=1}^{N}\begin{bmatrix}\mathbf{A}_n^T\mathbf{m}\mathbf{A}_n\\ \mathbf{A}_n^T\mathbf{c}\mathbf{A}_n\\ \mathbf{A}_n^T\mathbf{k}\mathbf{A}_n\\ \mathbf{A}_n^T\mathbf{k}_G\mathbf{A}_n\end{bmatrix} \tag{4.36}$$

then the following equilibrium condition at global level is obtained

$$\mathbf{M}\ddot{\mathbf{r}}(t) + \mathbf{C}\dot{\mathbf{r}}(t) + (\mathbf{K}_0+\mathbf{K}_G)\left[\bar{\mathbf{r}} + \mathbf{r}(t)\right] = \bar{\mathbf{R}} + \mathbf{R}(t) \tag{4.37}$$

Defining $\mathbf{K} = \mathbf{K}_0 + \mathbf{K}_G$, this may conveniently be split into a time invariant mean (static) part

$$\mathbf{K}\cdot\bar{\mathbf{r}} = \bar{\mathbf{R}} \tag{4.38}$$

and a purely dynamic part

$$\mathbf{M}\cdot\ddot{\mathbf{r}}(t) + \mathbf{C}\cdot\dot{\mathbf{r}}(t) + \mathbf{K}\cdot\mathbf{r}(t) = \mathbf{R}(t) \tag{4.39}$$

The solution to this equation is at the core of all structural dynamics. Various possible strategies will be presented in Chapters 5 and 6, while relevant solutions to the problems of ground motion and wind induced dynamic forces will be presented in Chapters 7 and 8. The special problem of structural damping will be dealt with in Chapter 9.

It is in the following taken for granted that $\mathbf{M}$, $\mathbf{C}$ and $\mathbf{K}$ are all symmetric, quadratic and invertible matrices, and that they are real and positive definite. Their dimension is N_r by N_r, where N_r is the number of degrees of freedom in the system ($N_r = 6 \cdot N_p$ where N_p is the number of nodes in the system).

Elaboration 4.4: Condensation of Obsolete Degrees of Freedom

Any degree of freedom r_n that due to structural restrictions or boundary conditions are equal to zero may be discarded simply by deleting the row and column associated with r_n. If two degrees of freedom r_n and r_{n+j} are linearly dependent of each other, then the relevant rows and columns associated with r_n and r_{n+j} may simply be added linearly, or as expressed in a mathematical way

$$\text{if} \quad \mathbf{Ar} = \mathbf{B} \quad \text{and} \quad \mathbf{r} = \mathbf{Dr}_1 \quad \text{then} \quad \left(\mathbf{D}^T \mathbf{AD}\right)\mathbf{r}_1 = \mathbf{D}^T\mathbf{B}$$

In some cases it may also be a convenient solution strategy to reduce the number of degrees of freedom in a system from $N_r = N_{r_1} + N_{r_2}$ to N_{r_1} simply by not allocating mass and damping to the N_{r_2} degrees of freedom that are considered obsolete. The system reduction may then be obtained by rewriting the original equilibrium condition

$$\mathbf{M}\ddot{\mathbf{r}} + \mathbf{C}\dot{\mathbf{r}} + \mathbf{Kr} = \mathbf{R} \text{ into } \begin{bmatrix} \mathbf{M}_{11} & \mathbf{0} \\ \mathbf{0} & \mathbf{0} \end{bmatrix} \begin{bmatrix} \ddot{\mathbf{r}}_1 \\ \ddot{\mathbf{r}}_2 \end{bmatrix} + \begin{bmatrix} \mathbf{C}_{11} & \mathbf{0} \\ \mathbf{0} & \mathbf{0} \end{bmatrix} \begin{bmatrix} \dot{\mathbf{r}}_1 \\ \dot{\mathbf{r}}_2 \end{bmatrix} + \begin{bmatrix} \mathbf{K}_{11} & \mathbf{K}_{12} \\ \mathbf{K}_{21} & \mathbf{K}_{22} \end{bmatrix} \begin{bmatrix} \mathbf{r}_1 \\ \mathbf{r}_2 \end{bmatrix} = \begin{bmatrix} \mathbf{R}_1 \\ \mathbf{R}_2 \end{bmatrix}$$

from which $\mathbf{r}_2 = \mathbf{K}_{22}^{-1}\mathbf{R}_2 - \mathbf{K}_{22}^{-1}\mathbf{K}_{21}\mathbf{r}_1$ and thus the reduced equilibrium condition is given by

$$\mathbf{M}_{11}\ddot{\mathbf{r}}_1 + \mathbf{C}_{11}\dot{\mathbf{r}}_1 + \left(\mathbf{K}_{11} - \mathbf{K}_{12}\mathbf{K}_{22}^{-1}\mathbf{K}_{21}\right)\mathbf{r}_1 = \mathbf{R}_1 - \mathbf{K}_{12}\mathbf{K}_{22}^{-1}\mathbf{R}_2$$

Example 4.1

A simple bridge type of structural system is illustrated in Fig. 4.7. It has five elements and two relevant degrees of freedom. For simplicity, the effect of axial force in the columns is disregarded. Let us assume that $m_2 = m_1/3$ *and* $EI_2 = EI_1/2$. *We shall then establish the mass and stiffness matrices of the system.*

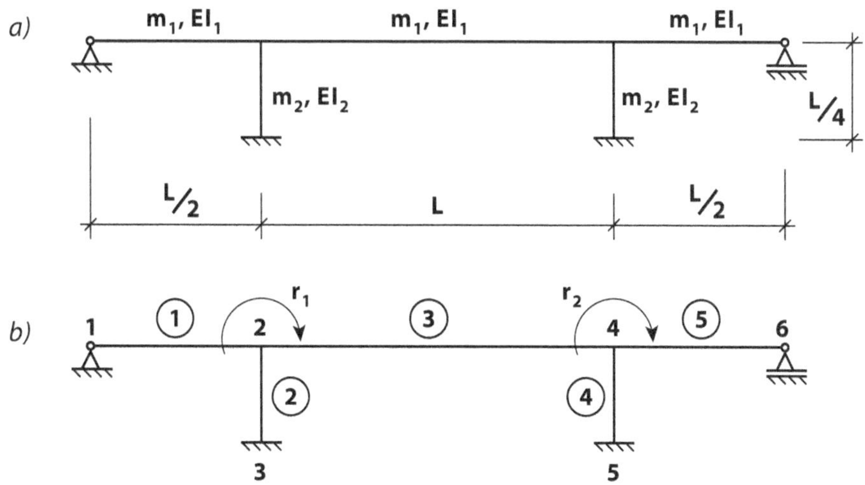

Fig. 4.7 Bridge type of structure; a) structural system, b) relevant degrees of freedom, node and element numbering

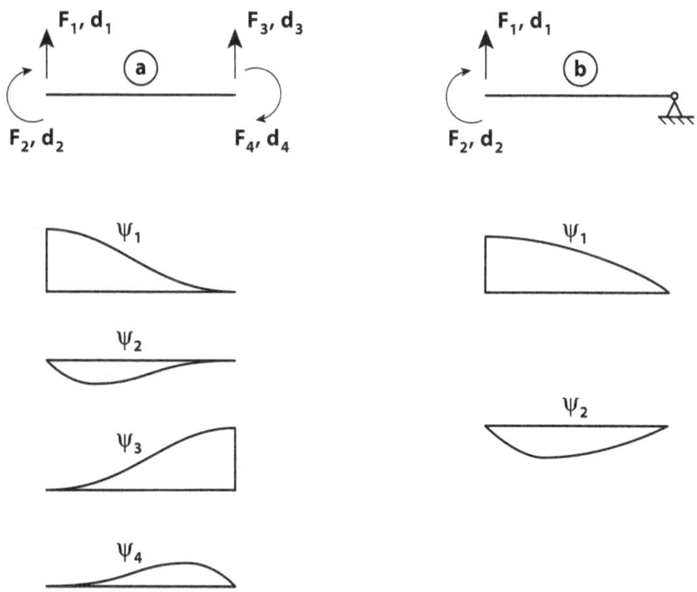

Fig. 4.8 Elements types a and b

The system contains two types of elements, i.e. elements types a *and* b *as illustrated in Fig. 4.8. The properties of element type* a *may be extracted from the properties of the six degrees of freedom element in Appendix C.2, simply by omitting the first and the fourth degrees of freedom. Thus:*

$$\mathbf{k}_a = \frac{EI}{L^3}\begin{bmatrix} 12 & -6L & -12 & -6L \\ -6L & 4L^2 & 6L & 2L^2 \\ -12 & 6L & 12 & 6L \\ -6L & 2L^2 & 6L & 4L^2 \end{bmatrix} \qquad \mathbf{m}_a = \frac{mL}{420}\begin{bmatrix} 156 & -22L & 54 & 13L \\ -22L & 4L^2 & -13L & -3L^2 \\ 54 & -13L & 156 & 22L \\ 13L & -3L^2 & 22L & 4L^2 \end{bmatrix}$$

The properties of element type b may also be extracted from the six degrees of freedom element in Appendix C.2. However, for this element type it is more consistent to use the following shape functions (see right hand side of Fig. 4.8):

$$\psi_1 = 1 - \frac{3}{2}\left(\frac{x}{L}\right)^2 + \frac{1}{2}\left(\frac{x}{L}\right)^3 \quad \textit{whose properties are such that} \quad \begin{cases} x = 0 \Rightarrow \begin{cases} \psi_1 = 1 \\ \psi_1' = 0 \end{cases} \\ x = L \Rightarrow \begin{cases} \psi_1 = 0 \\ \psi_1'' = 0 \end{cases} \end{cases}$$

$$\psi_2 = -x\left[1 - \frac{3}{2}\frac{x}{L} + \frac{1}{2}\left(\frac{x}{L}\right)^2\right] \quad \textit{whose properties are such that} \quad \begin{cases} x = 0 \Rightarrow \begin{cases} \psi_1 = 0 \\ \psi_1' = -1 \end{cases} \\ x = L \Rightarrow \begin{cases} \psi_1 = 0 \\ \psi_1'' = 0 \end{cases} \end{cases}$$

Defining $\boldsymbol{\Psi} = \begin{bmatrix} \psi_1 & \psi_2 \end{bmatrix}$ *and* $\hat{\hat{\boldsymbol{\Psi}}} = \begin{bmatrix} \psi_1'' & \psi_2'' \end{bmatrix}$ *then (see Eq. 4.24)*

$$\mathbf{k}_b = EI\int_0^L \hat{\hat{\boldsymbol{\Psi}}}^T \hat{\hat{\boldsymbol{\Psi}}} dx = EI\int_0^L \begin{bmatrix} (\psi_1'')^2 & \psi_1''\psi_2'' \\ \psi_2''\psi_1'' & (\psi_2'')^2 \end{bmatrix} dx = \frac{3EI}{L^3}\begin{bmatrix} 1 & -L \\ -L & L^2 \end{bmatrix}$$

$$\mathbf{m}_b = m\int_0^L \boldsymbol{\Psi}^T \boldsymbol{\Psi} dx = m\int_0^L \begin{bmatrix} (\psi_1)^2 & \psi_1\psi_2 \\ \psi_2\psi_1 & (\psi_2)^2 \end{bmatrix} dx = \frac{mL}{210}\begin{bmatrix} 102 & -18L \\ -18L & 4L^2 \end{bmatrix}$$

As mentioned above, the system has two global degrees of freedom r_1 and r_2. Hence, the stiffness and mass matrices will be two by two

$$\mathbf{K} = \begin{bmatrix} K_{11} & K_{12} \\ K_{21} & K_{22} \end{bmatrix} \quad \textit{and} \quad \mathbf{M} = \begin{bmatrix} M_{11} & M_{12} \\ M_{21} & M_{22} \end{bmatrix}$$

The contribution from element number 1 is then given by

$$\mathbf{K}_1 = \frac{EI_1}{(L/2)^3}\begin{bmatrix} 3(L/2)^2 & 0 \\ 0 & 0 \end{bmatrix} \quad \textit{and} \quad \mathbf{M}_1 = \frac{m_1 L/2}{210}\begin{bmatrix} 4(L/2)^2 & 0 \\ 0 & 0 \end{bmatrix}$$

The contribution from element number 2 is given by

$$\mathbf{K}_2 = \frac{EI_1/2}{(L/4)^3}\begin{bmatrix} 4(L/4)^2 & 0 \\ 0 & 0 \end{bmatrix} \quad \textit{and} \quad \mathbf{M}_2 = \frac{(m_1/3)L/4}{420}\begin{bmatrix} 4(L/4)^2 & 0 \\ 0 & 0 \end{bmatrix}$$

The contribution from element number 3 is given by

$$\mathbf{K}_3 = \frac{EI_1}{L^3}\begin{bmatrix} 4L^2 & 2L^2 \\ 2L^2 & 4L^2 \end{bmatrix} \quad \textit{and} \quad \mathbf{M}_3 = \frac{m_1 L}{420}\begin{bmatrix} 4L^2 & -3L^2 \\ -3L^2 & 4L^2 \end{bmatrix}$$

The contribution from element number 4 is given by

$$\mathbf{K}_4 = \frac{EI_1/2}{(L/4)^3}\begin{bmatrix} 0 & 0 \\ 0 & 4(L/4)^2 \end{bmatrix} \quad \textit{and} \quad \mathbf{M}_4 = \frac{(m_1/3)L/4}{420}\begin{bmatrix} 0 & 0 \\ 0 & 4(L/4)^2 \end{bmatrix}$$

The contribution from element number 5 is given by

$$\mathbf{K}_5 = \frac{EI_1}{(L/2)^3}\begin{bmatrix} 0 & 0 \\ 0 & 3(L/2)^2 \end{bmatrix} \quad \textit{and} \quad \mathbf{M}_5 = \frac{m_1 L/2}{210}\begin{bmatrix} 0 & 0 \\ 0 & 4(L/2)^2 \end{bmatrix}$$

Thus:

$$\mathbf{K} = \sum_{n=1}^{5} K_n = \frac{2EI_1}{L}\begin{bmatrix} 9 & 1 \\ 1 & 9 \end{bmatrix} \quad \textit{and} \quad \mathbf{M} = \sum_{n=1}^{5} M_n = \frac{m_1 L^3}{420}\begin{bmatrix} 241/48 & -3 \\ -3 & 241/48 \end{bmatrix}$$

This is a two degree of freedom system. The solution to its eigenvalue problem

$$\left(\mathbf{K} - \omega^2 \mathbf{M}\right)\boldsymbol{\varphi} = \left(\frac{2EI_1}{L}\begin{bmatrix} 9 & 1 \\ 1 & 9 \end{bmatrix} - \omega^2 \frac{m_1 L^3}{420}\begin{bmatrix} 241/48 & -3 \\ -3 & 241/48 \end{bmatrix} \right)\begin{bmatrix} a_1 \\ a_2 \end{bmatrix} = \begin{bmatrix} 0 \\ 0 \end{bmatrix}$$

is given by $\begin{cases} \omega_1 \approx 29\sqrt{\dfrac{EI_1}{m_1 L^4}} \\ \omega_2 \approx 64.5\sqrt{\dfrac{EI_1}{m_1 L^4}} \end{cases}$ *with corresponding eigenmodes* $\begin{cases} \boldsymbol{\varphi}_1 = [1 \quad -1]^T \\ \boldsymbol{\varphi}_2 = [1 \quad 1]^T \end{cases}$

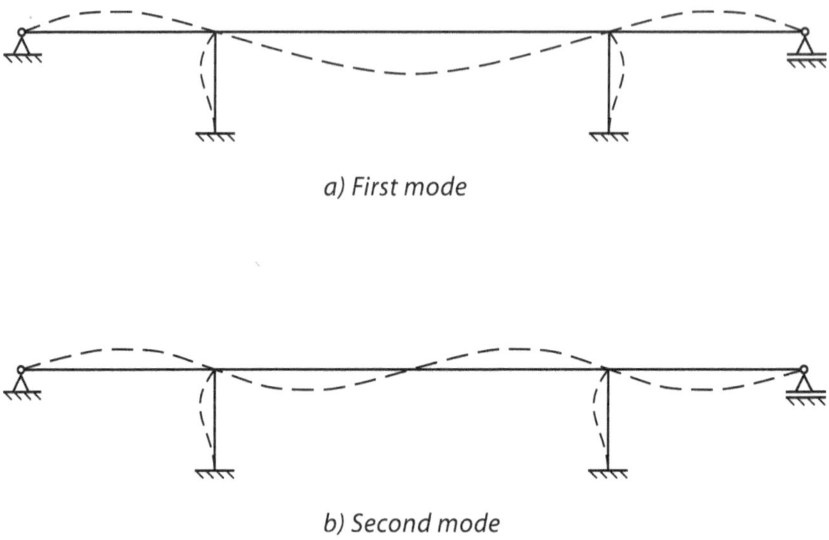

Fig. 4.9 Eigenmodes $\boldsymbol{\varphi}_1$ and $\boldsymbol{\varphi}_2$

4.4 The Numeric Eigenvalue Problem

Most structural systems have very low damping. Thus, they possess a distinct set of preference frequencies and associated shapes of motion which largely depend on the mass and stiffness distribution within the system. We call them eigenfrequencies and associated normal modes (or simply modeshapes). These properties play an important role in the evaluation of potential danger of excitation and to establish consistent solutions that focus on the most relevant aspects of what can be expected to occur (e.g. the normal mode method as presented in Chapter 5). In most cases they may be determined by solving the undamped and unloaded version of Eq. 4.39, rendering a classic eigenvalue problem which is covered in this chapter. (However, in some cases is it necessary to include damping, and that case is briefly covered in Chapter 4.5 below.) The solution to an eigenvalue problem plays a significant part in mathematics, and there are numerous effective routines for its numerical solution (see e.g. Bell [12]). It is beyond the scope of this book to cover the subject in great detail. However, in Chapters 2 and 3, where continuous and no more than two degrees of freedom systems have been covered, it has been demonstrated that the subject is a necessary part of the identification of dynamic properties to structural systems. Thus, for the sake of completeness a brief coverage is given below.

General Background

The steady state solution to the undamped and unloaded version of Eq. 4.39

$$\mathbf{M} \cdot \ddot{\mathbf{r}}(t) + \mathbf{K} \cdot \mathbf{r}(t) = \mathbf{0} \tag{4.40}$$

is given by $\mathbf{r}(t) = \mathrm{Re}\left(\boldsymbol{\varphi} \cdot e^{i\omega t}\right)$, where $\boldsymbol{\varphi} = \begin{bmatrix} \varphi_1 & \cdots & \varphi_k & \cdots & \varphi_{N_r} \end{bmatrix}^T$ is a vector with length equal to the number of degrees of freedom (N_r) in the system, and containing the displacement (or rotation) components associated with the relevant degrees of freedom at the structural position assigned to node p (see Fig. 4.1). Introducing this into Eq. 4.40, then the following is obtained

$$\left(\mathbf{K} - \omega^2 \mathbf{M}\right) \boldsymbol{\varphi} = \mathbf{0} \tag{4.41}$$

The solution to this equation is in mathematics known as the general eigenvalue problem. By pre-multiplication with $\mathbf{M}^{-1}$ then the special eigenvalue problem is obtained

$$\left(\mathbf{A} - \lambda \mathbf{I}\right) \boldsymbol{\varphi} = \mathbf{0} \tag{4.42}$$

where $\mathbf{I}$ is the identity matrix, $\lambda = \omega^2$ and $\mathbf{A} = \mathbf{M}^{-1}\mathbf{K}$. Similarly, by pre-multiplication with $-\omega^{-2}\mathbf{K}^{-1}$ then the inverse version of the special eigenvalue problem is obtained, i.e.

$$\left(\mathbf{B} - \beta \mathbf{I}\right) \boldsymbol{\varphi} = \mathbf{0} \tag{4.43}$$

where $\beta = \omega^{-2}$ and $\mathbf{B} = \mathbf{K}^{-1}\mathbf{M}$. It is readily seen that any of Eqs. 4.42 and 4.43 may be chosen for the determination of ω and $\boldsymbol{\varphi}$. In general there will be N_r possible solutions, as a nontrivial $\boldsymbol{\varphi}$ will require that the determinant to $\mathbf{A} - \lambda \mathbf{I}$ must be equal to zero, and this may be expressed by expanding $\det\left(\mathbf{A} - \lambda \mathbf{I}\right)$ into an N_r dimensional polynomial

$$f(\lambda) = \det\left(\mathbf{A} - \lambda \mathbf{I}\right) = \left(\lambda_1 - \lambda\right)\left(\lambda_2 - \lambda\right) \cdots \left(\lambda_i - \lambda\right) \cdots \left(\lambda_{N_r} - \lambda\right) \tag{4.44}$$

whose roots $\lambda_1, \lambda_2, \ldots, \lambda_{N_r}$ (see Fig. 4.9) all represents a possible solution to the eigenvalue problem.

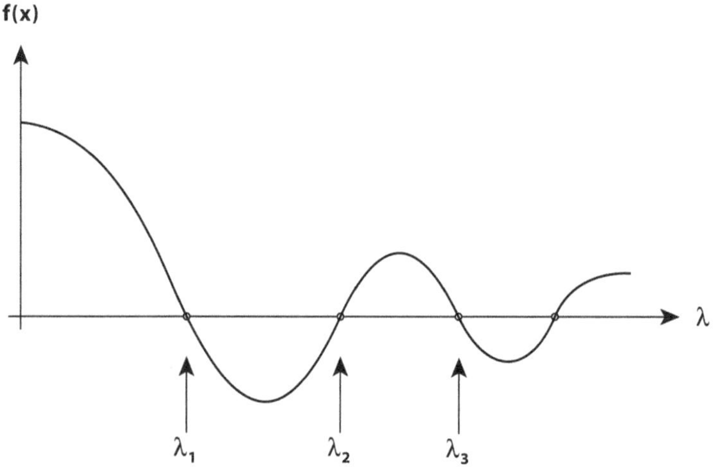

Fig. 4.10 N_r dimensional polynomial $f(\lambda)$

It is customary to arrange the eigenvalues in ascending order, i.e. $\lambda_1 \le \lambda_2 \cdots \le \lambda_i \cdots \le \lambda_{N_r}$, because it is the lowest that requires the least amount of energy to excite. The corresponding set of eigenmodes $\boldsymbol{\varphi}_i$, $i = 1, 2, \ldots, N_r$, may then be obtained by re-introducing any of the solutions λ_i back into Eq. 4.42 (or 4.43)

$$(\mathbf{A} - \lambda_i \mathbf{I})\boldsymbol{\varphi}_i = \mathbf{0} \tag{4.45}$$

Thus, it is seen that $\boldsymbol{\varphi}_i$ is scalable, i.e. that it may be multiplied or divided by any constant number. It only represents a characteristic shape of structural displacements. The actual displacements cannot be quantified unless a full version of Eq. 4.39 is solved. Re-writing Eq. 4.45 into

$$\mathbf{A}\boldsymbol{\varphi}_i = \lambda_i \boldsymbol{\varphi}_i \tag{4.46}$$

it is seen that an eigenmode $\boldsymbol{\varphi}_i$ of a matrix $\mathbf{A}$ is a non-zero vector which multiplied by $\mathbf{A}$ is the vector itself multiplied by a constant λ_i called the eigenvalue of $\mathbf{A}$. For a two by two matrix

$$\mathbf{A} = \begin{bmatrix} a_{11} & a_{12} \\ a_{21} & a_{22} \end{bmatrix} \tag{4.47}$$

and corresponding eigenmode $\boldsymbol{\varphi}_i = \begin{bmatrix} \varphi_1 & \varphi_2 \end{bmatrix}_i^T$ this is illustrated in Fig. 4.11.

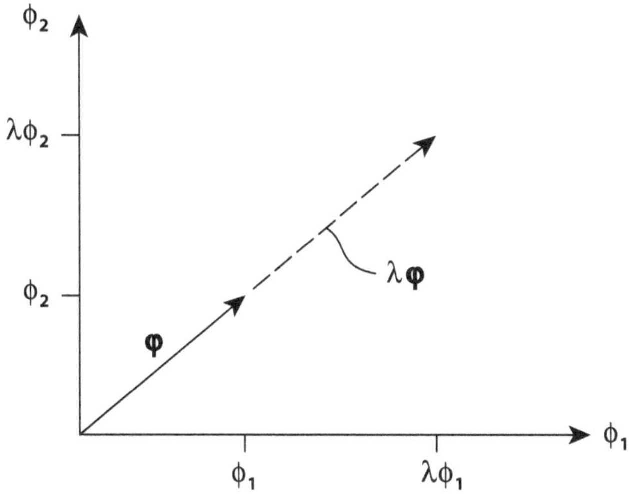

Fig. 4.11 Eigenmode and eigenvalue in two-dimensional vector space

Example 4.2: Polynomial Solution to Simple Two by Two Systems

Let $\mathbf{A} = \begin{bmatrix} a_{11} & a_{12} \\ a_{21} & a_{22} \end{bmatrix}$. *Then the solution to the eigenvalue problem*

$$(\mathbf{A} - \lambda\mathbf{I})\boldsymbol{\varphi} = \left(\begin{bmatrix} a_{11} & a_{12} \\ a_{21} & a_{22} \end{bmatrix} - \lambda\begin{bmatrix} 1 & 0 \\ 0 & 1 \end{bmatrix}\right)\begin{bmatrix} \varphi_1 \\ \varphi_2 \end{bmatrix} = \begin{bmatrix} a_{11}-\lambda & a_{12} \\ a_{21} & a_{22}-\lambda \end{bmatrix}\begin{bmatrix} \varphi_1 \\ \varphi_2 \end{bmatrix} = \begin{bmatrix} 0 \\ 0 \end{bmatrix}$$

is given by

$$\det\begin{bmatrix} a_{11}-\lambda & a_{12} \\ a_{21} & a_{22}-\lambda \end{bmatrix} = (a_{11}-\lambda)(a_{22}-\lambda) - a_{12}a_{21}$$
$$= \lambda^2 - (a_{11}+a_{22})\lambda + a_{11}a_{22} - a_{12}a_{21} = 0$$

rendering $$\lambda_{1,2} = \frac{a_{11}+a_{22}}{2} \pm \sqrt{\left(\frac{a_{11}+a_{22}}{2}\right)^2 - (a_{11}a_{22} - a_{12}a_{21})}$$

and $$(a_{11} - \lambda_{1,2})\varphi_1 + a_{12}\varphi_2 = 0 \qquad \Rightarrow \qquad \frac{\varphi_2}{\varphi_1} = \frac{\lambda_{1,2} - a_{11}}{a_{12}}$$

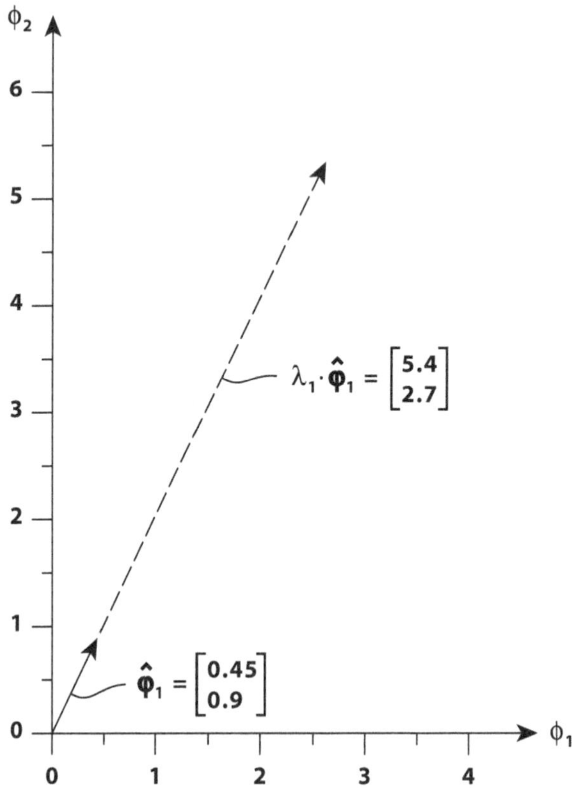

Fig. 4.12 Graphic illustration

Let for instance $\mathbf{A} = \begin{bmatrix} 2 & 2 \\ 2 & 5 \end{bmatrix}$. *Then* $\begin{cases} \lambda_1 = 6 \\ \lambda_2 = 1 \end{cases}$ *while* $\boldsymbol{\varphi}_1 = \begin{bmatrix} \varphi_1 \\ 2\varphi_1 \end{bmatrix}$ *and* $\boldsymbol{\varphi}_2 = \begin{bmatrix} \varphi_1 \\ -0.5\varphi_1 \end{bmatrix}$.

It is seen that they are orthogonal, i.e. that $\boldsymbol{\varphi}_1^T \boldsymbol{\varphi}_2 = 0$. *Since they are arbitrarily scalable, it is often convenient to scale them such that* $\hat{\boldsymbol{\varphi}}_i^T \hat{\boldsymbol{\varphi}}_i = 1$, *i.e.*

$$\hat{\boldsymbol{\varphi}}_1 = \boldsymbol{\varphi}_1 \Big/ \sqrt{\boldsymbol{\varphi}_1^T \boldsymbol{\varphi}_1} \approx \begin{bmatrix} 0.45 \\ 0.9 \end{bmatrix} \quad \textit{and} \quad \hat{\boldsymbol{\varphi}}_2 = \boldsymbol{\varphi}_2 \Big/ \sqrt{\boldsymbol{\varphi}_2^T \boldsymbol{\varphi}_2} \approx \begin{bmatrix} 0.9 \\ -0.45 \end{bmatrix}$$

The normalised eigenmode $\hat{\boldsymbol{\varphi}}_1$ *is illustrated in Fig. 4.12 above. As can be seen*

$$\lambda_1 \hat{\boldsymbol{\varphi}}_1 = 6\begin{bmatrix} 0.45 \\ 0.9 \end{bmatrix} = \begin{bmatrix} 2.7 \\ 5.4 \end{bmatrix} \quad \textit{is identical to} \quad \mathbf{A}\hat{\boldsymbol{\varphi}}_1 = \begin{bmatrix} 2 & 2 \\ 2 & 5 \end{bmatrix}\begin{bmatrix} 0.45 \\ 0.9 \end{bmatrix} = \begin{bmatrix} 2.7 \\ 5.4 \end{bmatrix}$$

The same holds for $\hat{\boldsymbol{\varphi}}_2$.

Eigenmode Orthogonality

Let ω_i^2 and ω_j^2, and corresponding $\boldsymbol{\varphi}_i$ and $\boldsymbol{\varphi}_j$, be two arbitrary but independent and non-trivial solutions (different from zero) to the general eigenvalue problem $\left(\mathbf{K} - \omega^2\mathbf{M}\right)\boldsymbol{\varphi} = \mathbf{0}$, i.e.

$$\left(\mathbf{K} - \omega_i^2\mathbf{M}\right)\boldsymbol{\varphi}_i = \mathbf{0} \quad \text{and} \quad \left(\mathbf{K} - \omega_j^2\mathbf{M}\right)\boldsymbol{\varphi}_j = \mathbf{0} \tag{4.48}$$

Pre-multiplying the first with $\boldsymbol{\varphi}_j^T$ and the second by $\boldsymbol{\varphi}_i^T$

$$\boldsymbol{\varphi}_j^T\mathbf{K}\boldsymbol{\varphi}_i - \omega_i^2\boldsymbol{\varphi}_j^T\mathbf{M}\boldsymbol{\varphi}_i = 0 \quad \text{and} \quad \boldsymbol{\varphi}_i^T\mathbf{K}\boldsymbol{\varphi}_j - \omega_j^2\boldsymbol{\varphi}_i^T\mathbf{M}\boldsymbol{\varphi}_j = 0 \tag{4.49}$$

transpose throughout the second (recalling that $\mathbf{K}$ and $\mathbf{M}$ are symmetric)

$$\boldsymbol{\varphi}_j^T\mathbf{K}\boldsymbol{\varphi}_i - \omega_j^2\boldsymbol{\varphi}_j^T\mathbf{M}\boldsymbol{\varphi}_i = 0 \tag{4.50}$$

and subtracting the first, then the following is obtained

$$\left(\omega_i^2 - \omega_j^2\right)\boldsymbol{\varphi}_j^T\mathbf{M}\boldsymbol{\varphi}_i = 0 \tag{4.51}$$

Thus, since $\omega_i \neq \omega_j$ we must conclude that

$$\boldsymbol{\varphi}_j^T\mathbf{M}\boldsymbol{\varphi}_i = 0 \tag{4.52}$$

It is seen from the first part of Eq. 4.49, $\boldsymbol{\varphi}_j^T\mathbf{K}\boldsymbol{\varphi}_i - \omega_i^2\boldsymbol{\varphi}_j^T\mathbf{M}\boldsymbol{\varphi}_i = 0$, that since ω_i is a non-trivial solution different from zero, we must also have that

$$\boldsymbol{\varphi}_j^T\mathbf{K}\boldsymbol{\varphi}_i = 0 \tag{4.53}$$

Thus, the eigenmode solutions to the general eigenvalue problem are $\mathbf{M}$ and $\mathbf{K}$ orthonormal. Similarly, let λ_i and λ_j, and corresponding $\boldsymbol{\varphi}_i$ and $\boldsymbol{\varphi}_j$, be two arbitrary but independent and non-trivial solutions different from zero to the special eigenvalue problem $(\mathbf{A} - \lambda\mathbf{I})\boldsymbol{\varphi} = \mathbf{0}$, i.e.

$$(\mathbf{A} - \lambda_i\mathbf{I})\boldsymbol{\varphi}_i = \mathbf{0} \quad \text{and} \quad (\mathbf{A} - \lambda_j\mathbf{I})\boldsymbol{\varphi}_j = \mathbf{0} \tag{4.54}$$

Pre-multiplying the first with $\boldsymbol{\varphi}_j^T$ and the second by $\boldsymbol{\varphi}_i^T$

$$\boldsymbol{\varphi}_j^T \mathbf{A} \boldsymbol{\varphi}_i - \lambda_i \boldsymbol{\varphi}_j^T \boldsymbol{\varphi}_i = 0 \quad \text{and} \quad \boldsymbol{\varphi}_i^T \mathbf{A} \boldsymbol{\varphi}_j - \lambda_j \boldsymbol{\varphi}_i^T \boldsymbol{\varphi}_j = 0 \tag{4.55}$$

transpose throughout the second (recalling that because $\mathbf{K}$ and $\mathbf{M}$ are symmetric, so is $\mathbf{A}$)

$$\boldsymbol{\varphi}_j^T \mathbf{A} \boldsymbol{\varphi}_i - \lambda_j \boldsymbol{\varphi}_j^T \boldsymbol{\varphi}_i = 0 \tag{4.56}$$

and subtracting the first, then the following is obtained

$$\left(\lambda_i - \lambda_j\right) \boldsymbol{\varphi}_j^T \boldsymbol{\varphi}_i = 0 \tag{4.57}$$

Again, since $\lambda_i \neq \lambda_j$ we must conclude that

$$\boldsymbol{\varphi}_j^T \boldsymbol{\varphi}_i = 0 \tag{4.58}$$

Thus, the eigenmode solutions to the special eigenvalue problem are orthogonal. Since $\boldsymbol{\varphi}_j^T \boldsymbol{\varphi}_i = 0$ and $\lambda_j \neq 0$ it is seen from Eq. 4.56 that as long as $i \neq j$ also

$$\boldsymbol{\varphi}_j^T \mathbf{A} \boldsymbol{\varphi}_i = 0 \tag{4.59}$$

The Rayleigh Quotient

Pre-multiplying $\left(\mathbf{K} - \omega_i^2 \mathbf{M}\right) \boldsymbol{\varphi}_i = \mathbf{0}$ by $\boldsymbol{\varphi}_i^T$ it is seen that

$$\omega_i^2 = \frac{\boldsymbol{\varphi}_i^T \mathbf{K} \boldsymbol{\varphi}_i}{\boldsymbol{\varphi}_i^T \mathbf{M} \boldsymbol{\varphi}_i} \tag{4.60}$$

which in structural dynamics is called the Rayleigh quotient. It is also seen that if the mode shapes are scaled such that

$$\hat{\boldsymbol{\varphi}}_i = \boldsymbol{\varphi}_i \Big/ \sqrt{\boldsymbol{\varphi}_i^T \mathbf{M} \boldsymbol{\varphi}_i} \tag{4.61}$$

then

$$\hat{\boldsymbol{\varphi}}_i^T \mathbf{M} \hat{\boldsymbol{\varphi}}_i = 1 \tag{4.62}$$

and thus

$$\omega_i^2 = \hat{\boldsymbol{\varphi}}_i^T \mathbf{K} \hat{\boldsymbol{\varphi}}_i \tag{4.63}$$

Similarly, pre-multiplying $\left(\mathbf{A} - \lambda_i \mathbf{I}\right) \boldsymbol{\varphi}_i = \mathbf{0}$ by $\boldsymbol{\varphi}_i^T$ it is seen that

$$\lambda_i = \frac{\boldsymbol{\varphi}_i^T \mathbf{A} \boldsymbol{\varphi}_i}{\boldsymbol{\varphi}_i^T \boldsymbol{\varphi}_i} \tag{4.64}$$

and if the mode shapes are scaled (by the norm of $\boldsymbol{\varphi}_i$) such that

$$\hat{\boldsymbol{\varphi}}_i = \boldsymbol{\varphi}_i \Big/ \sqrt{\boldsymbol{\varphi}_i^T \boldsymbol{\varphi}_i} \tag{4.65}$$

then

$$\hat{\boldsymbol{\varphi}}_i^T \hat{\boldsymbol{\varphi}}_i = 1 \tag{4.66}$$

and thus

$$\lambda_i = \hat{\boldsymbol{\varphi}}_i^T \mathbf{A} \hat{\boldsymbol{\varphi}}_i \tag{4.67}$$

Recalling that $(\mathbf{A} - \lambda_i \mathbf{I})\boldsymbol{\varphi}_i = \mathbf{0}$ where $\boldsymbol{\varphi}_i = \begin{bmatrix} \varphi_1 & \cdots & \varphi_k & \cdots & \varphi_{N_r} \end{bmatrix}_i^T$ and that there will be as many eigenmodes as there are degrees of freedom, then this may be expanded to an overall level by defining

$$\hat{\boldsymbol{\Phi}} = \begin{bmatrix} \hat{\boldsymbol{\varphi}}_1 & \cdots & \hat{\boldsymbol{\varphi}}_i & \cdots & \hat{\boldsymbol{\varphi}}_{N_r} \end{bmatrix} \text{ and } \boldsymbol{\lambda} = diag\begin{bmatrix} \lambda_1 & \cdots & \lambda_i & \cdots & \lambda_{N_r} \end{bmatrix} \tag{4.68}$$

in which case the equation covering the entire eigenvalue vector space is given by

$$\mathbf{A}\hat{\boldsymbol{\Phi}} - \hat{\boldsymbol{\Phi}}\boldsymbol{\lambda} = \mathbf{0} \tag{4.69}$$

and since

$$\hat{\boldsymbol{\Phi}}^T \hat{\boldsymbol{\Phi}} = \mathbf{I} \tag{4.70}$$

then pre-multiplication with $\hat{\boldsymbol{\Phi}}^T$ renders

$$\boldsymbol{\lambda} = \hat{\boldsymbol{\Phi}}^T \mathbf{A} \hat{\boldsymbol{\Phi}} \tag{4.71}$$

Elaboration 4.5: The Similarity Transformation

Similarly we may define the eigenmode vector space $\boldsymbol{\Phi} = \begin{bmatrix} \boldsymbol{\varphi}_1 & \cdots & \boldsymbol{\varphi}_i & \cdots & \boldsymbol{\varphi}_{N_r} \end{bmatrix}$ and correspondingly $\boldsymbol{\lambda} = diag\begin{bmatrix} \lambda_1 & \cdots & \lambda_i & \cdots & \lambda_{N_r} \end{bmatrix}$, then

$$\mathbf{A}\boldsymbol{\Phi} - \boldsymbol{\Phi}\boldsymbol{\lambda} = \mathbf{0}$$

Pre-multiplication by the inverse of a non-singular N_r by N_r matrix $\mathbf{Y}$

$$\mathbf{Y}^{-1}\mathbf{A}\boldsymbol{\Phi} - \mathbf{Y}^{-1}\boldsymbol{\Phi}\boldsymbol{\lambda} = \mathbf{0}$$

and defining a new unknown matrix $\mathbf{Z} = \mathbf{Y}^{-1}\boldsymbol{\Phi}$, then

$$\mathbf{Y}^{-1}\mathbf{AYZ} - \mathbf{Z}\boldsymbol{\lambda} = \mathbf{0}$$

from which the following transformed eigenvalue problem is obtained

$$\mathbf{BZ} - \mathbf{Z}\boldsymbol{\lambda} = \mathbf{0}$$

where $\mathbf{B}=\mathbf{Y}^{-1}\mathbf{A}\mathbf{Y}$. It is seen that $\mathbf{A}$ and $\mathbf{B}$ have the same eigenvalues $\boldsymbol{\lambda}$, but different eigenmodes $\boldsymbol{\Phi}$ and $\mathbf{Z}$. This is a linear similarity transformation. The connection between $\boldsymbol{\Phi}$ and $\mathbf{Z}$ is given by $\mathbf{Z}=\mathbf{Y}^{-1}\boldsymbol{\Phi}$, and thus

$$\boldsymbol{\Phi}=\mathbf{Y}\mathbf{Z}$$

Elaboration 4.6: The Cholesky Decomposition

The Cholesky decomposition is presented in appendix B.4. The reason why it has been included is that apart from being used in the simulation of non-coherent time series from spectra, it is in some cases convenient to perform a Cholesky decomposition of the mass matrix $\mathbf{M}$ (which in our case is always real) to obtain a solution to the general eigenvalue problem

$$\left(\mathbf{K}-\omega^2\mathbf{M}\right)\boldsymbol{\varphi}=\mathbf{0}$$

Thus, it is written as the product of a lower triangular matrix $\mathbf{Y}$ and its transposed

$$\mathbf{M}=\mathbf{Y}\mathbf{Y}^T$$

Introducing this into the equation above, defining $\mathbf{z}=\mathbf{Y}^T\boldsymbol{\varphi}$ and pr-multiplying by $\mathbf{Y}^{-1}$

$$\mathbf{Y}^{-1}\mathbf{K}\left(\mathbf{Y}^T\right)^{-1}\mathbf{z}-\omega^2\mathbf{Y}^{-1}\mathbf{Y}\mathbf{Y}^T\left(\mathbf{Y}^T\right)^{-1}\mathbf{z}=\mathbf{0}$$

will render the equivalent eigenvalue problem

$$\left(\mathbf{B}-\omega^2\mathbf{I}\right)\mathbf{z}=\mathbf{0}$$

where $\mathbf{B}=\mathbf{Y}^{-1}\mathbf{K}\left(\mathbf{Y}^T\right)^{-1}$. The original eigenmodes may then be retrieved from

$$\boldsymbol{\varphi}=\left(\mathbf{Y}^T\right)^{-1}\mathbf{z}$$

The advantage of this transformation becomes particularly apparent if

$$\mathbf{M}=diag\left(M_1 \quad \cdots \quad M_i \quad \cdots \quad M_{N_r}\right)$$

in which case

$$\mathbf{Y}=\mathbf{Y}^T=\mathbf{M}^{1/2}=diag\left(M_1^{1/2} \quad \cdots \quad M_i^{1/2} \quad \cdots \quad M_{N_r}^{1/2}\right)$$

$$\mathbf{Y}^{-1}=\left(\mathbf{Y}^T\right)^{-1}=\mathbf{M}^{-1/2}=diag\left(M_1^{-1/2} \quad \cdots \quad M_i^{-1/2} \quad \cdots \quad M_{N_r}^{-1/2}\right)$$

and thus

$$\mathbf{B} = \mathbf{M}^{-1/2}\mathbf{K}\mathbf{M}^{-1/2} \qquad \text{and} \qquad \boldsymbol{\varphi} = \mathbf{M}^{-1/2}\mathbf{z}$$

Below, the numeric routines of direct and inverse vector iterations as well as the Jacobi method are included, not because they are more important than many of the alternative routines available, but because they represent two main classes of approaches, namely iterations on $\mathbf{A}$ itself and iterations on a transformed version of $\mathbf{A}$. They all require a opening vector $\boldsymbol{\psi}_0$. Obviously, the closer $\boldsymbol{\psi}_0$ is to the exact solution the faster convergence is obtained.

Direct (Power) Vector Iteration

Starting with sound engineering guess $\boldsymbol{\psi}_0$ as an initial solution to

$$(\mathbf{A} - \lambda_k \mathbf{I})\boldsymbol{\psi}_k = \mathbf{0} \tag{4.72}$$

where $\mathbf{A} = \mathbf{M}^{-1}\mathbf{K}$, may iteratively be expanded into an improved solution by

$$\boldsymbol{\psi}_k = \frac{\mathbf{A}\boldsymbol{\psi}_{k-1}}{\left\| \mathbf{A}\boldsymbol{\psi}_{k-1} \right\|} \qquad k = 1, 2, \ldots, N \tag{4.73}$$

where $\left\| \mathbf{A}\boldsymbol{\psi}_{k-1} \right\| = \sqrt{(\mathbf{A}\boldsymbol{\psi}_{k-1})^T (\mathbf{A}\boldsymbol{\psi}_{k-1})}$ is the norm of the previous solution. The corresponding eigenvalue may be determined by using the Rayleigh quotient

$$\lambda_k = \frac{\boldsymbol{\psi}_k^T \mathbf{A} \boldsymbol{\psi}_k}{\boldsymbol{\psi}_k^T \boldsymbol{\psi}_k} \tag{4.74}$$

(Normalisation of $\boldsymbol{\psi}_k$ such that after each step $\hat{\boldsymbol{\psi}}_k = \boldsymbol{\psi}_k \Big/ \sqrt{\boldsymbol{\psi}_k^T \boldsymbol{\psi}_k}$ is optional.) The method is not widely in use because the solution converges towards a larger dominant eigenvalue, provided $\boldsymbol{\psi}_0$ has a non-zero component in the direction of the corresponding eigenmode.

Inverse Vector Iteration

The inverse vector iteration method is in principal equivalent to the direct method presented above, only it operates on the inverse eigenvalue problem (see Eq. 4.43)

$$(\mathbf{B} - \beta_k \mathbf{I})\boldsymbol{\psi}_k = \mathbf{0} \tag{4.75}$$

where $\mathbf{B} = \mathbf{K}^{-1}\mathbf{M}$ and $\beta_k = \omega_k^{-2}$. Let $\hat{\boldsymbol{\psi}}_1 = \boldsymbol{\psi}_0 \Big/ \sqrt{\boldsymbol{\psi}_0^T \boldsymbol{\psi}_0}$ be the normalised vector of an initial guess $\boldsymbol{\psi}_0$. Then

$$\boldsymbol{\psi}_k = \mathbf{B}\hat{\boldsymbol{\psi}}_k \qquad \text{where} \qquad \hat{\boldsymbol{\psi}}_k = \frac{\boldsymbol{\psi}_{k-1}}{\sqrt{\boldsymbol{\psi}_{k-1}^T \boldsymbol{\psi}_{k-1}}} \tag{4.76}$$

and because $\hat{\boldsymbol{\psi}}_k^T \left(\mathbf{B} - \beta_k \mathbf{I}\right)\hat{\boldsymbol{\psi}}_k = \hat{\boldsymbol{\psi}}_k^T \mathbf{B}\hat{\boldsymbol{\psi}}_k - \beta_k \hat{\boldsymbol{\psi}}_k^T \hat{\boldsymbol{\psi}}_k = \hat{\boldsymbol{\psi}}_k^T \boldsymbol{\psi}_k - \beta_k = 0$, then

$$\beta_k = \hat{\boldsymbol{\psi}}_k^T \boldsymbol{\psi}_k \tag{4.77}$$

The main advantage with this method is that with a choice of $\boldsymbol{\psi}_0$ that is not unduly off target it will always converge towards the solution with numerically lowest eigenfrequency ω_1 and corresponding eigenmode $\hat{\boldsymbol{\varphi}}_1$, i.e.

$$\beta_k \xrightarrow[k\to\infty]{} \beta_1 = \omega_1^{-2} \qquad \text{and} \qquad \hat{\boldsymbol{\psi}}_k \xrightarrow[k\to\infty]{} \hat{\boldsymbol{\varphi}}_1 \tag{4.78}$$

The second or any higher eigenvalue may be determined by introducing a shift (see Fig. 4.13 below) and rewrite the original eigenvalue problem into

$$\left[\mathbf{K} - \mu\mathbf{M} - \left(\omega^2 - \mu\right)\mathbf{M}\right]\boldsymbol{\varphi} = \mathbf{0} \tag{4.79}$$

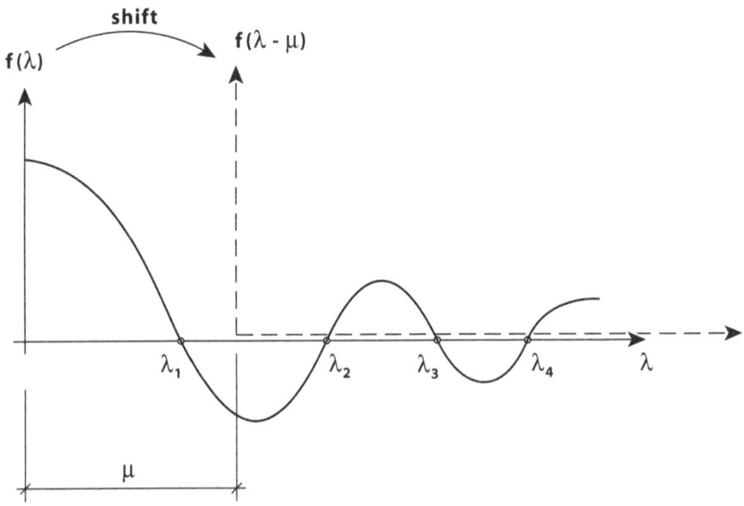

Fig. 4.13 Eigenvalue polynomial $f\left(\lambda\right)$ with shift

Defining $\beta = \left(\omega^2 - \mu\right)^{-1}$ and pre-multiplication by $-\beta\left(\mathbf{K} - \mu\mathbf{M}\right)^{-1}$ will then render the following equivalent eigenvalue problem

$$\left(\mathbf{B}_\mu - \beta\mathbf{I}\right)\boldsymbol{\varphi} = \mathbf{0} \tag{4.80}$$

where $\mathbf{B}_\mu = \left(\mathbf{K} - \mu\mathbf{M}\right)^{-1}\mathbf{M}$. Thus, iterations may be performed in an equivalent procedure to that which has been shown above, and the solution will converge towards the eigenvalue immediately above μ, while $\omega_k = \sqrt{\mu + 1/\beta_k}$.

Example 4.3: Inverse Vector Iteration

Let us consider the simple shear frame shown in Fig. 4.13.a and let

$$\left(\frac{12EI}{H^3}\right)_1 = 9 \cdot 10^4 \text{ N/m} \qquad \left(\frac{12EI}{H^3}\right)_2 = 6 \cdot 10^4 \text{ N/m} \qquad \left(\frac{12EI}{H^3}\right)_3 = 3 \cdot 10^4 \text{ N/m}$$
$$\mathrm{M}_1 = 6 \cdot 10^3 \text{ kg} \qquad \mathrm{M}_2 = 6 \cdot 10^3 \text{ kg} \qquad \mathrm{M}_3 = 3 \cdot 10^3 \text{ kg}$$

The stiffness of a shear beam is shown in Fig. 4.10.b. The necessary mass and stiffness properties of the system may then be established by successive unit displacements as illustrated in Fig. 4.10.c, from which the following is obtained

$$\mathbf{K} = 6 \cdot 10^4 \begin{bmatrix} 5 & -2 & 0 \\ -2 & 3 & -1 \\ 0 & -1 & 1 \end{bmatrix} \quad \textit{and} \quad \mathbf{M} = 3 \cdot 10^3 \begin{bmatrix} 2 & 0 & 0 \\ 0 & 2 & 0 \\ 0 & 0 & 1 \end{bmatrix}$$

Then
$$\mathbf{B}(\mu = 0) = \mathbf{K}^{-1}\mathbf{M} = \begin{bmatrix} 1/30 & 1/30 & 1/60 \\ 1/30 & 1/12 & 1/24 \\ 1/30 & 1/12 & 11/120 \end{bmatrix}$$

and the successive inverse iterations are shown in Table 4.1 below, starting with

$$\boldsymbol{\psi}_0 = \begin{bmatrix} 1 & 1 & 1 \end{bmatrix}^T \quad \Rightarrow \quad \hat{\boldsymbol{\varphi}}_1 = \boldsymbol{\psi}_0 \Big/ \sqrt{\boldsymbol{\psi}_0^T \boldsymbol{\psi}_0} \approx \begin{bmatrix} 0.5774 & 0.5774 & 0.5774 \end{bmatrix}^T$$

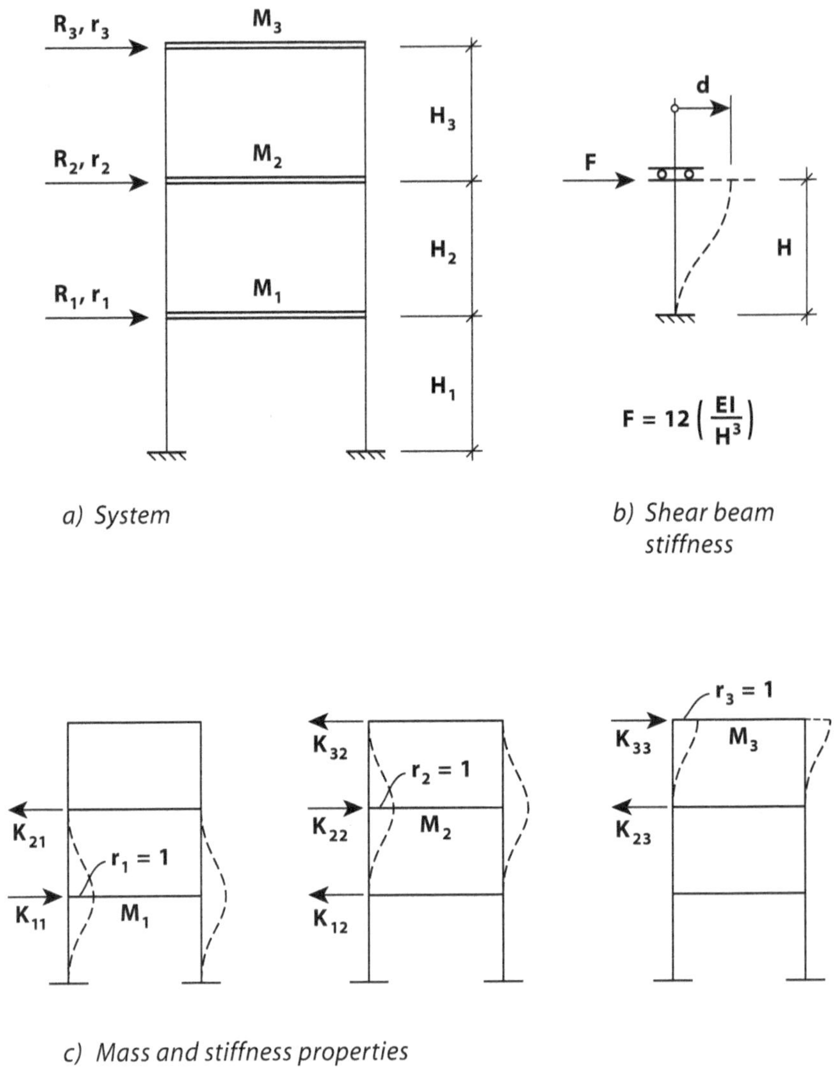

Fig. 4.14 Three degree of freedom shear frame

The iterative value of ω *vs.* k *is shown in Fig. 4.14. As can be seen the convergence is approaching its limit value* $\omega_1 \approx 2.5055$ rad/s *from below. It represents the first (lowest) eigenvalue of the system. The corresponding normalised eigenmode is given by*

$$\boldsymbol{\varphi}_1 \approx \hat{\boldsymbol{\psi}}_6 = \boldsymbol{\psi}_6 \Big/ \sqrt{\boldsymbol{\psi}_6^T \boldsymbol{\psi}_6} = [0.2504 \quad 0.5478 \quad 0.7982]^T$$

Table 4.1 Inverse vector iteration

k	1	2	3	4	5	6
$\mathbf{B}(\mu=0)$	$\hat{\boldsymbol{\psi}}_k = \boldsymbol{\psi}_{k-1} \Big/ \sqrt{\boldsymbol{\psi}_{k-1}^T \boldsymbol{\psi}_{k-1}}$					
0.0333 0.0333 0.0167	0.5774	0.3034	0.2597	0.2523	0.2509	0.2506
0.0333 0.0833 0.0417	0.5774	0.5766	0.5554	0.5495	0.5481	0.5478
0.0333 0.0833 0.0914	0.5774	0.7586	0.7900	0.7965	0.7979	0.7982
$\boldsymbol{\psi}_k = \mathbf{B}\hat{\boldsymbol{\psi}}_k$	0.0481	0.0420	0.0403	0.0400	0.0399	0.0399
	0.0914	0.0898	0.0879	0.0874	0.0873	0.0873
	0.1203	0.1277	0.1274	0.1272	0.1272	0.1272
$\beta_k = \hat{\boldsymbol{\psi}}_k^T \boldsymbol{\psi}_k$	0.15	0.1614	0.1599	0.1594	0.1593	0.1593
$\omega = \beta_k^{-1/2}$	2.582	2.4891	2.5008	2.5047	2.5055	2.5055

To obtain the second eigenvalue trial and error has shown that $\mu = 24$ *is a suitable choice of shift. Then*

$$\mathbf{B}_\mu = (\mathbf{K} - \mu\mathbf{M})^{-1}\mathbf{M} = 10^{-2}\begin{bmatrix} 5.303 & 1.894 & -4.735 \\ 1.894 & 2.462 & -6.155 \\ -9.47 & -12.31 & 5.777 \end{bmatrix}$$

Table 4.2 Inverse vector iteration with shift

k	1	2	3	4	5	6
$10^2 \cdot \mathbf{B}(\mu=24)$	$\hat{\boldsymbol{\psi}}_k = \boldsymbol{\psi}_{k-1} \Big/ \sqrt{\boldsymbol{\psi}_{k-1}^T \boldsymbol{\psi}_{k-1}}$					
5.303 1.894 -4.735	0.5774	0.1511	0.5317	0.3669	0.42299	0.4034
1.894 2.462 -6.155	0.5774	-0.1104	0.6146	0.3323	0.4338	0.3995
-9.47 -12.31 5.777	0.5774	-0.9823	-0.5827	-0.8688	-0.7956	-0.8232
$\boldsymbol{\psi}_k = \mathbf{B}\hat{\boldsymbol{\psi}}_k$	0.0142	0.0524	0.0674	0.0689	0.06831	0.0679
	-0.0104	0.0606	0.0611	0.0689	0.0677	0.0681
	-0.0924	-0.0575	-0.1597	-1.1258	-0.1394	-0.1349
$\beta_k = \hat{\boldsymbol{\psi}}_k^T \boldsymbol{\psi}_k$		0.0577	0.1664	0.1567	0.1692	0.1657
$\omega = (1/\beta_k + \mu)^{1/2}$		6.4289	5.4781	5.5119	5.469	5.480

The successive inverse iterations, again starting with

$$\boldsymbol{\psi}_0 = \begin{bmatrix} 1 & 1 & 1 \end{bmatrix}^T \quad \Rightarrow \quad \hat{\boldsymbol{\psi}}_1 = \boldsymbol{\psi}_0 \Big/ \sqrt{\boldsymbol{\psi}_0^T \boldsymbol{\psi}_0} \approx \begin{bmatrix} 0.5774 & 0.5774 & 0.5774 \end{bmatrix}^T$$

are shown in Table 4.2 above. Thus $\omega_2 \approx 5.48$ rad/s *and*

$$\boldsymbol{\varphi}_2 \approx \hat{\boldsymbol{\psi}}_6 = \boldsymbol{\psi}_6 \Big/ \sqrt{\boldsymbol{\psi}_6^T \boldsymbol{\psi}_6} = \begin{bmatrix} 0.4099 & 0.4111 & -0.8143 \end{bmatrix}^T.$$

The Jacobi Method

(The method is named after the German mathematician Carl Gustav Jacob Jacobi, [41].) Let the overall eigenvalue problem

$$(\mathbf{A} - \boldsymbol{\lambda})\boldsymbol{\Phi} = \mathbf{0} \tag{4.81}$$

be defined by

$$\mathbf{A} = \begin{bmatrix} a_{11} & \cdots & a_{1i} & \cdots & a_{1N} \\ \vdots & \ddots & \vdots & \iddots & \vdots \\ a_{i1} & \cdots & a_{ii} & \cdots & a_{iN} \\ \vdots & \iddots & \vdots & \ddots & \vdots \\ a_{N1} & \cdots & a_{Ni} & \cdots & a_{NN} \end{bmatrix} \quad \text{and} \quad \boldsymbol{\lambda} = \begin{bmatrix} \lambda_1 & & & & \\ & \ddots & & \mathbf{0} & \\ & & \lambda_i & & \\ & \mathbf{0} & & \ddots & \\ & & & & \lambda_N \end{bmatrix}$$

$$\text{and } \boldsymbol{\Phi} = \begin{bmatrix} \boldsymbol{\varphi}_1 & \cdots & \boldsymbol{\varphi}_2 & \cdots & \boldsymbol{\varphi}_N \end{bmatrix} = \begin{bmatrix} \begin{bmatrix} \varphi_1 \\ \vdots \\ \varphi_i \\ \vdots \\ \varphi_N \end{bmatrix}_1 & \cdots & \begin{bmatrix} \varphi_1 \\ \vdots \\ \varphi_i \\ \vdots \\ \varphi_N \end{bmatrix}_i & \cdots & \begin{bmatrix} \varphi_1 \\ \vdots \\ \varphi_i \\ \vdots \\ \varphi_N \end{bmatrix}_N \end{bmatrix} \tag{4.82}$$

The Jacobi method is then based on a successive similarity transformation

$$\mathbf{A}_k = \mathbf{Y}_{k-1}^T \mathbf{A}_{k-1} \mathbf{Y}_{k-1} \qquad k = 1, 2, 3, \ldots, N_k \tag{4.83}$$

where $\mathbf{Y}$ is an N by N orthogonal ($\mathbf{Y}^{-1} = \mathbf{Y}^T$) vector rotation matrix

$$\mathbf{Y} = \begin{bmatrix} 1 & & \overset{i}{\downarrow} & & & & \overset{j}{\downarrow} & & & \\ & \ddots & & & & & & & & \\ & & \cos\alpha & & & & -\sin\alpha & & \leftarrow i & \\ & & & \ddots & & \cdot^{\cdot^{\cdot}} & & & & \\ & & & & 1 & & & & & \\ & & & \cdot^{\cdot^{\cdot}} & & \ddots & & & & \\ & & \sin\alpha & & & & \cos\alpha & & \leftarrow j & \\ & & & & & & & \ddots & & \\ & & & & & & & & 1 & \end{bmatrix} \tag{4.84}$$

and where α is determined such that all other off-diagonal terms become zero,

$$\text{i.e.} \qquad \left. \begin{aligned} &\tan 2\alpha = \frac{2a_{ij}}{a_{ii} - a_{jj}} && \text{for } i \ne j \\ &\alpha = \pi/4 && \text{for } i = j \end{aligned} \right\} \tag{4.85}$$

It is seen that by successive expansion of $\mathbf{A}_{N_k}$

$$\begin{aligned} \mathbf{A}_{N_k} &= \mathbf{Y}_{N_k-1}^T \mathbf{A}_{N_k-1} \mathbf{Y}_{N_k-1} \\ &= \mathbf{Y}_{N_k-1}^T \mathbf{Y}_{N_k-2}^T \mathbf{A}_{N_k-2} \mathbf{Y}_{N_k-2} \mathbf{Y}_{N_k-1} \\ &= \mathbf{Y}_{N_k-1}^T \mathbf{Y}_{N_k-2}^T \mathbf{Y}_{N_k-3}^T \mathbf{A}_{N_k-3} \mathbf{Y}_{N_k-3} \mathbf{Y}_{N_k-2} \mathbf{Y}_{N_k-1} \\ &= \mathbf{Y}_{N_k-1}^T \quad \cdots \quad \mathbf{Y}_k^T \quad \cdots \quad \mathbf{Y}_1^T \mathbf{A} \mathbf{Y}_1 \quad \cdots \quad \mathbf{Y}_k \quad \cdots \quad \mathbf{Y}_{N_k-1} \end{aligned} \tag{4.86}$$

the solution is equivalent to a vector rotation of $\mathbf{A}_k$ into a position of $\boldsymbol{\Psi}_k^T \mathbf{A}_k \boldsymbol{\Psi}_k \xrightarrow[k \to \infty]{} \boldsymbol{\Psi}_k^T \boldsymbol{\lambda}_k \boldsymbol{\Psi}_k$, where $\boldsymbol{\Psi}_k = \mathbf{Y}_1 \cdot \mathbf{Y}_2 \cdots \mathbf{Y}_{k-1} \cdot \mathbf{Y}_k = \boldsymbol{\Psi}_{k-1} \cdot \mathbf{Y}_k$, i.e. a vector rotation which will converge towards $\boldsymbol{\lambda\Phi}$, and thus

$$\left. \begin{aligned} \mathbf{A}_{N_k} &\xrightarrow[N_k \to \infty]{} \boldsymbol{\lambda} \\ \mathbf{Y}_1 \cdot \mathbf{Y}_2 \cdots \mathbf{Y}_{N_k-1} &\xrightarrow[N_k \to \infty]{} \boldsymbol{\Phi} \end{aligned} \right\} \tag{4.87}$$

The advantage with the Jacobi method is that all the relevant eigenvalue solutions are obtained in one go.

Example 4.4: Jacobi Vector Iterations

Let $\mathbf{K}=6\cdot 10^4\begin{bmatrix} 5 & -2 & 0 \\ -2 & 3 & -1 \\ 0 & -1 & 1 \end{bmatrix}$ *and* $\mathbf{M}=3\cdot 10^3\begin{bmatrix} 2 & 0 & 0 \\ 0 & 2 & 0 \\ 0 & 0 & 1 \end{bmatrix}$

(i.e. identical to that which has been investigated in Example 4.3 above). As shown in Elaborations 4.3 and 4.4 it is greatly advantageous if $\mathbf{M}$ *is diagonal, in which case an overall solution may be obtained by defining* $\mathbf{\Phi}=\begin{bmatrix} \boldsymbol{\varphi}_1 & \cdots & \boldsymbol{\varphi}_i & \cdots & \boldsymbol{\varphi}_{N_r} \end{bmatrix}$ *and* $\boldsymbol{\lambda}=diag\left(\lambda_1 \quad \cdots \quad \lambda_i \quad \cdots \quad \lambda_{N_r}\right)$

$$(\mathbf{K}-\boldsymbol{\lambda}\mathbf{M})\mathbf{\Phi}=\mathbf{0}$$

Introducing $\mathbf{Z}=\begin{bmatrix} \mathbf{z}_1 & \cdots & \mathbf{z}_i & \cdots & \mathbf{z}_{N_r} \end{bmatrix}=\mathbf{M}^{1/2}\mathbf{\Phi}$ *(i.e.* $\mathbf{\Phi}=\mathbf{M}^{-1/2}\mathbf{Z}$ *) and pre-multiplying by* $\mathbf{M}^{-1/2}$ $\Rightarrow$ $\mathbf{M}^{-1/2}(\mathbf{K}-\boldsymbol{\lambda}\mathbf{M})\mathbf{M}^{-1/2}\mathbf{Z}=\mathbf{0}$ *then the following is obtained* $(\mathbf{A}-\boldsymbol{\lambda})\mathbf{Z}=\mathbf{0}$ *where* $\mathbf{A}=\mathbf{M}^{-1/2}\mathbf{K}\mathbf{M}^{-1/2}$

$$\Rightarrow \qquad \mathbf{A}=\begin{bmatrix} 50 & -20 & 0 \\ -20 & 30 & -14.1421 \\ 0 & -14.1421 & 20 \end{bmatrix}$$

The development of $\mathbf{Y}_k$ *and corresponding iterations are shown in Table 4.3.*

Thus $\boldsymbol{\lambda}\approx\begin{bmatrix} 63.72 & 0 & 0 \\ 0 & 6.28 & 0 \\ 0 & 0 & 30 \end{bmatrix} \quad \Rightarrow \quad \boldsymbol{\omega}=\boldsymbol{\lambda}^{1/2}=diag\begin{bmatrix} 7.982 & 2.506 & 5.477 \end{bmatrix}$

As can be seen, the first and the second eigenvalues comes out as the second and the third entry. Their numerical values comply well with that which was obtained in Example 4.3 above. The eigenmode matrix is given by

$$\mathbf{\Phi}\approx\mathbf{M}^{-1/2}\mathbf{Z}=\begin{bmatrix} 0.0105 & 0.0039 & -0.0065 \\ -0.0072 & 0.0086 & -0.0065 \\ 0.0033 & 0.0125 & 0.0129 \end{bmatrix}$$

Table 4.3 Jacobi vector iteration

k	i, j $\tan 2\alpha$ α	$\mathbf{Y}_k$	$\mathbf{A}_k = \mathbf{Y}_k^T \mathbf{A}_{k-1} \mathbf{Y}_k$	$\mathbf{Z}_k = \mathbf{Z}_{k-1} \mathbf{Y}_k$
1	$i=1, j=2$ -2 -0.5536	0.851 0.526 0 −0.526 0.851 0 0 0 1	62.36 0 7.43 0 17.64 −12.03 7.43 −12.03 20	0.851 0.526 0 −0.526 0.851 0 0 0 1
2	$i=2, j=3$ 10.192 0.7365	1 *o* 0 0 0.741 −0.672 0 0.672 0.741	62.36 4.99 5.51 4.99 6.73 0 5.51 0 30.92	0.851 0.389 −0.353 −0.526 0.630 −0.571 0 0.672 0.741
3	$i=1, j=3$ 0.3502 0.1684	0.986 0 −0.168 0 1 0 0.168 0 0.986	63.30 4.92 0 4.92 6.73 −0.84 0 −0.84 29.97	0.779 0.389 −0.491 −0.614 0.630 −0.475 0.124 0.672 0.730
4	$i=1, j=2$ 0.1741 0.0862	0.996 −0.861 0 0.861 0.996 0 0 0 1	63.72 0 −0.072 0 6.31 −0.83 −0.072 −0.83 29.97	0.81 0.321 −0.491 −0.558 0.681 −0.475 0.182 0.659 0.730
5	$i=2, j=3$ 0.0705 0.0352	1 0 0 0 0.999 −0.035 0 0.035 0.999	63.72 −0.003 −0.072 −0.003 6.28 0 −0.072 0 30	0.81 0.303 −0.502 −0.558 0.663 −0.499 0.182 0.684 0.707
6	$i=1, j=3$ −0.0043 −0.0021	1 0 0002 0 1 0 −0.002 0 1	63.72 −0.003 0 0.003 6.28 0 0 0 30	0.811 0.303 −0.500 −0.556 0.664 −0.500 0.180 0.684 0.707

It is seen that $\mathbf{\Phi}^T\mathbf{\Phi} \approx 10^{-3} \cdot diag\begin{bmatrix} 0.1719 & 0.2447 & 0.2499 \end{bmatrix}$ *and thus, the normalised version of the mode shape matrix is given by*

$$\hat{\mathbf{\Phi}} = \mathbf{\Phi} \Big/ \left(\mathbf{\Phi}^T \mathbf{\Phi}\right)^{1/2} = \begin{bmatrix} 0.7986 & 0.2501 & -0.4083 \\ -0.5475 & 0.5480 & -0.4083 \\ 0.2507 & 0.7983 & 0.8165 \end{bmatrix}$$

As can be seen, also the mode shapes complies well with that which was obtained in Example 4.3 above. Convergence of eigenvalues as obtained by the inverse method in Example 4.3 and from the Jacobi method in Example 4.4 are shown in Fig. 4.15.

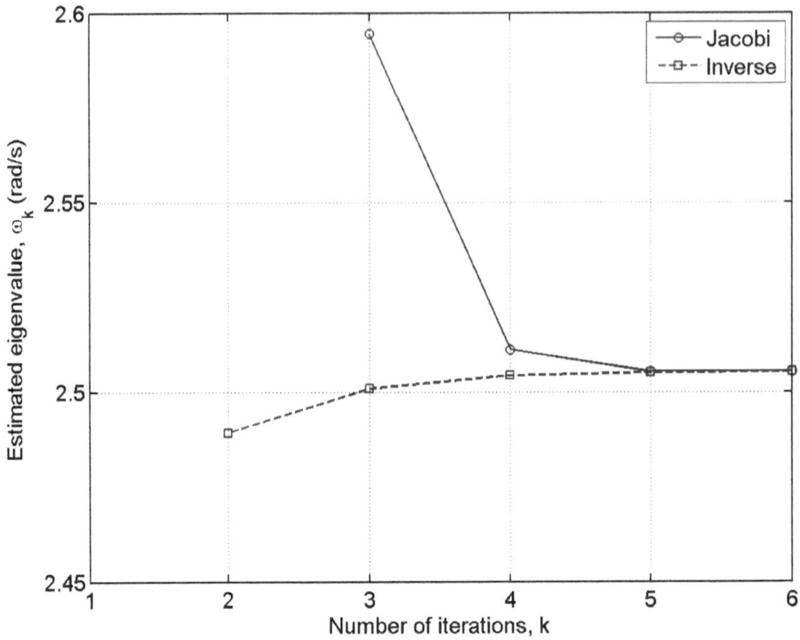

Fig. 4.15 Convergence of inverse and Jacobi iterations (Examples 4.3 and 4.4)

Elaboration 4.7: Damped Eigenvalues

For a damped but unloaded system the solution to $\mathbf{M\ddot{r}} + \mathbf{C\dot{r}} + \mathbf{Kr} = \mathbf{0}$

may be obtained by adding a dummy equation, e.g. $\mathbf{I\dot{r}} - \mathbf{I\dot{r}} = \mathbf{0}$ (where $\mathbf{I}$ is the identity matrix), in which case the problem may be re-stated into

$$\begin{bmatrix} \mathbf{I} & \mathbf{0} \\ \mathbf{0} & \mathbf{M} \end{bmatrix} \begin{bmatrix} \mathbf{\dot{r}} \\ \mathbf{\ddot{r}} \end{bmatrix} + \begin{bmatrix} \mathbf{0} & -\mathbf{I} \\ \mathbf{K} & \mathbf{C} \end{bmatrix} \begin{bmatrix} \mathbf{r} \\ \mathbf{\dot{r}} \end{bmatrix} = \mathbf{0}$$

Defining a new variable $\mathbf{z} = \begin{bmatrix} \mathbf{r} \\ \mathbf{\dot{r}} \end{bmatrix}$ and the matrices $\mathbf{B} = \begin{bmatrix} \mathbf{I} & \mathbf{0} \\ \mathbf{0} & \mathbf{M} \end{bmatrix}$ and $\mathbf{A} = \begin{bmatrix} \mathbf{0} & -\mathbf{I} \\ \mathbf{K} & \mathbf{C} \end{bmatrix}$, then the equation above is $\mathbf{B\dot{z}} + \mathbf{Az} = \mathbf{0}$

Since the damping affected solution is given by $\mathbf{r} = \mathrm{Re}\left(\boldsymbol{\varphi} e^{\lambda t}\right)$ then

$$\mathbf{z} = \mathrm{Re}\left(\begin{bmatrix} \boldsymbol{\varphi} \\ \lambda \boldsymbol{\varphi} \end{bmatrix} e^{\lambda t} \right)$$

Thus, the solution may be obtained from the eigenvalue problem

$$(\mathbf{A} + \lambda \mathbf{B}) \boldsymbol{\Phi}_z = \mathbf{0} \qquad \text{where} \qquad \boldsymbol{\Phi}_z = [\boldsymbol{\varphi} \quad \lambda \boldsymbol{\varphi}]^T$$

If $\mathbf{C} = \mathbf{0}$, then we are back to the ordinary eigenvalue problem previously described in this chapter, and as we have seen, λ is negative and purely imaginary (i.e. $\lambda = i\omega$) while $\boldsymbol{\varphi}$ only contains real quantities. This means that the motion is a simple harmonic and its size and shape remains unchanged with time, and all the degrees of freedom are in phase. If $\mathbf{C} \neq \mathbf{0}$, then λ will be complex (and not necessarily purely imaginary) while $\boldsymbol{\varphi}$ may also contain complex quantities. This means that the motion is not a simple harmonic, it is decaying with time (as long as all entries in $\mathbf{C}$ are positive), while its shape is time dependent as there is a phase between the motion of the various degrees of freedom in the system. Let us consider the most simple case of a single degree of freedom system

$$M\ddot{r} + C\dot{r} + Kr = 0 \;\Rightarrow\; \mathbf{B} = \begin{bmatrix} 1 & 0 \\ 0 & M \end{bmatrix}, \; \mathbf{A} = \begin{bmatrix} 0 & -1 \\ K & C \end{bmatrix} \text{ and } \mathbf{z} = \begin{bmatrix} r \\ \dot{r} \end{bmatrix} = \begin{bmatrix} \varphi \\ \lambda \varphi \end{bmatrix} e^{\lambda t}$$

Then the eigenvalue problem is given by $\begin{bmatrix} \lambda & -1 \\ K & C + \lambda M \end{bmatrix} \begin{bmatrix} \varphi \\ \lambda \varphi \end{bmatrix} = \mathbf{0}$, from which a non-trivial solution is obtained from $\lambda(C + \lambda M) + K = 0$, which may alternatively be written

$$\left(\frac{\lambda}{\omega_0}\right)^2 + 2\zeta_0 \left(\frac{\lambda}{\omega_0}\right) + 1 = 0, \text{ where } \omega_0 = \sqrt{K/M} \text{ and } \zeta_0 = C/(2M\omega_0)$$

Its solution is given by $\dfrac{\lambda}{\omega_0} = -\zeta_0 \pm \sqrt{\zeta_0^2 - 1}$. There are three possible scenarios:

1) $\zeta_0 > 1 \qquad \Rightarrow \qquad r(t) = e^{-\zeta_0 \omega_0 t} \left[a_1 e^{\omega_0 t \sqrt{\zeta_0^2 - 1}} + a_2 e^{-\omega_0 t \sqrt{\zeta_0^2 - 1}} \right]$

2) $\zeta_0 = 1 \qquad \Rightarrow \qquad r(t) = a \cdot e^{-\omega_0 t}$

$$3)\quad 0<\zeta_0<1 \quad \Rightarrow \quad \begin{cases} r(t)=\operatorname{Re}\left[e^{-\zeta_0\omega_0 t}\left(a_1 e^{i\omega_0 t\sqrt{1-\zeta_0^2}}+a_2 e^{-i\omega_0 t\sqrt{1-\zeta_0^2}}\right)\right] \\ \quad =a\cdot e^{-\zeta_0\omega_0 t}\cos\left(\omega_0 t\sqrt{1-\zeta_0^2}\right) \end{cases}$$

In the first case where $\zeta_0>1$ then the system is over-critically damped and the motion is consistently decaying without any oscillations, i.e. there is no such thing as an eigenvalue solution. In the second case where $\zeta_0=1$ the system is critically damped and the motion is still consistently decaying without any oscillations. In the third case where $0<\zeta_0<1$ the system is under-critically damped and the motion is oscillating with a frequency of $\omega_d=\omega_0\sqrt{1-\zeta_0^2}$, but still decaying. (This is identical to that which was obtained in Chapter 2.2, where the various types of motions are illustrated in Fig. 2.3 and a more comprehensive solution to the case that $\zeta_0=1$ is given in Eq. 2.18.)

Let us then consider the more demanding case of a two degree of freedom system

$$\mathbf{M}=M\begin{bmatrix}1&0\\0&1\end{bmatrix} \qquad \mathbf{C}=C\begin{bmatrix}1&0\\0&1\end{bmatrix} \qquad \text{and} \qquad \mathbf{K}=K\begin{bmatrix}2&-1\\-1&1\end{bmatrix}$$

It is convenient first to solve the eigenvalue problem $(\mathbf{K}-\lambda_0\mathbf{M})\boldsymbol{\varphi}_0=\mathbf{0}$, rendering

$$\begin{bmatrix}2-\lambda_0\dfrac{M}{K} & -1\\ -1 & 1-\lambda_0\dfrac{M}{K}\end{bmatrix}\begin{bmatrix}\varphi_1\\ \varphi_2\end{bmatrix}_0=\mathbf{0} \quad\Rightarrow\quad \begin{cases}\lambda_0\dfrac{M}{K}=\dfrac{3}{2}\pm\dfrac{\sqrt{5}}{2}\approx\begin{cases}2.618\\0.382\end{cases}\\ \dfrac{\varphi_{20}}{\varphi_{10}}\approx\begin{cases}-0.618\\1.618\end{cases}\end{cases}$$

$$\text{i.e.}\quad \begin{cases}\omega_{0_1}=0.618\sqrt{\dfrac{K}{M}}\ \text{rad/s}\ \&\ \boldsymbol{\varphi}_{0_1}=\begin{bmatrix}1\\1.618\end{bmatrix} \Rightarrow \hat{\boldsymbol{\varphi}}_{0_1}=\begin{bmatrix}0.526\\0.851\end{bmatrix}\\ \omega_{0_2}=1.618\sqrt{\dfrac{K}{M}}\ \text{rad/s}\ \&\ \boldsymbol{\varphi}_{0_2}=\begin{bmatrix}1\\-0.618\end{bmatrix} \Rightarrow \hat{\boldsymbol{\varphi}}_{0_2}=\begin{bmatrix}0.851\\-0.526\end{bmatrix}\end{cases}$$

Then, by adding $\mathbf{C}$ the eigenvalue problem is given by $(\mathbf{A}+\lambda\mathbf{B})\mathbf{\Phi}_z=\mathbf{0}$ where $\mathbf{\Phi}_z=[\boldsymbol{\varphi} \quad \lambda\boldsymbol{\varphi}]^T$, and where

$$\mathbf{A}=\begin{bmatrix} \mathbf{0} & -\mathbf{I} \\ \mathbf{K} & \mathbf{C} \end{bmatrix}=\begin{bmatrix} 0 & 0 & -1 & 0 \\ 0 & 0 & 0 & -1 \\ 2K & -K & C & 0 \\ -K & K & 0 & C \end{bmatrix} \qquad \mathbf{B}=\begin{bmatrix} \mathbf{I} & 0 \\ 0 & \mathbf{M} \end{bmatrix}=\begin{bmatrix} 1 & 0 & 0 & 0 \\ 0 & 1 & 0 & 0 \\ 0 & 0 & M & 0 \\ 0 & 0 & 0 & M \end{bmatrix}$$

$$\Rightarrow \qquad \begin{bmatrix} \lambda & 0 & -1 & 0 \\ 0 & \lambda & 0 & -1 \\ 2 & -1 & (C/K+\lambda M/K) & 0 \\ -1 & 1 & 0 & (C/K+\lambda M/K) \end{bmatrix}\begin{bmatrix} \varphi_1 \\ \phi_2 \\ \lambda\varphi_1 \\ \lambda\varphi_2 \end{bmatrix}=\mathbf{0}$$

The non-trivial damped eigenvalues may then be determined from the requirement that the determinant to the coefficient matrix is zero, i.e. that

$$\left[\lambda\left(\frac{C}{K}+\lambda\frac{M}{K}\right)\right]^2+3\left[\lambda\left(\frac{C}{K}+\lambda\frac{M}{K}\right)\right]+1=0$$

and thus $$\lambda\left(\frac{C}{K}+\lambda\frac{M}{K}\right)=-\frac{3}{2}\pm\frac{\sqrt{5}}{2}=-\lambda_0\frac{M}{K}$$

where $\lambda_0=\omega_{0_n}^2$, $n=1,2$, and where ω_{0_1} and ω_{0_2} are the un-damped eigenfrequencies of the system.

Thus $$\lambda^2+(C/M)\lambda+\omega_{0_n}^2=0$$

$$\Rightarrow \qquad \lambda=-\frac{C}{2M}\pm\sqrt{\left(\frac{C}{2M}\right)^2-\omega_{0_n}^2}=-\frac{C}{2M}\pm i\sqrt{\omega_{0_n}^2-\left(\frac{C}{2M}\right)^2}$$

Introducing $\zeta_{0_n}=C/(2M\omega_{0_n})$, $n=1,2$, then $\lambda/\omega_{0_n}=-\zeta_{0_n}\pm i\sqrt{1-\zeta_{0_n}^2}$.

Thus, the following roots are obtained:

$$\begin{cases} \lambda_1/\omega_{0_1}=-\zeta_{0_1}+i\sqrt{1-\zeta_{0_1}^2} \\ \lambda_2/\omega_{0_1}=-\zeta_{0_1}-i\sqrt{1-\zeta_{0_1}^2} \end{cases} \qquad \begin{cases} \lambda_3/\omega_{0_2}=-\zeta_{0_2}+i\sqrt{1-\zeta_{0_2}^2} \\ \lambda_4/\omega_{0_2}=-\zeta_{0_2}-i\sqrt{1-\zeta_{0_2}^2} \end{cases}$$

It is seen that the eigenvalues come in complex conjugate pairs. The corresponding eigenmodes may be obtained from the third row of the equation

$$2\varphi_1 - \varphi_2 + \lambda\left(\frac{C}{K} + \lambda\frac{M}{K}\right)\varphi_1 = 2\varphi_1 - \varphi_2 + \left(-\frac{3}{2} \pm \frac{\sqrt{5}}{2}\right)\varphi_1 = 0 \Rightarrow \frac{\varphi_2}{\varphi_1} = \begin{cases} 1.618 \\ -0.618 \end{cases}$$

I.e.:

$$\begin{cases} \boldsymbol{\varphi}_1 = \boldsymbol{\varphi}_2 = \begin{bmatrix} 1 \\ 1.618 \end{bmatrix} \Rightarrow \hat{\boldsymbol{\varphi}}_1 = \hat{\boldsymbol{\varphi}}_2 = \begin{bmatrix} 0.526 \\ 0.851 \end{bmatrix} \\ \boldsymbol{\varphi}_3 = \boldsymbol{\varphi}_4 = \begin{bmatrix} 1 \\ -0.618 \end{bmatrix} \Rightarrow \hat{\boldsymbol{\varphi}}_3 = \hat{\boldsymbol{\varphi}}_4 = \begin{bmatrix} 0.851 \\ -0.526 \end{bmatrix} \end{cases}$$

which is identical to that which was obtained for the un-damped case. The reason for this comes from the choice of damping matrix which contains no coupling between the two degrees of freedom. In a general case the damped eigenmodes will also be complex, implying there is a phase (time lag) between the occurrences of the various degrees of freedom.

Chapter 5
The Normal Mode Method

5.1 Introduction

In Chapter 4 we developed the equilibrium condition

$$\mathbf{M}\cdot\ddot{\mathbf{r}}(t)+\mathbf{C}\cdot\dot{\mathbf{r}}(t)+\mathbf{K}\cdot\mathbf{r}(t)=\mathbf{R}(t) \tag{5.1}$$

for a general discrete system with a chosen set of degrees of freedom $\mathbf{r}=\begin{bmatrix}\mathbf{r}_1 & \cdots & \mathbf{r}_p & \cdots & \mathbf{r}_{N_r}\end{bmatrix}^T$ and subject to a corresponding set of external loads $\mathbf{R}=\begin{bmatrix}\mathbf{R}_1 & \cdots & \mathbf{R}_p & \cdots & \mathbf{R}_{N_r}\end{bmatrix}^T$. Because structural systems have very low damping, we demonstrated that they have distinct sets of preference frequencies ω_n and associated shapes of motion $\boldsymbol{\varphi}_n$, which, under the assumption of a general harmonic and stationary motion $\mathbf{r}=\mathrm{Re}\left(\boldsymbol{\varphi}e^{i\omega t}\right)$, emerged from the eigenvalue solution of the undamped and unloaded version of Eq. 5.1

$$\left(\mathbf{K}-\omega^2\mathbf{M}\right)\boldsymbol{\varphi}=\mathbf{0} \tag{5.2}$$

These are all equations in what we define as being expressed in the physical degrees of freedom, or alternatively, in what we call the original coordinates. Frequency ω_n and associated shape vectors $\boldsymbol{\varphi}_n$ are called the eigenfrequencies and eiegenmodes of the system. As can be seen, they largely depend on the mass and stiffness distribution within the system. We have seen (Eqs. 4.48 – 4.59) that the eigenmodes $\boldsymbol{\varphi}_n$ are mass orthogonal, i.e. they are linearly independent vectors with respect to the mass distribution. If the system is given an arbitrary perturbation and it is left to oscillate by itself without any further external influence, then what will occur is an eigenmode oscillation at the corresponding eigenfrequency, depending on the type of perturbation that was applied. Thus, it

E.N. Strømmen, *Structural Dynamics*, Springer Series in Solid and Structural Mechanics 2,
DOI: 10.1007/978-3-319-01802-7_5,

has been conceived that if the system is subject to any external load an approximate solution may be obtained by separation of variables (position and time) and a linear combination of a limited set of known (or chosen) eigenmodes, i.e.

$$\mathbf{r}(t) \approx \sum_{i=1}^{N_{\text{mod}}} \boldsymbol{\varphi}_n \eta_n(t) \tag{5.3}$$

where $\eta_n(t)$ are the new unknown variables in the system. This is an approach that in most cases of structural engineering has proved to be reliable and accurate. It has the advantage that it enables a direct focus on that which is important, and it should be noted that while N_r may be an uncomfortably large number, N_{mod} may with sufficient accuracy be chosen at a very low number (in some special cases it may even suffice to set $N_{\text{mod}} = 1$). Thus, computational advantage may be considerable. As indicated above, it should also be noted that in some cases it may render sufficient accuracy to apply approximate shape functions, i.e. to replace $\boldsymbol{\varphi}_n$ by $\boldsymbol{\psi}_n \approx \boldsymbol{\varphi}_n$, where $\boldsymbol{\psi}_n$ is based on sound engineering judgement of what can be expected.

As we have seen in Chapters 1 and 3, in some cases it is convenient to establish the dynamic equilibrium conditions in a continuous format. For instance, the differential equilibrium condition for a simple beam whose motion is exclusively in the z-direction is given by

$$m_z \cdot \ddot{r}_z + c_z \cdot \dot{r}_z + EI_y \cdot r_z'''' = q_z \tag{5.4}$$

in which case the shape functions are continuous $\varphi_{z_n}(x)$ and thus

$$r_z(x,t) \approx \sum_{n=1}^{N_{\text{mod}}} \varphi_{z_n}(x) \cdot \eta_n(t) \tag{5.5}$$

The concept is known as the normal mode method. The method is intended for the calculation of dynamic load effects. For low frequency (quasi-static) type of problems it may be advisable to replace the eigenmodes by static shape functions, or alternatively to pursue a solution in original degrees of freedom.

5.2 The Discrete Normal Mode Approach

In a discrete format it is convenient to define the mode shape matrix

$$\boldsymbol{\Phi} = \begin{bmatrix} \boldsymbol{\varphi}_1 & \boldsymbol{\varphi}_2 & \cdots & \boldsymbol{\varphi}_n & \cdots & \boldsymbol{\varphi}_{N_{\text{mod}}} \end{bmatrix} \tag{5.6}$$

where $\boldsymbol{\varphi}_n$ $(n = 1, 2, \ldots, N_{\text{mod}})$ contains the mode shape numerical values, each associated with the corresponding global degree of freedom number $p = 1, 2, \ldots, N_r$

$$\boldsymbol{\varphi}_n = \begin{bmatrix} \phi_1 & \phi_2 & \cdots & \phi_p & \cdots & \phi_{N_r} \end{bmatrix}^T \tag{5.7}$$

and a time dependent unknown vector

$$\boldsymbol{\eta}(t) = \begin{bmatrix} \eta_1 & \eta_2 & \cdots & \eta_n & \cdots & \eta_{N_{\text{mod}}} \end{bmatrix}^T \tag{5.8}$$

(The content of $\boldsymbol{\eta}$ is often called generalised coordinates.) Thus, Eq. 5.3 may be written

$$\mathbf{r}(t) \approx \sum_{i=1}^{N_{\text{mod}}} \boldsymbol{\varphi}_n \eta_n(t) = \boldsymbol{\Phi} \cdot \boldsymbol{\eta}(t) \tag{5.9}$$

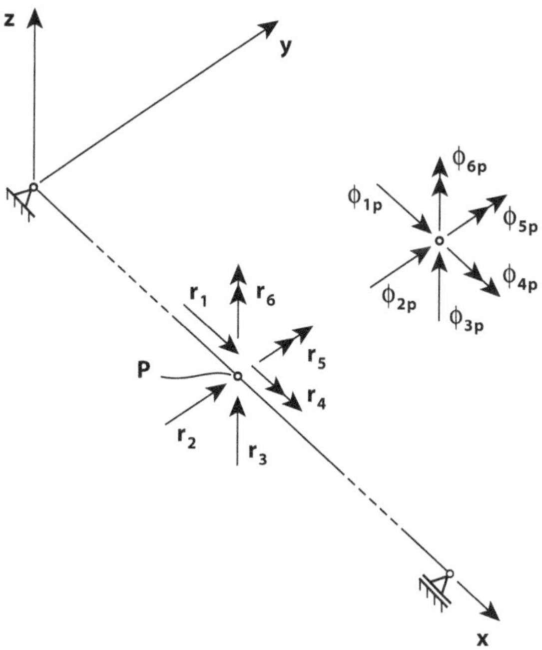

Fig. 5.1 Discrete system

Introducing Eq. 5.9 into Eq. 5.1 and pre-multiplying the entire equation by $\mathbf{\Phi}^T$ will then render

$$\mathbf{\Phi}^T\mathbf{M}\mathbf{\Phi}\ddot{\mathbf{\eta}}(t)+\mathbf{\Phi}^T\mathbf{C}\mathbf{\Phi}\dot{\mathbf{\eta}}(t)+\mathbf{\Phi}^T\mathbf{K}\mathbf{\Phi}\mathbf{\eta}(t)=\mathbf{\Phi}^T\mathbf{R}(t) \tag{5.10}$$

Defining the modally equivalent structural properties

$$\left.\begin{array}{ll} \bullet \text{ modal mass matrix:} & \tilde{\mathbf{M}}=\mathbf{\Phi}^T\mathbf{M}\mathbf{\Phi} \\ \bullet \text{ modal damping matrix:} & \tilde{\mathbf{C}}=\mathbf{\Phi}^T\mathbf{C}\mathbf{\Phi} \\ \bullet \text{ modal stiffness matrix:} & \tilde{\mathbf{K}}=\mathbf{\Phi}^T\mathbf{K}\mathbf{\Phi} \end{array}\right\} \tag{5.11}$$

and the modal load vector

$$\tilde{\mathbf{R}}(t)=\mathbf{\Phi}^T\mathbf{R}(t) \tag{5.12}$$

then the following modal dynamic equilibrium condition is obtained

$$\tilde{\mathbf{M}}\ddot{\mathbf{\eta}}(t)+\tilde{\mathbf{C}}\dot{\mathbf{\eta}}(t)+\tilde{\mathbf{K}}\mathbf{\eta}(t)=\tilde{\mathbf{R}}(t) \tag{5.13}$$

Due to the orthogonal properties of the mode shapes all the off diagonal terms in $\tilde{\mathbf{M}}$ and $\tilde{\mathbf{K}}$ are zeros. Thus

$$\left.\begin{array}{l} \tilde{\mathbf{M}}=diag\left[\tilde{M}_n\right] \\ \tilde{\mathbf{K}}=diag\left[\tilde{K}_n\right] \end{array}\right\} \quad \text{where} \quad \left.\begin{array}{l} \tilde{M}_n=\mathbf{\varphi}_n^T\mathbf{M}\mathbf{\varphi}_n \\ \tilde{K}_n=\mathbf{\varphi}_n^T\mathbf{K}\mathbf{\varphi}_n \end{array}\right\} \tag{5.14}$$

However, introducing an arbitrary mode shape $\mathbf{\varphi}_n$ and its corresponding eigenfrequency ω_n into Eq. 5.2

$$\left(\mathbf{K}-\omega_n^2\mathbf{M}\right)\mathbf{\varphi}_n=\mathbf{0} \tag{5.15}$$

and pre-multiplying by $\mathbf{\varphi}_n^T$ it is readily seen that $\tilde{K}_n$ may more conveniently be determined from

$$\mathbf{\varphi}_n^T\mathbf{K}\mathbf{\varphi}_n=\omega_n^2\mathbf{\varphi}_n^T\mathbf{M}\mathbf{\varphi}_n \quad \Rightarrow \quad \tilde{K}_n=\omega_n^2\tilde{M}_n \tag{5.16}$$

Furthermore, the damping properties of a structural system are usually obtained from unloaded decay motions or ambient more or less resonant recordings, i.e. damping data are in general associated with a particular mode shape. Therefore, it is common practice to introduce N_{mod} modal damping ratios ζ_n, each associated with its corresponding mode shape and critical modal damping $2\tilde{M}_n\omega_n$. Thus

$$\tilde{\mathbf{C}} = diag\left[\tilde{C}_n\right] \quad \text{where} \qquad \tilde{C}_n = 2\tilde{M}_n\omega_n\zeta_n \tag{5.17}$$

The modal equilibrium condition in Eq. 5.13 together with the basic coordinate transformation in Eq. 5.9 may then be used to solve complex dynamic load effect cases. A solution strategy in time or frequency domain is optional. It should be noted that, apart from loading time series $\mathbf{R}(t)$, it is in a modal format sufficient to know the content of the mass matrix $\mathbf{M}$, the mode shape matrix $\mathbf{\Phi} = \left[\boldsymbol{\varphi}_1 \ \cdots \ \boldsymbol{\varphi}_n \ \cdots \ \boldsymbol{\varphi}_{N_{\text{mod}}}\right]$, corresponding eigenfrequencies ω_n and damping ratios ζ_n. (I.e., as soon as a necessary set of $\boldsymbol{\varphi}_n$ and corresponding ω_n have been determined then the stiffness matrix $\mathbf{K}$ is no longer needed.) Modal damping properties are further discussed in Chapter 9.

Example 5.1

The three storey shear frame shown in Fig. 5.2 has original degrees of freedom

$$\mathbf{r}(t) = \begin{bmatrix} r_1 & r_2 & r_3 \end{bmatrix}^T$$

Associated with each of these degrees of freedom it is subject to the external load

$$\mathbf{R} = \text{Re}\left[\mathbf{a}_R \cdot e^{i\omega_2 t}\right] \quad \textit{where} \quad \begin{cases} \omega_2 \text{ is its second lowest eigenfrequency} \\ \mathbf{a}_R = \begin{bmatrix} 1800 & 1800 & 900 \end{bmatrix}^T \text{ (with unit } N\text{)} \end{cases}$$

The three storey shear frame has previously been dealt with in Example 4.3, where the two lowest eigenfrequencies and corresponding eigenmodes

$$\omega_1 \approx 2.5 \text{ rad/s} \qquad \textit{and} \qquad \boldsymbol{\varphi}_1 = \begin{bmatrix} 0.25 & 0.55 & 0.8 \end{bmatrix}^T$$

$$\omega_2 \approx 5.5 \text{ rad/s} \qquad \textit{and} \qquad \boldsymbol{\varphi}_2 = \begin{bmatrix} 0.41 & 0.41 & -0.81 \end{bmatrix}^T$$

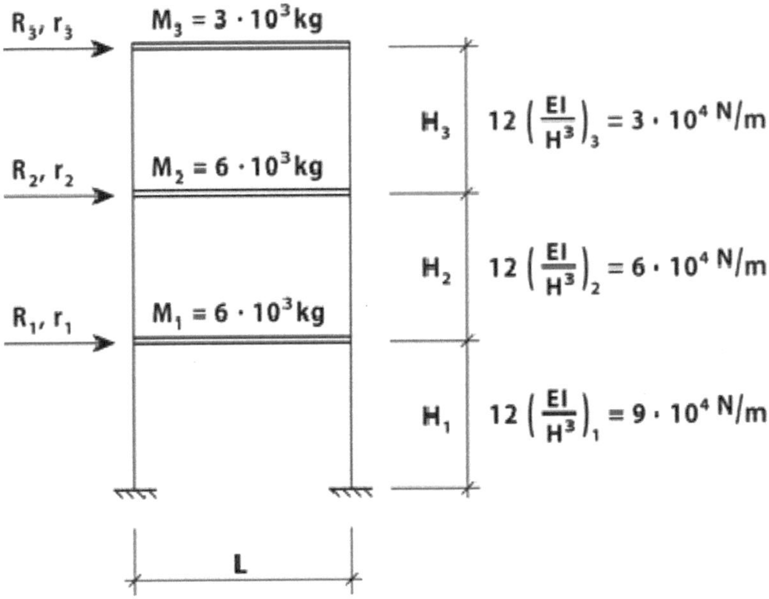

Fig. 5.2 Three storey shear frame

were obtained by an inverse vector iteration procedure. The mode shapes (including the third) are illustrated in Fig. 5.3.a. It is assumed that the corresponding modal damping ratios are given by $\zeta_1 = 0.02$, $\zeta_2 = 0.025$ *and* $\zeta_3 = 0.03$. *In addition to this we need the mass matrix*

$$\mathbf{M} = \begin{bmatrix} M_1 & 0 & 0 \\ 0 & M_2 & 0 \\ 0 & 0 & M_3 \end{bmatrix} = 3 \cdot 10^3 \begin{bmatrix} 2 & 0 & 0 \\ 0 & 2 & 0 \\ 0 & 0 & 1 \end{bmatrix} \text{ with unit } kg$$

Since the load frequency is exclusively identical to ω_2 *it is assumed that a sufficiently accurate solution may be obtained in a modal approach where only the two first eigenmodes are included, i.e. that*

$$\mathbf{r}(t) = \begin{bmatrix} r_1 \\ r_2 \\ r_3 \end{bmatrix} = \mathbf{\Phi}\boldsymbol{\eta}(t) = [\boldsymbol{\varphi}_1 \quad \boldsymbol{\varphi}_2] \begin{bmatrix} \eta_1 \\ \eta_2 \end{bmatrix} = \begin{bmatrix} \varphi_{11} & \varphi_{21} \\ \varphi_{12} & \varphi_{22} \\ \varphi_{13} & \varphi_{23} \end{bmatrix} \begin{bmatrix} \eta_1 \\ \eta_2 \end{bmatrix} = \begin{bmatrix} 0.25 & 0.41 \\ 0.55 & 0.41 \\ 0.8 & -0.81 \end{bmatrix} \begin{bmatrix} \eta_1 \\ \eta_2 \end{bmatrix}$$

Then the modal equilibrium condition is given by $\tilde{\mathbf{M}}\ddot{\boldsymbol{\eta}} + \tilde{\mathbf{C}}\dot{\boldsymbol{\eta}} + \tilde{\mathbf{K}}\boldsymbol{\eta} = \tilde{\mathbf{R}}$

where

$$\tilde{\mathbf{M}} = \begin{bmatrix} \tilde{M}_1 & 0 \\ 0 & \tilde{M}_2 \end{bmatrix} = \mathbf{\Phi}^T \mathbf{M} \mathbf{\Phi} = \begin{bmatrix} 0.25 & 0.41 \\ 0.55 & 0.41 \\ 0.8 & -0.81 \end{bmatrix}^T 3 \cdot 10^3 \begin{bmatrix} 2 & 0 & 0 \\ 0 & 2 & 0 \\ 0 & 0 & 1 \end{bmatrix} \begin{bmatrix} 0.25 & 0.41 \\ 0.55 & 0.41 \\ 0.8 & -0.81 \end{bmatrix}$$

$$\Rightarrow \qquad \tilde{\mathbf{M}} = \begin{bmatrix} \tilde{M}_1 & 0 \\ 0 & \tilde{M}_2 \end{bmatrix} \approx \begin{bmatrix} 4110 & 0 \\ 0 & 3986 \end{bmatrix}$$

$$\tilde{\mathbf{C}} = \begin{bmatrix} \tilde{C}_1 & 0 \\ 0 & \tilde{C}_2 \end{bmatrix} = \begin{bmatrix} 2\tilde{M}_1 \omega_1 \zeta_1 & 0 \\ 0 & 2\tilde{M}_2 \omega_2 \zeta_2 \end{bmatrix} = \begin{bmatrix} 411 & 0 \\ 0 & 1096 \end{bmatrix}$$

$$\tilde{\mathbf{K}} = \begin{bmatrix} \tilde{K}_1 & 0 \\ 0 & \tilde{K}_2 \end{bmatrix} = \begin{bmatrix} \omega_1^2 \tilde{M}_1 & 0 \\ 0 & \omega_2^2 \tilde{M}_2 \end{bmatrix} = 10^5 \begin{bmatrix} 0.26 & 0 \\ 0 & 1.2 \end{bmatrix}$$

and $\tilde{\mathbf{R}} = \mathbf{\Phi}^T \mathbf{R} = \begin{bmatrix} \tilde{R}_1 \\ \tilde{R}_2 \end{bmatrix} = \begin{bmatrix} \boldsymbol{\varphi}_1 & \boldsymbol{\varphi}_2 \end{bmatrix}^T \operatorname{Re}\left(\mathbf{a}_R \cdot e^{i\omega t}\right) = \operatorname{Re}\left(\begin{bmatrix} \boldsymbol{\varphi}_1^T \mathbf{a}_R \\ \boldsymbol{\varphi}_2^T \mathbf{a}_R \end{bmatrix} e^{i\omega t} \right)$

$$\Rightarrow \qquad \tilde{\mathbf{R}}(t) = \begin{bmatrix} 0.25 & 0.41 \\ 0.55 & 0.41 \\ 0.8 & -0.81 \end{bmatrix}^T \cdot \operatorname{Re}\left(\begin{bmatrix} 1800 \\ 1800 \\ 900 \end{bmatrix} e^{i\omega t} \right) \approx \operatorname{Re}\left(\begin{bmatrix} 2160 \\ 747 \end{bmatrix} e^{i\omega t} \right)$$

The solution to the modal equilibrium equation

$$\begin{bmatrix} \tilde{M}_1 & 0 \\ 0 & \tilde{M}_2 \end{bmatrix} \begin{bmatrix} \ddot{\eta}_1 \\ \ddot{\eta}_2 \end{bmatrix} + \begin{bmatrix} \tilde{C}_1 & 0 \\ 0 & \tilde{C}_2 \end{bmatrix} \begin{bmatrix} \dot{\eta}_1 \\ \dot{\eta}_2 \end{bmatrix} + \begin{bmatrix} \tilde{K}_1 & 0 \\ 0 & \tilde{K}_2 \end{bmatrix} \begin{bmatrix} \eta_1 \\ \eta_2 \end{bmatrix} = \begin{bmatrix} \tilde{R}_1 \\ \tilde{R}_2 \end{bmatrix}$$

is given by $\qquad \boldsymbol{\eta}(t) = \operatorname{Re}\left(\mathbf{a}_\eta \cdot e^{i\omega t}\right) \qquad$ *where* $\qquad \mathbf{a}_\eta = \begin{bmatrix} a_{\eta_1} & a_{\eta_2} \end{bmatrix}^T$,

which introduced into the equilibrium equation

$$\operatorname{Re}\left\{ \left(-\omega^2 \begin{bmatrix} \tilde{M}_1 & 0 \\ 0 & \tilde{M}_2 \end{bmatrix} + i\omega \begin{bmatrix} \tilde{C}_1 & 0 \\ 0 & \tilde{C}_2 \end{bmatrix} + \begin{bmatrix} \tilde{K}_1 & 0 \\ 0 & K_2 \end{bmatrix} \right) \begin{bmatrix} a_{\eta_1} \\ a_{\eta_2} \end{bmatrix} \right\} = \operatorname{Re} \begin{bmatrix} \tilde{R}_1 \\ \tilde{R}_2 \end{bmatrix}$$

and pre-multiplication by $\tilde{\mathbf{K}}^{-1}$ *renders* $\mathbf{a}_\eta = \begin{bmatrix} a_{\eta_1} \\ a_{\eta_2} \end{bmatrix} = \operatorname{Re}\left\{ \hat{\mathbf{H}}(\omega) \begin{bmatrix} \tilde{R}_1 / \tilde{K}_1 \\ \tilde{R}_2 / \tilde{K}_2 \end{bmatrix} \right\}$

where $\qquad \hat{\mathbf{H}}(\omega) = \begin{bmatrix} \hat{H}_1 & 0 \\ 0 & \hat{H}_2 \end{bmatrix} \qquad$ *and*

$$\hat{H}_1 = \frac{1}{1 - (\omega/\omega_1)^2 + 2i(\omega/\omega_1)\zeta_1} = \frac{1 - (\omega/\omega_1)^2 - i2\zeta_1(\omega/\omega_1)}{\left[1 - (\omega/\omega_1)^2\right]^2 + (2\zeta_1 \omega/\omega_1)^2}$$

$$\hat{H}_2 = \frac{1}{1-(\omega/\omega_2)^2 + 2i(\omega/\omega_2)\zeta_2} = \frac{1-(\omega/\omega_2)^2 - i2\zeta_2(\omega/\omega_2)}{\left[1-(\omega/\omega_2)^2\right]^2 + (2\zeta_2\omega/\omega_2)^2}$$

The trace of the absolute value of the frequency response matrix is illustrated in Fig. 5.3.b. Thus

$$\boldsymbol{\eta}(t) = \begin{bmatrix} \eta_1 \\ \eta_2 \end{bmatrix} = \mathrm{Re}\left(\mathbf{a}_\eta \cdot e^{i\omega t}\right) = \mathrm{Re}\left(\begin{bmatrix} \hat{H}_1(\omega)\cdot \tilde{R}_1/\tilde{K}_1 \\ \hat{H}_2(\omega)\cdot \tilde{R}_2/\tilde{K}_2 \end{bmatrix} \cdot e^{i\omega t}\right)$$

and since $\mathbf{r}(t) = \begin{bmatrix} r_1 \\ r_2 \\ r_3 \end{bmatrix} = \boldsymbol{\Phi}\boldsymbol{\eta}(t) = \begin{bmatrix} \varphi_{11} & \varphi_{21} \\ \varphi_{12} & \varphi_{22} \\ \varphi_{13} & \varphi_{23} \end{bmatrix} \begin{bmatrix} \eta_1 \\ \eta_2 \end{bmatrix}$ *then*

$$\mathbf{r}(t) = \mathrm{Re}\left(\begin{bmatrix} \phi_{11}\hat{H}_1(\omega)\tilde{R}_1/\tilde{K}_1 + \phi_{21}\hat{H}_2(\omega)\tilde{R}_2/\tilde{K}_2 \\ \phi_{12}\hat{H}_1(\omega)\tilde{R}_1/\tilde{K}_1 + \phi_{22}\hat{H}_2(\omega)\tilde{R}_2/\tilde{K}_2 \\ \phi_{13}\hat{H}_1(\omega)\tilde{R}_1/\tilde{K}_1 + \phi_{23}\hat{H}_2(\omega)\tilde{R}_2/\tilde{K}_2 \end{bmatrix}\right)$$

$$= \mathrm{Re}\left(\begin{bmatrix} \phi_{11}\left|\hat{H}_1(\omega)\right|e^{-i\beta_1}\, a_{\tilde{R}_1}e^{i\omega t}/\tilde{K}_1 + \phi_{21}\left|\hat{H}_2(\omega)\right|e^{-i\beta_2}\, a_{\tilde{R}_2}e^{i\omega t}/\tilde{K}_2 \\ \phi_{12}\left|\hat{H}_1(\omega)\right|e^{-i\beta_1}\, a_{\tilde{R}_1}e^{i\omega t}/\tilde{K}_1 + \phi_{22}\left|\hat{H}_2(\omega)\right|e^{-i\beta_2}\, a_{\tilde{R}_2}e^{i\omega t}/\tilde{K}_2 \\ \phi_{13}\left|\hat{H}_1(\omega)\right|e^{-i\beta_1}\, a_{\tilde{R}_1}e^{i\omega t}/\tilde{K}_1 + \phi_{23}\left|\hat{H}_2(\omega)\right|e^{-i\beta_2}\, a_{\tilde{R}_2}e^{i\omega t}/\tilde{K}_2 \end{bmatrix}\right)$$

where

$$\left|\hat{H}_1\right| = \left\{\left[1-(\omega/\omega_1)^2\right]^2 + (2\zeta_1\,\omega/\omega_1)^2\right\}^{-1/2} \qquad \tan\beta_1 = \frac{2\zeta_1\,\omega/\omega_1}{1-(\omega/\omega_1)^2}$$

and

$$\left|\hat{H}_2\right| = \left\{\left[1-(\omega/\omega_2)^2\right]^2 + (2\zeta_2\,\omega/\omega_2)^2\right\}^{-1/2} \qquad \tan\beta_2 = \frac{2\zeta_2\,\omega/\omega_2}{1-(\omega/\omega_2)^2}$$

Thus

$$r_1(t) = \phi_{11}\left|\hat{H}_1(\omega)\right|\left(a_{\tilde{R}_1}/\tilde{K}_1\right)\cos(\omega t-\beta_1) + \phi_{21}\left|\hat{H}_2(\omega)\right|\left(a_{\tilde{R}_2}/\tilde{K}_2\right)\cos(\omega t-\beta_2)$$

$$r_2(t) = \phi_{12}\left|\hat{H}_1(\omega)\right|\left(a_{\tilde{R}_1}/\tilde{K}_1\right)\cos(\omega t-\beta_1) + \phi_{22}\left|\hat{H}_2(\omega)\right|\left(a_{\tilde{R}_2}/\tilde{K}_2\right)\cos(\omega t-\beta_2)$$

$$r_3(t) = \phi_{13}\left|\hat{H}_1(\omega)\right|\left(a_{\tilde{R}_1}/\tilde{K}_1\right)\cos(\omega t-\beta_1) + \phi_{23}\left|\hat{H}_2(\omega)\right|\left(a_{\tilde{R}_2}/\tilde{K}_2\right)\cos(\omega t-\beta_2)$$

A plot of the response displacements are shown in Fig. 5.4.

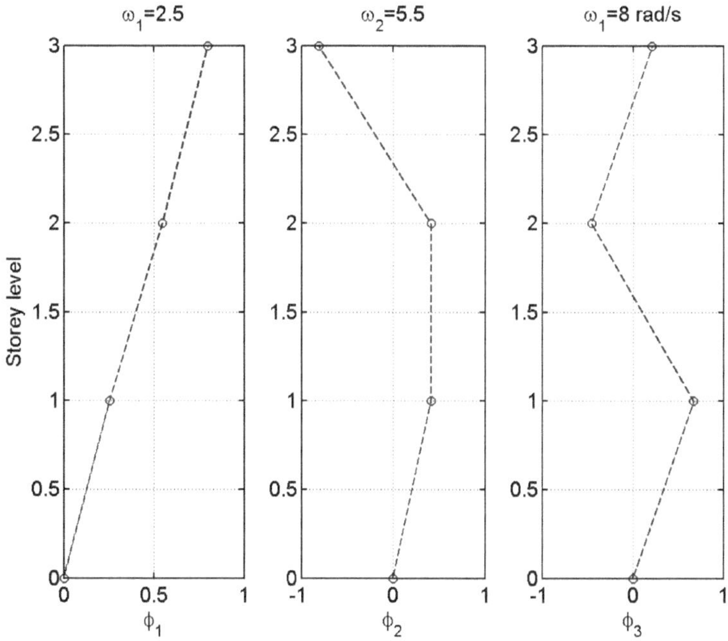

a) Mode shapes

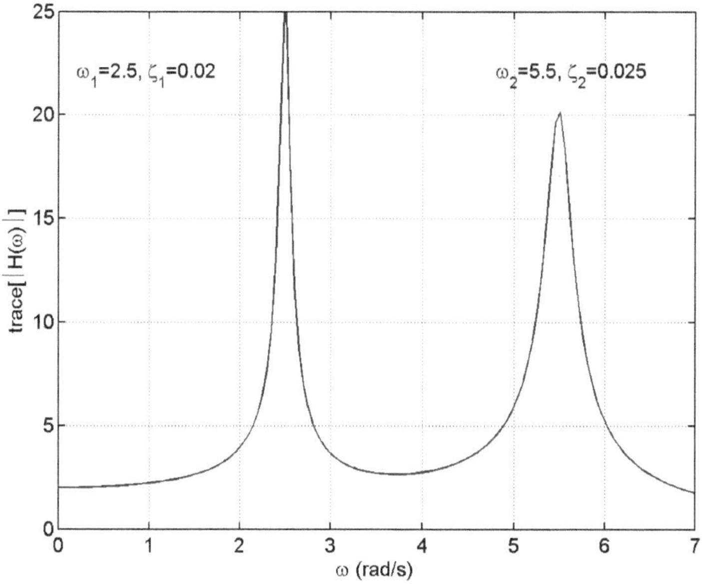

b) Trace of the absolute vale of the frequency response matrix

Fig. 5.3 Mode shapes and modal frequency response function

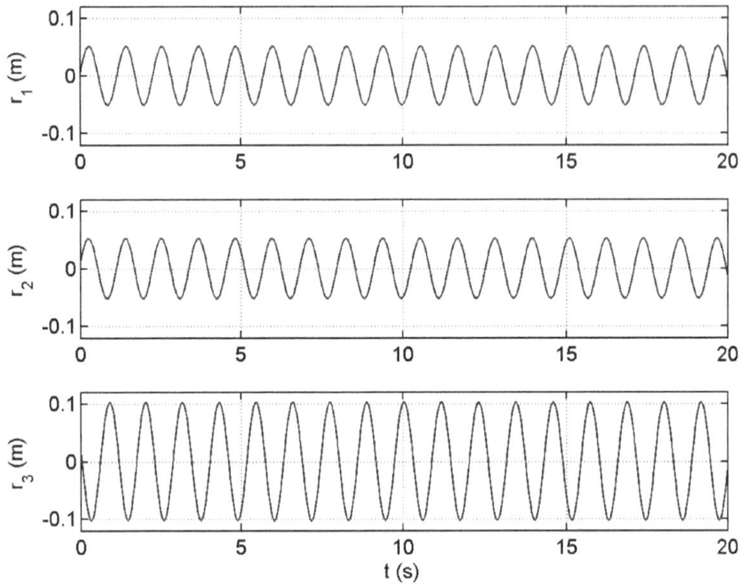

Fig. 5.4 Response displacements, $\omega = 5.5\ rad/s$, $\beta_1 = -0.023$ and $\beta_2 = \pi/2$

5.3 The Normal Mode Approach in a Continuous Format

The main idea behind the normal mode approach is to enable the designer to focus on the important aspects of the problem and to weed out superfluous computational efforts. For a continuous line-like system it is favourable to establish the necessary equilibrium conditions in accordance to the cross sectional three component load and displacement vectors illustrated in Fig. 5.5. Thus, the mode shape for such a system will contain three components ϕ_y , ϕ_z and ϕ_θ as shown in Fig. 5.6, and correspondingly, for a multi-mode approach it is convenient to organise the mode shape in cross sectional components as shown in Fig. 5.7, i.e.

$$\mathbf{r}(x,t) = \begin{bmatrix} r_y(x,t) \\ r_z(x,t) \\ r_\theta(x,t) \end{bmatrix} \approx \sum_{n=1}^{N_{\text{mod}}} \begin{bmatrix} \phi_y(x) \\ \phi_z(x) \\ \phi_\theta(x) \end{bmatrix}_n \eta_n(t) = \sum_{n=1}^{N_{\text{mod}}} \boldsymbol{\varphi}_n \cdot \eta_n(t) = \boldsymbol{\Phi}(x) \cdot \boldsymbol{\eta}(t) \quad (5.18)$$

where $$\boldsymbol{\Phi}(x) = \begin{bmatrix} \boldsymbol{\varphi}_1 & \boldsymbol{\varphi}_2 & \cdots & \boldsymbol{\varphi}_n & \cdots & \boldsymbol{\varphi}_{N_{\text{mod}}} \end{bmatrix} \quad (5.19)$$

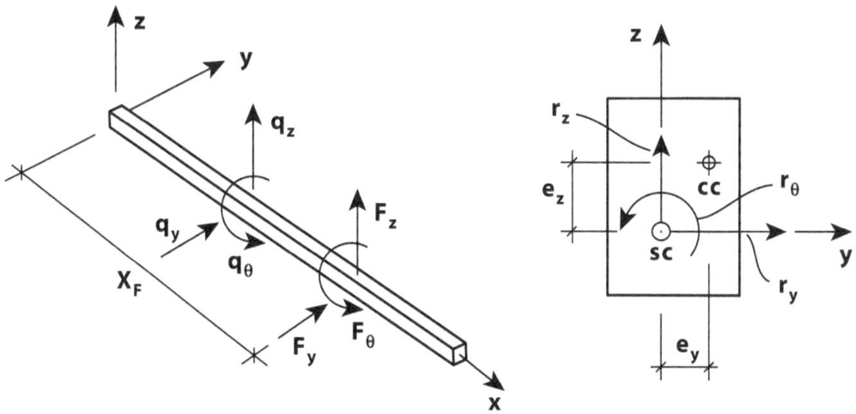

Fig. 5.5 Cross sectional loads and displacements

$$\boldsymbol{\varphi}_n(x) = \begin{bmatrix} \phi_y & \phi_z & \phi_\theta \end{bmatrix}^T \tag{5.20}$$

and
$$\boldsymbol{\eta}(t) = \begin{bmatrix} \eta_1 & \eta_2 & \cdots & \eta_n & \cdots & \eta_{N_{\text{mod}}} \end{bmatrix}^T \tag{5.21}$$

(Return to a discrete description may readily be performed on final equations.) Similarly (see Fig. 5.5), the distributed and concentrated load vectors are defined by

$$\mathbf{q}(x,t) = \begin{bmatrix} q_y & q_z & q_\theta \end{bmatrix}^T \quad \text{and} \quad \mathbf{F}_j(x_F,t) = \begin{bmatrix} F_y & F_z & F_\theta \end{bmatrix}_j^T \tag{5.22}$$

Let us for simplicity also assume that the system contains no concentrated mass set into dynamic motion. The dynamic equilibrium condition may then be obtained by applying the principle of virtual work as obtained in Eq. 1.119, i.e.

$$\begin{aligned} &\sum_{j=1}^{N_F} \left\{ \delta\mathbf{r}(x_F) \right\}^T \mathbf{F}_j + \int_L \delta\mathbf{r}^T \mathbf{q} dx = \int_L \delta\mathbf{r}^T \mathbf{m}_g \ddot{\mathbf{r}} dx + \int_L \delta\mathbf{r}^T \mathbf{c}_0 \dot{\mathbf{r}} dx \\ &+ \int_L \left[\delta r_y' \left(\bar{N} r_y' - \bar{M}_y r_\theta' \right) + \delta r_z' \left(\bar{N} r_z' - \bar{M}_z r_\theta' \right) + \delta r_y'' E I_z r_y'' + \delta r_z'' E I_y r_z'' \right. \\ &\left. \quad + \delta r_\theta' \left(G I_t r_\theta' + \bar{N} e_0^2 r_\theta' - \bar{M}_y r_y' - \bar{M}_z r_z' \right) \right] dx \end{aligned} \tag{5.23}$$

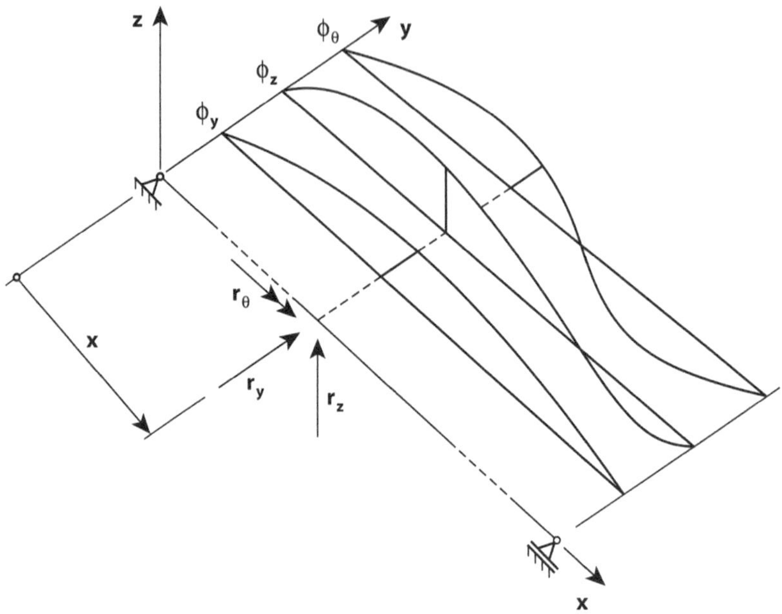

Fig. 5.6 Continuous systems

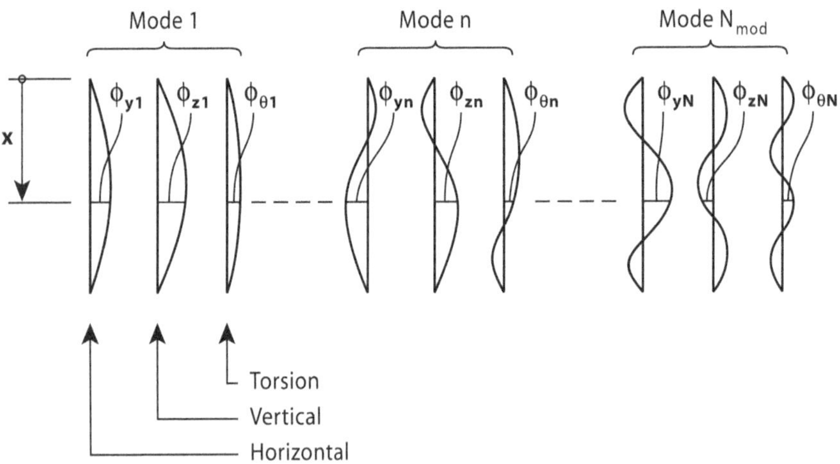

Fig. 5.7 Multiple modes for a continuous system

where $$\delta\mathbf{r}(x) = \begin{bmatrix} \delta r_y(x) \\ \delta r_z(x) \\ \delta r_\theta(x) \end{bmatrix} = \sum_{n=1}^{N_{\text{mod}}} \begin{bmatrix} \phi_y(x) \\ \phi_z(x) \\ \phi_\theta(x) \end{bmatrix}_n \delta\eta_n = \mathbf{\Phi}(x) \cdot \delta\boldsymbol{\eta} \tag{5.24}$$

and $$\delta\boldsymbol{\eta} = \begin{bmatrix} \delta\eta_1 & \cdots & \delta\eta_n & \cdots & \delta\eta_{N_{\text{mod}}} \end{bmatrix}^T \tag{5.25}$$

and where $$\mathbf{m}_g = \begin{bmatrix} m_y & 0 & -m_y e_z \\ & m_z & m_z e_y \\ Sym. & & m_\theta \end{bmatrix} \text{ and } \mathbf{c}_0 = \begin{bmatrix} c_y & 0 & 0 \\ & c_z & 0 \\ Sym & & c_\theta \end{bmatrix} \tag{5.26}$$

see Eqs. 1.106 and 1.107 (excluding contributions from the motion in the x direction). Defining

$$\left.\begin{aligned} &\begin{bmatrix} r_y' \\ r_z' \\ r_\theta' \end{bmatrix} = \hat{\mathbf{\Phi}}(x)\boldsymbol{\eta}(t) \text{ where } \hat{\mathbf{\Phi}} = \begin{bmatrix} \hat{\boldsymbol{\varphi}}_1 & \cdots & \hat{\boldsymbol{\varphi}}_n & \cdots & \hat{\boldsymbol{\varphi}}_{N_{\text{mod}}} \end{bmatrix} \text{ and } \hat{\boldsymbol{\varphi}}_n = \begin{bmatrix} \phi_y' \\ \phi_z' \\ \phi_\theta' \end{bmatrix} \\ &\begin{bmatrix} r_y'' \\ r_z'' \\ r_\theta' \end{bmatrix} = \hat{\hat{\mathbf{\Phi}}}(x)\boldsymbol{\eta}(t) \text{ where } \hat{\hat{\mathbf{\Phi}}} = \begin{bmatrix} \hat{\hat{\boldsymbol{\varphi}}}_1 & \cdots & \hat{\hat{\boldsymbol{\varphi}}}_n & \cdots & \hat{\hat{\boldsymbol{\varphi}}}_{N_{\text{mod}}} \end{bmatrix} \text{ and } \hat{\hat{\boldsymbol{\varphi}}}_n = \begin{bmatrix} \phi_y'' \\ \phi_z'' \\ \phi_\theta' \end{bmatrix} \end{aligned}\right\} \tag{5.27}$$

and, similarly

$$\begin{bmatrix} \delta r_y' \\ \delta r_z' \\ \delta r_\theta' \end{bmatrix} = \hat{\mathbf{\Phi}}(x)\delta\boldsymbol{\eta}(x) \quad \text{and} \quad \begin{bmatrix} \delta r_y'' \\ \delta r_z'' \\ \delta r_\theta' \end{bmatrix} = \hat{\hat{\mathbf{\Phi}}}(x)\delta\boldsymbol{\eta}(x) \tag{5.28}$$

then Eq. 5.23 may be written

$$\begin{aligned} &\sum_{j=1}^{N_F} \left\{\mathbf{\Phi}(x_F)\delta\boldsymbol{\eta}\right\}^T \mathbf{F}_j + \int_L (\mathbf{\Phi}\delta\boldsymbol{\eta})^T \mathbf{q}\,dx = \int_L (\mathbf{\Phi}\delta\boldsymbol{\eta})^T \mathbf{m}_g (\mathbf{\Phi}\ddot{\boldsymbol{\eta}})\,dx + \\ &\int_L (\mathbf{\Phi}\delta\boldsymbol{\eta})^T \mathbf{c}_0 (\mathbf{\Phi}\dot{\boldsymbol{\eta}})\,dx + \int_L \left(\hat{\hat{\mathbf{\Phi}}}\delta\boldsymbol{\eta}\right)^T \mathbf{k}_0 \left(\hat{\hat{\mathbf{\Phi}}}\boldsymbol{\eta}\right) dx + \int_L \left(\hat{\mathbf{\Phi}}\delta\boldsymbol{\eta}\right)^T \mathbf{k}_G \left(\hat{\mathbf{\Phi}}\boldsymbol{\eta}\right) dx \end{aligned} \tag{5.29}$$

where

$$\mathbf{k}_0 = \begin{bmatrix} EI_z & 0 & 0 \\ 0 & EI_y & 0 \\ 0 & 0 & GI_t \end{bmatrix} \quad \text{and} \quad \mathbf{k}_G = \begin{bmatrix} \bar{N} & 0 & -\bar{M}_y \\ 0 & \bar{N} & -\bar{M}_z \\ -\bar{M}_y & -\bar{M}_z & \bar{N}e_0^2 \end{bmatrix} \tag{5.30}$$

Performing the transposed operations

$$\begin{aligned} &\delta\boldsymbol{\eta}^T \sum_{j=1}^{N_F} \{\boldsymbol{\Phi}(x_F)\}^T \mathbf{F}_j + \delta\boldsymbol{\eta}^T \int_L \boldsymbol{\Phi}^T \mathbf{q} dx = \delta\boldsymbol{\eta}^T \int_L \boldsymbol{\Phi}^T \mathbf{m}_g \boldsymbol{\Phi} dx \cdot \ddot{\boldsymbol{\eta}} + \\ &\delta\boldsymbol{\eta}^T \int_L \boldsymbol{\Phi}^T \mathbf{c}_0 \boldsymbol{\Phi} dx \cdot \dot{\boldsymbol{\eta}} + \delta\boldsymbol{\eta}^T \int_L \hat{\hat{\boldsymbol{\Phi}}}^T \mathbf{k}_0 \hat{\hat{\boldsymbol{\Phi}}} dx \cdot \boldsymbol{\eta} + \delta\boldsymbol{\eta}^T \int_L \hat{\boldsymbol{\Phi}}^T \mathbf{k}_G \hat{\boldsymbol{\Phi}} dx \cdot \boldsymbol{\eta} \end{aligned} \tag{5.31}$$

it is seen that pre-multiplication by $\delta\boldsymbol{\eta}^T$ may be omitted, and thus, the dynamic equilibrium equation in modal coordinates is given by

$$\tilde{\mathbf{M}} \cdot \ddot{\boldsymbol{\eta}}(t) + \tilde{\mathbf{C}} \cdot \dot{\boldsymbol{\eta}}(t) + \tilde{\mathbf{K}} \cdot \boldsymbol{\eta}(t) = \tilde{\mathbf{R}}(t) \tag{5.32}$$

where

$$\left. \begin{aligned} \tilde{\mathbf{M}} &= \int_L \boldsymbol{\Phi}^T \mathbf{m}_g \boldsymbol{\Phi} dx \\ \tilde{\mathbf{C}} &= \int_L \boldsymbol{\Phi}^T \mathbf{c}_0 \boldsymbol{\Phi} dx \end{aligned} \right\} \tag{5.33}$$

$$\tilde{\mathbf{K}} = \tilde{\mathbf{K}}_0 + \tilde{\mathbf{K}}_G \qquad \text{where} \qquad \left\{ \begin{aligned} \tilde{\mathbf{K}}_0 &= \int_L \hat{\hat{\boldsymbol{\Phi}}}^T \mathbf{k}_0 \hat{\hat{\boldsymbol{\Phi}}} dx \\ \tilde{\mathbf{K}}_G &= \int_L \hat{\boldsymbol{\Phi}}^T \mathbf{k}_G \hat{\boldsymbol{\Phi}} dx \end{aligned} \right. \tag{5.34}$$

and

$$\tilde{\mathbf{R}} = \tilde{\mathbf{R}}_F + \tilde{\mathbf{R}}_q \qquad \text{where} \qquad \left\{ \begin{aligned} \tilde{\mathbf{R}}_F &= \sum_j^{N_F} \{\boldsymbol{\Phi}(x_F)\}^T \mathbf{F}_j \\ \tilde{\mathbf{R}}_q &= \int_L \boldsymbol{\Phi}^T \mathbf{q} dx \end{aligned} \right. \tag{5.35}$$

[If there are concentrated masses $\mathbf{M}_j = \begin{bmatrix} M_y & M_z & M_\theta \end{bmatrix}_j^T$ in motion at positions x_j, $j = 1, 2, 3, \ldots, N_M$ then the modal mass calculation in Eq. 5.33 is expanded into $\tilde{\mathbf{M}} = \int_L \boldsymbol{\Phi}^T \mathbf{m}_g \boldsymbol{\Phi} dx + \sum_j^{N_M} \left\{\boldsymbol{\Phi}(x_j)\right\}^T \mathbf{M}_j$.]

Elaboration 5.1: Mode Shape Orthogonality in a Continuous Format

The organisation of the mode shapes in Eqs. 5.19 and 5.20 is different (but usually more compact) than that which is obtained from a finite element solution of the system in original degrees of freedom, defined in Eqs. 5.6 and 5.7. However, the orthogonal properties of the system still apply. This is readily seen by considering the undamped and unloaded case to Eq. 5.32, in which case the relevant harmonic solution is given by

$$\boldsymbol{\eta}(t) = \mathbf{a} \cdot e^{i\omega t}$$

where $\mathbf{a} = \begin{bmatrix} a_1 & \cdots & a_n & \cdots & a_{N_{\text{mod}}} \end{bmatrix}^T$ is a vector containing the amplitude contributions associated with each of the modes that participates in the motion. Introduced into Eq. 5.32 (with $\tilde{\mathbf{C}}$ and $\tilde{\mathbf{R}}$ at zero) the following is obtained

$$\left(\tilde{\mathbf{K}} - \omega^2 \tilde{\mathbf{M}}\right)\mathbf{a} = \mathbf{0}$$

from which a non-trivial solution $\mathbf{a} \neq 0$ can only be obtained if $\tilde{\mathbf{K}}$ and $\tilde{\mathbf{M}}$ are the solution to the eigenvalue problem of the system, i.e. they are diagonal

$$\tilde{\mathbf{K}} = diag\left[\tilde{K}_n\right] \qquad \text{and} \qquad \tilde{\mathbf{M}} = diag\left[\tilde{M}_n\right]$$

while $\omega = \omega_n$ is the corresponding eigenvalue of the system.

The orthogonal properties may also be shown by return to the basic differential equations of unloaded and undamped equilibrium shown for the case of continuous systems in Chapter 1 (see Eqs. 1.29 and 1.40)

- y-direction equilibrium $\quad m_y \ddot{r}_y + EI_z r_y'''' = 0$
- z-direction equilibrium $\quad m_z \ddot{r}_z + EI_y r_z'''' = 0$
- torsion equilibrium $\quad m_\theta \ddot{r}_\theta + GI_t r_\theta'' = 0$

Let us for instance consider the differential equation of vertical equilibrium and introduce two equally valid harmonic eigenvalue solutions

$$r_z(x,t) = \begin{cases} \phi_{z_n} e^{i\omega_n t} \\ \phi_{z_m} e^{i\omega_m t} \end{cases} \quad \Rightarrow \quad \begin{aligned} EI_y \phi_{z_n}'''' &= \omega_n^2 m_z \phi_{z_n} \quad (1) \\ EI_y \phi_{z_m}'''' &= \omega_m^2 m_z \phi_{z_m} \quad (2) \end{aligned}$$

Let us then pre-multiply equations (1) and (2) by ϕ_{z_m} and ϕ_{z_n}, respectively, and integrate over the span of the system

$$\Rightarrow \begin{array}{ll} (1) & \int_L \phi_{z_m} EI_y \phi''''_{z_n} \mathrm{dx} = \omega_n^2 \int_L \phi_{z_m} m_z \phi_{z_n} dx \\ (2) & \int_L \phi_{z_n} EI_y \phi''''_{z_m} \mathrm{dx} = \omega_m^2 \int_L \phi_{z_n} m_z \phi_{z_m} dx \end{array}$$

Taking it for granted that the system has simply supported, fixed or free ends, and integrating the left hand side of these expressions by parts twice

$$\int_L \phi_{z_n} EI_y \phi''''_{z_m} \mathrm{dx} = \phi_{z_n} EI_y \phi'''_{z_m} \Big|_0^L - \int_L \phi'_{z_n} EI_y \phi'''_{z_m} \mathrm{dx}$$

$$= -\phi'_{z_n} EI_y \phi''_{z_m} \Big|_0^L + \int_L \phi''_{z_n} EI_y \phi''_{z_m} \mathrm{dx} = \int_L \phi''_{z_n} EI_y \phi''_{z_m} \mathrm{dx}$$

it is seen that

$$\begin{array}{ll} (1) & \int_L \phi''_{z_m} EI_y \phi''_{z_n} \mathrm{dx} = \omega_n^2 \int_L \phi_{z_m} m_z \phi_{z_n} dx \\ (2) & \int_L \phi''_{z_n} EI_y \phi''_{z_m} \mathrm{dx} = \omega_m^2 \int_L \phi_{z_n} m_z \phi_{z_m} dx \end{array}$$

Subtraction (1) – (2) will then render: $\left(\omega_n^2 - \omega_m^2\right) \int_L \phi_{z_n} m_z \phi_{z_m} dx = 0$

and since $\omega_n^2 - \omega_m^2 \neq 0$ for $n \neq m$ then $\int_L \phi_{z_n} m_z \phi_{z_m} dx = 0$

which proves the mass orthogonal properties of the mode shapes associated with vertical motion. For the other displacement components an equivalent proof may readily be developed.

Thus,

$$\tilde{\mathbf{M}} = \int_L \mathbf{\Phi}^T \mathbf{m}_g \mathbf{\Phi} dx = diag\left[\tilde{M}_n\right] \quad \text{where} \quad \tilde{M}_n = \int_L \boldsymbol{\varphi}_n^T \mathbf{m}_g \boldsymbol{\varphi}_n dx \tag{5.36}$$

$$\tilde{\mathbf{K}} = \tilde{\mathbf{K}}_0 + \tilde{\mathbf{K}}_G = diag\left[\tilde{K}_n\right] \quad \text{where} \quad \tilde{K}_n = \omega_n^2 \tilde{M}_n \tag{5.37}$$

and, similarly to that which was adopted in the discrete modal approach, we define the modal damping matrix by (see Eq. 5.17)

$$\tilde{\mathbf{C}} = \int_L \mathbf{\Phi}^T \mathbf{c} \mathbf{\Phi} dx = diag\left[\tilde{C}_n\right] \qquad \text{where} \qquad \tilde{C}_n = 2\tilde{M}_n \omega_n \zeta_n \tag{5.38}$$

where ζ_n is the modal damping ratio associated with mode shape $\boldsymbol{\varphi}_n$.

Case 1: Multi-mode Approach

In case of a multi-mode approach the equilibrium condition in modal degrees of freedom is given by Eq. 5.32 and the necessary calculation of modal mass may readily be obtained by combining Eqs. 5.26 and 5.36, rendering

$$\begin{aligned} \tilde{M}_n &= \int_L \boldsymbol{\varphi}_n^T \mathbf{m}_g \boldsymbol{\varphi}_n dx \\ &= \int_L \left[\phi_{y_n} m_y \left(\phi_{y_n} - 2e_z \phi_{\theta_n} \right) + \phi_{z_n} m_z \left(\phi_{z_n} + 2e_y \phi_{\theta_n} \right) + \phi_{\theta_n}^2 m_\theta \right] dx \end{aligned} \tag{5.39}$$

The modal load $\tilde{\mathbf{R}} = \tilde{\mathbf{R}}_F + \tilde{\mathbf{R}}_q$ is obtained from Eq. 5.35, rendering

$$\begin{aligned} &\tilde{\mathbf{R}}_F = \sum_j^{N_F} \left\{ \mathbf{\Phi}(x_F) \right\}^T \mathbf{F}_j = \begin{bmatrix} \tilde{R}_{F_1} & \cdots & \tilde{R}_{F_n} & \cdots & \tilde{R}_{F_{N_{\text{mod}}}} \end{bmatrix}^T \\ &\text{where} \quad \tilde{R}_{F_n} = \sum_{j=1}^{N_F} \left(\phi_{y_n} F_{y_j} + \phi_{z_n} F_{z_j} + \phi_{\theta_n} F_{\theta_j} \right) \end{aligned} \tag{5.40}$$

and

$$\begin{aligned} &\tilde{\mathbf{R}}_q = \int_L \mathbf{\Phi}^T \mathbf{q} dx = \begin{bmatrix} \tilde{R}_{q_1} & \cdots & \tilde{R}_{q_n} & \cdots & \tilde{R}_{q_{N_{\text{mod}}}} \end{bmatrix}^T \\ &\text{where} \quad \tilde{R}_{q_n} = \int_L \left(\phi_{y_n} q_y + \phi_{z_n} q_z + \phi_{\theta_n} q_\theta \right) dx \end{aligned} \tag{5.41}$$

Case 2: Single-Mode Approach

In case of a single-mode approach the equilibrium condition is reduced to

$$\tilde{M}_n \cdot \ddot{\eta}(t) + \tilde{C}_n \cdot \dot{\eta}(t) + \tilde{K}_n \cdot \eta(t) = \tilde{R}_n(t) \tag{5.42}$$

where $\tilde{R}_n = \tilde{R}_{F_n} + \tilde{R}_{q_n}$ and all other quantities are defined in Eqs. 5.37 – 5.41 above.

Elaboration 5.2: The Rayleigh Quotient in a Continuous Format

It has been shown above that $\tilde{K}_n = \omega_n^2 \tilde{M}_n$ i.e. that $\omega_n^2 = \tilde{K}_n / \tilde{M}_n$

Thus: $$\omega_n^2 = \frac{\tilde{K}_n}{\tilde{M}_n} = \frac{\tilde{K}_{0_n} + \tilde{K}_{G_n}}{\tilde{M}_n} = \frac{\int_L \hat{\hat{\boldsymbol{\varphi}}}_n^T \mathbf{k}_0 \hat{\hat{\boldsymbol{\varphi}}}_n dx + \int_L \hat{\boldsymbol{\varphi}}_n^T \mathbf{k}_G \hat{\boldsymbol{\varphi}}_n dx}{\int_L \boldsymbol{\varphi}_n^T \mathbf{m}_g \boldsymbol{\varphi}_n dx}$$

This is the so-called Rayleigh quotient, here expanded to apply to any line-like system with non-symmetric cross section and the possible presence of time invariant forces $\bar{N}$, $\bar{M}_y$ and $\bar{M}_z$. Using sound engineering judgement and setting

$$\boldsymbol{\varphi}_n = \begin{bmatrix} \phi_{y_n} & \phi_{z_n} & \phi_{\theta_n} \end{bmatrix}^T \approx \boldsymbol{\psi}_n = \begin{bmatrix} \psi_{y_n} & \psi_{z_n} & \psi_{\theta_n} \end{bmatrix}^T$$

then the Rayleigh quotient may be used to make approximate calculations of the eigenfrequency

$$\omega_n^2 \approx \left(\int_L \hat{\hat{\boldsymbol{\psi}}}_n^T \mathbf{k}_0 \hat{\hat{\boldsymbol{\psi}}}_n dx + \int_L \hat{\boldsymbol{\psi}}_n^T \mathbf{k}_G \hat{\boldsymbol{\psi}}_n dx \right) \Big/ \int_L \boldsymbol{\psi}_n^T \mathbf{m}_g \boldsymbol{\psi}_n dx$$

Introducing $\mathbf{k}_0$, $\mathbf{k}_G$ and $\mathbf{m}_g$ from Eqs. 5.26 and 5.30 then the following is obtained

$$\omega_n^2 \approx \frac{\begin{gathered}\int_L \Big[\psi_{y_n}''^2 EI_z + \psi_{z_n}''^2 EI_y + \psi_{\theta_n}'^2 GI_t + \bar{N}\left(\psi_{y_n}'^2 + \psi_{z_n}'^2 + \psi_{\theta_n}'^2 e_0^2 \right) \\ -2\psi_{\theta_n}' \left(\psi_{y_n}' \bar{M}_y + \psi_{z_n}' \bar{M}_z \right) \Big] dx\end{gathered}}{\int_L \left[\psi_{y_n} m_y \left(\psi_{y_n} - 2e_z \psi_{\theta_n} \right) + \psi_{z_n} m_z \left(\psi_{z_n} + 2e_y \psi_{\theta_n} \right) + \psi_{\theta_n}^2 m_\theta \right] dx}$$

It should be noted that for simplicity warping torsion effects have not been included above. It may readily be inferred from Eqs. 1.39 – 1.42 that the stiffness contribution from this effect is given by $\int_L \psi_\theta'''^2 EI_w dx$. It may simply be added to the numerator above.

Example 5.2

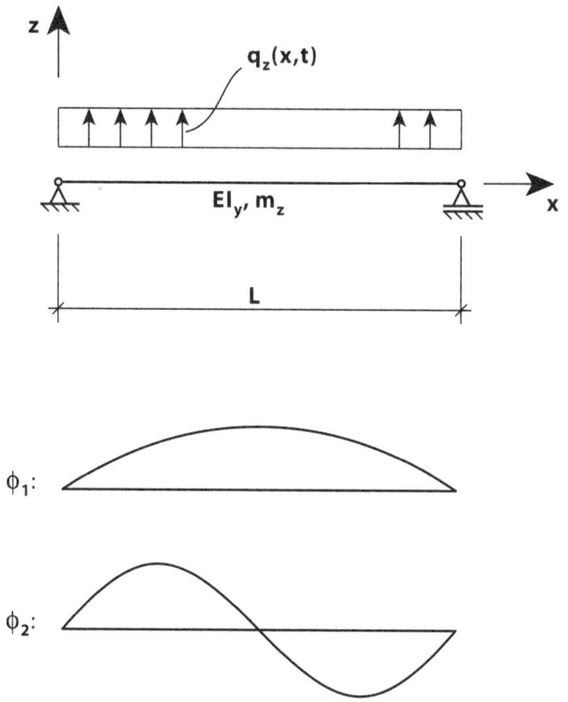

Fig. 5.8 Simple symmetric beam subject to distributed load $q_z(x,t)$

Let us consider the beam shown in Fig. 5.8. It is subject to a distributed and dynamic load

$$q_z(x,t) = \mathrm{Re}\left(q_0 e^{i\psi t}\right) = q_0 \cdot \mathrm{Re}\left(e^{i\psi t}\right) = q_0 \cdot \cos(\omega t)$$

Its cross section is for simplicity symmetric about y *and* z *axes, such that its shear centre coincides with the centroid. It is taken for granted that cross sectional properties are constant along its span. We know from Chapter 1.2 (see Example 1.6) that the two first mode shapes of such beams are given by*

$$\left.\begin{aligned} \varphi_1 &= \sin(\pi\hat{x}) \\ \varphi_2 &= \sin(2\pi\hat{x}) \end{aligned}\right\} \qquad \textit{where} \qquad \hat{x} = x/L$$

Using the Rayleigh- quotient (see Elaboration 5.2 above) then the following eigenfrequencies are obtained:

$$\omega_1^2 = \frac{\int_0^L EI_y (\varphi_1'')^2 dx}{\int_0^L m_z \varphi_1^2 dx} = \frac{\int_0^L EI_y \left[-(\pi/L)^2 \sin(\pi x/L)\right]^2 dx}{\int_0^L m_z \left[\sin(\pi x/L)\right]^2 dx} = \pi^4 \frac{EI_y}{m_z L^4}$$

$$\omega_2^2 = \frac{\int_0^L EI_y (\varphi_2'')^2 dx}{\int_0^L m_z \varphi_2^2 dx} = \frac{\int_0^L EI_y \left[-(2\pi/L)^2 \sin(2\pi x/L)\right]^2 dx}{\int_0^L m_z \left[\sin(2\pi x/L)\right]^2 dx} = (2\pi)^4 \frac{EI_y}{m_z L^4}$$

The system has the following modal properties:

$$\tilde{M}_1 = \int_0^L m_z \varphi_1^2 dx = m_z \int_0^L \sin\left(\frac{\pi x}{L}\right) dx = \frac{m_z L}{2}$$

and

$$\tilde{M}_2 = \int_0^L m_z \varphi_2^2 dx = m_z \int_0^L \sin\left(\frac{2\pi x}{L}\right) dx = \frac{m_z L}{2}$$

while

$$\tilde{K}_1 = \omega_1^2 \tilde{M}_1 \qquad \tilde{C}_1 = 2\tilde{M}_1 \omega_1 \zeta_1$$

and

$$\tilde{K}_2 = \omega_2^2 \tilde{M}_2 \qquad \tilde{C}_2 = 2\tilde{M}_2 \omega_2 \zeta_2$$

Similarly, the modal loads associated with mode shapes φ_1 *and* φ_2 *are given by*

$$\tilde{R}_1 = \int_0^L \varphi_1 q_z dx = \mathrm{Re}\left[q_0 e^{i\psi t} \int_0^L \sin\left(\frac{\pi x}{L}\right) dx\right] = \frac{2Lq_0}{\pi} \cdot \mathrm{Re}\left(e^{i\omega t}\right)$$

$$\tilde{R}_2 = \int_0^L \varphi_2 q_z dx = \mathrm{Re}\left[q_0 e^{i\psi t} \int_0^L \sin\left(\frac{2\pi x}{L}\right) dx\right] = 0$$

Thus, the modal equilibrium condition is given by $\tilde{\mathbf{M}} \cdot \ddot{\boldsymbol{\eta}}(t) + \tilde{\mathbf{C}} \cdot \dot{\boldsymbol{\eta}}(t) + \tilde{\mathbf{K}} \cdot \boldsymbol{\eta}(t) = \tilde{\mathbf{R}}(t)$ *where*

$$\tilde{\mathbf{M}} = \begin{bmatrix} \tilde{M}_1 & 0 \\ 0 & \tilde{M}_2 \end{bmatrix} = \frac{mL}{2} \begin{bmatrix} 1 & 0 \\ 0 & 1 \end{bmatrix}$$

$$\tilde{\mathbf{C}} = \begin{bmatrix} \tilde{C}_1 & 0 \\ 0 & \tilde{C}_2 \end{bmatrix} = \begin{bmatrix} 2\tilde{M}_1 \omega_1 \zeta_1 & 0 \\ 0 & 2\tilde{M}_2 \omega_2 \zeta_2 \end{bmatrix} = \frac{mL}{2} \begin{bmatrix} 2\omega_1 \zeta_1 & 0 \\ 0 & 2\omega_2 \zeta_2 \end{bmatrix}$$

$$\tilde{\mathbf{K}} = \begin{bmatrix} \tilde{K}_1 & 0 \\ 0 & \tilde{K}_2 \end{bmatrix} = \begin{bmatrix} \omega_1^2 \tilde{M}_1 & 0 \\ 0 & \omega_2^2 \tilde{M}_2 \end{bmatrix} = \frac{mL}{2} \begin{bmatrix} \omega_1^2 & 0 \\ 0 & \omega_2^2 \end{bmatrix}$$

and $$\tilde{\mathbf{R}} = \begin{bmatrix} \tilde{R}_1 \\ \tilde{R}_2 \end{bmatrix} = \mathrm{Re}\left(\frac{2Lq_0}{\pi} \begin{bmatrix} 1 \\ 0 \end{bmatrix} \cdot e^{i\omega t} \right).$$

I.e.: $$\begin{bmatrix} 1 & 0 \\ 0 & 1 \end{bmatrix} \begin{bmatrix} \ddot{\eta}_1 \\ \ddot{\eta}_2 \end{bmatrix} + \begin{bmatrix} 2\omega_1\zeta_1 & 0 \\ 0 & 2\omega_2\zeta_2 \end{bmatrix} \begin{bmatrix} \dot{\eta}_1 \\ \dot{\eta}_2 \end{bmatrix} + \begin{bmatrix} \omega_1^2 & 0 \\ 0 & \omega_2^2 \end{bmatrix} \begin{bmatrix} \eta_1 \\ \eta_2 \end{bmatrix} = \frac{4q_0}{\pi m} \mathrm{Re}\left(\begin{bmatrix} 1 \\ 0 \end{bmatrix} e^{i\omega t} \right)$$

The steady state solution to this equation is given by: $\boldsymbol{\eta}(t) = \begin{bmatrix} \eta_1 \\ \eta_2 \end{bmatrix} = \mathrm{Re}\left(\begin{bmatrix} a_1 \\ a_2 \end{bmatrix} \cdot e^{i\omega t} \right)$.

Thus

$$\mathrm{Re}\left\{ -\begin{bmatrix} 1 & 0 \\ 0 & 1 \end{bmatrix} \omega^2 + \begin{bmatrix} 2\omega_1\zeta_1 & 0 \\ 0 & 2\omega_2\zeta_2 \end{bmatrix} i\omega + \begin{bmatrix} \omega_1^2 & 0 \\ 0 & \omega_2^2 \end{bmatrix} \right\} \begin{bmatrix} a_1 \\ a_2 \end{bmatrix} = \mathrm{Re}\left\{ \frac{4q_0}{\pi m} \begin{bmatrix} 1 \\ 0 \end{bmatrix} e^{i\omega t} \right\}$$

$$\Rightarrow \qquad \mathrm{Re}\left\{ \hat{\mathbf{E}}(\omega) \right\} \begin{bmatrix} a_1 \\ a_2 \end{bmatrix} = \mathrm{Re}\left\{ \frac{4q_0}{\pi m \omega_1^2} \begin{bmatrix} 1 \\ 0 \end{bmatrix} e^{i\omega t} \right\}$$

where $$\hat{\mathbf{E}}(\omega) = \begin{bmatrix} \left\{ 1 - (\omega/\omega_1)^2 + 2i(\omega/\omega_1)\zeta_1 \right\} & 0 \\ 0 & \left\{ 1 - (\omega/\omega_2)^2 + 2i(\omega/\omega_2)\zeta_2 \right\} \end{bmatrix}$$

is the non-dimensional impedance matrix of the system. The unknown amplitude vector $\mathbf{a} = [a_1 \quad a_2]^T$ *is then given by*

$$\mathbf{a} = \begin{bmatrix} a_1 \\ a_2 \end{bmatrix} = \mathrm{Re}\left\{ \frac{4q_0}{\pi m \omega_1^2} \cdot \hat{\mathbf{H}}(\omega) \begin{bmatrix} 1 \\ 0 \end{bmatrix} e^{i\omega t} \right\} \quad \textit{where}$$

$$\hat{\mathbf{H}}(\omega) = \hat{\mathbf{E}}^{-1} = \begin{bmatrix} \hat{H}_{11} & 0 \\ 0 & \hat{H}_{22} \end{bmatrix}$$

where

$$\hat{H}_{11} = \left[1 - \left(\frac{\omega}{\omega_1} \right)^2 + 2i \left(\frac{\omega}{\omega_1} \right) \zeta_1 \right]^{-1} \quad \textit{and} \quad \hat{H}_{22} = \left[1 - \left(\frac{\omega}{\omega_2} \right)^2 + 2i \left(\frac{\omega}{\omega_2} \right) \zeta_2 \right]^{-1}$$

$\hat{\mathbf{H}}(\omega)$ *is what we call the modal non-dimensional frequency response matrix. Thus,*

$$r_z(x,t)=\mathbf{\Phi}(x)\cdot\boldsymbol{\eta}(t)=[\varphi_1 \quad \varphi_2]\begin{bmatrix}\eta_1\\ \eta_2\end{bmatrix}=[\varphi_1 \quad \varphi_2]\cdot\mathrm{Re}\left(\begin{bmatrix}a_1\\ a_2\end{bmatrix}e^{i\omega t}\right)$$

$$=[\varphi_1 \quad \varphi_2]\cdot\mathrm{Re}\left\{\frac{4q_0}{\pi m\omega_1^2}\cdot\hat{\mathbf{H}}(\omega)\cdot\begin{bmatrix}1\\ 0\end{bmatrix}e^{i\omega t}\right\}=\frac{4q_0\varphi_1}{\pi m\omega_1^2}\cdot\mathrm{Re}\left\{\hat{H}_1e^{i\omega t}\right\}$$

Since $$\hat{H}_{11}=\frac{1}{1-(\omega/\omega_1)^2+2i(\omega/\omega_1)\zeta_1}=\frac{1-(\omega/\omega_1)^2-i2\zeta_1(\omega/\omega_1)}{\left[1-(\omega/\omega_1)^2\right]^2+(2\zeta_1\omega/\omega_1)^2}$$

then

$$\hat{H}_{11}=\left|\hat{H}_{11}\right|\cdot e^{-i\phi} \text{ where } \begin{cases}\left|\hat{H}_{11}\right|=\sqrt{\hat{H}_{11}^*\hat{H}_{11}}=\dfrac{1}{\sqrt{\left[1-(\omega/\omega_1)^2\right]^2+(2\zeta_1\,\omega/\omega_1)^2}}\\[2ex] \tan\phi=\dfrac{\mathrm{Im}\left(\hat{H}_{11}\right)}{\mathrm{Re}\left(\hat{H}_{11}\right)}=\dfrac{2\zeta_1\,\omega/\omega_1}{1-(\omega/\omega_1)^2}\end{cases}$$

Thus

$$r_z(x,t)=\frac{4q_0\varphi_1}{\pi m\omega_1^2}\cdot\mathrm{Re}\left\{\left|\hat{H}_1\right|e^{-i\phi}e^{i\omega t}\right\}=\frac{4q_0\varphi_1}{\pi m\omega_1^2}\cdot\left|\hat{H}_1\right|\cos(\omega t-\phi)$$

$$=\frac{2}{\pi}\cdot\frac{q_0L}{\tilde{K}_1}\cdot\frac{\sin(\pi x/L)\cdot\cos(\omega t-\phi)}{\sqrt{\left[1-(\omega/\omega_1)^2\right]^2+(2\zeta_1\,\omega/\omega_1)^2}}$$

As can be seen, the 2nd mode shape is not excited at all. The reason for this is that the load is evenly distributed and constant along its span while the spanwise integration of the 2nd mode is zero. In general, a load that is constant along the span will only excite symmetric modes.

Example 5.3

Let us consider the case of an identical beam to that which was investigated in Example 5.2, except in this case it is subject to a concentrated dynamic load $F(t)=\mathrm{Re}\left(F_0\cdot e^{i\omega t}\right)$ at $x_F=L/3$ as shown in Fig. 5.9. Still $\varphi_1=\sin(\pi\hat{x})$ and $\varphi_2=\sin(2\pi\hat{x})$, $\hat{x}=x/L$ and, as shown in Example 5.2, then $\omega_1^2=\pi^4EI_y/m_zL^4$ and $\omega_2^2=(2\pi)^4EI_y/m_zL^4$ while $\tilde{M}_1=\tilde{M}_2=m_zL/2$, $\tilde{K}_1=\omega_1^2\tilde{M}_1$, $\tilde{K}_2=\omega_2^2\tilde{M}_2$, $\tilde{C}_1=2\tilde{M}_1\omega_1\zeta_1$ and $\tilde{C}_2=2\tilde{M}_2\omega_2\zeta_2$.

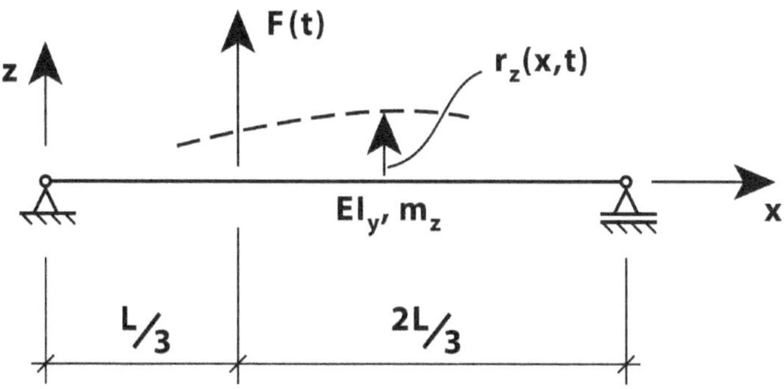

Fig. 5.9 Simple beam with concentrated dynamic load $F(t) = \mathrm{Re}\left(F_0 \cdot e^{i\omega t}\right)$

However, in this case

$$\tilde{R}_1 = \varphi_1(x = L/3) \cdot F(t) = \sin(\pi/3) \cdot \mathrm{Re}\left[F_0 e^{i\omega t}\right] = \frac{\sqrt{3}}{2} \cdot \mathrm{Re}\left[F_0 e^{i\omega t}\right]$$

and $\tilde{R}_2 = \varphi_2(x = L/3) \cdot F(t) = \sin(2\pi/3) \cdot \mathrm{Re}\left[F_0 e^{i\omega t}\right] = \dfrac{\sqrt{3}}{2} \cdot \mathrm{Re}\left[F_0 e^{i\omega t}\right]$

and thus the modal equilibrium condition $\tilde{\mathbf{M}} \cdot \ddot{\boldsymbol{\eta}}(t) + \tilde{\mathbf{C}} \cdot \dot{\boldsymbol{\eta}}(t) + \tilde{\mathbf{K}} \cdot \boldsymbol{\eta}(t) = \tilde{\mathbf{R}}(t)$ *with* $\boldsymbol{\eta} = \mathrm{Re}\left(\begin{bmatrix} a_1 & a_2 \end{bmatrix}^T e^{i\omega t}\right)$ *given by*

$$\mathrm{Re}\left\{\left(-\begin{bmatrix} 1 & 0 \\ 0 & 1 \end{bmatrix}\omega^2 + \begin{bmatrix} 2\omega_1\zeta_1 & 0 \\ 0 & 2\omega_2\zeta_2 \end{bmatrix} i\omega + \begin{bmatrix} \omega_1^2 & 0 \\ 0 & \omega_2^2 \end{bmatrix}\right)\begin{bmatrix} a_1 \\ a_2 \end{bmatrix} = \frac{\sqrt{3}F_0}{mL}\begin{bmatrix} 1 \\ 1 \end{bmatrix} e^{i\omega t}\right\}$$

Thus

$$\boldsymbol{\eta}(t) = \frac{\sqrt{3}F_0}{mL} \mathrm{Re}\left\{\hat{\mathbf{H}}(\omega)\begin{bmatrix} 1/\omega_1^2 \\ 1/\omega_2^2 \end{bmatrix} e^{i\omega t}\right\}$$

where

$$\hat{\mathbf{H}}(\omega) = \begin{bmatrix} \hat{H}_{11} & 0 \\ 0 & \hat{H}_{22} \end{bmatrix} \quad \text{and} \quad \begin{cases} \hat{H}_{11} = \left[1 - (\omega/\omega_1)^2 + 2i(\omega/\omega_1)\zeta_1\right]^{-1} \\ \hat{H}_{22} = \left[1 - (\omega/\omega_2)^2 + 2i(\omega/\omega_2)\zeta_2\right]^{-1} \end{cases}$$

Thus, (see Example 5.2 above)

$$\boldsymbol{\eta}(t) = \begin{bmatrix} \eta_1 \\ \eta_2 \end{bmatrix} = \frac{\sqrt{3}}{2} F_0 \begin{bmatrix} \tilde{K}_1^{-1} \cdot \left|\hat{H}_{11}(\omega)\right| \cdot \cos(\omega t - \phi_1) \\ \tilde{K}_2^{-1} \cdot \left|\hat{H}_{22}(\omega)\right| \cdot \cos(\omega t - \phi_2) \end{bmatrix} \qquad \textit{where}$$

$$\left|\hat{H}_{11}\right| = \left\{\left[1-(\omega/\omega_1)^2\right]^2 + (2\zeta_1\,\omega/\omega_1)^2\right\}^{-1/2}$$
$$\left|\hat{H}_{22}\right| = \left\{\left[1-(\omega/\omega_2)^2\right]^2 + (2\zeta_2\,\omega/\omega_2)^2\right\}^{-1/2}$$

and

$$\begin{cases} \tan\phi_1 = \dfrac{2\zeta_1\,\omega/\omega_1}{1-(\omega/\omega_1)^2} \\ \tan\phi_2 = \dfrac{2\zeta_2\,\omega/\omega_2}{1-(\omega/\omega_2)^2} \end{cases}$$

The dynamic response at an arbitrary point x *is then given by*

$$r_z(x,t) = \mathbf{\Phi}(x)\cdot\boldsymbol{\eta}(t) = \begin{bmatrix}\varphi_1(x) & \varphi_2(x)\end{bmatrix}\begin{bmatrix}\eta_1(t)\\ \eta_2(t)\end{bmatrix} = \varphi_1(x)\cdot\eta_1(t) + \varphi_2(x)\cdot\eta_2(t) =$$

$$\frac{\sqrt{3}}{2}\left\{\frac{F_0}{\tilde{K}_1}\left|\hat{H}_{11}(\omega)\right|\sin\left(\frac{\pi x}{L}\right)\cos(\omega t-\phi_1) + \frac{F_0}{\tilde{K}_2}\left|\hat{H}_{22}(\omega)\right|\sin\left(\frac{2\pi x}{L}\right)\cos(\omega t-\phi_2)\right\}$$

Elaboration 5.3: Determination of Cross Sectional Forces

In the case of a discrete normal mode approach the results of the calculations are the displacements in original finite element coordinates $\mathbf{r}(t) = \mathbf{\Phi}\boldsymbol{\eta}(t)$. Thus, the element end forces may be calculated the way it would be done in a regular finite element approach, i.e. $\mathbf{F}_n(t) = \mathbf{m}\ddot{\mathbf{d}}_n + \mathbf{c}\dot{\mathbf{d}}_n + \mathbf{k}\mathbf{d}_n$ where $\mathbf{d}_n(t) = \mathbf{A}_n\mathbf{r}(t)$, see Chapter 4.

In the case of a continuous normal mode approach

$$\mathbf{r}(x,t) = \begin{bmatrix} r_y \\ r_z \\ r_\theta \end{bmatrix} = \mathbf{\Phi}(x)\boldsymbol{\eta}(t) \quad \text{where} \quad \begin{cases} \mathbf{\Phi} = \begin{bmatrix}\boldsymbol{\varphi}_1 & \cdots & \boldsymbol{\varphi}_n & \cdots & \boldsymbol{\varphi}_{N_{\text{mod}}}\end{bmatrix} \\ \boldsymbol{\varphi}_n = \begin{bmatrix}\phi_y & \phi_z & \phi_\theta\end{bmatrix}^T \end{cases}$$

Since (see Chapter 1) $M_z = EI_z r_y''$, $M_y = -EI_y r_z''$ and $M_\theta = GI_t r_\theta'$ it is seen that

$$\begin{bmatrix} M_z(x,t) \\ -M_y(x,t) \\ M_\theta(x,t) \end{bmatrix} = \begin{bmatrix} EI_z(x) & 0 & 0 \\ 0 & EI_y(x) & 0 \\ 0 & 0 & GI_t(x) \end{bmatrix}\begin{bmatrix} r_y''(x,t) \\ r_z''(x,t) \\ r_\theta'(x,t) \end{bmatrix} = \mathbf{k}_0(x)\cdot\hat{\hat{\mathbf{\Phi}}}(x)\cdot\boldsymbol{\eta}(t)$$

where
$$\begin{cases} \hat{\hat{\mathbf{\Phi}}}(x) = \begin{bmatrix}\hat{\hat{\boldsymbol{\varphi}}}_1 & \cdots & \hat{\hat{\boldsymbol{\varphi}}}_n & \cdots & \hat{\hat{\boldsymbol{\varphi}}}_{N_{\text{mod}}}\end{bmatrix} \\ \hat{\hat{\boldsymbol{\varphi}}}_n(x) = \begin{bmatrix}\phi_y'' & \phi_z'' & \phi_\theta'\end{bmatrix}^T \end{cases} \quad \text{and } \mathbf{k}_0(x) = \begin{bmatrix} EI_z & 0 & 0 \\ 0 & -EI_y & 0 \\ 0 & 0 & GI_t \end{bmatrix}$$

Chapter 6
Frequency and Time Domain Response Calculations

6.1 Introduction

The relevant equilibrium equations that are necessary for the pursuit of a solution to the problem of dynamic load effects have been developed above, in Chapters 4 with respect to the original degrees of freedom (see Eq. 4.39) and in Chapter 5 with respect to the modal degrees of freedom (see Eq. 5.13 for a discrete system description and Eq. 5.32 for a continuous system description). In this chapter we shall present possible solution strategies. Basically, one is free to choose whichever approach is deemed most suitable solution strategy, e.g. with respect to efficiency or accuracy. In any case, there are three alternatives:

- a solution in time domain where the load and corresponding response development is pursued stepwise for a sufficiently long period of time, in which case time series of the structural response is obtained,
- an incremental stepwise state-space solution in time domain based on the Duhamel integral and applying the fluctuating load as a consecutive sequence of short impulses, or
- a solution where a Fourier transform is applied throughout the equilibrium equation and the problem is transferred into a frequency domain description, in which case a frequency domain spectral representation of the response is obtained.

Below, the time domain approach is presented in Chapter 6.3, while the frequency domain approach is presented in Chapters 6.4 and 6.5. The state-space solution and the Duhamel integral are presented in Chapter 6.6. For the sake of completeness the time invariant mean (static) as well as the quasi-static solutions are presented in Chapter 6.2. The quasi-static solution is only applicable if the there are no significant dynamic effects in the system, i.e. if the effects of

E.N. Strømmen, *Structural Dynamics*, Springer Series in Solid and Structural Mechanics 2,
DOI: 10.1007/978-3-319-01802-7_6, © Springer International Publishing Switzerland 2014

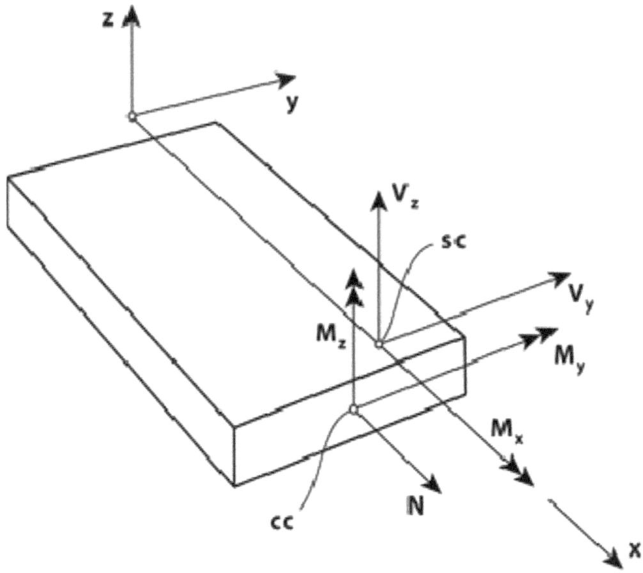

Fig. 6.1 Cross sectional force components

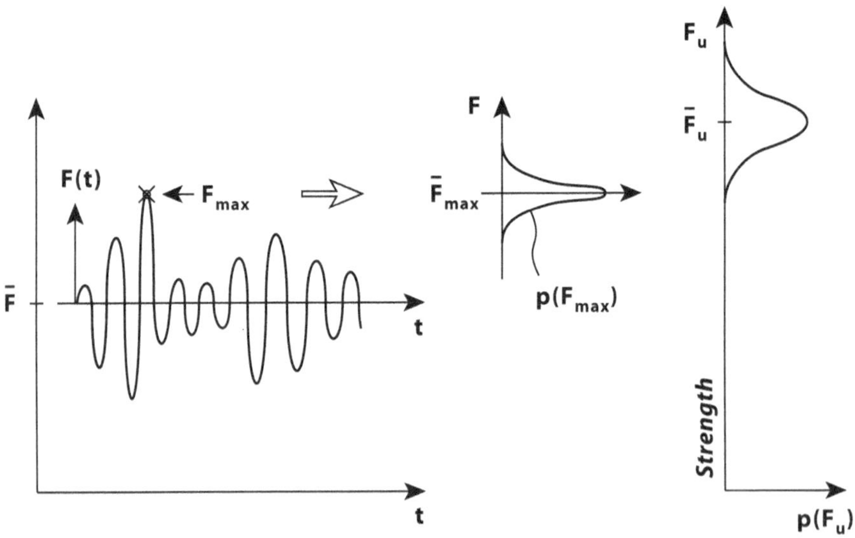

Fig. 6.2 Necessary cross sectional strength considerations

structural velocity and accelerations are negligible. This may be assumed to occur if the load is slowly fluctuating at low frequencies or if the lowest eigenfrequency of the system is beyond say 5 Hz. (However, it must not be inferred from this that beyond this limit there are no fluctuating load effects, as the limit is only an indication of where inertia and damping effects are negligible. Beyond this limit there may still be considerable quasi-static fluctuating stresses.) Otherwise, the solution strategy must be based on the methods shown in Chapters 6.3 – 6.5, whichever is most suitable. In any case, it is taken for granted that the total displacement response may be obtained as a sum of the time invariant solution $\bar{\mathbf{r}}$ and a fluctuating part $\mathbf{r}(t)$, i.e. $\mathbf{r}_{tot}(t)=\bar{\mathbf{r}}+\mathbf{r}(t)$. Having determined the fluctuating total dynamic displacement response it is in all design cases necessary to determine the corresponding cross sectional design forces $\mathbf{F}_{tot}=\bar{\mathbf{F}}+\mathbf{F}(t)$ for all elements n in order to enable a safety assessment of the system. In a time domain solution extreme values may be taken directly from time series as illustrated in Fig. 6.2. In a frequency domain solution where it is the spectral densities of the response that has been determined, $\mathbf{S}_{rr}(\omega)$, a similar approach may be adopted together with a time domain simulation of the relevant response quantities (see Appendix B), or from a probabilistic approach where the extreme values are obtained from the standard deviation of the dynamic response multiplied by a statistically appropriate peak factor (see Appendix A and Elaborations 6.3 and 6.4).

6.2 The Time Invariant and Quasi-static Solutions

It is generally recommended that the time invariant and quasi-static solutions are pursued in original degrees of freedom. The reason for this is that a solution in modal degrees of freedom is associated with displacement functions obtained from the eigenvalue problem, i.e. with the mode shapes of the system, and these functions may render a solution whose derivatives deviates considerably from the more relevant static solution.

The Time Invariant Solution

The time invariant mean (static) solution in original degrees of freedom may be obtained from Eq. 4.38, i.e.:

$$\bar{\mathbf{r}}=\mathbf{K}^{-1}\cdot\bar{\mathbf{R}} \tag{6.1}$$

The corresponding cross sectional forces on an arbitrary element n may be obtained from Eq. 4.26, i.e.

$$\bar{\mathbf{F}}_n = \mathbf{k}_n \bar{\mathbf{d}}_n = \mathbf{k}_n \mathbf{A}_n \bar{\mathbf{r}} \tag{6.2}$$

where $\mathbf{k}_n = \left(\mathbf{k}_0 + \mathbf{k}_G \right)_n$.

The Quasi-static Solution

If the lowest eigenfrequency of the structure is high, and the structural behaviour is quasi-static, then the solution may be obtained as a sum of the time invariant solution $\bar{\mathbf{r}}$ (given in Eq. 6.1 above) and a fluctuating part

$$\mathbf{r}(t) = \mathbf{K}^{-1} \cdot \mathbf{R}(t) \tag{6.3}$$

If $\mathbf{R}(t)$ is known for a sufficiently long period of time, then $\mathbf{r}(t)$ may be obtained directly form Eq. 6.3. If only the stochastic properties of the load process are known quantities, i.e. if only the cross spectral densities of all load components are known

$$\mathbf{S}_{RR}(\omega) = \begin{bmatrix} S_{R_1R_1} & \cdots & S_{R_1R_i} & \cdots & S_{R_1R_j} & \cdots & S_{R_1R_{N_r}} \\ \vdots & \ddots & \vdots & & \vdots & .\cdot^{\cdot} & \vdots \\ S_{R_iR_1} & \cdots & S_{R_iR_i} & \cdots & S_{R_iR_j} & \cdots & S_{R_iR_{N_r}} \\ \vdots & & \vdots & \ddots & \vdots & & \vdots \\ S_{R_jR_1} & \cdots & S_{R_jR_i} & \cdots & S_{R_jR_j} & \cdots & S_{R_jR_{N_r}} \\ \vdots & .\cdot^{\cdot} & \vdots & & \vdots & \ddots & \vdots \\ S_{R_{N_r}R_1} & \cdots & S_{R_{N_r}R_i} & \cdots & S_{R_{N_r}R_j} & \cdots & S_{R_{N_r}R_{N_r}} \end{bmatrix} \tag{6.4}$$

where $N_r = 6 \cdot N_p$ and N_p is the number of nodes in the system, then a time domain solution may still be pursued, but this will require the time domain simulations of $\mathbf{R}$, e.g. as shown in Appendix B. Alternatively, a stochastic solution may be directly obtained by developing the corresponding covariance properties of the displacement components, i.e.

$$\mathbf{Cov}_{rr} = \begin{bmatrix} \sigma^2_{r_1 r_1} & \cdots & Cov_{r_1 r_i} & \cdots & Cov_{r_1 r_j} & \cdots & Cov_{r_1 r_{N_r}} \\ \vdots & \ddots & \vdots & & \vdots & ⋰ & \vdots \\ Cov_{r_i r_1} & \cdots & \sigma^2_{r_i r_i} & \cdots & Cov_{r_i r_j} & \cdots & Cov_{r_i r_{N_r}} \\ \vdots & & \vdots & \ddots & \vdots & & \vdots \\ Cov_{r_j r_1} & \cdots & Cov_{r_j r_i} & \cdots & \sigma^2_{r_j r_j} & \cdots & Cov_{r_j r_{N_r}} \\ \vdots & ⋰ & \vdots & & \vdots & \ddots & \vdots \\ Cov_{r_{N_r} r_1} & \cdots & Cov_{r_{N_r} r_i} & \cdots & Cov_{r_{N_r} r_j} & \cdots & \sigma^2_{r_{N_r} r_{N_r}} \end{bmatrix} \tag{6.5}$$

$$= E\left[\{\mathbf{r}(t)\}\cdot\{\mathbf{r}(t)\}^T\right] = E\left[\{\mathbf{K}^{-1}\mathbf{R}(t)\}\cdot\{\mathbf{K}^{-1}\mathbf{R}(t)\}^T\right]$$

$$= \mathbf{K}^{-1}\cdot E\left[\{\mathbf{R}(t)\}\cdot\{\mathbf{R}(t)\}^T\right]\cdot\left(\mathbf{K}^{-1}\right)^T = \mathbf{K}^{-1}\cdot\mathbf{Cov}_{RR}\cdot\left(\mathbf{K}^{-1}\right)^T$$

where $\mathbf{Cov}_{RR}$ may be obtained from a frequency domain integration of Eq. 6.4:

$$\mathbf{Cov}_{RR} = \int_0^\infty \mathbf{S}_{RR}(\omega)\,d\omega \tag{6.6}$$

Assuming Gaussian and stationary stochastic properties, then for design purposes the extreme response values may be obtained from

$$\mathbf{r}_{tot_{\max}} = \bar{\mathbf{r}} + \mathbf{r}_{\max} = \bar{\mathbf{r}} + k_p \boldsymbol{\sigma}_r \tag{6.7}$$

where k_p is a peak factor defined in Eq. A.45 and, and where $\boldsymbol{\sigma}_r$ is a vector containing the square root of all the entries on the diagonal of $\mathbf{Cov}_{rr}$ in Eq. 6.5.

Elaboration 6.1: The Spectra of the Quasi Static Solution

An alternative approach may be chosen if it is considered necessary or advantageous to obtain time domain plots of the quasi-static response. It is seen that by taking the Fourier transform throughout Eq. 6.3

$$\Rightarrow \qquad \mathbf{a}_r(t) = \mathbf{K}^{-1}\cdot\mathbf{a}_R(t)$$

where $\mathbf{a}_r$ and $\mathbf{a}_R$ are vectors containing the Fourier amplitudes of the response and load, respectively, and thus the cross spectral density matrix of the response components is given by

$$\mathbf{S}_{rr}(\omega) = \begin{bmatrix} S_{r_1r_1} & \cdots & S_{r_1r_i} & \cdots & S_{r_1r_j} & \cdots & S_{r_1r_{N_r}} \\ \vdots & \ddots & \vdots & & \vdots & \iddots & \vdots \\ S_{r_ir_1} & \cdots & S_{r_ir_i} & \cdots & S_{r_ir_j} & \cdots & S_{r_ir_{N_r}} \\ \vdots & & \vdots & \ddots & \vdots & & \vdots \\ S_{r_jr_1} & \cdots & S_{r_jr_i} & \cdots & S_{r_jr_j} & \cdots & S_{r_jr_{N_r}} \\ \vdots & \iddots & \vdots & & \vdots & \ddots & \vdots \\ S_{r_{N_r}r_1} & \cdots & S_{r_{N_r}r_i} & \cdots & S_{r_{N_r}r_j} & \cdots & S_{r_{N_r}r_{N_r}} \end{bmatrix}$$

$$= \lim_{T\to\infty} \frac{1}{\pi T} \{\mathbf{a}_r(\omega)\}^* \{\mathbf{a}_r(\omega)\}^T = \lim_{T\to\infty} \frac{1}{\pi T} \{\mathbf{K}^{-1}\mathbf{a}_R(\omega)\}^* \{\mathbf{K}^{-1}\mathbf{a}_R(\omega)\}^T$$

$$= \left(\mathbf{K}^{-1}\right)^* \lim_{T\to\infty} \frac{1}{\pi T} \{\mathbf{a}_R(\omega)\}^* \{\mathbf{a}_R(\omega)\}^T \left(\mathbf{K}^{-1}\right)^T = \left(\mathbf{K}^{-1}\right)^* \mathbf{S}_{RR}(\omega) \left(\mathbf{K}^{-1}\right)^T$$

where $\mathbf{S}_{RR}(\omega) = \lim_{T\to\infty} \frac{1}{\pi T} \{\mathbf{a}_R(\omega)\}^* \cdot \{\mathbf{a}_R(\omega)\}^T$. From $\mathbf{S}_{rr}(\omega)$ a time domain simulation (see Appendix B) may be performed to obtain time series of the instantaneous values of the response. The corresponding response covariance matrix may readily be obtained by frequency domain integration

$$\mathbf{Cov}_{rr} = \int_0^\infty \mathbf{S}_{rr}(\omega)\, d\omega$$

6.3 Response Calculations in Time Domain

In a time domain solution the total displacement response may be obtained as a sum of the time invariant solution $\overline{\mathbf{r}}$ (given in Eq. 6.1) and a purely dynamic part $\mathbf{r}(t)$, i.e.

$$\mathbf{r}_{tot}(t) = \overline{\mathbf{r}} + \mathbf{r}(t) \tag{6.8}$$

A solution strategy may be pursued in the original finite element degrees of freedom (developed in Chapter 4.3) or in modal coordinates (see Chapters 5.2 and

5.3). If a solution in original degrees of freedom is pursued, then the calculation of $\mathbf{r}(t)$ will require the solution of the dynamic equation given in Eq. 4.39: $\mathbf{M}\cdot\ddot{\mathbf{r}}(t)+\mathbf{C}\cdot\dot{\mathbf{r}}(t)+\mathbf{K}\cdot\mathbf{r}(t)=\mathbf{R}(t)$. If a solution in modal degrees of freedom is pursued, then the calculation of $\mathbf{r}(t)$ will first require the solution of the dynamic equation given in Eq. 5.13 or 5.32: $\tilde{\mathbf{M}}\ddot{\boldsymbol{\eta}}(t)+\tilde{\mathbf{C}}\dot{\boldsymbol{\eta}}(t)+\tilde{\mathbf{K}}\boldsymbol{\eta}(t)=\tilde{\mathbf{R}}(t)$, after which the response may be obtained from Eq. 5.9 or 5.18: $\mathbf{r}(t)=\boldsymbol{\Phi}\cdot\boldsymbol{\eta}(t)$. Since the size of the relevant matrices in a modal format is generally considerably smaller than those in original coordinates, it is recommended to think twice before an approach in original coordinates is chosen. However, for the sake of generality, the symbolism shown below is that of an approach in original coordinates. There are a number of iteration procedures available for a time domain solution strategy. Only a selected few are included below. In any case, as illustrated in Fig. 6.3, a time domain solution will involve some discrete representation of the load processes $\mathbf{R}$ or $\tilde{\mathbf{R}}$ at time steps t_k ($k=1,2,\ldots,N_k$), followed by a stepwise calculation of the corresponding response ($\mathbf{r}$ or $\boldsymbol{\eta}$). Based on knowledge of the response at time step t_k and discrete load values, the task at hand is to calculate the response at time step t_{k+1}.

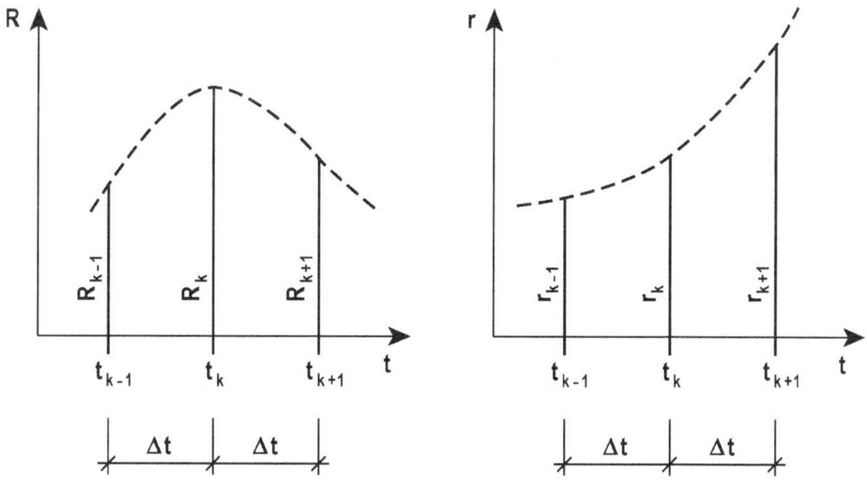

Fig. 6.3 Time domain

Such a forward prediction routine is called explicit if it is based on the known response history alone. It is called implicit if it contains assumptions about the response development or equilibrium condition in the unknown future of the system. I.e., in an explicit routine $\mathbf{r}(t_k + \Delta t)$ is a function of $\mathbf{r}$, $\dot{\mathbf{r}}$ and $\ddot{\mathbf{r}}$ at $t = t_k$, while an implicit routine contains some assumptions about the development of the motion in the time step between t_k and $t_k + \Delta t$. Obviously, the shorter time step, the easier it is to obtain a good solution. There are two main classes of forward iteration strategies, the direct iteration procedures and the numeric integration methods. Some of these are presented below.

The Second Central Difference Method

Consider the situation at time step t_{k-1}, t_k and t_{k+1}. A Taylor series expansion of $\mathbf{r}_{k+1}$ and $\mathbf{r}_{k-1}$ is given by:

$$\left.\begin{aligned} \mathbf{r}_{k+1} &= \mathbf{r}_k + \Delta t \cdot \dot{\mathbf{r}}_k + \frac{\Delta t^2}{2} \cdot \ddot{\mathbf{r}}_k + \cdots \\ \mathbf{r}_{k-1} &= \mathbf{r}_k - \Delta t \cdot \dot{\mathbf{r}}_k + \frac{\Delta t^2}{2} \cdot \ddot{\mathbf{r}}_k - \cdots \end{aligned}\right\} \tag{6.9}$$

Thus, considering only the three first terms of the Taylor series expansion, $\mathbf{r}_{k+1} - \mathbf{r}_{k-1}$ renders

$$\dot{\mathbf{r}}_k \approx \frac{1}{2\Delta t} \cdot \left(\mathbf{r}_{k+1} - \mathbf{r}_{k-1}\right) \tag{6.10}$$

while $\mathbf{r}_{k+1} + \mathbf{r}_{k-1}$ renders

$$\ddot{\mathbf{r}}_k \approx \frac{1}{\Delta t^2} \cdot \left(\mathbf{r}_{k+1} - 2\mathbf{r}_k + \mathbf{r}_{k-1}\right) \tag{6.11}$$

Dynamic equilibrium at t_k is given by

$$\mathbf{M}\,\ddot{\mathbf{r}}_k + \mathbf{C}\dot{\mathbf{r}}_k + \mathbf{K}\mathbf{r}_k = \mathbf{R}_k \tag{6.12}$$

Introducing $\mathbf{r}_k$, $\dot{\mathbf{r}}_k$ and $\ddot{\mathbf{r}}_k$ from Eqs. 6.10 and 6.11

$$\mathbf{M}\,\frac{1}{\Delta t^2} \cdot \left(\mathbf{r}_{k+1} - 2\mathbf{r}_k + \mathbf{r}_{k-1}\right) + \mathbf{C}\frac{1}{2\Delta t} \cdot \left(\mathbf{r}_{k+1} - \mathbf{r}_{k-1}\right) + \mathbf{K}\mathbf{r}_k = \mathbf{R}_k \tag{6.13}$$

and solving for $\mathbf{r}_{k+1}$

$$\mathbf{r}_{k+1} = \left(\mathbf{M} + \frac{\Delta t}{2}\mathbf{C} \right)^{-1} \left[\Delta t^2 \mathbf{R}_k + \left(2\mathbf{M} - \Delta t^2 \mathbf{K} \right) \mathbf{r}_k - \left(\mathbf{M} - \frac{\Delta t}{2}\mathbf{C} \right) \mathbf{r}_{k-1} \right] \tag{6.14}$$

Thus, it is seen that $\mathbf{r}_{k+1}$ may be estimated based on knowledge about the load and response quantise at t_k and t_{k-1}, i.e. this approach belongs among the explicit routines. For the establishment of initial conditions at $t = 0$ before the iteration procedure can start it is necessary to define (choose) $\mathbf{r}_0$ and $\dot{\mathbf{r}}_0$. From dynamic equilibrium (Eq. 6.12) at $t = 0$ the corresponding acceleration

$$\ddot{\mathbf{r}}_0 = \mathbf{M}^{-1} \cdot \left(\mathbf{R}_0 - \mathbf{C}\dot{\mathbf{r}}_0 - \mathbf{K}\mathbf{r}_0 \right) \tag{6.15}$$

is obtained, while eliminating $\mathbf{r}_1$ from (see Eqs. 6.10 and 6.11)

$$\left. \begin{aligned} \dot{\mathbf{r}}_0 &\approx \frac{1}{2\Delta t} \cdot \left(\mathbf{r}_1 - \mathbf{r}_{-1} \right) \\ \ddot{\mathbf{r}}_0 &\approx \frac{1}{\Delta t^2} \cdot \left(\mathbf{r}_1 - 2\mathbf{r}_0 + \mathbf{r}_{-1} \right) \end{aligned} \right\} \tag{6.16}$$

renders

$$\mathbf{r}_{-1} = \mathbf{r}_0 - \Delta t \cdot \dot{\mathbf{r}}_0 + \frac{\Delta t^2}{2} \ddot{\mathbf{r}}_0 \tag{6.17}$$

Then, introducing this back into Eq. 6.15, the following is obtained

$$\mathbf{r}_1 = \left(\mathbf{M} + \frac{\Delta t}{2}\mathbf{C} \right)^{-1} \left[\Delta t^2 \mathbf{R}_0 + \left(2\mathbf{M} - \Delta t^2 \mathbf{K} \right) \mathbf{r}_0 - \left(\mathbf{M} - \frac{\Delta t}{2}\mathbf{C} \right) \mathbf{r}_{-1} \right] \tag{6.18}$$

The stability of the second central difference method may be evaluated by considering an undamped and unloaded single mode system (se Eqs. 6.13) with modal mass $\tilde{M}_n$, stiffness $\tilde{K}_n$ and eigenfrequency $\omega_n^2 = \tilde{K}_n / \tilde{M}_n$:

$$\left. \begin{aligned} & \tilde{M}_n \frac{1}{\Delta t^2} \left(r_{k+1} - 2r_k + r_{k-1} \right) + \tilde{K}_n r_k = 0 \\ & \Rightarrow r_{k+1} + \left(\Delta t^2 \omega_n^2 - 2 \right) r_k + r_{k-1} = 0 \end{aligned} \right\} \tag{6.19}$$

which, for a simple harmonic motion $r = \mathrm{Re}\left(a \cdot e^{\lambda t}\right)$ (where a is the amplitude), will render

$$\left.\begin{aligned} & a e^{\lambda(t_k+\Delta t)} + \left(\Delta t^2 \omega_n^2 - 2\right) a e^{\lambda t_k} + a e^{\lambda(t_k-\Delta t)} = 0 \\ \Rightarrow\quad & \left(e^{\lambda\Delta t}\right)^2 + \left(\Delta t^2 \omega_n^2 - 2\right) e^{\lambda\Delta t} + 1 = 0 \\ \Rightarrow\quad & \left.\begin{matrix} \left(e^{\lambda\Delta t}\right)_1 \\ \left(e^{\lambda\Delta t}\right)_2 \end{matrix}\right\} = \frac{1}{2}\left[2 - \Delta t^2 \omega_n^2 \pm \Delta t \omega_n \sqrt{\Delta t^2 \omega_i^2 - 4}\right] \end{aligned}\right\} \tag{6.20}$$

and thus the response at t_{k+1} is given by

$$\begin{aligned} r_{k+1} &= \mathrm{Re}\left\{a e^{\lambda(t_k+\Delta t)}\right\} = \mathrm{Re}\left\{a e^{\lambda t_k} e^{\lambda\Delta t}\right\} \\ &= \mathrm{Re}\left\{a e^{\lambda t_k}\left[c_1\left(e^{\lambda\Delta t}\right)_1 + c_1\left(e^{\lambda\Delta t}\right)_2\right]\right\} \end{aligned} \tag{6.21}$$

where c_1 and c_2 are constants dependant on initial conditions. In a second order equation $\alpha x^2 + \beta x + \gamma = 0$ the product of the roots $x_1 \cdot x_2 = \gamma/\alpha$. Thus, it is seen that $\left(e^{\lambda\Delta t}\right)_1 \cdot \left(e^{\lambda\Delta t}\right)_2 = 1$, and it may be taken for granted that both roots are distinct. A positive radicand in the solution in Eq. 6.20 will render two real roots, and since the product of the two roots is unity one of them must be larger than one, and thus, the solution is consistently growing, i.e. it is unstable. A negative radicand on the other hand will render complex roots, and the product of the two roots can only be unity if they are complex conjugates and both has an absolute value equal to one. Thus, the solution is numerically stable if

$$\Delta t^2 \omega_n^2 - 4 \le 0 \qquad \Rightarrow \qquad \Delta t \le \frac{2}{\omega_n} \tag{6.22}$$

where (obviously), ω_n is the largest eigenfrequency expected to play any significant role in the response behaviour of the system. In any case Δt should not be chosen larger than about $1/\left(2\omega_{R_{\max}}\right)$ where $\omega_{R_{\max}}$ is the largest frequency contained in the load.

Numeric Integration Methods

The numeric integration methods belong among the implicit routines. They are based on the assumption that future response quantities may be obtained from the situation at the onset of the time step and an integration of the approximate higher order development between time steps, i.e. that

$$\left.\begin{aligned}\dot{\mathbf{r}}_{k+1} &= \dot{\mathbf{r}}_k + \int_0^{\Delta t} \ddot{\mathbf{r}}(\tau)\,d\tau \\ \mathbf{r}_{k+1} &= \mathbf{r}_k + \int_0^{\Delta t} \dot{\mathbf{r}}(\tau)\,d\tau\end{aligned}\right\} \quad \text{where} \quad 0 \le \tau \le \Delta t \tag{6.23}$$

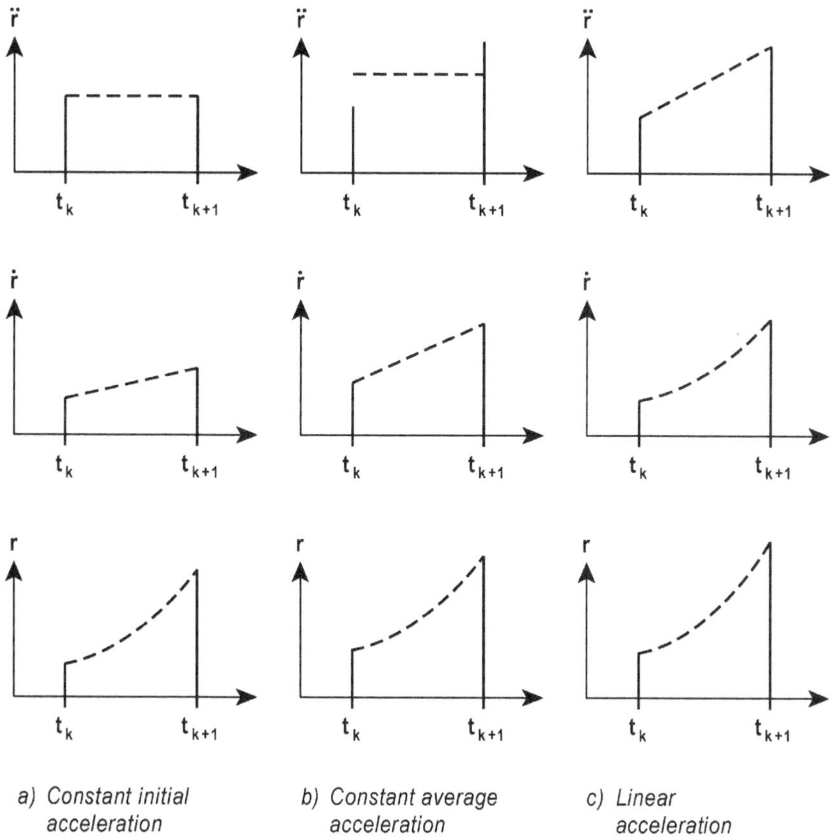

Fig. 6.4 Numeric integration assumptions

As illustrated in Fig. 6.4 the approximation is that the acceleration variation within the time step is either assumed constant and equal to its initial value, it is assumed equal to the average acceleration over the time step, or it is assumed linear across the time step. If constant initial acceleration is adopted (see Fig. 6.4.a), then

$$\ddot{\mathbf{r}}(\tau)=\ddot{\mathbf{r}}_k \qquad \Rightarrow \qquad \dot{\mathbf{r}}(\tau)=\dot{\mathbf{r}}_k+\int_0^{\tau}\ddot{\mathbf{r}}_k d\tau=\dot{\mathbf{r}}_k+\ddot{\mathbf{r}}_k\cdot\tau \tag{6.24}$$

and thus

$$\left.\begin{aligned}\dot{\mathbf{r}}_{k+1}&=\dot{\mathbf{r}}_k+\int_0^{\Delta t}\ddot{\mathbf{r}}_k d\tau=\dot{\mathbf{r}}_k+\Delta t\ddot{\mathbf{r}}_k\\ \mathbf{r}_{k+1}&=\mathbf{r}_k+\int_0^{\Delta t}\left(\dot{\mathbf{r}}_k+\tau\cdot\ddot{\mathbf{r}}_k\right)d\tau=\mathbf{r}_k+\Delta t\cdot\dot{\mathbf{r}}_k+\Delta t^2\ddot{\mathbf{r}}_k/2\end{aligned}\right\} \tag{6.25}$$

If the concept of a constant average acceleration is adopted (see Fig. 6.4.b), then

$$\ddot{\mathbf{r}}(\tau)=\frac{1}{2}(\ddot{\mathbf{r}}_k+\ddot{\mathbf{r}}_{k+1}) \Rightarrow \dot{\mathbf{r}}(\tau)=\dot{\mathbf{r}}_k+\int_0^{\tau}\frac{1}{2}(\ddot{\mathbf{r}}_k+\ddot{\mathbf{r}}_{k+1})d\tau=\dot{\mathbf{r}}_k+\frac{\tau}{2}(\ddot{\mathbf{r}}_k+\ddot{\mathbf{r}}_{k+1}) \tag{6.26}$$

and thus

$$\left.\begin{aligned}\dot{\mathbf{r}}_{k+1}&=\dot{\mathbf{r}}_k+\int_0^{\Delta t}\frac{1}{2}(\ddot{\mathbf{r}}_k+\ddot{\mathbf{r}}_{k+1})d\tau=\dot{\mathbf{r}}_k+\frac{\Delta t}{2}(\ddot{\mathbf{r}}_k+\ddot{\mathbf{r}}_{k+1})\\ \mathbf{r}_{k+1}&=\mathbf{r}_k+\int_0^{\Delta t}\left[\dot{\mathbf{r}}_k+\frac{\tau}{2}(\ddot{\mathbf{r}}_k+\ddot{\mathbf{r}}_{k+1})\right]d\tau=\mathbf{r}_k+\Delta t\cdot\dot{\mathbf{r}}_k+\frac{\Delta t^2}{4}(\ddot{\mathbf{r}}_k+\ddot{\mathbf{r}}_{k+1})\end{aligned}\right\} \tag{6.27}$$

If the concept of a linear acceleration is adopted (see Fig. 6.4.c), then

$$\ddot{\mathbf{r}}(\tau)=\ddot{\mathbf{r}}_k+(\ddot{\mathbf{r}}_{k+1}-\ddot{\mathbf{r}}_k)\tau/\Delta t \tag{6.28}$$

in which case

$$\left.\begin{aligned}\dot{\mathbf{r}}(\tau)&=\dot{\mathbf{r}}_k+\int_0^{\tau}\left[\ddot{\mathbf{r}}_k+(\ddot{\mathbf{r}}_{k+1}-\ddot{\mathbf{r}}_k)\frac{\tau}{\Delta t}\right]d\tau=\dot{\mathbf{r}}_k+\ddot{\mathbf{r}}_k\tau+(\ddot{\mathbf{r}}_{k+1}-\ddot{\mathbf{r}}_k)\frac{\tau^2}{2\Delta t}\\ \mathbf{r}(\tau)&=\mathbf{r}_k+\int_0^{\tau}\left[\dot{\mathbf{r}}_k+\ddot{\mathbf{r}}_k\tau+(\ddot{\mathbf{r}}_{k+1}-\ddot{\mathbf{r}}_k)\frac{\tau^2}{2\Delta t}\right]d\tau=\mathbf{r}_k+\dot{\mathbf{r}}_k\tau+\ddot{\mathbf{r}}_k\frac{\tau^2}{2}+(\ddot{\mathbf{r}}_{k+1}-\ddot{\mathbf{r}}_k)\frac{\tau^3}{6\Delta t}\end{aligned}\right\} \tag{6.29)}$$

and thus

$$\left.\begin{aligned} \dot{\mathbf{r}}_{k+1} &= \dot{\mathbf{r}}_k + \Delta t \ddot{\mathbf{r}}_k + \frac{\Delta t}{2}(\ddot{\mathbf{r}}_{k+1} - \ddot{\mathbf{r}}_k) = \dot{\mathbf{r}}_k + \frac{\Delta t}{2}(\ddot{\mathbf{r}}_k + \ddot{\mathbf{r}}_{k+1}) \\ \mathbf{r}_{k+1} &= \mathbf{r}_k + \Delta t \cdot \dot{\mathbf{r}}_k + \frac{\Delta t^2}{2}\ddot{\mathbf{r}}_k + \frac{\Delta t^2}{6}(\ddot{\mathbf{r}}_{k+1} - \ddot{\mathbf{r}}_k) = \mathbf{r}_k + \Delta t \cdot \dot{\mathbf{r}}_k + \frac{\Delta t^2}{3}\ddot{\mathbf{r}}_k + \frac{\Delta t^2}{6}\ddot{\mathbf{r}}_{k+1} \end{aligned}\right\} \tag{6.30}$$

The concept of integrating an assumed variation of the acceleration between t_k and t_{k+1} presented above may all be generalised into the following formulation (first suggested by Newmark [32]):

$$\left.\begin{aligned} \dot{\mathbf{r}}_{k+1} &= \dot{\mathbf{r}}_k + (1-\gamma) \cdot \Delta t \cdot \ddot{\mathbf{r}}_k + \gamma \cdot \Delta t \cdot \ddot{\mathbf{r}}_{k+1} \\ \mathbf{r}_{k+1} &= \mathbf{r}_k + \Delta t \cdot \dot{\mathbf{r}}_k + (1/2 - \beta) \cdot \Delta t^2 \cdot \ddot{\mathbf{r}}_k + \beta \cdot \Delta t^2 \cdot \ddot{\mathbf{r}}_{k+1} \end{aligned}\right\} \tag{6.31}$$

where γ and β are weighting parameters, each to be chosen according to prescribed requirements regarding numerical stability and accuracy. From the second expression in Eq. 6.31 the acceleration at t_{k+1}

$$\ddot{\mathbf{r}}_{k+1} = \frac{1}{\beta \Delta t^2}(\mathbf{r}_{k+1} - \mathbf{r}_k) - \left[\frac{1}{\beta \Delta t}\dot{\mathbf{r}}_k + \left(\frac{1}{2\beta} - 1\right)\ddot{\mathbf{r}}_k\right] \tag{6.32}$$

is obtained, which, combined with the first expression in Eq. 6.31, renders

$$\dot{\mathbf{r}}_{k+1} = \frac{\gamma}{\beta \Delta t}(\mathbf{r}_{k+1} - \mathbf{r}_k) - \left(\frac{\gamma}{\beta} - 1\right) \cdot \dot{\mathbf{r}}_k - \left(\frac{\gamma}{2\beta} - 1\right)\Delta t \cdot \ddot{\mathbf{r}}_k \tag{6.33}$$

For convenience the conditions at t_k may be defined by

$$\left.\begin{aligned} \mathbf{a}_k &= \frac{1}{\beta \Delta t^2}\mathbf{r}_k + \frac{1}{\beta \Delta t}\dot{\mathbf{r}}_k + \left(\frac{1}{2\beta} - 1\right)\ddot{\mathbf{r}}_k \\ \mathbf{b}_k &= \frac{\gamma}{\beta \Delta t}\mathbf{r}_k + \left(\frac{\gamma}{\beta} - 1\right) \cdot \dot{\mathbf{r}}_k + \left(\frac{\gamma}{2\beta} - 1\right)\Delta t \ddot{\mathbf{r}}_k \end{aligned}\right\} \tag{6.34}$$

in which case $\ddot{\mathbf{r}}_{k+1}$ and $\dot{\mathbf{r}}_{k+1}$ simplifies into

$$\left.\begin{aligned} \ddot{\mathbf{r}}_{k+1} &= (\mathbf{r}_{k+1} - \mathbf{a}_k) / (\beta \Delta t^2) \\ \dot{\mathbf{r}}_{k+1} &= \gamma(\mathbf{r}_{k+1} - \mathbf{b}_k) / (\beta \Delta t) \end{aligned}\right\} \tag{6.35}$$

Introducing this into the dynamic equilibrium equation at t_{k+1}

$$\mathbf{M}\,\ddot{\mathbf{r}}_{k+1} + \mathbf{C}\dot{\mathbf{r}}_{k+1} + \mathbf{K}\mathbf{r}_{k+1} = \mathbf{R}_{k+1} \tag{6.36}$$

will then render

$$\left(\frac{1}{\beta \Delta t^2}\mathbf{M} + \frac{\gamma}{\beta \Delta t}\mathbf{C} + \mathbf{K}\right)\mathbf{r}_{k+1} = \mathbf{R}_{k+1} + \mathbf{M}\cdot\mathbf{a}_k + \mathbf{C}\cdot\mathbf{b}_k \tag{6.37}$$

Defining

$$\left.\begin{aligned} \mathbf{K}_{eff_{k+1}} &= \frac{1}{\beta \Delta t^2}\mathbf{M} + \frac{\gamma}{\beta \Delta t}\mathbf{C} + \mathbf{K} \\ \mathbf{R}_{eff_{k+1}} &= \mathbf{R}_{k+1} + \mathbf{M}\cdot\mathbf{a}_k + \mathbf{C}\cdot\mathbf{b}_k \end{aligned}\right\} \tag{6.38}$$

and thus

$$\mathbf{r}_{k+1} = \mathbf{K}_{eff_{k+1}}^{-1}\cdot\mathbf{R}_{eff_{k+1}} \tag{6.39}$$

It is seen that the response at time step t_{k+1} is calculated from the load at t_{k+1} as well as the displacement, velocity and acceleration response at t_k. If the system is entirely linear, then $\mathbf{K}_{eff}$ remains constant throughout the iterations.

Hilber, Hughes & Taylor [33] have suggested an extension of Newmark's method by the introduction of the numerical dampening coefficient $\alpha \le 0$ into the dynamic equilibrium condition at t_{k+1}

$$\mathbf{M}\,\ddot{\mathbf{r}}_{k+1} + (1+\alpha)\mathbf{C}\dot{\mathbf{r}}_{k+1} - \alpha\mathbf{C}\dot{\mathbf{r}}_k + (1+\alpha)\mathbf{K}\mathbf{r}_{k+1} - \alpha\mathbf{K}\mathbf{r}_k = \mathbf{R}_\alpha \tag{6.40}$$

and accordingly, evaluate the dynamic load at $(1+\alpha)t_{k+1} - \alpha t_k = t_{k+1} + \alpha\Delta t$. I.e., if load linearity within the time step is adopted, then

$$\mathbf{R}_\alpha = (1+\alpha)\mathbf{R}_{k+1} - \alpha\mathbf{R}_k \tag{6.41}$$

Combining Eqs. 6.35, 6.40 and 6.41 and solving for $\mathbf{r}_{k+1}$ will then again render $\mathbf{r}_{k+1} = \mathbf{K}_{eff_{k+1}}^{-1}\cdot\mathbf{R}_{eff_{k+1}}$, but now $\mathbf{K}_{eff_{k+1}}$ and $\mathbf{R}_{eff_{k+1}}$ are extended into

$$\left.\begin{aligned} \mathbf{K}_{eff_{k+1}} &= \frac{1}{\beta \Delta t^2}\mathbf{M} + (1+\alpha)\frac{\gamma}{\beta \Delta t}\mathbf{C} + (1+\alpha)\mathbf{K} \\ \mathbf{R}_{eff_{k+1}} &= (1+\alpha)\mathbf{R}_{k+1} - \alpha\mathbf{R}_k + \mathbf{M}\cdot\mathbf{a}_k + \mathbf{C}\cdot\mathbf{b}_k + \alpha\mathbf{c}_k \end{aligned}\right\} \tag{6.42}$$

where

$$\mathbf{c}_k = \left(\frac{\gamma}{\beta \Delta t} \mathbf{C} + \mathbf{K} \right) \cdot \mathbf{r}_k + \mathbf{C} \frac{\gamma}{\beta} \dot{\mathbf{r}}_k + \mathbf{C} \left(\frac{\gamma}{2\beta} - 1 \right) \Delta t \cdot \ddot{\mathbf{r}}_k \tag{6.43}$$

For all of the numeric integration methods the establishment of initial conditions at $t = 0$ before the iteration procedure can start requires the choice of $\mathbf{r}_0$ and $\dot{\mathbf{r}}_0$. From dynamic equilibrium at $t = 0$ the corresponding acceleration

$$\ddot{\mathbf{r}}_0 = \mathbf{M}^{-1} \cdot \left(\mathbf{R}_0 - \mathbf{C}\dot{\mathbf{r}}_0 - \mathbf{K}\mathbf{r}_0 \right) \tag{6.44}$$

is obtained, and thus, iteration may commence.

Stability may be evaluated from the properties of a single degree of freedom system (or a modal approach) similar to that which has been shown for the central difference method above. In general, the Newmark method is unconditionally stable if

$$\gamma \geq \gamma_0 = 1/2 \qquad \text{and} \qquad \beta \geq \beta_0 = \frac{1}{4} \left(\gamma + \frac{1}{2} \right)^2 \tag{6.45}$$

For β-values below β_0 it is only conditionally stable. The stability limit is then given by

$$\Delta t \leq \Delta t_{cr} = \frac{1}{\omega_n \sqrt{\beta_0 - \beta}} \tag{6.46}$$

Positive or negative numeric damping is introduced into the system depending on $\gamma > \gamma_0$ or $\gamma < \gamma_0$. Positive numeric damping may be used as an effective tool to dampen out undesirable effects of higher modes in the system (which may also be obtained by adopting Hilber, Hughes & Taylor method with $-1/3 < \alpha < 0$). With $\gamma = 1$ and $\beta = 0$ Newmark's method becomes identical to a numeric integration method based on the assumption of constant initial acceleration which is only conditionally stable. If $\gamma = 1/2$ and $\beta = 1/4$ then Newmark's method becomes identical to a numerical integration method based on the assumption of a constant average acceleration, which is unconditionally stable. If $\gamma = 1/2$ and $\beta = 1/6$ then Newmark's method becomes identical to a numerical integration method based on the assumption of a linear variation of the acceleration, in which case the stability limit is given by $\Delta t_{cr} = \sqrt{12}/\omega_n$. As previously mentioned Δt should never be chosen larger than about $1/(2\omega_{R_{max}})$ where $\omega_{R_{max}}$ is the largest frequency contained in the load.

Tangent-Stiffness Approach

For large displacements or heavily non-linear material problems the stiffness may change considerably throughout the response process, in which case a sufficient level of accuracy may be obtained by updating the stiffness from one time step to the next. In such cases a tangent-stiffness approach may be adopted. Assuming a system of only short elements and sufficiently short time steps (such that linearity within each time step may be justifiable), then the change of internal forces from t_k to t_{k+1} is given by

$$\Delta \mathbf{R}_k^{\text{int}} = \mathbf{K}_k^{\tan} \cdot \Delta \mathbf{r} \tag{6.47}$$

where $\mathbf{K}_k^{\tan}$ is the updated tangent stiffness at t_k and $\Delta \mathbf{r} = \mathbf{r}_{k+1} - \mathbf{r}_k$. Thus, the internal force vector at t_{k+1} is

$$\mathbf{R}_{k+1}^{\text{int}} = \mathbf{R}_k^{\text{int}} + \mathbf{K}_k^{\tan} \cdot \Delta \mathbf{r} \tag{6.48}$$

The dynamic equilibrium condition at t_{k+1} is then given by (see Eq. 6.36)

$$\mathbf{M}\,\ddot{\mathbf{r}}_{k+1} + \mathbf{C}\dot{\mathbf{r}}_{k+1} + \mathbf{R}_k^{\text{int}} + \mathbf{K}_k^{\tan} \Delta \mathbf{r} = \mathbf{R}_{k+1} \tag{6.49}$$

By introducing the Newmark iteration scheme given in Eqs. 6.32 and 6.33 (and that $\mathbf{r}_{k+1} - \mathbf{r}_k = \Delta \mathbf{r}$) into Eq. 6.49, then the following is obtained

$$\begin{aligned}
&\left(\frac{1}{\beta \Delta t^2} \mathbf{M} + \frac{\gamma}{\beta \Delta t} \mathbf{C} + \mathbf{K}_k^{\tan} \right) \cdot \Delta \mathbf{r} = \mathbf{R}_{k+1} - \mathbf{R}_k^{\text{int}} \\
&+\mathbf{M} \cdot \left[\frac{1}{\beta \Delta t} \dot{\mathbf{r}}_k + \left(\frac{1}{2\beta} - 1 \right) \cdot \ddot{\mathbf{r}}_k \right] + \mathbf{C} \cdot \left[\left(\frac{\gamma}{\beta} - 1 \right) \cdot \dot{\mathbf{r}}_k + \left(\frac{\gamma}{2\beta} - 1 \right) \cdot \Delta t \cdot \ddot{\mathbf{r}}_k \right]
\end{aligned} \tag{6.50}$$

Thus

$$\Delta \mathbf{r} = \mathbf{K}_{eff_k}{}^{-1} \cdot \left(\mathbf{R}_{k+1} - \mathbf{R}_k^{\text{int}} + \mathbf{M} \cdot \mathbf{a}_{eff_k} + \mathbf{C} \cdot \mathbf{b}_{eff_k} \right) \tag{6.51}$$

where

$$\left.\begin{aligned} \mathbf{K}_{eff_k} &= \frac{1}{\beta \Delta t^2}\mathbf{M} + \frac{\gamma}{\beta \Delta t}\mathbf{C} + \mathbf{K}_k^{\tan} \\ \mathbf{a}_{eff_k} &= \frac{1}{\beta \Delta t}\dot{\mathbf{r}}_k + \left(\frac{1}{2\beta} - 1\right)\cdot \ddot{\mathbf{r}}_k \\ \mathbf{b}_{eff_k} &= \left(\frac{\gamma}{\beta} - 1\right)\cdot \dot{\mathbf{r}}_k + \left(\frac{\lambda}{2\beta} - 1\right)\cdot \Delta t \cdot \ddot{\mathbf{r}}_k \end{aligned}\right\} \tag{6.52}$$

Such a procedure will generally require error control. This may be obtained by minimising the estimated external load error $\Delta\mathbf{R}_{k+1}^{err}$, defined as the difference between the actual load at t_{k+1} and the corresponding load which can be calculated from the estimated displacements

$$\Delta\mathbf{R}_{k+1}^{err} = \mathbf{R}_{k+1} - \left(\mathbf{M}\,\ddot{\mathbf{r}}_{k+1} + \mathbf{C}\dot{\mathbf{r}}_{k+1} + \mathbf{R}_k^{\text{int}}\right)_{est} \tag{6.53}$$

Thus, iterations until $\Delta\mathbf{R}_{k+1}^{err}$ is less than a specified limit will be required within each time step. Initial conditions and stability criteria are identical to those presented above for the numeric integration methods.

Elaboration 6.2: Determination of Cross Sectional Forces

Having determined a sufficiently long time window of the response vector $\mathbf{r}_k$ the corresponding element end forces associated with element number n

$$\mathbf{F}_{n_k} = \begin{bmatrix} \mathbf{F}_1 \\ \mathbf{F}_2 \end{bmatrix}_{n_k} \quad \text{where} \quad \begin{cases} \mathbf{F}_{1_{n_k}} = \begin{bmatrix} F_1 & F_2 & F_3 & F_4 & F_5 & F_6 \end{bmatrix}_{n_k}^T \\ \mathbf{F}_{2_{n_k}} = \begin{bmatrix} F_7 & F_8 & F_9 & F_{10} & F_{11} & F_{12} \end{bmatrix}_{n_k}^T \end{cases}$$

may subsequently be obtained by (see Eq. 4.27):

$$\mathbf{F}_{n_k} = \mathbf{m}\ddot{\mathbf{d}}_{n_k} + \mathbf{c}\dot{\mathbf{d}}_{n_k} + \mathbf{k}\mathbf{d}_{n_k} \qquad \text{where} \qquad \mathbf{d}_{n_k} = \mathbf{A}_n \mathbf{r}_k$$

6.4 The Frequency Domain Solution in Original Coordinates

If the load is given in the form of its stochastic properties (mean values, covariance and cross spectral densities) a time domain solution may still be pursued, but this will then require the time domain simulation of the load components at all nodes (e.g. as suggested in Appendix B.3). However, in many cases it may be more appropriate to pursue a stochastic solution in frequency domain. Such a solution may be obtained in original degrees of freedom, or alternatively, and often far more convenient, in modal degrees of freedom. The solution strategy in original degrees of freedom is shown below, while the corresponding solution in modal degrees of freedom is shown in Chapter 6.5.

From a stochastic solution in frequency domain it is the corresponding statistical properties of the response that will emerge, i.e. the result of the response calculation is the covariance matrix

$$\mathbf{Cov}_{rr} = E\left[\mathbf{r}\cdot\mathbf{r}^T\right] = \begin{bmatrix} \sigma_1^2 & \cdots & Cov_{1i} & \cdots & Cov_{1j} & \cdots & Cov_{1N_r} \\ \vdots & \ddots & \vdots & & \vdots & & \vdots \\ Cov_{i1} & \cdots & \sigma_i^2 & \cdots & Cov_{ij} & & \vdots \\ \vdots & & \vdots & \ddots & \vdots & & \vdots \\ Cov_{j1} & \cdots & Cov_{ji} & \cdots & \sigma_j^2 & & \vdots \\ \vdots & & & & & \ddots & \vdots \\ Cov_{N_r1} & \cdots & \cdots & \cdots & \cdots & \cdots & \sigma_{N_r}^2 \end{bmatrix} \quad (6.54)$$

where N_r is the number of degrees of freedom in the system. Assuming stationary and Gaussian probability density distributions (of load and hence, also of the response), then extreme values of displacement events are given by

$$\mathbf{r}_{tot_{\max}} = \overline{\mathbf{r}} + \mathbf{r}_{\max} = \overline{\mathbf{r}} + k_p \boldsymbol{\sigma}_r \quad (6.55)$$

where $\overline{\mathbf{r}}$ is given in Eq. 6.1, k_p is a peak factor defined in Appendix A and $\boldsymbol{\sigma}_r$ is a vector containing all the standard deviations of the chosen set of displacement degrees of freedom in the system. $\boldsymbol{\sigma}_r$ may be extracted from the square root of the elements contained on the diagonal of the covariance matrix in Eq. 6.54. Since it is taken for granted that all load and response quantities are stationary a Fourier transform will render predictable coefficients throughout the entire time window of the process. Thus, a response calculation in frequency domain in original coordinates is based on the dynamic equilibrium equation given in Eq. 4.39. $\mathbf{M}\cdot\ddot{\mathbf{r}}(t)+\mathbf{C}\cdot\dot{\mathbf{r}}(t)+\mathbf{K}\cdot\mathbf{r}(t)=\mathbf{R}(t)$. Taking the Fourier transform throughout this equation, i.e. setting

$$\mathbf{r}(t)=\mathrm{Re}\sum_{\omega}\mathbf{a}_r(\omega)\cdot e^{i\omega t} \quad \text{and} \quad \mathbf{R}(t)=\mathrm{Re}\sum_{\omega}\mathbf{a}_R(\omega)\cdot e^{i\omega t} \tag{6.56}$$

where $\mathbf{a}_r(\omega)$ and $\mathbf{a}_R(\omega)$ are N_r by 1 vectors containing the Fourier coefficients of the displacement and load processes, then the dynamic equilibrium equation is satisfied for each ω-setting if

$$\left(-\mathbf{M}\omega^2+\mathbf{C}i\omega+\mathbf{K}\right)\cdot\mathbf{a}_r=\mathbf{a}_R \tag{6.57}$$

Thus,

$$\mathbf{a}_r=\mathbf{H}_r(\omega)\cdot\mathbf{a}_R \tag{6.58}$$

where

$$\mathbf{H}_r(\omega)=\left(-\mathbf{M}\omega^2+\mathbf{C}i\omega+\mathbf{K}\right)^{-1} \tag{6.59}$$

The cross spectral density matrix of the response quantities corresponding to the chosen degrees of freedom

$$\mathbf{S}_{rr}(\omega)=\lim_{T\to\infty}\frac{\left(\mathbf{a}_r^*\mathbf{a}_r^T\right)}{\pi T}=\begin{bmatrix} S_{r_1r_1} & \cdots & S_{r_1r_i} & \cdots & S_{r_1r_j} & \cdots & S_{r_1N_r} \\ \vdots & \ddots & \vdots & & \vdots & ⋰ & \vdots \\ S_{r_ir_1} & \cdots & S_{r_ir_i} & \cdots & S_{r_ir_j} & \cdots & S_{r_iN_r} \\ \vdots & & \vdots & \ddots & \vdots & & \vdots \\ S_{r_jr_1} & \cdots & S_{r_jr_i} & \cdots & S_{r_jr_j} & \cdots & S_{r_jN_r} \\ \vdots & & \vdots & & \vdots & \ddots & \vdots \\ S_{N_rr_1} & \cdots & S_{N_rr_i} & \cdots & S_{N_rr_j} & \cdots & S_{r_{N_r}r_{N_r}} \end{bmatrix} \tag{6.60}$$

may be obtained by

$$\begin{aligned}\mathbf{S}_{rr}(\omega)&=\lim_{T\to\infty}\frac{1}{\pi T}\left(\mathbf{a}_r^*\cdot\mathbf{a}_r^T\right)=\lim_{T\to\infty}\frac{1}{\pi T}\left[\left(\mathbf{H}_r\mathbf{a}_R\right)^*\cdot\left(\mathbf{H}_r\mathbf{a}_R\right)^T\right]\\&=\mathbf{H}_r^*\cdot\lim_{T\to\infty}\frac{1}{\pi T}\left(\mathbf{a}_R^*\cdot\mathbf{a}_R^T\right)\cdot\mathbf{H}_r^T=\mathbf{H}_r^*\cdot\mathbf{S}_{RR}\cdot\mathbf{H}_r^T\end{aligned} \tag{6.61}$$

where $\mathbf{S}_{RR}(\omega) = \lim_{T\to\infty} \frac{1}{\pi T}\left(\mathbf{a}_R^* \cdot \mathbf{a}_R^T\right)$ is the cross spectral density matrix of the load. As illustrated in Figs. 4.1 and 4.2 it is assumed that the load vector comprise contributions $\mathbf{R}_p$ from global external forces as well as contributions $\mathbf{R}_n$ from distributed element loads $\mathbf{q}_n(x,t)$, i.e. that

$$\mathbf{R} = \mathbf{R}_p(t) + \sum_{n=1}^{N} \mathbf{A}_n^T \cdot \mathbf{R}_n(t) \tag{6.62}$$

where N is the number of elements in the system. Defining the Fourier amplitude vectors $\mathbf{a}_{R_p}$ and $\mathbf{a}_{R_n}$ of $\mathbf{R}_p$ and $\mathbf{R}_n$, and assuming that the cross coherence between externally added load contributions $\mathbf{R}_p$ and those defined at element level $\mathbf{R}_n$ is zero or negligible, i.e. that $\mathbf{S}_{R_pR_n}(\omega) \approx 0$, then

$$\begin{aligned}
\mathbf{S}_{RR} &= \lim_{T\to\infty} \frac{1}{\pi T}\mathbf{a}_R^*\mathbf{a}_R^T = \lim_{T\to\infty} \frac{1}{\pi T}\left(\mathbf{a}_{R_p} + \sum_{n=1}^{N}\mathbf{A}_n^T\mathbf{a}_{R_n}\right)^*\left(\mathbf{a}_p + \sum_{n=1}^{N}\mathbf{A}_n^T\mathbf{a}_{R_n}\right)^T \\
&\approx \lim_{T\to\infty} \frac{1}{\pi T}\left(\mathbf{a}_{R_p}^*\mathbf{a}_{R_p}^T\right) + \lim_{T\to\infty} \frac{1}{\pi T}\left(\sum_{n=1}^{N}\mathbf{A}_n^T\mathbf{a}_{R_n}\right)^*\left(\sum_{n=1}^{N}\mathbf{A}_n^T\mathbf{a}_{R_n}\right)^T \\
&= \lim_{T\to\infty} \frac{1}{\pi T}\left(\mathbf{a}_{R_p}^*\mathbf{a}_{R_p}^T\right) + \sum_{n=1}^{N}\sum_{m=1}^{N}\mathbf{A}_n^T \lim_{T\to\infty} \frac{1}{\pi T}\left(\mathbf{a}_{R_n}^*\mathbf{a}_{R_m}^T\right)\mathbf{A}_m \\
&= \mathbf{S}_{R_pR_p}(\omega) + \sum_{n=1}^{N}\sum_{m=1}^{N}\mathbf{A}_n^T\mathbf{S}_{R_nR_m}(\omega)\mathbf{A}_m
\end{aligned} \tag{6.63}$$

where
$$\begin{cases}
\mathbf{S}_{R_pR_p}(\omega) = \lim_{T\to\infty} \frac{1}{\pi T}\left(\mathbf{a}_{R_p}^*\mathbf{a}_{R_p}^T\right) \\
\mathbf{S}_{R_nR_m}(\omega) = \lim_{T\to\infty} \frac{1}{\pi T}\left(\mathbf{a}_{R_n}^*\mathbf{a}_{R_m}^T\right)
\end{cases} \tag{6.64}$$

and where $\mathbf{S}_{R_pR_p}$ is an N_r by N_r matrix containing the cross spectral density of the R_p forces while $\mathbf{S}_{R_nR_m}$ is an N_r by N_r matrix containing the cross spectra between integrated element load forces of elements n and m. The response covariance matrix may then be obtained simply by integration throughout the frequency domain, i.e.

$$\mathbf{Cov}_{rr} = \begin{bmatrix} \sigma_1^2 & \cdots & Cov_{1i} & \cdots & Cov_{1j} & \cdots & Cov_{1N_r} \\ \vdots & \ddots & \vdots & & \vdots & & \vdots \\ Cov_{i1} & \cdots & \sigma_i^2 & \cdots & Cov_{ij} & & \vdots \\ \vdots & & \vdots & \ddots & \vdots & & \vdots \\ Cov_{j1} & \cdots & Cov_{ji} & \cdots & \sigma_j^2 & & \vdots \\ \vdots & & & & & \ddots & \vdots \\ Cov_{N_r 1} & \cdots & \cdots & \cdots & \cdots & \cdots & \sigma_{N_r}^2 \end{bmatrix} = \int_0^\infty \mathbf{S}_{rr}(\omega)\, d\omega \tag{6.65}$$

Elaboration 6.3: Covariance between Element End Forces

For the calculations of the ensuing stochastic properties of cross sectional response forces at element end points it is necessary also to determine the covariance between the displacement response and its derivatives. The general solution to the problem of determining the covariance between a stationary process $x(t)$ and its derivatives $\dot{x}(t)$ and $\ddot{x}(t)$ is given in Appendix A (see Eq. A.96), where it has been shown that for a stationary process $E[x \cdot \dot{x}] = E[\dot{x} \cdot \ddot{x}] = 0$. Then

$$\begin{bmatrix} \mathbf{Cov}_{rr} & \mathbf{Cov}_{r\dot{r}} & \mathbf{Cov}_{r\ddot{r}} \\ \mathbf{Cov}_{\dot{r}r} & \mathbf{Cov}_{\dot{r}\dot{r}} & \mathbf{Cov}_{\dot{r}\ddot{r}} \\ \mathbf{Cov}_{\ddot{r}r} & \mathbf{Cov}_{\ddot{r}\dot{r}} & \mathbf{Cov}_{\ddot{r}\ddot{r}} \end{bmatrix} = \int_0^\infty \begin{bmatrix} 1 & 0 & -\omega^2 \\ 0 & \omega^2 & 0 \\ -\omega^2 & 0 & \omega^4 \end{bmatrix} \cdot \mathbf{S}_{rr}\, d\omega$$

Since the displacement response vector associated with element number n is given by $\mathbf{d}_n = \mathbf{A}_n \cdot \mathbf{r}$ it is seen that

$$\begin{bmatrix} \mathbf{Cov}_{d_n \dot{d}_n} \\ \mathbf{Cov}_{\dot{d}_n d_n} \\ \mathbf{Cov}_{\dot{d}_n \ddot{d}_n} \\ \mathbf{Cov}_{\ddot{d}_n \dot{d}_n} \end{bmatrix} = E \begin{bmatrix} \mathbf{d} \cdot \dot{\mathbf{d}}^T \\ \dot{\mathbf{d}} \cdot \mathbf{d}^T \\ \dot{\mathbf{d}} \cdot \ddot{\mathbf{d}}^T \\ \ddot{\mathbf{d}} \cdot \dot{\mathbf{d}}^T \end{bmatrix}_n = E \begin{bmatrix} (\mathbf{A}_n \mathbf{r}) \cdot (\mathbf{A}_n \dot{\mathbf{r}})^T \\ (\mathbf{A}_n \dot{\mathbf{r}}) \cdot (\mathbf{A}_n \mathbf{r})^T \\ (\mathbf{A}_n \dot{\mathbf{r}}) \cdot (\mathbf{A}_n \ddot{\mathbf{r}})^T \\ (\mathbf{A}_n \ddot{\mathbf{r}}) \cdot (\mathbf{A}_n \dot{\mathbf{r}})^T \end{bmatrix} = \mathbf{A}_n \cdot E \begin{bmatrix} \mathbf{r} \cdot \dot{\mathbf{r}}^T \\ \dot{\mathbf{r}} \cdot \mathbf{r}^T \\ \dot{\mathbf{r}} \cdot \ddot{\mathbf{r}}^T \\ \ddot{\mathbf{r}} \cdot \dot{\mathbf{r}}^T \end{bmatrix} \cdot \mathbf{A}_n^T = \begin{bmatrix} \mathbf{0} \\ \mathbf{0} \\ \mathbf{0} \\ \mathbf{0} \end{bmatrix}$$

while

$$\begin{bmatrix} \mathbf{Cov}_{d_n d_n} \\ \mathbf{Cov}_{d_n \ddot{d}_n} \\ \mathbf{Cov}_{\ddot{d}_n d_n} \\ \mathbf{Cov}_{\dot{d}_n \dot{d}_n} \\ \mathbf{Cov}_{\ddot{d}_n \ddot{d}_n} \end{bmatrix} = E \begin{bmatrix} \mathbf{d}\mathbf{d}^T \\ \mathbf{d}\ddot{\mathbf{d}}^T \\ \ddot{\mathbf{d}}\mathbf{d}^T \\ \dot{\mathbf{d}}\dot{\mathbf{d}}^T \\ \ddot{\mathbf{d}}\ddot{\mathbf{d}}^T \end{bmatrix}_n = E \begin{bmatrix} (\mathbf{A}_n\mathbf{r})(\mathbf{A}_n\mathbf{r})^T \\ (\mathbf{A}_n\mathbf{r})(\mathbf{A}_n\ddot{\mathbf{r}})^T \\ (\mathbf{A}_n\ddot{\mathbf{r}})(\mathbf{A}_n\mathbf{r})^T \\ (\mathbf{A}_n\dot{\mathbf{r}})(\mathbf{A}_n\dot{\mathbf{r}})^T \\ (\mathbf{A}_n\ddot{\mathbf{r}})(\mathbf{A}_n\ddot{\mathbf{r}})^T \end{bmatrix} = \mathbf{A}_n E \begin{bmatrix} \mathbf{r}\mathbf{r}^T \\ \mathbf{r}\ddot{\mathbf{r}}^T \\ \ddot{\mathbf{r}}\mathbf{r}^T \\ \dot{\mathbf{r}}\dot{\mathbf{r}}^T \\ \ddot{\mathbf{r}}\ddot{\mathbf{r}}^T \end{bmatrix} \mathbf{A}_n^T = \mathbf{A}_n \begin{bmatrix} \mathbf{Cov}_{rr} \\ \mathbf{Cov}_{r\ddot{r}} \\ \mathbf{Cov}_{\ddot{r}r} \\ \mathbf{Cov}_{\dot{r}\dot{r}} \\ \mathbf{Cov}_{\ddot{r}\ddot{r}} \end{bmatrix} \mathbf{A}_n^T$$

Thus, the following is obtained
$$\begin{bmatrix} \mathbf{Cov}_{d_n d_n} \\ \mathbf{Cov}_{d_n \ddot{d}_n} \\ \mathbf{Cov}_{\ddot{d}_n d_n} \\ \mathbf{Cov}_{\dot{d}_n \dot{d}_n} \\ \mathbf{Cov}_{\ddot{d}_n \ddot{d}_n} \end{bmatrix} = \mathbf{A}_n \cdot \left(\int_0^\infty \begin{bmatrix} 1 \\ -\omega^2 \\ -\omega^2 \\ \omega^2 \\ \omega^4 \end{bmatrix} \cdot \mathbf{S}_{rr}(\omega) d\omega \right) \cdot \mathbf{A}_n^T$$

The response force vector

$$\mathbf{F}_n(t) = \begin{bmatrix} \mathbf{F}_1 \\ \mathbf{F}_2 \end{bmatrix}_n \quad \text{where} \quad \begin{cases} \mathbf{F}_{1_n} = [F_1 \quad F_2 \quad F_3 \quad F_4 \quad F_5 \quad F_6]_n^T \\ \mathbf{F}_{2_n} = [F_7 \quad F_8 \quad F_9 \quad F_{10} \quad F_{11} \quad F_{12}]_n^T \end{cases}$$

associated with element number n is defined by the local element dynamic equilibrium condition (see Eq. 4.27)

$$\mathbf{F}_n = \mathbf{m}_n \ddot{\mathbf{d}}_n + \mathbf{c}_n \dot{\mathbf{d}}_n + \mathbf{k}_n \mathbf{d}_n$$

and thus:

$$\begin{aligned} \mathbf{Cov}_{F_n F_n} &= E\left[\mathbf{F}_n \cdot \mathbf{F}_n^T\right] = E\left[\left(\mathbf{m}_n\ddot{\mathbf{d}}_n + \mathbf{c}_n\dot{\mathbf{d}}_n + \mathbf{k}_n\mathbf{d}_n\right) \cdot \left(\mathbf{m}_n\ddot{\mathbf{d}}_n + \mathbf{c}_n\dot{\mathbf{d}}_n + \mathbf{k}_n\mathbf{d}_n\right)^T\right] \\ &= \mathbf{m}_n \cdot E\left[\ddot{\mathbf{d}}_n \cdot \ddot{\mathbf{d}}_n^T\right] \cdot \mathbf{m}_n^T + \mathbf{c}_n \cdot E\left[\dot{\mathbf{d}}_n \cdot \dot{\mathbf{d}}_n^T\right] \cdot \mathbf{c}_n^T + \mathbf{k}_n \cdot E\left[\mathbf{d}_n \cdot \mathbf{d}_n^T\right] \cdot \mathbf{k}_n^T \\ &\quad + \mathbf{k}_n \cdot E\left[\mathbf{d}_n \cdot \ddot{\mathbf{d}}_n^T\right] \cdot \mathbf{m}_n^T + \mathbf{m}_n \cdot E\left[\ddot{\mathbf{d}}_n \cdot \mathbf{d}_n^T\right] \cdot \mathbf{k}_n^T \end{aligned}$$

from which the following is obtained:

$$\mathbf{Cov}_{F_nF_n} = E\left[\mathbf{F}_n \cdot \mathbf{F}_n^T\right] = \begin{bmatrix} \sigma_{F_1}^2 & \cdots & Cov_{F_1F_i} & \cdots & Cov_{F_1F_{12}} \\ \vdots & \ddots & \vdots & \therefore & \vdots \\ Cov_{F_iF_1} & \cdots & \sigma_{F_i}^2 & \cdots & Cov_{F_iF_{12}} \\ \vdots & \therefore & \vdots & \ddots & \vdots \\ Cov_{F_{12}F_1} & \cdots & Cov_{F_{12}F_i} & \cdots & \sigma_{F_{12}}^2 \end{bmatrix}$$

$$= \mathbf{m}_n \cdot \mathbf{Cov}_{\ddot{d}_n\ddot{d}_n} \mathbf{m}_n^T + \mathbf{c}_n \cdot \mathbf{Cov}_{\dot{d}_n\dot{d}_n} \mathbf{c}_n^T + \mathbf{k}_n \cdot \mathbf{Cov}_{d_nd_n} \mathbf{k}_n^T$$

$$+ \mathbf{m}_n \cdot \mathbf{Cov}_{\ddot{d}_nd_n} \mathbf{k}_n^T + \mathbf{k}_n \cdot \mathbf{Cov}_{d_n\ddot{d}_n} \mathbf{m}_n^T$$

where $\mathbf{Cov}_{\ddot{d}_n\ddot{d}_n}$, $\mathbf{Cov}_{\dot{d}_n\dot{d}_n}$, $\mathbf{Cov}_{d_nd_n}$, $\mathbf{Cov}_{\ddot{d}_nd_n}$ and $\mathbf{Cov}_{d_n\ddot{d}_n}$ are defined above. It should be noted that if damping has been defined at a global level (e.g. in the form of Rayleigh damping $\mathbf{C} = \alpha\mathbf{M} + \beta\mathbf{K}$), then the damping properties at element level should comply with the same choices of damping properties (i.e. $\mathbf{c}_n = \alpha\mathbf{m}_n + \beta\mathbf{k}_n$). It is also worth noting that if the chosen element length L_n is sufficiently small then the mass and damping terms above will be small, and hence $\mathbf{Cov}_{F_nF_n} \approx \mathbf{k}_n \cdot \mathbf{Cov}_{d_nd_n} \cdot \mathbf{k}_n^T$.

Elaboration 6.4: The Variance of Stress Components

Design calculations are intended to cover a certain unfavourable loading condition, e.g. an extreme wind or earthquake excitation that is characteristic to the particular place where the structure is located, and whose probability of occurrence is suitably small. In this situation it is the comparison of structural strength or capacity to the extreme value of some critical stress or stress resultant that is of interest. The situation is illustrated in Fig. 6.2. Since structural behaviour is assumed linear elastic, these quantities may in general be obtained from the extreme values of the displacements as shown above.

However, in this situation mean values are time invariants, and the response calculations have inevitably been based on predetermined values taken from standards or other design specifications. They have been established from authoritative sources to represent the characteristic values within a certain short term load condition chosen for the special purpose of design safety considerations. Therefore, in a particular design situation time invariant quantities may be considered as deterministic quantities, and thus, the mean values of displacements or stress resultants may be obtained directly from simple linear static calculations. I.e., it is only the fluctuating part of the response quantities that requires treatment as stochastic or transient processes. However, in a design situation it is necessary to consider the combined effects of stresses or stress resultants, and therefore, it is not only the standard deviation of processes that are of interest but also the covariance between fluctuating components. For instance, let a fluctuating (dynamic) displacement response at arbitrary position x

$$\begin{bmatrix} r_y \\ r_z \\ r_\theta \end{bmatrix} = \begin{bmatrix} \bar{r}_y(x) \\ \bar{r}_z(x) \\ \bar{r}_\theta(x) \end{bmatrix} + \begin{bmatrix} r_y(x,t) \\ r_z(x,t) \\ r_\theta(x,t) \end{bmatrix}$$

be associated with corresponding cross sectional moment and shear force components

$$\begin{bmatrix} M_y \\ M_z \\ M_x \\ V_y \\ V_z \end{bmatrix} = \begin{bmatrix} \bar{M}_y(x) \\ \bar{M}_z(x) \\ \bar{M}_x(x) \\ \bar{V}_y(x) \\ \bar{V}_z(x) \end{bmatrix} + \begin{bmatrix} M_y(x,t) \\ M_z(x,t) \\ M_x(x,t) \\ V_y(x,t) \\ V_z(x,t) \end{bmatrix}$$

Then the normal stress σ_x and shear stress τ_{yz} components at cross sectional position x are given by

$$\begin{bmatrix} \sigma_x \\ \tau_{yz} \end{bmatrix} = \begin{bmatrix} \bar{\sigma}_x \\ \bar{\tau}_{yz} \end{bmatrix} + \begin{bmatrix} \sigma_x(t) \\ \tau_{yz}(t) \end{bmatrix} = \begin{bmatrix} \dfrac{\bar{M}_y}{I_y} z + \dfrac{\bar{M}_z}{I_z} y \\ \dfrac{\bar{V}_y}{A_y} + \dfrac{\bar{V}_z}{A_z} + \dfrac{\bar{M}_x}{2A_m t_0} \end{bmatrix} + \begin{bmatrix} \dfrac{M_y}{I_y} z + \dfrac{M_z}{I_z} y \\ \dfrac{V_y}{A_y} + \dfrac{V_z}{A_z} + \dfrac{M_x}{2A_m t_0} \end{bmatrix}$$

where I_y and I_z are moments of inertia associated with bending about y and z axis, A_y and A_z are the cross sectional shear areas (rendering averaged values of shear stresses) and, for simplicity assuming that we are dealing with a closed box type of cross section, A_m is the sector area inscribed by the cross section and t_0 is the material thickness at position x. The variance of the normal stress is then given by

$$\begin{aligned} \sigma_{\sigma_x}^2 &= E\left[\sigma_x^2(t)\right] = E\left[\left(\frac{M_y}{I_y} z + \frac{M_z}{I_z} y\right)^2\right] \\ &= E\left[\left(\frac{M_y}{I_y} z\right)^2 + 2\left(\frac{M_y}{I_y} z\right)\cdot\left(\frac{M_z}{I_z} y\right) + \left(\frac{M_z}{I_z} y\right)^2\right] \\ &= \left(\frac{\sigma_{M_y}}{I_y} z\right)^2 + 2\frac{Cov_{M_y M_z}}{I_y I_z} yz + \left(\frac{\sigma_{M_z}}{I_z} y\right)^2 \end{aligned}$$

Where $\sigma_{M_y}^2 = E\left[M_y^2\right]$, $\sigma_{M_z}^2 = E\left[M_z^2\right]$ and $Cov_{M_y M_z} = E\left[M_y M_z\right]$. This may be further developed into

$$\sigma_{\sigma_x}^2 = \left(\frac{\sigma_{M_y}}{I_y} z\right)^2 + 2\left(\frac{\sigma_{M_y}}{I_y} z\right)\left(\frac{\sigma_{M_z}}{I_z} y\right)\rho_{M_y M_z} + \left(\frac{\sigma_{M_z}}{I_z} y\right)^2$$

where $\rho_{M_y M_z} = Cov_{M_y M_z} \Big/ \left(\sigma_{M_y}\sigma_{M_z}\right)$ is the covariance coefficient between M_y and M_z fluctuations. Similarly, the variance of the shear stress is given by

$$\sigma_{\tau_{yz}}^2 = E\left[\tau_{yz}^2(t)\right] = E\left[\left(\frac{V_y}{A_y} + \frac{V_z}{A_z} + \frac{M_x}{2A_m t_0}\right)^2\right]$$

$$= E\left[\left(\frac{V_y}{A_y}\right)^2 + \left(\frac{V_z}{A_z}\right)^2 + \left(\frac{M_x}{2A_m t_0}\right)^2 + 2\left(\frac{V_y}{A_y}\frac{V_z}{A_z}\right) + 2\left(\frac{V_y}{A_y}\frac{M_x}{2A_m t_0}\right) + 2\left(\frac{V_z}{A_z}\frac{M_x}{2A_m t_0}\right)\right]$$

$$= \left(\frac{\sigma_{V_y}}{A_y}\right)^2 + \left(\frac{\sigma_{V_z}}{A_z}\right)^2 + \left(\frac{\sigma_{M_x}}{2A_m t_0}\right)^2 + 2\frac{Cov_{V_y V_z}}{A_y A_z} + 2\frac{Cov_{V_y M_x}}{A_y\left(2A_m t_0\right)} + 2\frac{Cov_{V_z M_x}}{A_z\left(2A_m t_0\right)}$$

where $\begin{bmatrix} \sigma_{V_y}^2 \\ \sigma_{V_z}^2 \\ \sigma_{M_x}^2 \end{bmatrix} = E\begin{bmatrix} V_y^2 \\ V_z^2 \\ M_x^2 \end{bmatrix}$ and $\begin{bmatrix} Cov_{V_y V_z} \\ Cov_{V_y M_x} \\ Cov_{V_z M_x} \end{bmatrix} = E\begin{bmatrix} V_y V_z \\ V_y M_x \\ V_z M_x \end{bmatrix}$

I.e.

$$\sigma_{\tau_{yz}}^2 = \left(\frac{\sigma_{V_y}}{A_y}\right)^2 + \left(\frac{\sigma_{V_z}}{A_z}\right)^2 + \left(\frac{\sigma_{M_x}}{2A_m t_0}\right)^2$$
$$+2\frac{\sigma_{V_y}\sigma_{V_z}}{A_y A_z}\rho_{V_y V_z} + 2\frac{\sigma_{V_y}\sigma_{M_x}}{A_y\left(2A_m t_0\right)}\rho_{V_y M_x} + 2\frac{\sigma_{V_z}\sigma_{M_x}}{A_z\left(2A_m t_0\right)}\rho_{V_z M_x}$$

where $\begin{bmatrix} \rho_{V_y V_z} \\ \rho_{V_y M_x} \\ \rho_{V_z M_x} \end{bmatrix} = \begin{bmatrix} Cov_{V_y V_z} \Big/ \left(\sigma_{V_y}\sigma_{V_z}\right) \\ Cov_{V_y M_x} \Big/ \left(\sigma_{V_y}\sigma_{M_x}\right) \\ Cov_{V_z M_x} \Big/ \left(\sigma_{V_z}\sigma_{M_x}\right) \end{bmatrix}$

6.5 The Frequency Domain Solution in Modal Coordinates

By defining the mode shape matrix $\mathbf{\Phi} = \begin{bmatrix} \boldsymbol{\varphi}_1 & \cdots & \boldsymbol{\varphi}_n & \cdots & \boldsymbol{\varphi}_{N_{\text{mod}}} \end{bmatrix}$ and modal coordinates $\boldsymbol{\eta}(t) = \begin{bmatrix} \eta_1 & \cdots & \eta_n & \cdots & \eta_{N_{\text{mod}}} \end{bmatrix}^T$ such that

$$\mathbf{r}(t) = \mathbf{\Phi} \cdot \boldsymbol{\eta}(t) \tag{6.66}$$

then it was shown in Chapter 5 that the dynamic equilibrium condition in original discrete coordinates

$$\mathbf{M}\,\ddot{\mathbf{r}}(t) + \mathbf{C}\dot{\mathbf{r}}(t) + \mathbf{K}\mathbf{r}(t) = \mathbf{R}(t) \tag{6.67}$$

may be transformed into an equivalent equilibrium condition in modal coordinates

$$\tilde{\mathbf{M}}\ddot{\boldsymbol{\eta}}(t) + \tilde{\mathbf{C}}\dot{\boldsymbol{\eta}}(t) + \tilde{\mathbf{K}}\boldsymbol{\eta}(t) = \tilde{\mathbf{R}}(t) \tag{6.68}$$

(see Eqs. 4.39 and 5.13 or 5.32). Taking the Fourier transform throughout this equation, i.e. setting

$$\boldsymbol{\eta}(t) = \operatorname{Re}\sum_{\omega} \mathbf{a}_{\eta}(\omega) \cdot e^{i\omega t} \quad \text{and} \quad \tilde{\mathbf{R}}(t) = \operatorname{Re}\sum_{\omega} \mathbf{a}_{\tilde{R}}(\omega) \cdot e^{i\omega t} \tag{6.69}$$

where $\mathbf{a}_{\eta}(\omega)$ and $\mathbf{a}_{\tilde{R}}(\omega)$ are N_{mod} by 1 vectors containing the Fourier coefficients of the modal coordinates and the modal load, and pre-multiplying by $\tilde{\mathbf{K}}^{-1}$, then the modal dynamic equilibrium equation is satisfied for each ω-setting if

$$\left(-\tilde{\mathbf{K}}^{-1}\tilde{\mathbf{M}}\omega^2 + \tilde{\mathbf{K}}^{-1}\tilde{\mathbf{C}}i\omega + \mathbf{I}\right) \cdot \mathbf{a}_{\eta} = \tilde{\mathbf{K}}^{-1}\mathbf{a}_{\tilde{R}} \tag{6.70}$$

Recalling that

$$\tilde{\mathbf{M}} = diag\left[\tilde{M}_n\right] \qquad \tilde{\mathbf{K}} = diag\left[\tilde{K}_n\right] \qquad \text{and} \qquad \tilde{\mathbf{C}} = diag\left[\tilde{C}_n\right] \tag{6.71}$$

where $\tilde{M}_n = \boldsymbol{\varphi}_n^T \mathbf{M} \boldsymbol{\varphi}_n$, $\tilde{K}_n = \omega_n^2 \tilde{M}_n$ and $\tilde{C}_n = 2\tilde{M}_n \omega_n \zeta_n$ (see Eqs. 5.11, 5.14, 5.16 and 5.17), then

$$\mathbf{a}_{\eta} = \hat{\mathbf{H}}_{\eta}(\omega) \cdot \tilde{\mathbf{K}}^{-1} \cdot \mathbf{a}_{\tilde{R}} \tag{6.72}$$

where

$$\left.\begin{aligned}&\hat{\mathbf{H}}_\eta(\omega)=\left\{\mathbf{I}-diag\left[(\omega/\omega_n)^2\right]+2i\cdot diag(\zeta_n\omega/\omega_n)\right\}^{-1}=diag\left[\hat{H}_{\eta_n}\right]\\&\text{where}\quad \hat{H}_{\eta_n}=\left[1-(\omega/\omega_n)^2+2i\zeta_n\omega/\omega_n\right]^{-1}\end{aligned}\right\}\qquad(6.73)$$

The cross spectral density matrix of the modal coordinates is defined by

$$\mathbf{S}_{\eta\eta}(\omega)=\lim_{T\to\infty}\frac{\mathbf{a}_\eta^*\cdot\mathbf{a}_\eta^T}{\pi T}=\begin{bmatrix}S_{\eta_1\eta_1} & \cdots & S_{\eta_1\eta_m} & \cdots & S_{\eta_1\eta_{N_{\text{mod}}}}\\ \vdots & \ddots & \vdots & \iddots & \vdots\\ S_{\eta_n\eta_1} & \cdots & S_{\eta_n\eta_m} & \cdots & S_{\eta_n\eta_{N_{\text{mod}}}}\\ \vdots & \iddots & \vdots & \ddots & \vdots\\ S_{\eta_{N_{\text{mod}}}\eta_1} & \cdots & S_{\eta_{N_{\text{mod}}}\eta_m} & \cdots & S_{\eta_{N_{\text{mod}}}\eta_{N_{\text{mod}}}}\end{bmatrix}\qquad(6.74)$$

Introducing Eq. 6.72, then

$$\begin{aligned}\mathbf{S}_{\eta\eta}(\omega)&=\lim_{T\to\infty}\frac{1}{\pi T}\left(\mathbf{a}_\eta^*\cdot\mathbf{a}_\eta^T\right)=\lim_{T\to\infty}\frac{1}{\pi T}\left[\left(\hat{\mathbf{H}}_\eta\tilde{\mathbf{K}}^{-1}\mathbf{a}_{\tilde{R}}\right)^*\cdot\left(\hat{\mathbf{H}}_\eta\tilde{\mathbf{K}}^{-1}\mathbf{a}_{\tilde{R}}\right)^T\right]\\&=\hat{\mathbf{H}}_\eta^*\tilde{\mathbf{K}}^{-1}\cdot\lim_{T\to\infty}\frac{1}{\pi T}\left(\mathbf{a}_{\tilde{R}}^*\cdot\mathbf{a}_{\tilde{R}}^T\right)\cdot\left(\tilde{\mathbf{K}}^{-1}\right)^T\hat{\mathbf{H}}_\eta^T=\hat{\mathbf{H}}_\eta^*\tilde{\mathbf{K}}^{-1}\cdot\mathbf{S}_{\tilde{R}\tilde{R}}\cdot\left(\tilde{\mathbf{K}}^{-1}\right)^T\hat{\mathbf{H}}_\eta^T\end{aligned}\qquad(6.75)$$

where $\mathbf{S}_{\tilde{R}\tilde{R}}(\omega)=\lim_{T\to\infty}\frac{1}{\pi T}\left(\mathbf{a}_{\tilde{R}}^*\cdot\mathbf{a}_{\tilde{R}}^T\right)$ is the cross spectral density matrix of the modal load. Furthermore, since $\mathbf{r}(t)=\mathbf{\Phi}\boldsymbol{\eta}(t)$ then $\mathbf{a}_r=\mathbf{\Phi}\cdot\mathbf{a}_\eta$, and thus

$$\begin{aligned}\mathbf{S}_{rr}(\omega)&=\lim_{T\to\infty}\frac{1}{\pi T}\left(\mathbf{a}_r^*\mathbf{a}_r^T\right)=\lim_{T\to\infty}\frac{1}{\pi T}\left[\left(\mathbf{\Phi}\mathbf{a}_\eta\right)^*\left(\mathbf{\Phi}\mathbf{a}_\eta\right)^T\right]\\&=\mathbf{\Phi}\lim_{T\to\infty}\frac{1}{\pi T}\left(\mathbf{a}_\eta^*\mathbf{a}_\eta^T\right)\mathbf{\Phi}^T=\mathbf{\Phi}\mathbf{S}_{\eta\eta}(\omega)\mathbf{\Phi}^T=\mathbf{\Phi}\hat{\mathbf{H}}_\eta^*\tilde{\mathbf{K}}^{-1}\mathbf{S}_{\tilde{R}\tilde{R}}\left(\tilde{\mathbf{K}}^{-1}\right)^T\hat{\mathbf{H}}_\eta^T\mathbf{\Phi}^T\end{aligned}\qquad(6.76)$$

A Discrete Format

Since $\tilde{\mathbf{R}}(t)=\mathbf{\Phi}^T\mathbf{R}(t)$ then $\mathbf{a}_{\tilde{R}}=\mathbf{\Phi}^T\cdot\mathbf{a}_R$, and thus

$$\begin{aligned}\mathbf{S}_{\tilde{R}\tilde{R}}(\omega) &= \lim_{T\to\infty}\frac{1}{\pi T}\left(\mathbf{a}_{\tilde{R}}^{*}\cdot\mathbf{a}_{\tilde{R}}^{T}\right) = \lim_{T\to\infty}\frac{1}{\pi T}\left[\left(\mathbf{\Phi}^{T}\cdot\mathbf{a}_{R}\right)^{*}\cdot\left(\mathbf{\Phi}^{T}\cdot\mathbf{a}_{R}\right)^{T}\right]\\ &= \mathbf{\Phi}^{T}\cdot\lim_{T\to\infty}\frac{1}{\pi T}\left(\mathbf{a}_{R}^{*}\cdot\mathbf{a}_{R}^{T}\right)\cdot\mathbf{\Phi} = \mathbf{\Phi}^{T}\cdot\mathbf{S}_{RR}(\omega)\cdot\mathbf{\Phi}\end{aligned} \tag{6.77}$$

By combination of Eqs. 6.75 – 6.77, then the following is obtained:

$$\Rightarrow \begin{aligned}\mathbf{S}_{rr}(\omega) &= \mathbf{\Phi}\mathbf{S}_{\eta\eta}(\omega)\mathbf{\Phi}^{T} = \mathbf{\Phi}\left(\hat{\mathbf{H}}_{\eta}^{*}\tilde{\mathbf{K}}^{-1}\mathbf{S}_{\tilde{R}\tilde{R}}\cdot\left(\tilde{\mathbf{K}}^{-1}\right)^{T}\hat{\mathbf{H}}_{\eta}^{T}\right)\mathbf{\Phi}^{T}\\ &= \mathbf{\Phi}\left\{\hat{\mathbf{H}}_{\eta}^{*}(\omega)\tilde{\mathbf{K}}^{-1}\left[\mathbf{\Phi}^{T}\mathbf{S}_{RR}(\omega)\mathbf{\Phi}\right]\left(\tilde{\mathbf{K}}^{-1}\right)^{T}\hat{\mathbf{H}}_{\eta}^{T}(\omega)\right\}\mathbf{\Phi}^{T}\end{aligned} \tag{6.78}$$

where $\mathbf{S}_{RR}(\omega)$ is defined in Eq. 6.63.

A Continuous Format

Let us also consider a continuous format (see Chapter 5.3, Eqs. 5.18 – 5.22 and 5.32) and for simplicity assume that the entire load is defined by $\mathbf{q}(x,t) = \begin{bmatrix} q_y & q_z & q_\theta \end{bmatrix}^T$ (i.e. $\mathbf{R}_p = \mathbf{0}$ and $q_x = 0$), then

$$\tilde{\mathbf{R}} = \int_L \mathbf{\Phi}^T \mathbf{q}\, dx = \begin{bmatrix} \tilde{R}_1 & \cdots & \tilde{R}_n & \cdots & \tilde{R}_{N_{\text{mod}}} \end{bmatrix}^T \tag{6.79}$$

The Fourier transform of $\tilde{\mathbf{R}}(t)$ is then

$$\mathbf{a}_{\tilde{R}}(\omega) = \int_L \mathbf{\Phi}^T(x)\cdot\mathbf{a}_q(x,\omega)\, dx \quad \text{where } \mathbf{a}_q(x,\omega) = \begin{bmatrix} a_{q_y} & a_{q_z} & a_{q\theta} \end{bmatrix}^T \tag{6.80}$$

Thus

$$\begin{aligned}\mathbf{S}_{\tilde{R}\tilde{R}}(\omega) &= \lim_{T\to\infty}\frac{1}{\pi T}\mathbf{a}_{\tilde{R}}^{*}\mathbf{a}_{\tilde{R}}^{T} = \lim_{T\to\infty}\frac{1}{\pi T}\left(\int_L \mathbf{\Phi}^T\mathbf{a}_q dx\right)^{*}\left(\int_L \mathbf{\Phi}^T\mathbf{a}_q dx\right)^{T}\\ &= \int_L\int_L \mathbf{\Phi}^T(x_1)\cdot\lim_{T\to\infty}\frac{1}{\pi T}\mathbf{a}_q^{*}(x_1,\omega)\cdot\mathbf{a}_q^{T}(x_2,\omega)\cdot\mathbf{\Phi}(x_2)\,dx_1 dx_2\\ &= \int_L\int_L \mathbf{\Phi}^T(x_1)\cdot\mathbf{S}_{qq}(x_1,x_2,\omega)\cdot\mathbf{\Phi}(x_2)\,dx_1 dx_2\end{aligned} \tag{6.81}$$

where $$\mathbf{S}_{qq}(x_1,x_2,\omega)=\begin{bmatrix} S_{q_y q_y} & S_{q_y q_z} & S_{q_y q_\theta} \\ S_{q_z q_y} & S_{q_z q_z} & S_{q_z q_\theta} \\ S_{q_\theta q_y} & S_{q_\theta q_z} & S_{q_\theta q_\theta} \end{bmatrix} \tag{6.82}$$

Thus

$$\mathbf{S}_{rr}(x,\omega)=\begin{bmatrix} S_{r_y r_y} & S_{r_y r_z} & S_{r_y r_\theta} \\ S_{r_z r_y} & S_{r_z r_z} & S_{r_z r_\theta} \\ S_{r_\theta r_y} & S_{r_\theta r_z} & S_{r_\theta r_\theta} \end{bmatrix}= \\ \mathbf{\Phi}(x)\left\{\hat{\mathbf{H}}_\eta^*(\omega)\tilde{\mathbf{K}}^{-1}\left[\int\limits_L\int\limits_L \mathbf{\Phi}^T(x_1)\mathbf{S}_{qq}(x_1,x_2,\omega)\mathbf{\Phi}(x_2)\,dx_1 dx_2\right]\left(\tilde{\mathbf{K}}^{-1}\right)^T\hat{\mathbf{H}}_\eta^T(\omega)\right\}\mathbf{\Phi}^T(x) \tag{6.83}$$

As can be seen, if the load is stochastic and non-coherent in space then the calculations of the dynamic effects will inevitably involve spatial averaging.

6.6 The State-Space Equation and the Duhamel Integral

In elaboration 4.5 the damped eigenvalue problem was solved by the introduction of an additional dummy equation $\mathbf{I}\dot{\mathbf{r}} - \mathbf{I}\dot{\mathbf{r}} = \mathbf{0}$ and a substitute variable

$$\mathbf{z}(t)=\begin{bmatrix}\mathbf{r} & \dot{\mathbf{r}}\end{bmatrix}^T \tag{6.84}$$

Thus, the equilibrium equation $\mathbf{M}\ddot{\mathbf{r}}(t)+\mathbf{C}\dot{\mathbf{r}}(t)+\mathbf{K}\mathbf{r}(t)=\mathbf{R}(t)$ may be expanded into

$$\begin{bmatrix}\mathbf{I} & \mathbf{0}\\ \mathbf{0} & \mathbf{M}\end{bmatrix}\begin{bmatrix}\dot{\mathbf{r}}\\ \ddot{\mathbf{r}}\end{bmatrix}+\begin{bmatrix}\mathbf{0} & -\mathbf{I}\\ \mathbf{K} & \mathbf{C}\end{bmatrix}\begin{bmatrix}\mathbf{r}\\ \dot{\mathbf{r}}\end{bmatrix}=\begin{bmatrix}\mathbf{0}\\ \mathbf{R}\end{bmatrix} \tag{6.85}$$

Pre-multiplying by

$$\begin{bmatrix}\mathbf{I} & \mathbf{0}\\ \mathbf{0} & \mathbf{M}^{-1}\end{bmatrix} \tag{6.86}$$

$$\Rightarrow \qquad \begin{bmatrix}\mathbf{I} & \mathbf{0}\\ \mathbf{0} & \mathbf{I}\end{bmatrix}\begin{bmatrix}\dot{\mathbf{r}}\\ \ddot{\mathbf{r}}\end{bmatrix}+\begin{bmatrix}\mathbf{0} & -\mathbf{I}\\ \mathbf{M}^{-1}\mathbf{K} & \mathbf{M}^{-1}\mathbf{C}\end{bmatrix}\begin{bmatrix}\mathbf{r}\\ \dot{\mathbf{r}}\end{bmatrix}=\begin{bmatrix}\mathbf{0}\\ \mathbf{M}^{-1}\mathbf{R}\end{bmatrix} \tag{6.87}$$

and introducing

$$\mathbf{D} = -\begin{bmatrix} \mathbf{0} & -\mathbf{I} \\ \mathbf{M}^{-1}\mathbf{K} & \mathbf{M}^{-1}\mathbf{C} \end{bmatrix} \quad \text{and} \quad \mathbf{Q} = \begin{bmatrix} \mathbf{0} \\ \mathbf{M}^{-1}\mathbf{R} \end{bmatrix} \tag{6.88}$$

then
$$\dot{\mathbf{z}}(t) = \mathbf{D} \cdot \mathbf{z}(t) + \mathbf{Q}(t) \tag{6.89}$$

This is the state-space equation. Its general solution is given by

$$\mathbf{z}(t) = \text{Re}\left\{ \mathbf{z}(t_0) \cdot e^{\mathbf{D} \cdot s_0} + \int_{t_0}^{t} e^{\mathbf{D} \cdot s} \cdot \mathbf{Q}(\tau) d\tau \right\} \tag{6.90}$$

where $s_0 = t - t_0$ and $s = t - \tau$, and where the mathematical operation $e^{\mathbf{D}s} = \left(e^{\mathbf{D}}\right)^s = \left(\sum_{k=0}^{\infty} \frac{1}{k!} \mathbf{D}^k \right)^s$. Let us for simplicity consider a single degree of freedom system, whose starting condition is defined by

$$\mathbf{z}(t_0 = 0) = \begin{bmatrix} r(t_0 = 0) \\ \dot{r}(t_0 = 0) \end{bmatrix} = \begin{bmatrix} 0 \\ 0 \end{bmatrix} \tag{6.91}$$

Then Eq. 6.89 is reduced to

$$\dot{\mathbf{z}}(t) = \mathbf{D}\mathbf{z}(t) + \mathbf{Q}(t) \tag{6.92}$$

where
$$\mathbf{z} = \begin{bmatrix} r \\ \dot{r} \end{bmatrix}, \quad \mathbf{D} = \begin{bmatrix} 0 & 1 \\ -K/M & -C/M \end{bmatrix} \quad \text{and} \quad \mathbf{Q} = \begin{bmatrix} 0 \\ R/M \end{bmatrix} \tag{6.93}$$

The general solution is given by

$$\mathbf{z}(t) = \text{Re}\left\{ \int_{0}^{t} e^{\mathbf{D} \cdot s} \cdot \mathbf{Q}(\tau) d\tau \right\} \tag{6.94}$$

where $e^{\mathbf{D}s} = f(s) \cdot \mathbf{I} + g(s) \cdot \mathbf{D}$ and

$$f(s) = \begin{cases} (1 - \alpha_1 s) e^{\alpha_1 s} & \text{if } \alpha_1 = \alpha_2 \\ \dfrac{\alpha_1 e^{\alpha_2 s} - \alpha_2 e^{\alpha_1 s}}{\alpha_1 - \alpha_2} & \text{if } \alpha_1 \neq \alpha_2 \end{cases}, \quad g(s) = \begin{cases} s e^{\alpha_1 s} & \text{if } \alpha_1 = \alpha_2 \\ \dfrac{e^{\alpha_1 s} - e^{\alpha_2 s}}{\alpha_1 - \alpha_2} & \text{if } \alpha_1 \neq \alpha_2 \end{cases} \tag{6.95}$$

and where α_1 and α_2 are the roots in the damped eigenvalue problem

$$M\alpha^2 + C\alpha + K = 0 \tag{6.96}$$

I.e.:

$$\frac{\alpha}{\omega_0} = -\zeta_0 \pm \sqrt{\zeta_0^2 - 1} \quad \text{where} \quad \begin{cases} \omega_0 = \sqrt{K/M} \\ \zeta_0 = C/(2M\omega_0) \end{cases} \tag{6.97}$$

Assuming $\zeta_0 < 1 \quad \Rightarrow \quad \begin{cases} \alpha_1 = -\zeta_0\omega_0 + i\omega_d \\ \alpha_2 = -\zeta_0\omega_0 - i\omega_d \end{cases}$ where $\omega_d = \omega_0\sqrt{1-\zeta_0^2}$

Thus

$$\left.\begin{aligned} &\alpha_1 - \alpha_2 = 2i\omega_d \\ &e^{\alpha_1 s} = e^{-\zeta_0\omega_0 s}\cdot e^{i\omega_d s} = e^{-\zeta_0\omega_0 s}\cdot(\cos\omega_d s + i\sin\omega_d s) \\ &e^{\alpha_2 s} = e^{-\zeta_0\omega_0 s}\cdot e^{-i\omega_d s} = e^{-\zeta_0\omega_0 s}\cdot(\cos\omega_d s - i\sin\omega_d s) \end{aligned}\right\} \tag{6.98}$$

rendering

$$\left.\begin{aligned} &f(s) = e^{-\zeta_0\omega_0 s}\cdot\left[\cos\omega_d s + \zeta_0(\omega_0/\omega_d)\sin\omega_d s\right] \\ &g(s) = (1/\omega_d)e^{-\zeta_0\omega_0 s}\cdot\sin\omega_d s \end{aligned}\right\} \tag{6.99}$$

from which it is seen that

$$\begin{aligned} e^{\mathbf{D}s}\mathbf{Q} &= \{f(s)\cdot\mathbf{I} + g(s)\cdot\mathbf{D}\}\mathbf{Q} = \begin{bmatrix} f(s) & g(s) \\ -\omega_0^2 g(s) & f(s) - 2\zeta_0\omega_0 g(s) \end{bmatrix}\begin{bmatrix} 0 \\ R/M \end{bmatrix} \\ &= \frac{R}{M}\begin{bmatrix} g(s) \\ f(s) - 2\zeta_0\omega_0 g(s) \end{bmatrix} = \frac{R}{M}e^{-\zeta_0\omega_0 s}\begin{bmatrix} (1/\omega_d)\cdot\sin\omega_d s \\ \cos\omega_d s + \zeta_0(\omega_0/\omega_d)\sin\omega_d s \end{bmatrix} \end{aligned} \tag{6.100}$$

Thus, the following solution is obtained

$$\mathbf{z}(t) = \begin{bmatrix} r(t) \\ \dot{r}(t) \end{bmatrix} = \int_0^t \frac{R(\tau)}{M} e^{-\zeta_0\omega_0(t-\tau)} \begin{bmatrix} \dfrac{1}{\omega_d}\cdot\sin\omega_d(t-\tau) \\ \cos\omega_d(t-\tau) + \zeta_0\dfrac{\omega_0}{\omega_d}\sin\omega_d(t-\tau) \end{bmatrix} d\tau \tag{6.101}$$

The expression

$$h(t-\tau) = \frac{1}{M\omega_d}e^{-\zeta_0\omega_0(t-\tau)}\sin\omega_d(t-\tau) \tag{6.102}$$

is the unit impulse response function, and

$$r(t) = \int_0^t R(\tau) \cdot h(t-\tau) d\tau \qquad (6.103)$$

is the Duhamel integral. As illustrated in Fig. 6.5, $h(t-\tau)$ is the linear relationship between the impulse load $R_0 \cdot \Delta t$ and the corresponding incremental displacement response $r(t) = (R_0 \Delta t) h(t-\tau)$ and thus, as illustrated in Fig. 6.6, the response to an arbitrary load impulse sequence $R_n \Delta t$ $(n = 1, 2, 3, \dots)$ may be perceived as a continuous succession of such impulses, and thus the response may be obtained by integration, i.e.

$$r(t) = \lim_{\Delta t \to 0} \sum_{n=1}^{N} (R \cdot \Delta t)_n \cdot h(t - N \cdot \Delta t) = \int_0^t R(\tau) \cdot h(t-\tau) d\tau \qquad (6.104)$$

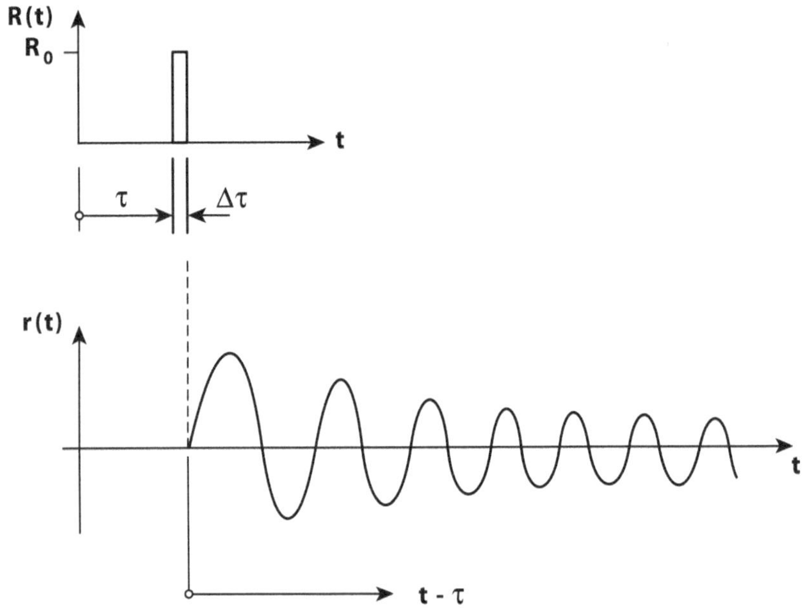

Fig. 6.5 Incremental displacement response using the Duhamel integral

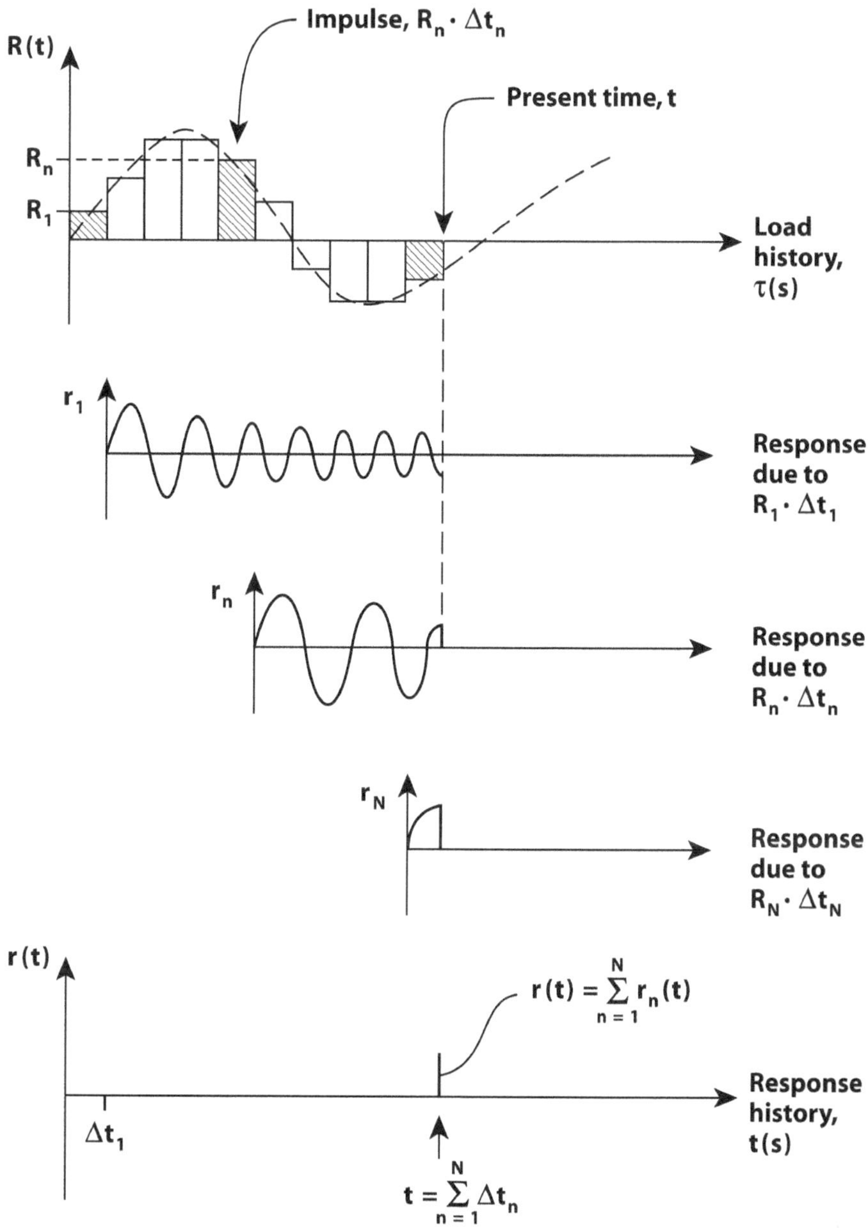

Fig. 6.6 The impulse response method

Equivalent expressions are obtained in modal coordinates. Let us substitute

$$\mathbf{r}(t) = \mathbf{\Phi}\boldsymbol{\eta}(t) \quad \text{where} \quad \begin{cases} \mathbf{\Phi} = \begin{bmatrix} \boldsymbol{\varphi}_1 & \cdots & \boldsymbol{\varphi}_n & \cdots & \boldsymbol{\varphi}_{N_{\text{mod}}} \end{bmatrix} \\ \boldsymbol{\varphi}_n = \begin{bmatrix} \phi_1 & \cdots & \phi_k & \cdots & \phi_{N_r} \end{bmatrix}_n^T \\ \boldsymbol{\eta}(t) = \begin{bmatrix} \eta_1 & \cdots & \eta_n & \cdots & \eta_{N_{\text{mod}}} \end{bmatrix}^T \end{cases} \tag{6.105}$$

into Eq. 6.85

$$\begin{bmatrix} \mathbf{I} & \mathbf{0} \\ \mathbf{0} & \mathbf{M} \end{bmatrix} \mathbf{\Phi} \begin{bmatrix} \dot{\boldsymbol{\eta}} \\ \ddot{\boldsymbol{\eta}} \end{bmatrix} + \begin{bmatrix} \mathbf{0} & -\mathbf{I} \\ \mathbf{K} & \mathbf{C} \end{bmatrix} \mathbf{\Phi} \begin{bmatrix} \boldsymbol{\eta} \\ \dot{\boldsymbol{\eta}} \end{bmatrix} = \begin{bmatrix} \mathbf{0} \\ \mathbf{R} \end{bmatrix} \tag{6.106}$$

and pre-multiply by $\begin{bmatrix} \mathbf{I} & \mathbf{0} \\ \mathbf{0} & \mathbf{\Phi}^T \end{bmatrix}$ will render the same in modal coordinates

$$\begin{bmatrix} \mathbf{I} & \mathbf{0} \\ \mathbf{0} & \tilde{\mathbf{M}} \end{bmatrix} \begin{bmatrix} \dot{\boldsymbol{\eta}} \\ \ddot{\boldsymbol{\eta}} \end{bmatrix} + \begin{bmatrix} \mathbf{0} & -\mathbf{I} \\ \tilde{\mathbf{K}} & \tilde{\mathbf{C}} \end{bmatrix} \begin{bmatrix} \boldsymbol{\eta} \\ \dot{\boldsymbol{\eta}} \end{bmatrix} = \begin{bmatrix} \mathbf{0} \\ \tilde{\mathbf{R}} \end{bmatrix} \quad \text{where} \quad \begin{cases} \tilde{\mathbf{M}} = \mathbf{\Phi}^T \mathbf{M} \mathbf{\Phi} \\ \tilde{\mathbf{K}} = \mathbf{\Phi}^T \mathbf{K} \mathbf{\Phi} \\ \tilde{\mathbf{C}} = \mathbf{\Phi}^T \mathbf{C} \mathbf{\Phi} \\ \tilde{\mathbf{R}} = \mathbf{\Phi}^T \mathbf{R} \end{cases} \tag{6.107}$$

Pre-multiplying by

$$\begin{bmatrix} \mathbf{I} & \mathbf{0} \\ \mathbf{0} & \tilde{\mathbf{M}}^{-1} \end{bmatrix} \tag{6.108}$$

$$\Rightarrow \quad \begin{bmatrix} \mathbf{I} & \mathbf{0} \\ \mathbf{0} & \mathbf{I} \end{bmatrix} \begin{bmatrix} \dot{\boldsymbol{\eta}} \\ \ddot{\boldsymbol{\eta}} \end{bmatrix} + \begin{bmatrix} \mathbf{0} & -\mathbf{I} \\ \tilde{\mathbf{M}}^{-1}\tilde{\mathbf{K}} & \tilde{\mathbf{M}}^{-1}\tilde{\mathbf{C}} \end{bmatrix} \begin{bmatrix} \boldsymbol{\eta} \\ \dot{\boldsymbol{\eta}} \end{bmatrix} = \begin{bmatrix} \mathbf{0} \\ \tilde{\mathbf{M}}^{-1}\tilde{\mathbf{R}} \end{bmatrix} \tag{6.109}$$

and introducing

$$\mathbf{y}(t) = \begin{bmatrix} \boldsymbol{\eta} \\ \dot{\boldsymbol{\eta}} \end{bmatrix}, \quad \tilde{\mathbf{D}} = -\begin{bmatrix} \mathbf{0} & -\mathbf{I} \\ \tilde{\mathbf{M}}^{-1}\tilde{\mathbf{K}} & \tilde{\mathbf{M}}^{-1}\tilde{\mathbf{C}} \end{bmatrix} \quad \text{and} \quad \tilde{\mathbf{Q}} = \begin{bmatrix} \mathbf{0} \\ \tilde{\mathbf{M}}^{-1}\tilde{\mathbf{R}} \end{bmatrix} \tag{6.110}$$

then

$$\dot{\mathbf{y}}(t) = \tilde{\mathbf{D}} \cdot \mathbf{y}(t) + \tilde{\mathbf{Q}}(t) \tag{6.111}$$

which is equivalent to that which was obtained in original coordinates in Eq. 6.89 above. Thus, the unit impulse response function and the Duhamel integrals for an arbitrary mode $\boldsymbol{\varphi}_n$ with corresponding eigenfrequency ω_n are given by

$$\left.\begin{aligned} \tilde{h}_n(t-\tau) &= \frac{1}{\tilde{M}_n \omega_{n_d}} e^{-\zeta_n \omega_n (t-\tau)} \sin \omega_{n_d}(t-\tau) \\ \eta_n(t) &= \int_0^t \tilde{R}_n(\tau) \cdot \tilde{h}_n(t-\tau) d\tau \end{aligned}\right\} \tag{6.112}$$

where $\omega_{n_d} = \omega_n \sqrt{1-\zeta_n^2}$. A direct solution strategy using the impulse response method in original degrees of freedom is not often pursued, because it involves the determination of an N_r by N_r transfer matrix containing the load effects in all p nodes ($p = 1, 2, \ldots, N_r$) due to a unit impulse in any arbitrary node number k ($k = 1, 2, \ldots, N_r$), which for a real system may be a formidable task. However, in a modal format, where all matrices in the equilibrium condition are diagonal, the method is effective and straight forward as long as the load is not unduly demanding. Defining an N_{mod} by 1 vector

$$\tilde{\mathbf{R}}(\tau) = \begin{bmatrix} \tilde{R}_1 & \cdots & \tilde{R}_n & \cdots & \tilde{R}_{N_{\text{mod}}} \end{bmatrix}^T \tag{6.113}$$

and an N_{mod} by N_{mod} diagonal matrix

$$\tilde{\mathbf{h}} = diag\left[\tilde{h}_n(t-\tau)\right] \tag{6.114}$$

then

$$\boldsymbol{\eta}(t) = \int_0^t \tilde{\mathbf{h}}(t-\tau) \cdot \tilde{\mathbf{R}}(\tau) d\tau \tag{6.115}$$

Elaboration 6.5: Impulse Loads

For a single degree of freedom system the impulse response function may be developed in a more physically direct way by defining a force $R(\tau)$ which has a largest value R_0 and is acting only during a short time period from $\tau = 0$ to $\tau = \Delta t$. It is assumed that during this period any build-up of elastic spring forces and viscous damping forces within the system may be ignored, such that instantly

$$M\ddot{r}(\tau) = R(\tau)$$

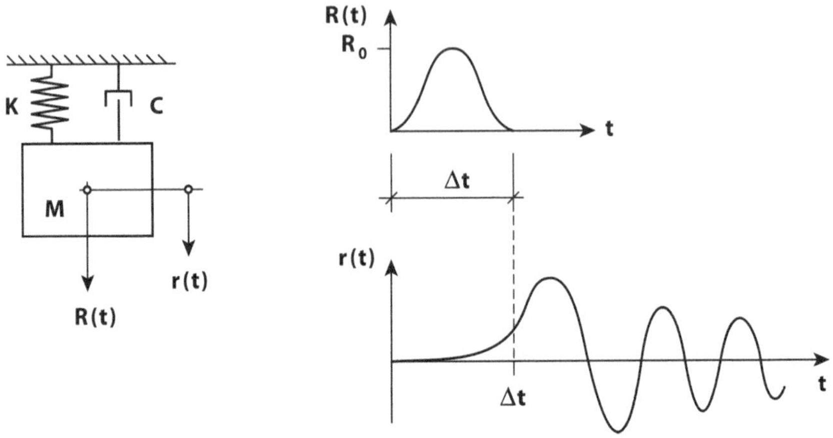

Fig. 6.7 The impulse response function

Integration will then render

$$\int_0^{\Delta t} M\ddot{r}(\tau)d\tau = M\left[\dot{r}(\tau)\right]_0^{\Delta t} = M\dot{r}(\Delta t) = \int_0^{\Delta t} R(\tau)dt \approx I \qquad \text{where} \qquad I = R_0\Delta t$$

I.e., any motion of the system during the impulse period Δt is ignored such that at the end of the impulse $r(\Delta t)=0$ and $\dot{r}(\Delta t)=I/M$. Thus, the response of the system from time $t=0$ and onwards is given by (see Eq. 2.30)

$$r(t) = \frac{\dot{r}_0}{\omega_d} e^{-\omega_n \zeta_n t} \sin(\omega_d t) \qquad \text{where} \qquad \dot{r}_0 = \dot{r}(\Delta t) = I/M$$

Thus, the impulse response function is given by

$$r(t) = \frac{I}{M\omega_d} e^{-\omega_n \zeta_n t} \sin(\omega_d t)$$

The link to the Duhamel integral is obvious as the introduction of the limiting situation that $I(\tau) = R(\tau)d\tau$ will render

$$dr = \frac{R(\tau)d\tau}{M\omega_d} e^{-\zeta_n \omega_n (t-\tau)} \sin \omega_d (t-\tau)$$

and thus

$$r(t) = \int_0^t dr = \int_0^t \frac{R(\tau)}{M\,\omega_d} e^{-\omega_n \zeta_n (t-\tau)} \sin \omega_d (t-\tau)\, d\tau$$

Example 6.1: The Step Load Case

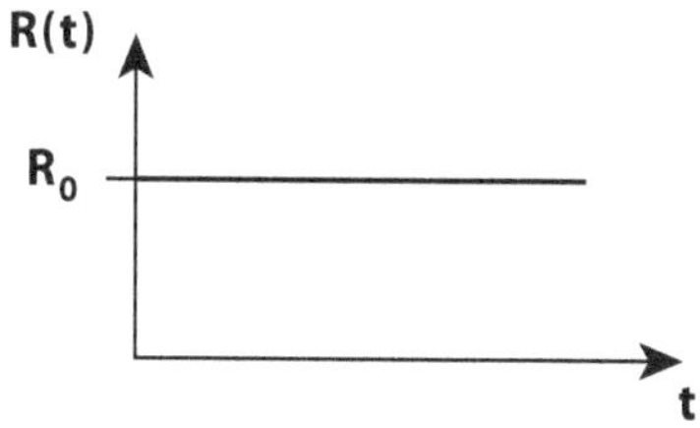

Fig. 6.8 The step load case

A solution to the step load case

$$r(t) = \frac{R_0}{M\,\omega_d} \int_0^t e^{-\omega_n \zeta_n (t-\tau)} \sin \omega_d (t-\tau)\, d\tau$$

may be obtained by the substitution $s = t - \tau$*, rendering*

$$r(t) = \frac{R_0}{M\,\omega_d} \int_t^0 e^{-\omega_n \zeta_n s} \sin(\omega_d s)(-ds) = \frac{R_0}{M\,\omega_d} \int_0^t e^{-\omega_n \zeta_n s} \sin(\omega_d s)\, ds$$

$$= \frac{R_0}{M\,\omega_d} \left\{ \frac{e^{-\omega_n \zeta_n s} \left[\omega_n \zeta_n \sin(\omega_d s) + \omega_d \cos(\omega_d s) \right]}{(-\omega_n \zeta_n)^2 + \omega_d^2} \right\}_0^t$$

Thus

$$r(t) = \frac{R_0}{K} \left\{ 1 - e^{-\omega_n \zeta_n t} \left[\frac{\omega_n}{\omega_d} \zeta_n \sin(\omega_d t) + \cos(\omega_d t) \right] \right\}$$

The solution is shown in the upper diagram in Fig. 6.10, in the special case that $\omega_n = 2\ rad/s$ *and* $\zeta_n = 0.02$.

Example 6.2: The Half Sinus Impulse Load Case

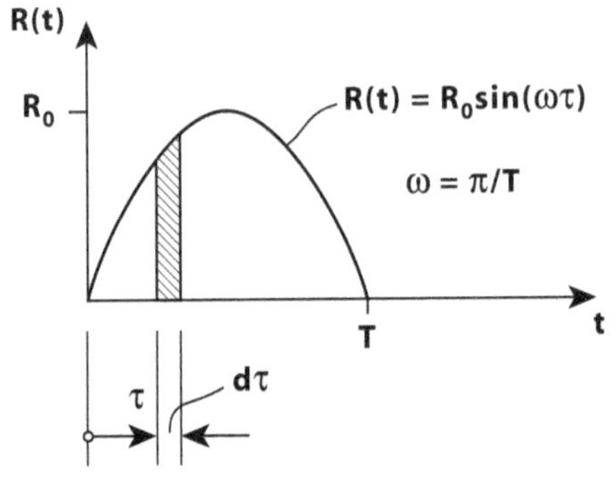

Fig. 6.9 The half sinus impulse load case

An approximate solution to the case of an impulse load in the shape of half a sine wave with duration T *(see Fig. 6.9 above) may be obtained by using the Duhamel integral. But, since an accurate solution to the problem of a sine load to a single degree of freedom system has been developed in Chapter 2.3, it is in this case possible to develop an accurate solution. First, at* $t \le T$ *the solution is given in Eq. 2.50, i.e.*

$$\frac{r(t)}{F_0/K} = \frac{\dfrac{\sin(\beta_p)}{\cos(\beta_h)} e^{-\omega_n \zeta_n t} \cos(\omega_d t - \beta_h) + \sin(\omega t - \beta_p)}{\sqrt{\left[1-(\omega/\omega_n)^2\right]^2 + (2\zeta_n \omega/\omega_n)^2}}$$

where $\tan(\beta_p) = \dfrac{2\zeta_n \omega/\omega_n}{1-(\omega/\omega_n)^2}$ *and* $\tan(\beta_h) = \dfrac{1-2\zeta_n^2 - (\omega/\omega_n)^2}{2\zeta_n\sqrt{1-\zeta_n^2}}$

and where the artificial load frequency $\omega = \pi/T$ *. Defining*

$$A_1 = \frac{\dfrac{\sin(\beta_p)}{\cos(\beta_h)} \cos(\omega_d T - \beta_h) + e^{\omega_n \zeta_n T} \sin(\beta_p)}{\sqrt{\left[1-(\omega/\omega_n)^2\right]^2 + (2\zeta_n \omega/\omega_n)^2}}$$

and

$$A_2 = \frac{-\dfrac{\sin(\beta_p)}{\cos(\beta_h)}\left[\sin(\omega_d T - \beta_h) + \zeta_n \dfrac{\omega_n}{\omega_d}\cos(\omega_d T - \beta_h)\right] + e^{\omega_n \zeta_n T}\dfrac{\omega}{\omega_d}\cos(\beta_p)}{\sqrt{\left[1-(\omega/\omega_n)^2\right]^2 + (2\zeta_n\,\omega/\omega_n)^2}}$$

it is seen that $\dfrac{r(t=T)}{R_0/K} = A_1 e^{-\omega_n \zeta_n T}$ *and* $\dfrac{\dot{r}(t=T)}{R_0/K} = A_2 \omega_d e^{-\omega_d \zeta_n T}$.

Second, at $t > T$ *then the system is free to oscillate without any loading, and thus, the solution is given in Eq. 2.26, i.e.*

$$r(t) = e^{-\omega_n \zeta_n t} \cdot \left[a_1 \sin(\omega_d t) + a_2 \cos(\omega_d t)\right]$$

The initial conditions to this freely decaying motion is that

$$r(t=T) = e^{-\omega_n \zeta_n T} \cdot \left[a_1 \sin(\omega_d T) + a_2 \cos(\omega_d T)\right] = \frac{R_0}{K} A_1 e^{-\omega_n \zeta_n T}$$

and

$$\dot{r}(t=T) = e^{-\omega_n \zeta_n T} \cdot \left\{a_1\left[\omega_d \cos(\omega_d T) - \omega_n \zeta_n \sin(\omega_d T)\right]\right.$$
$$\left. - a_2\left[\omega_d \sin(\omega_d T) + \omega_n \zeta_n \cos(\omega_d T)\right]\right\} = \frac{R_0}{K} A_2 \omega_d e^{-\omega_n \zeta_n T}$$

Thus, defining $\hat{a}_1 = a_1/(R_0/K)$ *and* $\hat{a}_2 = a_2/(R_0/K)$, *it is seen that*

$$\begin{bmatrix} \sin(\omega_d T) & \cos(\omega_d T) \\ \left\{\cos(\omega_d T) - \zeta_n \dfrac{\omega_n}{\omega_d}\sin(\omega_d T)\right\} & -\left\{\sin(\omega_d T) + \zeta_n \dfrac{\omega_n}{\omega_d}\cos(\omega_d T)\right\} \end{bmatrix} \begin{bmatrix} \hat{a}_1 \\ \hat{a}_2 \end{bmatrix} = \begin{bmatrix} A_1 \\ A_2 \end{bmatrix}$$

from which the following solution is obtained

$$\hat{a}_1 = A_1 \cdot \left[\sin(\omega_d T) + \zeta_n \frac{\omega_n}{\omega_d}\cos(\omega_d T)\right] + A_2 \cdot \cos(\omega_d T)$$

$$\hat{a}_2 = A_1 \cdot \left[\cos(\omega_d T) - \zeta_n \frac{\omega_n}{\omega_d}\sin(\omega_d T)\right] - A_2 \cdot \sin(\omega_d T)$$

Thus, at $t > T$

$$r(t) = (R_0/K) \cdot e^{-\omega_n \zeta_n t} \cdot \left[\hat{a}_1 \sin(\omega_d t) + \hat{a}_2 \cos(\omega_d t) \right]$$

The response will have its largest response $\left[r(t) \right]_{\max} \approx 1.75 \cdot (R_0/K)$ *if* $T/T_d \approx 0.8$, *where* $T_d = 2\pi/\omega_d$ *is the eigen-period of the system. This particular case is illustrated in the lower diagram in Fig. 6.10 below.*

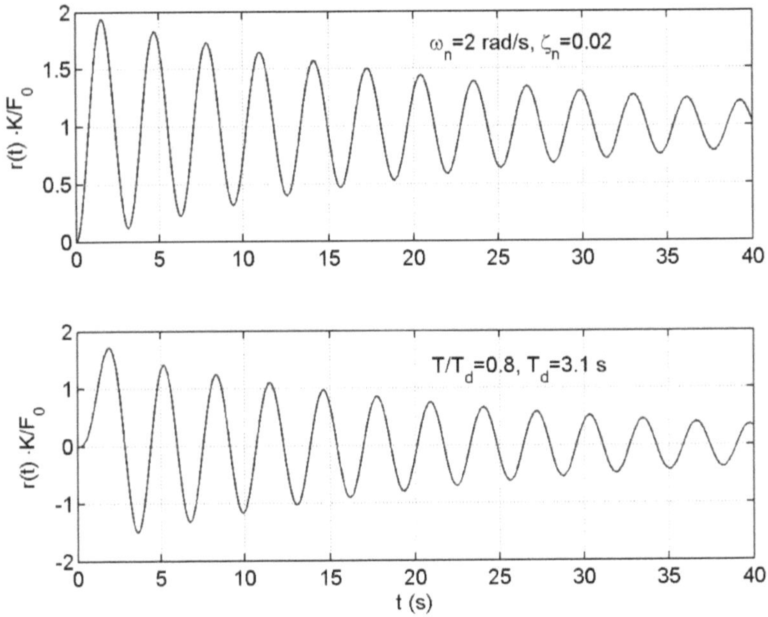

Fig. 6.10 Upper diagram: the step load case. Lower diagram: the half sinus impulse load case

Chapter 7
Dynamic Response to Earthquake Excitation

7.1 Introduction

Earthquake excitation of civil engineering structures is in general a complex process. However, an idealised situation where the structure is subject to representative single component horizontal ground acceleration will usually suffice for design purposes. The problem of dynamic response calculations under such excitation is pursued in the present chapter. For simplicity, the focus of the theory below is limited to structures with main extension in the vertical direction (e.g. vertical frames or cantilevered type of tower buildings). A typical time series of such a ground motion is illustrated in the upper diagram in Fig. 7.1. Its spectral

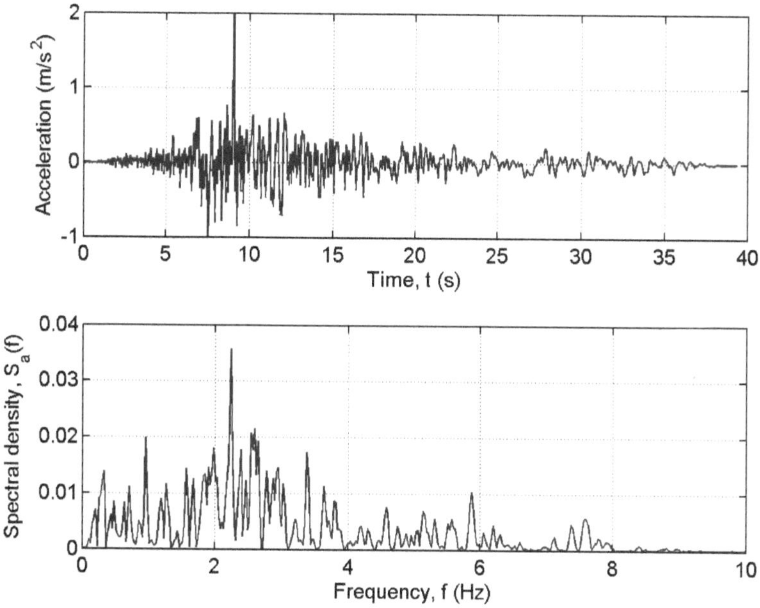

Fig. 7.1 Typical horizontal ground acceleration due to earthquake excitation (upper diagram: time domain, lower diagram: frequency domain)

E.N. Strømmen, *Structural Dynamics*, Springer Series in Solid and Structural Mechanics 2,
DOI: 10.1007/978-3-319-01802-7_7, © Springer International Publishing Switzerland 2014

density is shown in the lower diagram. As can be seen, a typical earthquake excitation is a non-periodic transient process. Hence, the perception of earthquake excitation as a random base acceleration process may not be suitable for a full frequency domain approach, as this may require some stationarity in the process. Thus, the dynamic response calculations for a system subject to earthquake excitation may often require a solution strategy in time domain. Nonetheless, in many cases a frequency domain solution may still render sufficiently accurate results, particularly if the response is close to quasi-static or narrow-banded resonant, i.e. if the system is either fairly stiff or else very slender.

7.2 Single Degree of Freedom Shear Frame

Let us first consider the simple case of a single storey shear frame, i.e. a single storey frame with an infinitely stiff beam and columns whose distributed mass is insignificant. The eigenvalue calculation for this case is shown in Example 2.2. The system is now subject to horizontal ground acceleration $\ddot{r}_g(t)$. The basic idea is that the total horizontal acceleration of the system is the ground acceleration plus the acceleration of the system itself, i.e. that $\ddot{r}_{tot} = \ddot{r}_g + \ddot{r}$, and thus, the equilibrium condition of the mass in motion is given by:

$$M\left(\ddot{r}_g + \ddot{r}\right) + C\dot{r} + Kr = 0 \quad \Rightarrow \quad M\ddot{r} + C\dot{r} + Kr = -M\ddot{r}_g \tag{7.1}$$

where $K = 2\left(12EI_y/L^3\right)$, M is the mass of the beam (assuming negligible column mass contribution) and C is the overall damping coefficient of the system. Taking the Fourier transform throughout Eq. 7.1, i.e. setting

$$\begin{bmatrix} r(t) \\ \ddot{r}_g(t) \end{bmatrix} = \mathrm{Re}\sum_{\omega} \begin{bmatrix} a_r(\omega) \\ a_{\ddot{r}_g}(\omega) \end{bmatrix} e^{i\omega t} \tag{7.2}$$

will then render

$$\left(-M\omega^2 + Ci\omega + K\right)a_r = -Ma_{\ddot{r}_g} \tag{7.3}$$

By pre-multiplication with K^{-1}, and the introduction of

$$\left.\begin{aligned} &\omega_n^2 = K/M \\ &\zeta_n = C/(2M\omega_n) = C\omega_n/(2K) \\ &\hat{H}(\omega) = \left[1 - (\omega/\omega_n)^2 + 2i\zeta_n\,\omega/\omega_n\right]^{-1} \end{aligned}\right\} \tag{7.4}$$

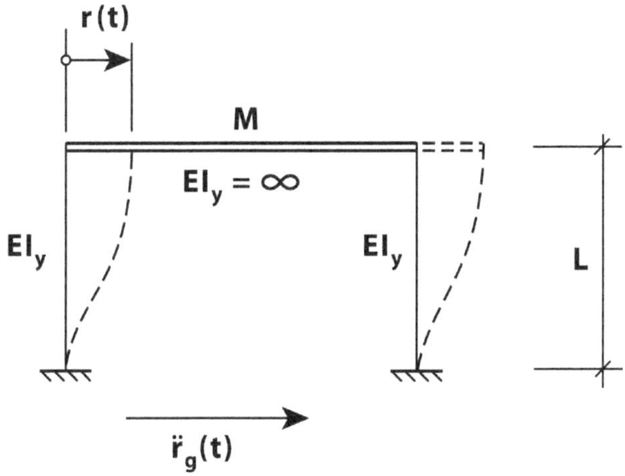

Fig. 7.2 Single storey shear frame

then the following is obtained

$$a_r(\omega) = -\hat{H}(\omega)\frac{M}{K}a_{\ddot{r}_g}(\omega) \tag{7.5}$$

The physical response $r(t)$ may then be obtained from the first row of Eq. 7.2. If the initial transient part of the response is disregarded, and the ground acceleration is narrow banded, i.e. it may be represented by a single harmonic component

$$\ddot{r}_g(t) \approx \mathrm{Re}\left[a_{\ddot{r}_g}(\omega_g)\cdot e^{i\omega_g t}\right] \tag{7.6}$$

then

$$\begin{aligned} r(t) &= \mathrm{Re}\left[a_r(\omega_g)\cdot e^{i\omega_g t}\right] = \mathrm{Re}\left[-\hat{H}(\omega_g)\frac{M}{K}a_{\ddot{r}_g}(\omega_g)\cdot e^{i\omega_g t}\right] \\ &= -\frac{M}{K}a_{\ddot{r}_g}\left|\hat{H}(\omega_g)\right|\cos(\omega_g t - \beta) = -\frac{a_{\ddot{r}_g}}{\omega_n^2}\left|\hat{H}(\omega_g)\right|\cos(\omega_g t - \beta) \end{aligned} \tag{7.7}$$

where $\tan\beta = (2\zeta_n\,\omega_g/\omega_n)\Big/\left[1-(\omega_g/\omega_n)^2\right]$. If, on the other hand, the ground acceleration may be represented by a stationary stochastic process whose spectral density is given by

$$S_{\ddot{r}_g}(\omega) = \lim_{T\to\infty} \frac{1}{\pi T} a^*_{\ddot{r}_g}(\omega) \cdot a_{\ddot{r}_g}(\omega) \tag{7.8}$$

Then, recalling that $K = \omega_n^2 M$, the spectral density of the corresponding dynamic response is given by

$$\begin{aligned} S_{rr}(\omega) &= \lim_{T\to\infty} \frac{1}{\pi T} a_r^* a_r = \lim_{T\to\infty} \frac{1}{\pi T} \left[-\hat{H}(\omega) \frac{M}{K} a_{\ddot{r}_g}(\omega) \right]^* \left[-\hat{H}(\omega) \frac{M}{K} a_{\ddot{r}_g}(\omega) \right] \\ &= \left(\frac{M}{K} \right)^2 \left| \hat{H}(\omega) \right|^2 \lim_{T\to\infty} \frac{1}{\pi T} a^*_{\ddot{r}_g}(\omega) \cdot a_{\ddot{r}_g}(\omega) = \omega_n^{-4} \left| \hat{H}(\omega) \right|^2 S_{\ddot{r}_g}(\omega) \end{aligned} \tag{7.9}$$

and the corresponding variance of the response $r(t)$ is given by

$$\sigma_r^2 = \int_0^\infty S_{rr}(\omega) d\omega = \omega_n^{-4} \int_0^\infty \left| \hat{H}(\omega) \right|^2 S_{\ddot{r}_g}(\omega) d\omega \tag{7.10}$$

σ_r^2 may most often with sufficient accuracy be split into a background part and a resonant part, i.e.

$$\begin{aligned} \sigma_r^2 &= \omega_n^{-4} \int_0^\infty \left| \hat{H}(\omega) \right|^2 S_{\ddot{r}_g}(\omega) d\omega \\ &\approx \omega_n^{-4} \left[\int_0^\infty \left| \hat{H}(\omega=0) \right|^2 S_{\ddot{r}_g}(\omega) d\omega + \int_0^\infty \left| \hat{H}(\omega) \right|^2 S_{\ddot{r}_g}(\omega=\omega_n) d\omega \right] \\ &= \omega_n^{-4} \left[\int_0^\infty S_{\ddot{r}_g}(\omega) d\omega + S_{\ddot{r}_g}(\omega=\omega_n) \int_0^\infty \left| \hat{H}(\omega) \right|^2 d\omega \right] \\ &= \omega_n^{-4} \left[\sigma_{\ddot{r}_g}^2 + \frac{\pi\omega_n}{4\zeta_n} S_{\ddot{r}_g}(\omega=\omega_n) \right] = \left(\frac{\sigma_{\ddot{r}_g}}{\omega_n^2} \right)^2 + \frac{\pi \cdot S_{\ddot{r}_g}(\omega=\omega_n)}{4\omega_n^3 \zeta_n} \end{aligned} \tag{7.11}$$

where $\sigma_{\ddot{r}_g}^2$ is the ground acceleration variance. It is seen that the background (quasi-static) part of the standard deviation of the response is $\sigma_{\ddot{r}_g} / \omega_n^2$ while the resonant part of the standard deviation is given by $\sqrt{\pi S_{\ddot{r}_g}(\omega_n) / (4\omega_n^3 \zeta_n)}$. (It should be noted that the peak factor of the background part is likely to be in the order of 3.5 – 4.5, see Eq. A.45, while the peak factor for the resonant part is likely to be slightly above $\sqrt{2}$).

If structural damping forces are small, then the shear force in the columns may be obtained by the equilibrium requirement that $2V_y(t) + M(\ddot{r} + \ddot{r}_g) = 0$, and since $M(\ddot{r} + \ddot{r}_g) + Kr = 0$, then

$$2V_y(t) = Kr(t) \tag{7.12}$$

Thus, if $\ddot{r}_g$ is narrow banded (close to harmonic), then

$$V_y(t) = \frac{K}{2}\left(-\frac{a_{\ddot{r}_g}}{\omega_n^2}\left|\hat{H}(\omega_g)\right|\cos(\omega_g t - \beta)\right) = -\frac{Ma_{\ddot{r}_g}}{2}\left|\hat{H}(\omega_g)\right|\cos(\omega_g t - \beta) \tag{7.13}$$

while, if $\ddot{r}_g$ is stochastic and broad banded, then

$$\sigma_{V_y} = \frac{K}{2}\sigma_r \approx \frac{K}{2}\sqrt{\left(\frac{\sigma_{\ddot{r}_g}}{\omega_n^2}\right)^2 + \frac{\pi S_{\ddot{r}_g}(\omega_n)}{4\omega_n^3\zeta_n}} = \frac{M}{2}\sqrt{\sigma_{\ddot{r}_g}^2 + \frac{\pi\omega_n S_{\ddot{r}_g}(\omega_n)}{4\zeta_n}} \tag{7.14}$$

7.3 Two Degrees of Freedom Shear Frame

A two storey frame with infinitely stiff beams and columns whose distributed mass is negligible is shown in Fig. 7.2. This is a two-degree of freedom system, $\mathbf{r} = \begin{bmatrix} r_1 & r_2 \end{bmatrix}^T$, and the eigenvalue calculations for this case is shown in Example 2.3. The system is now subject to a horizontal ground acceleration $\ddot{r}_g(t)$. Again, the basic idea is that the total horizontal acceleration is the ground acceleration plus the acceleration of the system itself, i.e. that

$$\ddot{\mathbf{r}}_{tot} = \begin{bmatrix} \ddot{r}_1 + \ddot{r}_g \\ \ddot{r}_2 + \ddot{r}_g \end{bmatrix} \tag{7.15}$$

and thus, the equilibrium condition in original degrees of freedom is given by

$$\mathbf{M}\ddot{\mathbf{r}}_{tot} + \mathbf{C}\dot{\mathbf{r}} + \mathbf{K}\mathbf{r} = \mathbf{0} \tag{7.16}$$

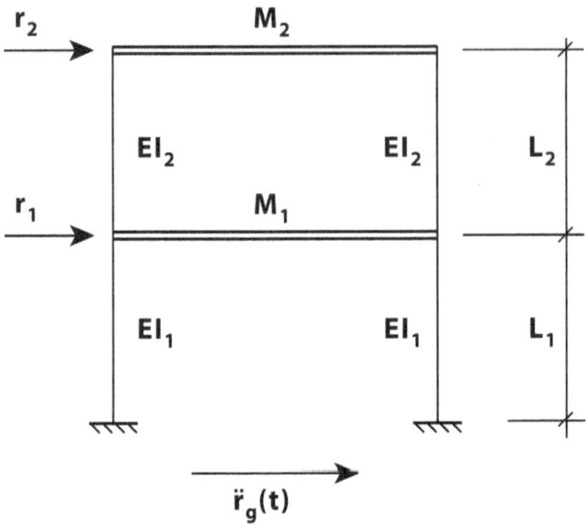

Fig. 7.3 Two storey shear frame

where $\mathbf{M} = diag(M_n)$, $n = 1$ or 2, and $\mathbf{C}$ and $\mathbf{K}$ are defined in Example 2.3. By defining

$$\mathbf{A}_g = [1 \quad 1]^T \tag{7.17}$$

this may also be written

$$\mathbf{M\ddot{r}} + \mathbf{C\dot{r}} + \mathbf{Kr} = -\mathbf{MA}_g \ddot{r}_g \tag{7.18}$$

By introducing modal coordinates

$$\mathbf{r} = \boldsymbol{\Phi} \cdot \boldsymbol{\eta} \qquad \text{where} \qquad \begin{cases} \boldsymbol{\eta} = [\eta_1 \quad \eta_2]^T \\ \boldsymbol{\Phi} = [\boldsymbol{\varphi}_1 \quad \boldsymbol{\varphi}_2] = \begin{bmatrix} \varphi_{11} & \varphi_{12} \\ \varphi_{21} & \varphi_{22} \end{bmatrix} \end{cases} \tag{7.19}$$

and pre-multiplication by $\boldsymbol{\Phi}^T$, then the modal dynamic equilibrium condition is given by

$$\tilde{\mathbf{M}}\ddot{\boldsymbol{\eta}} + \tilde{\mathbf{C}}\dot{\boldsymbol{\eta}} + \tilde{\mathbf{K}}\boldsymbol{\eta} = \tilde{\mathbf{R}} \tag{7.20}$$

where

$$\left.\begin{array}{l}\tilde{\mathbf{M}} = \mathbf{\Phi}^T \mathbf{M} \mathbf{\Phi} = \begin{bmatrix} \boldsymbol{\varphi}_1^T \mathbf{M} \boldsymbol{\varphi}_1 & 0 \\ 0 & \boldsymbol{\varphi}_2^T \mathbf{M} \boldsymbol{\varphi}_2 \end{bmatrix} = \begin{bmatrix} \tilde{M}_1 & 0 \\ 0 & \tilde{M}_2 \end{bmatrix} \\ \tilde{\mathbf{C}} = \begin{bmatrix} \tilde{C}_1 & 0 \\ 0 & \tilde{C}_2 \end{bmatrix} \quad \text{and} \quad \tilde{\mathbf{K}} = \begin{bmatrix} \tilde{K}_1 & 0 \\ 0 & \tilde{K}_2 \end{bmatrix}\end{array}\right\} \tag{7.21}$$

and

$$\left.\begin{array}{l}\tilde{M}_1 = \varphi_{11}^2 M_1 + \varphi_{21}^2 M_2 \\ \tilde{M}_2 = \varphi_{12}^2 M_1 + \varphi_{22}^2 M_2\end{array}\right\} \quad \left.\begin{array}{l}\tilde{C}_1 = 2\tilde{M}_1 \omega_1 \zeta_1 \\ \tilde{C}_2 = 2\tilde{M}_2 \omega_2 \zeta_2\end{array}\right\} \text{ and } \left.\begin{array}{l}\tilde{K}_1 = \omega_1^2 \tilde{M}_1 \\ \tilde{K}_2 = \omega_2^2 \tilde{M}_2\end{array}\right\} \tag{7.22}$$

and where the modal load vector is given by

$$\tilde{\mathbf{R}} = -\ddot{r}_g \tilde{\mathbf{M}}_g \qquad \text{where} \quad \tilde{\mathbf{M}}_g = \mathbf{\Phi}^T \mathbf{M} \mathbf{A}_g = \begin{bmatrix} \tilde{M}_{g1} \\ \tilde{M}_{g2} \end{bmatrix} = \begin{bmatrix} \varphi_{11} M_1 + \varphi_{21} M_2 \\ \varphi_{12} M_1 + \varphi_{22} M_2 \end{bmatrix} \tag{7.23}$$

Introducing this into Eq. 7.20 then the following is obtained:

$$\begin{bmatrix} \tilde{M}_1 & 0 \\ 0 & \tilde{M}_2 \end{bmatrix} \ddot{\boldsymbol{\eta}} + \begin{bmatrix} 2\tilde{M}_1 \omega_1 \zeta_1 & 0 \\ 0 & 2\tilde{M}_2 \omega_2 \zeta_2 \end{bmatrix} \dot{\boldsymbol{\eta}} + \begin{bmatrix} \omega_1^2 \tilde{M}_1 & 0 \\ 0 & \omega_2^2 \tilde{M}_2 \end{bmatrix} \boldsymbol{\eta} = -\ddot{r}_g \begin{bmatrix} \tilde{M}_{g1} \\ \tilde{M}_{g2} \end{bmatrix} \tag{7.24}$$

Taking the Fourier transform throughout Eq. 7.24, i.e. setting

$$\left.\begin{array}{l}\ddot{r}_g = \operatorname{Re} \sum\limits_{\omega} a_{\ddot{r}_g}(\omega) \cdot e^{i\omega t} \\ \boldsymbol{\eta}(t) = \begin{bmatrix} \eta_1 \\ \eta_2 \end{bmatrix} = \operatorname{Re} \sum\limits_{\omega} \begin{bmatrix} a_{\eta_1} \\ a_{\eta_2} \end{bmatrix} \cdot e^{i\omega t} = \operatorname{Re} \sum\limits_{\omega} \mathbf{a}_{\eta}(\omega) \cdot e^{i\omega t}\end{array}\right\} \tag{7.25}$$

and pre-multiplying by $\tilde{\mathbf{K}}^{-1}$, then the following is obtained

$$\begin{bmatrix} 1 - \left(\dfrac{\omega}{\omega_1}\right)^2 + 2i\dfrac{\omega}{\omega_1}\zeta_1 & 0 \\ 0 & 1 - \left(\dfrac{\omega}{\omega_2}\right)^2 + 2i\dfrac{\omega}{\omega_2}\zeta_2 \end{bmatrix} \mathbf{a}_{\eta} = -a_{\ddot{r}_g} \begin{bmatrix} \dfrac{\tilde{M}_{g1}}{\omega_1^2 \tilde{M}_1} \\ \dfrac{\tilde{M}_{g2}}{\omega_2^2 \tilde{M}_2} \end{bmatrix} \tag{7.26}$$

Introducing the modal frequency response matrix

$$\hat{\mathbf{H}}(\omega)=\begin{bmatrix}\hat{H}_1 & 0\\ 0 & \hat{H}_2\end{bmatrix} \quad \text{where} \quad \begin{cases}\hat{H}_1=\left[1-\left(\dfrac{\omega}{\omega_1}\right)^2+2i\dfrac{\omega}{\omega_1}\zeta_1\right]^{-1}\\ \hat{H}_2=\left[1-\left(\dfrac{\omega}{\omega_2}\right)^2+2i\dfrac{\omega}{\omega_2}\zeta_2\right]^{-1}\end{cases} \tag{7.27}$$

it is seen that

$$\mathbf{a}_\eta=-a_{\ddot{r}_g}(\omega)\cdot\hat{\mathbf{H}}(\omega)\cdot\begin{bmatrix}\dfrac{\tilde{M}_{g1}}{\omega_1^2\tilde{M}_1}\\ \dfrac{\tilde{M}_{g2}}{\omega_2^2\tilde{M}_2}\end{bmatrix}=-a_{\ddot{r}_g}(\omega)\cdot\begin{bmatrix}\hat{H}_1(\omega)\dfrac{\tilde{M}_{g1}}{\omega_1^2\tilde{M}_1}\\ \hat{H}_2(\omega)\dfrac{\tilde{M}_{g2}}{\omega_2^2\tilde{M}_2}\end{bmatrix} \tag{7.28}$$

For instance, if $\ddot{r}_g$ is stationary and narrow banded (close to harmonic), i.e.

$$\ddot{r}_g(t)\approx a_{\ddot{r}_g}\cdot\cos(\omega_g t)=\operatorname{Re}\left(a_{\ddot{r}_g}e^{i\omega_g t}\right) \tag{7.29}$$

then $$\boldsymbol{\eta}(t)=\operatorname{Re}\left(\mathbf{a}_\eta e^{i\omega_g t}\right)=\operatorname{Re}\{-a_{\ddot{r}_g}\begin{bmatrix}\hat{H}_1(\omega_g)\dfrac{\tilde{M}_{g1}}{\omega_1^2\tilde{M}_1}\\ \hat{H}_2(\omega_g)\dfrac{\tilde{M}_{g2}}{\omega_2^2\tilde{M}_2}\end{bmatrix}\cdot e^{i\omega_g t}\} \tag{7.30}$$

Since $\mathbf{r}(t)=\mathbf{\Phi}\boldsymbol{\eta}(t)$, it is seen that

$$\mathbf{r}(t)=\begin{bmatrix}r_1\\ r_2\end{bmatrix}=-a_{\ddot{r}_g}\cdot\operatorname{Re}\{\begin{bmatrix}\varphi_{11}\hat{H}_1(\omega_g)\dfrac{\tilde{M}_{g1}}{\omega_1^2\tilde{M}_1}+\varphi_{12}\hat{H}_2(\omega_g)\dfrac{\tilde{M}_{g2}}{\omega_2^2\tilde{M}_2}\\ \varphi_{21}\hat{H}_1(\omega_g)\dfrac{\tilde{M}_{g1}}{\omega_1^2\tilde{M}_1}+\varphi_{22}\hat{H}_2(\omega_g)\dfrac{\tilde{M}_{g2}}{\omega_2^2\tilde{M}_2}\end{bmatrix}\cdot e^{i\omega_g t}\}$$

$$=-a_{\ddot{r}_g}\{\left|\hat{H}_1(\omega_g)\right|\frac{\tilde{M}_{g1}}{\omega_1^2\tilde{M}_1}\begin{bmatrix}\varphi_{11}\\ \varphi_{21}\end{bmatrix}\cos(\omega_g t-\beta_1)+\left|\hat{H}_2(\omega_g)\right|\frac{\tilde{M}_{g2}}{\omega_2^2\tilde{M}_2}\begin{bmatrix}\varphi_{12}\\ \varphi_{22}\end{bmatrix}\cos(\omega_g t-\beta_2)\} \tag{7.31}$$

where $$\tan\beta_n=\left(2\zeta_n\frac{\omega_g}{\omega_n}\right)\Big/\left[1-\left(\frac{\omega_g}{\omega_n}\right)^2\right],\quad n=\begin{cases}1\\2\end{cases}$$

Example 7.1

Let us consider the two storey frame in Fig. 7.3 and assume:

$$K_1 = 120 \cdot 10^6 \ N/m \qquad M_1 = 4 \cdot 10^6 \ kg \quad \text{and} \qquad \zeta_1 = 0.01$$
$$K_2 = 100 \cdot 10^6 \ N/m \qquad M_2 = 2 \cdot 10^6 \ kg \quad \text{and} \qquad \zeta_2 = 0.02$$

Then (see Example 2.3),

$$\omega = \left\{ \frac{1}{2}\left(\frac{K_1 + K_2}{M_1} + \frac{K_2}{M_2}\right) \pm \left[\frac{1}{4}\left(\frac{K_1 + K_2}{M_1} + \frac{K_2}{M_2}\right)^2 - \frac{K_1}{M_1}\frac{K_2}{M_2} \right]^{1/2} \right\}^{1/2}$$

$$\Rightarrow \qquad \omega_1 \approx 4.1 \ rad/s \qquad \text{and} \qquad \omega_2 \approx 9.4 \ rad/s$$

$$\text{and} \qquad \frac{a_2}{a_1} = \frac{K_2 - \omega^2 M_2}{K_2} \quad \Rightarrow \quad \mathbf{\Phi} = \begin{bmatrix} 0.66 & -0.76 \\ 1 & 1 \end{bmatrix}$$

$$\mathbf{M} = 10^6 \begin{bmatrix} 4 & 0 \\ 0 & 2 \end{bmatrix} \quad \Rightarrow \quad \tilde{\mathbf{M}} = \mathbf{\Phi}^T \mathbf{M} \mathbf{\Phi} = \begin{bmatrix} \tilde{M}_1 & 0 \\ 0 & \tilde{M}_2 \end{bmatrix} = 10^3 \begin{bmatrix} 3737 & 0 \\ 0 & 4304 \end{bmatrix} kg$$

$$\mathbf{A}_g = \begin{bmatrix} 1 \\ 1 \end{bmatrix} \quad \Rightarrow \quad \tilde{\mathbf{M}}_g = \mathbf{\Phi}^T \mathbf{M} \mathbf{A}_g = \begin{bmatrix} \tilde{M}_{g1} \\ \tilde{M}_{g2} \end{bmatrix} = 10^3 \begin{bmatrix} 4636 \\ -1036 \end{bmatrix} kg$$

Let us assume $a_{\ddot{r}_g} = 1 \ m/s^2$ *and that* $\ddot{r}_g$ *has a constant period of* $T_g = 2 \ s$ *(i.e.* $\omega_g = \pi$ *). Then*

$$\left|\hat{H}_1(\omega_g)\right| = 2.37 \qquad \qquad \beta_1 = 0.036$$
$$\text{and}$$
$$\left|\hat{H}_2(\omega_g)\right| = 1.13 \qquad \qquad \beta_2 = 0.015$$

Thus

$$\begin{cases} r_1(t) = -0.1137 \cdot \cos(\pi t - 0.036) - 0.0023 \cdot \cos(\pi t - 0.015) \\ r_2(t) = -0.1725 \cdot \cos(\pi t - 0.036) + 0.0031 \cdot \cos(\pi t - 0.015) \end{cases}$$

The response is shown in Fig. 7.4.

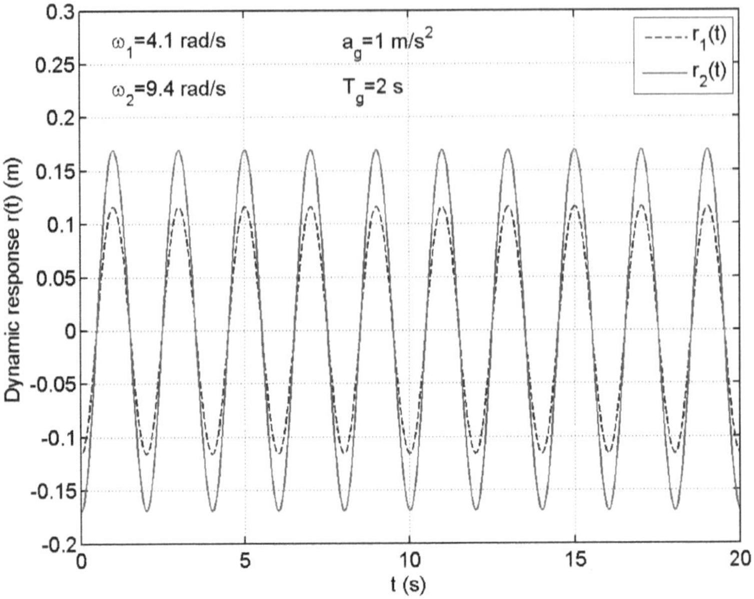

Fig. 7.4 Response to simple harmonic ground acceleration

On the other hand, if $\ddot{r}_g$ is a broad banded stochastic process defined by its spectral density $S_{\ddot{r}_g}(\omega)$, then (see Eq. 7.28)

$$\mathbf{S}_\eta(\omega) = \lim_{T\to\infty} \frac{1}{\pi T} \mathbf{a}_\eta^* \cdot \mathbf{a}_\eta^T$$

$$= \lim_{T\to\infty} \frac{1}{\pi T} \left(-a_{\ddot{r}_g}(\omega) \cdot \begin{bmatrix} \hat{H}_1(\omega) \dfrac{\tilde{M}_{g1}}{\omega_1^2 \tilde{M}_1} \\ \hat{H}_2(\omega) \dfrac{\tilde{M}_{g2}}{\omega_2^2 \tilde{M}_2} \end{bmatrix} \right)^* \cdot \left(-a_{\ddot{r}_g}(\omega) \cdot \begin{bmatrix} \hat{H}_1(\omega) \dfrac{\tilde{M}_{g1}}{\omega_1^2 \tilde{M}_1} \\ \hat{H}_2(\omega) \dfrac{\tilde{M}_{g2}}{\omega_2^2 \tilde{M}_2} \end{bmatrix} \right)^T \quad (7.32)$$

$$\Rightarrow \quad \mathbf{S}_\eta(\omega) = S_{\ddot{r}_g}(\omega) \begin{bmatrix} \hat{S}_{\eta_{11}}(\omega) & \hat{S}_{\eta_{11}}(\omega) \\ \hat{S}_{\eta_{11}}(\omega) & \hat{S}_{\eta_{11}}(\omega) \end{bmatrix} \quad (7.33)$$

where

$$\left.\begin{aligned}
\hat{S}_{\eta_{11}}(\omega) &= \left|\hat{H}_1(\omega)\right|^2 \left[\tilde{M}_{g1}\big/\left(\omega_1^2\tilde{M}_1\right)\right]^2 \\
\hat{S}_{\eta_{12}}(\omega) &= \hat{H}_1^*(\omega)\cdot\hat{H}_2(\omega)\cdot\left[\tilde{M}_{g1}\big/\left(\omega_1^2\tilde{M}_1\right)\right]\left[\tilde{M}_{g2}\big/\left(\omega_2^2\tilde{M}_2\right)\right] \\
\hat{S}_{\eta_{21}}(\omega) &= \hat{H}_2^*(\omega)\cdot\hat{H}_1(\omega)\cdot\left[\tilde{M}_{g2}\big/\left(\omega_2^2\tilde{M}_2\right)\right]\left[\tilde{M}_{g1}\big/\left(\omega_1^2\tilde{M}_1\right)\right] \\
\hat{S}_{\eta_{22}}(\omega) &= \left|\hat{H}_2(\omega)\right|^2 \left[\tilde{M}_{g2}\big/\left(\omega_2^2\tilde{M}_2\right)\right]^2
\end{aligned}\right\} \tag{7.34}$$

The dynamic response itself is given by $\mathbf{r} = \mathbf{\Phi}\cdot\boldsymbol{\eta}$, and thus

$$\begin{aligned}
\mathbf{S}_r(\omega) &= \lim_{T\to\infty}\frac{1}{\pi T}\mathbf{a}_r^*\cdot\mathbf{a}_r^T = \lim_{T\to\infty}\frac{1}{\pi T}\left(\mathbf{\Phi}\mathbf{a}_\eta\right)^*\cdot\left(\mathbf{\Phi}\mathbf{a}_\eta\right)^T \\
&= \mathbf{\Phi}\cdot\lim_{T\to\infty}\frac{1}{\pi T}\mathbf{a}_\eta^*\cdot\mathbf{a}_\eta^T\cdot\mathbf{\Phi}^T = \mathbf{\Phi}\cdot\mathbf{S}_\eta(\omega)\cdot\mathbf{\Phi}^T \\
&= S_{\ddot{r}_g}(\omega)\cdot\begin{bmatrix}\varphi_{11} & \varphi_{12}\\ \varphi_{21} & \varphi_{22}\end{bmatrix}\cdot\begin{bmatrix}\hat{S}_{\eta_{11}}(\omega) & \hat{S}_{\eta_{12}}(\omega)\\ \hat{S}_{\eta_{21}}(\omega) & \hat{S}_{\eta_{22}}(\omega)\end{bmatrix}\cdot\begin{bmatrix}\varphi_{11} & \varphi_{21}\\ \varphi_{12} & \varphi_{22}\end{bmatrix}
\end{aligned} \tag{7.35}$$

$$\Rightarrow \qquad \mathbf{S}_r(\omega) = \begin{bmatrix} S_{r_{11}} & S_{r_{12}} \\ S_{r_{21}} & S_{r_{22}} \end{bmatrix} = S_{\ddot{r}_g}(\omega)\begin{bmatrix}\hat{S}_{r_{11}}(\omega) & \hat{S}_{r_{12}}(\omega)\\ \hat{S}_{r_{21}}(\omega) & \hat{S}_{r_{22}}(\omega)\end{bmatrix} \tag{7.36}$$

where

$$\left.\begin{aligned}
\hat{S}_{r_{11}}(\omega) &= \varphi_{11}^2\hat{S}_{\eta_{11}} + \varphi_{11}\varphi_{12}\left[\hat{S}_{\eta_{12}} + \hat{S}_{\eta_{21}}\right] + \varphi_{12}^2\hat{S}_{\eta_{22}} \\
\hat{S}_{r_{12}}(\omega) &= \varphi_{11}\varphi_{21}\hat{S}_{\eta_{11}} + \varphi_{11}\varphi_{22}\hat{S}_{\eta_{12}} + \varphi_{12}\varphi_{21}\hat{S}_{\eta_{21}} + \varphi_{12}\varphi_{22}\hat{S}_{\eta_{22}} \\
\hat{S}_{r_{21}}(\omega) &= \varphi_{11}\varphi_{21}\hat{S}_{\eta_{11}} + \varphi_{21}\varphi_{12}\hat{S}_{\eta_{12}} + \varphi_{11}\varphi_{22}\hat{S}_{\eta_{21}} + \varphi_{12}\varphi_{22}\hat{S}_{\eta_{22}} \\
\hat{S}_{r_{22}}(\omega) &= \varphi_{21}^2\hat{S}_{\eta_{11}} + \varphi_{22}\varphi_{21}\left[\hat{S}_{\eta_{12}} + \hat{S}_{\eta_{21}}\right] + \varphi_{22}^2\hat{S}_{\eta_{22}}
\end{aligned}\right\} \tag{7.37}$$

The relevant variances and co-variances are given by

$$\int_0^\infty \mathbf{S}_r(\omega)\,d\omega = \begin{bmatrix}\sigma_{r_1}^2 & Cov_{r_1r_2}\\ Cov_{r_2r_1} & \sigma_{r_2}^2\end{bmatrix} \tag{7.38}$$

Elaboration 7.1: Quasi Static Response

In cases where $\omega_g << \omega_1$ then the response is primarily quasi-static, i.e.

$$\hat{S}_{\eta 11} = \left(\frac{\tilde{M}_{g1}}{\omega_1^2 \tilde{M}_1} \right)^2 \qquad \hat{S}_{\eta 12} = \hat{S}_{\eta 21} = \frac{\tilde{M}_{g1}}{\omega_1^2 \tilde{M}_1} \cdot \frac{\tilde{M}_{g2}}{\omega_2^2 \tilde{M}_2} \qquad \hat{S}_{\eta 22} = \left(\frac{\tilde{M}_{g2}}{\omega_2^2 \tilde{M}_2} \right)^2$$

and

$$\begin{bmatrix} \sigma_{r_1}^2 & Cov_{r_1 r_2} \\ Cov_{r_2 r_1} & \sigma_{r_2}^2 \end{bmatrix} = \int_0^\infty \mathbf{S}_r(\omega) d\omega = \int_0^\infty \begin{bmatrix} S_{r_1 1} & S_{r_1 2} \\ S_{r_2 1} & S_{r_2 2} \end{bmatrix} d\omega = \int_0^\infty S_{\ddot{r}_g}(\omega) \begin{bmatrix} \hat{S}_{r_1 1} & \hat{S}_{r_1 2} \\ \hat{S}_{r_2 1} & \hat{S}_{r_2 2} \end{bmatrix} d\omega$$

$$= \int_0^\infty S_{\ddot{r}_g}(\omega) d\omega \cdot \begin{bmatrix} \varphi_{11} & \varphi_{12} \\ \varphi_{21} & \varphi_{22} \end{bmatrix} \cdot \begin{bmatrix} \hat{S}_{\eta 11} & \hat{S}_{\eta 12} \\ \hat{S}_{\eta 21} & \hat{S}_{\eta 22} \end{bmatrix} \cdot \begin{bmatrix} \varphi_{11} & \varphi_{21} \\ \varphi_{12} & \varphi_{22} \end{bmatrix}$$

$$= \sigma_{\ddot{r}_g}^2 \cdot \begin{bmatrix} \varphi_{11} & \varphi_{12} \\ \varphi_{21} & \varphi_{22} \end{bmatrix} \cdot \begin{bmatrix} \hat{S}_{\eta 11} & \hat{S}_{\eta 12} \\ \hat{S}_{\eta 21} & \hat{S}_{\eta 22} \end{bmatrix} \cdot \begin{bmatrix} \varphi_{11} & \varphi_{21} \\ \varphi_{12} & \varphi_{22} \end{bmatrix}$$

Thus

$$\sigma_{r_1}^2 = \sigma_{\ddot{r}_g}^2 \left[\varphi_{11}^2 \left(\frac{\tilde{M}_{g1}}{\omega_1^2 \tilde{M}_1} \right)^2 + \varphi_{11}\varphi_{12} \frac{2\tilde{M}_{g1}\tilde{M}_{g2}}{\left(\omega_1^2 \tilde{M}_1\right)\left(\omega_2^2 \tilde{M}_2\right)} + \varphi_{12}^2 \left(\frac{\tilde{M}_{g2}}{\omega_2^2 \tilde{M}_2} \right)^2 \right]$$

$$\sigma_{r_2}^2 = \sigma_{\ddot{r}_g}^2 \left[\varphi_{21}^2 \left(\frac{\tilde{M}_{g1}}{\omega_1^2 \tilde{M}_1} \right)^2 + \varphi_{22}\varphi_{21} \frac{2\tilde{M}_{g1}\tilde{M}_{g2}}{\left(\omega_1^2 \tilde{M}_1\right)\left(\omega_2^2 \tilde{M}_2\right)} + \varphi_{22}^2 \left(\frac{\tilde{M}_{g2}}{\omega_2^2 \tilde{M}_2} \right)^2 \right]$$

$$Cov_{r_1 r_2} = Cov_{r_2 r_1}$$

$$= \sigma_{\ddot{r}_g}^2 \left[\varphi_{11}\varphi_{21} \left(\frac{\tilde{M}_{g1}}{\omega_1^2 \tilde{M}_1} \right)^2 + \left(\varphi_{11}\varphi_{22} + \varphi_{12}\varphi_{21}\right) \frac{\tilde{M}_{g1}\tilde{M}_{g2}}{\left(\omega_1^2 \tilde{M}_1\right)\left(\omega_2^2 \tilde{M}_2\right)} + \varphi_{12}\varphi_{22} \left(\frac{\tilde{M}_{g2}}{\omega_2^2 \tilde{M}_2} \right)^2 \right]$$

The shear forces in the columns may be obtained from simple equilibrium consideration. Thus

$$\left.\begin{aligned} 2V_{1_y} - 2V_{2_y} + M_1 \ddot{r}_{1_{tot}} &= 0 \\ 2V_{2_y} + M_2 \ddot{r}_{2_{tot}} &= 0 \end{aligned}\right\} \tag{7.39}$$

which may be written

$$\mathbf{B}\mathbf{V}_y + \mathbf{M}\ddot{\mathbf{r}}_{tot} = \mathbf{0} \tag{7.40}$$

where $\mathbf{M}$ and $\ddot{\mathbf{r}}_{tot}$ are defined in Eqs. 7.15 and 7.16, $\mathbf{V}_y = \begin{bmatrix} V_{1_y} & V_{2_y} \end{bmatrix}^T$ and

$$\mathbf{B} = \begin{bmatrix} 2 & -2 \\ 0 & 2 \end{bmatrix} \tag{7.41}$$

Since damping is disregarded, then $\mathbf{M}\ddot{\mathbf{r}}_{tot} + \mathbf{K}\mathbf{r} = \mathbf{0}$, and thus

$$\mathbf{V}_y = \begin{bmatrix} V_{1_y} \\ V_{2_y} \end{bmatrix} = \mathbf{B}^{-1}\mathbf{K}\mathbf{r} = \frac{1}{4}\begin{bmatrix} 2 & 2 \\ 0 & 2 \end{bmatrix}\begin{bmatrix} K_1 + K_2 & -K_2 \\ -K_2 & K_2 \end{bmatrix}\begin{bmatrix} r_1 \\ r_2 \end{bmatrix} = \frac{1}{2}\begin{bmatrix} K_1 r_1 \\ K_2 (r_2 - r_1) \end{bmatrix} \tag{7.42}$$

If $\ddot{r}_g$ is a broad banded stochastic process, then

$$\sigma_{V_{2_y}}^2 = E\left\{\left[V_{2_y}(t)\right]^2\right\} = E\left[\left(\frac{K_2}{2}\right)^2 (r_2 - r_1)^2\right] = \left(\frac{K_2}{2}\right)^2 E\left[r_2^2 - 2r_1 r_2 + r_1^2\right] \tag{7.43}$$

$$\Rightarrow \qquad \sigma_{V_{2_y}} = \frac{K_2}{2}\sqrt{\sigma_{r_1}^2 - 2Cov_{r_1 r_2} + \sigma_{r_2}^2} \tag{7.44}$$

and

$$\sigma_{V_{1_y}}^2 = E\left\{\left[V_{1_y}(t)\right]^2\right\} = E\left[\left(\frac{K_1}{2}\right)^2 r_1^2(t)\right] = \left(\frac{K_1}{2}\right)^2 E\left[r_1^2(t)\right] = \left(\frac{K_1}{2}\right)^2 \sigma_{r_1}^2 \tag{7.45}$$

$$\Rightarrow \qquad \sigma_{V_{1_y}} = \frac{K_1}{2}\sigma_{r_1} \tag{7.46}$$

7.4 The General Case of a Discrete System

The system shown in Fig. 7.5 is intended to represent a more general type of structural system, which in a finite element format has N_r degrees of freedom. As indicated in Fig. 7.5 it is taken for granted that the ground acceleration $\ddot{r}_g(t)$ is close to horizontal. In an arbitrary node p the system may have several degrees of freedom, of which one r_k may be associated with a displacement (more or less) parallel to the ground. In original degrees of freedom the equilibrium condition for this system is then given by

$$\mathbf{M}\ddot{\mathbf{r}}(t)+\mathbf{C}\dot{\mathbf{r}}(t)+\mathbf{K}\mathbf{r}(t)=\mathbf{R}(t) \tag{7.47}$$

where

$$\mathbf{R}(t)=-\mathbf{M}\mathbf{A}_g\ddot{r}_g(t) \tag{7.48}$$

and

$$\mathbf{A}_g=\begin{bmatrix} A_{g_1} & \cdots & A_{g_n} & \cdots & A_{g_{N_r}} \end{bmatrix}^T \quad \text{where} \quad A_{g_n}=\begin{cases} 1 & \text{at } n=k \\ 0 & \text{elsewhere} \end{cases} \tag{7.49}$$

Introducing the modal format

$$\mathbf{r}=\mathbf{\Phi}\boldsymbol{\eta} \qquad \text{where} \qquad \begin{cases} \mathbf{\Phi}=\begin{bmatrix} \boldsymbol{\varphi}_1 & \cdots & \boldsymbol{\varphi}_n & \cdots & \boldsymbol{\varphi}_{N_{\text{mod}}} \end{bmatrix} \\ \boldsymbol{\eta}=\begin{bmatrix} \eta_1 & \cdots & \eta_n & \cdots & \eta_{N_{\text{mod}}} \end{bmatrix}^T \end{cases} \tag{7.50}$$

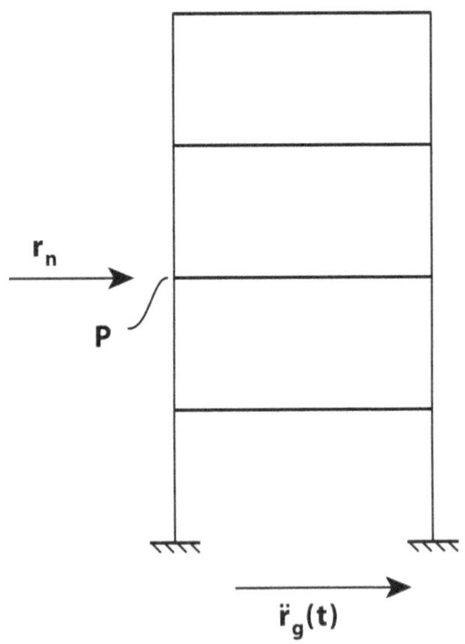

Fig. 7.5 A more general type of system

and pre-multiplying by $\mathbf{\Phi}^T$, then the modal equilibrium condition is given by

$$\tilde{\mathbf{M}}\ddot{\boldsymbol{\eta}} + \tilde{\mathbf{C}}\dot{\boldsymbol{\eta}} + \tilde{\mathbf{K}}\boldsymbol{\eta} = \tilde{\mathbf{R}}_g \tag{7.51}$$

where

$$\begin{cases} \tilde{\mathbf{M}} = \mathbf{\Phi}^T\mathbf{M}\mathbf{\Phi} = diag\left[\tilde{M}_n\right] \text{ where } \tilde{M}_n = \boldsymbol{\varphi}_n^T\mathbf{M}\boldsymbol{\varphi}_n \\ \tilde{\mathbf{C}} = \mathbf{\Phi}^T\mathbf{C}\mathbf{\Phi} = diag\left[\tilde{C}_n\right] \text{ where } \tilde{C}_n = 2\tilde{M}_n\omega_n\zeta_n \\ \tilde{\mathbf{K}} = \mathbf{\Phi}^T\mathbf{K}\mathbf{\Phi} = diag\left[\tilde{K}_n\right] \text{ where } \tilde{K}_n = \omega_n^2\tilde{M}_n \end{cases} \tag{7.52}$$

and

$$\tilde{\mathbf{R}}_g(t) = -\mathbf{\Phi}^T\mathbf{M}\mathbf{A}_g\ddot{r}_g(t) \tag{7.53}$$

Taking the Fourier transform throughout this equation, i.e. introducing

$$\left.\begin{aligned} \boldsymbol{\eta}(t) &= \sum_{\omega}\mathbf{a}_\eta(\omega)\cdot e^{i\omega t} \quad \text{where } \mathbf{a}_\eta(\omega) = \begin{bmatrix} a_{\eta_1} & \cdots & a_{\eta_n} & \cdots & a_{\eta_{N_{\text{mod}}}} \end{bmatrix}^T \\ \ddot{r}_g(t) &= \sum_{\omega} a_{\ddot{r}_g}(\omega)\cdot e^{i\omega t} \end{aligned}\right\} \tag{7.54}$$

and pre-multiplying by $\tilde{\mathbf{K}}^{-1}$, then

$$\left(-\omega^2\tilde{\mathbf{K}}^{-1}\tilde{\mathbf{M}} + i\omega\tilde{\mathbf{K}}^{-1}\tilde{\mathbf{C}} + \mathbf{I}\right)\mathbf{a}_\eta = -\tilde{\mathbf{K}}^{-1}\mathbf{\Phi}^T\mathbf{M}\mathbf{A}_g a_{\ddot{r}_g} \tag{7.55}$$

from which it is seen that

$$\mathbf{a}_\eta(\omega) = -\hat{\mathbf{H}}_\eta(\omega)\tilde{\mathbf{K}}^{-1}\mathbf{\Phi}^T\mathbf{M}\mathbf{A}_g a_{\ddot{r}_g}(\omega) \tag{7.56}$$

where

$$\hat{\mathbf{H}}_\eta(\omega) = \left\{\mathbf{I} - diag\left[\left(\frac{\omega}{\omega_n}\right)^2\right] + 2i\cdot diag\left[\zeta_n\frac{\omega}{\omega_n}\right]\right\}^{-1} = diag\left[\hat{H}_n\right] \tag{7.57}$$

and $\hat{H}_n = \left[1-(\omega/\omega_n)^2 + 2i\zeta_n(\omega/\omega_n)\right]^{-1}$. Thus, the spectral density matrix of the modal degrees of freedom is given by

$$\begin{aligned}\mathbf{S}_{\eta}(\omega) &= \lim_{T\to\infty}\frac{1}{\pi T}\mathbf{a}_{\eta}^{*}\mathbf{a}_{\eta}^{T}\\ &= \lim_{T\to\infty}\frac{1}{\pi T}\left(-\hat{\mathbf{H}}_{\eta}\tilde{\mathbf{K}}^{-1}\mathbf{\Phi}^{T}\mathbf{M}\mathbf{A}_{g}a_{\ddot{r}_{g}}\right)^{*}\left(-\tilde{\mathbf{H}}_{\eta}\tilde{\mathbf{K}}^{-1}\mathbf{\Phi}^{T}\mathbf{M}\mathbf{A}_{g}a_{\ddot{r}_{g}}\right)^{T}\\ &= \hat{\mathbf{H}}_{\eta}^{*}\tilde{\mathbf{K}}^{-1}\mathbf{\Phi}^{T}\mathbf{M}\mathbf{A}_{g}\left(\lim_{T\to\infty}\frac{1}{\pi T}a_{\ddot{r}_{g}}^{*}a_{\ddot{r}_{g}}\right)\mathbf{A}_{g}^{T}\mathbf{M}^{T}\mathbf{\Phi}\left(\tilde{\mathbf{K}}^{-1}\right)^{T}\hat{\mathbf{H}}_{\eta}^{T}\\ &= \hat{\mathbf{H}}_{\eta}^{*}(\omega)\tilde{\mathbf{K}}^{-1}\mathbf{\Phi}^{T}\mathbf{M}\mathbf{A}_{g}\mathbf{A}_{g}^{T}\mathbf{M}^{T}\mathbf{\Phi}\left(\tilde{\mathbf{K}}^{-1}\right)^{T}\hat{\mathbf{H}}_{\eta}^{T}(\omega)S_{\ddot{r}_{g}}(\omega)\end{aligned} \tag{7.58}$$

where $S_{\ddot{r}_{g}}(\omega) = \lim_{T\to\infty}\frac{1}{\pi T}a_{\ddot{r}_{g}}^{*}a_{\ddot{r}_{g}}$ is the spectral density of the ground motion acceleration. Since $\mathbf{r} = \mathbf{\Phi}\boldsymbol{\eta}$ then $\mathbf{a}_{r}(\omega) = \mathbf{\Phi}\mathbf{a}_{\eta}(\omega)$ and thus the spectral density response matrix is given by

$$\begin{aligned}&\mathbf{S}_{r}(\omega) = \lim_{T\to\infty}\frac{1}{\pi T}\mathbf{a}_{r}^{*}\mathbf{a}_{r}^{T} = \lim_{T\to\infty}\frac{1}{\pi T}\left(\mathbf{\Phi}\mathbf{a}_{\eta}\right)^{*}\left(\mathbf{\Phi}\mathbf{a}_{\eta}\right)^{T} = \mathbf{\Phi}\lim_{T\to\infty}\frac{1}{\pi T}\mathbf{a}_{\eta}^{*}\mathbf{a}_{\eta}^{T}\mathbf{\Phi}^{T}\\ &= \mathbf{\Phi}\mathbf{S}_{\eta}\mathbf{\Phi}^{T} = \mathbf{\Phi}\hat{\mathbf{H}}_{\eta}^{*}(\omega)\tilde{\mathbf{K}}^{-1}\mathbf{\Phi}^{T}\mathbf{M}\mathbf{A}_{g}\mathbf{A}_{g}^{T}\mathbf{M}^{T}\mathbf{\Phi}\left(\tilde{\mathbf{K}}^{-1}\right)^{T}\hat{\mathbf{H}}_{\eta}^{T}(\omega)\mathbf{\Phi}^{T}S_{\ddot{r}_{g}}(\omega)\end{aligned} \tag{7.59}$$

In the special case that the response is quasi-static, then the response covariance matrix is given by

$$\mathbf{Cov}_{rr} = \begin{bmatrix} \sigma_{1}^{2} & \cdots & Cov_{1i} & \cdots & Cov_{1j} & \cdots & Cov_{1N_r} \\ \vdots & \ddots & \vdots & & \vdots & & \vdots \\ Cov_{i1} & \cdots & \sigma_{i}^{2} & \cdots & Cov_{ij} & & \vdots \\ \vdots & & \vdots & \ddots & \vdots & & \vdots \\ Cov_{j1} & \cdots & Cov_{ji} & \cdots & \sigma_{j}^{2} & & \vdots \\ \vdots & & & & & \ddots & \vdots \\ Cov_{N_r 1} & \cdots & \cdots & \cdots & \cdots & \cdots & \sigma_{N_r}^{2} \end{bmatrix} \tag{7.60}$$

$$= \int_{0}^{\infty}\mathbf{S}_{r}(\omega)d\omega = \mathbf{\Phi}\tilde{\mathbf{K}}^{-1}\mathbf{\Phi}^{T}\mathbf{M}\mathbf{A}_{g}\mathbf{A}_{g}^{T}\mathbf{M}^{T}\mathbf{\Phi}\left(\tilde{\mathbf{K}}^{-1}\right)^{T}\mathbf{\Phi}^{T}\cdot\sigma_{\ddot{r}_{g}}^{2}$$

Elaboration 7.2: Non-symmetric Two Storey Shear Frame

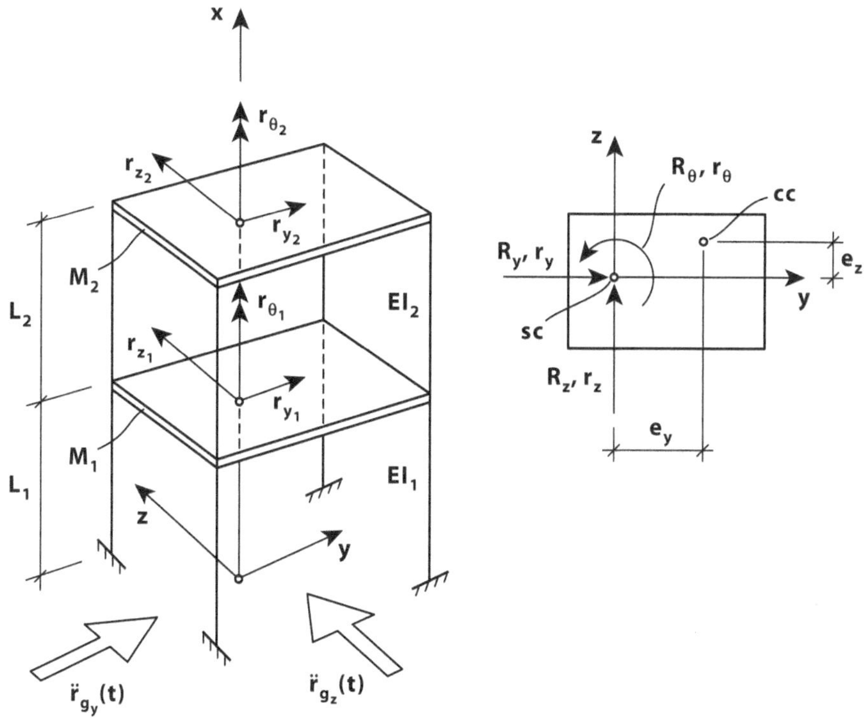

Fig. 7.6 Two storey shear frame with non-matching shear and mass centres

A two storey shear frame with non-matching shear and mass centres is illustrated on Fig. 7.6. The ground acceleration $\ddot{r}_g(t)$ has been split into a component $\ddot{r}_{g_y}(t)$ parallel to the global y-axis and a component $\ddot{r}_{g_z}(t)$ parallel to the global z-axis. The system has six degrees of freedom

$$\mathbf{r}(t) = [\mathbf{r}_1 \quad \mathbf{r}_2]^T \qquad \text{where} \qquad \begin{cases} \mathbf{r}_1 = \begin{bmatrix} r_y & r_z & r_\theta \end{bmatrix}_1^T \\ \mathbf{r}_2 = \begin{bmatrix} r_y & r_z & r_\theta \end{bmatrix}_2^T \end{cases}$$

and load components $\mathbf{R}(t) = [\mathbf{R}_1 \quad \mathbf{R}_2]^T$ where $\begin{cases} \mathbf{R}_1 = \begin{bmatrix} R_y & R_z & R_\theta \end{bmatrix}_1^T \\ \mathbf{R}_2 = \begin{bmatrix} R_y & R_z & R_\theta \end{bmatrix}_2^T \end{cases}$

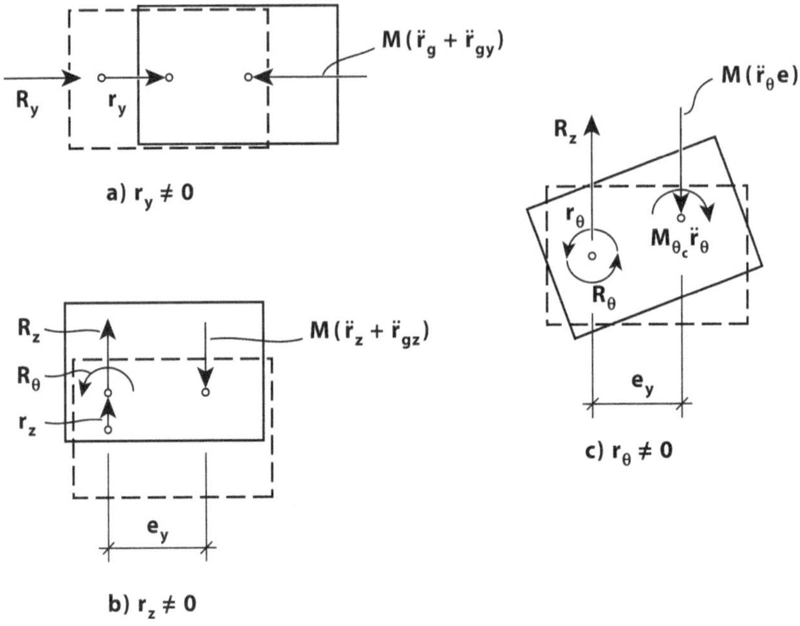

Fig. 7.7 Y , z and θ equilibrium, component by component

The content of $\mathbf{M}$, $\mathbf{C}$ and $\mathbf{K}$ may conveniently be obtained by considering the equilibrium condition for each component, one after the other, as illustrated in Fig. 7.7. Let for instance damping be associated with corresponding stiffness contributions, then:

$$r_{y_1} \neq 0 \Rightarrow \begin{cases} R_{y_1} = M_1\left(\ddot{r}_{y_1} + \ddot{r}_{g_y}\right) + \left(C_{y_1} + C_{y_1}\right)\dot{r}_{y_1} + \left(K_{y_1} + K_{y_1}\right) r_{y_1} \\ R_{z_1} = 0, \quad R_{\theta_1} = -M_1\left(\ddot{r}_{y_1} + \ddot{r}_{g_y}\right) e_z \\ R_{y_2} = -C_{y_2}\dot{r}_{y_1} - K_{y_2} r_{y_1}, \quad R_{z_2} = 0, \quad R_{\theta_2} = 0 \end{cases}$$

$$r_{y_2} \neq 0 \Rightarrow \begin{cases} R_{y_1} = -C_{y_2}\dot{r}_{y_2} - K_{y_2} r_{y_2}, \quad R_{z_1} = 0, \quad R_{\theta_1} = 0 \\ R_{y_2} = M_2\left(\ddot{r}_{y_2} + \ddot{r}_{g_y}\right) + C_{y_2}\dot{r}_{y_2} + K_{y_2} r_{y_2}, R_{z_2} = 0 \\ R_{\theta_2} = -M_2\left(\ddot{r}_{y_2} + \ddot{r}_{g_z}\right) e_z \end{cases}$$

$$r_{z_1} \neq 0 \Rightarrow \begin{cases} R_{y_1} = 0, \quad R_{z_1} = M_1\left(\ddot{r}_{z_1} + \ddot{r}_{g_z}\right) + \left(C_{z_1} + C_{z_1}\right)\dot{r}_{z_1} + \left(K_{z_1} + K_{z_1}\right)r_{z_1} \\ R_{\theta_1} = M_1\left(\ddot{r}_{z_1} + \ddot{r}_{g_z}\right)e_y \\ R_{y_2} = 0, \quad R_{z_2} = -C_{z_2}\dot{r}_{z_1} - K_{z_2}r_{z_1}, \quad R_{\theta_2} = 0 \end{cases}$$

$$r_{z_2} \neq 0 \Rightarrow \begin{cases} R_{y_1} = 0, \quad R_{z_1} = -C_{z_2}\dot{r}_{z_2} - K_{z_2}r_{z_2}, \quad R_{\theta_1} = 0, \quad R_{y_2} = 0 \\ R_{z_2} = M_2\left(\ddot{r}_{z_2} + \ddot{r}_{g_z}\right) + C_{z_2}\dot{r}_{z_2} + K_{z_2}r_{z_2}, \quad R_{\theta_2} = M_2\left(\ddot{r}_{z_2} + \ddot{r}_{g_z}\right)e_y \end{cases}$$

$$r_{\theta_1} \neq 0 \Rightarrow \begin{cases} R_{y_1} = -M_1\ddot{r}_{\theta_1}e_z, \quad R_{z_1} = M_1\ddot{r}_{\theta_1}e_y \\ R_{\theta_1} = M_{\theta_1}\ddot{r}_{\theta_1} + M_1\ddot{r}_{\theta_1}e_y^2 + M_1\ddot{r}_{\theta_1}e_z^2 + \left(C_{\theta_1} + C_{\theta_2}\right)\dot{r}_{\theta_1} + \left(K_{\theta_1} + K_{\theta_2}\right)r_{\theta_1} \\ R_{y_2} = 0, \quad R_{z_2} = 0, \quad R_{\theta_2} = -C_{\theta_2}\dot{r}_{\theta_1} - K_{\theta_2}r_{\theta_1} \end{cases}$$

$$r_{\theta_2} \neq 0 \Rightarrow \begin{cases} R_{y_1} = 0, \quad R_{z_1} = 0, \quad R_{\theta_1} = -C_{\theta_2}\dot{r}_{\theta 2} - K_{\theta_2}r_{\theta_2}, \quad R_{y_2} = -M_2\ddot{r}_{\theta_2}e_z \\ R_{z_2} = M_2\ddot{r}_{\theta_2}e_y, \quad R_{\theta_2} = M_{\theta_2}\ddot{r}_{\theta_2} + M_2\ddot{r}_{\theta_2}e_y^2 + M_2\ddot{r}_{\theta_2}e_z^2 + C_{\theta_2}\dot{r}_{\theta 2} + K_{\theta_2}r_{\theta_1} \end{cases}$$

Thus $\mathbf{M}\ddot{\mathbf{r}} + \mathbf{C}\dot{\mathbf{r}} + \mathbf{K}\mathbf{r} = \mathbf{R} - \mathbf{M}\mathbf{A}_g\ddot{\mathbf{r}}_g$ where

$$\ddot{\mathbf{r}}_g(t) = \begin{bmatrix} \ddot{r}_{y_g} \\ \ddot{r}_{z_g} \end{bmatrix} \text{ and } \mathbf{A}_g = \begin{bmatrix} \mathbf{A}_{g_y} & \mathbf{A}_{g_z} \end{bmatrix} \text{ where } \begin{cases} \mathbf{A}_{g_y} = [1 \quad 0 \quad 0 \quad 1 \quad 0 \quad 0]^T \\ \mathbf{A}_{g_z} = [0 \quad 1 \quad 0 \quad 0 \quad 1 \quad 0]^T \end{cases}$$

$$\mathbf{M} = \begin{bmatrix} \mathbf{M}_1 & \mathbf{0} \\ \mathbf{0} & \mathbf{M}_2 \end{bmatrix} \text{ where } \mathbf{M}_j = \begin{bmatrix} M_j & 0 & -M_j e_z \\ 0 & M_j & M_j e_y \\ -M_j e_z & M_j e_y & M_{\theta_{j_{tot}}} \end{bmatrix} \quad j = \begin{cases} 1 \\ 2 \end{cases}$$

$$M_{\theta_{1_{tot}}} = M_{\theta_1} + M_1\left(e_y^2 + e_z^2\right) \qquad M_{\theta_{2_{tot}}} = M_{\theta_2} + M_2\left(e_y^2 + e_z^2\right)$$

$$\mathbf{K} = \begin{bmatrix} \mathbf{K}_1 + \mathbf{K}_2 & -\mathbf{K}_2 \\ -\mathbf{K}_2 & \mathbf{K}_2 \end{bmatrix} \quad \text{where} \quad \mathbf{K}_j = \begin{bmatrix} K_{y_j} & 0 & 0 \\ 0 & K_{z_j} & 0 \\ 0 & 0 & K_{\theta_j} \end{bmatrix} \quad j = \begin{cases} 1 \\ 2 \end{cases}$$

$$\mathbf{C} = \begin{bmatrix} \mathbf{C}_1 + \mathbf{C}_2 & -\mathbf{C}_2 \\ -\mathbf{C}_2 & \mathbf{C}_2 \end{bmatrix} \quad \text{where} \quad \mathbf{C}_j = \begin{bmatrix} C_{y_j} & 0 & 0 \\ 0 & C_{z_j} & 0 \\ 0 & 0 & C_{\theta_j} \end{bmatrix} \quad j = \begin{cases} 1 \\ 2 \end{cases}$$

and $\mathbf{R} = \begin{bmatrix} \mathbf{R}_1 \\ \mathbf{R}_2 \end{bmatrix}$ where $\begin{cases} \mathbf{R}_1 = \begin{bmatrix} R_y & R_z & R_\theta \end{bmatrix}_1^T \\ \mathbf{R}_2 = \begin{bmatrix} R_y & R_z & R_\theta \end{bmatrix}_2^T \end{cases}$

Since the superposition principle applies, the load effects of $\mathbf{R}$ and $-\mathbf{MA}_g\ddot{\mathbf{r}}_g$ may be handled in separate calculations.

Example 7.2: Single Mode Solution with Torsion and z Direction Coupling

For the case above, let us for simplicity assume that $e_z = 0$. *Then the motion* r_y *in the* y *direction may be handled separately and identical to that which has been shown in Example 7.1. Accordingly, the response vector associated with motion* r_z *and* r_θ *at levels 1 and 2 may be reduces into* $\mathbf{r}(t) = \begin{bmatrix} r_{z_1} & r_{\theta_1} & r_{z_2} & r_{\theta_2} \end{bmatrix}^T$. *The corresponding mass matrix and load vector are given by*

$$\mathbf{M} = \begin{bmatrix} M_1 & e_y M_1 & 0 & 0 \\ e_y M_1 & M_{\theta_{1tot}} & 0 & 0 \\ 0 & 0 & M_2 & e_y M_2 \\ 0 & 0 & e_y M_2 & M_{\theta_{2tot}} \end{bmatrix} \quad \text{and} \quad -\mathbf{MA}_g\ddot{\mathbf{r}}_g(t) = -\ddot{r}_{g_z}(t) \begin{bmatrix} M_1 \\ e_y M_1 \\ M_2 \\ e_y M_2 \end{bmatrix}$$

Let us for simplicity only include a single mode $\boldsymbol{\varphi}_n = [\phi_1 \quad \phi_2 \quad \phi_3 \quad \phi_4]_n^T$ *and corresponding eigenfrequency* ω_n. *Then the modal mass is given by*

$$\tilde{M}_n = \boldsymbol{\varphi}_n^T \mathbf{M} \boldsymbol{\varphi}_n = \left(\phi_1^2 + 2e_y\phi_1\phi_2\right)M_1 + \left(\phi_3^2 + 2e_y\phi_3\phi_4\right)M_2 + \phi_2^2 M_{\theta_1} + \phi_4^2 M_{\theta_2}$$

Modal stiffness and damping are given by $\tilde{K}_n = \omega_n^2 \tilde{M}_n$ *and* $\tilde{C} = 2\zeta_n\omega_n\tilde{M}_n$, *while the modal load is defined by*

$$\tilde{R}_n = -\boldsymbol{\varphi}_n^T \mathbf{MA}_g\ddot{\mathbf{r}}_g(t) = -\ddot{r}_{g_z} A_n \quad \text{where } A_n = \left(\phi_1 + e_y\phi_2\right)M_1 + \left(\phi_3 + e_y\phi_4\right)M_2$$

Setting $\mathbf{r}(t) = \boldsymbol{\varphi}_n \cdot \eta_n(t)$ *then, as usual, the modal equilibrium condition is expressed by*

$$\tilde{M}_n\ddot{\eta}_n + \tilde{C}_n\dot{\eta}_n + \tilde{K}_n\eta_n = \tilde{R}_n$$

Taking the Fourier transform throughout this equation and pre multiplying by $\tilde{K}_n^{-1}$, *then* $$\left[1-(\omega/\omega_n)^2+2i\zeta_n\,\omega/\omega_n\right]a_{\eta_n}(\omega)=\frac{a_{\tilde{R}_n}}{\tilde{K}_n}=-\frac{A_n}{\tilde{K}_n}\cdot a_{\ddot{r}_{g_z}}(\omega)$$

where $a_{\eta_n}(\omega)$ *and* $a_{\ddot{r}_{g_z}}(\omega)$ *are the Fourier amplitudes of the modal degree of freedom and the ground acceleration. Thus*

$$a_{\eta_n}(\omega)=-\frac{A_n}{\tilde{K}_n}\hat{H}(\omega)a_{\ddot{r}_{g_z}}(\omega) \text{ where } \hat{H}(\omega)=\left[1-(\omega/\omega_n)^2+2i\zeta_n\,\omega/\omega_n\right]^{-1}$$

$$S_{\eta_n}(\omega)=\lim_{T\to\infty}\frac{1}{\pi T}\left(a_{\eta_n}^{*}a_{\eta_n}\right)=\lim_{T\to\infty}\frac{1}{\pi T}\left[\left(-\frac{A_n}{\tilde{K}_n}\hat{H}_n a_{\ddot{r}_{g_z}}\right)^{*}\left(-\frac{A_n}{\tilde{K}_n}\hat{H}_n a_{\ddot{r}_{g_z}}\right)\right]$$

$$=\left(\frac{A_n}{\tilde{K}_n}\right)^2\left|\hat{H}_n\right|^2\lim_{T\to\infty}\frac{1}{\pi T}\left(a_{\ddot{r}_{g_z}}^{*}a_{\ddot{r}_{g_z}}\right)=\left(\frac{A_n}{\omega_n^2\tilde{M}_n}\right)^2\left|\hat{H}_n(\omega)\right|^2 S_{\ddot{r}_{g_z}}(\omega)$$

Since $\mathbf{r}(t)=\boldsymbol{\varphi}_n\cdot\eta_n(t)$, *then* $\mathbf{a}_r(\omega)=\boldsymbol{\varphi}_n\cdot a_{\eta_n}(\omega)$, *and hence*

$$\mathbf{S}_{rr}(\omega)=\begin{bmatrix} S_{r_1r_1} & S_{r_1r_2} & S_{r_1r_3} & S_{r_1r_4} \\ & S_{r_2r_2} & S_{r_2r_3} & S_{r_2r_4} \\ sym. & & S_{r_3r_3} & S_{r_3r_4} \\ & & & S_{r_4r_4} \end{bmatrix}=\lim_{T\to\infty}\frac{1}{\pi T}\left(\mathbf{a}_r^{*}\mathbf{a}_r^{T}\right)$$

$$\mathbf{S}_{rr}(\omega)=\lim_{T\to\infty}\frac{1}{\pi T}\left[\left(\boldsymbol{\varphi}_n\cdot a_{\eta_n}\right)^{*}\left(\boldsymbol{\varphi}_n\cdot a_{\eta_n}\right)^{T}\right]=\boldsymbol{\varphi}_n\boldsymbol{\varphi}_n^{T}S_{\eta_n}(\omega)$$

$$\Rightarrow \quad =\boldsymbol{\varphi}_n\boldsymbol{\varphi}_n^{T}\left(\frac{A_n}{\omega_n^2\tilde{M}_n}\right)^2\left|\hat{H}_n(\omega)\right|^2 S_{\ddot{r}_{g_z}}(\omega)$$

In the special case that the response is quasi static, then

$$\mathbf{C}ov_{rr}=\begin{bmatrix} \sigma_{r_1}^2 & Cov_{r_1r_2} & Cov_{r_1r_3} & Cov_{r_1r_4} \\ & \sigma_{r_2}^2 & Cov_{r_2r_3} & Cov_{r_2r_4} \\ sym. & & \sigma_{r_3}^2 & Cov_{r_3r_4} \\ & & & \sigma_{r_4}^2 \end{bmatrix}=\int_0^\infty \mathbf{S}_{rr}(\omega)d\omega$$

$$\Rightarrow \mathbf{C}ov_{rr}\approx\boldsymbol{\varphi}_n\boldsymbol{\varphi}_n^{T}\left[A_n/\left(\omega_n^2\tilde{M}_n\right)\right]^2\int_0^\infty S_{\ddot{r}_{g_z}}(\omega)d\omega=\boldsymbol{\varphi}_n\boldsymbol{\varphi}_n^{T}\left[A_n\sigma_{\ddot{r}_{g_z}}/\left(\omega_n^2\tilde{M}_n\right)\right]^2$$

where $\sigma_{\ddot{r}_{g_z}}$ *is the standard deviation of the horizontal ground acceleration.*

7.5 The Case of Continuous Line-Like Systems

For a continuous line-like system, in Fig. 7.8 arbitrarily represented by a cantilevered beam, the instantaneous equilibrium condition of an incremental element dx is shown in the lower illustration of Fig. 7.8. Let, as usual introduce

$$\mathbf{r}(x,t)=\begin{bmatrix} r_y \\ r_z \\ r_\theta \end{bmatrix} = \sum_{n=1}^{N_{\text{mod}}} \begin{bmatrix} \phi_y(x) \\ \phi_z(x) \\ \phi_\theta(x) \end{bmatrix}_n \cdot \eta_n(t) = \sum_{n=1}^{N_{\text{mod}}} \boldsymbol{\varphi}_n(x)\cdot\eta_n(t) = \boldsymbol{\Phi}(x)\cdot\boldsymbol{\eta}(t) \tag{7.61}$$

where

$$\left\{\begin{aligned} &\boldsymbol{\Phi}(x)=\begin{bmatrix}\boldsymbol{\varphi}_1 & \cdots & \boldsymbol{\varphi}_n & \cdots & \boldsymbol{\varphi}_{N_{\text{mod}}}\end{bmatrix} \\ &\boldsymbol{\eta}(t)=\begin{bmatrix}\eta_1 & \cdots & \eta_n & \cdots & \eta_{N_{\text{mod}}}\end{bmatrix}^T \end{aligned}\right. \tag{7.62}$$

and where the N_{mod} mode shapes $\boldsymbol{\varphi}_n(x)=\begin{bmatrix}\phi_y & \phi_z & \phi_\theta\end{bmatrix}^T$ are assumed to be continuous functions of x. The solution has been developed in Chapter 6.5 (and 5.3), where it was shown that for a continuous system subject to a distributed stochastic load $\mathbf{q}(x,t)=\begin{bmatrix}q_y & q_z & q_\theta\end{bmatrix}^T$ then the response at position x_r is given by

$$\mathbf{S}_{rr}(x_r,\omega)=\begin{bmatrix} S_{r_y r_y} & S_{r_y r_z} & S_{r_y r_\theta} \\ S_{r_z r_y} & S_{r_z r_z} & S_{r_z r_\theta} \\ S_{r_\theta r_y} & S_{r_\theta r_z} & S_{r_\theta r_\theta} \end{bmatrix} = \boldsymbol{\Phi}(x_r)\cdot\mathbf{S}_{\eta\eta}(\omega)\cdot\boldsymbol{\Phi}^T(x_r) \tag{7.63}$$

where

$$\left.\begin{aligned} &\boldsymbol{\Phi}(x_r)=\begin{bmatrix} \phi_{y_1}(x_r) & & \phi_{y_n}(x_r) & & \phi_{yN_{\text{mod}}}(x_r) \\ \phi_{z_1}(x_r) & \cdots & \phi_{z_n}(x_r) & \cdots & \phi_{zN_{\text{mod}}}(x_r) \\ \phi_{\theta_1}(x_r) & & \phi_{\theta_n}(x_r) & & \phi_{\theta N_{\text{mod}}}(x_r) \end{bmatrix} \\ &\mathbf{S}_{\eta\eta}(\omega)=\begin{bmatrix} \ddots & & ⋰ \\ & S_{\eta_n\eta_m} & \\ ⋰ & & \ddots \end{bmatrix} \end{aligned}\right\} \tag{7.64}$$

and

$$\left.\begin{aligned} S_{\eta_n\eta_m} &= \frac{\hat{H}^*_{\eta_n}(\omega)\cdot\hat{H}_{\eta_m}(\omega)}{\left(\omega_n^2\tilde{M}_n\right)\cdot\left(\omega_m^2\tilde{M}_m\right)}\cdot\int\limits_L\int\limits_L \boldsymbol{\varphi}_n^T(x_1)\cdot\mathbf{S}_{qq}(x_1,x_2,\omega)\cdot\boldsymbol{\varphi}_m(x_2)\,dx_1dx_2 \\ &\left.\begin{aligned} \hat{H}_{\eta_j}(\omega) &= \left[1-\left(\omega/\omega_j\right)^2+2i\zeta_j\,\omega/\omega_j\right]^{-1} \\ \tilde{M}_j &= \int\limits_L \boldsymbol{\varphi}_j^T(x)\cdot\mathbf{m}_g(x)\cdot\boldsymbol{\varphi}_j(x)\,dx \end{aligned}\right\}\ \text{where } j=n \text{ or } m \end{aligned}\right\} \tag{7.65}$$

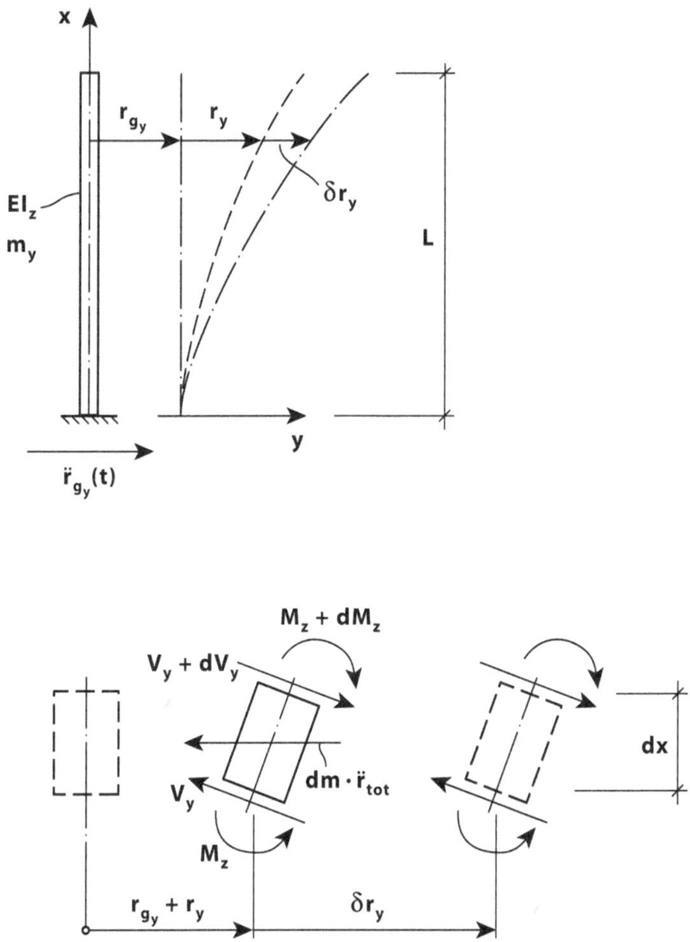

Fig. 7.8 Continuous line-like system

and

$$\mathbf{S}_{qq}(x_1,x_2,\omega)=\lim_{T\to\infty}\frac{1}{\pi T}\left[\mathbf{a}_q^*(x,\omega)\cdot\mathbf{a}^T(x,\omega)\right]=\begin{bmatrix} S_{q_yq_y} & S_{q_yq_z} & S_{q_yq_\theta} \\ S_{q_zq_y} & S_{q_zq_z} & S_{q_zq_\theta} \\ S_{q_\theta q_y} & S_{q_\theta q_z} & S_{q_\theta q_\theta} \end{bmatrix} \tag{7.66}$$

As illustrated in Fig. 7.8, the inertia force due to horizontal ground acceleration is given by

$$-d\mathbf{m}\cdot\ddot{\mathbf{r}}_{tot}=-\mathbf{m}_g dx\begin{bmatrix}\ddot{r}_y+\ddot{r}_g\\ \ddot{r}_z\\ \ddot{r}_\theta\end{bmatrix}=-\left(\mathbf{m}_g\ddot{\mathbf{r}}+\ddot{r}_g\mathbf{m}_g\mathbf{A}\right)dx \tag{7.67}$$

where $$\mathbf{m}_g=\begin{bmatrix} m_y & 0 & -m_ye_z\\ & m_z & m_ze_y\\ Sym. & & m_\theta\end{bmatrix} \quad\text{and}\quad \mathbf{A}=\begin{bmatrix}1\\0\\0\end{bmatrix} \tag{7.68}$$

Thus, it is seen that the horizontal ground acceleration is equivalent to an evenly distributed horizontal load

$$\mathbf{q}(x,t)=\begin{bmatrix}q_y & q_z & q_\theta\end{bmatrix}^T=-\ddot{r}_g\mathbf{m}_g\mathbf{A} \tag{7.69}$$

whose Fourier transform is defined by

$$\mathbf{a}_q(x,\omega)=\begin{bmatrix}a_{q_y} & a_{q_z} & a_{q\theta}\end{bmatrix}^T=-a_{\ddot{r}_g}\mathbf{m}_g\mathbf{A} \tag{7.70}$$

and thus, its cross spectral load effect is given by

$$\begin{aligned}\mathbf{S}_{qq}(x_1,x_2,\omega)&=\lim_{T\to\infty}\frac{1}{\pi T}\left[\mathbf{a}_q^*(x_1,\omega)\cdot\mathbf{a}_q^T(x_2,\omega)\right]\\ &=\lim_{T\to\infty}\frac{1}{\pi T}\left\{\left[-a_{\ddot{r}_g}(\omega)\mathbf{m}_g(x_1)\mathbf{A}\right]^*\left[-a_{\ddot{r}_g}(\omega)\mathbf{m}_g(x_2)\mathbf{A}\right]^T\right\}\\ &=\mathbf{m}_g(x_1)\mathbf{A}\mathbf{A}^T\mathbf{m}_g^T(x_2)S_{\ddot{r}_g}(\omega)\end{aligned} \tag{7.71}$$

where $S_{\ddot{r}_g}(\omega)=\lim_{T\to\infty}\frac{1}{\pi T}\left(a_{\ddot{r}_g}^* a_{\ddot{r}_g}\right)$ is the spectral density of the horizontal ground acceleration. Thus, the modal cross spectral density $S_{\eta_n\eta_m}(\omega)$ in Eq. 6.65 is given by

$$
\begin{aligned}
S_{\eta_n\eta_m} &= \frac{\hat{H}_{\eta_n}^*(\omega)\hat{H}_{\eta_m}(\omega)}{\left(\omega_n^2\tilde{M}_n\right)\left(\omega_m^2\tilde{M}_m\right)} S_{\ddot{r}_g} \int_L\int_L \boldsymbol{\varphi}_n^T(x_1)\mathbf{m}_g(x_1)\mathbf{A}\mathbf{A}^T\mathbf{m}_g^T(x_2)\boldsymbol{\varphi}_m(x_2)dx_1dx_2 \\
&= \frac{\hat{H}_{\eta_n}^*(\omega)\hat{H}_{\eta_m}(\omega)}{\left(\omega_n^2\tilde{M}_n\right)\left(\omega_m^2\tilde{M}_m\right)} S_{\ddot{r}_g}(\omega) \int_L\int_L m_y(x_1)m_y(x_2)\big[\phi_y(x_1)\phi_y(x_2) \\
&\quad -e_z\phi_\theta(x_1)\phi_y(x_2) - e_z\phi_y(x_1)\phi_\theta(x_2) + e_z^2\phi_\theta(x_1)\phi_\theta(x_2)\big]dx_1dx_2
\end{aligned}
\tag{7.72}
$$

The relevant variances and co-variances are then given by

$$
\begin{aligned}
&\begin{bmatrix} \sigma_{r_yr_y}^2 & Cov_{r_yr_z} & Cov_{r_yr_\theta} \\ Cov_{r_zr_y} & \sigma_{r_zr_z}^2 & Cov_{r_zr_\theta} \\ Cov_{r_\theta r_y} & Cov_{r_\theta r_z} & \sigma_{r_\theta r_\theta}^2 \end{bmatrix} = \int_0^\infty \mathbf{S}_{rr}(x_r,\omega)d\omega \\
&= \int_0^\infty \boldsymbol{\Phi}(x_r)\mathbf{S}_{\eta\eta}(\omega)\boldsymbol{\Phi}^T(x_r)d\omega = \int_0^\infty \begin{bmatrix} S_{r_yr_y} & S_{r_yr_z} & S_{r_yr_\theta} \\ S_{r_zr_y} & S_{r_zr_z} & S_{r_zr_\theta} \\ S_{r_\theta r_y} & S_{r_\theta r_z} & S_{r_\theta r_\theta} \end{bmatrix} d\omega
\end{aligned}
\tag{7.73}
$$

This will generally be applicable to any continuous system subject to horizontal ground acceleration. The actual response itself may be obtained by a time domain simulation from spectra (see Appendix B) or by a frequency domain integration, as shown in Eq. 7.73.

Elaboration 7.3: Single Mode Single Component Response

If a single mode approach is considered adequate, i.e. $N_{\text{mod}} = 1$, and there are no cross sectional asymmetry in the entire system, i.e. if

$$
\boldsymbol{\varphi}_n(x) = \begin{bmatrix} \phi_y \\ 0 \\ 0 \end{bmatrix}_n \qquad \text{and} \qquad \mathbf{m}_g(x) = \begin{bmatrix} m_y & 0 & 0 \\ & m_z & 0 \\ Sym. & & m_\theta \end{bmatrix}
$$

then the response calculation simplifies into $S_{r_yr_y}(x_r,\omega) = \phi_{y_n}^2(x_r)\cdot S_{\eta_n\eta_n}(\omega)$

where $$S_{\eta_n\eta_n}(\omega) = \left[\frac{\left|\hat{H}_n(\omega)\right|}{\omega_n^2 \tilde{M}_n} \int_0^L \phi_{y_n}(x) \cdot m_y(x)\, dx \right]^2 S_{\ddot{r}_g}(\omega)$$

Frequency domain integration may in most cases with sufficient accuracy be split into a background part and a resonant part, and thus

$$\sigma_{r_y}^2(x_r) = \int_0^\infty S_{r_y r_y}(\omega)\, d\omega = \left[\frac{\phi_{y_n}(x_r)}{\omega_n^2 \tilde{M}_n} \int_0^L \phi_{y_n}(x) m_y(x)\, dx \right]^2 \int_0^\infty \left|\hat{H}_n(\omega)\right|^2 S_{\ddot{r}_g}(\omega)\, d\omega$$

$$\approx \left[\frac{\phi_{y_n}(x_r)}{\omega_n^2 \tilde{M}_n} \int_0^L \phi_{y_n} m_y dx \right]^2 \left[\int_0^\infty \left|\hat{H}_n(\omega=0)\right|^2 S_{\ddot{r}_g}(\omega)\, d\omega + \int_0^\infty \left|\hat{H}_n(\omega)\right|^2 S_{\ddot{r}_g}(\omega=\omega_n)\, d\omega \right]$$

$$= \left[\frac{\phi_{y_n}(x_r)}{\omega_n^2 \tilde{M}_n} \int_0^L \phi_{y_n}(x) m_y(x)\, dx \right]^2 \left[\sigma_{\ddot{r}_g}^2 + \frac{\pi \omega_n}{4\zeta_n} S_{\ddot{r}_g}(\omega=\omega_n) \right]$$

where $\sigma_{\ddot{r}_g}^2 = \int_0^\infty S_{\ddot{r}_g}(\omega)\, d\omega$ is the variance of the horizontal ground acceleration.

Thus $$\sigma_{r_y}(x_r) \approx \left|\phi_{y_n}(x_r)\right| \cdot \frac{\int_0^L \phi_{y_n}(x) m_y(x)\, dx}{\int_0^L \phi_{y_n}^2(x) m_y(x)\, dx} \cdot \left[\left(\frac{\sigma_{\ddot{r}_g}}{\omega_n^2} \right)^2 + \frac{\pi S_{\ddot{r}_g}(\omega=\omega_n)}{4\omega_n^3 \zeta_n} \right]^{\frac{1}{2}}$$

Chapter 8
Wind Induced Dynamic Response Calculations

8.1 Introduction

It is in the following taken for granted that the main wind direction throughout the entire span of the structure is perpendicular to the direction of its span. The wind velocity vector is split into three fluctuating orthogonal components, U in the main wind direction, and v and w in the across wind horizontal and vertical directions. Typical full scale recordings of U, v and w are illustrated in Figs. 8.1 and 8.2 above. For a relevant structural design situation it is assumed that U may be split into a mean value V that only varies with height above ground and a fluctuating part u, i.e. $U=V+u$. V is commonly known as the mean wind velocity, and u, v and w are the turbulence components, created by friction between the terrain and the flow of the main weather system. It is taken for granted that the instantaneous wind velocity pressure is given by Bernoulli's equation $q(t)=\rho U^2/2$. Within the relevant time and space of response calculations it is also taken for granted that the variations of the wind velocity components are stationary and homogeneous.

If an air flow is met by the obstacle of a more or less solid line–like structure, the flow/structure interaction will give raise to forces acting on the system. Unless the body is extremely streamlined and the speed of the flow is very low and smooth, these forces will fluctuate. Firstly, because the oncoming flow in which the structure is submerged contains turbulence, i.e. it is itself fluctuating in time and space. Secondly, on the surface of the structure additional flow turbulence and vortices are created due to friction as well as flow separation, causing vortices to be shed in the wake of the body. And finally, if the structure is flexible the fluctuating forces will cause the body to oscillate, and the alternating flow and the oscillating body will interact and generate further forces. The first of these effects is known as buffeting, the second as vortex shedding, and the third is aerodynamic motion induced forces. In literature, the response calculations due to buffeting and vortex shedding are usually treated separately. The reason for this is that for most civil engineering structures they occur at their strongest in fairly separate wind velocity regions, i.e. vortex shedding is at its strongest at fairly low wind velocities, while buffeting occur at stronger wind velocities. An illustration of what can be expected is shown in Fig. 8.3. At large wind velocities, in the vicinity of a certain limiting (critical) wind velocity, the response curve may increase

E.N. Strømmen, *Structural Dynamics*, Springer Series in Solid and Structural Mechanics 2, DOI: 10.1007/978-3-319-01802-7_8, © Springer International Publishing Switzerland 2014

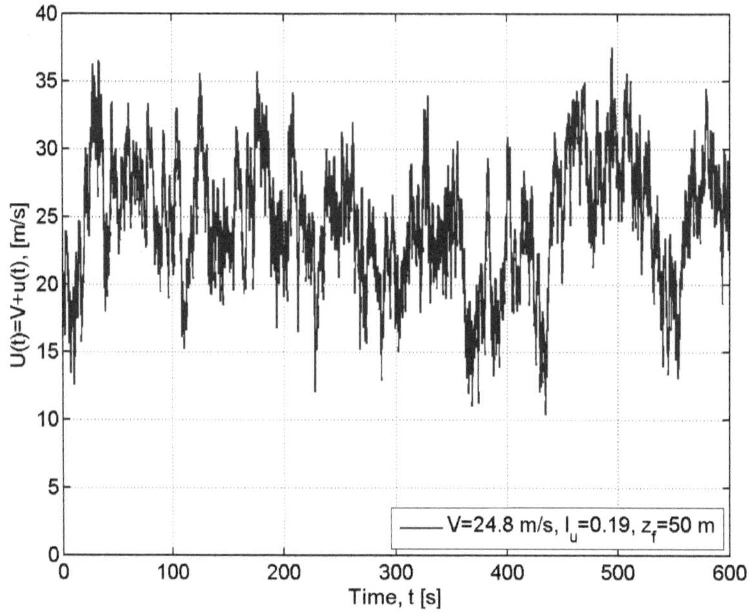

Fig. 8.1 Typical recording of along wind velocity U component

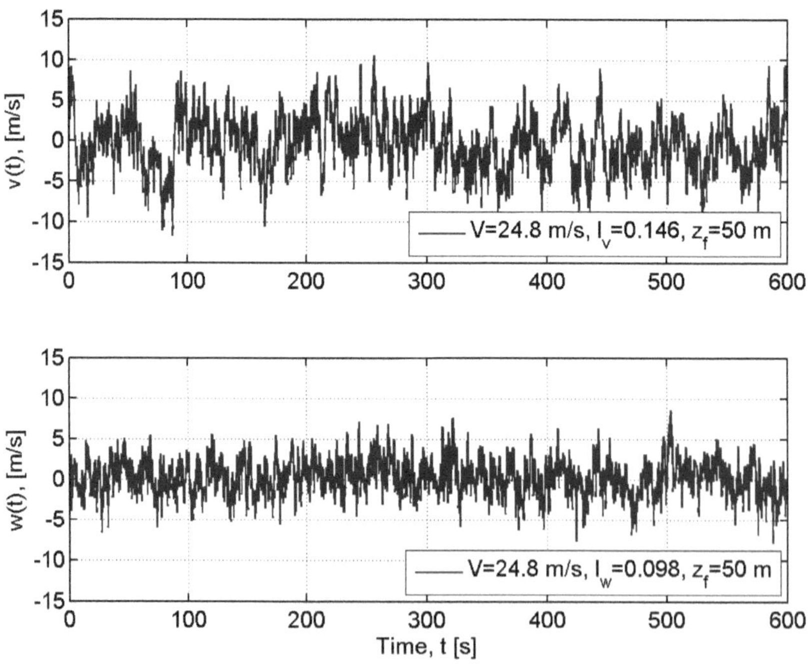

Fig. 8.2 Typical recording of across wind velocity v and w components

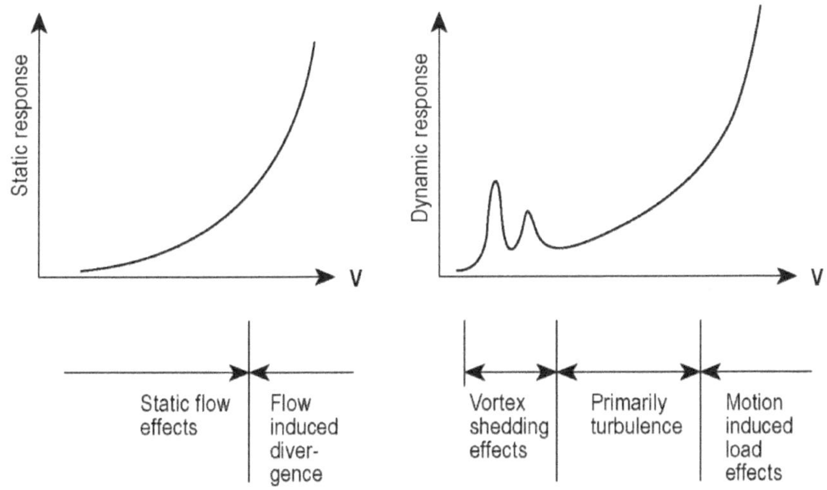

Fig. 8.3 Typical response behaviour of slender civil engineering structures

rapidly and the system will show signs of unstable behaviour in the sense that a small increase in V implies a large increase of static or dynamic response. It should be noted that in the vicinity of such a stability limit the problem is nonlinear. The reason for this is that the unstable behaviour is primarily caused by motion induced load effects, which contain contributions that will change the resonance frequency (as well as mode shapes) and the damping properties to the combined structure and flow system. Thus, the solution may require iterations.

8.2 The Dynamic Buffeting Load

The situation is illustrated in Fig. 8.4. As can be seen, the buffeting load from the wind is assumed to comprise distributed drag, lift and pitching moment forces (q_D, q_L and q_M), whose components in structural axis are denoted q_y, q_z and q_θ. (It should be noted that in aeronautics as well as in wind engineering it is customary to define the pitching moment positive in the opposite direction of that which is shown in Fig. 8.4, see Fig. 8.5.) The theory presented below was first developed by A.G. Davenport [42, 43].

The basic assumptions are that the load may be calculated from the instantaneous velocity pressure and appropriate load coefficients that have been obtained from static tests, and that linearization of any fluctuating parts will render results with sufficient accuracy. It is a requirement for linearization of load components that structural displacements and cross sectional rotations are small and that the turbulence components u, v and w are small as compared to V. It is taken for granted that the local element x axis is either horizontal or vertical (i.e. parallel to either z_f or y_f as illustrated in Fig. 8.4).

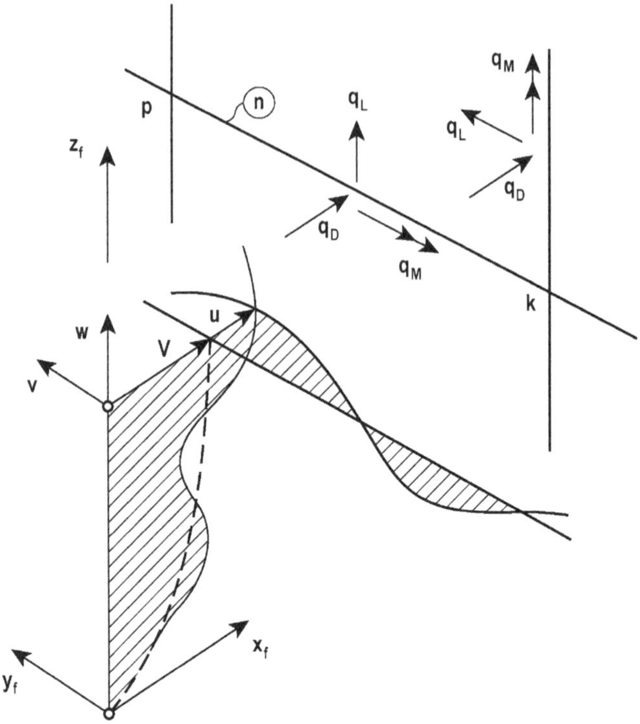

Fig. 8.4 Line-like structure in a turbulent wind field

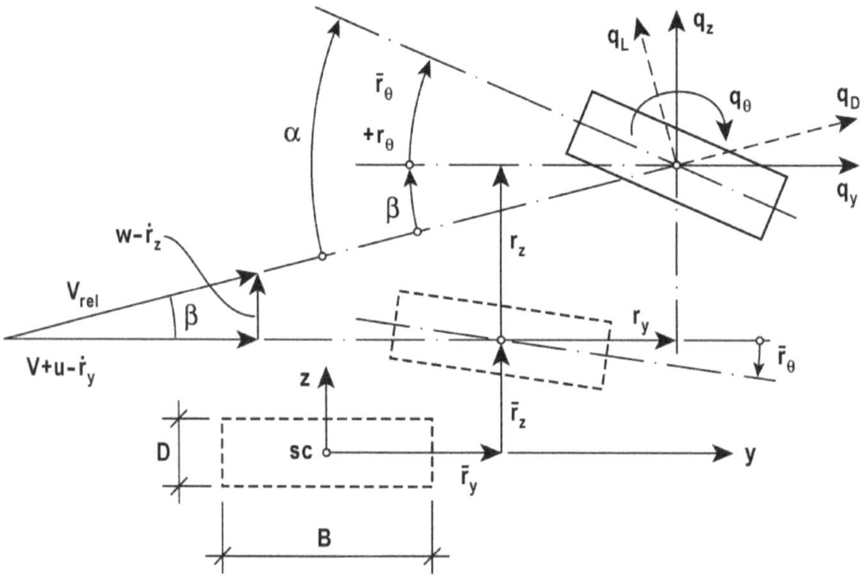

Fig. 8.5 Instantaneous flow and displacement quantities

The situation for a horizontal element is show in Fig. 8.5. (The situation for a vertical element is identical, except that w should be replaced by v .) The cross section at an arbitrary position x is first given the time invariant (mean) displacements $\bar{r}_y(x)$, $\bar{r}_z(x)$ and $\bar{r}_\theta(x)$. In this position the wind velocity vector is $V+u(x,t)$ in the along wind horizontal direction and $w(x,t)$ in the vertical across wind direction. It is about this position that the structure oscillates. Adopting the principle of d'Alambert, the cross section is given an additional dynamic displacements $r_y(x,t)$, $r_z(x,t)$ and $r_\theta(x,t)$. In this position the instantaneous cross sectional drag, lift and moment forces in flow axes are given by

$$\begin{bmatrix} q_D(x,t) \\ q_L(x,t) \\ q_M(x,t) \end{bmatrix} = \frac{1}{2}\rho V_{rel}^2 \cdot \begin{bmatrix} D \cdot C_D(\alpha) \\ B \cdot C_L(\alpha) \\ B^2 \cdot C_M(\alpha) \end{bmatrix} \tag{8.1}$$

where C_D, C_L and C_M are cross sectional characteristic load coefficients from static tests (see Fig. 8.6), V_{rel} is the instantaneous relative wind velocity and α is the angle of flow incidence. Transformation into structural axis is obtained by

$$\mathbf{q}_{tot}(x,t) = \begin{bmatrix} q_y \\ q_z \\ q_\theta \end{bmatrix}_{tot} = \begin{bmatrix} \cos\beta & -\sin\beta & 0 \\ \sin\beta & \cos\beta & 0 \\ 0 & 0 & 1 \end{bmatrix} \cdot \begin{bmatrix} q_D \\ q_L \\ q_M \end{bmatrix} \tag{8.2}$$

where:

$$\beta = \arctan\left(\frac{w - \dot{r}_z}{V + u - \dot{r}_y}\right) \tag{8.3}$$

The first linearization involves the assumption that the fluctuating flow components $u(x,t)$ and $w(x,t)$ are small as compared to V, and that structural displacements (as well as cross sectional rotation) are also small. Then $\cos\beta \approx 1$ and $\sin\beta \approx \tan\beta \approx \beta \approx (w-\dot{r}_z)/(V+u-\dot{r}_y) \approx (w-\dot{r}_z)/V$, and thus

$$\left.\begin{aligned} V_{rel}^2 &= (V+u-\dot{r}_y)^2 + (w-\dot{r}_z)^2 \approx V^2 + 2Vu - 2V\dot{r}_y \\ \alpha &= \bar{r}_\theta + r_\theta + \beta \approx \bar{r}_\theta + r_\theta + w/V - \dot{r}_z/V \end{aligned}\right\} \tag{8.4}$$

The second linearization involves the flow incidence dependent load coefficients. As illustrated in Fig. 8.6, the nonlinear variation of the load coefficient curves is replaced by the following linear approximation

$$\begin{bmatrix} C_D(\alpha) \\ C_L(\alpha) \\ C_M(\alpha) \end{bmatrix} = \begin{bmatrix} C_D(\bar{\alpha}) \\ C_L(\bar{\alpha}) \\ C_M(\bar{\alpha}) \end{bmatrix} + \alpha_f \cdot \begin{bmatrix} C'_D(\bar{\alpha}) \\ C'_L(\bar{\alpha}) \\ C'_M(\bar{\alpha}) \end{bmatrix} \tag{8.5}$$

where $\bar{\alpha}$ and α_f are the mean value and the fluctuating part of the angle of flow incidence, and where C'_D, C'_L and C'_M are the slopes of the load coefficient curves at $\bar{\alpha}$. It follows from Eq. 8.4 that $\bar{\alpha} = \bar{r}_\theta$ and $\alpha_f = r_\theta + w/V - \dot{r}_z/V$. For simplicity the following notation is introduced

$$\begin{bmatrix} C_D(\bar{\alpha}) \\ C_L(\bar{\alpha}) \\ C_M(\bar{\alpha}) \end{bmatrix} = \begin{bmatrix} \bar{C}_D \\ \bar{C}_L \\ \bar{C}_M \end{bmatrix} \quad \text{and} \quad \begin{bmatrix} C'_D(\bar{\alpha}) \\ C'_L(\bar{\alpha}) \\ C'_M(\bar{\alpha}) \end{bmatrix} = \begin{bmatrix} C'_D \\ C'_L \\ C'_M \end{bmatrix} \tag{8.6}$$

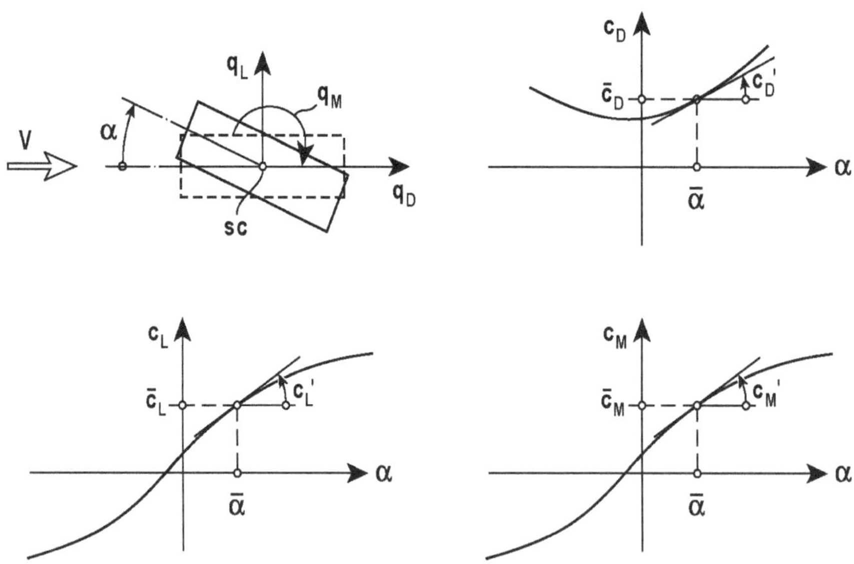

Fig. 8.6 Load coefficients obtained from static tests

Combining Eqs. 8.2 – 8.6

$$\begin{bmatrix} q_y \\ q_z \\ q_\theta \end{bmatrix}_{tot} = \rho V \left(\frac{V}{2} + u - \dot{r}_y \right) \left\{ \begin{bmatrix} D\bar{C}_D \\ B\bar{C}_L \\ B^2\bar{C}_M \end{bmatrix} + \left(r_\theta + \frac{w}{V} - \frac{\dot{r}_z}{V} \right) \begin{bmatrix} DC'_D \\ BC'_L \\ B^2C'_M \end{bmatrix} + \frac{w - \dot{r}_z}{V} \begin{bmatrix} -B\bar{C}_L \\ D\bar{C}_D \\ 0 \end{bmatrix} \right\} \tag{8.7}$$

and discarding higher order terms (i.e. terms containing the square or the product of quantities that have been assumed small) the following is obtained

$$\mathbf{q}_{tot}(x,t) = \begin{bmatrix} \bar{q}_y(x) \\ \bar{q}_z(x) \\ \bar{q}_\theta(x) \end{bmatrix} + \begin{bmatrix} q_y(x,t) \\ q_z(x,t) \\ q_\theta(x,t) \end{bmatrix} = \bar{\mathbf{q}} + \mathbf{b}_q \cdot \mathbf{v} + \mathbf{c}_{q_{ae}} \cdot \dot{\mathbf{r}} + \mathbf{k}_{q_{ae}} \cdot \mathbf{r} \tag{8.8}$$

Where

$$\mathbf{v}(x,t) = \begin{bmatrix} u & w \end{bmatrix}^T \text{ and } \mathbf{r}(x,t) = \begin{bmatrix} r_y & r_z & r_\theta \end{bmatrix}^T \tag{8.9}$$

$$\bar{\mathbf{q}}(x) = \begin{bmatrix} \bar{q}_y \\ \bar{q}_z \\ \bar{q}_\theta \end{bmatrix} = \frac{\rho V^2 B}{2} \begin{bmatrix} (D/B)\bar{C}_D \\ \bar{C}_L \\ B\bar{C}_M \end{bmatrix} \tag{8.10}$$

$$\mathbf{b}_q(x) = \frac{\rho V B}{2} \begin{bmatrix} 2(D/B)\bar{C}_D & \left((D/B)C'_D - \bar{C}_L\right) \\ 2\bar{C}_L & \left(C'_L + (D/B)\bar{C}_D\right) \\ 2B\bar{C}_M & BC'_M \end{bmatrix} = \frac{\rho V B}{2} \cdot \hat{\mathbf{b}}_q \tag{8.11}$$

$$\mathbf{c}_{q_{ae}}(x) = -\frac{\rho V B}{2} \begin{bmatrix} 2(D/B)\bar{C}_D & \left((D/B)C'_D - \bar{C}_L\right) & 0 \\ 2\bar{C}_L & \left(C'_L + (D/B)\bar{C}_D\right) & 0 \\ 2B\bar{C}_M & BC'_M & 0 \end{bmatrix} \tag{8.12}$$

$$\mathbf{k}_{q_{ae}}(x) = \frac{\rho V^2 B}{2} \begin{bmatrix} 0 & 0 & (D/B)C'_D \\ 0 & 0 & C'_L \\ 0 & 0 & BC'_M \end{bmatrix} \tag{8.13}$$

It is seen that the total load vector comprises a time invariant (mean static) part

$$\bar{\mathbf{q}}(x) = \begin{bmatrix} \bar{q}_y & \bar{q}_z & \bar{q}_\theta \end{bmatrix}^T \tag{8.14}$$

and a fluctuating (dynamic) part

$$\mathbf{q}(x,t) = \begin{bmatrix} q_y & q_z & q_\theta \end{bmatrix}^T = \mathbf{b}_q \cdot \mathbf{v} + \mathbf{c}_{q_{ae}} \cdot \dot{\mathbf{r}} + \mathbf{k}_{q_{ae}} \cdot \mathbf{r} \tag{8.15}$$

where $\mathbf{b}_q \cdot \mathbf{v}$ is the dynamic loading associated with turbulence (u and w) in the oncoming flow, while $\mathbf{c}_{q_{ae}} \cdot \dot{\mathbf{r}}$ and $\mathbf{k}_{q_{ae}} \cdot \mathbf{r}$ are motion induced loads associated with structural velocity and displacement. The theory is applicable in time domain as well as in frequency domain. For an element that is vertical in the flow, the local axis system is maintained and thus, the necessary load equations may simply be obtained by replacing w by v in Eqs. 8.3 – 8.15.

Used in a frequency domain approach $\mathbf{c}_{q_{ae}}$ and $\mathbf{k}_{q_{ae}}$ in Eqs. 8.12 and 8.13 may be expanded into the more general theory of aerodynamic derivatives, first developed in the field of aeronautics and later made applicable to line like civil engineering structures by Scanlan & Tomko [44]. Following their notations, the frequency domain versions of $\mathbf{c}_{q_{ae}}$ and $\mathbf{k}_{q_{ae}}$ are then given by

$$\mathbf{c}_{q_{ae}} = \frac{\rho B^2}{2} \cdot \omega_i(V) \cdot \hat{\mathbf{c}}_{q_{ae}} \text{ and } \mathbf{k}_{q_{ae}} = \frac{\rho B^2}{2} \cdot \left[\omega_i(V)\right]^2 \cdot \hat{\mathbf{k}}_{q_{ae}} \tag{8.16}$$

where

$$\hat{\mathbf{c}}_{q_{ae}} = \begin{bmatrix} P_1^* & P_5^* & BP_2^* \\ H_5^* & H_1^* & BH_2^* \\ BA_5^* & BA_1^* & B^2A_2^* \end{bmatrix} \text{ and } \hat{\mathbf{k}}_{q_{ae}} = \begin{bmatrix} P_4^* & P_6^* & BP_3^* \\ H_6^* & H_4^* & BH_3^* \\ BA_6^* & BA_4^* & B^2A_3^* \end{bmatrix} \tag{8.17}$$

The non-dimensional coefficients $P_k^*, H_k^*, A_k^*,\ k = 1-6$ contained in $\hat{\mathbf{c}}_{q_{ae}}$ and $\hat{\mathbf{k}}_{q_{ae}}$ are usually called aerodynamic derivatives. Usually, they have been experimentally determined in wind tunnel aeroelastic section model tests, where they are given as functions of the reduced mean wind velocity $\hat{V} = V/(B\omega_n)$, where B is the cross sectional width and ω_n is the in-wind resonance frequency associated with the modelled mode shape of the system. For the purpose of full scale calculations the similarity requirements between model scale and full scale conditions must be fulfilled, and thus, the aerodynamic derivatives will have to be extracted as functions of $\hat{V} = V/(B\omega_n)$, which itself is affected by stiffness contributions from $\mathbf{k}_{q_{ae}}$. Hence, iterations may be required.

8.3 Dynamic Response to Wind Buffeting

The buffeting theory in a general finite element format is presented in ref. [15]. In this chapter the theory is presented in a continuous version. It is tailored for simple line like bridges (as illustrated in Fig. 8.7) or tower type of structural systems.

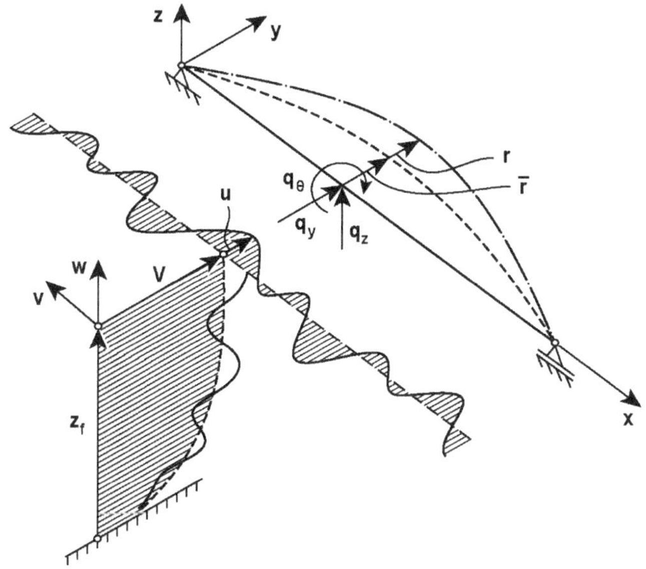

Fig. 8.7 Simple bridge type of structural system subject to wind turbulence

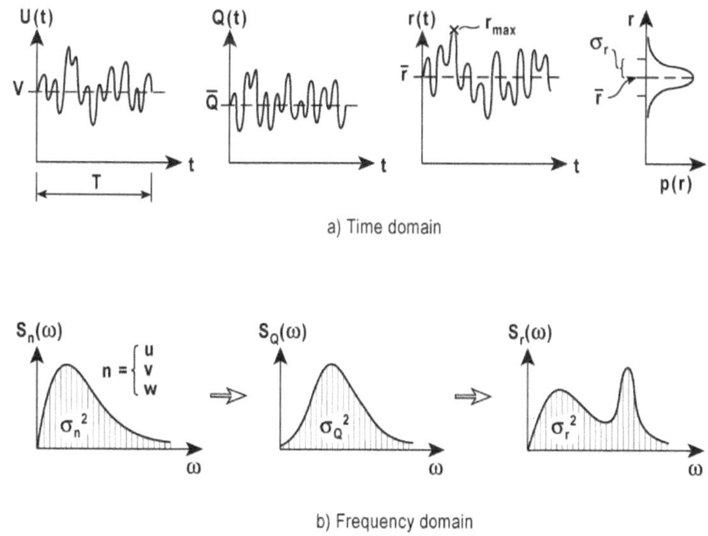

Fig. 8.8 Frequency domain representation

The solution is based on a modal frequency domain approach, as previously developed in Chapter 5.3 and further developed in Chapter 6.5. Thus, the entire focus is on the calculation of the variances and co-variances of the fluctuating displacement components see Fig. 8.8. Since the flow is assumed Gaussian, stationary and homogeneous and the structural system is linear elastic, then the same stochastic properties will apply to the response, and thus, extreme values of the response at an arbitrary position x_r along the span may be obtained from (see Appendix A)

$$\mathbf{r}_{tot_{\max}}(x_r) = \bar{\mathbf{r}} + k_p \boldsymbol{\sigma}_r \text{ where } \begin{cases} \bar{\mathbf{r}} = \begin{bmatrix} \bar{r}_y & \bar{r}_z & \bar{r}_\theta \end{bmatrix}^T \\ \boldsymbol{\sigma}_r = \begin{bmatrix} \sigma_y & \sigma_z & \sigma_\theta \end{bmatrix}^T \end{cases} \tag{8.18}$$

and where k_p is a peak factor defined in Eq. A.45, or else, $\mathbf{r}_{tot_{\max}}$ may be taken from a representative number of time domain simulations. In a modal format the cross sectional displacements $\mathbf{r}(x,t) = \begin{bmatrix} r_y & r_z & r_\theta \end{bmatrix}^T$ may be replaced by the sum of the products between natural eigenmodes $\boldsymbol{\varphi}_n(x) = \begin{bmatrix} \phi_y & \phi_z & \phi_\theta \end{bmatrix}_n^T$ (see Chapter 5.3) and unknown exclusively time dependent functions $\eta_n(t)$, i.e.

$$\mathbf{r}(x,t) = \sum_{n=1}^{N_{\text{mod}}} \begin{bmatrix} \phi_y(x) \\ \phi_z(x) \\ \phi_\theta(x) \end{bmatrix}_n \cdot \eta_n(t) = \boldsymbol{\Phi}(x) \cdot \boldsymbol{\eta}(t) \tag{8.19}$$

where N_{mod} is the number of modes that has been deemed necessary for a sufficiently accurate or representative solution (see Fig. 5.7). The mode shape matrix $\boldsymbol{\Phi}(x)$ and the generalised coordinate vector $\boldsymbol{\eta}(t)$ are defined by $\boldsymbol{\Phi}(x) = \begin{bmatrix} \boldsymbol{\varphi}_1 & \cdots & \boldsymbol{\varphi}_n & \cdots & \boldsymbol{\varphi}_{N_{\text{mod}}} \end{bmatrix}$ and $\boldsymbol{\eta}(t) = \begin{bmatrix} \eta_1 & \ldots & \eta_n & \ldots & \eta_{N_{\text{mod}}} \end{bmatrix}^T$. It was then shown in Chapter 5.3 that the introduction of Eq. 8.19 into the equilibrium equations, followed by consecutive weighing with each (orthogonal) modeshape and span-wise integration will render N_{mod} equivalent modal equilibrium conditions (see Eq. 5.32)

$$\tilde{\mathbf{M}} \cdot \ddot{\boldsymbol{\eta}}(t) + \tilde{\mathbf{C}} \cdot \dot{\boldsymbol{\eta}}(t) + \tilde{\mathbf{K}} \cdot \boldsymbol{\eta}(t) = \tilde{\mathbf{R}}_q(t) \tag{8.20}$$

where

$$\left.\begin{array}{l}\tilde{\mathbf{M}}=diag\left[\tilde{M}_n\right]\\ \tilde{\mathbf{C}}=diag\left[\tilde{C}_n\right]\\ \tilde{\mathbf{K}}=diag\left[\tilde{K}_n\right]\end{array}\right\}\text{ and }\left\{\begin{array}{l}\tilde{M}_n=\int\limits_L\left(\boldsymbol{\varphi}_n^T\cdot\mathbf{m}_g\cdot\boldsymbol{\varphi}_n\right)dx\\ \tilde{C}_n=2\tilde{M}_n\omega_n\zeta_n\\ \tilde{K}_n=\omega_n^2\tilde{M}_n\end{array}\right. \tag{8.21}$$

and where

$$\tilde{\mathbf{R}}_q=\int\limits_L\boldsymbol{\Phi}^T\mathbf{q}dx \tag{8.22}$$

What remains is to expand this theory to cover the load case of wind buffeting $\mathbf{q}(x,t)=\begin{bmatrix}q_y & q_z & q_\theta\end{bmatrix}^T=\mathbf{b}_q\mathbf{v}+\mathbf{c}_{q_{ae}}\cdot\dot{\mathbf{r}}+\mathbf{k}_{q_{ae}}\cdot\mathbf{r}$ (see Eq. 8.15) where

$$\left.\begin{array}{l}\mathbf{v}=\begin{bmatrix}u(x,t) & w(x,t)\end{bmatrix}^T\text{ if the system is horisontal}\\ \mathbf{v}=\begin{bmatrix}u(x,t) & v(x,t)\end{bmatrix}^T\text{ if the system is vertical}\end{array}\right\} \tag{8.23}$$

The modal load vector is then given by

$$\begin{aligned}\tilde{\mathbf{R}}_q(t)&=\int\limits_{L_{\exp}}\boldsymbol{\Phi}^T(x)\cdot\mathbf{q}(x,t)dx\\ &=\int\limits_{L_{\exp}}\begin{bmatrix}\boldsymbol{\varphi}_1 & \cdots & \boldsymbol{\varphi}_n & \cdots & \boldsymbol{\varphi}_{N_{\text{mod}}}\end{bmatrix}^T\left(\mathbf{b}_q\mathbf{v}+\mathbf{c}_{q_{ae}}\dot{\mathbf{r}}+\mathbf{k}_{q_{ae}}\mathbf{r}\right)dx\\ &=\int\limits_{L_{\exp}}\boldsymbol{\Phi}^T\mathbf{b}_q\mathbf{v}dx+\int\limits_{L_{\exp}}\boldsymbol{\Phi}^T\mathbf{c}_{q_{ae}}\boldsymbol{\Phi}dx\cdot\dot{\boldsymbol{\eta}}+\int\limits_{L_{\exp}}\boldsymbol{\Phi}^T\mathbf{k}_{q_{ae}}\boldsymbol{\Phi}dx\cdot\boldsymbol{\eta}\end{aligned} \tag{8.24}$$

and thus

$$\Rightarrow\ \tilde{\mathbf{R}}_q(t)=\tilde{\mathbf{R}}(t)+\tilde{\mathbf{C}}_{ae}\dot{\boldsymbol{\eta}}(t)+\tilde{\mathbf{K}}_{ae}\boldsymbol{\eta}(t) \tag{8.25}$$

where:

$$\left.\begin{array}{l}\tilde{\mathbf{R}}(t)=\begin{bmatrix}\tilde{R}_1 & \cdots & \tilde{R}_n & \cdots & \tilde{R}_{N_{\text{mod}}}\end{bmatrix}^T\\ \tilde{R}_n(t)=\int\limits_{L_{\exp}}\boldsymbol{\varphi}_n^T(x)\mathbf{b}_q\mathbf{v}(x,t)dx\end{array}\right\} \tag{8.26}$$

is the flow induced (buffeting) part of the modal load, and where

$$\left.\begin{aligned}\tilde{\mathbf{K}}_{ae} &= \int_{L_{\exp}} \mathbf{\Phi}^T \mathbf{k}_{q_{ae}} \mathbf{\Phi} dx \\ \tilde{\mathbf{C}}_{ae} &= \int_{L_{\exp}} \mathbf{\Phi}^T \mathbf{c}_{q_{ae}} \mathbf{\Phi} dx\end{aligned}\right\} \tag{8.27}$$

are aerodynamic modal stiffness and damping contributions. Moving these effects to the left hand side of Eq. 8.20 then the equilibrium condition in modal coordinates is given by

$$\tilde{\mathbf{M}} \cdot \ddot{\boldsymbol{\eta}}(t) + \left(\tilde{\mathbf{C}} - \tilde{\mathbf{C}}_{ae}\right) \cdot \dot{\boldsymbol{\eta}}(t) + \left(\tilde{\mathbf{K}} - \tilde{\mathbf{K}}_{ae}\right) \cdot \boldsymbol{\eta}(t) = \tilde{\mathbf{R}}(t) \tag{8.28}$$

Single Mode Response Calculations

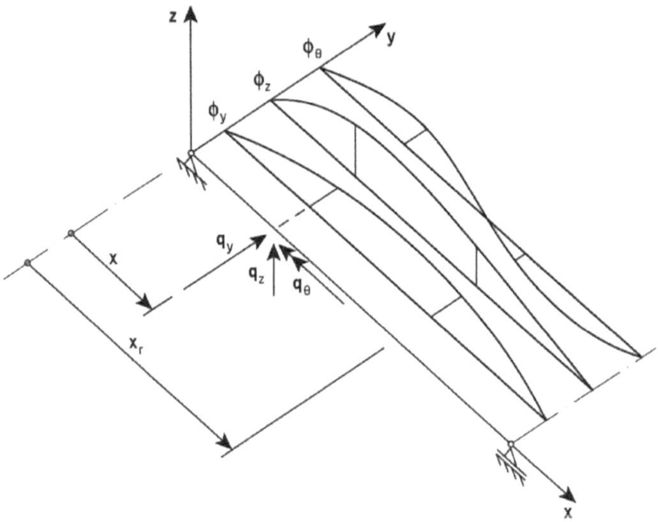

Fig. 8.9 Single mode shape containing three components

Let us first consider the simple case that a single mode approach may be adopted, i.e. that there are no significant sources of mechanical or flow induced coupling between modes and that they may with sufficient accuracy be handled as a sum of individual contributions. The response covariance between modes will then be zero, and thus, the variance of the total dynamic displacement response can be obtained as the sum of contributions from each mode, i.e. the variance of a displacement component is the sum of all variance contributions from excited modes

$$\begin{bmatrix}\sigma_y^2 & \sigma_z^2 & \sigma_\theta^2\end{bmatrix}^T = \sum_n \begin{bmatrix}\sigma_{y_n}^2 & \sigma_{z_n}^2 & \sigma_{\theta_n}^2\end{bmatrix}^T \tag{8.29}$$

Given an arbitrary mode shape (see Fig. 8.9)

$$\boldsymbol{\varphi}_n(x) = \begin{bmatrix}\phi_y & \phi_z & \phi_\theta\end{bmatrix}_n^T \tag{8.30}$$

with eigenfrequency ω_n and modal damping ratio ζ_n. The displacement response at a particular position x_r associated with this mode is simply $\mathbf{r}(x_r,t) = \boldsymbol{\varphi}_n(x_r)\cdot\eta_n(t)$. Since there is only a single mode then Eq. 8.28 is reduced to

$$\tilde{M}_n \cdot \ddot{\eta}_n(t) + \left(\tilde{C}_n - \tilde{C}_{ae_n}\right)\cdot\dot{\eta}_n(t) + \left(\tilde{K}_n - \tilde{K}_{ae_n}\right)\cdot\eta_n(t) = \tilde{R}_n(t) \tag{8.31}$$

where

$$\begin{bmatrix}\tilde{M}_n \\ \tilde{C}_n \\ \tilde{K}_n\end{bmatrix} = \begin{bmatrix}\int_L \boldsymbol{\varphi}_n^T \mathbf{m}_g \boldsymbol{\varphi}_n dx \\ 2\tilde{M}_n\omega_n\zeta_n \\ \omega_n^2 \tilde{M}_n\end{bmatrix} \tag{8.32}$$

$$\left.\begin{aligned}\tilde{K}_{ae_n} &= \int_{L_{\exp}} \boldsymbol{\varphi}_n^T \mathbf{k}_{q_{ae}} \boldsymbol{\varphi}_n dx \\ \tilde{C}_{ae_n} &= \int_{L_{\exp}} \boldsymbol{\varphi}_n^T \mathbf{c}_{q_{ae}} \boldsymbol{\varphi}_n dx\end{aligned}\right\} \tag{8.33}$$

and

$$\tilde{R}_n(t) = \int_{L_{\exp}} \boldsymbol{\varphi}_n^T(x)\mathbf{q}(x,t)dx = \int_{L_{\exp}} \boldsymbol{\varphi}_n^T(x)\mathbf{b}_q\mathbf{v}(x,t)dx \tag{8.34}$$

Transition into the frequency domain is obtained by taking the Fourier transform on either side of Eq. 8.31. Thus,

$$\left[-\tilde{M}_n\omega^2 + \left(\tilde{C}_n - \tilde{C}_{ae_n}\right)i\omega + \left(\tilde{K}_n - \tilde{K}_{ae_n}\right)\right]\cdot a_{\eta_n}(\omega) = a_{\tilde{R}_n}(\omega) \tag{8.35}$$

where a_{η_n} and $a_{\tilde{R}_n}$ are the Fourier amplitudes of $\eta_n(t)$ and $\tilde{R}_n(t)$,

$$a_{\tilde{R}_n}(\omega) = \int_{L_{\exp}} \boldsymbol{\varphi}_n^T(x)\mathbf{b}_q\mathbf{a}_v(x,\omega)dx \tag{8.36}$$

and

$$\left.\begin{array}{l} \mathbf{a}_v(x,\omega) = \begin{bmatrix} a_u & a_w \end{bmatrix}^T \text{ if the system is horisontal} \\ \mathbf{a}_v(x,\omega) = \begin{bmatrix} a_u & a_v \end{bmatrix}^T \text{ if the system is vertical} \end{array}\right\} \tag{8.37}$$

Pre-multiplying Eq. 8.35 by $\tilde{K}_n^{-1} = 1/\left(\omega_n^2 \tilde{M}_n\right)$, then the following is obtained

$$a_{\eta_n}(\omega) = \frac{\hat{H}_n(\omega)}{\tilde{K}_n} \cdot a_{\tilde{R}_n}(\omega) \tag{8.38}$$

where

$$\hat{H}_n(\omega) = \left[1 - \frac{\tilde{K}_{ae_n}}{\tilde{K}_n} - \left(\frac{\omega}{\omega_n}\right)^2 + 2i\left(\zeta_n - \frac{\tilde{C}_{ae_n}}{2\omega_n \tilde{M}_n}\right) \cdot \frac{\omega}{\omega_n}\right]^{-1} \tag{8.39}$$

is the non-dimensional modal frequency-response-function. Introducing $\kappa_{ae_n} = \tilde{K}_{ae_n} / \tilde{K}_n$ and $\zeta_{ae_n} = \tilde{C}_{ae_n} / \left(2\omega_n \tilde{M}_n\right)$, then

$$\hat{H}_n(\omega) = \left[1 - \kappa_{ae_n} - \left(\frac{\omega}{\omega_n}\right)^2 + 2i\left(\zeta_n - \zeta_{ae_n}\right) \cdot \frac{\omega}{\omega_n}\right]^{-1} \tag{8.40}$$

The single–sided spectrum of $\eta_n(t)$ is given by

$$S_{\eta_n}(\omega) = \lim_{T\to\infty} \frac{1}{\pi T} \cdot \left(a_{\eta_n}^* \cdot a_{\eta_n}\right) = \frac{\left|\hat{H}_n(\omega)\right|^2}{\tilde{K}_n^2} \cdot \lim_{T\to\infty} \frac{1}{\pi T} \cdot \left(a_{\tilde{R}_n}^* \cdot a_{\tilde{R}_n}\right) \tag{8.41}$$

$$\Rightarrow S_{\eta_n}(\omega) = \frac{\left|\hat{H}_n(\omega)\right|^2}{\tilde{K}_n^2} \cdot S_{\tilde{R}_n}(\omega) \tag{8.42}$$

where the single-sided spectrum of the modal loading is defined by (see Eq. 8.36)

$$
\begin{aligned}
S_{\tilde{R}_n}(\omega) &= \lim_{T\to\infty}\frac{1}{\pi T}\left(a_{\tilde{R}_n}^{*}\cdot a_{\tilde{R}_n}\right) \\
&= \lim_{T\to\infty}\frac{1}{\pi T}\left\{\left(\int_{L_{\exp}} \boldsymbol{\varphi}_n^T(x)\mathbf{b}_q\mathbf{a}_v(x,\omega)\,dx\right)^{*}\left(\int_{L_{\exp}} \boldsymbol{\varphi}_n^T(x)\mathbf{b}_q\mathbf{a}_v(x,\omega)\,dx\right)^{T}\right\} \\
&= \iint_{L_{\exp}} \boldsymbol{\varphi}_n^T(x_1)\mathbf{b}_q \lim_{T\to\infty}\frac{1}{\pi T}\left\{\mathbf{a}_v^{*}(x_1,\omega)\mathbf{a}_v^T(x_2,\omega)\right\}\mathbf{b}_q^T\boldsymbol{\varphi}_n(x_2)\,dx_1dx_2 \\
&= \iint_{L_{\exp}} \boldsymbol{\varphi}_n^T(x_1)\mathbf{b}_q\mathbf{S}_{vv}(x_1,x_2,\omega)\mathbf{b}_q^T\boldsymbol{\varphi}_n(x_2)\,dx_1dx_2
\end{aligned}
\tag{8.43}
$$

($L_{\exp}$ indicates integration over the wind exposed length of the system). In most cases cross spectra between turbulence components may be ignored, and thus

$$
\begin{aligned}
\mathbf{S}_{vv}(x_1,x_2,\omega) &= \lim_{T\to\infty}\frac{1}{\pi T}\left\{\mathbf{a}_v^{*}(x_1,\omega)\mathbf{a}_v^T(x_2,\omega)\right\} \\
&= \begin{bmatrix} S_{uu}(x_1,x_2,\omega) & S_{uw}(x_1,x_2,\omega) \\ S_{wu}(x_1,x_2,\omega) & S_{ww}(x_1,x_2,\omega)\end{bmatrix} \approx \begin{bmatrix} S_{uu}(x_1,x_2,\omega) & 0 \\ 0 & S_{ww}(x_1,x_2,\omega)\end{bmatrix}
\end{aligned}
\tag{8.44}
$$

if the system is horizontal, and

$$
\begin{aligned}
&\mathbf{S}_{vv}(x_1,x_2,\omega) = \lim_{T\to\infty}\frac{1}{\pi T}\left\{\mathbf{a}_v^{*}(x_1,\omega)\mathbf{a}_v^T(x_2,\omega)\right\} = \\
&\begin{bmatrix} S_{uu}(x_1,x_2,\omega) & S_{uv}(x_1,x_2,\omega) \\ S_{vu}(x_1,x_2,\omega) & S_{vv}(x_1,x_2,\omega)\end{bmatrix} \approx \begin{bmatrix} S_{uu}(x_1,x_2,\omega) & 0 \\ 0 & S_{vv}(x_1,x_2,\omega)\end{bmatrix}
\end{aligned}
\tag{8.45}
$$

if the system is vertical. Linearity implies that the Fourier amplitudes of the displacement components at an arbitrary position x_r are given by

$$
\mathbf{a}_{r_n}(x_r,\omega) = \begin{bmatrix} a_{r_y} \\ a_{r_z} \\ a_{r_\theta}\end{bmatrix}_n = \begin{bmatrix} \phi_y(x_r) \\ \phi_z(x_r) \\ \phi_\theta(x_r)\end{bmatrix}_n \cdot a_{\eta_n}(\omega) = \boldsymbol{\varphi}_n(x_r)\cdot a_{\eta_n}(\omega)
\tag{8.46}
$$

The cross spectral density matrix of the three displacement components is then

$$\begin{aligned}\mathbf{S}_{r_n}(x_r,\omega) &= \lim_{T\to\infty}\frac{1}{\pi T}\left(\mathbf{a}^*_{r_n}\cdot\mathbf{a}^T_{r_n}\right) = \\ \lim_{T\to\infty}\frac{1}{\pi T}&\left\{\left(\boldsymbol{\varphi}_n a_{\eta_n}\right)^*\left(\boldsymbol{\varphi}_n a_{\eta_n}\right)^T\right\} = \boldsymbol{\varphi}_n \lim_{T\to\infty}\frac{1}{\pi T}\left(a^*_{\eta_n}a_{\eta_n}\right)\boldsymbol{\varphi}^T_n\end{aligned} \tag{8.47}$$

from which the following is obtained:

$$\mathbf{S}_{r_n}(x_r,\omega) = \begin{bmatrix} S_{r_y r_y}(x_r,\omega) & S_{r_y r_z}(x_r,\omega) & S_{r_y r_\theta}(x_r,\omega) \\ S_{r_z r_y}(x_r,\omega) & S_{r_z r_z}(x_r,\omega) & S_{r_z r_\theta}(x_r,\omega) \\ S_{r_\theta r_y}(x_r,\omega) & S_{r_\theta r_z}(x_r,\omega) & S_{r_\theta r_\theta}(x_r,\omega) \end{bmatrix}_n = \boldsymbol{\varphi}_n(x_r)S_{\eta_n}(\omega)\boldsymbol{\varphi}^T_n(x_r) \tag{8.48}$$

where $S_{\eta_n}(\omega) = \lim_{T\to\infty}\frac{1}{\pi T}\left(a^*_{\eta_n}\cdot a_{\eta_n}\right)$ is given in Eq. 8.42. Thus (see also Eqs. 8.42, 8.43 and 8.48)

$$\begin{aligned}&\mathbf{S}_{r_n}(x_r,\omega) = \boldsymbol{\varphi}_n(x_r)\mathbf{S}_{\eta_n}(\omega)\boldsymbol{\varphi}^T_n(x_r) = \\ &\boldsymbol{\varphi}_n(x_r)\frac{\left|\hat{H}_n(\omega)\right|^2}{\tilde{K}^2_n}\left\{\iint_{L_{\exp}}\boldsymbol{\varphi}^T_n(x_1)\mathbf{b}_q\mathbf{S}_{vv}(x_1,x_2,\omega)\mathbf{b}^T_q\boldsymbol{\varphi}_n(x_2)\,dx_1dx_2\right\}\boldsymbol{\varphi}^T_n(x_r)\end{aligned} \tag{8.49}$$

The response covariance matrix is obtained by frequency domain integration of $\mathbf{S}_n(x_r,\omega)$, and thus

$$\begin{aligned}\mathbf{Cov}_{r_n}(x_r) &= \int_0^\infty \mathbf{S}_{r_n}(x_r,\omega)\,d\omega = \begin{bmatrix} \sigma^2_{r_y}(x_r) & Cov_{r_y r_z}(x_r) & Cov_{r_y r_\theta}(x_r) \\ Cov_{r_z r_y}(x_r) & \sigma^2_{r_z}(x_r) & Cov_{r_z r_\theta}(x_r) \\ Cov_{r_\theta r_y}(x_r) & Cov_{r_\theta r_z}(x_r) & \sigma^2_{r_\theta}(x_r) \end{bmatrix}_n \\ &= \boldsymbol{\varphi}_n(x_r)\sigma^2_{\eta_n}\boldsymbol{\varphi}^T_n(x_r) = \sigma^2_{\eta_n}\begin{bmatrix} \phi^2_y(x_r) & \phi_y(x_r)\cdot\phi_z(x_r) & \phi_y(x_r)\cdot\phi_\theta(x_r) \\ & \phi^2_z(x_r) & \phi_z(x_r)\cdot\phi_\theta(x_r) \\ sym. & & \phi^2_\theta(x_r) \end{bmatrix}_n\end{aligned} \tag{8.50}$$

where

$$\sigma_{\eta_n}^2 = \int_0^\infty S_{\eta_n}(\omega)\,d\omega =$$

$$\int_0^\infty \frac{\left|\hat{H}_n(\omega)\right|^2}{\tilde{K}_n^2} \left\{ \iint_{L_{\exp}} \boldsymbol{\varphi}_n^T(x_1)\mathbf{b}_q\mathbf{S}_{vv}(x_1,x_2,\omega)\mathbf{b}_q^T\boldsymbol{\varphi}_n(x_2)\,dx_1dx_2 \right\} d\omega \tag{8.51}$$

The total response covariance matrix may be obtained by adding up contributions from all modes deemed necessary for a sufficiently accurate solution, i.e.

$$\mathbf{Cov}(x_r) = \sum_{n=1}^{N_{\text{mod}}} \mathbf{Cov}_n(x_r) \tag{8.52}$$

Elaboration 8.1: Single Mode Single Component Horizontal System

For simplicity, let us consider a single mode single component and perfectly horizontal system, e.g. the case that the mode shape only contains a single horizontal y component

$$\boldsymbol{\varphi}_n \approx \begin{bmatrix} \phi_y & 0 & 0 \end{bmatrix}_n^T$$

Let us also take it for granted that V, σ_u, σ_w and all cross sectional quantities are constants, then

$$S_{\eta_n}(\omega) = \frac{\left|\hat{H}_n(\omega)\right|^2}{\tilde{K}_n^2} \left\{ \iint_{L_{\exp}} \boldsymbol{\varphi}_n^T(x_1)\mathbf{b}_q\mathbf{S}_{vv}(x_1,x_2,\omega)\mathbf{b}_q^T\boldsymbol{\varphi}_n(x_2)\,dx_1dx_2 \right\}$$

$$= \left[\frac{\rho V^2 B}{2} \cdot \frac{\left|\hat{H}_n(\omega)\right|}{\tilde{K}_n^2} \right]^2 \iint_{L_{\exp}} \phi_{y_n}(x_1) \cdot \phi_{y_n}(x_2) \cdot$$

$$\left\{ \left(2\frac{D}{B}\bar{C}_D I_u \right)^2 \frac{S_{uu}(x_1,x_2,\omega)}{\sigma_u^2} + \left[\left(\frac{D}{B}C_D' - \bar{C}_L \right) I_w \right]^2 \frac{S_{ww}(x_1,x_2,\omega)}{\sigma_w^2} \right\} dx_1dx_2$$

where $I_u = \sigma_u / V$ and $I_w = \sigma_w / V$ are the turbulence intensities, $\Delta x = |x_1 - x_2|$ is the spatial (span-wise) separation, and where x_1 and x_2 are two arbitrary positions. Introducing

$$\tilde{K}_y = \omega_y^2 \tilde{M}_y$$

and the modally equivalent and evenly distributed mass

$$\tilde{m}_n = \tilde{M}_n \Big/ \int_L \phi_{y_n}^2 dx = \int_L m_y \phi_{y_n}^2 dx \Big/ \int_L \phi_{y_n}^2 dx$$

then the following expression is obtained for the standard deviation of the dynamic response in the along wind y direction associated with mode shape n

$$\sigma_{r_{y_n}}(x_r) = \left|\phi_{y_n}(x_r)\right| \left\{ \int_0^\infty S_{\eta_n}(\omega) d\omega \right\}^{\frac{1}{2}}$$

$$= \frac{\left|\phi_{y_n}(x_r)\right|}{\int_L \phi_{y_n}^2 dx} \frac{\rho B^3}{2\tilde{m}_n} \left(\frac{V}{B\omega_n}\right)^2 \left(\int_0^\infty \left|\hat{H}_n(\omega)\right|^2 \left\{ \iint_{L_{\exp}} \phi_{y_n}(x_1) \phi_{y_n}(x_2) \cdot \right.\right.$$

$$\left.\left. \left[\left(2\frac{D}{B}\bar{C}_D I_u\right)^2 \frac{S_{uu}(\Delta x, \omega)}{\sigma_u^2} + \left(\left(\frac{D}{B}C_D' - \bar{C}_L\right) I_w\right)^2 \frac{S_{ww}(\Delta x, \omega)}{\sigma_w^2} \right] dx_1 dx_2 \right\} d\omega \right)^{\frac{1}{2}}$$

The non-dimensional frequency response function is given by

$$\hat{H}_n(\omega) = \left[1 - \kappa_{ae_n} - (\omega/\omega_n)^2 + 2i\left(\zeta_n - \zeta_{ae_n}\right)\cdot(\omega/\omega_n)\right]^{-1}$$

where

$$\begin{bmatrix} \kappa_{ae_n} \\ \zeta_{ae_n} \end{bmatrix} = \begin{bmatrix} \dfrac{\tilde{K}_{ae_n}}{\omega_n^2 \tilde{M}_n} \\ \dfrac{\tilde{C}_{ae_n}}{2\omega_n \tilde{M}_n} \end{bmatrix} = \begin{bmatrix} \dfrac{\dfrac{\rho B^2}{2} \omega_n^2 P_4^* \int_{L_{\exp}} \phi_{y_n}^2 dx}{\omega_n^2 \tilde{m}_n \int_L \phi_{y_n}^2 dx} \\ \dfrac{\dfrac{\rho B^2}{2} \omega_n P_1^* \int_{L_{\exp}} \phi_{y_n}^2 dx}{2\omega_n \tilde{m}_n \int_L \phi_{y_n}^2 dx} \end{bmatrix} = \frac{\rho B^2}{\tilde{m}_n} \cdot \frac{\int_{L_{\exp}} \phi_{y_n}^2 dx}{\int_L \phi_{y_n}^2 dx} \cdot \begin{bmatrix} \dfrac{1}{2} P_4^* \\ \dfrac{1}{4} P_1^* \end{bmatrix}$$

Similarly, if $\boldsymbol{\varphi}_n(x) \approx \begin{bmatrix} 0 & \phi_z & 0 \end{bmatrix}_n^T$ then

$$\sigma_{z_n}(x_r) = \frac{\left|\phi_{z_n}(x_r)\right|}{\int_L \phi_{z_n}^2 dx} \cdot \frac{\rho B^3}{2\tilde{m}_n} \cdot \left(\frac{V}{B\omega_n}\right)^2 \cdot \left(\int_0^\infty \left|\hat{H}_n(\omega)\right|^2 \cdot \left\{ \iint_{L_{\exp}} \phi_{z_n}(x_1) \cdot \phi_{z_n}(x_2) \cdot \right.\right.$$

$$\left.\left. \left[\left(2\bar{C}_L I_u\right)^2 \frac{S_{uu}(\Delta x, \omega)}{\sigma_u^2} \right] + \left[\left(C_L' + \frac{D}{B}\bar{C}_D\right) I_w \right]^2 \frac{S_{ww}(\Delta x, \omega)}{\sigma_w^2} dx_1 dx_2 \right\} d\omega \right)^{\frac{1}{2}}$$

where:

$$\tilde{m}_n = \frac{\tilde{M}_n}{\int_L \phi_{z_n}^2 dx} = \frac{\int_L m_n \phi_{z_n}^2 dx}{\int_L \phi_{z_n}^2 dx} \quad \text{and} \quad \begin{bmatrix} \kappa_{ae_n} \\ \zeta_{ae_n} \end{bmatrix} = \frac{\rho B^2}{\tilde{m}_n} \frac{\int_{L_{exp}} \phi_{z_n}^2 dx}{\int_L \phi_{z_n}^2 dx} \begin{bmatrix} \frac{1}{2} H_4^* \\ \frac{1}{4} H_1^* \end{bmatrix}$$

Or, if $\boldsymbol{\varphi}_n(x) \approx [0 \quad 0 \quad \phi_\theta]_n^T$ then

$$\sigma_{\theta_n}(x_r) = \frac{|\phi_\theta(x_r)|}{\int_L \phi_{\theta_n}^2 dx} \cdot \frac{\rho B^4}{2\tilde{m}_n} \cdot \left(\frac{V}{B\omega_n}\right)^2 \cdot \left(\int_0^\infty \left|\hat{H}_n(\omega)\right|^2 \cdot \left\{ \iint_{L_{exp}} \phi_{\theta_n}(x_1) \cdot \phi_{\theta_n}(x_2) \cdot \right. \right.$$

$$\left. \left. \left[\left(2\bar{C}_M I_u\right)^2 \frac{S_{uu}(\Delta x, \omega)}{\sigma_u^2} \right] + \left(C_M' I_w\right)^2 \frac{S_{ww}(\Delta x, \omega)}{\sigma_w^2} dx_1 dx_2 \right\} d\omega \right)^{\frac{1}{2}}$$

$$\text{where } \tilde{m}_n = \frac{\tilde{M}_n}{\int_L \phi_{\theta_n}^2 dx} = \frac{\int_L m_n \phi_{\theta_n}^2 dx}{\int_L \phi_{\theta_n}^2 dx} \quad \text{and} \quad \begin{bmatrix} \kappa_{ae_n} \\ \zeta_{ae_n} \end{bmatrix} = \frac{\rho B^4}{\tilde{m}_n} \frac{\int_{L_{exp}} \phi_{\theta_n}^2 dx}{\int_L \phi_{\theta_n}^2 dx} \begin{bmatrix} \frac{1}{2} A_3^* \\ \frac{1}{4} A_2^* \end{bmatrix}$$

Example 8.1

Let us consider a typical slender bridge type of system, where the three modes k,m,n

$$\boldsymbol{\varphi}_k = \begin{bmatrix} \phi_y & 0 & 0 \end{bmatrix}^T \qquad \boldsymbol{\varphi}_m = \begin{bmatrix} 0 & \phi_z & 0 \end{bmatrix}^T \qquad \boldsymbol{\varphi}_n = \begin{bmatrix} 0 & 0 & \phi_\theta \end{bmatrix}^T$$

with corresponding eigenfrequencies $\omega_y, \omega_z, \omega_\theta$ *have been singled out for a response calculation. Let us assume that the main girder cross section of the bridge is close to a flat plate, in which case the following load coefficient properties may be adopted*

$$\left.\begin{matrix} \bar{C}_D \\ C_L' \\ C_M' \end{matrix}\right\} \neq 0, \quad \left.\begin{matrix} C_D' \\ \bar{C}_L \\ \bar{C}_M \end{matrix}\right\} \approx 0 \quad \text{and} \quad \frac{D}{B}\bar{C}_D << C_L'$$

In that case

$$S_{r_y}(\omega, x_r) = \left[\phi_y(x_r) \cdot \frac{\rho B^2 D}{\tilde{m}_y} \cdot \left(\frac{V}{B\omega_y}\right)^2 \cdot \bar{C}_D I_u \cdot \left|\hat{H}_y(\omega)\right| \cdot \hat{J}_y(\omega) \right]^2 \cdot \frac{S_u(\omega)}{\sigma_u^2}$$

$$S_{r_z}(\omega, x_r) = \left[\phi_z(x_r) \frac{\rho B^3}{2\tilde{m}_z} \cdot \left(\frac{V}{B\omega_z} \right)^2 \cdot C'_L I_w \cdot \left| \hat{H}_z(\omega) \right| \cdot \hat{J}_z(\omega) \right]^2 \cdot \frac{S_w(\omega)}{\sigma_w^2}$$

$$S_{r_\theta}(\omega, x_r) = \left[\phi_\theta(x_r) \frac{\rho B^4}{2\tilde{m}_\theta} \cdot \left(\frac{V}{B\omega_\theta} \right)^2 \cdot C'_M I_w \cdot \left| \hat{H}_\theta(\omega) \right| \cdot \hat{J}_\theta(\omega) \right]^2 \cdot \frac{S_w(\omega)}{\sigma_w^2}$$

where it has been assumed that $\begin{cases} S_{uu}(\Delta x, \omega) = S_u(\omega) \cdot \hat{C}o_{uu}(\Delta x, \omega) \\ S_{ww}(\Delta x, \omega) = S_w(\omega) \cdot \hat{C}o_{ww}(\Delta x, \omega) \end{cases}$

and where

$$\hat{J}_y(\omega) = \left(\iint_{L_{\exp}} \phi_y(x_1) \cdot \phi_y(x_2) \cdot \hat{C}o_{uu}(\Delta x, \omega)\, dx_1 dx_2 \right)^{1/2} \Big/ \int_L \phi_y^2 dx$$

$$\hat{J}_z(\omega) = \left(\iint_{L_{\exp}} \phi_z(x_1) \cdot \phi_z(x_2) \cdot \hat{C}o_{ww}(\Delta x, \omega)\, dx_1 dx_2 \right)^{1/2} \Big/ \int_L \phi_z^2 dx$$

$$\hat{J}_\theta(\omega) = \left(\iint_{L_{\exp}} \phi_\theta(x_1) \cdot \phi_\theta(x_2) \cdot \hat{C}o_{ww}(\Delta x, \omega)\, dx_1 dx_2 \right)^{1/2} \Big/ \int_L \phi_\theta^2 dx$$

are the non–dimensional joint acceptance functions. Integrating across the entire frequency domain, then the following response standard deviations are obtained:

$$\sigma_{r_y}(x_r) = \left| \phi_y(x_r) \right| \cdot \frac{\rho B^2 D}{\tilde{m}_y} \cdot \left(\frac{V}{B\omega_y} \right)^2 \cdot \bar{C}_D I_u \cdot \left[\int_0^\infty \left| \hat{H}_y(\omega) \right|^2 \cdot \frac{S_u(\omega)}{\sigma_u^2} \cdot \hat{J}_y^2(\omega)\, d\omega \right]^{1/2}$$

$$\sigma_{r_z}(x_r) = \left| \phi_z(x_r) \right| \cdot \frac{\rho B^3}{2\tilde{m}_z} \cdot \left(\frac{V}{B\omega_z} \right)^2 \cdot C'_L I_w \cdot \left[\int_0^\infty \left| \hat{H}_z(\omega) \right|^2 \cdot \frac{S_w(\omega)}{\sigma_w^2} \cdot \hat{J}_z^2(\omega)\, d\omega \right]^{1/2}$$

$$\sigma_{r_\theta}(x_r) = \left| \phi_\theta(x_r) \right| \cdot \frac{\rho B^4}{2\tilde{m}_\theta} \cdot \left(\frac{V}{B\omega_\theta} \right)^2 \cdot C'_M I_w \cdot \left[\int_0^\infty \left| \hat{H}_\theta(\omega) \right|^2 \cdot \frac{S_w(\omega)}{\sigma_w^2} \cdot \hat{J}_\theta^2(\omega)\, d\omega \right]^{1/2}$$

Let us focus exclusively on the response in the y (drag) direction, and consider a suspension bridge with span $L = 1200m$ *that is elevated at a position* $z_f = 50m$. *Let us for simplicity assume that the relevant mode shape* $\phi_y(x) = \sin(\pi x / L)$ *and that* $x_r = L/2$, *in which case* $\phi_y(x_r) = 1$. *Let us also assume that the entire span is flow exposed, i.e.* $L_{\exp} = L$, *and adopt the following wind field properties:*

1) the turbulence intensity $I_u = \sigma_u / V = 0.15$

2) the integral length scale: ${}^{xf}L_u = 100 \cdot (z_f / 10)^{0.3} = 162m$

3) the auto spectral density: $\dfrac{S_u(\omega)}{\sigma_u^2} = \dfrac{1.08 \cdot {}^{xf}L_u / V}{\left(1 + 1.62 \cdot \omega \cdot {}^{xf}L_u / V\right)^{5/3}}$

4) the normalised co-spectrum: $\hat{C}o_{uu}(\omega, \Delta x) = \exp(-C_{ux} \cdot \omega \cdot \Delta x / V)$

where $C_{ux} = C_{uy_f} = 9/(2\pi) \approx 1.4$.

Let us allot the following values to the remaining constants that are necessary for a numerical calculation of $\sigma_{r_y}(x_r = L/2)$:

ρ (kg/m³)	$\bar{C}_D$	B (m)	D (m)	m_y (kg/m)	ω_y (rad/s)	ζ_y
1.25	0.7	20	4	10000	0.4	0.005

Since m_y *is constant along the span, then the modally equivalent and evenly distributed mass* $\tilde{m}_y = m_y$. *Finally, let us adopt quasi–static values to the aerodynamic derivatives, in which case* $\kappa_{ae_y} = 0$ *and the aerodynamic damping* ζ_{ae_y} *is given by*

$$\zeta_{ae_y} = \frac{\rho B^2}{4\tilde{m}_y} P_1^* = \frac{\rho B^2}{4\tilde{m}_y}\left(-2\bar{C}_D \frac{D}{B}\frac{V}{B\omega_y}\right) = -\frac{\rho D \bar{C}_D V}{2\tilde{m}_y \omega_y}$$

The standard deviation of the dynamic response at $x_r = L/2$ *is then given by*

$$\sigma_{r_y}(L/2) = 3.28 \cdot 10^{-4} \cdot V^2 \left[\int_0^\infty \left|\hat{H}_y(\omega)\right|^2 \cdot \frac{S_u(\omega)}{\sigma_u^2} \cdot \hat{J}_y^2(\hat{\omega})\, d\omega\right]^{1/2}$$

where $S_u(\omega)/\sigma_u^2$, $\hat{J}_y$ *are defined above, and where*

$$\hat{H}_y(\omega) = \left[1 - (\omega/\omega_y)^2 + 2i\left(\zeta_y - \zeta_{ae_y}\right) \cdot \omega/\omega_y\right]^{-1}$$

The chosen single point spectral density and corresponding normalised co–spectrum of the turbulent u *component are shown on the top left and right hand side diagrams in Fig. 8.10. The non-dimensional frequency response function and the squared normalised joint acceptance functions are shown on the lower left and right hand side diagrams in Fig. 8.10. The response spectrum of the along wind*

r_y component at $x_r = L/2$ and $V = 40\,m/s$ is shown in Fig. 8.11. As can be seen, it contains a broad banded background part and a narrow banded resonant part at $\omega = 0.4\,rad/s$. The standard deviation of the dynamic response at $x_r = L/2$ is plotted versus the mean wind velocity in Fig. 8.12. From the response spectrum in Fig. 8.11 a time domain simulation (see Appendix B) has been extracted, see Fig. 8.13.

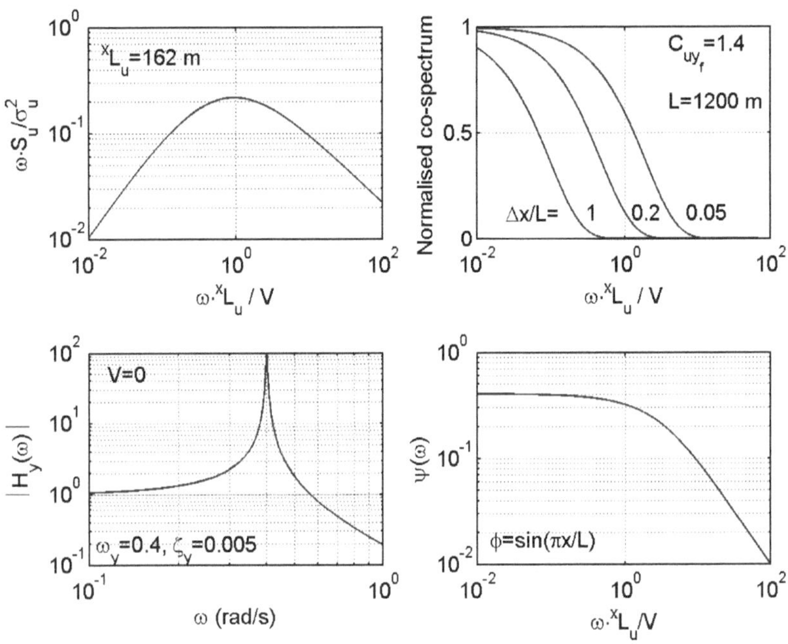

Fig. 8.10 Top left and right: single point u spectrum and corresponding normalised co-spectrum, lower left and right: frequency response function and joint acceptance function

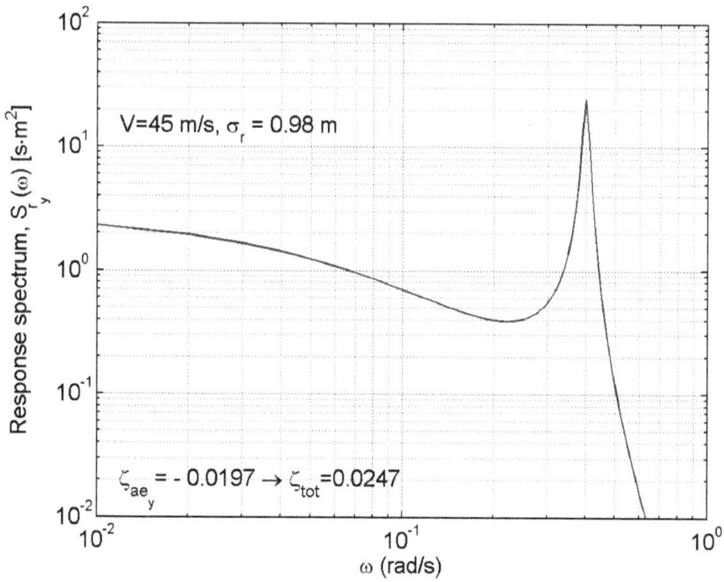

Fig. 8.11 Response spectrum of r_y displacements at $x_r = L/2$ and $V = 45\ m/s$

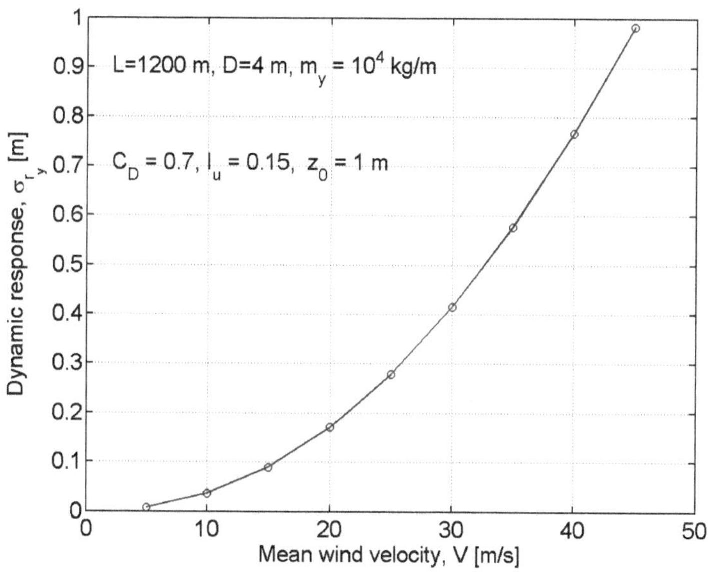

Fig. 8.12 The standard deviation of the dynamic response at $x_r = L/2$ versus the mean wind velocity

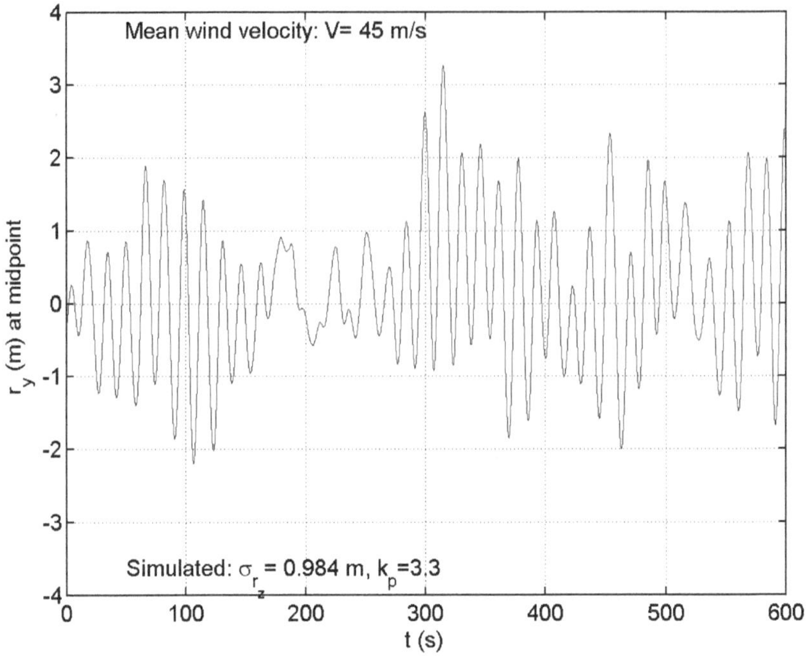

Fig. 8.13 Time domain response simulation at $x_r = L/2$ and $V = 45\ m/s$

Elaboration 8.2: Vertical System with Span-Wise Variable Properties
For a vertical tower type of system where cross sectional properties

$$B(x),\ D(x),\ \bar{C}_D(x),\ C'_D(x) \text{ and } \bar{C}_L(x)$$

as well as mean wind velocity $V(x)$ and turbulence intensities $I_u(x)$ and $I_v(x)$ are variables along the span, then a calculation of its single mode single component $\boldsymbol{\varphi}_n \approx \begin{bmatrix} \phi_y & 0 & 0 \end{bmatrix}_n^T$ response spectral density at position x_r is given by

$$S_{r_{y_n}}(\omega, x_r) = \left|\phi_{y_n}(x_r)\right|^2 \int_0^\infty S_{\eta_n}(\omega)\, d\omega$$

where

$$S_{\eta_n}(\omega) = \frac{\left|\hat{H}_n(\omega)\right|^2}{\tilde{K}_n^2} \left\{ \iint\limits_{L_{exp}} \boldsymbol{\varphi}_n^T(x_1) \mathbf{b}_q(x_1) \mathbf{S}_{vv}(x_1, x_2, \omega) \mathbf{b}_q^T(x_2) \boldsymbol{\varphi}_n(x_2) dx_1 dx_2 \right\}$$

$$= \left(\frac{\left|\hat{H}_n(\omega)\right|}{\tilde{K}_n^2} \right)^2 \iint\limits_{L_{exp}} \left(\frac{\rho V^2 B}{2} \right)_{x_1} \left(\frac{\rho V^2 B}{2} \right)_{x_2} \phi_{y_n}(x_1) \cdot \phi_{y_n}(x_2) \cdot$$

$$\left\{ \left(2\frac{D}{B}\bar{C}_D I_u \right)_{x_1} \left(2\frac{D}{B}\bar{C}_D I_u \right)_{x_2} \frac{S_{uu}(x_1, x_2, \omega)}{\sigma_u^2} + \right.$$

$$\left. \left[\left(\frac{D}{B} C_D' - \bar{C}_L \right) I_v \right]_{x_1} \left[\left(\frac{D}{B} C_D' - \bar{C}_L \right) I_v \right]_{x_2} \frac{S_{vv}(x_1, x_2, \omega)}{\sigma_v^2} \right\} dx_1 dx_2$$

and

$$\begin{bmatrix} \kappa_{ae_n} \\ \zeta_{ae_n} \end{bmatrix} = \begin{bmatrix} \dfrac{\tilde{K}_{ae_n}}{\omega_n^2 \tilde{M}_n} \\ \dfrac{\tilde{C}_{ae_n}}{2\omega_n \tilde{M}_n} \end{bmatrix} = \frac{1}{\int\limits_L \phi_{y_n}^2 dx} \cdot \begin{bmatrix} \int\limits_{L_{exp}} \dfrac{\rho B^2}{\tilde{m}_n} \cdot \dfrac{\phi_{y_n}^2}{2} P_4^* dx \\ \int\limits_{L_{exp}} \dfrac{\rho B^2}{\tilde{m}_n} \cdot \dfrac{\phi_{y_n}^2}{4} P_1^* dx \end{bmatrix}$$

General Multi-mode Response Calculations

In the final section of this chapter it is assumed that a full multi-mode approach is required (see also Chapters 5.3 and 6.5), i.e. that

$$\mathbf{r}(x,t) = \boldsymbol{\Phi}(x) \cdot \boldsymbol{\eta}(t) \tag{8.53}$$

where

$$\begin{cases} \mathbf{r}(x,t) = \begin{bmatrix} r_y & r_z & r_\theta \end{bmatrix}^T \\ \boldsymbol{\Phi}(x) = \begin{bmatrix} \boldsymbol{\varphi}_1 \ldots & \boldsymbol{\varphi}_n & \ldots \boldsymbol{\varphi}_{N_{mod}} \end{bmatrix} \\ \boldsymbol{\eta}(x) = \begin{bmatrix} \eta_1 \ldots & \eta_n & \ldots \eta_{N_{mod}} \end{bmatrix}^T \end{cases} \tag{8.54}$$

and that the modal equilibrium equation is given by (see Eq. 8.28)

$$\tilde{\mathbf{M}} \cdot \ddot{\boldsymbol{\eta}}(t) + \left(\tilde{\mathbf{C}} - \tilde{\mathbf{C}}_{ae} \right) \cdot \dot{\boldsymbol{\eta}}(t) + \left(\tilde{\mathbf{K}} - \tilde{\mathbf{K}}_{ae} \right) \cdot \boldsymbol{\eta}(t) = \tilde{\mathbf{R}}(t) \tag{8.55}$$

where $\tilde{\mathbf{M}}$, $\tilde{\mathbf{C}}$ and $\tilde{\mathbf{K}}$ are N_{mod} by N_{mod} diagonal matrices defined in Eq. 8.18, and the modal N_{mod} by one flow induced load vectors is given by

$$\tilde{\mathbf{R}}(t) = \left[\tilde{R}_1 \quad \dots \quad \tilde{R}_n \quad \dots \quad \tilde{R}_{N_{\text{mod}}} \right]^T$$

where

$$\tilde{R}_n = \int_{L_{\text{exp}}} \left(\boldsymbol{\varphi}_n^T \cdot \mathbf{q} \right) dx \tag{8.56}$$

and where $\tilde{\mathbf{C}}_{ae}$ and $\tilde{\mathbf{K}}_{ae}$ are N_{mod} by N_{mod} matrices

$$\tilde{\mathbf{C}}_{ae} = \begin{bmatrix} \ddots & & \cdot^{\cdot^{\cdot}} \\ & \tilde{C}_{ae_{nm}} & \\ \cdot^{\cdot^{\cdot}} & & \ddots \end{bmatrix} \quad \tilde{\mathbf{K}}_{ae} = \begin{bmatrix} \ddots & & \cdot^{\cdot^{\cdot}} \\ & \tilde{K}_{ae_{nm}} & \\ \cdot^{\cdot^{\cdot}} & & \ddots \end{bmatrix} \tag{8.57}$$

whose elements on row n column m are given by

$$\begin{bmatrix} \tilde{C}_{ae_{nm}} \\ \tilde{K}_{ae_{nm}} \end{bmatrix} = \int_{L_{\text{exp}}} \begin{bmatrix} \boldsymbol{\varphi}_n^T \cdot \mathbf{c}_{q_e} \cdot \boldsymbol{\varphi}_m \\ \boldsymbol{\varphi}_n^T \cdot \mathbf{k}_{q_{ae}} \cdot \boldsymbol{\varphi}_m \end{bmatrix} dx \tag{8.58}$$

where $\mathbf{c}_{q_{ae}}$ and $\mathbf{k}_{q_{ae}}$ are three by three motion dependent cross sectional load coefficient matrices, e.g. as given in Eqs. 8.12 and 8.13 or alternatively as given in Eq. 8.16, whichever is deemed most appropriate. First, the Fourier transform is taken throughout Eq. 8.55, i.e.

$$\boldsymbol{\eta}(t) = \sum_{\omega=0}^{\infty} \mathbf{a}_{\eta}(\omega) \cdot e^{i\omega t} \text{ and } \tilde{\mathbf{R}}(t) = \sum_{\omega=0}^{\infty} \mathbf{a}_{\tilde{R}}(\omega) \cdot e^{i\omega t} \tag{8.59}$$

and thus, Eq. 8.55 is satisfied at all frequencies if

$$\left[-\left(\tilde{\mathbf{M}} - \tilde{\mathbf{M}}_{ae} \right) \omega^2 + \left(\tilde{\mathbf{C}} - \tilde{\mathbf{C}}_{ae} \right) i\omega + \left(\tilde{\mathbf{K}} - \tilde{\mathbf{K}}_{ae} \right) \right] \cdot \mathbf{a}_{\eta}(\omega) = \mathbf{a}_{\tilde{R}}(\omega) \tag{8.60}$$

Pre-multiplication by $\tilde{\mathbf{K}}^{-1}$, recalling that

$$\left. \begin{aligned} \tilde{\mathbf{K}} &= diag\left[\omega_n^2 \tilde{M}_n \right] \\ \tilde{\mathbf{C}} &= diag\left[2\tilde{M}_n \omega_n \zeta_n \right] \end{aligned} \right\} \tag{8.61}$$

and introducing a reduced modal load vector

$$\mathbf{a}_{\hat{R}}(\omega) = \tilde{\mathbf{K}}^{-1} \cdot \mathbf{a}_{\hat{R}}(\omega) = \begin{bmatrix} \cdots & \dfrac{\int_{L_{\exp}} \boldsymbol{\varphi}_n^T(x) \cdot \mathbf{a}_q(x,\omega)\,dx}{\omega_n^2 \tilde{M}_n} & \cdots \end{bmatrix}^T \tag{8.62}$$

where $\mathbf{a}_q^{\cdot}(x,\omega) = \begin{bmatrix} a_{q_y} & a_{q_z} & a_{q_\theta} \end{bmatrix}^T$, then the following is obtained

$$\mathbf{a}_\eta(\omega) = \hat{\mathbf{H}}_\eta(\omega) \cdot \mathbf{a}_{\hat{R}}(\omega) \tag{8.63}$$

where

$$\hat{\mathbf{H}}_\eta(\omega) = \left\{ \mathbf{I} - \tilde{\mathbf{K}}^{-1}\tilde{\mathbf{K}}_{ae} - diag\left[\frac{1}{\omega_n^2}\right]\omega^2 + \left(diag\left[\frac{2\zeta_n}{\omega_n}\right] - \tilde{\mathbf{K}}^{-1}\tilde{\mathbf{C}}_{ae} \right) i\omega \right\}^{-1} \tag{8.64}$$

is the non-dimensional frequency-response-matrix, and $\mathbf{I}$ is the N_{mod} by N_{mod} identity matrix. It is convenient to define the following N_{mod} by N_{mod} matrices

$$\left. \begin{aligned} \boldsymbol{\kappa}_{ae} &= \tilde{\mathbf{K}}^{-1}\tilde{\mathbf{K}}_{ae} \\ \boldsymbol{\zeta}_{ae} &= \frac{1}{2} diag[\omega_n] \cdot \left(\tilde{\mathbf{K}}^{-1} \cdot \tilde{\mathbf{C}}_{ae} \right) \end{aligned} \right\} \tag{8.65}$$

as well as introducing $\boldsymbol{\zeta} = diag[\zeta_n]$. The non-dimensional frequency-response-matrix is then given by

$$\hat{\mathbf{H}}_\eta(\omega) = \left\{ \mathbf{I} - \boldsymbol{\kappa}_{ae} - \left(\omega \cdot diag\left[\frac{1}{\omega_n}\right] \right)^2 + 2i\omega \cdot diag\left[\frac{1}{\omega_n}\right] \cdot (\boldsymbol{\zeta} - \boldsymbol{\zeta}_{ae}) \right\}^{-1} \tag{8.66}$$

Recalling that $\tilde{\mathbf{K}}^{-1} = diag\left[1/\left(\omega_n^2 \tilde{M}_n\right) \right]$, then the content of

$$\boldsymbol{\kappa}_{ae} = \begin{bmatrix} \ddots & & \cdot^{\cdot^{\cdot}} \\ & \kappa_{ae_{nm}} & \\ \cdot^{\cdot^{\cdot}} & & \ddots \end{bmatrix} \text{ and } \boldsymbol{\zeta}_{ae} = \begin{bmatrix} \ddots & & \cdot^{\cdot^{\cdot}} \\ & \zeta_{ae_{nm}} & \\ \cdot^{\cdot^{\cdot}} & & \ddots \end{bmatrix} \tag{8.67}$$

are given by (see Eq. 8.16)

$$\kappa_{ae_{nm}} = \frac{\tilde{K}_{ae_{nm}}}{\omega_n^2 \tilde{M}_n} = \frac{1}{\omega_n^2 \tilde{m}_n} \cdot \frac{\int\limits_{L_{\exp}} \left(\boldsymbol{\varphi}_n^T \mathbf{k}_{q_{ae}} \boldsymbol{\varphi}_m \right) dx}{\int\limits_{L} \left(\boldsymbol{\varphi}_n^T \boldsymbol{\varphi}_n \right) dx} \tag{8.68}$$

$$\zeta_{ae_{nm}} = \frac{\omega_n}{2} \frac{\tilde{C}_{ae_{nm}}}{\omega_n^2 \tilde{M}_n} = \frac{1}{2\omega_n \tilde{m}_n} \cdot \frac{\int\limits_{L_{\exp}} \left(\boldsymbol{\varphi}_n^T \hat{\mathbf{c}}_{q_{ae}} \boldsymbol{\varphi}_m \right) dx}{\int\limits_{L} \left(\boldsymbol{\varphi}_n^T \boldsymbol{\varphi}_n \right) dx} \tag{8.69}$$

Elaboration 8.3: Aerodynamic Stiffness and Damping

Fully expanded versions of Eqs. 8.68 and 8.69 are given by

$$\kappa_{ae_{nm}} = \frac{1}{\omega_n^2 \tilde{m}_n} \int\limits_{L_{\exp}} \left(\phi_{y_n} \phi_{y_m} P_4^* + \phi_{z_n} \phi_{y_m} H_6^* + \phi_{\theta_n} \phi_{y_m} B A_6^* + \phi_{y_n} \phi_{z_m} P_6^* + \phi_{z_n} \phi_{z_m} H_4^* + \right.$$

$$\left. \phi_{\theta_n} \phi_{z_m} B A_4^* + \phi_{y_n} \phi_{\theta_m} B P_3^* + \phi_{z_n} \phi_{\theta_m} B H_3^* + \phi_{\theta_n} \phi_{\theta_m} B^2 A_3^* \right) dx \Big/ \int\limits_{L} \left(\phi_{y_n}^2 + \phi_{z_n}^2 + \phi_{\theta_n}^2 \right) dx$$

$$\zeta_{ae_{nm}} = \frac{1}{2\omega_n \tilde{m}_n} \int\limits_{L_{\exp}} \left(\phi_{y_n} \phi_{y_m} P_1^* + \phi_{z_n} \phi_{y_m} H_5^* + \phi_{\theta_n} \phi_{y_m} B A_5^* + \phi_{y_n} \phi_{z_m} P_5^* + \phi_{z_n} \phi_{z_m} H_1^* \right.$$

$$\left. + \phi_{\theta_n} \phi_{z_m} B A_1^* + \phi_{y_n} \phi_{\theta_m} B P_2^* + \phi_{z_n} \phi_{\theta_m} B H_2^* + \phi_{\theta_n} \phi_{\theta_m} B^2 A_2^* \right) dx \Big/ \int\limits_{L} \left(\phi_{y_n}^2 + \phi_{z_n}^2 + \phi_{\theta_n}^2 \right) dx$$

Returning to Eq. 8.63, the response spectral density matrix (N_{mod} by N_{mod} and containing single-sided spectra) is obtained from the basic definition of spectra as expressed from the Fourier amplitudes, and thus, the following development applies (see general development in Chapter 6.5):

$$\begin{aligned} \mathbf{S}_\eta(\omega) &= \lim_{T \to \infty} \frac{1}{\pi T} \left(\mathbf{a}_\eta^* \cdot \mathbf{a}_\eta^T \right) = \lim_{T \to \infty} \frac{1}{\pi T} \left[\left(\hat{\mathbf{H}}_\eta \mathbf{a}_{\hat{R}} \right)^* \cdot \left(\hat{\mathbf{H}}_\eta \mathbf{a}_{\hat{R}} \right)^T \right] \\ &= \hat{\mathbf{H}}_\eta^* \cdot \left[\lim_{T \to \infty} \frac{1}{\pi T} \left(\mathbf{a}_{\hat{R}}^* \cdot \mathbf{a}_{\hat{R}}^T \right) \right] \cdot \hat{\mathbf{H}}_\eta^T = \hat{\mathbf{H}}_\eta^* \cdot \mathbf{S}_{\hat{R}} \cdot \hat{\mathbf{H}}_\eta^T \end{aligned} \tag{8.70}$$

where $\mathbf{S}_{\hat{R}}$ is an N_{mod} by N_{mod} normalised modal load matrix

$$\mathbf{S}_{\hat{R}}(\omega) = \lim_{T\to\infty}\frac{1}{\pi T}\left(\mathbf{a}_{\hat{R}}^{*}\mathbf{a}_{\hat{R}}^{T}\right) =$$

$$\lim_{T\to\infty}\frac{1}{\pi T}\left(\begin{bmatrix} a_{\hat{R}_1}^{*} \\ \vdots \\ a_{\hat{R}_n}^{*} \\ \vdots \\ a_{\hat{R}_{N_{\text{mod}}}}^{*}\end{bmatrix}\cdot\begin{bmatrix} a_{\hat{R}_1} & \cdots & a_{\hat{R}_m} & \cdots & a_{\hat{R}_{N_{\text{mod}}}}\end{bmatrix}\right) = \begin{bmatrix} \ddots & & \cdot^{\cdot^{\cdot}} \\ & S_{\hat{R}_n\hat{R}_m}(\omega) & \\ \cdot^{\cdot^{\cdot}} & & \ddots \end{bmatrix} \tag{8.71}$$

whose elements on row n column m are given by

$$\begin{aligned} S_{\hat{R}_n\hat{R}_m}(\omega) &= \lim_{T\to\infty}\frac{1}{\pi T}\left[a_{\hat{R}_n}^{*}\,a_{\hat{R}_m}\right] \\ &= \lim_{T\to\infty}\frac{1}{\pi T}\left(\frac{\int_{L_{\text{exp}}}\boldsymbol{\varphi}_n^T(x)\mathbf{a}_q^{*}(x,\omega)\,dx}{\omega_n^2\tilde{M}_n}\cdot\frac{\int_{L_{\text{exp}}}\left[\boldsymbol{\varphi}_m^T(x)\mathbf{a}_q(x,\omega)\right]^T dx}{\omega_m^2\tilde{M}_m}\right) \\ &= \frac{\iint_{L_{\text{exp}}}\boldsymbol{\varphi}_n^T(x_1)\cdot\lim_{T\to\infty}\frac{1}{\pi T}\left[\mathbf{a}_q^{*}(x_1,\omega)\cdot\mathbf{a}_q^T(x_2,\omega)\right]\cdot\boldsymbol{\varphi}_m(x_2)\,dx_1dx_2}{\left(\omega_n^2\tilde{M}_n\right)\cdot\left(\omega_m^2\tilde{M}_m\right)} \end{aligned} \tag{8.72}$$

Thus, the elements of $\mathbf{S}_{\hat{R}}(\omega)$ are given by

$$S_{\hat{R}_n\hat{R}_m}(\omega) = \frac{\iint_{L_{\text{exp}}}\boldsymbol{\varphi}_n^T(x_1)\cdot\mathbf{S}_{qq}(x_1,x_2,\omega)\cdot\boldsymbol{\varphi}_m(x_2)\,dx_1dx_2}{\left(\omega_n^2\tilde{M}_n\right)\cdot\left(\omega_m^2\tilde{M}_m\right)} \tag{8.73}$$

where $\mathbf{S}_{qq}(x_1,x_2,\omega)$ is the spectral density matrix of cross sectional loads. Since $\begin{bmatrix} q_y & q_z & q_\theta\end{bmatrix}^T = \mathbf{b}_q(x)\mathbf{v}(x,t) = (\rho VB/2)\hat{\mathbf{b}}_q\mathbf{v}$, then its Fourier transform is

$$\mathbf{a}_q(x,\omega) = \begin{bmatrix} a_{q_y} \\ a_{q_z} \\ a_{q_\theta}\end{bmatrix} = (\rho VB/2)\cdot\hat{\mathbf{b}}_q(x)\cdot\mathbf{a}_v(x,\omega) \tag{8.74}$$

where $\mathbf{a}_v(x,\omega)=[a_u \quad a_w]^T$ or $\mathbf{a}_v(x,\omega)=[a_u \quad a_v]^T$ depending on whether the system is horizontal or vertical (see Eqs. 8.23). Thus

$$\begin{aligned}\mathbf{S}_{qq}(x_1,x_2,\omega) &= \lim_{T\to\infty}\frac{1}{\pi T}\left[\mathbf{a}_q^*(x_1,\omega)\cdot\mathbf{a}_q^T(x_2,\omega)\right]\\ &=\left(\frac{\rho VB}{2}\right)^2\cdot\hat{\mathbf{b}}_q(x_1)\cdot\lim_{T\to\infty}\frac{1}{\pi T}\left[\mathbf{a}_v^*(x_1,\omega)\cdot\mathbf{a}_v^T(x_2,\omega)\right]\cdot\hat{\mathbf{b}}_q^T(x_2)\\ &=\left(\frac{\rho VB}{2}\right)^2\cdot\hat{\mathbf{b}}_q(x_1)\cdot\mathbf{S}_{vv}(x_1,x_2,\omega)\cdot\hat{\mathbf{b}}_q^T(x_2)\end{aligned} \tag{8.75}$$

where $\mathbf{S}_{vv}(x_1,x_2,\omega)$ is defined in Eqs. 8.44 or 8.45 depending on whether the system is horizontal or vertical. Then the content of the N_{mod} by N_{mod} normalised modal load matrix

$$\mathbf{S}_{\hat{R}}(\omega)=\begin{bmatrix}\ddots & & \cdot^{\cdot^{\cdot}}\\ & S_{\hat{R}_n\hat{R}_m}(\omega) & \\ \cdot^{\cdot^{\cdot}} & & \ddots\end{bmatrix} \tag{8.76}$$

is given by

$$\begin{aligned}S_{\hat{R}_n\hat{R}_m} &= \left(\frac{\rho V^2 B}{2}\right)^2\frac{\iint\limits_{L_{\exp}}\boldsymbol{\varphi}_n^T(x_1)\left\{\hat{\mathbf{b}}_q(x_1)\left[\mathbf{I}_v^2\hat{\mathbf{S}}_v(x_1,x_2,\omega)\right]\hat{\mathbf{b}}_q^T(x_2)\right\}\boldsymbol{\varphi}_m(x_2)dx_1dx_2}{\left(\omega_n^2\tilde{M}_n\right)\left(\omega_m^2\tilde{M}_m\right)}\\ &=\frac{\rho B^3}{2\tilde{m}_n}\cdot\frac{\rho B^3}{2\tilde{m}_m}\cdot\left(\frac{V}{B\omega_n}\right)^2\cdot\left(\frac{V}{B\omega_m}\right)^2\cdot\hat{J}_{nm}^2\end{aligned} \tag{8.77}$$

where

$\mathbf{I}_v=diag[I_u \quad I_w]$ or $\mathbf{I}_v=diag[I_u \quad I_v]$ and $\mathbf{S}_v=diag\left[S_{uu}/\sigma_u^2 \quad S_{ww}/\sigma_w^2\right]$ or $\mathbf{S}_v=diag\left[S_{uu}/\sigma_u^2 \quad S_{vv}/\sigma_v^2\right]$ depending on whether the system is horizontal or vertical, and where the $N_{\text{mod}}\cdot N_{\text{mod}}$ joint acceptance functions $\hat{J}_{nm}^2$, are defined by

$$\hat{J}_{nm}^2=\frac{\iint\limits_{L_{\exp}}\boldsymbol{\varphi}_n^T(x_1)\left\{\hat{\mathbf{b}}_q(x_1)\left[\mathbf{I}_v^2\cdot\hat{\mathbf{S}}_v(x_1,x_2,\omega)\right]\hat{\mathbf{b}}_q^T(x_2)\right\}\boldsymbol{\varphi}_m(x_2)dx_1dx_2}{\left(\int_L\boldsymbol{\varphi}_n^T\boldsymbol{\varphi}_n dx\right)\left(\int_L\boldsymbol{\varphi}_m^T\boldsymbol{\varphi}_m dx\right)} \tag{8.78}$$

Elaboration 8.4: Joint Acceptance Function for Horizontal System
A fully expanded version of the numerator in Eq. 8.78

$$J_{nm}^2 = \iint_{L_{\exp}} \boldsymbol{\varphi}_n^T(x_1)\left\{\hat{\mathbf{b}}_q(x_1)\left[\mathbf{I}_v^2 \cdot \hat{\mathbf{S}}_v(x_1,x_2,\omega)\right]\hat{\mathbf{b}}_q^T(x_2)\right\}\boldsymbol{\varphi}_m(x_2)\,dx_1 dx_2$$

for a horizontal system whose cross sectional properties are constant along its span is given by

$$\begin{aligned} J_{nm}^2 = \iint_{L_{\exp}} \Bigg\{ & \phi_{y_n}(x_1)\phi_{y_m}(x_2)\left[\left(2\frac{D}{B}\bar{C}_D\right)^2 I_u^2\hat{S}_{uu} + \left(\frac{D}{B}C_D' - \bar{C}_L\right)^2 I_w^2\hat{S}_{ww}\right] \\ & +\phi_{z_n}(x_1)\phi_{z_m}(x_2)\left[\left(2\bar{C}_L\right)^2 I_u^2\hat{S}_{uu} + \left(C_L' + \frac{D}{B}\bar{C}_D\right)^2 I_w^2\hat{S}_{ww}\right] \\ & +\phi_{\theta_n}(x_1)\phi_{\theta_m}(x_2)\left[\left(2B\bar{C}_M\right)^2 I_u^2\hat{S}_{uu} + \left(BC_M'\right)^2 I_w^2\hat{S}_{ww}\right] \\ & +\left[\phi_{y_n}(x_1)\phi_{z_m}(x_2) + \phi_{z_n}(x_1)\phi_{y_m}(x_2)\right]\cdot \\ & \left[4\frac{D}{B}\bar{C}_D\bar{C}_L I_u^2\hat{S}_{uu} + \left(\frac{D}{B}C_D' - \bar{C}_L\right)\left(C_L' + \frac{D}{B}\bar{C}_D\right) I_w^2\hat{S}_{ww}\right] \\ & +\left[\phi_{y_n}(x_1)\phi_{\theta_m}(x_2) + \phi_{\theta_n}(x_1)\phi_{y_m}(x_2)\right]\cdot \\ & \left[4\frac{D}{B}\bar{C}_D B\bar{C}_M I_u^2\hat{S}_{uu} + \left(\frac{D}{B}C_D' - \bar{C}_L\right) BC_M' I_w^2\hat{S}_{ww}\right] \\ & +\left[\phi_{z_n}(x_1)\phi_{\theta_m}(x_2) + \phi_{\theta_n}(x_1)\phi_{z_m}(x_2)\right]\cdot \\ & \left[4\bar{C}_L B\bar{C}_M I_u^2\hat{S}_{uu} + \left(C_L' + \frac{D}{B}\bar{C}_D\right) BC_M' I_w^2\hat{S}_{ww}\right]\Bigg\} dx_1 dx_2 \end{aligned}$$

and the corresponding reduced version is given by

$$\hat{J}_{nm}^2 = \frac{J_{nm}^2}{\left(\int_L \left(\phi_{y_n}^2 + \phi_{z_n}^2 + \phi_{\theta n}^2\right)dx\right)\cdot\left(\int_L \left(\phi_{y_m}^2 + \phi_{z_m}^2 + \phi_{\theta_m}^2\right)dx\right)}$$

The reduced cross spectra $\hat{S}_{uu}$ and $\hat{S}_{ww}$ are defined by

$$\begin{cases} \hat{S}_{uu} = S_{uu}(x_1,x_2,\omega)/\sigma_u^2 \\ \hat{S}_{ww} = S_{ww}(x_1,x_2,\omega)/\sigma_w^2 \end{cases}$$

Since spatial averaging will eliminate any complex parts of the cross spectra, then

$$\hat{S}_{uu} = \text{Re}\left[S_{uu}(x_1,x_2,\omega)\right]/\sigma_u^2 = \frac{S_u(x_1,\omega)}{\sigma_u^2}\cdot \hat{C}o_{uu}(\Delta x,\omega)$$

$$\hat{S}_{ww} = \text{Re}\left[S_{ww}(x_1,x_2,\omega)\right]/\sigma_w^2 = \frac{S_w(x_1,\omega)}{\sigma_w^2}\cdot \hat{C}o_{ww}(\Delta x,\omega)$$

where $\Delta x = |x_1 - x_2|$, $\hat{C}o_{uu}$ and $\hat{C}o_{ww}$ are the reduced u- and w- component co-spectra (see Eq. A.89).

Extracting from the mode shape matrix $\mathbf{\Phi}=\left[\boldsymbol{\varphi}_1 \ \dots \ \boldsymbol{\varphi}_i \ \dots \ \boldsymbol{\varphi}_{N_{mod}}\right]$ a three by N_{mod} matrix associated with a chosen span-wise position x_r

$$\mathbf{\Phi}_r(x_r) = \left[\boldsymbol{\varphi}_1(x_r) \ \dots \ \boldsymbol{\varphi}_n(x_r) \ \dots \ \boldsymbol{\varphi}_{N_{\text{mod}}}(x_r)\right]$$
$$= \left[\begin{bmatrix}\phi_y(x_r)\\ \phi_z(x_r)\\ \phi_\theta(x_r)\end{bmatrix}_1 \ \dots \ \begin{bmatrix}\phi_y(x_r)\\ \phi_z(x_r)\\ \phi_\theta(x_r)\end{bmatrix}_n \ \dots \ \begin{bmatrix}\phi_y(x_r)\\ \phi_z(x_r)\\ \phi_\theta(x_r)\end{bmatrix}_{N_{\text{mod}}}\right] \tag{8.79}$$

then the three by three cross spectral density matrix of the unknown displacements r_y, r_z and r_θ at $x = x_r$

$$\mathbf{S}_{rr}(x_r,\omega) = \begin{bmatrix} S_{r_y r_y} & S_{r_y r_z} & S_{r_y r_\theta} \\ S_{r_z r_y} & S_{r_z r_z} & S_{r_z r_\theta} \\ S_{r_\theta r_y} & S_{r_\theta r_z} & S_{r_\theta r_\theta}\end{bmatrix} \tag{8.80}$$

is given by

$$\mathbf{S}_{rr}(x_r,\omega) = \lim_{T\to\infty}\frac{1}{\pi T}\left\{\left[\mathbf{\Phi}_r(x_r)\mathbf{a}_\eta(\omega)\right]^*\left[\mathbf{\Phi}_r(x_r)\mathbf{a}_\eta(\omega)\right]^T\right\}$$
$$= \mathbf{\Phi}_r(x_r)\lim_{T\to\infty}\frac{1}{\pi T}\left\{\mathbf{a}_\eta^*(\omega)\mathbf{a}_\eta^T(\omega)\right\}\mathbf{\Phi}_r^T(x_r) = \mathbf{\Phi}_r(x_r)\cdot\mathbf{S}_\eta(\omega)\cdot\mathbf{\Phi}_r^T(x_r) \tag{8.81}$$

where $\mathbf{S}_\eta(\omega)$ is given in Eq. 8.70, i.e.:

$$\mathbf{S}_{rr}(x_r,\omega) = \mathbf{\Phi}_r(x_r)\cdot\left[\hat{\mathbf{H}}_\eta^*(\omega)\cdot\mathbf{S}_{\hat{R}}(\omega)\cdot\hat{\mathbf{H}}_\eta^T(\omega)\right]\cdot\mathbf{\Phi}_r^T(x_r) \tag{8.82}$$

where $\mathbf{S}_{\tilde{R}}(\omega)$ is defined in Eqs. 8.76 – 8.78. The corresponding covariance matrix is simply obtained by frequency domain integration, i.e.

$$\mathbf{Cov}_{rr}(x_r) = \begin{bmatrix} \sigma_{r_y}^2 & Cov_{r_y r_z} & Cov_{r_y r_\theta} \\ Cov_{r_z r_y} & \sigma_{r_z}^2 & S_{r_z r_\theta} \\ Cov_{r_\theta r_y} & Cov_{r_\theta r_z} & \sigma_{r_\theta}^2 \end{bmatrix} = \int_0^\infty \mathbf{S}_{rr}(x_r,\omega)\, d\omega \qquad (8.83)$$

Example 8.2

Let us again (similar to example 8.1) consider a suspension bridge with a single span of $L = 1200m$ *that is elevated at a position* $z_f = 50m$*, but now we set out to calculate the dynamic response at* $x_r = L/2$ *associated with two mode shapes*

$$\boldsymbol{\varphi}_1 = \begin{bmatrix} 0 & \phi_{z_1} & 0 \end{bmatrix}^T \qquad \textit{and} \qquad \boldsymbol{\varphi}_2 = \begin{bmatrix} 0 & 0 & \phi_{\theta_2} \end{bmatrix}^T$$

with corresponding eigenfrequencies $\omega_1 = 0.7$ *and* $\omega_2 = 2.75\ rad/s$*. As can be seen,* $\boldsymbol{\varphi}_1$ *contains only the displacement component in the across wind vertical direction while* $\boldsymbol{\varphi}_2$ *only contains torsion. Thus*

$$\boldsymbol{\Phi} = \begin{bmatrix} \boldsymbol{\varphi}_1 & \boldsymbol{\varphi}_2 \end{bmatrix} = \begin{bmatrix} 0 & 0 \\ \phi_{z_1} & 0 \\ 0 & \phi_{\theta_2} \end{bmatrix} \quad \textit{which may be simplified into } \boldsymbol{\Phi} = \begin{bmatrix} \phi_{z_1} & 0 \\ 0 & \phi_{\theta_2} \end{bmatrix}$$

Let us assume that

$$\phi_{z_1} = \frac{1}{2}\left[\sin(\pi x/L) - \sin(3\pi x/L)\right] \qquad \textit{and} \qquad \phi_{\theta_2} = \sin(\pi x/L)$$

Thus, the aim of this example is to calculate the corresponding dynamic response quantities $\sigma_{r_z r_z}$ *and* $\sigma_{r_\theta r_\theta}$ *at* $x_r = L/2$ *and the covariance* $Cov_{r_\theta r_z}$ *between them. It is taken for granted that the chosen mean wind velocity will also be set in the vicinity of the instability limit of the system, such that in-wind changes to resonance frequencies may not be ignored. Again, it is assumed that the cross section has the following static load coefficient properties:*

$$\bar{C}_L = 0 \qquad C_L' = 2.5 \qquad \bar{C}_M = 0 \qquad C_M' = 0.8 \quad \textit{and}$$

$$\frac{D}{B}\bar{C}_D \ll C_L'$$

(Quantifying the drag coefficient is obsolete since y direction response is not excited.)

It is taken for granted that $L_{\exp} = L$. *Let us adopt the following wind field properties:*

1) *the turbulence intensity* $I_w = \sigma_w / V = 0.08$
2) *the integral length scales:*

$${}^{xf}L_u = 100 \cdot \left(z_f/10\right)^{0.3} = 162m,$$

$${}^{xf}L_w = {}^{xf}L_u / 12$$

3) *the normalised auto spectral density:*

$$\hat{S}_w(\omega) = \frac{S_w(\omega)}{\sigma_w^2} = \frac{1.5\,{}^{xf}L_w/V}{\left(1 + 2.25\,\omega\,{}^{xf}L_w/V\right)^{5/3}}$$

4) *the normalised co-spectrum:*

$$\begin{cases} \hat{C}o_{ww}(\omega, \Delta x) = \exp\left(-C_{wx}\,\omega\Delta x/V\right) \\ C_{wx} = C_{wy_f} = 6.5/(2\pi) \approx 1.0 \end{cases}$$

(where $\Delta x = |x_1 - x_2|$*) Let us allot the following values to the remaining constants that are necessary for a numerical calculation of the relevant dynamic response quantities at* $x_r = L/2$:

ρ (kg/m^3)	B (m)	D (m)	m_1 (kg/m)	m_2 (kgm^2/m)	ω_1 (rad/s)	ω_2 (rad/s)	ζ_1	ζ_2
1.25	20	4	12500	425000	0.7	2.75	0.005	0.01

Since m_1 *and* m_2 *are constant along the span, then the modally equivalent and evenly distributed masses* $\tilde{m}_1 = m_1$ *and* $\tilde{m}_2 = m_2$. *It should be noted that*

$$\int_0^L \phi_{z1}^2\,dx = \frac{L}{4} \qquad \int_0^L \phi_{\theta_2}^2\,dx = \frac{L}{2} \qquad \int_0^L \phi_{z1}\phi_{\theta_2}\,dx = \frac{L}{4}$$

Finally, let us for simplicity adopt quasi-static values to the aerodynamic derivatives, except for A_2^* *which is responsible for aerodynamic damping in torsion. Adopting* $A_2^* = -\beta_M C_M' \left(V/B\omega_n\right)^2$ *and* $\beta_M = 0.2$ *provides a good approximation to the flat plate properties. Thus, the aerodynamic derivatives associated with motion in the across wind vertical direction and torsion are given by:*

$$\begin{bmatrix} H_1^* \\ H_2^* \\ H_3^* \end{bmatrix} = -C_L' \cdot \begin{bmatrix} \hat{V} \\ 0 \\ \hat{V}^2 \end{bmatrix} \quad and \quad \begin{bmatrix} A_1^* \\ A_2^* \\ A_3^* \end{bmatrix} = C_M' \cdot \begin{bmatrix} -\hat{V} \\ -\beta_M \cdot \hat{V}^2 \\ \hat{V}^2 \end{bmatrix} \quad and$$

$$\begin{bmatrix} H_4^* & A_4^* \\ H_5^* & A_5^* \\ H_6^* & A_6^* \end{bmatrix} = \mathbf{0}$$

where: $\hat{V} = V / \left[B\omega_n(V) \right]$ *and where* $\omega_n(V)$, $n = 1 \text{ or } 2$, *is the in-wind resonance frequency. The aerodynamic coefficients associated with changes in stiffness and damping are then given by:*

$$\kappa_{ae_{nm}} = \frac{\rho B^2}{2\tilde{m}_n} \cdot \int_{L_{\exp}} \left(\phi_{z_n} \phi_{\theta_m} B H_3^* + \phi_{\theta_n} \phi_{\theta_m} B^2 A_3^* \right) dx \bigg/ \int_L \left(\phi_{z_n}^2 + \phi_{\theta_n}^2 \right) dx$$

$$\zeta_{ae_{nm}} = \frac{\rho B^2}{4\tilde{m}_n} \cdot \int_{L_{\exp}} \left(\phi_{z_n} \phi_{z_m} H_1^* + \phi_{\theta_n} \phi_{z_m} B A_1^* + \phi_{\theta_n} \phi_{\theta_m} B^2 A_2^* \right) dx \bigg/ \int_L \left(\phi_{z_n}^2 + \phi_{\theta_n}^2 \right) dx$$

where n *and* m *are equal to 1 or 2. Introducing the choice of aerodynamic derivatives given above, then:*

$$\kappa_{ae_{11}} = 0,$$

$$\kappa_{ae_{12}} = \frac{\rho B^2}{2\tilde{m}_1} \cdot \frac{\int_{L_{\exp}} \phi_{z_1} \phi_{\theta_2} B H_3^* dx}{\int_L \phi_{z_1}^2 dx} = \frac{\rho B^3}{2\tilde{m}_1} \cdot H_3^* = \frac{\rho B^3}{2\tilde{m}_1} \cdot C_L' \cdot \left(\frac{V}{B\omega_1} \right)^2$$

$$\kappa_{ae_{21}} = 0,$$

$$\kappa_{ae_{22}} = \frac{\rho B^2}{2\tilde{m}_2} \cdot \frac{\int_{L_{\exp}} \phi_{\theta_2}^2 B^2 A_3^* dx}{\int_L \phi_{\theta_2}^2 dx} = \frac{\rho B^4}{2\tilde{m}_2} \cdot A_3^* = \frac{\rho B^4}{2\tilde{m}_2} \cdot C_M' \cdot \left(\frac{V}{B\omega_2} \right)^2$$

$$\zeta_{ae_{11}} = \frac{\rho B^2}{4\tilde{m}_1} \cdot \frac{\int_{L_{\exp}} \phi_{z_1}^2 H_1^* dx}{\int_L \phi_{z_1}^2 dx} = \frac{\rho B^2}{4\tilde{m}_1} \cdot H_1^* = -\frac{\rho B^2}{4\tilde{m}_1} \cdot C_L' \cdot \frac{V}{B\omega_1},$$

$$\zeta_{ae_{12}} = 0$$

$$\zeta_{ae21} = \frac{\rho B^2}{4\tilde{m}_2} \cdot \frac{\int_{L_{\exp}} \phi_{\theta_2} \phi_{z_1} B A_1^* dx}{\int_L \phi_{\theta_2}^2 dx} = \frac{\rho B^3}{8\tilde{m}_2} \cdot A_1^* = -\frac{\rho B^3}{8\tilde{m}_2} \cdot C_M' \cdot \frac{V}{B\omega_2}$$

$$\zeta_{ae22} = \frac{\rho B^2}{4\tilde{m}_2} \cdot \frac{\int_{L_{\exp}} \phi_{\theta_2}^2 B^2 A_2^* dx}{\int_L \phi_{\theta_2}^2 dx} = \frac{\rho B^4}{4\tilde{m}_2} \cdot A_2^* = -\frac{\rho B^4}{4\tilde{m}_2} \cdot \beta_M C_M' \cdot \left(\frac{V}{B\omega_2}\right)^2$$

The non-dimensional frequency response function is then given by

$$\hat{\mathbf{H}}_\eta(\omega) = \left\{ \mathbf{I} - \boldsymbol{\kappa}_{ae} - \left(diag\left[\frac{\omega}{\omega_n}\right]\right)^2 + 2i \cdot diag\left[\frac{\omega}{\omega_n}\right] \cdot (\boldsymbol{\zeta} - \boldsymbol{\zeta}_{ae}) \right\}^{-1}$$

$$= \left\{ \begin{bmatrix} 1 & 0 \\ 0 & 1 \end{bmatrix} - \begin{bmatrix} 0 & \kappa_{ae12} \\ 0 & \kappa_{ae22} \end{bmatrix} - \omega^2 \begin{bmatrix} \omega_1^{-2} & 0 \\ 0 & \omega_2^{-2} \end{bmatrix} + \right.$$

$$\left. 2i\omega \begin{bmatrix} \omega_1^{-1} & 0 \\ 0 & \omega_2^{-1} \end{bmatrix} \left(\begin{bmatrix} \zeta_1 & 0 \\ 0 & \zeta_2 \end{bmatrix} - \begin{bmatrix} \zeta_{ae11} & 0 \\ \zeta_{ae21} & \zeta_{ae22} \end{bmatrix} \right) \right\}^{-1}$$

It is worth noting that since $\kappa_{ae11} = 0$ then ω_1 will remain unchanged with increasing mean wind velocities. The absolute value of the determinant of the non–dimensional frequency response function (at $V = 0$) is shown in Fig. 8.14. The content of the normalised modal load matrix

$$\mathbf{S}_{\hat{R}}(\omega) = \begin{bmatrix} S_{\hat{R}_1\hat{R}_1} & S_{\hat{R}_1\hat{R}_2} \\ S_{\hat{R}_2\hat{R}_1} & S_{\hat{R}_2\hat{R}_2} \end{bmatrix}$$

is given by: $S_{\hat{R}_n\hat{R}_m}(\omega) = \frac{\rho B^3}{2\tilde{m}_n} \cdot \frac{\rho B^3}{2\tilde{m}_m} \cdot \left(\frac{V}{B\omega_n}\right)^2 \cdot \left(\frac{V}{B\omega_m}\right)^2 \cdot \hat{J}_{nm}^2$

where the reduced joint acceptance function $\hat{J}_{nm}$ is given in Eq. 8.78.

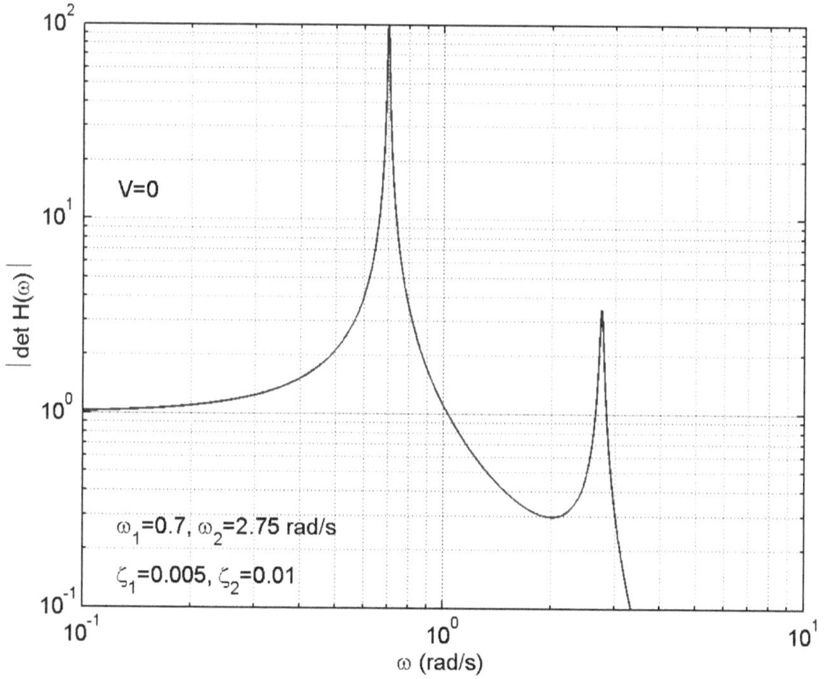

Fig. 8.14 The absolute value of the determinant of the frequency response matrix at $V = 0$

An expanded version of the joint acceptance function itself is given in Elaboration 8.4 above. Under the present circumstances it simplifies into

$$\hat{J}_{11}^2 = \iint\limits_{L_{\exp}} \phi_{z_1}(x_1) \cdot \phi_{z_1}(x_2) \cdot \left(C_L' I_w\right)^2 \cdot \frac{S_{ww}(\omega, \Delta x)}{\sigma_w^2} dx_1 dx_2 \Bigg/ \left(\int_0^L \phi_{z_1}^2 dx \right)^2$$

$$\hat{J}_{12}^2 = \iint\limits_{L_{\exp}} \phi_{z_1}(x_1) \phi_{\theta_2}(x_2) C_L' B C_M' I_w^2 \frac{S_{ww}(\omega, \Delta x)}{\sigma_w^2} dx_1 dx_2 \Bigg/ \left(\int_0^L \phi_{z_1}^2 dx \int_0^L \phi_{\theta_2}^2 dx \right)$$

$$\hat{J}_{21}^2 = \hat{J}_{12}^2$$

$$J_{22}^2 = \iint\limits_{L_{\exp}} \phi_{\theta_2}(x_1) \cdot \phi_{\theta_2}(x_2) \cdot \left(B C_M' I_w\right)^2 \cdot \frac{S_{ww}(\omega, \Delta x)}{\sigma_w^2} dx_1 dx_2 \Bigg/ \left(\int_0^L \phi_{\theta_2}^2 dx \right)^2$$

Introducing $S_{ww}(\omega, \Delta x) = S_w(\omega) \cdot \hat{C}o_{ww}(\omega, \Delta x)$ *and* $I_w = \sigma_w / V$ *, then the content of the normalised modal load matrix is given by*

$$S_{\hat{R}_1\hat{R}_1}(\omega) = \left(\frac{\rho VBC_L'}{2\omega_1^2\tilde{m}_1}\right)^2 \left(\frac{\psi_{11}(\omega)}{\int_0^L \phi_{z_1}^2 dx}\right)^2 S_w(\omega)$$

$$S_{\hat{R}_1\hat{R}_2}(\omega) = \left(\frac{\rho VB\sqrt{C_L' BC_M'}}{2\omega_1\omega_2\sqrt{\tilde{m}_1\tilde{m}_2}}\right)^2 \frac{\psi_{12}^2(\omega)}{\int_0^L \phi_{z_1}^2 dx \int_0^L \phi_{\theta_2}^2 dx} S_w(\omega)$$

$$S_{\hat{R}_2\hat{R}_1}(\omega) = S_{\hat{R}_1\hat{R}_2}(\omega)$$

and $S_{\hat{R}_2\hat{R}_2}(\omega) = \left(\frac{\rho VB^2 C_M'}{2\omega_2^2\tilde{m}_2}\right)^2 \left(\frac{\psi_{22}(\omega)}{\int_0^L \phi_{\theta_2}^2 dx}\right)^2 S_w(\omega)$

where:

$$\psi_{11} = \left[\int_0^L \int_0^L \phi_{z_1}(x_1)\cdot\phi_{z_1}(x_2)\cdot\hat{C}o_{ww}(\omega,\Delta x)\,dx_1 dx_2\right]^{1/2}$$

$$\psi_{12} = \left[\int_0^L \int_0^L \phi_{z_1}(x_1)\cdot\phi_{\theta_2}(x_2)\cdot\hat{C}o_{ww}(\omega,\Delta x)\,dx_1 dx_2\right]^{1/2}$$

$$\psi_{22} = \left[\int_0^L \int_0^L \phi_{\theta_2}(x_1)\cdot\phi_{\theta_2}(x_2)\cdot\hat{C}o_{ww}(\omega,\Delta x)\,dx_1 dx_2\right]^{1/2}$$

The joint acceptance functions ψ_{11}, ψ_{12} *and* ψ_{22} *are shown in Fig. 8.15. The normalised modal load matrix* $\mathbf{S}_{\hat{R}}(\omega)$ *is then given by*

$$\mathbf{S}_{\hat{R}}(\omega) = \begin{bmatrix} S_{\hat{R}_1\hat{R}_1} & S_{\hat{R}_1\hat{R}_2} \\ S_{\hat{R}_2\hat{R}_1} & S_{\hat{R}_2\hat{R}_2} \end{bmatrix}$$

$$= \frac{(\rho VB)^2 S_w(\omega)}{4(\omega_1^2\tilde{m}_1)(\omega_2^2\tilde{m}_2)} \begin{bmatrix} C_L'^2 \frac{\tilde{m}_2}{\tilde{m}_1}\left(\frac{\omega_2}{\omega_1}\frac{\psi_{11}(\omega)}{\int_0^L \phi_{z_1}^2 dx}\right)^2 & BC_L'C_M' \frac{\psi_{12}^2(\omega)}{\int_0^L \phi_{z_1}^2 dx \int_0^L \phi_{\theta_2}^2 dx} \\ Sym. & (BC_M')^2 \frac{\tilde{m}_1}{\tilde{m}_2}\left(\frac{\omega_1}{\omega_2}\frac{\psi_{22}(\omega)}{\int_0^L \phi_{\theta_2}^2 dx}\right)^2 \end{bmatrix}$$

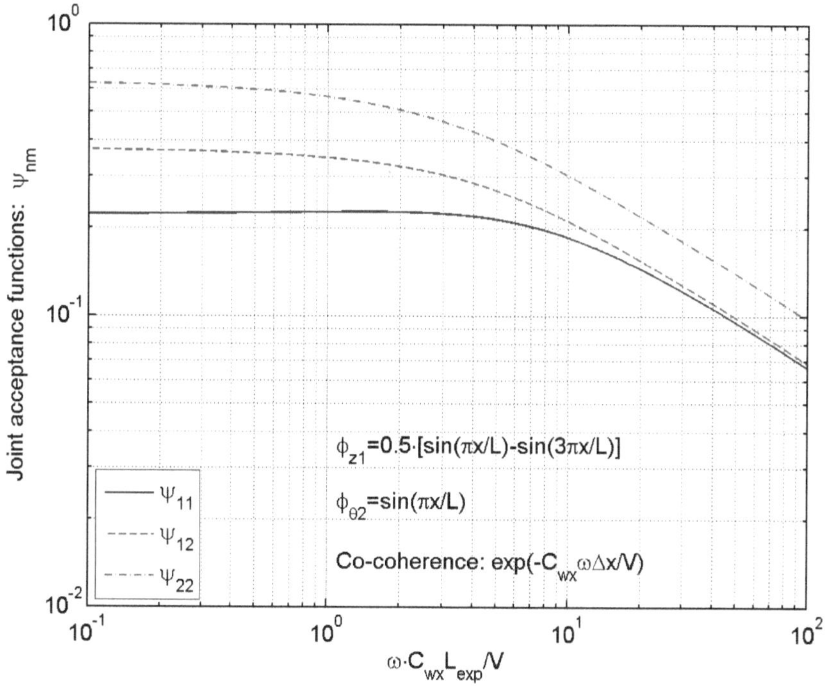

Fig. 8.15 Joint acceptance functions

And thus, the spectral density response matrix at $x_r = L/2$ *is given by (see Eq. 8.82)*

$$\mathbf{S}_{rr}(L/2,\omega) = \begin{bmatrix} S_{r_z r_z} & S_{r_z r_\theta} \\ S_{r_\theta r_z} & S_{r_\theta r_\theta} \end{bmatrix} = \mathbf{\Phi}_r(L/2) \cdot \mathbf{S}_\eta(\omega) \cdot \mathbf{\Phi}_r^T(L/2)$$

where: $\mathbf{S}_\eta(\omega) = \hat{\mathbf{H}}_\eta^*(\omega) \cdot \mathbf{S}_{\hat{R}}(\omega) \cdot \hat{\mathbf{H}}_\eta^T(\omega)$ *and* $\mathbf{\Phi}_r(L/2) = \begin{bmatrix} 1 & 0 \\ 0 & 1 \end{bmatrix}$

Introducing the impedance matrix $\mathbf{E}(\omega) = \begin{bmatrix} E_{11} & E_{12} \\ E_{21} & E_{22} \end{bmatrix}$ *where*

$$E_{11} = 1 - \left(\frac{\omega}{\omega_1}\right)^2 + 2i\frac{\omega}{\omega_1}\left(\zeta_1 - \zeta_{ae_{11}}\right) \qquad E_{12} = -\kappa_{ae_{12}}$$

$$E_{21} = -2i\frac{\omega}{\omega_2}\zeta_{ae_{21}} \qquad \textit{and}$$

$$E_{22} = 1 - \kappa_{ae_{22}} - \left(\frac{\omega}{\omega_2}\right)^2 + 2i\frac{\omega}{\omega_2}\left(\zeta_2 - \zeta_{ae_{22}}\right)$$

then $\hat{\mathbf{H}}_\eta(\omega) = \begin{bmatrix} \hat{H}_{11} & \hat{H}_{12} \\ \hat{H}_{21} & \hat{H}_{22} \end{bmatrix} = \mathbf{E}^{-1} = \dfrac{1}{\det \mathbf{E}} \begin{bmatrix} E_{22} & -E_{12} \\ -E_{21} & E_{11} \end{bmatrix}$

rendering the following expression for the spectral density response matrix at $x_r = L/2$

$$\mathbf{S}_{rr}(L/2,\omega) = \frac{\rho B^2}{2\tilde{m}_1}\frac{\rho B^4}{2\tilde{m}_2}\left(\frac{V}{B\omega_1}\right)^2\left(\frac{V}{B\omega_2}\right)^2 I_w^2 \frac{S_w(\omega)}{\sigma_w^2}\begin{bmatrix} \hat{S}_{\eta 11} & \hat{S}_{\eta 12} \\ \hat{S}_{\eta 21} & \hat{S}_{\eta 22} \end{bmatrix}$$

where:

$$\hat{S}_{\eta_{11}}(\omega) = \gamma_{LL}\cdot \hat{H}_{11}^*\hat{H}_{11} + 2\cdot\gamma_{LM}\cdot\hat{H}_{12}^*\cdot\hat{H}_{11} + \gamma_{MM}\cdot\hat{H}_{12}^*\hat{H}_{12}$$

$$\hat{S}_{\eta_{12}}(\omega) = \gamma_{LL}\cdot\hat{H}_{11}^*\cdot\hat{H}_{21} + \gamma_{LM}\cdot\left(\hat{H}_{12}^*\cdot\hat{H}_{21} + \hat{H}_{11}^*\cdot\hat{H}_{22}\right) + \gamma_{MM}\cdot\hat{H}_{12}^*\cdot\hat{H}_{22}$$

$$\hat{S}_{\eta_{21}}(\omega) = \gamma_{LL}\cdot\hat{H}_{11}^*\hat{H}_{21} + \gamma_{LM}\cdot\left(\hat{H}_{21}^*\hat{H}_{12} + \hat{H}_{22}^*\hat{H}_{11}\right) + \gamma_{MM}\cdot\hat{H}_{22}^*\cdot\hat{H}_{12}$$

$$\hat{S}_{\eta_{22}}(\omega) = \gamma_{LL}\cdot \hat{H}_{21}^*\hat{H}_{21} + 2\cdot\gamma_{LM}\cdot\hat{H}_{21}^*\hat{H}_{22} + \gamma_{MM}\cdot\hat{H}_{22}^*\hat{H}_{22}$$

where

$$\gamma_{LL} = C_L'^2\frac{\tilde{m}_2}{\tilde{m}_1}\left(\frac{\omega_2}{\omega_1}\frac{\psi_{11}(\omega)}{\int_0^L \phi_{z_1}^2 dx}\right)^2$$

$$\gamma_{LM} = BC_L'C_M'\frac{\psi_{12}^2(\omega)}{\int_0^L \phi_{z_1}^2 dx \int_0^L \phi_{\theta_2}^2 dx}$$

and

$$\gamma_{MM} = (BC_M')^2\frac{\tilde{m}_1}{\tilde{m}_2}\left(\frac{\omega_1}{\omega_2}\frac{\psi_{22}(\omega)}{\int_0^L \phi_{\theta_2}^2 dx}\right)^2$$

The corresponding covariance matrix is then obtained by frequency domain integration:

$$\mathbf{Cov}_{rr}(x_r = L/2) = \int_0^\infty \mathbf{S}_{rr}(L/2,\omega)\,d\omega = \begin{bmatrix} \sigma_{r_z r_z}^2 & Cov_{r_z r_\theta} \\ Cov_{r_\theta r_z} & \sigma_{r_\theta r_\theta}^2 \end{bmatrix}$$

The absolute value of the determinant of the non-dimensional frequency response function at a mean wind velocity of $V = 40\ m/s$ *is shown in the top left hand side diagram in Fig. 8.16. The top right hand side diagram shows the cross spectrum between* r_z *and* r_θ *while the two lower diagrams show the spectral densities of* r_z *and* r_θ *, both at* $x = L/2$ *and at a mean wind velocity of* $V = 40\ m/s$ *. As can be seen, there are traces of modal coupling. In this case the coupling effects are exclusively motion induced. Comparing* $\left|\det \hat{\mathbf{H}}(\omega)\right|$ *shown in the top left hand side diagram of Fig. 8.14 to that which is shown in Fig. 8.16 it is seen that the resonance frequency associated with the second mode shape (in torsion) is no longer precisely at* $2.75\ rad/s$*, but slightly below. It is also seen that the resonance peaks are reduced, and particularly the peak associated with* ϕ_{z1} *at* $\omega_1 = 0.7\ rad/s$ *. A time domain simulation of* r_z *and* r_θ *at* $x = L/2$ *and* $V = 40\ m/s$ *is shown in Fig. 8.17 (from spectra shown in Fig. 8.16). The standard deviation of the dynamic responses in the across wind direction (* r_z*) and in torsion (* r_θ *) at various mean wind velocities are shown on the two left hand*

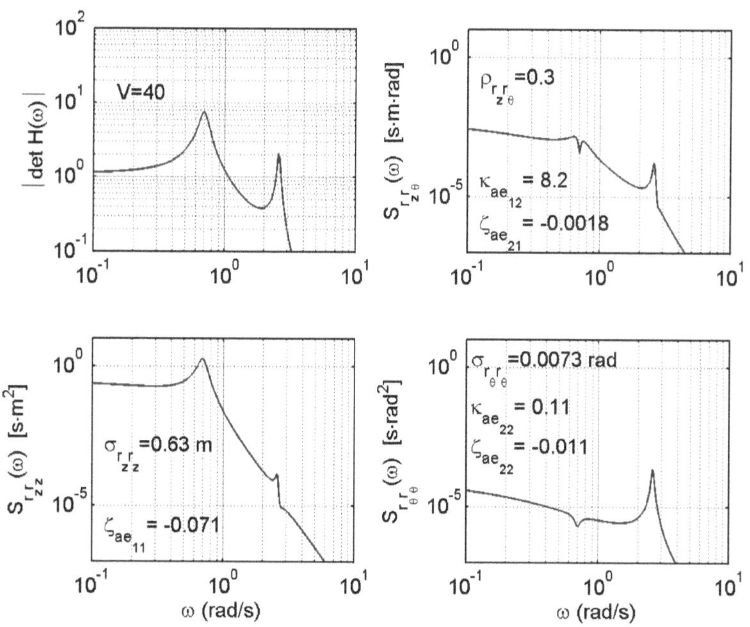

Fig. 8.16 Top left: absolute value of frequency response function. Top right: cross spectrum between vertical and torsion response components. Lower left and right: spectra of response components in vertical direction and torsion. $V = 40\ m/s$.

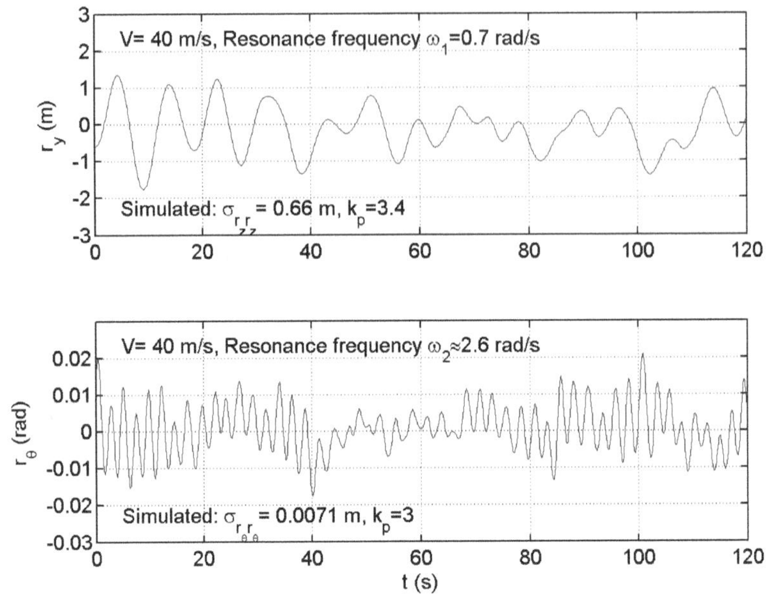

Fig. 8.17 Time domain simulation of dynamic response at $x = L/2$ and $V = 40\ m/s$

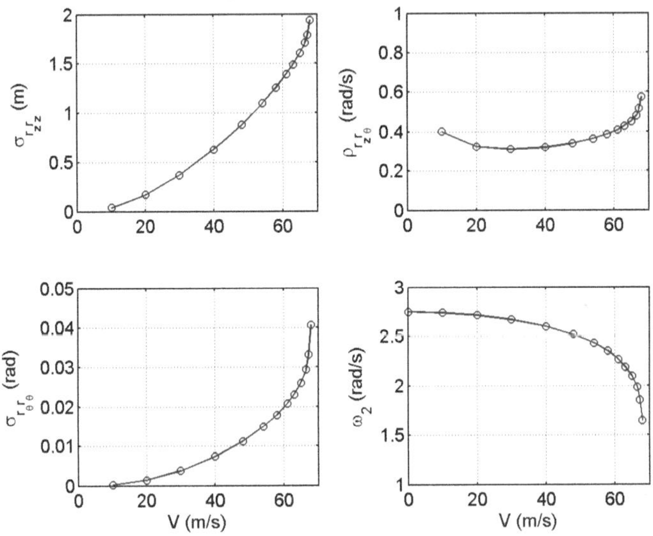

Fig. 8.18 Top and lower left: dynamic response in vertical direction and torsion. Top right: covariance coefficient. Lower right: resonance frequency associated with 2nd mode.

side diagrams in Fig. 8.18 As can be seen, the system has a stability limit slightly below a mean wind velocity of $V = 70\ m/s$ *. The covariance coefficient between the dynamic responses* r_z *and* r_θ *is shown on the top right hand side diagram, while the changes of the resonance frequency associated with the second mode shape (in torsion) at increasing mean wind velocities is shown on the lower right hand side diagram in Fig. 8.18. Since the instability is associated with coupling between* ϕ_{z1} *and* $\phi_{\theta 2}$ *the covariance coefficient will approach unity as* V *becomes closer and closer to the stability limit. At the same time* $\omega_2 \to \omega_1$ *.*

8.4 Dynamic Response to Vortex Shedding

When the air flow is met by a solid bridge or tower type of structure flow separation will occur on the surface of the structure causing vortices to be shed alternately on either side in the wake of the structure. Assuming that along wind load effects may be disregarded, these vortices give rise to fluctuating across wind forces q_z and cross sectional torsion moment q_θ, accompanied by corresponding dynamic displacements r_z or r_θ , as shown in Fig. 8.19. Harmful vortex induced vibrations may particularly occur in cases of resonance, but, even if no resonance occur, the possible long term fatigue effect of these fluctuating forces should not be underestimated. Experimental investigations on stiff models (where r_z and r_θ are zero) show that the single point (i.e. cross sectional) vortex induced forcing process is more or less narrow banded centred at a characteristic vortex shedding frequency f_s , as illustrated in Fig. 8.20.a. The shedding frequency is characteristic to the cross section of the line-like structure. It is proportional to the mean wind velocity V and inversely proportional to the across wind width D. Thus,

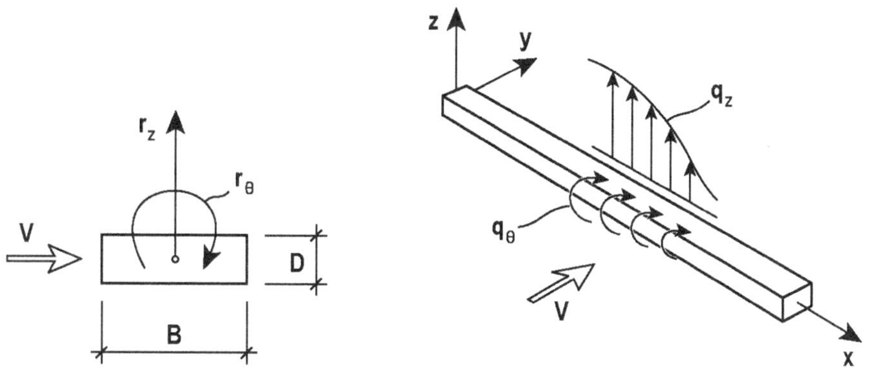

Fig. 8.19 Relevant displacement components and vortex shedding forces

$$f_s = St \cdot \frac{V}{D} \tag{8.84}$$

where St is the Strouhal number, which is available for a good number of typical structural cross sections in the literature.

Two-dimensional investigations also show that the forcing process has a more or less random distribution in the span-wise direction, as indicated by the decaying co-spectrum with increasing separation in Fig. 8.20.b.

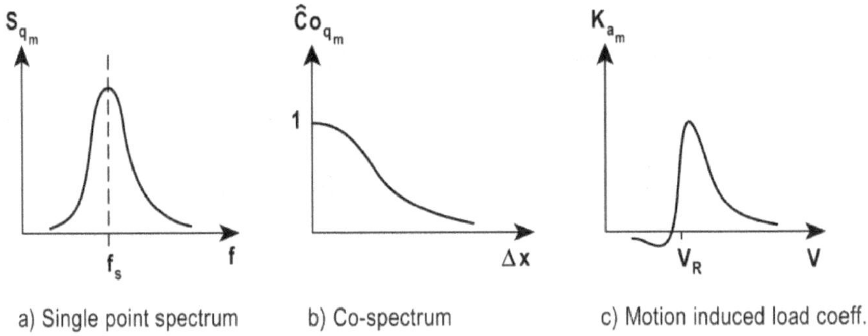

Fig. 8.20 Load characteristics associated with vortex shedding

Turning to a flexible structure it is assumed that the properties of f_s are maintained, i.e. that Eq. 8.84 still holds. The situation is illustrated in Fig. 8.21. Assuming that V is slowly increasing (from zero), then f_s will increase accordingly, and resonance will first occur when f_s becomes equal to the lowest eigenfrequency with respect to vibrations in the across wind direction or torsion. Further increase of V will cause resonance to occur when f_s is equal to the next eigenfrequency, and so on. Theoretically, resonance will occur when f_s is equal to any eigenfrequency $f_n = \omega_n / (2\pi)$. According to Eq. 8.84, the event that $f_s = f_n$ will occur when the mean wind velocity has a value given by

$$V_{R_n} = f_n D / St \tag{8.85}$$

Experiments show that when resonance occurs the flow and the oscillating structure will interact, and for a certain range of ensuing wind velocity settings f_s will deviate from Eq. 8.84 and stay equal or close to f_n, as shown on the upper right hand side of Fig. 8.21. This is what is usually called lock-in. Such vortex shedding induced interaction is accompanied by two important load effects. At lock-in the fluctuating load becomes better correlated in the span-wise direction, but what is more important is that a significant motion induced part is added. As V increase beyond V_R, resonant vibrations no longer occur, as illustrated on the diagram on the lower right hand side in Fig. 8.21.

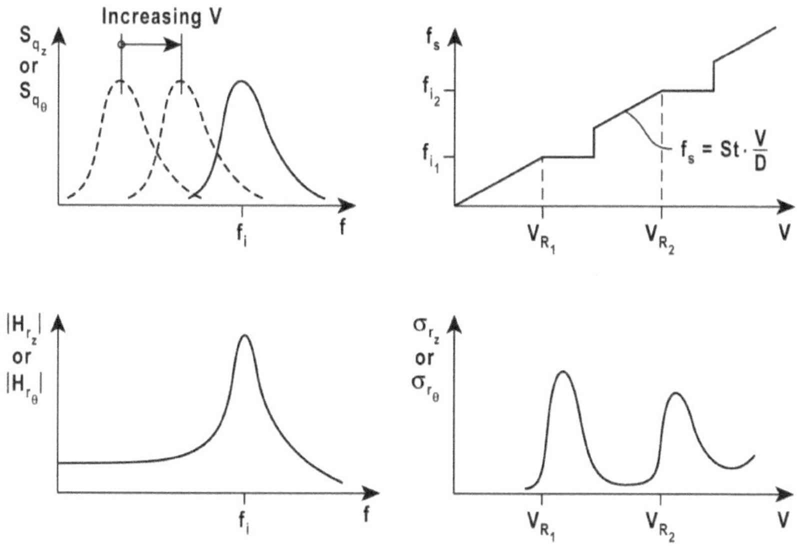

Fig. 8.21 Response characteristics associated with vortex shedding

It has been customary to ascribe the motion induced part of the vortex shedding induced load to the velocity of the system. Extensive research has been carried out on the phenomenon of vortex shedding induced vibrations. In the following it is the theory first developed by Vickery & Basu [20, 21] that will be presented, as it is the only comprehensive stochastic frequency domain theory currently available, rendering a solution at any setting of the mean wind velocity. In this theory the shape-wise description of the net motion–independent cross sectional load spectra and corresponding co-spectra are shown in Fig. 8.19. Mathematically they are given by

$$\begin{bmatrix} S_{q_z}(\omega) \\ S_{q\theta}(\omega) \end{bmatrix} = \frac{\left(\frac{1}{2}\rho V^2\right)^2}{\sqrt{\pi}\cdot\omega_s} \cdot \begin{bmatrix} \dfrac{\left(B\cdot\hat{\sigma}_{q_z}\right)^2}{b_z}\cdot\exp\left\{-\left(\dfrac{1-\omega/\omega_s}{b_z}\right)^2\right\} \\ \dfrac{\left(B^2\cdot\hat{\sigma}_{q\theta}\right)^2}{b_\theta}\cdot\exp\left\{-\left(\dfrac{1-\omega/\omega_s}{b_\theta}\right)^2\right\} \end{bmatrix} \tag{8.86}$$

and

$$\hat{C}o_{q_m}(\Delta x) = \cos\left(\frac{2}{3}\frac{\Delta x}{\lambda_m D}\right)\cdot \exp\left[-\left(\frac{\Delta x}{3\lambda_m D}\right)^2\right] \qquad m = z \text{ or } \theta \qquad (8.87)$$

where, $\omega_s = 2\pi f_s$, $\hat{\sigma}_{q_z} = \sigma_{q_z}/(\rho V^2 B/2)$ and $\hat{\sigma}_{q\theta} = \sigma_{q\theta}/(\rho V^2 B^2/2)$ are the non-dimensional root mean square lift or torsion moment coefficients, b_m is a non-dimensional load spectrum band width parameter, λ_m is a non-dimensional coherence length scale and Δx is span–wise separation, i.e. $\Delta x = |x_1 - x_2|$ where x_1 and x_2 are two arbitrary positions along the span of the system.

Elaboration 8.6

By substituting $\Delta x/(\lambda D) = \alpha$, $a = 1/3$ and $b = 2/3$, and using the known integral $\int_0^\infty \cos(b\alpha)\cdot \exp\left[-(a\alpha)^2\right] d\alpha = \left(\sqrt{\pi}/2a\right)\cdot \exp\left[-(b/2a)^2\right]$ it may readily be shown that $\int_0^\infty \hat{C}o_{q_m}(\Delta x)\, d(\Delta x) = \left(3\sqrt{\pi}/2\right)\cdot e^{-1}\cdot \lambda D \approx \lambda D$.

For the description of the characteristic motion induced load effects at lock-in Vickery & Basu [20, 21] have suggested that this may be accounted for by a negative motion dependent aerodynamic modal damping ratio, ζ_{ae_n}, such that the total modal damping ratio associated with mode n is given by

$$\zeta_{tot_n} = \zeta_n - \zeta_{ae_n} \qquad (8.88)$$

This is equivalent to the introduction of motion dependent aerodynamic derivatives as described in chapter 8.2 above. Let us for consistency adopt the notation given in Eqs. 8.16 and 8.17, and then it is the aerodynamic derivatives H_1^* and A_2^* that are responsible for aerodynamic damping effects in the vertical (z) direction or in torsion (θ). Assuming that in the vicinity of a distinct vortex shedding type of response all other motion induced effects may be ignored, then in this mean wind velocity region

$$\mathbf{C}_{ae} \approx \frac{\rho B^2}{2}\omega_n(V)\begin{bmatrix} 0 & 0 & 0 \\ 0 & H_1^* & 0 \\ 0 & 0 & B^2 A_2^* \end{bmatrix} \qquad \text{and} \qquad \mathbf{K}_{ae} \approx 0 \qquad (8.89)$$

where

$$H_1^* = K_{a_z}\left[1-\left(\frac{\sigma_z}{a_z D}\right)^2\right] \quad \text{and} \quad A_2^* = K_{a\theta}\left[1-\left(\frac{\sigma_\theta}{a_\theta}\right)^2\right] \tag{8.90}$$

and where K_{a_z} and $K_{a\theta}$ are velocity dependent damping coefficients equivalent to those defined by Vickery & Basu [20, 21]. Since $\mathbf{K}_{ae} \approx 0$ it is a reasonable assumption that $\omega_n(V) \approx \omega_n(V=0)$, and then the aerodynamic damping term in Eq. 8.88 may be taken from Eq. 8.69, i.e.:

$$\zeta_{ae_n} = \frac{\tilde{C}_{ae_n}}{2\omega_n \tilde{M}_n} = \frac{\int_{L_{\exp}} \boldsymbol{\varphi}_n^T \cdot \mathbf{C}_{ae} \cdot \boldsymbol{\varphi}_n dx}{2\omega_n \tilde{m}_n \int_L \boldsymbol{\varphi}_n^T \cdot \boldsymbol{\varphi}_n dx} = \frac{\rho B^2}{4\tilde{m}_n} \cdot \frac{\int_{L_{\exp}} \left(H_1^* \phi_{z_n}^2 + B^2 A_2^* \phi_{\theta_n}^2\right) dx}{\int_L \left(\phi_{y_n}^2 + \phi_{z_n}^2 + \phi_{\theta_n}^2\right) dx} \tag{8.91}$$

where

$$\tilde{m}_n = \frac{\tilde{M}_n}{\int_L \boldsymbol{\varphi}_n^T \cdot \boldsymbol{\varphi}_n dx} = \frac{\tilde{M}_n}{\int_L \left(\phi_{y_n}^2 + \phi_{z_n}^2 + \phi_{\theta_n}^2\right) dx} \tag{8.92}$$

is the evenly distributed and modally equivalent mass associated with mode n. K_{a_m} ($m = z$ or θ) are the coefficients that account for the accelerating part of the motion induced load when V is close to V_{R_n}. Apart from being cross sectional characteristics, they are functions of V and the resonance frequency of the mode in question (see right hand side diagram in Fig. 8.20). $a_z D$ and a_θ are quantities associated with the self-limiting nature of vortex shedding, i.e. they represent upper displacement or rotation limits at which the aerodynamic damping becomes insignificant. {It should be noted that in Eq. 8.89 the damping coefficients are defined such that consistency is obtained with the general definition of aerodynamic derivatives in Eqs. 8.16 and 8.17 rather than the definition adopted by Vickery & Basu. Thus, the K_{a_z} values given by Vickery & Basu in references [20, 21] are applicable in the expressions given above if they are multiplied by $4(D/B)^2$ }. Vortex shedding induced load effects at or in the vicinity of lock-in is dependent on the dynamic response of the structure, because the total damping in each mode is unknown prior to any knowledge about the actual structural displacements. Thus, response calculations will inevitably involve iterations. It should be acknowledged that the peak factor for vortex shedding response does not comply with the theory behind what may be obtained from Eq. A.45. For an ultra-narrow-banded vortex shedding response the peak factor is close to 1.5 (theoretically $\sqrt{2}$). For broad-banded response Eq. A.45 will most often render conservative results. A few time domain simulations of response spectra (see Appendix B) will give a good indication on what peak factor should be chosen.

Multi-mode Response Calculations

The general solution of a multi-mode approach to the problem of calculating vortex shedding induced dynamic response is identical to that which has been presented above for buffeting response calculations. I.e., the general solution to the calculation of the three by three cross spectra response matrix $\mathbf{S}_{rr}(x_r,\omega)$ is given in Eq. 8.82, while the corresponding covariance matrix is given in Eq. 8.83. The N_{mod} by N_{mod} frequency response matrix $\hat{\mathbf{H}}_{\eta}(\omega)$ is given in Eq. 8.66 while the modal load matrix $\mathbf{S}_{\hat{R}}(\omega)$ is given in Eqs. 8.76 – 8.78, except that for vortex shedding the motion induced load is assumed exclusively related to structural velocity, and its effect applies to the actual modal response and not to the individual Fourier components. As shown in Eq. 8.89, this implies that $\mathbf{K}_{ae}=0$ and $\mathbf{C}_{ae}=\left(\rho B^2/2\right)\cdot\omega_n\cdot diag\left[0 \quad H_1^* \quad B^2A_2^*\right]$, and thus

$$\hat{\mathbf{H}}_{\eta}(\omega)=\left\{\mathbf{I}-\left(\omega\cdot diag\left[1/\omega_n\right]\right)^2+2i\omega\cdot diag\left[1/\omega_n\right]\cdot\left(\boldsymbol{\zeta}-\boldsymbol{\zeta}_{ae}\right)\right\}^{-1} \tag{8.93}$$

where $\boldsymbol{\zeta}=diag\left[\zeta_n\right]$ and the content of $\boldsymbol{\zeta}_{ae}$ is given by

$$\begin{aligned}\zeta_{ae_{nm}}&=\frac{\omega_n}{2}\frac{\tilde{C}_{ae_{nm}}}{\omega_n^2\tilde{M}_n}=\frac{1}{2\omega_n\tilde{m}_n}\cdot\frac{\int_{L_{\exp}}\left(\boldsymbol{\varphi}_n^T\cdot\mathbf{C}_{ae}\cdot\boldsymbol{\varphi}_m\right)dx}{\int_L\left(\boldsymbol{\varphi}_n^T\cdot\boldsymbol{\varphi}_n\right)dx}\\&=\frac{\rho B^2}{4\tilde{m}_n}\cdot\frac{\int_{L_{\exp}}\phi_{z_n}\phi_{z_m}H_1^*dx+B^2\int\phi_{\theta_n}\phi_{\theta_m}A_2^*dx}{\int_L\left(\phi_{y_n}^2+\phi_{z_n}^2+\phi_{\theta_n}^2\right)dx}\end{aligned} \tag{8.94}$$

where H_1^* and A_2^*, are given in Eq. 8.90 and where $\tilde{m}_n$ is defined in Eq. 8.92. If H_1^* and A_2^* are taken as modal constants and independent of span-wise position, then, due to the orthogonal properties of the mode shapes, $\boldsymbol{\zeta}_{ae}$ becomes diagonal, i.e.

$$\boldsymbol{\zeta}_{ae} = diag\left[\zeta_{ae_n}\right]$$

where

$$\zeta_{ae_n} = \frac{\rho B^2}{4\tilde{m}_n} \cdot \frac{H_1^* \int\limits_{L_{exp}} \phi_{z_n}^2 dx + B^2 A_2^* \int\limits_{L_{exp}} \phi_{\theta_n}^2 dx}{\int\limits_L \left(\phi_{y_n}^2 + \phi_{z_n}^2 + \phi_{\theta_n}^2\right) dx} \tag{8.95}$$

This implies that $\hat{\mathbf{H}}_\eta(\omega)$ is an N_{mod} by N_{mod} diagonal matrix. In vortex shedding induced vibration problems it is usually not essential to include the along wind load effects. The load vector may then be reduced to

$$\mathbf{q}(x,t) = \begin{bmatrix} 0 & q_z & q_\theta \end{bmatrix}^T \tag{8.96}$$

and the corresponding Fourier transform is

$$\mathbf{a}_q(x,\omega) = \begin{bmatrix} 0 & a_{q_z} & a_{q\theta} \end{bmatrix}^T \tag{8.97}$$

The cross sectional load spectrum is defined by

$$\mathbf{S}_{qq}(\Delta x,\omega) = \lim_{T\to\infty} \frac{\left(\mathbf{a}_q^* \mathbf{a}_q^T\right)}{\pi T} = \lim_{T\to\infty} \frac{1}{\pi T} \begin{bmatrix} 0 & 0 & 0 \\ 0 & a_{q_z}^* a_{q_z} & a_{q_z}^* a_{q\theta} \\ 0 & a_{q\theta}^* a_{q_z} & a_{q\theta}^* a_{q\theta} \end{bmatrix} = \begin{bmatrix} 0 & 0 & 0 \\ 0 & S_{q_z q_z} & S_{q_z q_\theta} \\ 0 & S_{q_\theta q_z} & S_{q_\theta q_\theta} \end{bmatrix} \tag{8.98}$$

The problem is greatly simplified if the cross coupling between q_z and q_θ may be disregarded (i.e. they occur at distinctly different mean wind velocities), in which case

$$\mathbf{S}_{qq}(\Delta x,\omega) \approx \begin{bmatrix} 0 & 0 & 0 \\ 0 & S_{q_z q_z} & 0 \\ 0 & 0 & S_{q_\theta q_\theta} \end{bmatrix} \tag{8.99}$$

where $S_{q_z q_z}$ and $S_{q_\theta q_\theta}$ are given by

$$\left.\begin{aligned} S_{q_z q_z} &= S_{q_z}(\omega) \cdot \hat{C}o_{q_z}(\Delta x) \\ S_{q_\theta q_\theta} &= S_{q_\theta}(\omega) \cdot \hat{C}o_{q_\theta}(\Delta x) \end{aligned}\right\} \tag{8.100}$$

The single point spectra S_{q_z} and $S_{q\theta}$ are defined in Eq. 8.86, while the reduced co–spectra $\hat{C}o_{q_z}$ and $\hat{C}o_{q\theta}$ are defined in Eq. 8.87. Thus, the elements of $\mathbf{S}_{\hat{R}}$ (see Eq. 8.73) are reduced to

$$S_{\hat{R}_n\hat{R}_m}(\omega)=\frac{\iint\limits_{L_{\exp}} \boldsymbol{\varphi}_n^T(x_1)\cdot\mathbf{S}_{qq}(\Delta x,\omega)\cdot\boldsymbol{\varphi}_m(x_2)\,dx_1dx_2}{\left(\omega_n^2\tilde{M}_n\right)\cdot\left(\omega_m^2\tilde{M}_m\right)}=$$

$$\frac{\iint\limits_{L_{\exp}}\left\{\phi_{z_n}(x_1)\phi_{z_m}(x_2)S_{q_zq_z}+\phi_{\theta_n}(x_1)\phi_{\theta_m}(x_2)S_{q_\theta q_\theta}\right\}dx_1dx_2}{\left(\omega_n^2\tilde{M}_n\right)\cdot\left(\omega_m^2\tilde{M}_m\right)}= \tag{8.101}$$

$$\frac{S_{q_z}\iint\limits_{L_{\exp}}\phi_{z_n}(x_1)\phi_{z_m}(x_2)\hat{C}o_{q_z}dx_1dx_2+S_{q\theta}\iint\limits_{L_{\exp}}\phi_{\theta_n}(x_1)\phi_{\theta_m}(x_2)\hat{C}o_{q\theta}dx_1dx_2}{\left(\omega_n^2\tilde{M}_n\right)\cdot\left(\omega_m^2\tilde{M}_m\right)}$$

Furthermore, it is a reasonable assumption that the integral length–scale of the vortices λD is small as compared to the flow exposed length $L_{\exp}$ of the structure, and since q_z and q_θ are caused by the same vortices their coherence properties are likely to be identical, in which case [recalling that $\int_0^\infty \hat{C}o_{q_m}(\Delta x)d(\Delta x)\approx\lambda D$ (see Elaboration 8.6)] the following is obtained:

$$S_{\hat{R}_n\hat{R}_m}(\omega)\approx\frac{2\lambda D\left[S_{q_z}\int\limits_{L_{\exp}}\phi_{z_n}(x)\phi_{z_m}(x)dx+S_{q\theta}\int\limits_{L_{\exp}}\phi_{\theta_n}(x)\phi_{\theta_m}(x)dx\right]}{\left(\omega_n^2\tilde{M}_n\right)\cdot\left(\omega_m^2\tilde{M}_m\right)} \tag{8.102}$$

Again, due to the orthogonal properties of the mode shapes this implies that $\mathbf{S}_{\hat{R}}(\omega)$ becomes diagonal, i.e.

$$\mathbf{S}_{\hat{R}}(\omega)=diag\left[S_{\hat{R}_n}\right] \tag{8.103}$$

where

$$S_{\hat{R}_n}(\omega)=\frac{2\lambda D}{\left(\omega_n^2\tilde{M}_n\right)^2}\left[S_{q_z}(\omega)\int\limits_{L_{\exp}}\phi_{z_n}^2dx+S_{q\theta}(\omega)\int\limits_{L_{\exp}}\phi_{\theta_n}^2dx\right] \tag{8.104}$$

While the calculation of the spectral response matrix is given in Eq. 8.82, it should be noted that if the simplifications above hold then both $\hat{\mathbf{H}}_\eta$ and $\mathbf{S}_{\hat{R}}$ are diagonal, in which case

$$\begin{aligned}\mathbf{S}_{rr}(x_r,\omega) &= \mathbf{\Phi}_r(x_r)\,diag\left[S_{\eta_n}(\omega)\right]\mathbf{\Phi}_r^T(x_r) = \sum_{n=1}^{N_{\text{mod}}} \boldsymbol{\varphi}_n(x_r)\boldsymbol{\varphi}_n^T(x_r)S_{\eta_n}(\omega)\\ &= \sum_{n=1}^{N_{\text{mod}}} \begin{bmatrix} \phi_y^2(x_r) & \phi_y(x_r)\phi_z(x_r) & \phi_y(x_r)\phi_\theta(x_r)\\ & \phi_z^2(x_r) & \phi_z(x_r)\phi_\theta(x_r)\\ Sym. & & \phi_\theta^2(x_r)\end{bmatrix}_n S_{\eta_n}(\omega)\end{aligned} \tag{8.105}$$

where

$$S_{\eta_n}(\omega) = \left|\hat{H}_{\eta_n}(\omega)\right|^2 \cdot S_{\hat{R}_n}(\omega) \tag{8.106}$$

and $\hat{H}_{\eta_n}$ is given by (see Eq. 8.93)

$$\hat{H}_{\eta_n}(\omega) = \left[1-\left(\frac{\omega}{\omega_n}\right)^2 + 2i\cdot\left(\zeta_n - \zeta_{ae_n}\right)\cdot\frac{\omega}{\omega_n}\right]^{-1} \tag{8.107}$$

and ζ_{ae_n} is given in Eq. 8.95. The corresponding covariance response matrix $\mathbf{Cov}_{rr}(x_r)$ for the dynamic response at span-wise position x_r is then given by

$$\begin{aligned}\mathbf{Cov}_{rr}(x_r) &= \int_0^\infty \mathbf{S}_{rr}(x_r,\omega)\,d\omega = \begin{bmatrix} \sigma_{r_yr_y}^2 & Cov_{r_yr_z} & Cov_{r_yr_\theta}\\ & \sigma_{r_zr_z}^2 & Cov_{r_zr_\theta}\\ Sym. & & \sigma_{r_\theta r_\theta}^2\end{bmatrix}\\ &= \sum_{n=1}^{N_{\text{mod}}} \begin{bmatrix} \phi_y^2(x_r) & \phi_y(x_r)\cdot\phi_z(x_r) & \phi_y(x_r)\cdot\phi_\theta(x_r)\\ & \phi_z^2(x_r) & \phi_z(x_r)\cdot\phi_\theta(x_r)\\ Sym. & & \phi_\theta^2(x_r)\end{bmatrix}_n \sigma_{\eta_n}^2\end{aligned} \tag{8.108}$$

where

$$\sigma_{\eta_n}^2 = \int_0^\infty S_{\eta_n}\,d\omega \tag{8.109}$$

is the variance contribution from an arbitrary mode n. Usually, vortex shedding induced dynamic response is largely resonant and narrow-banded. It will then suffice to only consider the resonant part of the frequency domain integration in Eq. 8.109, and discard the background part. Thus,

$$\begin{aligned}\sigma_{\eta_n}^2 &= \int_0^\infty S_{\eta_n} d\omega = \int_0^\infty \left|\hat{H}_{\eta_n}(\omega)\right|^2 \cdot S_{\hat{R}_n}(\omega) d\omega \\ &\approx \int_0^\infty \left|\hat{H}_{\eta_n}(\omega)\right|^2 d\omega \cdot S_{\hat{R}_n}(\omega_n) = \frac{\pi\omega_n \cdot S_{\hat{R}_n}(\omega_n)}{4\left(\zeta_n - \zeta_{ae_n}\right)}\end{aligned} \tag{8.110}$$

where (see Eqs. 8.104 and 8.86)

$$\begin{aligned}S_{\hat{R}_n}(\omega_n) &= \frac{2\lambda D\left[S_{q_z}(\omega_n)\int_{L_{\exp}}\phi_{z_n}^2 dx + S_{q\theta}(\omega_n)\int_{L_{\exp}}\phi_{\theta_n}^2 dx\right]}{\left(\omega_n^2 \tilde{M}_n\right)^2} \\ &= \frac{2\lambda D}{\left(\omega_n^2 \tilde{M}_n\right)^2} \cdot \frac{\left(\rho V^2 B/2\right)^2}{\sqrt{\pi}\cdot\omega_s} \cdot \left\{\frac{\sigma_{q_z}^2}{b_z}\int_{L_{\exp}}\phi_{z_n}^2 dx \cdot \exp\left[-\left(\frac{1-\omega_n/\omega_s}{b_z}\right)^2\right]\right. \\ &\quad \left. + \frac{\left(B\sigma_{q\theta}\right)^2}{b_\theta}\int_{L_{\exp}}\phi_{\theta_n}^2 dx \cdot \exp\left[-\left(\frac{1-\omega_n/\omega_s}{b_\theta}\right)^2\right]\right\}\end{aligned} \tag{8.111}$$

and $\omega_s = 2\pi f_s$. As mentioned above, the calculations will inevitably demand iterations, because H_1^* and A_2^* are functions of $\sigma_{r_z r_z}$ and $\sigma_{r_\theta r_\theta}$. The iterations may be demanding as they will take place on the difference between ζ_n and ζ_{ae_n}, which in general is a small quantity.

Example 8.3

Let us consider a single span suspension bridge with span $L = L_{\exp} = 1200m$ *and set out to calculate the vortex shedding induced dynamic response at* $x_r = L/2$ *which is associated with the three mode shapes*

$$\boldsymbol{\varphi}_1 = \begin{bmatrix} 0 \\ \phi_{z_1} \\ 0 \end{bmatrix} = \begin{bmatrix} 0 \\ \sin\left(\dfrac{\pi x}{L}\right) \\ 0 \end{bmatrix}$$

$$\boldsymbol{\varphi}_2 = \begin{bmatrix} 0 \\ \phi_{z_2} \\ 0 \end{bmatrix} = \begin{bmatrix} 0 \\ \sin\left(\frac{3\pi x}{L}\right) \\ 0 \end{bmatrix}$$

$$\boldsymbol{\varphi}_3 = \begin{bmatrix} 0 \\ 0 \\ \phi_{\theta_3} \end{bmatrix} = \begin{bmatrix} 0 \\ 0 \\ \sin\left(\frac{\pi x}{L}\right) \end{bmatrix}$$

with corresponding eigenfrequencies 0.7 , 1.6 *and* 2.75 rad/s. *As can be seen,* $\boldsymbol{\varphi}_1$ *and* $\boldsymbol{\varphi}_2$ *contain only the displacement component in the across wind vertical direction while* $\boldsymbol{\varphi}_3$ *only contains torsion. Let us adopt the following structural properties:*

ρ $\frac{kg}{m^3}$	B m	D m	m_z $\frac{kg}{m}$	m_θ $\frac{kgm^2}{m}$	ω_1 $\frac{rad}{s}$	ω_2 $\frac{rad}{s}$	ω_3 $\frac{rad}{s}$	$\zeta_1 = \zeta_2$ %	ζ_3 %
1.25	20	4	12500	425000	0.7	1.6	2.75	0.5	1

and the following vortex induced wind load properties:

St	$\hat{\sigma}_{q_z}$	$\hat{\sigma}_{q\theta}$	b_z	b_θ	a_z	a_θ	$\lambda_z = \lambda_\theta$	K_{az0}	$K_{a\theta0}$
0.1	0.9	0.3	0.15	0.1	0.4	0.1	1.2	0.2	0.02

where $\hat{\sigma}_{q_z} = \sigma_{q_z} \big/ \left(\rho V^2 B/2\right)$ *and* $\hat{\sigma}_{q\theta} = \sigma_{q\theta} \big/ \left(\rho V^2 B/2\right)$. *Since* m_z *and* m_θ *are constant along the span, then the modally equivalent and evenly distributed masses* $\tilde{m}_1 = \tilde{m}_2 = m_z$ *and* $\tilde{m}_3 = m_\theta$. *Finally, let us adopt the following wind velocity variation of the relative aerodynamic damping coefficient*

$$K_a(V)/K_{a0} = 2.6 \cdot \left(V/V_{R_i}\right)^{-3} \cdot \exp\left[-\left(V/V_{R_i}\right)^{-4}\right]$$

where $$V_{R_i} = \frac{\omega_i}{2\pi} \cdot \frac{D}{St}$$

In this case (see Eq. 8.108)

$$\mathbf{Cov}_{rr}(x_r) = \int_0^\infty \mathbf{S}_{rr}(x_r,\omega)\,d\omega = \begin{bmatrix} \sigma_{r_y r_y}^2 & Cov_{r_y r_z} & Cov_{r_y r_\theta} \\ & \sigma_{r_z r_z}^2 & Cov_{r_z r_\theta} \\ Sym. & & \sigma_{r_\theta r_\theta}^2 \end{bmatrix}$$

$$= \begin{bmatrix} 0 & 0 & 0 \\ 0 & \phi_{z1}^2(x_r) & 0 \\ 0 & 0 & 0 \end{bmatrix} \cdot \sigma_{\eta 1}^2 + \begin{bmatrix} 0 & 0 & 0 \\ 0 & \phi_{z2}^2(x_r) & 0 \\ 0 & 0 & 0 \end{bmatrix} \cdot \sigma_{\eta 2}^2 + \begin{bmatrix} 0 & 0 & 0 \\ 0 & 0 & 0 \\ 0 & 0 & \phi_{\theta 3}^2(x_r) \end{bmatrix} \cdot \sigma_{\eta 3}^2$$

and thus:

$$\mathbf{Cov}_{rr}(x_r) = \begin{bmatrix} 0 & 0 & 0 \\ 0 & \sigma_{r_z r_z}^2 & 0 \\ 0 & 0 & \sigma_{r_\theta r_\theta}^2 \end{bmatrix} = \begin{bmatrix} 0 & 0 & 0 \\ 0 & \phi_{z1}^2(x_r)\cdot\sigma_{\eta 1}^2 + \phi_{z2}^2(x_r)\cdot\sigma_{\eta 2}^2 & 0 \\ 0 & 0 & \phi_{\theta 3}^2(x_r)\cdot\sigma_{\eta 3}^2 \end{bmatrix}$$

From Eqs. 8.110 and 8.111 (and taking it for granted that $\lambda_z = \lambda_\theta = \lambda$ *) the following variance contributions are obtained*

$$\sigma_{\eta 1}^2 = \left(\frac{D}{2^{7/2}\pi^{7/4}} \cdot \frac{\rho BD}{\tilde{m}_1} \cdot \frac{\hat{\sigma}_{q_z}}{St^2} \right)^2 \cdot \frac{\lambda}{b_z} \cdot \frac{1}{\zeta_1 - \zeta_{ae_1}} \cdot \frac{D \int_{L_{\exp}} \phi_{1_z}^2\,dx}{\left(\int_L \phi_{1_z}^2\,dx \right)^2} \cdot g_1^2\left(V_{R_1}, V\right)$$

$$\sigma_{\eta 2}^2 = \left(\frac{D}{2^{7/2}\pi^{7/4}} \cdot \frac{\rho BD}{\tilde{m}_2} \cdot \frac{\hat{\sigma}_{q_z}}{St^2} \right)^2 \cdot \frac{\lambda}{b_z} \cdot \frac{1}{\zeta_2 - \zeta_{ae_2}} \cdot \frac{D \int_{L_{\exp}} \phi_{2_z}^2\,dx}{\left(\int_L \phi_{2_z}^2\,dx \right)^2} \cdot g_2^2\left(V_{R_2}, V\right)$$

$$\sigma_{\eta 3}^2 = \left(\frac{1}{2^{7/2}\pi^{7/4}} \cdot \frac{\rho (BD)^2}{\tilde{m}_3} \cdot \frac{\hat{\sigma}_{q\theta}}{St^2} \right)^2 \cdot \frac{\lambda}{b_\theta} \cdot \frac{1}{\zeta_3 - \zeta_{ae_3}} \cdot \frac{D \int_{L_{\exp}} \phi_{3_\theta}^2\,dx}{\left(\int_L \phi_{3_\theta}^2\,dx \right)^2} \cdot g_3^2\left(V_{R_3}, V\right)$$

where

$$\left.\begin{aligned}
g_1\left(V_{R_1},V\right) &= \left(\frac{V}{V_{R_1}}\right)^{3/2} \exp\left[-\frac{1}{2}\left(\frac{1-V_{R_1}/V}{b_z}\right)^2\right] \\
g_2\left(V_{R_2},V\right) &= \left(\frac{V}{V_{R_2}}\right)^{3/2} \exp\left[-\frac{1}{2}\left(\frac{1-V_{R_2}/V}{b_z}\right)^2\right] \\
g_3\left(V_{R_3},V\right) &= \left(\frac{V}{V_{R_3}}\right)^{3/2} \exp\left[-\frac{1}{2}\left(\frac{1-V_{R_3}/V}{b_\theta}\right)^2\right]
\end{aligned}\right\} \quad \textit{and}$$

$$V_{R_n} = \frac{\omega_n}{2\pi} \cdot \frac{D}{St} \quad \text{where } n = \begin{cases} 1 \\ 2 \\ 3 \end{cases}$$

What then remains are the aerodynamic damping contributions given in Eq. 8.95, from which the following is obtained:

$$\zeta_{ae_1}(V) = \frac{\rho B^2}{4\tilde{m}_1} \cdot H_1^* \cdot \frac{\int_{L_{\exp}} \phi_{z_1}^2 dx}{\int_L \phi_{z_1}^2 dx} = \frac{\rho B^2}{4\tilde{m}_1} \cdot K_{a_z}\left(V_{R_1},V\right) \cdot \left\{1 - \left[\frac{\sigma_{r_z r_z}(V)}{a_z D}\right]^2\right\}$$

$$\zeta_{ae_2}(V) = \frac{\rho B^2}{4\tilde{m}_2} \cdot H_1^* \cdot \frac{\int_{L_{\exp}} \phi_{z_2}^2 dx}{\int_L \phi_{z_2}^2 dx} = \frac{\rho B^2}{4\tilde{m}_2} \cdot K_{a_z}\left(V_{R_2},V\right) \cdot \left\{1 - \left[\frac{\sigma_{r_z r_z}(V)}{a_z D}\right]^2\right\}$$

$$\zeta_{ae_3}(V) = \frac{\rho B^4}{4\tilde{m}_3} \cdot A_2^* \cdot \frac{\int_{L_{\exp}} \phi_{\theta_3}^2 dx}{\int_L \phi_{\theta_3}^2 dx} = \frac{\rho B^2}{4\tilde{m}_3} \cdot K_{a\theta}\left(V_{R_3},V\right) \cdot \left\{1 - \left[\frac{\sigma_{r_\theta r_\theta}(V)}{a_\theta}\right]^2\right\}$$

The relevant response diagrams are shown in Fig. 8.22 below.

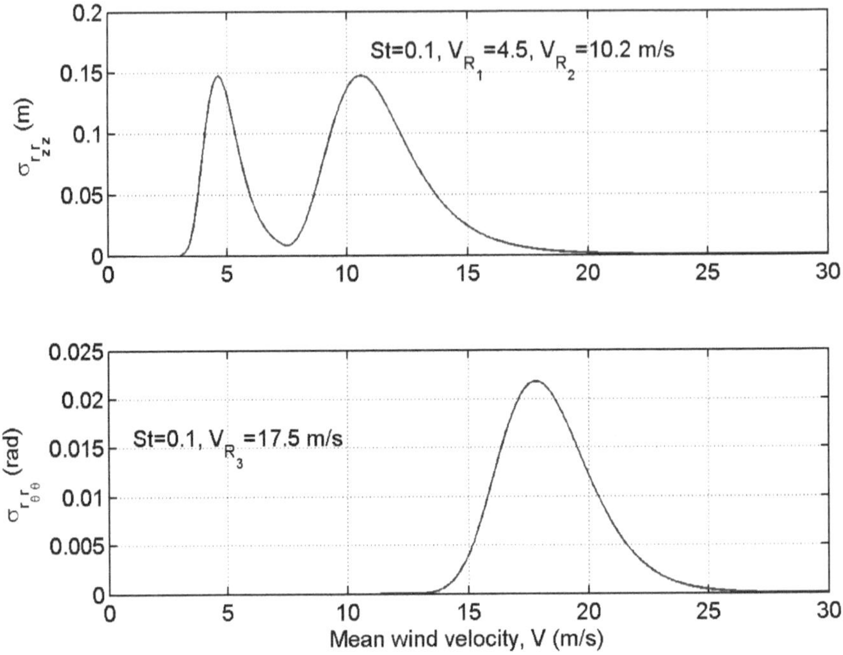

Fig. 8.22 Vortex shedding induced across wind and torsion response

Single Mode Single Component Response Calculations

A single mode single component response calculation is obtained by assuming that any of the following two conditions apply

$$\left.\begin{aligned} \boldsymbol{\varphi}_n(x) &\approx \begin{bmatrix} 0 & \phi_z & 0 \end{bmatrix}_n^T \\ \boldsymbol{\varphi}_n(x) &\approx \begin{bmatrix} 0 & 0 & \phi_\theta \end{bmatrix}_n^T \end{aligned}\right\} \tag{8.112}$$

Mode shape summation and off diagonal terms in Eq. 8.108 will then vanish, rendering all covariance quantities obsolete, and $\mathbf{S}_{rr}$ will simply contain the response variances of a single mode excitation on its diagonal. Thus, the response spectrum and the displacement variance associated with the excitation of an arbitrary mode n are given by

$$\left.\begin{aligned} S_{r_n}(\omega) &= \phi_n^2(x_r) \cdot \left|\hat{H}_{\eta_n}(\omega)\right|^2 \cdot S_{\hat{R}_n}(\omega) \\ \sigma_{r_n}^2 &= \int_0^\infty S_{r_n}(\omega)\,d\omega \end{aligned}\right\} n = \begin{cases} z \\ \theta \end{cases} \tag{8.113}$$

where

$$\left.\begin{aligned} \hat{H}_{\eta_n}(\omega) &= \left[1-\left(\frac{\omega}{\omega_n}\right)^2 + 2i\cdot\left(\zeta_n-\zeta_{ae_n}\right)\cdot\frac{\omega}{\omega_n}\right]^{-1} \\ S_{\hat{R}_n}(\omega) &= 2\lambda D\cdot\frac{S_{q_n}(\omega)\cdot\int\limits_{L_{\exp}}\phi_n^2(x)\,dx}{\left(\omega_n^2\tilde{M}_n\right)^2} \end{aligned}\right\} \tag{8.114}$$

and where aerodynamic damping properties may be extracted from Eq. 8.95, rendering

$$\left.\begin{aligned} \zeta_{ae_z} &= \frac{\tilde{C}_{ae_{zz}}}{2\omega_z\tilde{M}_z} = \frac{\rho B^2 H_1^*}{4\tilde{m}_z}\frac{\int\limits_{L_{\exp}}\phi_z^2dx}{\int\limits_L\phi_z^2dx} = \frac{\rho B^2}{4\tilde{m}_z}K_{a_z}\left[1-\left(\frac{\sigma_z}{a_zD}\right)^2\right]\frac{\int\limits_{L_{\exp}}\phi_z^2dx}{\int\limits_L\phi_z^2dx} \\ \zeta_{ae\theta} &= \frac{\tilde{C}_{ae\theta\theta}}{2\omega_\theta\tilde{M}_\theta} = \frac{\rho B^4 A_2^*}{4\tilde{m}_\theta}\frac{\int\limits_{L_{\exp}}\phi_\theta^2dx}{\int\limits_L\phi_\theta^2dx} = \frac{\rho B^4}{4\tilde{m}_\theta}K_{a\theta}\left[1-\left(\frac{\sigma_\theta}{a_\theta}\right)^2\right]\frac{\int\limits_{L_{\exp}}\phi_\theta^2dx}{\int\limits_L\phi_\theta^2dx} \end{aligned}\right\} \tag{8.115}$$

Again, vortex shedding induced dynamic response is largely resonant and narrow-banded. It will then suffice to only consider the resonant part of the frequency domain integration in Eq. 8.113, and discard the background part. Thus,

$$\begin{aligned} \sigma_{r_n}^2 &= \int_0^\infty S_{r_n}d\omega \approx \phi_n^2(x_r)\cdot\int_0^\infty\left|\hat{H}_{\eta_n}(\omega)\right|^2 d\omega\cdot S_{\hat{R}_n}(\omega_n) \\ &= \phi_n^2(x_r)\cdot\frac{\pi\omega_n S_{\hat{R}_n}(\omega_n)}{4\left(\zeta_n-\zeta_{ae_n}\right)} \end{aligned} \qquad n = \begin{cases} z \\ \theta \end{cases} \tag{8.116}$$

As mentioned above, it is also a reasonable assumption that the integral length–scale λD for q_z and q_θ are identical. Adopting the convenient notation

$$\tilde{M}_n = \tilde{m}_n\int\limits_L\phi_n^2dx \qquad n = \begin{cases} z \\ \theta \end{cases} \tag{8.117}$$

and introducing $S_{\hat{R}_n}$ from Eqs. 8.111, then the following is obtained

$$\frac{\sigma_{r_z}}{D} = \frac{\left|\phi_z(x_r)\right|}{2^{7/2}\pi^{7/4}}\frac{\rho BD}{\tilde{m}_z}\frac{\hat{\sigma}_{q_z}}{St^2}\left[\frac{\lambda}{b_z\left(\zeta_z-\zeta_{ae_z}\right)}\right]^{1/2}\frac{\left(D\int\limits_{L_{\exp}}\phi_z^2dx\right)^{1/2}}{\int\limits_L\phi_z^2dx}g_z\left(V_{R_z},V\right) \tag{8.118}$$

$$\sigma_{r\theta} = \frac{|\phi_\theta(x_r)|}{2^{7/2}\pi^{7/4}} \frac{\rho(BD)^2}{\tilde{m}_\theta} \frac{\hat{\sigma}_{q\theta}}{St^2} \left[\frac{\lambda}{b_\theta(\zeta_\theta - \zeta_{ae\theta})} \right]^{1/2} \frac{\left(D \int\limits_{L_{exp}} \phi_\theta^2 dx \right)^{1/2}}{\int\limits_L \phi_\theta^2 dx} g_\theta(V_{R\theta}, V) \tag{8.119}$$

where

$$g_n(V_{R_n}, V) = \left(\frac{V}{V_{R_n}} \right)^{3/2} \cdot \exp\left[-\frac{1}{2} \left(\frac{1 - V_{R_n}/V}{b_n} \right)^2 \right] \quad n = \begin{cases} z \\ \theta \end{cases} \tag{8.120}$$

and where $V_{R_n} = D\omega_n/(2\pi \cdot St)$.

Elaboration 8.7

For the simple case of a single mode across wind response calculation, the standard deviation of the dynamic response at $x = x_r$ is given by:

$$\frac{\sigma_{r_z}(x_r)}{D} = \frac{|\phi_z(x_r)|}{2^{7/2}\pi^{7/4}} \frac{\rho BD}{\tilde{m}_z} \frac{\hat{\sigma}_{q_z}}{St^2} \left[\frac{\lambda}{b_z(\zeta_z - \zeta_{ae_z})} \right]^{\frac{1}{2}} \frac{\left(D \int\limits_{L_{exp}} \phi_z^2 dx \right)^{\frac{1}{2}}}{\int\limits_L \phi_z^2 dx} g_z(V_{R_z}, V)$$

where

$$g_z(V_{R_z}, V) = \left(\frac{V}{V_{R_z}} \right)^{3/2} \cdot \exp\left[-\frac{1}{2} \left(\frac{1 - V_{R_z}/V}{b_z} \right)^2 \right]$$

and $\zeta_{ae_z} = \dfrac{\rho B^2}{4\tilde{m}_z} K_{a_z} \left[1 - \left(\dfrac{\sigma_z}{a_z D} \right)^2 \right] \dfrac{\int\limits_{L_{exp}} \phi_z^2 dx}{\int\limits_L \phi_z^2 dx}$

and where the resonance mean wind velocity $V_{R_z} = \omega_z D/(2\pi St)$. Under these circumstances the equation above may be rewritten into the following fourth order polynomial

$$\hat{\sigma}_{r_z}^4 - (1 - \hat{\zeta})\hat{\sigma}_{r_z}^2 - \hat{\beta}^2 = 0$$

where

$$\hat{\zeta} = \frac{4\tilde{m}_z}{\rho B^2} \frac{\zeta_z}{K_{a_z}} \frac{\int_L \phi_z^2 dx}{\int_{L_{\exp}} \phi_z^2 dx} \quad \text{and} \quad \hat{\beta} = \frac{|\phi_z(x_r)|}{2^{5/2}\pi^{7/4}} \left(\frac{\rho D^3}{\tilde{m}_z \int_L \phi_z^2 dx} \frac{\lambda}{b_z K_{a_z}} \right)^{1/2} \frac{\hat{\sigma}_{q_z}}{St^2} \frac{g_z}{a_z}$$

and where $\hat{\sigma}_{r_z} = \sigma_{r_z} / (a_z D)$. Thus, the reduced standard deviation of the vortex shedding induced dynamic response is given by

$$\hat{\sigma}_{r_z} = \left\{ \frac{1-\hat{\zeta}}{2} + \left[\left(\frac{1-\hat{\zeta}}{2} \right)^2 + \hat{\beta}^2 \right]^{1/2} \right\}^{1/2}$$

Thus, no iterations are required. Similarly, for the simple case of a single mode torsion response calculation, then

$$\hat{\sigma}_{r_\theta} = \left\{ \frac{1-\hat{\zeta}}{2} + \left[\left(\frac{1-\hat{\zeta}}{2} \right)^2 + \hat{\beta}^2 \right]^{1/2} \right\}^{1/2}$$

where

$$\hat{\sigma}_{r_\theta} = \sigma_{r_\theta} / a_\theta$$

and

$$\begin{cases} \hat{\zeta} = \dfrac{4\tilde{m}_\theta}{\rho B^4} \cdot \dfrac{\zeta_\theta}{K_{a\theta}} \cdot \dfrac{\int_L \phi_\theta^2 dx}{\int_{L_{\exp}} \phi_\theta^2 dx} \\ \hat{\beta} = \dfrac{|\phi_\theta(x_r)|}{2^{5/2}\pi^{7/4}} \cdot \left(\dfrac{\rho D^5}{\tilde{m}_\theta \int_L \phi_\theta^2 dx} \cdot \dfrac{\lambda}{b_\theta K_{a\theta}} \right)^{1/2} \cdot \dfrac{\hat{\sigma}_{q\theta}}{St^2} \cdot \dfrac{g_\theta}{a_\theta} \end{cases}$$

Chapter 9
Damping

9.1 Introduction

The introduction of small or moderate damping forces into the theory of structural dynamics is based on the simple observation that any linear elastic system set into a free unloaded motion will harmonically oscillate in a characteristic modeshape which gradually decays until the system again is at rest. The preference of a harmonic modeshape is usually associated with the characteristic largest period (i.e. the lowest eigenfrequency) of the system, as this is its slowest possible free motion, and therefore it requires the least amount of energy exchange between inertia and stress fluctuations. The reason for the decay of the motion (diminishing of kinetic energy) is ascribed to damping forces within the system or between the system and its surrounding air or water. There are several possible sources to these forces within a structural system:

- there are material stress-strain fluctuations causing yielding and cracking in structural elements (at micro as well as at macro level),
- there are sliding (and hence, friction) in support connections, in inner joints and in joints between structural elements and secondary elements such as cladding, inner walls etc.,
- there are stress-strain fluctuations and radiation in adjacent soil and structure interaction,
- there are resistance from surrounding air or water, and
- in earthquake prone areas there may be buildings with inner joints or supports deliberately designed to dissipate energy during large structural motions.

The magnitude of these damping forces will in general depend on

- the relevant static and dynamic stress-strain condition (e.g. bending and shear vs. pure axial strain),
- the type of motion (e.g. at support connections and inner joints), as well as
- temperature and humidity (e.g. in concrete and wood).

Even for the simplest types of structural systems the damping sources are by and large unidentifiable, and for normal buildings it is necessary to rely on overall behaviour. Thus, the choice of properties that may be ascribed to structural damping must in general be based on observations or recordings of decaying

E.N. Strømmen, *Structural Dynamics*, Springer Series in Solid and Structural Mechanics 2,
DOI: 10.1007/978-3-319-01802-7_9,

model experiments or preferably from full scale motion. Since it is very difficult to set a structural system into motion in anything beyond its first eigenmode the few observations that are available are almost exclusively associated with this motion, i.e. most data is strictly spoken only relevant for a motion in the mode shape associated with the first eigenfrequency of the system. Hence, what can be found in the literature are usually the modal damping properties associated with the modal degrees of freedom and not the damping properties associated with the original degrees of freedom in the system. It is considered a reasonable assumption that the damping effects are higher at higher modeshapes.

9.2 Damping Models

In structural dynamics there are two main models of damping, i.e. viscous and friction type of damping forces. The single degree of freedom mathematics of these models is described below. It is usually assumed that the overall physical behaviour of the inner forces of a single degree of freedom system may be replaced by a single material model comprising elastic and damping forces that are additive. Thus, possible material models may be illustrated as shown in Fig. 9.1. In both cases, the material model comprise a spring and a damper in parallel (because they are assumed additive). It is taken for granted that the spring is linear, i.e. that the spring force is proportional to its elongation. To the left in Fig. 9.1 is shown a viscous type of damper. The characteristic property of this damper is that the damper force is proportional to the velocity of its elongation. The origin of this type of material model is usually accredited to William Thomson, 1st Baron Kelvin (1824 – 1907) and Voldemar Voigt (1850 – 1919). To the right in Fig. 9.1 is shown a friction type of damper. Its characteristic property is that the damper elongation is zero until the force has exceeded a certain minimum value, after which the damping force is proportional to a constant friction coefficient, and hence, independent of any further damper elongation. This type of dry friction force is usually accredited to Charles-Augustine de Coulomb (1736 - 1806). It is in the following taken for granted that damping is a small quantity.

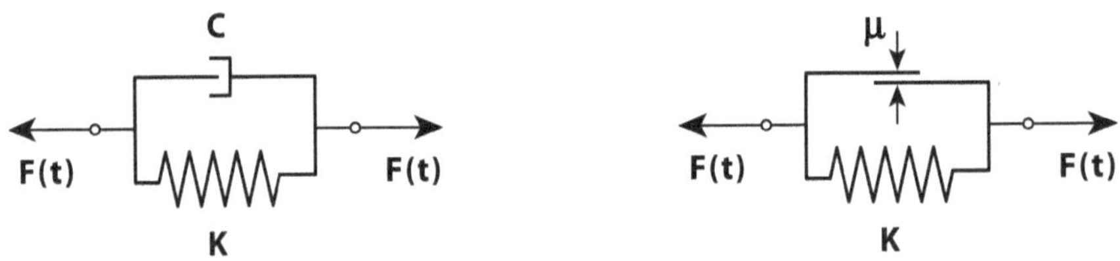

Fig. 9.1 Viscous and friction damping models

Viscous Damping:
Throughout this book it has been taken for granted that it is a viscous type of damping that prevails. It was first touched upon in Chapter 1.6 for a continuous beam (see Fig. 1.21), while a complete solution was developed in Chapter 2.2 for

a freely oscillating single degree of freedom system (see Figs. 2.2 and 2.3). For a real continuous system this solution applies to the case of unloaded free oscillations in any of its mode shapes $\boldsymbol{\varphi}_n(x)$ with corresponding modal properties $\tilde{M}_n$, $\tilde{K}_n$ and $\tilde{C}_n$. Let us consider a simple system (e.g. a beam) whose eigenmodes $\phi_n(x)$ are containing only a single y, z or θ component. Let us for simplicity also assume that at the position of maximum displacement, x_r, the mode shape has been normalised to unity. At this position the system is given an initial displacement r_0 and a velocity $\dot{r}_0$. Then the decaying displacements at x_r is given by $r(x_r,t) = \phi_n(x_r) \cdot \eta_n(t) = \eta_n(t)$, where $\eta_n(t)$ is the solution to the modal equilibrium condition

$$\tilde{M}_n \ddot{\eta}_n(t) + \tilde{C}_n \dot{\eta}_n(t) + \tilde{K}_n \eta_n(t) = 0 \tag{9.1}$$

which is given by (see Eq. 2.28)

$$\eta(t) = a_\eta \cdot e^{-\omega_n \zeta_n t} \cdot \cos(\omega_d t - \beta_n) \tag{9.2}$$

where $\tilde{C}/(2\tilde{M}\omega_n) = \zeta_n$ is the damping ratio (in the following it is taken for granted that $\zeta_n << 1$), $\tilde{K}/\tilde{M} = \omega_n^2$, $\omega_d = \omega_n \sqrt{1-\zeta_n^2}$, and where

$$a_\eta = \sqrt{r_0^2 + \left(\frac{\dot{r}_0}{\omega_d} + \frac{r_0 \zeta_n}{\sqrt{1-\zeta_n^2}} \right)^2} \quad \text{and} \quad \tan\beta_n = \frac{\dot{r}_0}{r_0 \omega_d} + \frac{\zeta_n}{\sqrt{1-\zeta_n^2}} \tag{9.3}$$

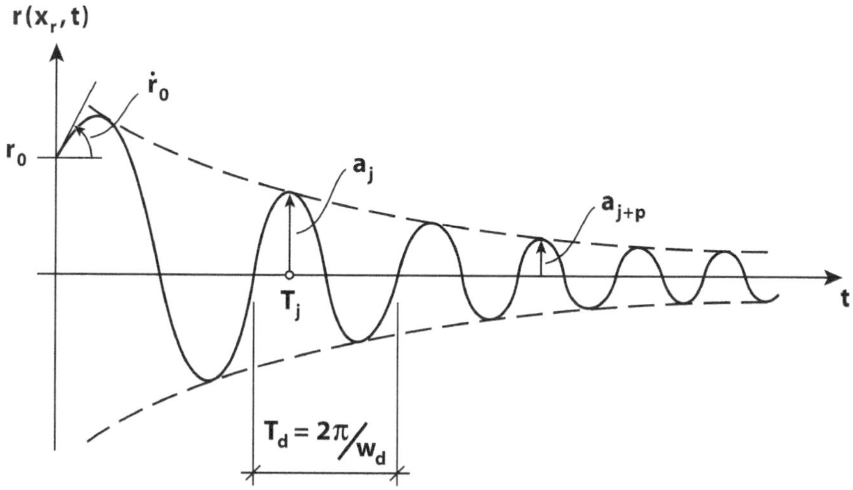

Fig. 9.2 Free decaying motion at x_r

The decaying oscillation at x_r is illustrated in Fig. 9.2 above. The amplitude of motion at $t = T_j$ is given by

$$a_j = a_\eta \cdot e^{-\omega_n \zeta_n T_j} \cdot \cos\left(\omega_d T_j - \beta_n\right) \tag{9.4}$$

The amplitude of motion p periods later, i.e. at $t = T_j + pT_d = T_j + p(2\pi/\omega_d)$ is given by

$$\begin{aligned} a_{j+p} &= a_\eta \cdot e^{-\omega_n \zeta_n (T_j + pT_d)} \cdot \cos\left[\omega_d \left(T_j + p\frac{2\pi}{\omega_d}\right) - \beta_n\right] \\ &= a_\eta \cdot e^{-\omega_n \zeta_n T_j} \cdot e^{-p\omega_n \zeta_n T_d} \cdot \cos\left(\omega_d T_j - \beta_n\right) \end{aligned} \tag{9.5}$$

Recalling that $\omega_d = \omega_n \sqrt{1 - \zeta_n^2}$ it is seen that the natural logarithm to the ratio between a_j and a_{j+p} is given by

$$\ln\left(\frac{a_j}{a_{j+p}}\right) = \ln\left(e^{p\omega_n \zeta_n T_d}\right) = p\omega_n \zeta_n \frac{2\pi}{\omega_d} = p \cdot \frac{2\pi\zeta_n}{\sqrt{1 - \zeta_n^2}} \tag{9.6}$$

The logarithmic decrement of damping associated with mode shape $\phi_n(x)$ is a measure of the decay of the process from one amplitude to the next, i.e. it is defined by

$$\lambda_n = \ln\left(\frac{a_j}{a_{j+1}}\right) = \frac{2\pi\zeta_n}{\sqrt{1 - \zeta_n^2}} \tag{9.7}$$

Since it has been taken for granted that $\zeta_n << 1$, then

$$\lambda_n \approx 2\pi\zeta_n \tag{9.8}$$

The loss of energy during a load cycle due to viscous damping may be illustrated by considering the single degree of freedom system illustrated in Fig. 9.3. The system is subject to a harmonic and perfectly resonant load $F(t) = F_0 \sin \omega_n t$, rendering the steady state harmonic response (see Chapter 2.3)

$$r(t) = r_0 \sin\left(\omega_n t - \pi/2\right) = r_0 \cos \omega_n t \tag{9.9}$$

where $r_0 = F_0 / (2K\zeta_n)$. Introducing $K = \omega_n^2 M$ and $\zeta_n = C/(2M\omega_n)$ it is seen that $r_0 = F_0/(C\omega_n)$.

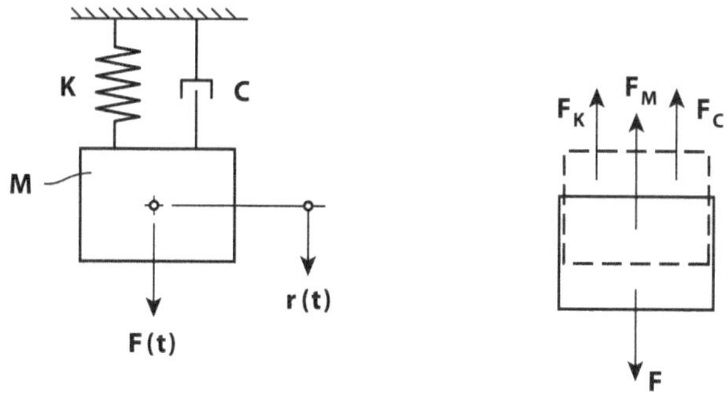

Fig. 9.3 Single degree of freedom system subject to resonant load

The inner elastic force is $F_K = Kr$, while the inner damping force is $F_C = C\dot{r} = C \cdot (-r_0 \omega_n \sin \omega_n t) = -Cr_0 \omega_n \cdot \sqrt{1 - \cos^2 \omega_n t}$. Thus, it is seen that the total inner force $F_i = F_K + F_C$ in the system is given by

$$F_i = Kr - Cr_0 \omega_n \sqrt{1 - \cos^2 \omega_n t} \tag{9.10}$$

which may also be written

$$\left(\frac{F_i - Kr}{Cr_0 \omega_n} \right)^2 + \left(\frac{r}{r_0} \right)^2 = 1 \tag{9.11}$$

or alternatively

$$\left(\frac{F_C}{Cr_0 \omega_n} \right)^2 + \left(\frac{r}{r_0} \right)^2 = 1 \Rightarrow F_C = Cr_0 \omega_n \sqrt{1 - (r/r_0)^2} \tag{9.12}$$

The elliptic function of F_i versus r (Eq. 9.11) is illustrated in Fig. 9.4.a.

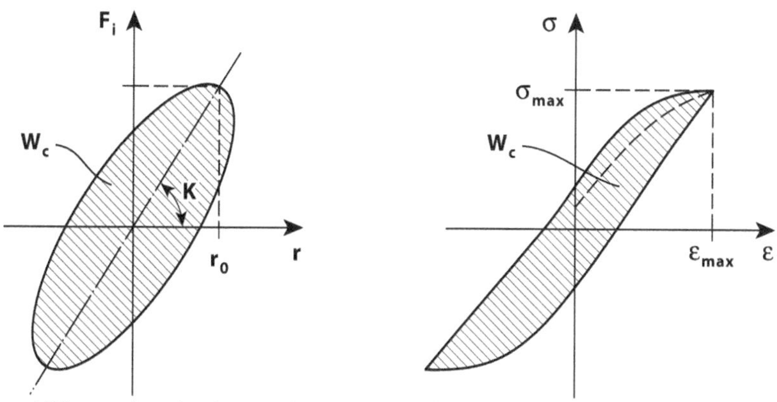

a) Viscous damping hysteresis *b) Material stress-strain hysteresis*

Fig. 9.4 Typical viscous damping and material stress-strain hysteresis

The area W_C within the curve is the viscous energy dissipation within one cycle of motion. For a linearly elastic system W_C may in general be defined by

$$W_C = \oint_s F_C dr \tag{9.13}$$

(where $\oint_s$ indicates a single load cycle integration). In the case above

$$W_C = \oint_s F_C dr = 4\int_0^{r_0} C r_0 \omega_n \sqrt{1-(r/r_0)^2}\, dr = \pi C r_0^2 \omega_n \tag{9.14}$$

The corresponding linear elastic strain energy per cycle of motion is given by

$$W_K = \int_r F_K dr = \int_0^{r_0} Kr dr = \frac{1}{2} K r_0^2 \tag{9.15}$$

The energy dissipation coefficient is then defined by

$$e_D = W_C / (2\pi W_K) \tag{9.16}$$

which, for the case above, may be further developed into

$$e_D = \frac{1}{2\pi}\frac{W_C}{W_K} = \frac{1}{2\pi}\frac{\pi C r_0^2 \omega_n}{K r_0^2/2} = \frac{C\omega_n}{K} = \frac{C\omega_n}{\omega_n^2 M} = \frac{C}{\omega_n M} = 2\zeta_n \tag{9.17}$$

Thus

$$\zeta_n = W_C / (4\pi W_K) \tag{9.18}$$

A typical test recording of material stress-strain hysteresis is illustrated in Fig. 9.4.b. The damping energy dissipation is the area W_C within the hysteresis, while the corresponding linear elastic strain energy is given by

$$W_K = \int_\varepsilon \sigma d\varepsilon = \int_0^{\varepsilon_0} E\varepsilon d\varepsilon = \frac{1}{2} E\varepsilon_0^2 \tag{9.19}$$

The equivalent viscous damping ratio as obtained from such a test recording is then given by

$$\zeta_{eq} = W_C / (4\pi W_K) \tag{9.20}$$

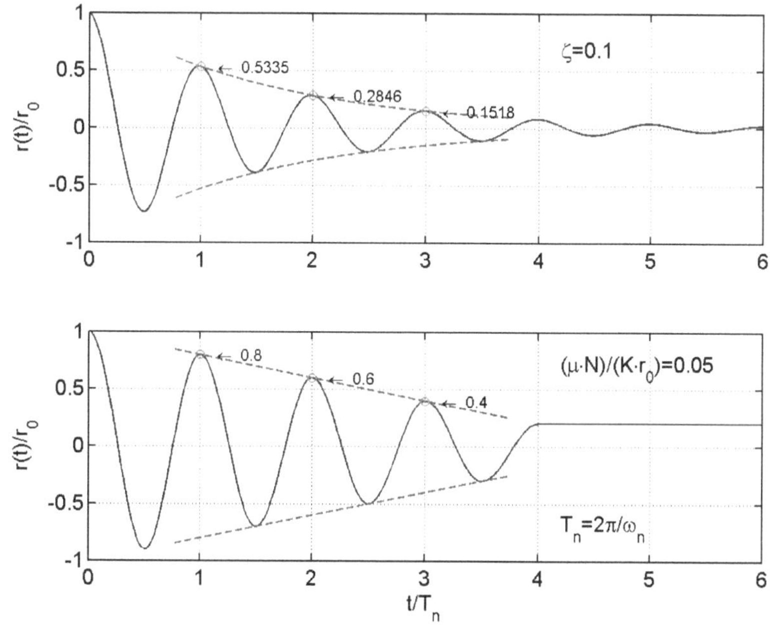

Fig. 9.5 Decaying unloaded motion after initial displacement r_0 and velocity $\dot{r}_0 = 0$; top diagram: viscous damping, lower diagram: friction damping

Setting $\dot{r}_0 = 0$ and $\zeta_n = 0.1$ then a normalised version of the decaying unloaded motion in Fig. 9.2 may be quantified, as shown in the upper curve in Fig. 9.5. Reading off the amplitudes at arbitrary positions it may readily be shown that the rate of decay coincides with $\zeta_n = 0.1$.

Dry Friction (Coulomb Damping):

A single degree of freedom system with a dry friction type of damper is illustrated in Fig. 9.6. The basic idea behind such a damper is that the damping force F_C is proportional to a friction coefficient μ and to the normal force N applied perpendicular to the friction surfaces, i.e.

$$F_C = sign(\dot{r}) \cdot \mu N \text{ where } sign(\dot{r}) = \begin{cases} +1 & \text{if } \dot{r} \geq 0 \\ -1 & \text{if } \dot{r} < 0 \end{cases} \tag{9.21}$$

and that it is independent of the displacement $r(t)$.

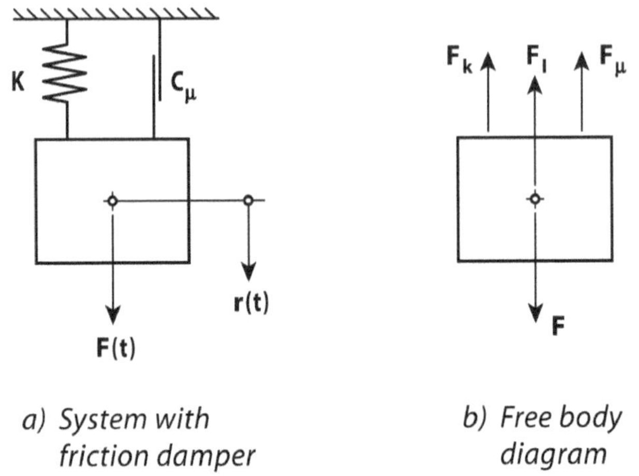

Fig. 9.6 Single degree of freedom system with dry friction type of damper

(It should be noted that a real friction damper designed as indicated above will have a static friction coefficient when the body is at rest which is higher than the friction coefficient it will exhibit when it is in motion. This effect is disregarded below. It is also taken for granted that the friction surface is parallel to the motion of the system.) Thus, the instantaneous equilibrium condition for the system in free unloaded motion ($F = 0$) is given by

$$\left.\begin{aligned} M\ddot{r}(t)+Kr(t)&=-F_C \quad \text{if } \dot{r}\geq 0 \\ M\ddot{r}(t)+Kr(t)&=+F_C \quad \text{if } \dot{r}<0 \end{aligned}\right\} \tag{9.22}$$

The homogeneous solution ($F_C = 0$) in a resonant motion ($\omega = \omega_n$) is given by

$$r = a \cdot \cos\left(\omega_n t - \beta_n\right) \tag{9.23}$$

where the amplitude a and the phase angle β_n will depend on the initial conditions $r(t=0) = r_0$ and $\dot{r}(t=0) = \dot{r}_0$. The particular solution is

$$\left.\begin{aligned} r&=-F_C/K \quad \text{if } \dot{r}\geq 0 \\ r&=+F_C/K \quad \text{if } \dot{r}<0 \end{aligned}\right\} \tag{9.24}$$

Thus,

$$\left.\begin{aligned} r&=a_1\cos\left(\omega_n t-\beta_{n_1}\right)-F_C/K \quad \text{if } \dot{r}\geq 0 \\ r&=a_2\cos\left(\omega_n t-\beta_{n_2}\right)+F_C/K \quad \text{if } \dot{r}<0 \end{aligned}\right\} \tag{9.25}$$

I.e., it is necessary to follow the process as $\dot{r}(t)$ varies between positive and negative values. Let us for instance assume that the system is initially moved at a positive distance $r_0 \gg \mu N/K$ and then let loose to oscillate freely and without any further external forces until it stops all by itself (i.e. $\dot{r}_0 = 0$). The motion is then as illustrated in the lower diagram in Fig. 9.5 [with $\mu N/(K r_0) = 0.05$]. It may be determined the following way:

$$0 \le t \le T_n/2 = \pi/\omega_n : \quad \dot{r} \le 0$$

$$\Rightarrow \begin{cases} r(t=0) = a_2 + F_C/K = r_0 \;\Rightarrow\; a_2 = r_0 - F_C/K \;\&\; \beta_{n2} = 0 \\ r(t) = (r_0 - F_C/K)\cos\omega_n t + F_C/K \\ r(t = T_n/2) = -r_0 + 2F_C/K \end{cases} \tag{9.26}$$

$$T_n/2 \le t \le T_n = 2\pi/\omega_n : \; \dot{r} > 0$$

$$\Rightarrow \begin{cases} r(t = T_n/2) = -a_2 - F_C/K = -r_0 + 2F_C/K \;\Rightarrow\; a_2 = r_0 - 3F_C/K \\ r(t) = (r_0 - 3F_C/K)\cos\omega_n t - F_C/K \\ r(t = T_n) = r_0 - 4F_C/K \end{cases} \tag{9.27}$$

and so on until the spring force is less than the friction force, after which the displacement remains constant. (It may readily be shown that the peak to peak amplitude reduction is in general given by $4F_C/K$.) The hysteresis F_i and F_C versus $r(t)$ is illustrated in Fig. 9.7. It is seen that while the elastic work performed by the spring is still $W_K = K r_0^2/2$, the work performed by the friction damper is given by

$$W_C = 4 r_0 \mu N \tag{9.28}$$

and thus, the equivalent viscous damping ratio is given by

$$\zeta_{eq} = \frac{1}{4\pi}\frac{4 r_0 \mu N}{K r_0^2/2} = \frac{2}{\pi}\frac{\mu N}{K r_0} \tag{9.29}$$

Thus it is seen that ζ_{eq} for a dry friction type of damper is inversely proportional to the amplitude of motion r_0 (i.e., it is likely to be large at small amplitudes of motion).

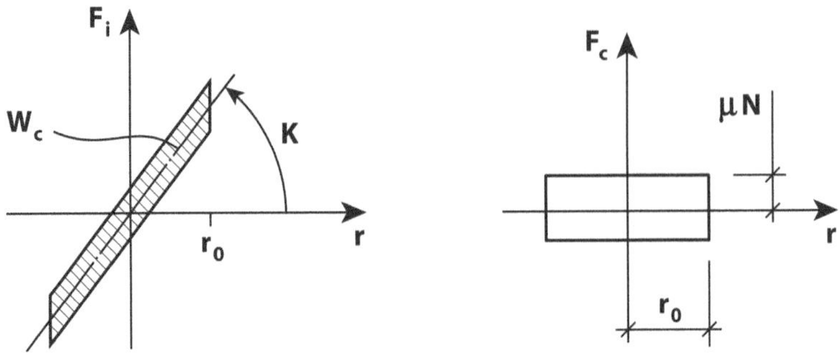

Fig. 9.7 Friction force hysteresis

9.3 Structural Damping

There are several more or less easily identifiable sources to the damping forces that sums up to resist the oscillating motion of a real structure. Throughout this book it is taken for granted that the overall properties of these forces are viscous, i.e. they are linearly proportional to structural velocity. The limited number of sources makes it difficult to give a simple description of what will be representative in a mathematical model of the dynamic properties of real systems. By and large it is necessary to rely on model experiments or full scale observations, e.g. ambient vibration recordings or deliberate tests where the structure is set into a more or less freely oscillating motion. As illustrated in Fig. 9.8 the statistical scatter of the data available for concrete, steel or wooden civil engineering structures is large. The main bulk of available data is related to small amplitude of motion in the first eigenmode of the system. The obvious reason for this is that in higher modes the energy input is rapidly increasing, rendering it unduly demanding to be excited. One of the first major publications regarding structural damping was presented by Lazan [30], who provided basic data for several relevant material damping properties. Full scale observations have been recorded among others by Lagomarsino [31], Çelebi [32], Jeary [33] and Satake et.al. [34]. It is generally agreed that the damping ratio is increasing with increasing eigenfrequency and also with increasing amplitude of motion.

Lagomarsino took his data from fairly small amplitudes of motion, suggesting

$\zeta_1 = \alpha_{-1}\omega_1^{-1} + \alpha_1\omega_1$ where

$$\begin{cases} \alpha_{-1} \approx 0.045 \;\&\; \alpha_1 \approx 0.0011 \text{ for Concrete} \\ \alpha_{-1} \approx 0.02 \;\;\;\&\; \alpha_1 \approx 0.0012 \text{ for Steel} \end{cases} \tag{9.30}$$

while Satake et.al. had data more relevant for earthquake induced large amplitude of motion, suggesting

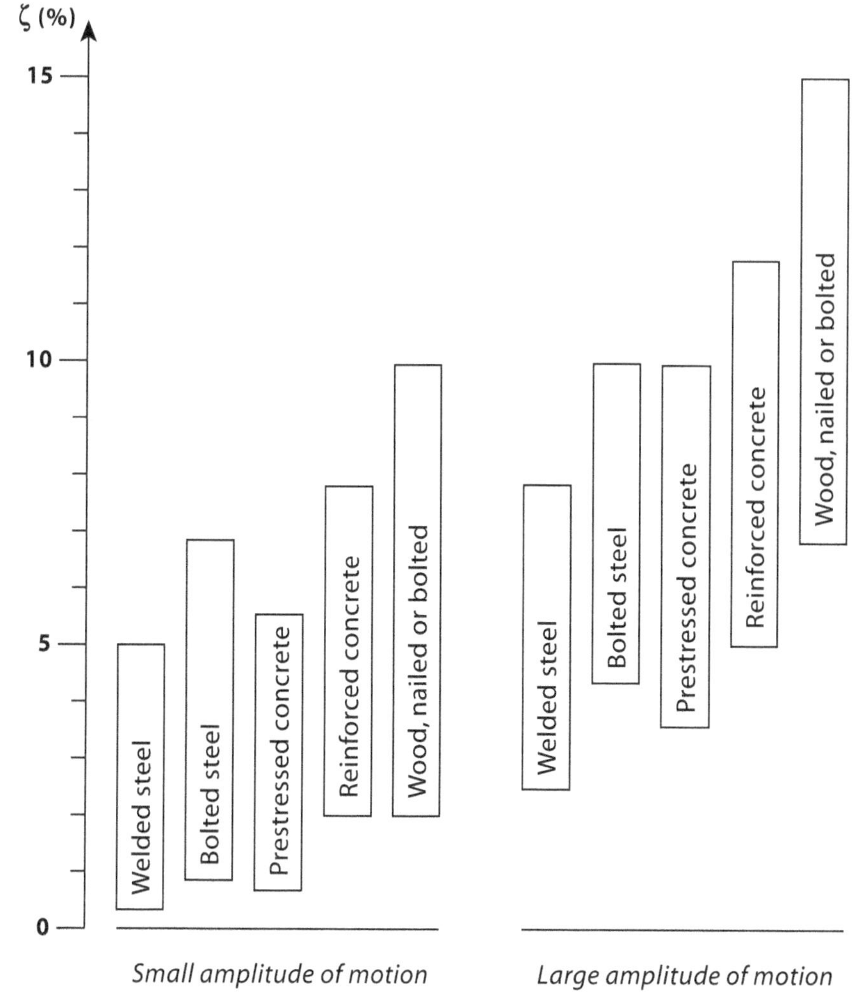

Fig. 9.8 Bandwidth variation of the damping ratio for typical civil engineering structures

$$\left.\begin{aligned} &\zeta_1 = \alpha_0 + \alpha_1\omega_1 + \beta r_{max}/L \\ &r_{max}/L \leq 2\cdot 10^{-5} \end{aligned}\right\} \text{where} \begin{cases} \left.\begin{aligned} &\alpha_0 \approx 0.0018 \\ &\alpha_1 \approx 0.0022 \\ &\beta \approx 470 \end{aligned}\right\} \text{for Concrete} \\ \left.\begin{aligned} &\alpha_0 \approx 0.0029 \\ &\alpha_1 \approx 0.0021 \\ &\beta \approx 400 \end{aligned}\right\} \text{for Steel} \end{cases} \tag{9.31}$$

and where r_{max} is the peak displacement at the top of the building and L is the building height. A similar amplitude dependency was observed by Jeary [33].

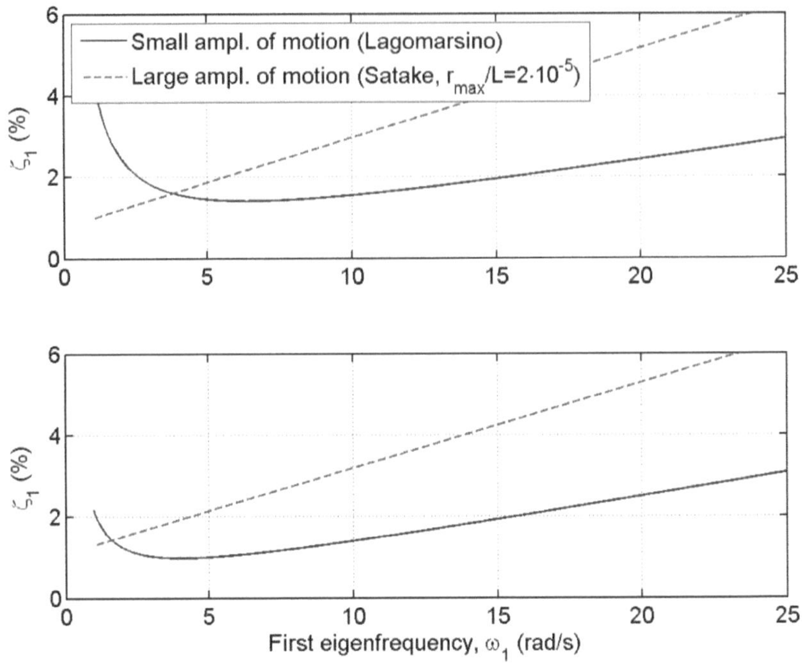

Fig. 9.9 Damping ratio associated with first eigenfrequency (Lagomarsino [31] and Satake et.al. [34]). Upper diagram: reinforced concrete buildings, lower diagram: steel buildings.

Eqs. 9.30 and 9.31 have been plotted in Fig. 9.9. Quantification of the relevant damping ratio should in general be obtained from national or international standards, keeping in mind that what can usually be obtained is only that which is associated with the first or the lower mode shapes of the system. As indicated above, it is usually the lower mode damping properties that are known from experiments and full scale observations. However, in structural dynamics it is not enough to establish the damping ratio for the lowest or the lowest few eigenmodes of the system. In fact, the type of damping coefficients that is required will depend on the type of solution strategy that has been chosen. If the chosen solution strategy is in modal degrees of freedom $\boldsymbol{\eta}$, then

$$\tilde{\mathbf{C}} = \mathbf{\Phi}^T \mathbf{C} \mathbf{\Phi} = diag\left[\tilde{C}_n\right] \text{ where } \tilde{C}_n = 2\tilde{M}_n \omega_n \zeta_n \tag{9.32}$$

and thus, it is necessary to quantify all ζ_n, $n = 1,\ldots,N_{\text{mod}}$, where N_{mod} is the number of modes deemed necessary to be included in the response prediction. If the chosen solution strategy is in original degrees of freedom $\mathbf{r}$, then it is the entire content of the N_r by N_r damping matrix $\mathbf{C}$ (where N_r is the total number of degrees of freedom) that needs to be filled. Apart from choosing a diagonal more or less conservative version (e.g. based on ζ_n), there are two main options:

- to perform a direct inversion from the modal damping matrix and the mass matrix as suggested by Wilson & Penzien [35], or
- to adopt the so-called theory of Rayleigh damping (as suggested by Lord Rayleigh [36]).

The direct development suggested by Wilson & Penzien is based on Eq. 9.32 and that

$$\tilde{\mathbf{M}} = \mathbf{\Phi}^T \mathbf{M} \mathbf{\Phi} = diag\left[\tilde{M}_n\right] \text{ where } \tilde{M}_n = \boldsymbol{\varphi}_n^T \mathbf{M} \boldsymbol{\varphi}_n \tag{9.33}$$

From Eq. 9.32 it is seen that

$$\mathbf{C} = \left(\mathbf{\Phi}^{-1}\right)^T \tilde{\mathbf{C}} \mathbf{\Phi}^{-1} \tag{9.34}$$

But, since $\mathbf{\Phi}$ is non-quadratic, its inversion has no meaning, and hence, it is more convenient to observe from Eq. 9.33 that

$$\mathbf{I} = \tilde{\mathbf{M}}^{-1} \tilde{\mathbf{M}} = \tilde{\mathbf{M}}^{-1} \mathbf{\Phi}^T \mathbf{M} \mathbf{\Phi} \tag{9.35}$$

Thus, since $\mathbf{I} = \mathbf{\Phi}^{-1} \mathbf{\Phi}$ and $\mathbf{M}$ is symmetric it is seen that

$$\mathbf{\Phi}^{-1} = \tilde{\mathbf{M}}^{-1} \mathbf{\Phi}^T \mathbf{M} \Rightarrow \left(\mathbf{\Phi}^{-1}\right)^T = \mathbf{M}^T \mathbf{\Phi} \left(\tilde{\mathbf{M}}^{-1}\right)^T = \mathbf{M} \mathbf{\Phi} \tilde{\mathbf{M}}^{-1} \tag{9.36}$$

Introducing this into Eq. 9.34 and observing that $\tilde{\mathbf{M}} = diag\left[\tilde{M}_n\right]$ and $\tilde{\mathbf{C}} = diag\left[2\tilde{M}_n \omega_n \zeta_n\right]$, then the following is obtained

$$\mathbf{C} = \mathbf{M} \mathbf{\Phi} \tilde{\mathbf{M}}^{-1} \tilde{\mathbf{C}} \tilde{\mathbf{M}}^{-1} \mathbf{\Phi}^T \mathbf{M} = \mathbf{M} \mathbf{\Phi} \tilde{\mathbf{D}} \mathbf{\Phi}^T \mathbf{M} \tag{9.37}$$

where $\tilde{\mathbf{D}} = \tilde{\mathbf{M}}^{-1} \tilde{\mathbf{C}} \tilde{\mathbf{M}}^{-1} = diag\left[2\omega_n \zeta_n / \tilde{M}_n\right]$. Thus, it is necessary to quantify all ζ_n within the frequency domain that is relevant for a sufficiently accurate solution.

Rayleigh damping is based on the simple hypothesis that wherever there is a mass or stiffness contribution to the equilibrium condition in original degrees of freedom (i.e. a mass in motion or elastic stress fluctuations) then there will also be energy dissipation due to what may be covered by a viscous damping model where

$$\mathbf{C} = \alpha \mathbf{M} + \beta \mathbf{K} \tag{9.38}$$

By using the orthogonality properties of the mode shapes and developing $\tilde{\mathbf{C}}$ from this hypothesis, it is seen that

$$\begin{aligned} \tilde{\mathbf{C}} &= \mathbf{\Phi}^T \mathbf{C} \mathbf{\Phi} = \alpha \mathbf{\Phi}^T \mathbf{M} \mathbf{\Phi} + \beta \mathbf{\Phi}^T \mathbf{K} \mathbf{\Phi} = \alpha \tilde{\mathbf{M}} + \beta \tilde{\mathbf{K}} \\ &= diag\left[\alpha \boldsymbol{\varphi}_n^T \mathbf{M} \boldsymbol{\varphi}_n + \beta \boldsymbol{\varphi}_n^T \mathbf{K} \boldsymbol{\varphi}_n \right] = diag\left[\tilde{M}_n \left(\alpha + \beta \omega_n^2 \right) \right] \end{aligned} \tag{9.39}$$

Since $\tilde{\mathbf{C}} = diag\left[2\tilde{M}_n \omega_n \zeta_n \right]$ then $2\tilde{M}_n \omega_n \zeta_n = \tilde{M}_n \left(\alpha + \beta \omega_n^2 \right)$, and thus

$$\zeta_n = \left(\alpha \omega_n^{-1} + \beta \omega_n \right) / 2 \tag{9.40}$$

Since it is usually only first mode data ζ_1 associated with ω_1 that may be found in the literature, it is a common strategy to determine α and β such that this pair will constitute the minimum point of the curve $\zeta_n(\omega_n) = \left(\alpha \omega_n^{-1} + \beta \omega_n \right) / 2$. This may readily be obtained by setting

$$\frac{\partial \zeta_n}{\partial \omega_n} = \frac{1}{2}\left(-\alpha \omega_n^{-2} + \beta \right) = 0 \Rightarrow \alpha = \omega_n^2 \beta \tag{9.41}$$

from which the corresponding β may be obtained from

$$\zeta_n = \left(\omega_n^2 \beta \omega_n^{-1} + \beta \omega_n \right) / 2 \Rightarrow \beta = \zeta_n / \omega_n \tag{9.42}$$

Thus, by choosing $\alpha = \omega_1 \zeta_1$ and $\beta = \zeta_1 / \omega_1$ then

$$\zeta_n = \frac{\zeta_1}{2}\left(\frac{\omega_1}{\omega_n} + \frac{\omega_n}{\omega_1} \right) \tag{9..43}$$

and the coordinates (ζ_1, ω_1) will always be at the minimum point, as illustrated in Fig. 9.10 below.

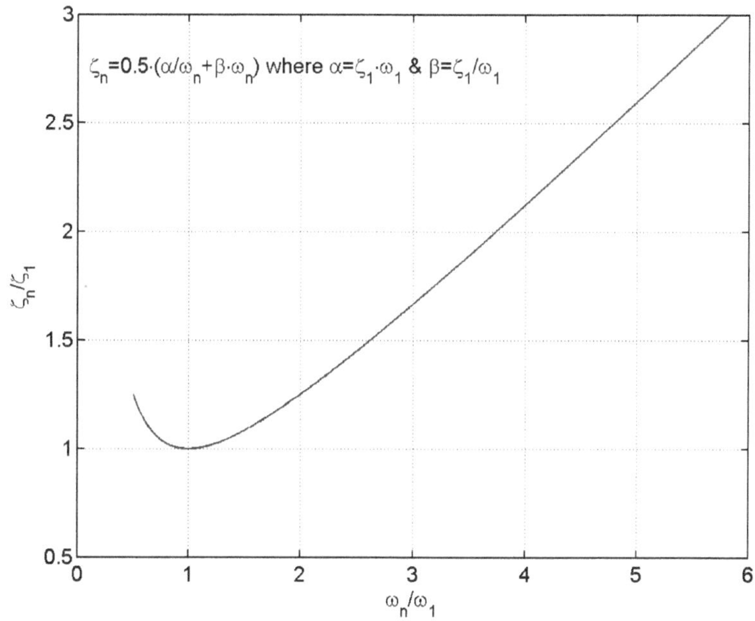

Fig. 9.10 Rayleigh damping with coordinates ζ_1, ω_1 at its minimum

In some cases this strategy may render a too steep or too moderate rate of increase at the upper end of the relevant band of eigenfrequencies. In that case it may be a better strategy to choose two points on the curve, e.g. the coordinates (ω_1, ζ_1) and (ω_2, ζ_2), in which case

$$\left.\begin{aligned}\zeta_1 &= \left(\alpha\omega_1^{-1} + \beta\omega_1\right)/2\\ \zeta_2 &= \left(\alpha\omega_2^{-1} + \beta\omega_2\right)/2\end{aligned}\right\} \Rightarrow \left.\begin{aligned}\alpha &= \frac{2\omega_1\omega_2\left(\omega_2\zeta_1 - \omega_1\zeta_2\right)}{\omega_2^2 - \omega_1^2}\\ \beta &= \frac{2\left(\omega_2\zeta_2 - \omega_1\zeta_1\right)}{\omega_2^2 - \omega_1^2}\end{aligned}\right\} \tag{9.44}$$

With this strategy it is to some extent possible to steer the upper tail of the curve such that the effects of eigenmodes associated with large eigenfrequencies may either fully contribute or effectively be damped out. Though, care must be taken to secure an acceptable position of the minimum point of the curve.

A more general type of damping variation may be obtained by setting

$$\zeta_n = \sum_j a_j \omega_n^j \tag{9.45}$$

From this general expression it is seen that the Rayleigh damping model is obtained by

$$j=-1,1 \Rightarrow \zeta_n = a_{-1}\omega_n^{-1} + a_1\omega_n \tag{9.46}$$

while the so called Caughey damping model [37] is obtained by

$$j=-1,0,1 \Rightarrow \zeta_n = a_{-1}\omega_n^{-1} + a_0 + a_1\omega_n \tag{9.47}$$

This type of damping may be chosen if there are additional damping dependencies, e.g. an increase with increasing amplitude of motion.

9.4 The Tuned Mass Damper

The tuned mass damper (see Elaboration 2.2) may be regarded as an effective way of adding artificial damping into an otherwise lightly damped system. It is in particular a most effective way of damping out resonant or near resonant oscillations. It should be noted that a tuned mass damper will only affect the effective damping properties of the system, it will not affect its stiffness, and thus, it is not helpful to reduce a problem of p quasi static behaviour.

The Tuned Mass Damper for a Single Degree of Freedom System:
The idea of a tuned mass damper may most easily be understood by considering a single degree of freedom system with an additional much smaller mass as illustrated in Fig. 9.11 below.

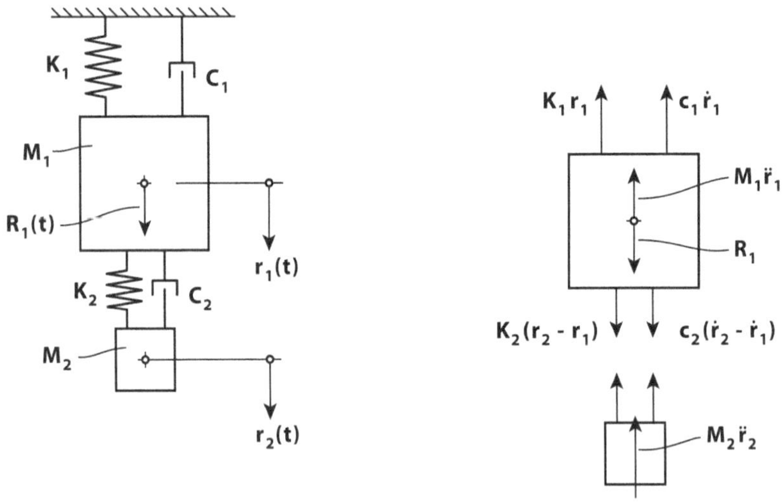

Fig. 9.11 Single degree of freedom system with tuned mass damper

The equilibrium conditions of the two bodies are then given by

$$\left.\begin{aligned} M_1\ddot{r}_1 + C_1\dot{r}_1 - C_2\left(\dot{r}_2 - \dot{r}_1\right) + K_1 r_1 - K_2\left(r_2 - r_1\right) - R_1 = 0 \\ M_2\ddot{r}_2 + C_2\left(\dot{r}_2 - \dot{r}_1\right) + K_2\left(r_2 - r_1\right) = 0 \end{aligned}\right\} \tag{9.48}$$

which may also be written

$$\begin{bmatrix} M_1 & 0 \\ 0 & M_2 \end{bmatrix}\begin{bmatrix} \ddot{r}_1 \\ \ddot{r}_2 \end{bmatrix} + \begin{bmatrix} C_1 + C_2 & -C_2 \\ -C_2 & C_2 \end{bmatrix}\begin{bmatrix} \dot{r}_1 \\ \dot{r}_2 \end{bmatrix} + \begin{bmatrix} K_1 + K_2 & -K_2 \\ -K_2 & K_2 \end{bmatrix}\begin{bmatrix} r_1 \\ r_2 \end{bmatrix} = \begin{bmatrix} R_1 \\ 0 \end{bmatrix} \tag{9.49}$$

or

$$\mathbf{M}_0\ddot{\mathbf{r}}_0 + \mathbf{C}_0\dot{\mathbf{r}}_0 + \mathbf{K}_0\mathbf{r}_0 = \mathbf{R}_0 \tag{9.50}$$

where: $\mathbf{r}_0 = \begin{bmatrix} r_1 & r_2 \end{bmatrix}^T$, $\mathbf{R}_0 = \begin{bmatrix} R_1 & 0 \end{bmatrix}^T$, and

$$\mathbf{M}_0 = \begin{bmatrix} M_1 & 0 \\ 0 & M_2 \end{bmatrix} \quad \mathbf{C}_0 = \begin{bmatrix} C_1 + C_2 & -C_2 \\ -C_2 & C_2 \end{bmatrix} \text{ and } \mathbf{K}_0 = \begin{bmatrix} K_1 + K_2 & -K_2 \\ -K_2 & K_2 \end{bmatrix}$$

For simplicity, let us introduce $r_1 = r$, and, rather that operating on the total displacements r_2 of the additional mass, it is convenient to introduce the relative displacement $\Delta r = r_2 - r_1$, i.e. that

$$\begin{bmatrix} r_1 \\ r_2 \end{bmatrix} = \begin{bmatrix} r \\ r + \Delta r \end{bmatrix} = \begin{bmatrix} 1 & 0 \\ 1 & 1 \end{bmatrix}\begin{bmatrix} r \\ \Delta r \end{bmatrix} \tag{9.51}$$

which may alternatively be written

$$\mathbf{r}_0 = \mathbf{\Psi r} \text{ where } \mathbf{\Psi} = \begin{bmatrix} 1 & 0 \\ 1 & 1 \end{bmatrix} \text{ and } \mathbf{r} = \begin{bmatrix} r \\ \Delta r \end{bmatrix} \tag{9.52}$$

Introducing this into Eq. 9.50 and pre-multiplying by $\mathbf{\Psi}^T$

$$\mathbf{\Psi}^T\mathbf{M}_0\mathbf{\Psi}\ddot{\mathbf{r}} + \mathbf{\Psi}^T\mathbf{C}_0\mathbf{\Psi}\dot{\mathbf{r}} + \mathbf{\Psi}^T\mathbf{K}_0\mathbf{\Psi}\mathbf{r} = \mathbf{\Psi}^T\mathbf{R}_0 \tag{9.53}$$

then the following is obtained

$$\mathbf{M}\ddot{\mathbf{r}} + \mathbf{C}\dot{\mathbf{r}} + \mathbf{K}\mathbf{r} = \mathbf{R} \tag{9.54}$$

where

$$\mathbf{M} = \mathbf{\Psi}^T \mathbf{M}_0 \mathbf{\Psi} = \begin{bmatrix} M_1 + M_2 & M_2 \\ M_2 & M_2 \end{bmatrix}, \quad \mathbf{K} = \mathbf{\Psi}^T \mathbf{K}_0 \mathbf{\Psi} = \begin{bmatrix} K_1 & 0 \\ 0 & K_2 \end{bmatrix}$$

$$\mathbf{C} = \mathbf{\Psi}^T \mathbf{C}_0 \mathbf{\Psi} = \begin{bmatrix} C_1 & 0 \\ 0 & C_2 \end{bmatrix} \quad \text{and} \quad \mathbf{R} = \mathbf{\Psi}^T \mathbf{R}_0 = \begin{bmatrix} R_1(t) \\ 0 \end{bmatrix}$$

Basically, this is a two degrees of freedom system identical to that which has been dealt with in Chapter 2.6 and whose eigenvalue problem in original degrees of freedom $\mathbf{r}_0 = \begin{bmatrix} r_1 & r_2 \end{bmatrix}^T$ was solved in Chapter 1.2, see Eqs. 1.20 – 1.22. In the present case of relative degrees of freedom the eigenvalue problem is obtained from Eq. 9.54 by setting $\mathbf{C} = \mathbf{0}$, $\mathbf{R} = \mathbf{0}$ and $\mathbf{r} = \boldsymbol{\varphi} e^{i\omega t}$, where $\boldsymbol{\varphi} = \begin{bmatrix} \phi_1 & \phi_2 \end{bmatrix}^T$, and after pre-multiplication by $\mathbf{K}^{-1}$ then the following is obtained

$$\left(\mathbf{I} - \omega^2 \mathbf{K}^{-1} \mathbf{M} \right) \boldsymbol{\varphi} = \mathbf{0} \tag{9.55}$$

where

$$\mathbf{K}^{-1}\mathbf{M} = \begin{bmatrix} 1/K_1 & 0 \\ 0 & 1/K_2 \end{bmatrix} \begin{bmatrix} M_1 + M_2 & M_2 \\ M_2 & M_2 \end{bmatrix} \tag{9.56}$$

Thus

$$\begin{bmatrix} 1 - \omega^2 \dfrac{M_1 + M_2}{K_1} & -\omega^2 \dfrac{M_2}{K_1} \\ -\omega^2 \dfrac{M_2}{K_2} & 1 - \omega^2 \dfrac{M_2}{K_2} \end{bmatrix} \begin{bmatrix} \phi_1 \\ \phi_2 \end{bmatrix} = \mathbf{0} \tag{9.57}$$

rendering

$$\omega^2 = \frac{\dfrac{M_1 + M_2}{K_1} + \dfrac{M_2}{K_2} \pm \sqrt{\left(\dfrac{M_1 + M_2}{K_1} + \dfrac{M_2}{K_2} \right)^2 - 4 \dfrac{M_1}{K_1} \dfrac{M_2}{K_2}}}{2 \dfrac{M_1}{K_1} \dfrac{M_2}{K_2}} \tag{9.58}$$

(which is identical to that which was obtained in Eq. 1.20). For a tuned mass damper it may be taken for granted that $M_2 \ll M_1$, and thus

$$\omega^2 \approx \frac{\frac{M_1}{K_1}+\frac{M_2}{K_2} \pm \left(\frac{M_1}{K_1}-\frac{M_2}{K_2}\right)}{2\frac{M_1}{K_1}\frac{M_2}{K_2}} \Rightarrow \begin{cases} \omega_1 \approx \sqrt{K_1/M_1} \\ \omega_2 \approx \sqrt{K_2/M_2} \end{cases} \tag{9.59}$$

Pre-multiplication by $\mathbf{K}^{-1}$ and taking the Fourier transform throughout Eq. 9.54, i.e. setting

$$\mathbf{r}(t) = \sum_{\omega} \mathbf{a}_r(\omega) \cdot e^{i\omega t} \text{ and } \mathbf{R} = \sum_{\omega} \mathbf{a}_R(\omega) \cdot e^{i\omega t} \tag{9.60}$$

where

$$\mathbf{a}_r(\omega) = \begin{bmatrix} a_r & a_{\Delta r} \end{bmatrix}^T \text{ and } \mathbf{a}_R(\omega) = \begin{bmatrix} a_{R_1} & 0 \end{bmatrix}^T \tag{9.61}$$

are the Fourier coefficient vectors of $\mathbf{r}$ and $\mathbf{R}$, then the following is obtained

$$\mathbf{a}_r(\omega) = \hat{\mathbf{H}}(\omega) \cdot \begin{bmatrix} a_{R_1}(\omega)/K_1 \\ 0 \end{bmatrix} \tag{9.62}$$

where

$$\hat{\mathbf{H}}^{-1} = \mathbf{I} + i\omega\mathbf{K}^{-1}\mathbf{C} - \omega^2\mathbf{K}^{-1}\mathbf{M} = \begin{bmatrix} D_1 & -\mu(\omega/\omega_1)^2 \\ -(\omega/\omega_2)^2 & D_2 \end{bmatrix} \tag{9.63}$$

and

$$\left.\begin{aligned} D_1(\omega) &= 1-(1+\mu)(\omega/\omega_1)^2 + 2i\zeta_1\,\omega/\omega_1 \\ D_2(\omega) &= 1-(\omega/\omega_2)^2 + 2i\zeta_2\,\omega/\omega_2 \end{aligned}\right\}$$

where

$$\left.\begin{aligned} \mu &= M_2/M_1 \\ \zeta_1 &= C_1/(2M_1\omega_1) \\ \zeta_2 &= C_2/(2M_2\omega_2) \end{aligned}\right\} \tag{9.64}$$

Thus

$$\hat{\mathbf{H}}(\omega)=\frac{1}{D_1D_2-\mu(\omega/\omega_1)^2(\omega/\omega_2)^2}\begin{bmatrix} D_2 & \mu(\omega/\omega_1)^2 \\ (\omega/\omega_2)^2 & D_1 \end{bmatrix} \tag{9.65}$$

Defining

$$\left.\begin{aligned} \hat{\omega} &= \omega/\omega_1 \\ \alpha &= \omega_1/\omega_2 \end{aligned}\right\} \tag{9.66}$$

and

$$\left.\begin{aligned} &a_1=2(\zeta_1+\alpha\zeta_2) && b_1=2\alpha\zeta_2 && \\ &a_2=1+\mu+\alpha^2+4\alpha\zeta_1\zeta_2 && b_2=\alpha^2 && d_2=-\alpha^2 \\ &a_3=2\alpha\left[\alpha\zeta_1+(1+\mu)\zeta_2\right] && c_1=2\zeta_1 && e_2=-\mu \\ &a_4=\alpha^2 && c_2=1+\mu && \end{aligned}\right\} \tag{9.67}$$

then Eq. 9.65 may be written in the following more convenient way

$$\hat{\mathbf{H}}(\hat{\omega})=\begin{bmatrix} \hat{H}_{11}(\hat{\omega}) & \hat{H}_{12}(\hat{\omega}) \\ \hat{H}_{21}(\hat{\omega}) & \hat{H}_{22}(\hat{\omega}) \end{bmatrix} \tag{9.68}$$

where

$$\left.\begin{aligned} \hat{H}_{11}(\hat{\omega}) &= \frac{1+b_1(i\hat{\omega})+b_2(i\hat{\omega})^2}{1+a_1(i\hat{\omega})+a_2(i\hat{\omega})^2+a_3(i\hat{\omega})^3+a_4(i\hat{\omega})^4} \\ \hat{H}_{22}(\hat{\omega}) &= \frac{1+c_1(i\hat{\omega})+c_2(i\hat{\omega})^2}{1+a_1(i\hat{\omega})+a_2(i\hat{\omega})^2+a_3(i\hat{\omega})^3+a_4(i\hat{\omega})^4} \\ \hat{H}_{21}(\hat{\omega}) &= \frac{d_2(i\hat{\omega})^2}{1+a_1(i\hat{\omega})+a_2(i\hat{\omega})^2+a_3(i\hat{\omega})^3+a_4(i\hat{\omega})^4} \\ \hat{H}_{12}(\hat{\omega}) &= \frac{e_2(i\hat{\omega})^2}{1+a_1(i\hat{\omega})+a_2(i\hat{\omega})^2+a_3(i\hat{\omega})^3+a_4(i\hat{\omega})^4} \end{aligned}\right\} \tag{9.69}$$

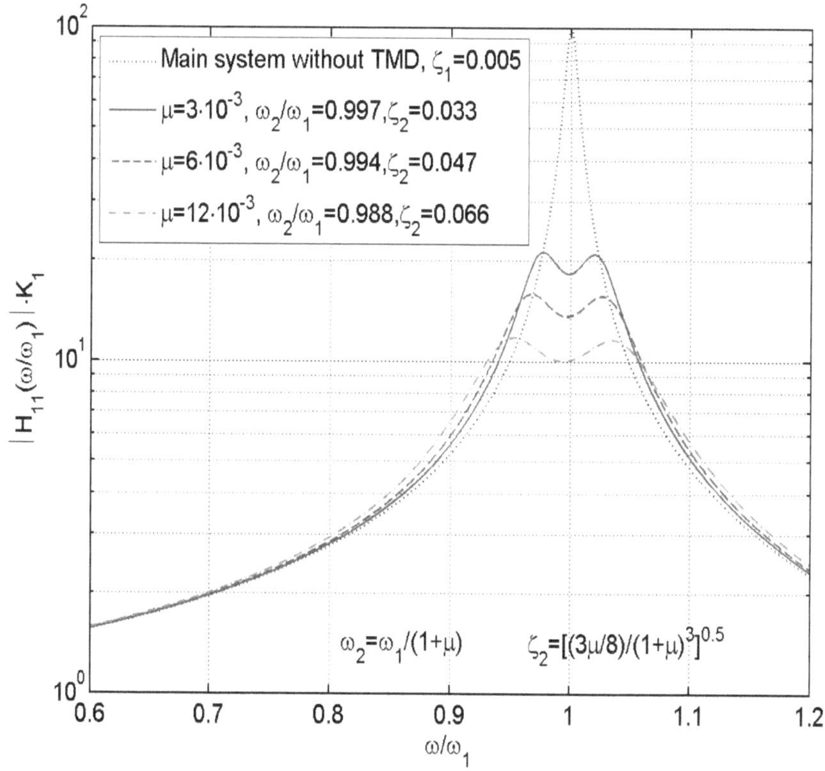

Fig. 9.12 The frequency response function for a single degree of freedom system with a tuned mass damper

Then the idea behind the tuned mass damper reveals itself by looking at a plot of the absolute values of the non-dimensional frequency response function $\hat{H}_{11}(\hat{\omega})$, which is directly associated with the dynamic response of the main system r, see Fig. 9.12 above. As can be seen, the addition of a small damper is equivalent to adding damping into the main system, and even when the mass ratio $\mu = M_2/M_1$ is only one percent or below its effect is significant. The response spectral density matrix $\mathbf{S}_r(\omega)$, containing the spectral density of $r(t)$ and $\Delta r(t)$ on its diagonal and the cross spectral densities between $r(t)$ and $\Delta r(t)$ on its off-diagonal terms, is defined by

$$\mathbf{S}_r(\omega)=\begin{bmatrix} S_{rr} & S_{r\Delta r} \\ S_{\Delta rr} & S_{\Delta r\Delta r}\end{bmatrix}=\lim_{T\to\infty}\frac{1}{\pi T}\mathbf{a}_r^*\mathbf{a}_r^T=\lim_{T\to\infty}\frac{1}{\pi T}\begin{bmatrix} a_r^* a_r & a_r^* a_{\Delta r} \\ a_{\Delta r}^* a_r & a_{\Delta r}^* a_{\Delta r}\end{bmatrix}$$
$$=\lim_{T\to\infty}\frac{1}{\pi T}\left(\hat{\mathbf{H}}(\omega)\cdot\begin{bmatrix} a_{R_1}(\omega)/K_1 \\ 0\end{bmatrix}\right)^*\left(\hat{\mathbf{H}}(\omega)\cdot\begin{bmatrix} a_{R_1}(\omega)/K_1 \\ 0\end{bmatrix}\right)^T \tag{9.70}$$
$$=\frac{S_{R_1}(\omega)}{K_1^2}\cdot\begin{bmatrix} \left|\hat{H}_{11}(\hat{\omega})\right|^2 & \hat{H}_{11}^*(\hat{\omega})\cdot\hat{H}_{21}(\hat{\omega}) \\ \hat{H}_{21}^*(\hat{\omega})\cdot\hat{H}_{11}(\hat{\omega}) & \left|\hat{H}_{21}(\hat{\omega})\right|^2\end{bmatrix}$$

where

$$S_{R_1}(\omega)=\lim_{T\to\infty}\frac{1}{\pi T}a_{R_1}^*(\omega)\cdot a_{R_1}(\omega) \tag{9.71}$$

is the spectral density of the load acting on the main system. Let us for simplicity assume a fairly broad banded load and a near to resonant response, in which case the response variances σ_r^2 and $\sigma_{\Delta r}^2$ are given by (see Newland, [38])

$$\sigma_r^2=\int_0^\infty S_{rr}(\omega)d\omega\approx\frac{S_{R_1}(\omega_1)}{K_1^2}\cdot\int_0^\infty\left|\hat{H}_{11}(\hat{\omega})\right|^2 d\omega$$
$$=\frac{\pi S_{R_1}(\omega_1)}{2K_1^2}\cdot\frac{a_2a_3-a_1a_4+a_1b_2^2+a_3\left(b_1^2-2b_2\right)}{a_1a_2a_3-a_1^2a_4-a_3^2} \tag{9.72}$$

and

$$\sigma_{\Delta r}^2=\int_0^\infty S_{\Delta r\Delta r}(\omega)d\omega\approx\frac{S_{R_1}(\omega_1)}{K_1^2}\int_0^\infty\left|\hat{H}_{21}(\hat{\omega})\right|^2 d\omega=\frac{\pi S_{R_1}(\omega_1)}{2K_1^2}\frac{a_1d_2^2}{a_1a_2a_3-a_1^2a_4-a_3^2} \tag{9.73}$$

where all the constants are defined in Eqs. 9.67 – 9.69.

Elaboration 9.1: Optimum Choice of Damper Properties
In general there is no mathematical optimum choice of damper properties. In most cases it is a matter of how large a damper it is convenient (or possible) to include into the design of the main system (what can usually be expected is μ in the range between 0.005 and 0.05). However,

Den Hartog [26] recommends $\omega_2 = \dfrac{\omega_1}{\alpha} = \dfrac{\omega_1}{1+\mu}$ and

$$\zeta_2 = \sqrt{\frac{3\mu}{8(1+\mu)^3}}$$

while R. Luft [39] recommends $\omega_2 = \dfrac{\omega_1}{\alpha} = \dfrac{\omega_1}{\sqrt{1+3\mu/2}}$ and $\zeta_2 = \sqrt{\dfrac{\mu}{4}\left(1-\dfrac{3\mu}{4}\right)}$

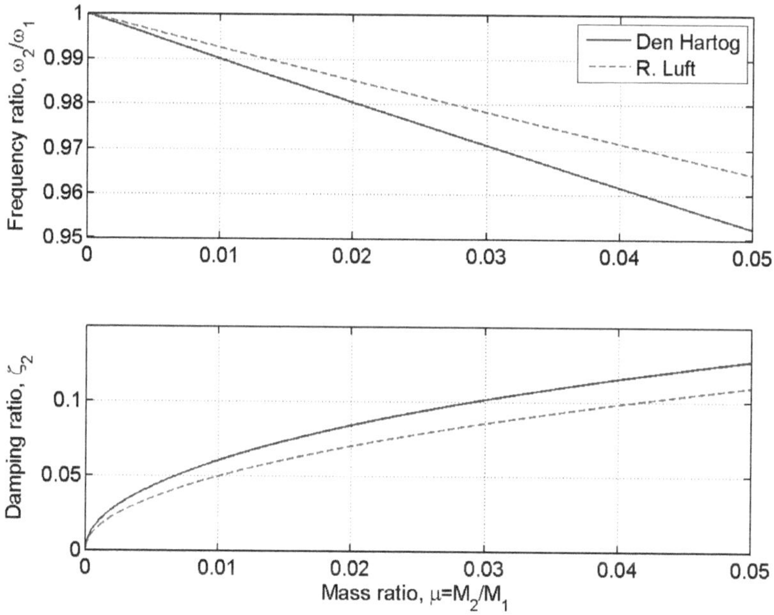

Fig. 9.13 Recommended damper properties

The Tuned Mass Damper in a Continuous Line-Like System:

For a real system it may be necessary to install dampers specially designed to provide artificial damping to several modes. Let us first consider the case of a continuous beam or column with dampers M_j, $j = 1, 2, \ldots, N_j$, at corresponding positions x_j, as shown in Fig. 9.14. The beam itself is subject to a distributed load $q_z(x,t)$ while it is taken for granted that the tuned mass dampers are unloaded. For simplicity, it is in the following also taken for granted that the motion of the beam is a single component displacement in the z-direction. It is also taken for granted that the damper masses are small, and that the motion of the system itself may modally be described by $r_z(x,t) = \mathbf{\Phi}(x) \cdot \boldsymbol{\eta}(t)$, where $\mathbf{\Phi}$ contains mode shapes from an eigenvalue solution in original coordinates (with or without the dampers, assuming $\mu_j = M_j / \tilde{M}_n \ll 1$).

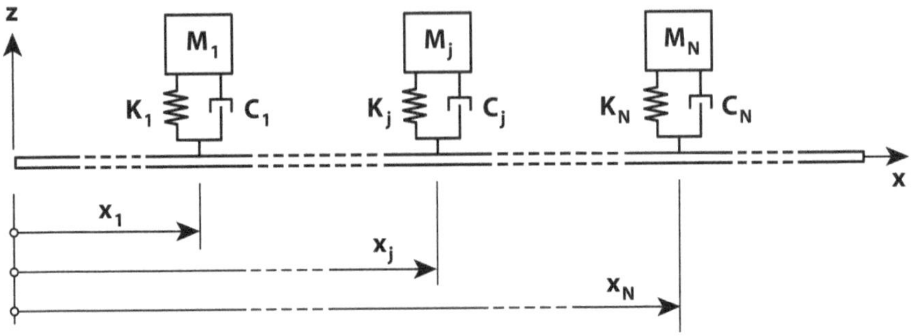

Fig. 9.14 Beam (or column) with dampers M_j at positions x_j

At arbitrary position x and time t the system is then in motion $r_z(x,t)$, see Fig. 9.15.a, while at the same time, an arbitrary mass damper M_j at position x_j is in motion $r_{j_z}(t)$, see Fig. 9.15.b. At this particular time the internal forces in the beam as well as in the mass damper are illustrated in Fig. 9.16.a and b, and the instantaneous equilibrium condition of the entire system may then be established by adopting d'Alambert's principle and the principle of virtual displacements, see Chapter 1.6. I.e., as illustrated in Fig. 9.15 the beam is given a virtual displacement $\delta r_z(x)$ and the damper is given a virtual displacement δr_{j_z}. During this virtual displacement the total energy of the system has not been changed, and thus, external work must equal internal work, i.e.

$$\begin{aligned} &\int_L \left[q_z(x,t) - m_z(x)\ddot{r}_z(x,t) - c_z(x)\dot{r}_z(x,t) \right] \delta r_z(x)\,dx + \sum_j F_j(t)\,\delta r_z(x_j) \\ &- \sum_j \left[M_j \ddot{r}_j(t) + F_j(t) \right] \delta r_{j_z} = \int_L \int_A \sigma_x(x,z,t)\,\delta\varepsilon_x(x,z,t)\,dA\,dx \end{aligned} \tag{9.74}$$

where q_z, m_z and c_z are distributed load, mass and viscous damping coefficient (i.e. per unit length), σ_x is normal stress associated with r_z, $\delta\varepsilon_x$ is normal strain associated with δr_z and

$$\begin{aligned} F_j(t) &= C_j\left[\dot{r}_{j_z}(t) - \dot{r}_z(x_j,t) \right] + K_j\left[r_{j_z}(t) - r_z(x_j,t) \right] \\ &= C_j \dot{r}_{j_z}(t) + K_j r_{j_z}(t) - \left[C_j \dot{r}_z(x_j,t) + K_j r_z(x_j,t) \right] \end{aligned} \tag{9.75}$$

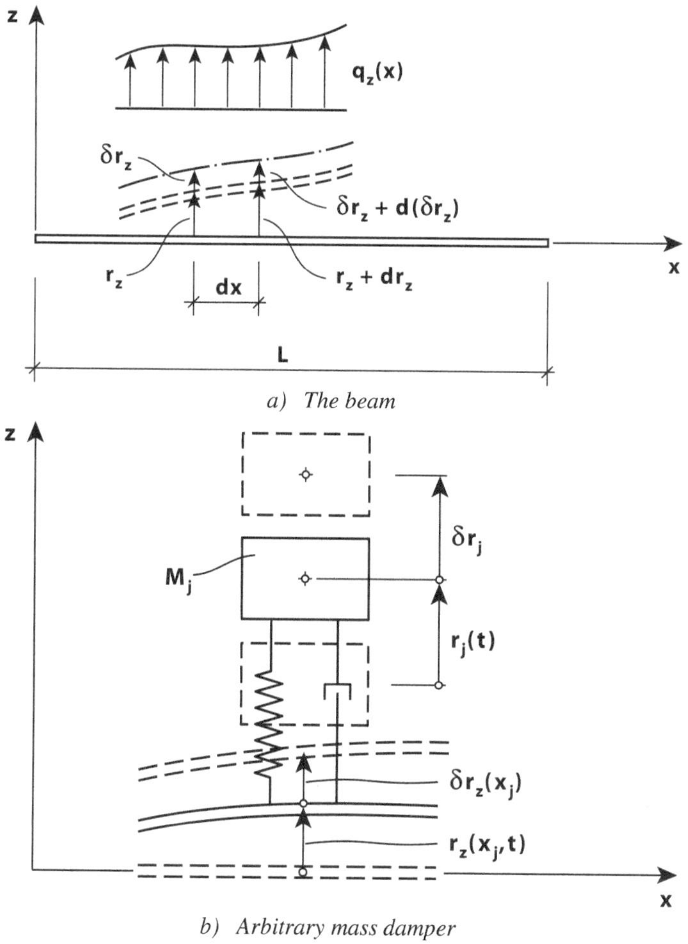

Fig. 9.15 Beam and mass damper motion and virtual displacement

where C_j and K_j are the damping and stiffness properties of the jth mass damper. From Chapter 1.2 (see Eqs. 1.26 – 1.27) we know that

$$\left.\begin{aligned} \sigma_x &= \frac{M_y(x,t)}{I_y} z = \frac{-EI_y r_z''(x,t)}{I_y} z = -Er_z''(x,t)\cdot z \\ \delta\varepsilon_x &= -\delta r_z''(x)\cdot z \end{aligned}\right\} \tag{9.76}$$

and thus

$$\begin{aligned}&\int_L\int_A \sigma_x(x,z,t)\,\delta\varepsilon_x(x,z,t)\,dA\,dx = \int_L\int_A \left[-Er_z''(x,t)\cdot z\right]\left[-\delta r_z''(x)\cdot z\right]dA\,dx\\&= \int_L Er_z''(x,t)\cdot\delta r_z''(x)\int_A z^2 dA\,dx = \int_L EI_y r_z''(x,t)\cdot\delta r_z''(x)\,dx\end{aligned} \tag{9.77}$$

Introducing this into Eq. 9.74, then the following is obtained (see Eq. 1.119):

$$\begin{aligned}&\int_L m_z(x)\cdot\ddot{r}_z(x,t)\cdot\delta r_z(x)\,dx + \int_L c_z(x)\cdot\dot{r}_z(x,t)\cdot\delta r_z(x)\,dx\\&+\int_L EI_y\cdot r_z''(x,t)\cdot\delta r_z''(x)\,dx - \sum_j F_j(t)\cdot\delta r_z(x_j)\\&+\sum_j\left[M_j\cdot\ddot{r}_{j_z}(t)+F_j(t)\right]\cdot\delta r_{j_z} = \int_L q_z(x,t)\cdot\delta r_z(x)\,dx\end{aligned} \tag{9.78}$$

Let us first consider the most simple case of a single mode (and single component) $\phi_z(x)$ approach with only one mass damper (M_1, C_1, K_1) at position x_1 within the span of the beam. Then (as the motion of the damper alone will represent a mode shape of its own)

$$\left.\begin{aligned}r_z(x,t)&=\phi_z(x)\cdot\eta_z(t)\\r_1(t)&=1\cdot\eta_1(t)\end{aligned}\right\} \tag{9.79}$$

and correspondingly, we choose virtual displacements

$$\left.\begin{aligned}\delta r_z(x)&=\phi_z(x)\cdot\delta\eta_z\\\delta r_1&=1\cdot\delta\eta_1\end{aligned}\right\} \tag{9.80}$$

Introducing this into Eq. 9.78, then the following is obtained

$$\begin{aligned}&\int_L m_z(x)\phi_z(x)\ddot{\eta}_z(t)\phi_z(x)\delta\eta_z dx + \int_L c_z(x)\phi_z(x)\dot{\eta}_z(t)\phi_z(x)\delta\eta_z dx\\&+\int_L EI_y\phi_z''(x)\eta_z(t)\phi_z''(x)\delta\eta_z dx\\&-\left\{C_1\dot{\eta}_1(t)+K_1\eta_1(t)-\left[C_1\phi_z(x_1)\dot{\eta}_z(t)+K_1\phi_z(x_1)\eta_z(t)\right]\right\}\phi_z(x_1)\delta\eta_z\\&+\left\{M_1\ddot{\eta}_1+C_1\dot{\eta}_1(t)+K_1\eta_1(t)-\left[C_1\phi_z(x_1)\dot{\eta}_z(t)+K_1\phi_z(x_1)\eta_z(t)\right]\right\}\delta\eta_1\\&=\int_L q_z(x,t)\phi_z(x)\delta\eta_z dx\end{aligned} \tag{9.81}$$

which may more conveniently be written

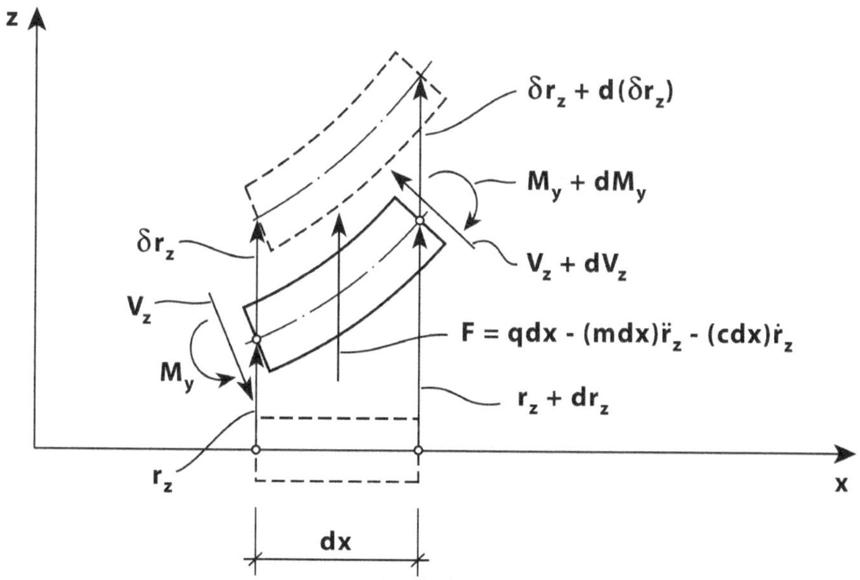

a) The beam

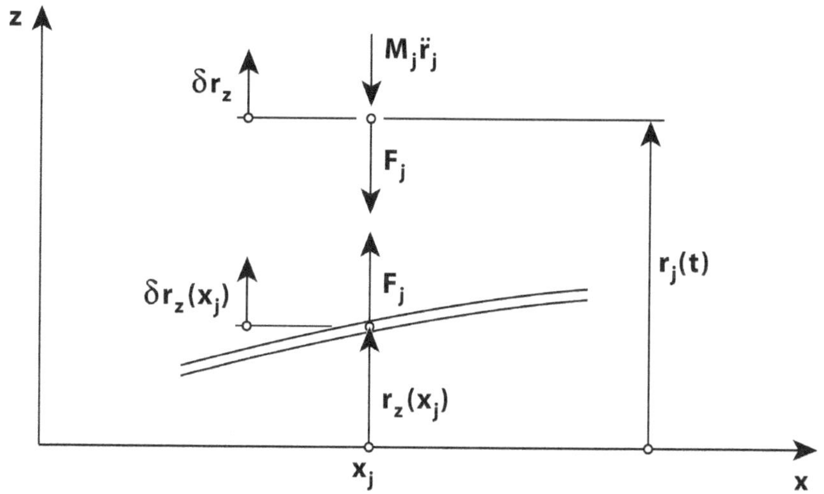

b) Arbitrary mass damper

Fig. 9.16 Internal beam and mass damper forces

$$\delta\boldsymbol{\eta}_z^T\left[\tilde{\mathbf{M}}_{z0}\ddot{\boldsymbol{\eta}}_{z0}(t)+\tilde{\mathbf{C}}_{z0}\dot{\boldsymbol{\eta}}_{z0}(t)+\tilde{\mathbf{K}}_{z0}\boldsymbol{\eta}_{z0}(t)\right]=\delta\boldsymbol{\eta}_z^T\tilde{\mathbf{R}}_{z0}(t) \tag{9.82}$$

where

$$\boldsymbol{\eta}_{z0}=\begin{bmatrix}\eta_z & \eta_1\end{bmatrix}^T \quad \delta\boldsymbol{\eta}_z=\begin{bmatrix}\delta\eta_z & \delta\eta_1\end{bmatrix}^T \quad \tilde{\mathbf{R}}_{z0}=\begin{bmatrix}\tilde{R}_z & 0\end{bmatrix}^T$$

$$\tilde{\mathbf{M}}_{z0} = \begin{bmatrix} \tilde{M}_z & 0 \\ 0 & M_1 \end{bmatrix} \qquad \left. \begin{aligned} \tilde{\mathbf{C}}_{z0} &= \begin{bmatrix} \tilde{C}_z + \phi_z^2(x_1)C_1 & -\phi_z(x_1)C_1 \\ -\varphi_z(x_1)C_1 & C_1 \end{bmatrix} \\ \tilde{\mathbf{K}}_{z0} &= \begin{bmatrix} \tilde{K}_z + \phi_z^2(x_1)K_1 & -\phi_z(x_1)K_1 \\ -\phi_z(x_1)K_1 & K_1 \end{bmatrix} \end{aligned} \right\}$$

$$\text{and} \quad \begin{bmatrix} \tilde{M}_z \\ \tilde{C}_z = 2\zeta_n\omega_n\tilde{M}_z \\ \tilde{K}_z = \omega_n^2\tilde{M}_z \\ \tilde{R}_z \end{bmatrix} = \int_L \begin{bmatrix} \phi_z^2(x)\cdot m_z(x) \\ \phi_z^2(x)\cdot c_z(x) \\ \phi_z''^2(x)\cdot EI_y(x) \\ \phi_z(x)\cdot q_z(x,t) \end{bmatrix} dx$$

The pre-multiplication by $\delta\boldsymbol{\eta}_z^T$ may obviously be omitted, and thus the equilibrium condition in total modal coordinates is given by

$$\tilde{\mathbf{M}}_{z0}\ddot{\boldsymbol{\eta}}_{z0}(t) + \tilde{\mathbf{C}}_{z0}\dot{\boldsymbol{\eta}}_{z0}(t) + \tilde{\mathbf{K}}_{z0}\boldsymbol{\eta}_{z0}(t) = \tilde{\mathbf{R}}_{z0}(t) \tag{9.83}$$

Defining the relative damper motion and corresponding modal quantity

$$\Delta r_1(t) = 1\cdot\Delta\eta_1(t) = r_1(t) - r_z(x_1,t) = 1\cdot\eta_1(t) - \phi_z(x_1)\cdot\eta(t) \tag{9.84}$$

it is readily seen that

$$\begin{bmatrix} r_z(x,t) \\ r_1(t) \end{bmatrix} = \begin{bmatrix} \phi_z(x) & 0 \\ 0 & 1 \end{bmatrix} \begin{bmatrix} \eta_z(t) \\ \eta_1(t) \end{bmatrix} = \begin{bmatrix} r_z(x) \\ r_z(x_1) + \Delta r_1 \end{bmatrix} = \begin{bmatrix} \phi_z(x) & 0 \\ \phi_z(x_1) & 1 \end{bmatrix} \begin{bmatrix} \eta_z(t) \\ \Delta\eta_1(t) \end{bmatrix} \tag{9.85}$$

from which the following is obtained:

$$\boldsymbol{\eta}_{z0}(t) = \boldsymbol{\Psi}_z(x)\cdot\boldsymbol{\eta}_z(t) \tag{9.86}$$

where $\boldsymbol{\eta}_z = \begin{bmatrix} \eta_z & \Delta\eta_1 \end{bmatrix}^T$ are the new relative degrees of freedom and

$$\boldsymbol{\Psi}_z = \begin{bmatrix} \phi_z(x) & 0 \\ 0 & 1 \end{bmatrix}^{-1} \cdot \begin{bmatrix} \phi_z(x) & 0 \\ \phi_z(x_1) & 1 \end{bmatrix} = \begin{bmatrix} 1 & 0 \\ \phi_z(x_1) & 1 \end{bmatrix} \tag{9.87}$$

Introducing this into Eq. 9.83 and pre-multiplication by $\boldsymbol{\Psi}_z^T$ will then render the following equilibrium condition in relative modal coordinates:

$$\tilde{\mathbf{M}}_z\ddot{\boldsymbol{\eta}}_z(t) + \tilde{\mathbf{C}}_z\dot{\boldsymbol{\eta}}_z(t) + \tilde{\mathbf{K}}_z\boldsymbol{\eta}_z(t) = \tilde{\mathbf{R}}_z(t) \tag{9.88}$$

where

$$\tilde{\mathbf{M}}_z = \mathbf{\Psi}_z^T \tilde{\mathbf{M}}_{z0} \mathbf{\Psi}_z = \begin{bmatrix} \tilde{M}_z + \phi_z^2(x_1) \cdot M_1 & \phi_z(x_1) \cdot M_1 \\ \phi_z(x_1) \cdot M_1 & M_1 \end{bmatrix} \tag{9.89}$$

$$\tilde{\mathbf{C}}_z = \mathbf{\Psi}_z^T \tilde{\mathbf{C}}_{z0} \mathbf{\Psi}_z = \begin{bmatrix} \tilde{C}_z & 0 \\ 0 & C_1 \end{bmatrix} \quad \tilde{\mathbf{K}}_z = \mathbf{\Psi}_z^T \tilde{\mathbf{K}}_{z0} \mathbf{\Psi}_z = \begin{bmatrix} \tilde{K}_z & 0 \\ 0 & K_1 \end{bmatrix} \tag{9.90}$$

and

$$\tilde{\mathbf{R}}_z = \mathbf{\Psi}_z^T \tilde{\mathbf{R}}_{z0} = \begin{bmatrix} \tilde{R}_z & 0 \end{bmatrix}^T \tag{9.91}$$

Eqs. 9.88 – 9.91 are mathematically identical to that which was developed for the case of a single degree of freedom system with an additional small mass in Eqs. 9.48 – 9.59, only with the properties of the single degree of freedom system replaced by the modal properties of the continuous system in a single mode and single component approach. Thus, the entire developments of a frequency domain approach in Eqs. 9.60 – 9.69 will also apply to the present case, only by replacing single degree of freedom properties by their equivalent modal quantities. Thus:

$$\begin{aligned} \mathbf{S}_\eta(\omega) &= \lim_{T\to\infty} \frac{1}{\pi T} \mathbf{a}_\eta^* \mathbf{a}_\eta^T = \lim_{T\to\infty} \frac{1}{\pi T} \begin{bmatrix} a_\eta^* a_\eta & a_\eta^* a_{\Delta\eta} \\ a_{\Delta\eta}^* a_\eta & a_{\Delta\eta}^* a_{\Delta\eta} \end{bmatrix} = \begin{bmatrix} S_{\eta_z\eta_z} & S_{\eta_z\Delta\eta} \\ S_{\Delta\eta\eta_z} & S_{\Delta\eta\Delta\eta} \end{bmatrix} \\ &= \lim_{T\to\infty} \frac{1}{\pi T} \left(\hat{\mathbf{H}}(\omega) \cdot \begin{bmatrix} a_{\tilde{R}_z}(\omega)/\tilde{K}_z \\ 0 \end{bmatrix} \right)^* \left(\hat{\mathbf{H}}(\omega) \cdot \begin{bmatrix} a_{\tilde{R}_z}(\omega)/\tilde{K}_z \\ 0 \end{bmatrix} \right)^T \\ &= \frac{S_{\tilde{R}_z}(\omega)}{\tilde{K}_z^2} \begin{bmatrix} \left|\hat{H}_{zz}(\hat{\omega})\right|^2 & \hat{H}_{zz}^*(\hat{\omega}) \cdot \hat{H}_{1z}(\hat{\omega}) \\ \hat{H}_{1z}^*(\hat{\omega}) \cdot \hat{H}_{zz}(\hat{\omega}) & \left|\hat{H}_{1z}(\hat{\omega})\right|^2 \end{bmatrix} \end{aligned} \tag{9.92}$$

where $S_{\tilde{R}_z}(\omega)$ is the spectral density of the modal load on the main system, and

$$\left.\begin{aligned} \hat{H}_{zz}(\hat{\omega}) &= \frac{1 + \tilde{b}_1(i\hat{\omega}) + \tilde{b}_2(i\hat{\omega})^2}{1 + \tilde{a}_1(i\hat{\omega}) + \tilde{a}_2(i\hat{\omega})^2 + \tilde{a}_3(i\hat{\omega})^3 + \tilde{a}_4(i\hat{\omega})^4} \\ \hat{H}_{1z}(\hat{\omega}) &= \frac{\tilde{d}_2(i\hat{\omega})^2}{1 + \tilde{a}_1(i\hat{\omega}) + \tilde{a}_2(i\hat{\omega})^2 + \tilde{a}_3(i\hat{\omega})^3 + \tilde{a}_4(i\hat{\omega})^4} \end{aligned}\right\} \tag{9.93}$$

$$\left.\begin{aligned} &\hat{\omega} = \omega / \omega_z \\ &\tilde{\alpha} = \omega_z / \omega_1 \text{ where } \omega_1 = \sqrt{K_1/M_1} \\ &\tilde{\mu} = M_1 / \tilde{M}_z \end{aligned}\right\} \tag{9.94}$$

$$\left.\begin{aligned} &\tilde{a}_1 = 2(\zeta_z + \tilde{\alpha}\zeta_1) && \tilde{b}_1 = 2\tilde{\alpha}\zeta_1 && \\ &\tilde{a}_2 = 1 + \tilde{\mu} + \tilde{\alpha}^2 + 4\tilde{\alpha}\zeta_z\zeta_1 && \tilde{b}_2 = \tilde{\alpha}^2 && \tilde{d}_2 = -\tilde{\alpha}^2 \\ &\tilde{a}_3 = 2\tilde{\alpha}\left[\tilde{\alpha}\zeta_z + (1+\tilde{\mu})\zeta_1\right] && \tilde{c}_1 = 2\zeta_z && \tilde{e}_2 = -\tilde{\mu} \\ &\tilde{a}_4 = \tilde{\alpha}^2 && \tilde{c}_2 = 1 + \tilde{\mu} && \end{aligned}\right\} \tag{9.95}$$

and where ζ_z and ζ_1 are the damping ratios of the beam and the mass damper, respectively. Assuming that the beam is subject to an evenly distributed stochastic load $q_z(x,t)$ with a cross spectral density $S_{q_z}(\omega,\Delta x)$, where $\Delta x = |x_a - x_b|$ is the absolute value of spatial separation between arbitrary positions x_a and x_b, then

$$\begin{aligned} S_{\tilde{R}_z}(\omega) &= \lim_{T\to\infty} \frac{1}{\pi T} a^*_{\tilde{R}_z}(\omega) \cdot a^T_{\tilde{R}_z}(\omega) \\ &= \lim_{T\to\infty} \frac{1}{\pi T} \left[\int_L \phi_z(x) \cdot a_{q_z}(\omega,x)\,dx \right]^* \left[\int_L \phi_z(x) \cdot a_{q_z}(\omega,x)\,dx \right] \\ &= \int_L \int_L \phi_z(x_a) \cdot \phi_z(x_b) \cdot \lim_{T\to\infty} \frac{1}{\pi T} a^*_{q_z}(\omega,x_a) \cdot a_{q_z}(\omega,x_b)\,dx_a dx_b \\ &= \int_L \int_L \phi_z(x_a) \cdot \phi_z(x_b) \cdot S_{q_z}(\omega,\Delta x)\,dx_a dx_b \end{aligned} \tag{9.96}$$

The physical response quantities may be obtained by acknowledging that

$$\begin{bmatrix} r_z(x,t) \\ \Delta r_1(t) \end{bmatrix} = \mathbf{\Psi}_0 \begin{bmatrix} \eta_z(t) \\ \Delta\eta_1(t) \end{bmatrix} \text{ where } \mathbf{\Psi}_0 = \begin{bmatrix} \phi_z(x) & 0 \\ 0 & 1 \end{bmatrix} \tag{9.97}$$

from which it follows that

$$\mathbf{S}_{r_z}(x,\omega)=\lim_{T\to\infty}\frac{1}{\pi T}\left(\begin{bmatrix}a_{r_z}(x,\omega)\\a_{\Delta\eta}(\omega)\end{bmatrix}^*\begin{bmatrix}a_{r_z}(x,\omega)\\a_{\Delta\eta}(\omega)\end{bmatrix}^T\right)=\begin{bmatrix}S_{r_zr_z} & S_{r_z\Delta\eta}\\S_{\Delta\eta r_z} & S_{\Delta\eta\Delta\eta}\end{bmatrix}$$

$$=\lim_{T\to\infty}\frac{1}{\pi T}\left(\mathbf{\Psi}_0\begin{bmatrix}a_{\eta_z}(\omega)\\a_{\Delta\eta_1}(\omega)\end{bmatrix}\right)^*\left(\mathbf{\Psi}_0\begin{bmatrix}a_{\eta_z}(\omega)\\a_{\Delta\eta_1}(\omega)\end{bmatrix}\right)^T=\mathbf{\Psi}_0\mathbf{S}_{\eta_z}(\omega)\mathbf{\Psi}_0^T \tag{9.98}$$

$$=\frac{S_{\tilde{R}_z}(\omega)}{\tilde{K}_z^2}\begin{bmatrix}\phi_z^2(x)\left|\hat{H}_{zz}(\hat{\omega})\right|^2 & \phi_z(x)\hat{H}_{zz}^*(\hat{\omega})\hat{H}_{1z}(\hat{\omega})\\\phi_z(x)\hat{H}_{1z}^*(\hat{\omega})\hat{H}_{zz}(\hat{\omega}) & \left|\hat{H}_{1z}(\hat{\omega})\right|^2\end{bmatrix}$$

Again, if $S_{\tilde{R}_z}(\omega)$ is sufficiently broad banded, then a near to resonant response will occur, in which case the response variances are given by (see Newland [38])

$$\sigma_{r_z}^2(x_r)=\int_0^\infty S_{r_zr_z}(\omega)d\omega\approx\phi_z^2(x_r)\frac{S_{\tilde{R}_z}(\omega_z)}{\tilde{K}_z^2}\cdot\int_0^\infty\left|\hat{H}_{zz}(\hat{\omega})\right|^2d\omega$$
$$=\phi_z^2(x_r)\frac{\pi S_{\tilde{R}_z}(\omega_z)}{2\tilde{K}_z^2}\cdot\frac{\tilde{a}_2\tilde{a}_3-\tilde{a}_1\tilde{a}_4+\tilde{a}_1\tilde{b}_2^2+\tilde{a}_3\left(\tilde{b}_1^2-2\tilde{b}_2\right)}{\tilde{a}_1\tilde{a}_2\tilde{a}_3-\tilde{a}_1^2\tilde{a}_4-\tilde{a}_3^2} \tag{9.99}$$

and

$$\sigma_{\Delta r}^2=\int_0^\infty S_{\Delta r\Delta r}(\omega)d\omega\approx\frac{S_{\tilde{R}_z}(\omega_z)}{K_z^2}\cdot\int_0^\infty\left|\hat{H}_{1z}(\hat{\omega})\right|^2d\omega$$
$$=\frac{\pi S_{\tilde{R}_z}(\omega_z)}{2\tilde{K}_z^2}\cdot\frac{\tilde{a}_1\tilde{d}_2^2}{\tilde{a}_1\tilde{a}_2\tilde{a}_3-\tilde{a}_1^2\tilde{a}_4-\tilde{a}_3^2} \tag{9.100}$$

Example 9.1:

A suspension bridge is subject to wind induced vortex shedding oscillations in its second vertical mode

$$\phi_{z_2}=0.4\cdot\sin(\pi\hat{x})-0.6\cdot\sin(3\pi\hat{x}),\ \hat{x}=x/L$$

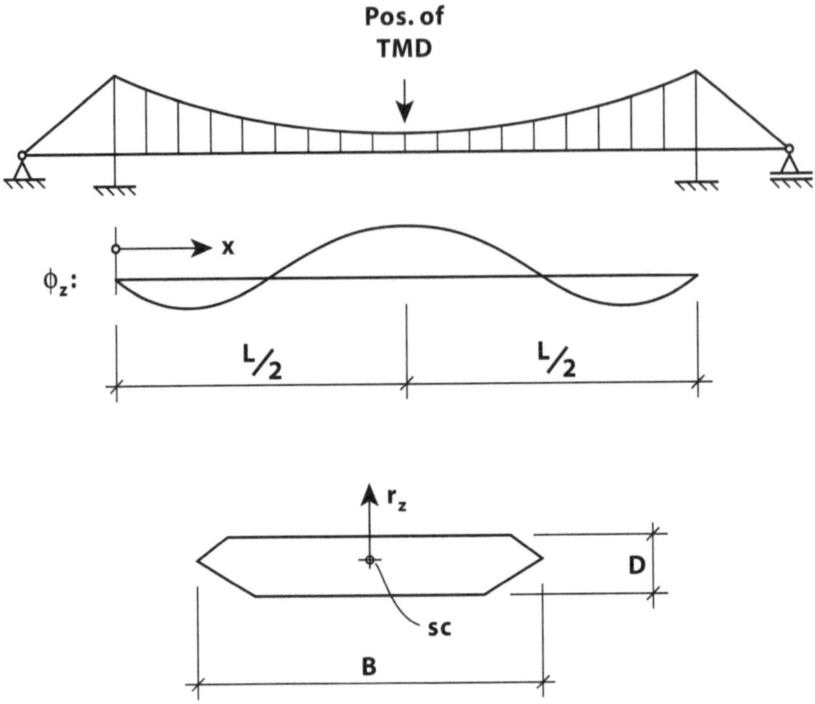

Fig. 9.17 Vortex shedding excitation of second vertical mode of suspension bridge

Basic structural data are defined by:

ρ (kg/m^3)	L (m)	B (m)	D (m)	m_z (kg/m)	ω_{z2} (rad/s)	ζ_{z2}
1.25	1200	20	3.5	9000	0.9	0.005

The necessary data for the description of the vortex shedding load is the following:

$\hat{\sigma}_{q_z} = \sigma_{q_z} / \left(\rho V^2 B/2\right)$	b_z	St	λ	K_{a_z}	a_L
0.63	0.1	0.16	3.5	0.294	0.3

The modal mass associated with this mode is

$$\tilde{M}_z = \int_0^L m_z \phi_{z2}^2 \, dx = 2.808 \cdot 10^6 \ kg \ \ while \ \ \tilde{m}_z = \tilde{M}_z \Big/ \int_0^L \phi_{z2}^2 \, dx = m_z$$

The spectral density of the cross sectional dynamic vortex shedding load has previously been described in Chapter 8.4

$$S_{q_z}(\omega) = \frac{\sigma_{q_z}^2}{\sqrt{\pi}\omega_s b_z} \exp\left[-\left(\frac{1-\omega/\omega_s}{b_z} \right)^2 \right]$$

where $\sigma_{q_z}^2$ *is the variance of the cross sectional vortex shedding load,* b_z *is its band width and* $\omega_s = 2\pi V St/D$ *is the shedding frequency,* V *is the mean wind velocity and* St *is the relevant Strouhal number. At resonance*

$$\omega_s = \omega_{z2} \Rightarrow V = V_{R_2} = \frac{D\omega_{z2}}{2\pi St} \approx 3.1 \text{ m/s}$$

Assuming that the coherence length λD *of the vortices is small as compared to the length* L *of the system, then, with sufficient accuracy, the modal load is defined by*

$$S_{\tilde{R}_z}(\omega) \approx 2\lambda D S_{q_z}(\omega) \int_0^L \phi_{z2}^2(x)\,dx = \frac{\sigma_{\tilde{R}_z}^2}{\pi^{1/2} b_z \omega_s} \exp\left[-\left(\frac{1-\omega/\omega_s}{b_z} \right)^2 \right]$$

where $\sigma_{\tilde{R}_z}^2 = \int_0^\infty S_{\tilde{R}_z}(\omega)\,d\omega = 2\lambda D \sigma_{q_z}^2 \int_0^L \phi_{z2}^2(x)\,dx$

It is taken for granted that the vortex shedding induced aerodynamic damping is defined by

$$\zeta_{ae_z} = \frac{\rho B^2}{4m_z} K_{a_z} \left[1 - \left(\frac{\sigma_{r_z}^2}{a_L D} \right)^2 \right]$$

where $\sigma_{r_z}^2$ *is the variance of the dynamic response,* K_{a_z} *is an aerodynamic damping coefficient and* a_L *is a limiting motion parameter. The spectral density of the dynamic response at mid-span is then given by*

$$S_{r_z}(x_r = L/2, \omega) = \left[\phi_{z2}(x_r = L/2) / \tilde{K}_{z2} \right]^2 \left| \hat{H}_z(\omega) \right|^2 S_{\tilde{R}_z}(\omega)$$

where $\hat{H}_z(\omega) = \left[1 - (\omega/\omega_{z2})^2 + 2i(\zeta_{z2} - \zeta_{ae_z})\omega/\omega_{z2} \right]^{-1}$ *is the frequency response function and* $\tilde{K}_z = \omega_{z2}^2 \cdot m_z \int_0^L \phi_{z2}^2(x)\,dx$ *is the modal stiffness associated with* ϕ_{z2}. *The modal load spectrum* $S_{\tilde{R}_z}(\omega)$ *at resonance (* $\omega_s = \omega_{z2}$ *) and the absolute value of the frequency response function are shown in Fig. 9.18. Due to the motion dependent aerodynamic damping any response calculations will involve iterations.*

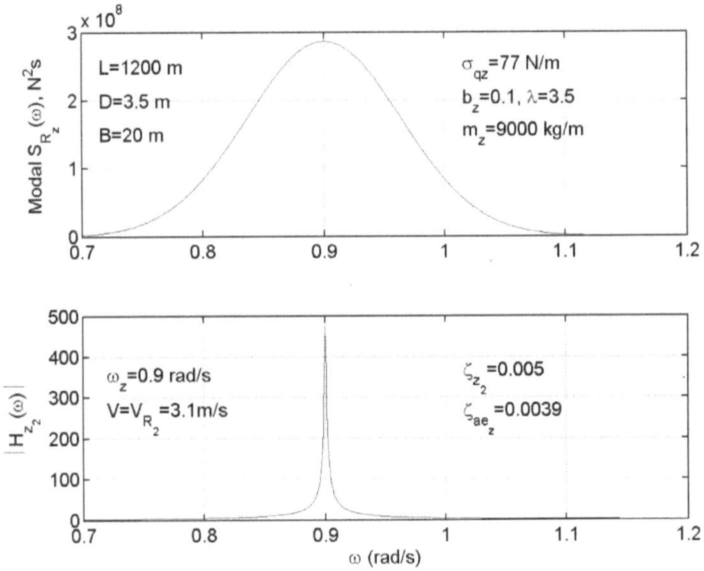

Fig. 9.18 Modal load spectrum (upper curve) and modal frequency response function at $x = L/2$ (lower curve), $V = V_{R_2} = 3.1$ m/s

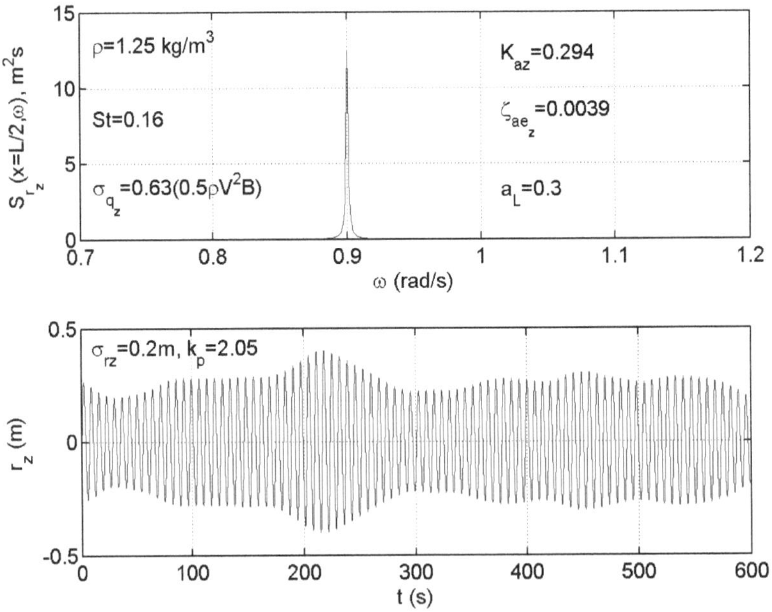

Fig. 9.19 Response spectrum of displacement r_z at $x = L/2$ (upper curve) and corresponding time domain version (lower curve) when the system has no additional tuned mass damper, $V = V_{R_2} = 3.1$ m/s

The response spectrum due to vortex shedding prior to any tuned mass damper is shown in the upper diagram in Fig. 9.19. A corresponding time domain simulation is illustrated in the lower diagram. As can be seen, the maximum dynamic displacements are in the order of about 0.4 m. This is deemed unduly large. In an attempt to quench the problem it has been decided to install a mass damper at mid-span (see Fig. 9.17). Choosing

$$\mu = 0.006,\ \omega_1/\omega_{z_2} = 1/(1+\mu) = 0.994 \text{ and } \zeta_1 = \sqrt{\frac{3\mu}{8(1+\mu)^3}} = 0.047$$

then (See Eqs. 9.93 and 9.98) the spectral densities of the dynamic response at mid-span and the corresponding relative displacements of the damper are given by

$$S_{r_z r_z} = \left[\frac{\phi_z(x_r = L/2)}{\tilde{K}_z}\right]^2 \left|\hat{H}_{zz}(\hat{\omega})\right|^2 S_{\tilde{R}_z}(\omega) \text{ and } S_{\Delta r \Delta r} = \frac{\left|\hat{H}_{1z}(\hat{\omega})\right|^2}{\tilde{K}_z^2} S_{\tilde{R}_z}(\omega)$$

where

$$\begin{cases} \hat{H}_{zz} = \dfrac{1+\tilde{b}_1(i\hat{\omega})+\tilde{b}_2(i\hat{\omega})^2}{1+\tilde{a}_1(i\hat{\omega})+\tilde{a}_2(i\hat{\omega})^2+\tilde{a}_3(i\hat{\omega})^3+\tilde{a}_4(i\hat{\omega})^4} \\ \hat{H}_{1z} = \dfrac{\tilde{d}_2(i\hat{\omega})^2}{1+\tilde{a}_1(i\hat{\omega})+\tilde{a}_2(i\hat{\omega})^2+\tilde{a}_3(i\hat{\omega})^3+\tilde{a}_4(i\hat{\omega})^4} \end{cases}$$

and

$$\begin{array}{llll} & \tilde{a}_1 = 2(\zeta_z + \tilde{\alpha}\zeta_1) & \tilde{b}_1 = 2\tilde{\alpha}\zeta_1 & \\ \hat{\omega} = \omega/\omega_z & \tilde{a}_2 = 1+\tilde{\mu}+\tilde{\alpha}^2+4\tilde{\alpha}\zeta_z\zeta_1 & \tilde{b}_2 = \tilde{\alpha}^2 & \zeta_z = \zeta_{z2} - \zeta_{ae_z} \\ \tilde{\alpha} = \omega_z/\omega_1 & \tilde{a}_3 = 2\tilde{\alpha}\left[\tilde{\alpha}\zeta_z + (1+\tilde{\mu})\zeta_1\right] & & \zeta_{ae_z} \approx \dfrac{\rho B^2}{4m_z} K_{a_z} \\ \tilde{\mu} = M_1/\tilde{M}_z & \tilde{a}_4 = \tilde{\alpha}^2 & \tilde{d}_2 = -\tilde{\alpha}^2 & \end{array}$$

Plots of $S_{r_z r_z}(x_r = L/2, \omega)$ *and* $S_{\Delta r \Delta r}(\omega)$ *are shown in Fig. 9.20, while corresponding time domain simulations of* $r_z(x_r = L/2, t)$ *and* $\Delta r(t)$ *are shown in Fig. 9.21.*

The variance of the displacement response of the bridge beam and the variance of the relative displacement of the damper may readily be obtained from the simulations, or more directly, from integration of spectra, rendering

$$\sigma_{r_z} = \sqrt{\int_0^\infty S_{r_z r_z}(\omega)\, d\omega} = 0.0386 \text{ m}$$

and

$$\sigma_{\Delta r} = \sqrt{\int_0^\infty S_{\Delta r \Delta r}(\omega)\, d\omega} = 0.3443 \text{ m}$$

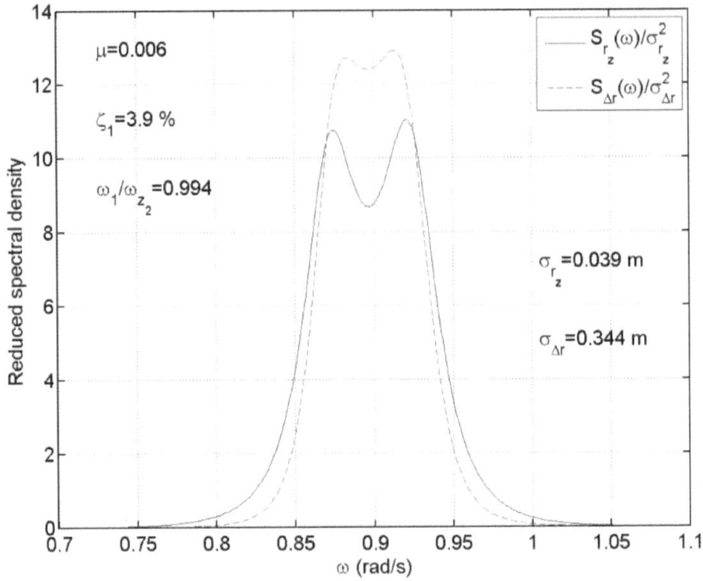

Fig. 9.20 Reduced response spectral densities of r_z at $x = L/2$ (solid line) and of tuned mass damper Δr relative to r_z (broken line)

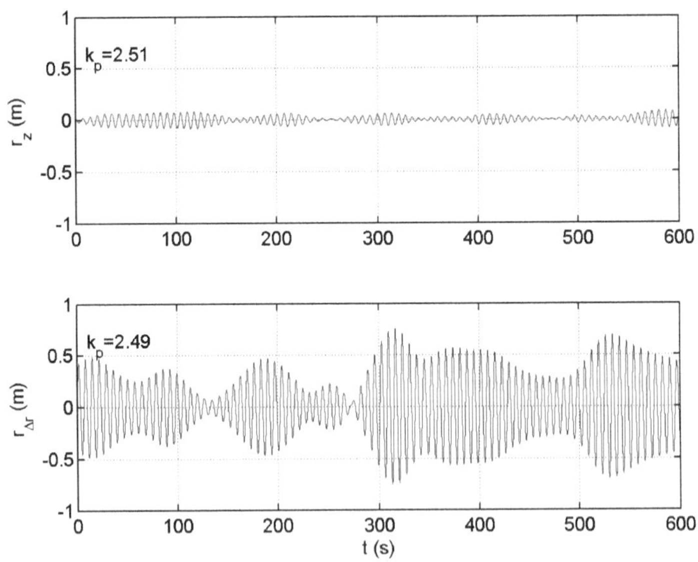

Fig. 9.21 Time domain simulations of $r_z(t)$ and $\Delta r(t)$

Under the present conditions this seems manageable. However, it should be noted that a suspension bridge similar to the one illustrated in Fig. 9.17 will have an asymmetric vertical eigenmode at a lower eigenfrequency than the symmetric case which has been dealt with above. Therefore, an extended case study including both of these eigenmodes is given further consideration in Example 9.2 below.

The case shown above is only adequate if the problem is limited to a case with only one mass damper (M_1, C_1, K_1) intended to quench unwanted response in a single mode (and single component) oscillations. Let us therefore also consider the case of a multi-mode system with tuned mass dampers at positions x_j, $j=1,2,\ldots,N_j$, i.e. the case that there are installed N_j dampers defined by the properties

$$\left.\begin{aligned}\mathbf{M}_d &= diag\left[M_j\right]\\ \mathbf{C}_d &= diag\left[C_j\right]\\ \mathbf{K}_d &= diag\left[K_j\right]\end{aligned}\right\} j=1,2,\ldots,N_j \tag{9.101}$$

intended to quench oscillations associated with N_j (or less) modes. Still, it is taken for granted that all damper masses are small as compared to the modal mass of the system itself. The displacement of the main system may then be described in modal coordinates by

$$r_z(x,t)=\sum_{n=1}^{N_{\text{mod}}} \phi_{z_n}(x)\cdot\eta_{z_n}(t)=\mathbf{\Phi}_z(x)\cdot\mathbf{\eta}_z(t) \tag{9.102}$$

where

$$\left.\begin{aligned}\mathbf{\Phi}_z(x) &= \begin{bmatrix}\phi_{z_1} & \cdots & \phi_{z_n} & \cdots & \phi_{zN_{\text{mod}}}\end{bmatrix}\\ \mathbf{\eta}_z(t) &= \begin{bmatrix}\eta_{z_1} & \cdots & \eta_{z_n} & \cdots & \eta_{zN_{\text{mod}}}\end{bmatrix}^T\end{aligned}\right\} \tag{9.103}$$

The mass damper displacements in original coordinates are defined by

$$\mathbf{r}_d(t)=\begin{bmatrix}r_1 & \cdots & r_j & \cdots & r_{N_j}\end{bmatrix}^T \tag{9.104}$$

Since the displacement of each damper alone will represent a mode shape on its own, i.e. $r_j=1\cdot\eta_j$, then $\mathbf{r}_d$ in original coordinates is equivalent to $\mathbf{\eta}_d$ in modal coordinates, i.e.:

$$\mathbf{r}_d(t)=1\cdot\mathbf{\eta}_d(t)=\begin{bmatrix}\eta_1 & \cdots & \eta_j & \cdots & \eta_{N_j}\end{bmatrix}^T \tag{9.105}$$

Thus, our real and modal degrees of freedom are defined by

$$\mathbf{r} = \begin{bmatrix} r_z(x,t) \\ \mathbf{r}_d(t) \end{bmatrix} \text{ and } \boldsymbol{\eta} = \begin{bmatrix} \eta_z(t) \\ \boldsymbol{\eta}_d(t) \end{bmatrix} \tag{9.106}$$

On these degrees of freedom we impose a corresponding set of time invariant virtual displacements

$$\delta\mathbf{r} = \begin{bmatrix} \delta r_z(x) \\ \delta\mathbf{r}_d \end{bmatrix} \tag{9.107}$$

where

$$\delta\mathbf{r}_d = \begin{bmatrix} \delta r_1 & \cdots & \delta r_j & \cdots & \delta r_{N_j} \end{bmatrix}^T \tag{9.108}$$

Introducing this into Eq. 9.78, then the following is obtained

$$\begin{aligned}
&\int_L m_z \ddot{r}_z(x,t)\delta r_z(x)dx + \int_L c_z \dot{r}_z(x,t)\delta r_z(x)dx + \int_L EI_y r_z''(x,t)\delta r_z''(x)dx \\
&-\sum_{j=1}^{N_j}\left[C_j\dot{r}_j(t) + K_j r_j(t) - C_j\dot{r}_z(x_j,t) - K_j r_z(x_j,t)\right]\delta r_z(x_j) \\
&+\sum_{j=1}^{N_j}\left[M_j\ddot{r}_j(t) + C_j\dot{r}_j(t) + K_j r_j(t) - C_j\dot{r}_z(x_j,t) - K_j r_z(x_j,t)\right]\delta r_j \\
&= \int_L q_z(x,t)\delta r_z(x)dx
\end{aligned} \tag{9.109}$$

The strategy is then to use the orthogonality properties of the mode shapes

$$\int_L \begin{bmatrix} m_z(x)\phi_n(x)\phi_m(x) \\ c_z(x)\phi_n(x)\phi_m(x) \\ EI_y(x)\phi_n''(x)\phi_m''(x) \end{bmatrix} dx = \mathbf{0} \tag{9.110}$$

and successively introduce

$$\left.\begin{array}{lll}
1) & \delta r_z = \phi_{z_1}\delta r_{z_1} \text{ and } \delta\mathbf{r}_d = \begin{bmatrix} \delta r_1 & \cdots & \delta r_j & \cdots & \delta r_{N_j} \end{bmatrix}^T \\
\vdots & \qquad\qquad\qquad \vdots \\
n) & \delta r_z = \phi_{z_n}\delta r_{z_n} \text{ and } \delta\mathbf{r}_d = \begin{bmatrix} \delta r_1 & \cdots & \delta r_j & \cdots & \delta r_{N_j} \end{bmatrix}^T \\
\vdots & \qquad\qquad\qquad \vdots \\
N_{\text{mod}}) & \delta r_z = \phi_{zN_{\text{mod}}}\delta r_{zN_{\text{mod}}} \text{ and } \delta\mathbf{r}_d = \begin{bmatrix} \delta r_1 & \cdots & \delta r_j & \cdots & \delta r_{N_j} \end{bmatrix}^T
\end{array}\right\} \tag{9.111}$$

Replacing $r_z(x,t)$ by $\boldsymbol{\Phi}_z(x)\cdot\boldsymbol{\eta}_z(t)$ and $r_j(t)$ by $\eta_j(t)$ this will yield $N_{\text{mod}}+N_j$ equations, where the equation associated with $\delta r_z(x)=\phi_{z_n}(x)\cdot\delta r_{z_n}$ and $\delta\mathbf{r}_d=\begin{bmatrix}\delta r_1 & \cdots & \delta r_j & \cdots & \delta r_{N_j}\end{bmatrix}^T$ is given by

$$\begin{aligned}
&\delta r_{z_n}\Big\{\tilde{M}_{z_n}\ddot{\eta}_{z_n}(t)+\tilde{C}_{z_n}\dot{\eta}_{z_n}(t)+\tilde{K}_{z_n}\eta_{z_n}(t)+\sum_{j=1}^{N_j}C_j\phi_{z_n}^2(x_j)\dot{\eta}_{z_n}(t)\\
&+\sum_{j=1}^{N}K_j\phi_{z_n}^2(x_j)\eta_{z_n}(t)-\sum_{j=1}^{N_j}C_j\phi_{z_n}(x_j)\dot{\eta}_j(t)-\sum_{j=1}^{N}K_j\phi_{z_n}(x_j)\eta_j(t)\Big\}\\
&+\sum_{j=1}^{N_j}\delta r_j\Big\{M_j\ddot{\eta}_j(t)+C_j\dot{\eta}_j(t)+K\eta_j(t)-C_j\phi_{z_n}(x_j)\dot{\eta}_{z_n}(t)-K_j\phi_{z_n}(x_j)\eta_{z_n}(t)\Big\}\\
&=\delta r_{z_n}\tilde{R}_{z_n}(t)
\end{aligned}\tag{9.112}$$

where

$$\begin{bmatrix}\tilde{M}_{z_n}\\ \tilde{C}_{z_n}\\ \tilde{K}_{z_n}\end{bmatrix}=\int_L\begin{bmatrix}m_z\phi_{z_n}^2\\ c_z\phi_{z_n}^2\\ EI_y\phi_{z_n}''^2\end{bmatrix}dx \quad\text{and}\quad \tilde{R}_{z_n}(t)=\int_L\phi_{z_n}(x)q_z(x,t)dx \tag{9.113}$$

Like usual, it is convenient to define the modal property matrices:

$$\left.\begin{aligned}\tilde{\mathbf{M}}_z&=diag\left[\tilde{M}_{z_n}\right]\\ \tilde{\mathbf{C}}_z&=diag\left[\tilde{C}_{z_n}\right]\\ \tilde{\mathbf{K}}_z&=diag\left[\tilde{K}_{z_n}\right]\end{aligned}\right\}\quad\text{and}\quad \tilde{\mathbf{R}}_z(t)=\begin{bmatrix}\tilde{R}_{z_1} & \cdots & \tilde{R}_{z_n} & \cdots & \tilde{R}_{z_{N_{\text{mod}}}}\end{bmatrix}^T \tag{9.114}$$

In addition to this it is convenient to define (see also Eq. 9.101):

$$\boldsymbol{\Phi}_d=\begin{bmatrix}\boldsymbol{\Phi}_z(x_1)\\ \vdots\\ \boldsymbol{\Phi}_z(x_j)\\ \vdots\\ \boldsymbol{\Phi}_z(x_{N_j})\end{bmatrix}=\begin{bmatrix}\phi_{z_1}(x_1) & \cdots & \phi_{z_n}(x_1) & \cdots & \phi_{z_{N_{\text{mod}}}}(x_1)\\ \vdots & & \vdots & & \vdots\\ \phi_{z_1}(x_j) & \cdots & \phi_{z_n}(x_j) & \cdots & \phi_{z_{N_{\text{mod}}}}(x_j)\\ \vdots & & \vdots & & \vdots\\ \phi_{z_1}(x_{N_j}) & \cdots & \phi_{z_n}(x_{N_j}) & \cdots & \phi_{z_{N_{\text{mod}}}}(x_{N_j})\end{bmatrix} \tag{9.115}$$

and

$$\left.\begin{aligned}\tilde{\mathbf{K}}_d &= \mathbf{\Phi}_d^T \mathbf{K}_d \mathbf{\Phi}\\ \tilde{\mathbf{C}}_d &= \mathbf{\Phi}_d^T \mathbf{C}_d \mathbf{\Phi}\end{aligned}\right\} \tag{9.116}$$

The total $N_{\mathrm{mod}} + N_j$ equations contained in Eq. 9.112 may then conveniently be written

$$\begin{aligned}\delta\mathbf{r}^T &\left\{\begin{bmatrix}\tilde{\mathbf{M}}_z & \mathbf{0}\\ \mathbf{0} & \mathbf{M}_d\end{bmatrix}\begin{bmatrix}\ddot{\boldsymbol{\eta}}_z\\ \ddot{\boldsymbol{\eta}}_d\end{bmatrix} + \begin{bmatrix}\tilde{\mathbf{C}}_z + \tilde{\mathbf{C}}_d & -\mathbf{\Phi}_d^T\mathbf{C}_d\\ -\mathbf{C}_d\mathbf{\Phi}_d & \mathbf{C}_d\end{bmatrix}\begin{bmatrix}\dot{\boldsymbol{\eta}}_z\\ \dot{\boldsymbol{\eta}}_d\end{bmatrix}\right.\\ &\left. + \begin{bmatrix}\tilde{\mathbf{K}}_z + \tilde{\mathbf{K}}_d & -\mathbf{\Phi}_d^T\mathbf{K}_d\\ -\mathbf{K}_d\mathbf{\Phi}_d & \mathbf{K}_d\end{bmatrix}\begin{bmatrix}\boldsymbol{\eta}_z\\ \boldsymbol{\eta}_d\end{bmatrix}\right\} = \delta\mathbf{r}^T\begin{bmatrix}\tilde{\mathbf{R}}_z\\ \mathbf{0}\end{bmatrix}\end{aligned} \tag{9.117}$$

As can be seen, the pre-multiplication by $\delta\mathbf{r}^T$ may be omitted, and thus the equilibrium condition in total modal coordinates is given by

$$\begin{aligned}&\begin{bmatrix}\tilde{\mathbf{M}}_z & \mathbf{0}\\ \mathbf{0} & \mathbf{M}_d\end{bmatrix}\begin{bmatrix}\ddot{\boldsymbol{\eta}}_z\\ \ddot{\boldsymbol{\eta}}_d\end{bmatrix} + \begin{bmatrix}\tilde{\mathbf{C}}_z + \tilde{\mathbf{C}}_d & -\mathbf{\Phi}_d^T\mathbf{C}_d\\ -\mathbf{C}_d\mathbf{\Phi}_d & \mathbf{C}_d\end{bmatrix}\begin{bmatrix}\dot{\boldsymbol{\eta}}_z\\ \dot{\boldsymbol{\eta}}_d\end{bmatrix}\\ &\quad + \begin{bmatrix}\tilde{\mathbf{K}}_z + \tilde{\mathbf{K}}_d & -\mathbf{\Phi}_d^T\mathbf{K}_d\\ -\mathbf{K}_d\mathbf{\Phi}_d & \mathbf{K}_d\end{bmatrix}\begin{bmatrix}\boldsymbol{\eta}_z\\ \boldsymbol{\eta}_d\end{bmatrix} = \begin{bmatrix}\tilde{\mathbf{R}}_z\\ \mathbf{0}\end{bmatrix}\end{aligned} \tag{9.118}$$

It is readily seen that the total degrees of freedom (see Eq. 9.106)

$$\begin{bmatrix}r_z(x,t)\\ \mathbf{r}_d(t)\end{bmatrix} = \begin{bmatrix}\mathbf{\Phi}_z(x)\cdot\boldsymbol{\eta}_z(t)\\ \boldsymbol{\eta}_d\end{bmatrix} = \begin{bmatrix}\mathbf{\Phi}_z(x) & \mathbf{0}\\ \mathbf{0} & \mathbf{I}\end{bmatrix}\begin{bmatrix}\boldsymbol{\eta}_z(t)\\ \boldsymbol{\eta}_d\end{bmatrix} \tag{9.119}$$

may alternatively be expressed in relative displacement degrees of freedom

$$\begin{bmatrix}r_z(x,t)\\ \mathbf{r}_d(t)\end{bmatrix} = \begin{bmatrix}r_z(x,t)\\ \mathbf{r}_z(x_j,t) + \Delta\mathbf{r}_d\end{bmatrix} = \begin{bmatrix}\mathbf{\Phi}_z(x) & \mathbf{0}\\ \mathbf{\Phi}_d & \mathbf{I}\end{bmatrix}\begin{bmatrix}\boldsymbol{\eta}_z(t)\\ \Delta\boldsymbol{\eta}_d(t)\end{bmatrix} \tag{9.120}$$

or

$$\mathbf{r}_{rel} = \begin{bmatrix}r_z(x,t)\\ \Delta\mathbf{r}_d(t)\end{bmatrix} = \begin{bmatrix}\mathbf{\Phi}_z(x)\boldsymbol{\eta}_z(t)\\ \Delta\boldsymbol{\eta}_d(t)\end{bmatrix} = \begin{bmatrix}\mathbf{\Phi}_z(x) & \mathbf{0}\\ \mathbf{0} & \mathbf{I}\end{bmatrix}\boldsymbol{\eta}_{rel}(t) \tag{9.121}$$

where the relative modal coordinate vector is given by

$$\boldsymbol{\eta}_{rel}(t) = \begin{bmatrix}\boldsymbol{\eta}_z & \Delta\boldsymbol{\eta}_d\end{bmatrix}^T \tag{9.122}$$

(and where $\mathbf{I}$ is an N_j by N_j identity matrix, while the size of $\mathbf{0}$ depend on its first or second row position, i.e. and $\mathbf{0}$ is a 1 by N_j vector when it is located on the first row while $\mathbf{0}$ is a 1 by N_{mod} vector when it is located on the second row). Thus, by combining Eqs. 9.119 and 9.120 it is seen that

$$\begin{bmatrix} \mathbf{\Phi}_z(x) & \mathbf{0} \\ \mathbf{0} & \mathbf{I} \end{bmatrix} \begin{bmatrix} \boldsymbol{\eta}_z(t) \\ \boldsymbol{\eta}_d \end{bmatrix} = \begin{bmatrix} \mathbf{\Phi}_z(x) & \mathbf{0} \\ \mathbf{\Phi}_d & \mathbf{I} \end{bmatrix} \begin{bmatrix} \boldsymbol{\eta}_z(t) \\ \Delta\boldsymbol{\eta}_d(t) \end{bmatrix} \tag{9.123}$$

rendering

$$\begin{bmatrix} \boldsymbol{\eta}_z(t) \\ \boldsymbol{\eta}_d \end{bmatrix} = \begin{bmatrix} \mathbf{I} & \mathbf{0} \\ \mathbf{\Phi}_d & \mathbf{I} \end{bmatrix} \begin{bmatrix} \boldsymbol{\eta}_z(t) \\ \Delta\boldsymbol{\eta}_d(t) \end{bmatrix} \tag{9.124}$$

Defining the transformation matrix

$$\mathbf{\Psi} = \begin{bmatrix} \mathbf{I} & \mathbf{0} \\ \mathbf{\Phi}_d & \mathbf{I} \end{bmatrix} \tag{9.125}$$

then

$$\boldsymbol{\eta}(t) = \mathbf{\Psi}\boldsymbol{\eta}_{rel}(t)$$

Introducing this into Eq. 9.118 and pre-multiplying by $\mathbf{\Psi}^T$ then the equilibrium condition in relative modal degrees of freedom is given by

$$\tilde{\mathbf{M}}\ddot{\boldsymbol{\eta}}_{rel}(t) + \tilde{\mathbf{C}}\dot{\boldsymbol{\eta}}_{rel}(t) + \tilde{\mathbf{K}}\boldsymbol{\eta}_{rel}(t) = \tilde{\mathbf{R}}(t) \tag{9.126}$$

where

$$\tilde{\mathbf{M}} = \mathbf{\Psi}^T \begin{bmatrix} \tilde{\mathbf{M}}_z & \mathbf{0} \\ \mathbf{0} & \mathbf{M}_d \end{bmatrix} \mathbf{\Psi} = \begin{bmatrix} \tilde{\mathbf{M}}_z + \mathbf{\Phi}_d^T\mathbf{M}_d\mathbf{\Phi}_d & \mathbf{\Phi}_d^T\mathbf{M}_d \\ \mathbf{M}_d\mathbf{\Phi}_d & \mathbf{M}_d \end{bmatrix} \tag{9.127}$$

$$\tilde{\mathbf{C}} = \mathbf{\Psi}^T \begin{bmatrix} \tilde{\mathbf{C}}_z + \mathbf{\Phi}_d^T\mathbf{C}_d\mathbf{\Phi}_d & -\mathbf{\Phi}_d^T\mathbf{C}_d \\ -\mathbf{C}_d\mathbf{\Phi}_d & \mathbf{C}_d \end{bmatrix} \mathbf{\Psi}_z = \begin{bmatrix} \tilde{\mathbf{C}}_z & \mathbf{0} \\ \mathbf{0} & \mathbf{C}_d \end{bmatrix} \tag{9.128}$$

$$\tilde{\mathbf{K}} = \mathbf{\Psi}^T \begin{bmatrix} \tilde{\mathbf{K}}_z + \mathbf{\Phi}_d^T\mathbf{K}_d\mathbf{\Phi}_d & -\mathbf{\Phi}_d^T\mathbf{K}_d \\ -\mathbf{K}_d\mathbf{\Phi}_d & \mathbf{K}_d \end{bmatrix} \mathbf{\Psi}_z = \begin{bmatrix} \tilde{\mathbf{K}}_z & \mathbf{0} \\ \mathbf{0} & \mathbf{K}_d \end{bmatrix} \tag{9.129}$$

and

$$\tilde{\mathbf{R}} = \mathbf{\Psi}^T \begin{bmatrix} \tilde{\mathbf{R}}_z \\ \mathbf{0} \end{bmatrix} = \begin{bmatrix} \tilde{\mathbf{R}}_z \\ \mathbf{0} \end{bmatrix} \tag{9.130}$$

A general solution to the modal equilibrium condition in Eq. 9.126 may be obtained in a frequency domain approach. It is then convenient first to split $\tilde{\mathbf{M}}$ into

$$\tilde{\mathbf{M}} = \begin{bmatrix} \tilde{\mathbf{M}}_z & \mathbf{0} \\ \mathbf{0} & \mathbf{M}_d \end{bmatrix} + \begin{bmatrix} \tilde{\mathbf{M}}_z \hat{\mathbf{D}} \mathbf{\Phi}_d & \tilde{\mathbf{M}}_z \hat{\mathbf{D}} \\ \mathbf{M}_d \mathbf{\Phi}_d & \mathbf{0} \end{bmatrix} \tag{9.131}$$

where $\hat{\mathbf{D}} = \tilde{\mathbf{M}}_z^{-1} \mathbf{\Phi}_d^T \mathbf{M}_d$, after which the entire equation is pre-multiplied by $\tilde{\mathbf{K}}^{-1}$. Since

$$\tilde{\mathbf{K}}^{-1}\tilde{\mathbf{M}} = \begin{bmatrix} \tilde{\mathbf{K}}_z^{-1} & \mathbf{0} \\ \mathbf{0} & \mathbf{K}_d^{-1} \end{bmatrix} \begin{bmatrix} \tilde{\mathbf{M}}_z & \mathbf{0} \\ \mathbf{0} & \mathbf{M}_d \end{bmatrix} \left(\begin{bmatrix} \mathbf{I}_{N_{mod}} & \mathbf{0} \\ \mathbf{0} & \mathbf{I}_{N_j} \end{bmatrix} + \begin{bmatrix} \hat{\mathbf{D}} \mathbf{\Phi}_d & \hat{\mathbf{D}} \\ \mathbf{\Phi}_d & \mathbf{0} \end{bmatrix} \right) \tag{9.132}$$

it is seen that the following is obtained:

$$\begin{aligned} &\begin{bmatrix} \mathbf{I}_{N_{\text{mod}}} & \mathbf{0}_{N_{\text{mod}}N_j} \\ \mathbf{0}_{N_j N_{\text{mod}}} & \mathbf{I}_{N_j} \end{bmatrix} \begin{bmatrix} \boldsymbol{\eta}_z \\ \Delta\boldsymbol{\eta}_d \end{bmatrix} \\ &+ \begin{bmatrix} \boldsymbol{\omega}_z^{-2} & \mathbf{0}_{N_{\text{mod}}N_j} \\ \mathbf{0}_{N_j N_{\text{mod}}} & \boldsymbol{\omega}_d^{-2} \end{bmatrix} \begin{bmatrix} \mathbf{I}_{N_{\text{mod}}} + \hat{\mathbf{D}}\mathbf{\Phi}_d & \hat{\mathbf{D}} \\ \mathbf{\Phi}_d & \mathbf{I}_{N_j} \end{bmatrix} \begin{bmatrix} \ddot{\boldsymbol{\eta}}_z \\ \Delta\ddot{\boldsymbol{\eta}}_d \end{bmatrix} \\ &+2 \begin{bmatrix} \boldsymbol{\omega}_z^{-1} & \mathbf{0}_{N_{\text{mod}}N_j} \\ \mathbf{0}_{N_j N_{\text{mod}}} & \boldsymbol{\omega}_d^{-1} \end{bmatrix} \begin{bmatrix} \boldsymbol{\zeta}_z & \mathbf{0}_{N_{\text{mod}}N_j} \\ \mathbf{0}_{N_j N_{\text{mod}}} & \boldsymbol{\zeta}_d \end{bmatrix} \begin{bmatrix} \dot{\boldsymbol{\eta}}_z \\ \Delta\dot{\boldsymbol{\eta}}_d \end{bmatrix} = \begin{bmatrix} \tilde{\mathbf{K}}_z^{-1}\tilde{\mathbf{R}}_z \\ \mathbf{0}_{N_j 1} \end{bmatrix} \end{aligned} \tag{9.133}$$

where $\begin{cases} \boldsymbol{\omega}_z = diag\left[\omega_{z_n}\right] \\ \boldsymbol{\omega}_d = diag\left[\omega_j\right] \end{cases}$ and $\begin{cases} \boldsymbol{\zeta}_z = diag\left[\zeta_{z_n}\right] \\ \boldsymbol{\zeta}_d = diag\left[\zeta_j\right] \end{cases}$

and where the indices on the identity and zero matrices indicate their size, i.e. the number of rows and columns (a single index means that the matrix is square). Taking the Fourier transform

$$\begin{bmatrix} \boldsymbol{\eta}_z \\ \Delta\boldsymbol{\eta}_d \end{bmatrix} = \sum_\omega \begin{bmatrix} \mathbf{a}_{\eta_z}(\omega) \\ \mathbf{a}_{\Delta\eta_d}(\omega) \end{bmatrix} e^{i\omega t} \quad \text{and} \quad \begin{bmatrix} \tilde{\mathbf{K}}_z^{-1}\tilde{\mathbf{R}}_z \\ \mathbf{0}_{N_j 1} \end{bmatrix} = \sum_\omega \begin{bmatrix} \tilde{\mathbf{K}}_z^{-1}\mathbf{a}_{\tilde{R}_z}(\omega) \\ \mathbf{0}_{N_j 1} \end{bmatrix} e^{i\omega t}$$

throughout Eq. 9.133 will then require that for every ω-setting

$$\begin{bmatrix} \mathbf{a}_{\eta_z}(\omega) \\ \mathbf{a}_{\Delta\eta_d}(\omega) \end{bmatrix} = \hat{\mathbf{H}}(\omega) \cdot \begin{bmatrix} \tilde{\mathbf{K}}_z^{-1} \mathbf{a}_{\tilde{R}_z}(\omega) \\ \mathbf{0}_{N_j 1} \end{bmatrix} \tag{9.134}$$

where

$$\hat{\mathbf{H}}(\omega) = \begin{bmatrix} \left\{ \mathbf{I}_{N_{\text{mod}}} - \hat{\boldsymbol{\omega}}_z^2 \left(\mathbf{I}_{N_{\text{mod}}} + \hat{\mathbf{D}} \boldsymbol{\Phi}_d \right) + 2i \hat{\boldsymbol{\omega}}_z \boldsymbol{\zeta}_z \right\} & -\hat{\boldsymbol{\omega}}_z^2 \hat{\mathbf{D}} \\ -\hat{\boldsymbol{\omega}}_d^2 \boldsymbol{\Phi}_d & \left\{ \mathbf{I}_{N_j} - \hat{\boldsymbol{\omega}}_d^2 + 2i \hat{\boldsymbol{\omega}}_d \boldsymbol{\zeta}_d \right\} \end{bmatrix}^{-1} \tag{9.135}$$

and where $\hat{\boldsymbol{\omega}}_z = diag\left[\omega/\omega_{z_n}\right]$ and $\hat{\boldsymbol{\omega}}_d = diag\left[\omega/\omega_j\right]$. As shown in Eq. 9.121, at a particular position x_r where we wish to determine the structural displacement response $r_z(x_r,t)$ and the corresponding relative displacement response of all the N_j mass dampers $\Delta\mathbf{r}_d(t)$

$$\begin{bmatrix} r_z(x_r,t) \\ \Delta\mathbf{r}_d(t) \end{bmatrix} = \boldsymbol{\psi}_r(x_r) \begin{bmatrix} \boldsymbol{\eta}_z(t) \\ \Delta\boldsymbol{\eta}_d(t) \end{bmatrix}$$

where

$$\boldsymbol{\psi}_r = \begin{bmatrix} \boldsymbol{\Phi}_z(x_r) & \mathbf{0}_{1N_j} \\ \mathbf{0}_{N_j N_{\text{mod}}} & \mathbf{I}_{N_j} \end{bmatrix} \tag{9.136}$$

A Fourier transform will then render

$$\begin{bmatrix} a_{r_z}(x_r,\omega) \\ \mathbf{a}_{\Delta r}(\omega) \end{bmatrix} = \boldsymbol{\psi}_r(x_r) \begin{bmatrix} \mathbf{a}_{\eta_z}(\omega) \\ \mathbf{a}_{\Delta\eta}(\omega) \end{bmatrix} \tag{9.137}$$

Thus

$$\begin{aligned} \mathbf{S}_r(x_r,\omega) &= \lim_{T\to\infty} \frac{1}{\pi T} \begin{bmatrix} a_{r_z} \\ \mathbf{a}_{\Delta r} \end{bmatrix}^* \begin{bmatrix} a_{r_z} \\ \mathbf{a}_{\Delta r} \end{bmatrix}^T = \\ & \lim_{T\to\infty} \frac{1}{\pi T} \left\{ \boldsymbol{\psi}_r(x_r) \begin{bmatrix} \mathbf{a}_{\eta_z}(\omega) \\ \mathbf{a}_{\Delta\eta_d}(\omega) \end{bmatrix} \right\}^* \left\{ \boldsymbol{\psi}_r(x_r) \begin{bmatrix} \mathbf{a}_{\eta_z}(\omega) \\ \mathbf{a}_{\Delta\eta_d}(\omega) \end{bmatrix} \right\}^T = \\ & \lim_{T\to\infty} \frac{1}{\pi T} \left\{ \boldsymbol{\psi}_r(x_r) \hat{\mathbf{H}}(\omega) \begin{bmatrix} \tilde{\mathbf{K}}_z^{-1} \mathbf{a}_{\tilde{R}_z}(\omega) \\ \mathbf{0}_{N_j 1} \end{bmatrix} \right\}^* \left\{ \boldsymbol{\psi}_r(x_r) \hat{\mathbf{H}}(\omega) \begin{bmatrix} \tilde{\mathbf{K}}_z^{-1} \mathbf{a}_{\tilde{R}_z}(\omega) \\ \mathbf{0}_{N_j 1} \end{bmatrix} \right\}^T \end{aligned} \tag{9.138}$$

from which the following is obtained

$$\mathbf{S}_r(x_r,\omega)=\mathbf{\Psi}_r(x_r)\cdot\hat{\mathbf{H}}^*(\omega)\cdot\hat{\mathbf{S}}_{\tilde{R}}(\omega)\cdot\hat{\mathbf{H}}^T(\omega)\cdot\mathbf{\Psi}_r^T(x_r) \tag{9.139}$$

where (recalling that $\tilde{\mathbf{K}}_z$ is diagonal and real)

$$\hat{\mathbf{S}}_{\tilde{R}}(\omega)=\begin{bmatrix}\tilde{\mathbf{K}}_z^{-1}\cdot\mathbf{S}_{\tilde{R}_z}(\omega)\cdot\left(\tilde{\mathbf{K}}_z^{-1}\right)^T & \mathbf{0}_{N_{\text{mod}}N_j}\\ \mathbf{0}_{N_jN_{\text{mod}}} & \mathbf{0}_{N_j}\end{bmatrix} \tag{9.140}$$

$$\text{and}\begin{cases}\mathbf{S}_{\tilde{R}_z}(\omega)=\lim\limits_{T\to\infty}\dfrac{1}{\pi T}\mathbf{a}_{\tilde{R}_z}^*(\omega)\cdot\mathbf{a}_{\tilde{R}_z}^T(\omega)\\ \text{where } \mathbf{a}_{\tilde{R}_z}(\omega)=\begin{bmatrix}a_{\tilde{R}_{z1}} & \cdots & a_{\tilde{R}_{zn}} & \cdots & a_{\tilde{R}_{zN_{\text{mod}}}}\end{bmatrix}^T\\ \text{and } a_{\tilde{R}_{zn}}(t)=\displaystyle\int_L a_{q_z}(x,\omega)\phi_{z_n}(x)\,dx\end{cases}$$

(see also Eq. 9.113), and where $a_{q_z}(x,\omega)$ is the Fourier amplitude of the distributed load $q_z(x,t)$.

Example 9.2:

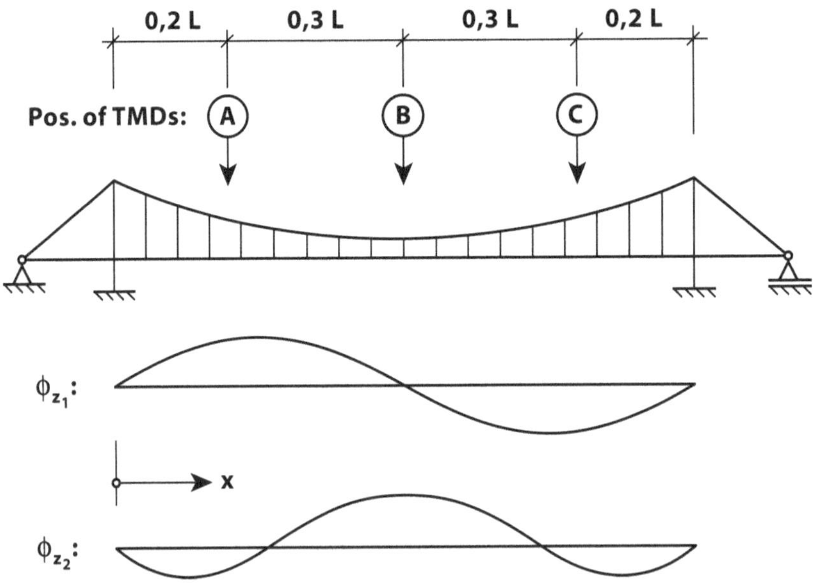

Fig. 9.22 Suspension bridge vertical modes susceptible to vortex shedding

The suspension bridge in Example 9.1 has two onerous vertical eigenmodes and corresponding eigenfrequencies and damping ratios

$$\left.\begin{aligned}\phi_{z_1} &= \sin(2\pi\hat{x}) \\ \phi_{z_2} &= 0.4\sin(\pi\hat{x}) - 0.6\sin(3\pi\hat{x})\end{aligned}\right\} \quad \hat{x} = \frac{x}{L} \quad \left.\begin{aligned}\omega_{z_1} &= 0.7 \\ \omega_{z_2} &= 0.9\end{aligned}\right\} rad/s$$

$$\left.\begin{aligned}\zeta_{z1} \\ \zeta_{z2}\end{aligned}\right\} = 0.005$$

which are vulnerable to wind induced vortex shedding oscillations, see Fig. 9.22 above. Therefore, it has been decided to install three mass dampers (A, B and C) with the mass properties $\mu_A = \mu_C = 0.0025$ *and* $\mu_B = 0.005$ *at positions* $x_A = 0.2L$, $x_B = 0.5L$ *and* $x_C = 0.8L$. *Since mass dampers A and C are primarily intended to quench possible oscillations in the first vertical mode while the mass damper B is intended to quench oscillations in the second, it is taken for granted that*

$$\left.\begin{aligned}\omega_A &= \omega_C = \omega_{z_1} / (1 + \mu_A) \\ \omega_B &= \omega_{z_2} / (1 + \mu_B)\end{aligned}\right\} \quad \text{and} \quad \begin{aligned}\zeta_A &= \zeta_C = \sqrt{\frac{3\mu_A}{8(1+\mu_A)^3}} \\ \zeta_B &= \sqrt{\frac{3\mu_B}{8(1+\mu_B)^3}}\end{aligned}$$

Basic structural data and load properties are given in Example 9.1 above. Since there are two modes that may be excited, there will be two critical wind velocities

$$\omega_s = \omega_{z_1} \quad \Rightarrow \quad V_{R_1} = \frac{D\omega_{z_1}}{2\pi St} \approx 2.4 \text{ m/s}$$

$$\omega_s = \omega_{z_2} \quad \Rightarrow \quad V_{R_2} = \frac{D\omega_{z_2}}{2\pi St} \approx 3.1 \text{ m/s}$$

for the onset of resonant vortex shedding induced oscillations. There are two mode shapes and three mass dampers, and hence, the size of the system is 5 by 5. It is necessary to establish the following matrices:

$$\mathbf{\Phi}_z = \begin{bmatrix}\phi_{z_1}(x) & \phi_{z_2}(x)\end{bmatrix} \qquad \mathbf{\Phi}_d = \begin{bmatrix}\mathbf{\Phi}_z(x_A) \\ \mathbf{\Phi}_z(x_B) \\ \mathbf{\Phi}_z(x_C)\end{bmatrix} = \begin{bmatrix}\phi_{z_1}(x_A) & \phi_{z_2}(x_A) \\ \phi_{z_1}(x_B) & \phi_{z_2}(x_B) \\ \phi_{z_1}(x_C) & \phi_{z_2}(x_C)\end{bmatrix}$$

$$\tilde{\mathbf{M}}_z = \begin{bmatrix}\tilde{M}_{z1} & 0 \\ 0 & \tilde{M}_{z2}\end{bmatrix} \quad \text{where} \quad \begin{bmatrix}\tilde{M}_{z1} \\ \tilde{M}_{z2}\end{bmatrix} = \int_0^L m_z \begin{bmatrix}\phi_{z1}^2 \\ \phi_{z2}^2\end{bmatrix} dx$$

$$\mathbf{M}_d = \begin{bmatrix} M_A & 0 & 0 \\ 0 & M_B & 0 \\ 0 & 0 & M_C \end{bmatrix} = \begin{bmatrix} \mu_A \tilde{M}_{z1} & 0 & 0 \\ 0 & \mu_B \tilde{M}_{z2} & 0 \\ 0 & 0 & \mu_C \tilde{M}_{z1} \end{bmatrix}$$

$$\hat{\boldsymbol{\omega}}_z = \omega \cdot \begin{bmatrix} \omega_{z1}^{-1} & 0 \\ 0 & \omega_{z2}^{-1} \end{bmatrix} \qquad \hat{\boldsymbol{\omega}}_d = \omega \cdot \begin{bmatrix} \omega_A^{-1} & 0 & 0 \\ 0 & \omega_B^{-1} & 0 \\ 0 & 0 & \omega_C^{-1} \end{bmatrix}$$

$$\boldsymbol{\zeta}_z = \begin{bmatrix} \left(\zeta_{z1} - \zeta_{ae_z}\right) & 0 \\ 0 & \left(\zeta_{z2} - \zeta_{ae_z}\right) \end{bmatrix} \qquad \boldsymbol{\zeta}_d = \begin{bmatrix} \zeta_A & 0 & 0 \\ 0 & \zeta_B & 0 \\ 0 & 0 & \zeta_C \end{bmatrix}$$

$$\tilde{\mathbf{K}}_z^{-1} = \begin{bmatrix} \left(\omega_{z1}^2 \tilde{M}_{z1}\right)^{-1} & 0 \\ 0 & \left(\omega_{z2}^2 \tilde{M}_{z2}\right)^{-1} \end{bmatrix} \; \textit{and} \; \hat{\mathbf{S}}_{\tilde{R}}(\omega) = \begin{bmatrix} \tilde{\mathbf{K}}_z^{-1} \mathbf{S}_{\tilde{R}_z}(\omega) \tilde{\mathbf{K}}_z^{-1} & \mathbf{0}_{2 \cdot 3} \\ \mathbf{0}_{3 \cdot 2} & \mathbf{0}_{3 \cdot 3} \end{bmatrix}$$

$$\mathbf{S}_{\tilde{R}_z} = \lim_{T \to \infty} \frac{1}{\pi T} \mathbf{a}_{\tilde{R}_z}^* \mathbf{a}_{\tilde{R}_z}^T = \lim_{T \to \infty} \frac{1}{\pi T} \begin{bmatrix} a_{\tilde{R}_{z1}} \\ a_{\tilde{R}_{z2}} \end{bmatrix}^* \begin{bmatrix} a_{\tilde{R}_{z1}} \\ a_{\tilde{R}_{z2}} \end{bmatrix}^T = \begin{bmatrix} S_{\tilde{R}_{z1}\tilde{R}_{z1}} & S_{\tilde{R}_{z1}\tilde{R}_{z2}} \\ S_{\tilde{R}_{z2}\tilde{R}_{z1}} & S_{\tilde{R}_{z2}\tilde{R}_{z2}} \end{bmatrix}$$

Assuming that the covariance between vortex shedding forces at V_{R_1} *and* V_{R_2} *are negligible, assuming that the resonance cases* V_{R_1} *and* V_{R_2} *are most relevant cases, i.e.* $\omega = \omega_s = 2\pi V St / D$ *and taking it for granted that the tuned mass dampers are effective such that* $r_z(x,t)$ *is small, in which case* $\zeta_{ae_z} \approx 0$, *then*

$$\hat{\mathbf{S}}_{\tilde{R}}(\omega) = \begin{bmatrix} \left(\omega_{z1}^2 \tilde{M}_{z1}\right)^{-2} S_{\tilde{R}_{z1}\tilde{R}_{z1}}(\omega) & 0 & 0 & 0 & 0 \\ 0 & \left(\omega_{z2}^2 \tilde{M}_{z2}\right)^{-2} S_{\tilde{R}_{z2}\tilde{R}_{z2}}(\omega) & 0 & 0 & 0 \\ 0 & 0 & 0 & 0 & 0 \\ 0 & 0 & 0 & 0 & 0 \\ 0 & 0 & 0 & 0 & 0 \end{bmatrix}$$

where

$$\begin{bmatrix} S_{\tilde{R}_{z1}\tilde{R}_{z1}} \\ S_{\tilde{R}_{z2}\tilde{R}_{z2}} \end{bmatrix} = 2\lambda D S_{q_z} \int_0^L \begin{bmatrix} \phi_{z1}^2 \\ \phi_{z2}^2 \end{bmatrix} dx \quad \textit{where}$$

$$S_{q_z} = \frac{\left(\frac{1}{2}\rho V^2 B \hat{\sigma}_{q_z}\right)^2}{\sqrt{\pi}\omega_s b_z} \exp\left[-\left(\frac{1-\frac{\omega}{\omega_s}}{b_z}\right)^2\right]$$

In this case it will suffice to

$$\boldsymbol{\Psi}_r = \begin{bmatrix} \boldsymbol{\Phi}_z(x_r) & \mathbf{0}_{1\cdot 3} \\ \mathbf{0}_{3\cdot 2} & \mathbf{I}_{3\cdot 3} \end{bmatrix} = \begin{bmatrix} \phi_{z1}(x_r) & \phi_{z2}(x_r) & 0 & 0 & 0 \\ 0 & 0 & 1 & 0 & 0 \\ 0 & 0 & 0 & 1 & 0 \\ 0 & 0 & 0 & 0 & 1 \end{bmatrix} \qquad \textit{where}$$

$$x_r = \begin{cases} x_A \\ \text{or } x_B \end{cases}$$

because due to symmetry or asymmetry the response at x_C *will be equal to or opposite to the response at* x_A. *What then remains is to calculate* $\hat{\mathbf{D}} = \tilde{\mathbf{M}}_z^{-1}\boldsymbol{\Phi}_d^T\mathbf{M}_d$ *and the frequency response function*

$$\hat{\mathbf{H}}(\omega) = \begin{bmatrix} \left\{\mathbf{I}_2 - \hat{\boldsymbol{\omega}}_z^2\left(\mathbf{I}_2 + \hat{\mathbf{D}}\boldsymbol{\Phi}_d\right) + 2i\hat{\boldsymbol{\omega}}_z\boldsymbol{\zeta}_z\right\} & -\hat{\boldsymbol{\omega}}_z^2\hat{\mathbf{D}} \\ -\hat{\boldsymbol{\omega}}_d^2\boldsymbol{\Phi}_d & \left\{\mathbf{I}_3 - \hat{\boldsymbol{\omega}}_d^2 + 2i\hat{\boldsymbol{\omega}}_d\boldsymbol{\zeta}_d\right\} \end{bmatrix}^{-1}$$

Thus, introducing $V = \begin{cases} V_{R1} \\ \text{or } V_{R2} \end{cases}$ *and* $x_r = \begin{cases} x_A \\ \text{or } x_B \end{cases}$ *and then the spectral density of the response components are given by*

$$\mathbf{S}_r = \begin{bmatrix} S_{r_z} & S_{r_z\Delta r_A} & S_{r_z\Delta r_B} & S_{r_z\Delta r_C} \\ & S_{\Delta r_A} & S_{\Delta r_A\Delta r_B} & S_{\Delta r_A\Delta r_C} \\ & & S_{\Delta r_B} & S_{\Delta r_B\Delta r_C} \\ Sym. & & & S_{\Delta r_C} \end{bmatrix} = \boldsymbol{\Psi}_r(x_r)\hat{\mathbf{H}}^*(\omega)\hat{\mathbf{S}}_{\tilde{R}}(\omega)\hat{\mathbf{H}}^T(\omega)\boldsymbol{\Psi}_r^T(x_r)$$

Setting $V = V_{R1}$, *then the spectral densities of the dynamic response of the main system at* $x_r = x_A = 0.2L$ *and of the relative dynamic response of the mass damper A are shown in Fig. 9.23. As could be expected, they are centred on the resonance frequency* $\omega_{z1} = 0.7\ rad/s$. *Corresponding time domain plots are shown in Fig. 9.24. The response is not particularly narrow banded, and hence, a fairly high peak factor should be adopted. As can be seen, the excitation of the main system at* x_B *and mass damper B are small. Similarly, setting* $V = V_{R_2}$, *then the spectral densities of the dynamic response of the main system at* $x_r = x_B = 0.5L$ *and of the relative dynamic response of the mass damper B are shown in Fig. 9.25. This time they are centred on the resonance frequency* $\omega_{z_2} = 0.9\ rad/s$. *Corresponding time domain plots are shown in Fig. 9.26. Again the response is not particularly narrow banded, while the excitation of the main system at* x_A *and mass damper A are small. However, as to whether or not relative damper displacements in the order of* $\pm 1\ m$ *are manageable is a practical design question.*

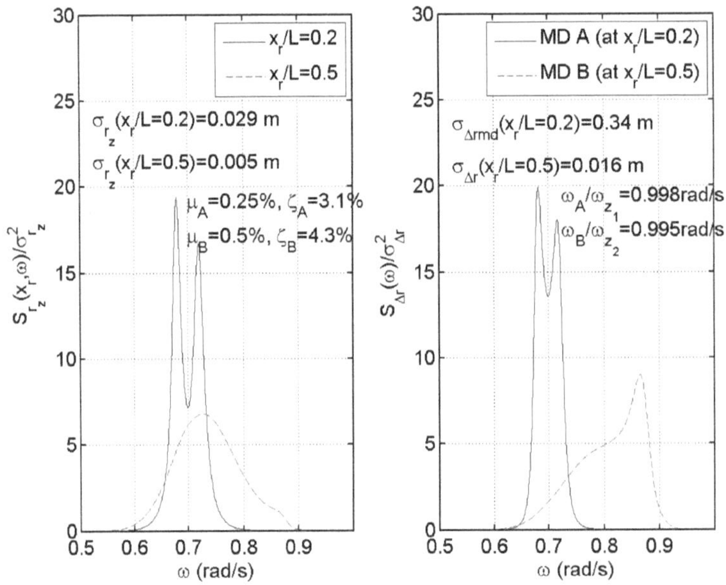

Fig. 9.23 $V = V_{R_1}$, response of main system to the left, response of mass dampers A and B to the right

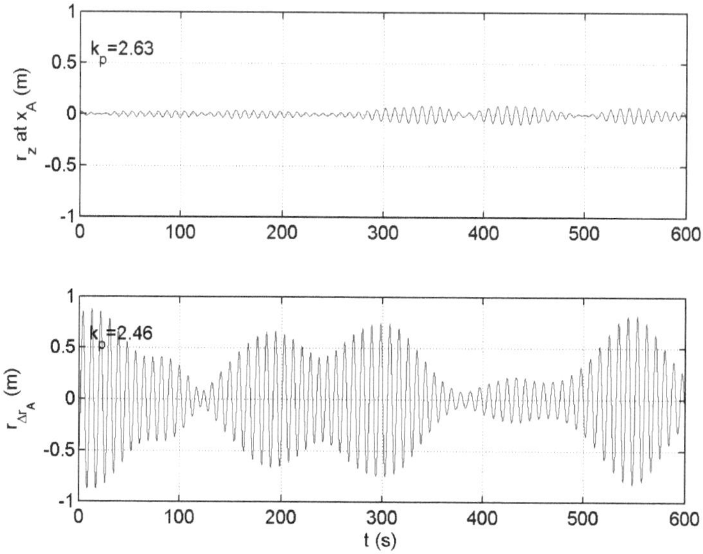

Fig. 9.24 $V = V_{R_1}$, response of main system top diagram, response of mass damper A in lower diagram

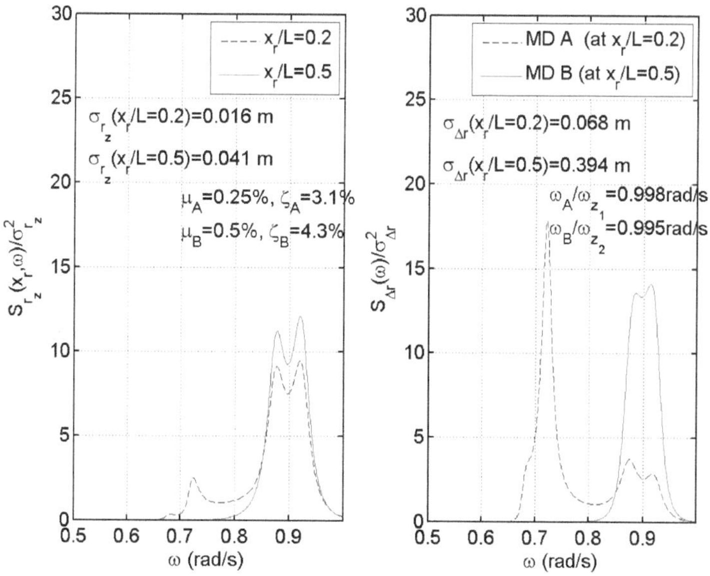

Fig. 9.25 $V = V_{R_2}$, response of main system to the left, response of mass dampers A and B to the right

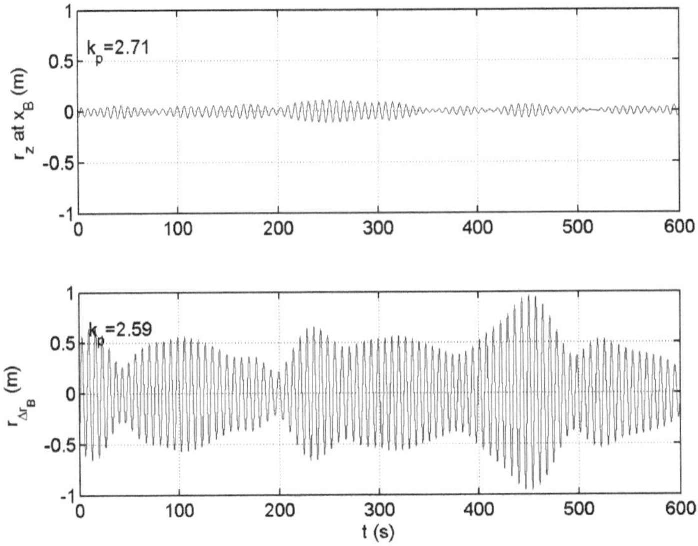

Fig. 9.26 $V = V_{R_2}$, response of main system top diagram, response of mass damper B in lower diagram

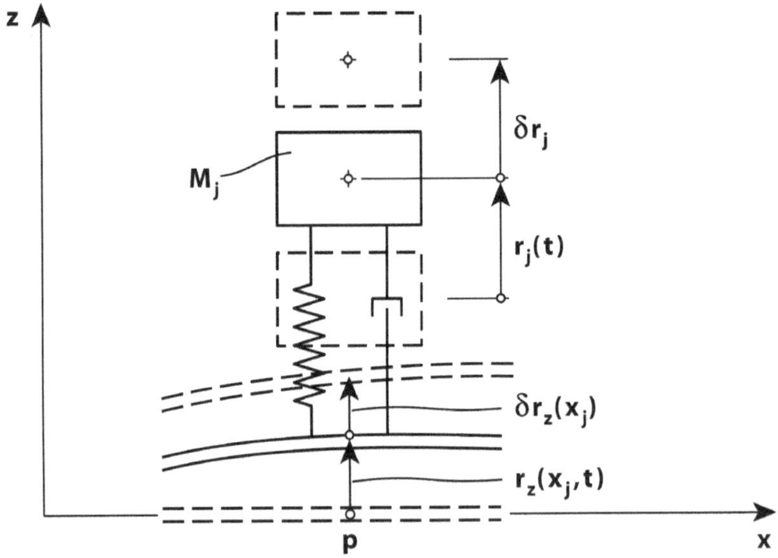

Fig. 9.27 Arbitrary mass damper number j attached to node p

The Mass Damper in a Finite Element Format:

The theory above may also be formulated in a finite element format. It is in the following assumed that the system contains N elements, N_p nodes and N_j

mass dampers. As illustrated in Fig. 9.27 it is taken for granted that the mass dampers are all attached to a node, i.e. that an arbitrary mass damper j with properties M_j, C_j and K_j is attached to node p.

Let the damper properties be defined by the diagonal matrices

$$\left.\begin{aligned} \mathbf{M}_d &= diag\left[M_j\right] \\ \mathbf{C}_d &= diag\left[C_j\right] \\ \mathbf{K}_d &= diag\left[K_j\right] \end{aligned}\right\} j = 1,2,\ldots,N_j \tag{9.141}$$

and, at global level, the N_r physical degrees of freedom of the system and corresponding load components are defined by

$$\left.\begin{aligned} \mathbf{r} &= \begin{bmatrix} r_1 & \cdots & r_p & \cdots & r_{N_r} \end{bmatrix}^T \\ \mathbf{R} &= \begin{bmatrix} R_1 & \cdots & R_p & \cdots & R_{N_r} \end{bmatrix}^T \end{aligned}\right\} \tag{9.142}$$

At element level the physical degrees of freedom and corresponding force components are as usual defined by

$$\left.\begin{aligned} \mathbf{d}_n &= \begin{bmatrix} \mathbf{d}_1 \\ \mathbf{d}_2 \end{bmatrix}_n \quad \text{where} \begin{cases} \mathbf{d}_1 = \begin{bmatrix} d_1 & d_2 & d_3 & d_4 & d_5 & d_6 \end{bmatrix}^T \\ \mathbf{d}_2 = \begin{bmatrix} d_7 & d_8 & d_9 & d_{10} & d_{11} & d_{12} \end{bmatrix}^T \end{cases} \\ \mathbf{F}_n &= \begin{bmatrix} \mathbf{F}_1 \\ \mathbf{F}_2 \end{bmatrix}_n \quad \text{where} \begin{cases} \mathbf{F}_1 = \begin{bmatrix} F_1 & F_2 & F_3 & F_4 & F_5 & F_6 \end{bmatrix}^T \\ \mathbf{F}_2 = \begin{bmatrix} F_7 & F_8 & F_9 & F_{10} & F_{11} & F_{12} \end{bmatrix}^T \end{cases} \end{aligned}\right\} \tag{9.143}$$

As shown in Chapter 4, the equilibrium condition at element level is given by

$$\mathbf{F}_n(t) = \mathbf{m}_n \ddot{\mathbf{d}}_n(t) + \mathbf{c}_n \dot{\mathbf{d}}_n(t) + \mathbf{k}_n \mathbf{d}_n(t) \tag{9.144}$$

where $\mathbf{m}_n$, $\mathbf{c}_n$ and $\mathbf{k}_n$ are defined in Chapter 4.2. The connection between local and global degrees of freedom is defined by the connectivity matrix $\mathbf{A}_n$ such that

$$\mathbf{d}_n = \mathbf{A}_n \mathbf{r} \tag{9.145}$$

Let the N_j damper degrees of freedom be defined by

$$\mathbf{r}_d = \begin{bmatrix} r_1 & \cdots & r_j & \cdots & r_{N_j} \end{bmatrix}^T \tag{9.146}$$

In addition to this, it is necessary to define a vector $\mathbf{r}_p$ with length N_j, containing all the global degrees of freedom associated with a mass damper (i.e. the global degrees of freedom at a node where there is attached a mass damper and whose degree of freedom is parallel to that response motion)

$$\mathbf{r}_p = \begin{bmatrix} \cdots & \cdots & \left(r_{p_j} = r_p \text{ at mass damper number } j \right) & \cdots & \cdots \end{bmatrix}^T \tag{9.147}$$

Furthermore, it is convenient to define a N_r by 1 vector $\mathbf{a}_j$

$$\mathbf{a}_j = \begin{bmatrix} 0 & \cdots & \left(1 \text{ at the } r_p \text{ position of mass damper number } j \right) & \cdots & 0 \end{bmatrix}^T \tag{9.148}$$

such that an arbitrary r_{p_j} associated with mass damper number j is identified by $r_{p_j} = \mathbf{a}_j^T \mathbf{r}$. Thus, the N_j by N_r connectivity matrix $\mathbf{A}_p$ between $\mathbf{r}_p$ and $\mathbf{r}$ is defined by

$$\mathbf{r}_p = \mathbf{A}_p \mathbf{r} \text{ where } \mathbf{A}_p = \begin{bmatrix} \mathbf{a}_1 & \cdots & \mathbf{a}_j & \cdots & \mathbf{a}_{N_j} \end{bmatrix}^T \tag{9.149}$$

Finally, it is convenient to define the vectors

$$\left. \begin{aligned} \mathbf{F}_d &= \begin{bmatrix} F_1 & \cdots & F_j & \cdots & F_{N_j} \end{bmatrix}^T \\ \mathbf{Q}_d &= \begin{bmatrix} Q_1 & \cdots & Q_j & \cdots & Q_{N_j} \end{bmatrix}^T \end{aligned} \right\} \tag{9.150}$$

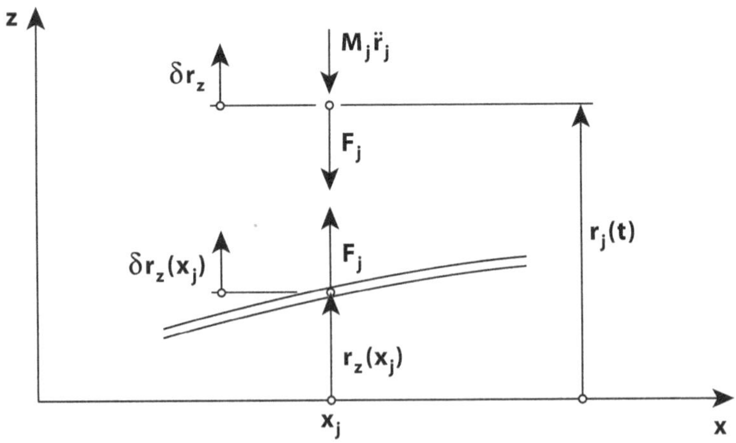

Fig. 9.28 Internal forces in mass damper number j

where F_j is the sum of internal spring and damper forces associated with mass damper j and Q_j is the corresponding inertia force, see Fig. 9.28, from which it is seen that

$$\mathbf{F}_d = \mathbf{C}_d\left(\dot{\mathbf{r}}_d - \dot{\mathbf{r}}_p\right) + \mathbf{K}_d\left(\mathbf{r}_d - \mathbf{r}_p\right) = \mathbf{C}_d\left(\dot{\mathbf{r}}_d - \mathbf{A}_p\dot{\mathbf{r}}\right) + \mathbf{K}_d\left(\mathbf{r}_d - \mathbf{A}_p\mathbf{r}\right) \quad (9.151)$$

and

$$\mathbf{Q}_d = \mathbf{M}_d\ddot{\mathbf{r}}_d \quad (9.152)$$

To the equilibrium condition of this system there is imposed a set of virtual displacements

$$\delta\mathbf{r} = \begin{bmatrix} \delta r_1 & \cdots & \delta r_p & \cdots & \delta r_{N_r} \end{bmatrix}^T \quad (9.153)$$

on the N_r degrees of freedom of the system itself, and

$$\delta\mathbf{r}_d = \begin{bmatrix} \delta r_1 & \cdots & \delta r_j & \cdots & \delta r_{N_j} \end{bmatrix}^T \quad (9.154)$$

on the N_j mass damper degrees of freedom. Accordingly $\delta\mathbf{d}_n = \mathbf{A}_n\delta\mathbf{r}$ and $\delta\mathbf{r}_p = \mathbf{A}_p\delta\mathbf{r}$. Thus

$$\delta\mathbf{r}^T\mathbf{R} - \sum_{n=1}^{N}\delta\mathbf{d}_n^T\mathbf{F}_n + \delta\mathbf{r}_p^T\mathbf{F}_d - \delta\mathbf{r}_d^T\left(\mathbf{F}_d + \mathbf{Q}_d\right) = 0 \quad (9.155)$$

from which the following is obtained

$$\begin{aligned} &\delta\mathbf{r}^T\left(\mathbf{M}\ddot{\mathbf{r}} + \mathbf{C}\dot{\mathbf{r}} + \mathbf{K}\mathbf{r} + \mathbf{A}_p^T\mathbf{C}_d\mathbf{A}_p\dot{\mathbf{r}} + \mathbf{A}_p^T\mathbf{K}_d\mathbf{A}_p\mathbf{r} - \mathbf{A}_p^T\mathbf{C}_d\dot{\mathbf{r}}_d - \mathbf{A}_p^T\mathbf{K}_d\mathbf{r}_d\right) \\ &+\delta\mathbf{r}_d^T\left(-\mathbf{C}_d\mathbf{A}_p\dot{\mathbf{r}} - \mathbf{K}_d\mathbf{A}_p\mathbf{r} + \mathbf{M}_d\ddot{\mathbf{r}}_d + \mathbf{C}_d\dot{\mathbf{r}}_d + \mathbf{K}_d\mathbf{r}_d\right) = \delta\mathbf{r}^T\mathbf{R} \end{aligned} \quad (9.156)$$

where

$$\begin{bmatrix} \mathbf{M} \\ \mathbf{C} \\ \mathbf{K} \end{bmatrix} = \sum_{n=1}^{N}\mathbf{A}_n^T \begin{bmatrix} \mathbf{m}_n \\ \mathbf{c}_n \\ \mathbf{k}_n \end{bmatrix} \mathbf{A}_n \quad (9.157)$$

and $\mathbf{m}_n$, $\mathbf{c}_n$ and $\mathbf{k}_n$ are cross sectional mass , damping and stiffness matrices of element number n. This may more conveniently be written

$$\begin{bmatrix} \delta \mathbf{r} \\ \delta \mathbf{r}_d \end{bmatrix}^T \left(\begin{bmatrix} \mathbf{M} & \mathbf{0} \\ \mathbf{0} & \mathbf{M}_d \end{bmatrix} \begin{bmatrix} \ddot{\mathbf{r}} \\ \ddot{\mathbf{r}}_d \end{bmatrix} + \begin{bmatrix} \mathbf{C} + \mathbf{A}_p^T \mathbf{C}_d \mathbf{A}_p & -\mathbf{A}_p^T \mathbf{C}_d \\ -\mathbf{C}_d \mathbf{A}_p & \mathbf{C}_d \end{bmatrix} \begin{bmatrix} \dot{\mathbf{r}} \\ \dot{\mathbf{r}}_d \end{bmatrix} + \begin{bmatrix} \mathbf{K} + \mathbf{A}_p^T \mathbf{K}_d \mathbf{A}_p & -\mathbf{A}_p^T \mathbf{K}_d \\ -\mathbf{K}_d \mathbf{A}_p & \mathbf{K}_d \end{bmatrix} \begin{bmatrix} \mathbf{r} \\ \mathbf{r}_d \end{bmatrix} \right) = \begin{bmatrix} \delta \mathbf{r} \\ \delta \mathbf{r}_d \end{bmatrix}^T \begin{bmatrix} \mathbf{R} \\ \mathbf{0} \end{bmatrix} \tag{9.158}$$

As can be seen, the pre-multiplication by the virtual displacement vectors may be omitted, and thus the equilibrium condition of a discrete N_r by N_r system with a set of N_j mass dampers is given by

$$\begin{bmatrix} \mathbf{M} & \mathbf{0} \\ \mathbf{0} & \mathbf{M}_d \end{bmatrix} \begin{bmatrix} \ddot{\mathbf{r}} \\ \ddot{\mathbf{r}}_d \end{bmatrix} + \begin{bmatrix} \mathbf{C} + \mathbf{A}_p^T \mathbf{C}_d \mathbf{A}_p & -\mathbf{A}_p^T \mathbf{C}_d \\ -\mathbf{C}_d \mathbf{A}_p & \mathbf{C}_d \end{bmatrix} \begin{bmatrix} \dot{\mathbf{r}} \\ \dot{\mathbf{r}}_d \end{bmatrix} + \begin{bmatrix} \mathbf{K} + \mathbf{A}_p^T \mathbf{K}_d \mathbf{A}_p & -\mathbf{A}_p^T \mathbf{K}_d \\ -\mathbf{K}_d \mathbf{A}_p & \mathbf{K}_d \end{bmatrix} \begin{bmatrix} \mathbf{r} \\ \mathbf{r}_d \end{bmatrix} = \begin{bmatrix} \mathbf{R} \\ \mathbf{0} \end{bmatrix} \tag{9.159}$$

which may also be transformed into relative coordinates

$$\begin{bmatrix} \mathbf{r} \\ \mathbf{r}_d \end{bmatrix} = \begin{bmatrix} \mathbf{r} \\ \mathbf{r}_p + \mathbf{r}_d \end{bmatrix} = \begin{bmatrix} \mathbf{r} \\ \mathbf{A}_p \mathbf{r} + \Delta \mathbf{r}_d \end{bmatrix} = \begin{bmatrix} \mathbf{I}_{N_r} & \mathbf{0} \\ \mathbf{A}_p & \mathbf{I}_{N_j} \end{bmatrix} \begin{bmatrix} \mathbf{r} \\ \Delta \mathbf{r}_d \end{bmatrix} \tag{9.160}$$

where $\mathbf{I}_{N_r}$ and $\mathbf{I}_{N_j}$ are N_r by N_r and N_j by N_j identity matrices. Thus

$$\begin{bmatrix} \mathbf{r} \\ \mathbf{r}_d \end{bmatrix} = \mathbf{\Psi}_{rel} \begin{bmatrix} \mathbf{r} \\ \Delta \mathbf{r}_d \end{bmatrix} \quad \text{where} \quad \mathbf{\Psi}_{rel} = \begin{bmatrix} \mathbf{I}_{N_r} & \mathbf{0} \\ \mathbf{A}_p & \mathbf{I}_{N_j} \end{bmatrix} \tag{9.161}$$

By introducing this into the equilibrium equation above and pre-multiplication by $\mathbf{\Psi}_{rel}^T$ then the following equilibrium condition in relative degrees of freedom is obtained

$$\begin{bmatrix} \mathbf{M} + \mathbf{A}_p^T \mathbf{M}_d \mathbf{A}_p & \mathbf{A}_p^T \mathbf{M}_d \\ \mathbf{M}_d \mathbf{A}_p & \mathbf{M}_d \end{bmatrix} \begin{bmatrix} \ddot{\mathbf{r}} \\ \Delta \ddot{\mathbf{r}}_d \end{bmatrix} + \begin{bmatrix} \mathbf{C} & \mathbf{0} \\ \mathbf{0} & \mathbf{C}_d \end{bmatrix} \begin{bmatrix} \dot{\mathbf{r}} \\ \Delta \dot{\mathbf{r}}_d \end{bmatrix} + \begin{bmatrix} \mathbf{K} & \mathbf{0} \\ \mathbf{0} & \mathbf{K}_d \end{bmatrix} \begin{bmatrix} \mathbf{r} \\ \Delta \mathbf{r}_d \end{bmatrix} = \begin{bmatrix} \mathbf{R} \\ \mathbf{0} \end{bmatrix} \tag{9.162}$$

Chapter 10
Rectangular Plates

10.1 Introduction

The development below is limited to cover the theory of rectangular thin and plane plates. It is often referred to as the Kirchhoff-Love theory, as it was first presented by A.E.H. Love [45] based on basic assumptions outline by G.R. Kirchhoff (1824 – 1887). Perpendicular to the plate it is subject to a fluctuating and distributed load $q_z(x, y, t)$ (with unit N/m^2). In the plane of the plate it may be subject to time invariant evenly distributed axial loads N_x and N_y (with units N/m). It is taken for granted that the plate is homogeneous and isotropic, that it is linear elastic, and, although this is not a general requirement, it is most often assumed that the plate thickness h is constant. Since the plate is thin as compared to its overall dimension ($h \ll L_x$ and L_y) it is assumed that stresses perpendicular to the plate plane may be ignored. For the same reason Navier's

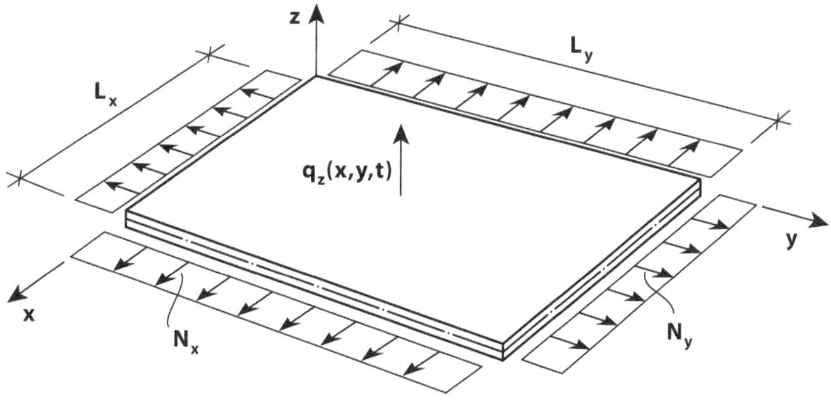

Fig. 10.1 Rectangular, plane and isotropic plate

E.N. Strømmen, *Structural Dynamics*, Springer Series in Solid and Structural Mechanics 2,
DOI: 10.1007/978-3-319-01802-7_10, © Springer International Publishing Switzerland 2014

hypothesis [4] is adopted, implying that a straight section which is normal to the middle surface before any plate deformations will remain straight after deformations, i.e. there are no cross sectional distortion. Finally, the theory is limited to plate deformations $r_z(x,y,t)$ which are small, such that for any cross sectional rotation α it will be sufficiently accurate to assume that $\cos\alpha \approx 1$ and that $\sin\alpha \approx \tan\alpha \approx \partial r_z/\partial j$, $j = x$ or y. Finally, in establishing the relevant equilibrium conditions it is taken for granted that the principle of d'Alambert applies. Thus, it is seen that the Kirchhoff-Love theory is an extension of the beam theory first presented in Chapter 1.2.

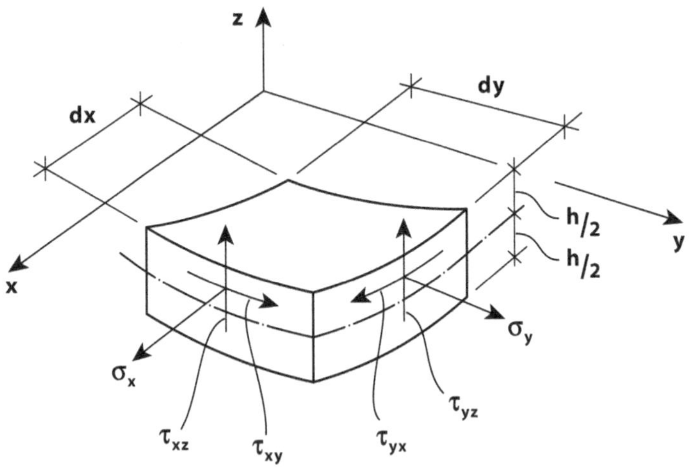

Fig. 10.2 Definition of stress components

The relevant normal and shear stress components σ_x, σ_y, τ_{xy}, τ_{yx}, τ_{xz} and τ_{yz} are shown in Fig. 10.2. Considering an infinitesimal element $dx \cdot dy \cdot dz$ then the requirements with respect to force equilibrium in the x and y directions

$$\left.\begin{aligned}(\sigma_x + d\sigma_x)dydz - \sigma_x dydz + (\tau_{yx} + d\tau_{yx})dxdz - \tau_{yx}dxdz = 0\\(\sigma_y + d\sigma_y)dxdz - \sigma_y dxdz + (\tau_{xy} + d\tau_{xy})dydz - \tau_{xy}dydz = 0\end{aligned}\right\} \tag{10.1}$$

and moment equilibrium about a vertical axis z_s through the middle of the element

$$\left(\tau_{xy}+d\tau_{xy}\right)dydz\frac{dx}{2}+\tau_{xy}dydz\frac{dx}{2}-\left(\tau_{yx}+d\tau_{yx}\right)dxdz\frac{dy}{2}-\tau_{yx}dxdz\frac{dy}{2}=0 \quad (10.2)$$

will render

$$\left.\begin{aligned}\partial\sigma_x/\partial x+\partial\tau_{yx}/\partial y=0\\ \partial\sigma_y/\partial y+\partial\tau_{xy}/\partial x=0\end{aligned}\right\} \quad \text{and} \quad \tau_{xy}=\tau_{yx} \quad (10.3)$$

Thus, for the relevant in plane stress components it will suffice to focus on σ_x, σ_y and τ_{xy}.

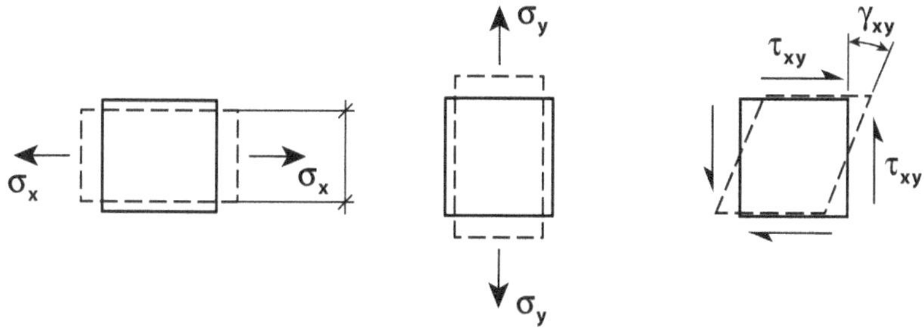

Fig. 10.3 Superposition of in-plane strain components

The relationship between stresses σ_x, σ_y, τ_{xy} and corresponding strain components ε_x, ε_y and γ_{xy} may, as illustrated in Fig. 10.3, be obtained by using the principle of superposition, i.e. by adding the effects of each strain component separately:

$$\left.\begin{aligned}&\sigma_x \text{ alone} \Rightarrow \begin{cases}\varepsilon_x=\sigma_x/E\\ \varepsilon_y=-\upsilon\varepsilon_x\end{cases}\\ &\sigma_y \text{ alone} \Rightarrow \begin{cases}\varepsilon_y=\sigma_y/E\\ \varepsilon_x=-\upsilon\varepsilon_y\end{cases}\\ &\tau_{xy} \text{ alone} \Rightarrow \gamma_{xy}=\tau_{xy}/G\end{aligned}\right\} \Rightarrow \begin{cases}\varepsilon_x=\dfrac{\sigma_x}{E}-\upsilon\dfrac{\sigma_y}{E}\\ \varepsilon_y=\dfrac{\sigma_y}{E}-\upsilon\dfrac{\sigma_x}{E}\\ \gamma_{xy}=\dfrac{\tau_{xy}}{G}\end{cases} \quad (10.4)$$

where E is the elastic normal stress modulus, G is the corresponding shear stress modulus and υ is the Poisson ratio. (For an elastic homogeneous and isotropic material the shear stress modulus is given by $G=E/\left[2(1+\upsilon)\right]$, see e.g. Timoshenko & Goodier [5].) Thus,

$$\begin{bmatrix} \varepsilon_x \\ \varepsilon_y \\ \gamma_{xy} \end{bmatrix} = \begin{bmatrix} E^{-1} & -\upsilon E^{-1} & 0 \\ -\upsilon E^{-1} & E^{-1} & 0 \\ 0 & 0 & G^{-1} \end{bmatrix} \begin{bmatrix} \sigma_x \\ \sigma_y \\ \tau_{xy} \end{bmatrix} \tag{10.5}$$

from which the following in-plane stress-strain relationship is obtained

$$\begin{bmatrix} \sigma_x \\ \sigma_y \\ \tau_{xy} \end{bmatrix} = \frac{E}{1-\upsilon^2} \begin{bmatrix} 1 & \upsilon & 0 \\ \upsilon & 1 & 0 \\ 0 & 0 & (1-\upsilon)/2 \end{bmatrix} \begin{bmatrix} \varepsilon_x \\ \varepsilon_y \\ \gamma_{xy} \end{bmatrix} \tag{10.6}$$

First, it is for simplicity assumed that the displacements $r_z(x, y, t)$ are so small that they will only cause insignificant stretching, i.e. that

$$\left.\begin{aligned} \left(dx^2 + dr_z^2\right)^{1/2} &\approx dx\left[1 + (\partial r_z/\partial x)^2/2\right] \approx dx \\ \left(dy^2 + dr_z^2\right)^{1/2} &\approx dy\left[1 + (\partial r_z/\partial y)^2/2\right] \approx dy \end{aligned}\right\} \tag{10.7}$$

and thus, the normal stresses σ_x and σ_y that are associated with the deformation r_z will only create pure plate bending, as illustrated in Fig. 10.4. It is seen that for an infinitesimal element $dx \cdot dy \cdot dz$

$$\left.\begin{aligned} \varepsilon_x &= \frac{-\left(\frac{\partial r_z}{\partial x} + \frac{\partial^2 r_z}{\partial x^2} dx\right)\cdot z + \frac{\partial r_z}{\partial x}\cdot z}{dx} = -\frac{\partial^2 r_z}{\partial x^2}\cdot z \\ \varepsilon_y &= \frac{-\left(\frac{\partial r_z}{\partial y} + \frac{\partial^2 r_z}{\partial y^2} dy\right)\cdot z + \frac{\partial r_z}{\partial y}\cdot z}{dy} = -\frac{\partial^2 r_z}{\partial x^2}\cdot z \\ \gamma_{xy} &= \frac{-\left(\frac{\partial r_z}{\partial x} + \frac{\partial^2 r_z}{\partial y \partial x} dx\right)\cdot z + \frac{\partial r_z}{\partial x}\cdot z}{dx} + \frac{-\left(\frac{\partial r_z}{\partial y} + \frac{\partial^2 r_z}{\partial x \partial y} dx\right)\cdot z + \frac{\partial r_z}{\partial y}\cdot z}{dx} = -2\frac{\partial^2 r_z}{\partial x \partial y}\cdot z \end{aligned}\right\} \tag{10.8}$$

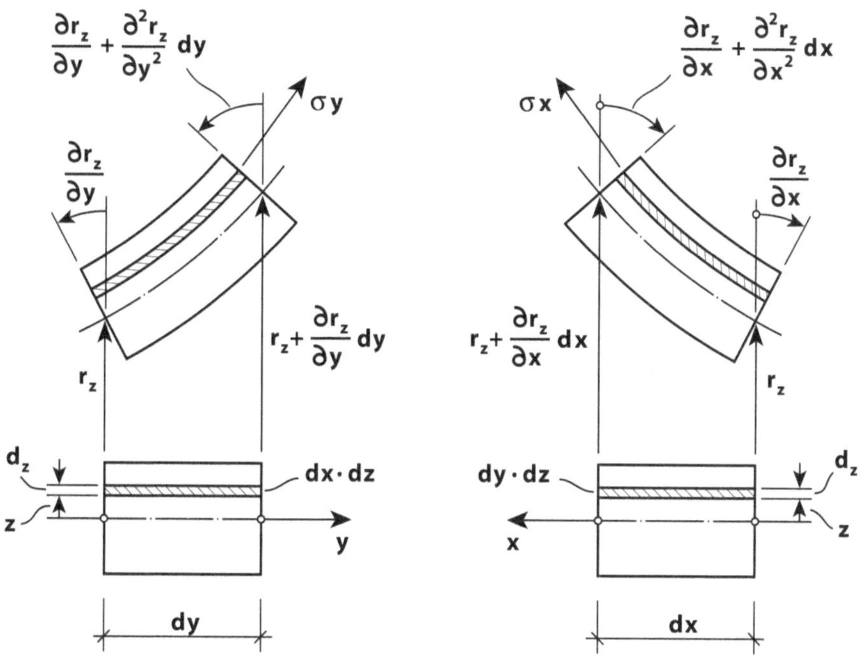

Fig. 10.4 Pure plate bending

Thus
$$\begin{bmatrix} \varepsilon_x \\ \varepsilon_y \\ \gamma_{xy} \end{bmatrix} = -z \cdot \begin{bmatrix} \partial^2 r_z / \partial x^2 \\ \partial^2 r_z / \partial y^2 \\ 2 \cdot \partial^2 r_z / \partial x \partial y \end{bmatrix} \tag{10.9}$$

from which the following connection between the instantaneous displacement $r_z(x, y, t)$ and the 'cross sectional' stress resultants M_x, M_y and M_{xy} (moments per unit length), see Fig. 10.5, is obtained:

$$\begin{bmatrix} M_x \\ M_y \\ M_{xy} \\ M_{yx} \end{bmatrix} = \int_{-h/2}^{h/2} \begin{bmatrix} -z\sigma_y \\ z\sigma_x \\ z\tau_{yx} \\ -z\tau_{xy} \end{bmatrix} dx = D \begin{bmatrix} \upsilon & 1 & 0 \\ -1 & -\upsilon & 0 \\ 0 & 0 & -(1-\upsilon) \\ 0 & 0 & (1-\upsilon) \end{bmatrix} \begin{bmatrix} \partial^2 r_z / \partial x^2 \\ \partial^2 r_z / \partial y^2 \\ \partial^2 r_z / \partial x \partial y \end{bmatrix} \tag{10.10}$$

where $$D = \frac{E}{1-\upsilon^2} \int_{-h/2}^{h/2} z^2 dz = \frac{Eh^3}{12\left(1-\upsilon^2\right)} \tag{10.11}$$

is the plate stiffness (from which it is also seen that $M_{yx} = -M_{xy}$).

10.2 The Differential Equation of Motion

What then remains is to establish the equilibrium conditions for an infinitesimal element $dx \cdot dy$. For such an element the stress resultants (per unit length) are defined in Fig. 10.5. In addition to these forces, the element is subject to external forces $q_z dxdy$, $N_x dy$ and $N_y dx$ (N_x and N_y assumed constants) as well as resisting inertia and damping forces $(m\ddot{r}_z)dxdy$ and $(c\dot{r}_z)dxdy$, where m is the plate mass (kg/m^2) and c is its damping coefficient (with unit Ns/m^3). The relevant plate deformations and the variation of stress resultants in the y direction are illustrated in Fig. 10.6. A similar variation will occur in the x direction. The normal force variation (in both x and y directions) is illustrated in Fig. 10.7. Thus, the following equilibrium requirements are obtained:

1) Force equilibrium in x-direction:

$$\left(V_{yx} + dV_{yx}\right)dx - V_{yx}dx = 0 \;\Rightarrow\; dV_{yx} = 0 \Rightarrow V_{yx} = \int_{-h/2}^{h/2} \tau_{yx} dz = 0 \tag{10.12}$$

2) Force equilibrium in y-direction:

$$\left(V_{xy} + dV_{xy}\right)dy - V_{xy}dy = 0 \;\Rightarrow\; dV_{xy} = 0 \Rightarrow V_{xy} = \int_{-h/2}^{h/2} \tau_{xy} dz = 0 \tag{10.13}$$

3) Force equilibrium in z-direction:

$$q_z dxdy - \left(m\ddot{r}_z + c\dot{r}_z\right)dxdy + \left(V_{xz} + dV_{xz}\right)dy - V_{xz}dy + \left(V_{yz} + dV_{yz}\right)dx - V_{yz}dx$$

$$+N_x\left(\frac{\partial r_z}{\partial x} + \frac{\partial^2 r_z}{\partial x^2}dx\right)dy - N_x\frac{\partial r_z}{\partial x}dy + N_y\left(\frac{\partial r_z}{\partial y} + \frac{\partial^2 r_z}{\partial y^2}dy\right)dx - N_y\frac{\partial r_z}{\partial x}dx = 0$$

$$\Rightarrow \quad m\ddot{r}_z + c\dot{r}_z - \frac{\partial V_{xz}}{\partial x} - \frac{\partial V_{yz}}{\partial y} - N_x\frac{\partial^2 r_z}{\partial x^2} - N_y\frac{\partial^2 r_z}{\partial y^2} = q_z \tag{10.14}$$

4) Moment equilibrium about an axis through the mid-point p and parallel to x:

$$\left.\begin{array}{l}\left(V_{yz}+dV_{yz}\right)dx\,dy/2+V_{yz}dx\,dy/2+\\ \left(M_x+dM_x\right)dx-M_x dx+\left(M_{yx}+dM_{yx}\right)dy-M_{yx}dy=0\end{array}\right\}$$

$$\Rightarrow \qquad V_{yz}=-\frac{\partial M_x}{\partial y}-\frac{\partial M_{yx}}{\partial x} \tag{10.15}$$

5) Moment equilibrium about an axis through the mid-point p and parallel to y:

$$\left.\begin{array}{l}-\left(V_{xz}+dV_{xz}\right)dy\,dx/2-V_{xz}dy\,dx/2+\\ \left(M_y+dM_y\right)dy-M_y dy+\left(M_{xy}+dM_{xy}\right)dx-M_{xy}dx=0\end{array}\right\}$$

$$\Rightarrow \qquad V_{xz}=\frac{\partial M_y}{\partial x}+\frac{\partial M_{xy}}{\partial y} \tag{10.16}$$

6) Moment equilibrium about an axis through the mid-point p and parallel to z:

$$\left(V_{xy}+dV_{xy}\right)dy\frac{dx}{2}+V_{xy}dy\frac{dx}{2}-\left(V_{yx}+dV_{yx}\right)dx\frac{dy}{2}-V_{yx}dx\frac{dy}{2}=0$$

$$\Rightarrow \qquad V_{xy}=V_{yx} \tag{10.17}$$

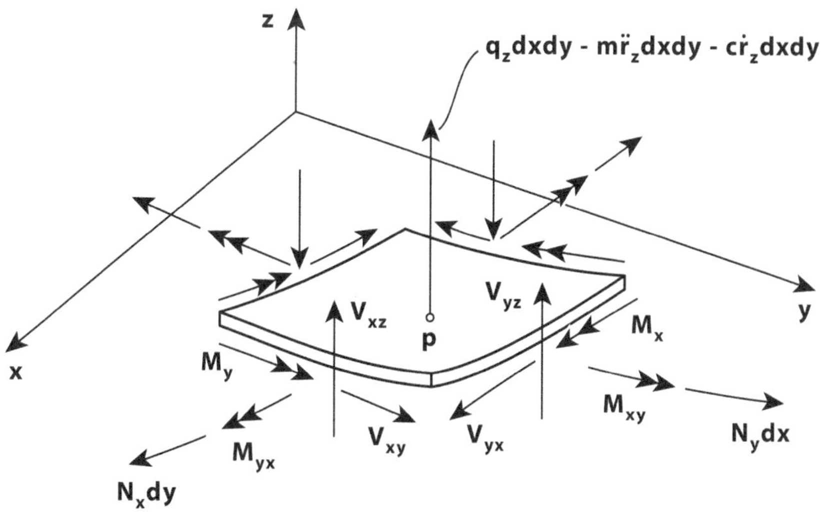

Fig. 10.5 Definition of stress resultants

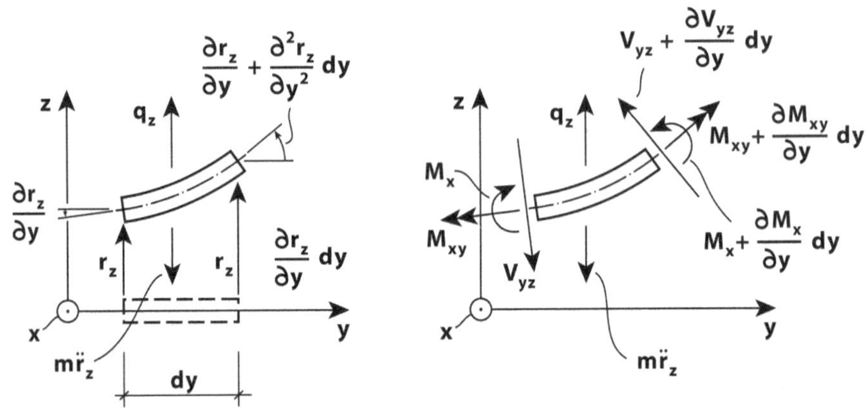

Fig. 10.6 Deformation and cross sectional stress resultant variation in y -direction

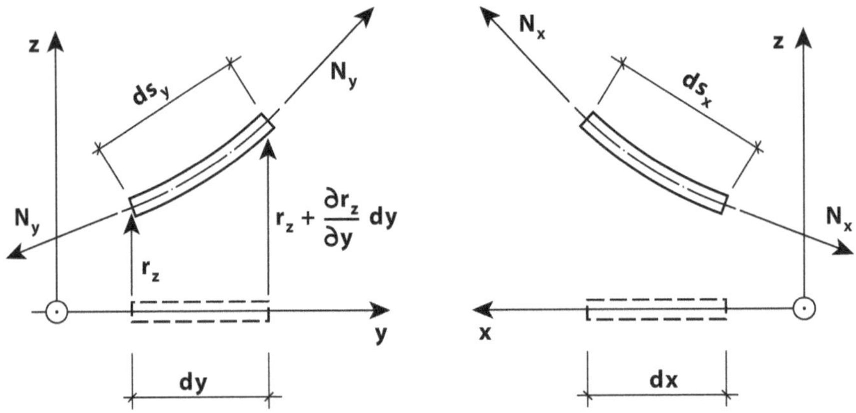

Fig. 10.7 The effects of in-plane (membrane) axial forces

Thus, Eq. 10.14 may be further developed into (see Eq. 10.10):

$$
\begin{aligned}
& m\ddot{r}_z + c\dot{r}_z - \frac{\partial V_{xz}}{\partial x} - \frac{\partial V_{yz}}{\partial y} - N_x \frac{\partial^2 r_z}{\partial x^2} - N_y \frac{\partial^2 r_z}{\partial y^2} \\
& = m\ddot{r}_z + c\dot{r}_z + \frac{\partial^2 M_x}{\partial y^2} - \frac{\partial^2 M_{xy}}{\partial x \partial y} + \frac{\partial^2 M_{yx}}{\partial x \partial y} - \frac{\partial^2 M_y}{\partial x^2} - N_x \frac{\partial^2 r_z}{\partial x^2} - N_y \frac{\partial^2 r_z}{\partial y^2} \\
& = m\ddot{r}_z + c\dot{r}_z + D\frac{\partial^2}{\partial x^2}\left(\frac{\partial^2 r_z}{\partial x^2} + \upsilon \frac{\partial^2 r_z}{\partial y^2}\right) + 2D(1-\upsilon)\frac{\partial^2}{\partial x \partial y}\left(\frac{\partial^2 r_z}{\partial x \partial y}\right) \\
& + D\frac{\partial^2}{\partial y^2}\left(\upsilon \frac{\partial^2 r_z}{\partial x^2} + \frac{\partial^2 r_z}{\partial y^2}\right) - N_x \frac{\partial^2 r_z}{\partial x^2} - N_y \frac{\partial^2 r_z}{\partial y^2} = q_z
\end{aligned}
\tag{10.18}
$$

from which the following differential equation is obtained:

$$m\ddot{r}_z + c\dot{r}_z + D\left(\frac{\partial^4 r_z}{\partial x^4} + 2\frac{\partial^4 r_z}{\partial x^2 \partial y^2} + \frac{\partial^4 r_z}{\partial y^4}\right) - N_x \frac{\partial^2 r_z}{\partial x^2} - N_y \frac{\partial^2 r_z}{\partial y^2} = q_z \tag{10.19}$$

In general eigenvalues may be obtained under the condition that c and q_z are zero, in which case the plate is free oscillate in an unknown eigenfrequency ω_n and a corresponding eigenmode $\varphi_n(x, y)$, i.e.

$$r_z = \mathrm{Re}\left[\varphi_n(x, y) \cdot e^{i\omega_n t}\right] \tag{10.20}$$

Thus, Eq. 10.19 is transformed into the following eigenvalue problem:

$$D\left(\frac{\partial^4 \varphi_n}{\partial x^4} + 2\frac{\partial^4 \varphi_n}{\partial x^2 \partial y^2} + \frac{\partial^4 \varphi_n}{\partial y^4}\right) - N_x \frac{\partial^2 \varphi_n}{\partial x^2} - N_y \frac{\partial^2 \varphi_n}{\partial y^2} - \omega_n^2 m \varphi_n = 0 \tag{10.21}$$

10.3 Solution to the Eigenvalue Problem

There are three possible solution strategies: to search for an exact solution to the differential equation throughout the space of the system, to apply an approximate solution to an energy formulation (e.g. Rayleigh-Ritz) or to apply an approximate solution to the weighted residuals method of Galerkin.

The choice of an exact solution will render a transcendental equation whose roots may be obtained numerically. This solution strategy is in general rather cumbersome. Nonetheless, the method, including a few examples, is briefly presented below. Many of the classic solutions have been obtained by an energy formulation. This method is not included below, where the main focus is on the Galerkin method of weighted residuals, as this approach is far more effective and well suited for practical applications (and computer programming) than any other analytical method.

Exact Eigenvalue Solution to the Differential Equation

It is seen from Eq. 10.21 that the second and fourth derivatives of $\varphi_n(x, y)$ must be congruent to $\varphi_n(x, y)$ itself, i.e. the solution is given by a sum of harmonic and hyperbolic functions. Thus, for an isotropic rectangular plate with constant mass and thickness, the general solution to the eigenvalue problem is given by

$$\varphi_n(x,y)=\varphi_p(x)\cdot\varphi_k(y) \tag{10.22}$$

where

$$\left.\begin{array}{l}\varphi_j(\hat{s})=c_1\sin(\lambda_j\hat{s})+c_2\sinh(\lambda_j\hat{s})+c_3\cos(\lambda_j\hat{s})+c_4\cosh(\lambda_j\hat{s})\\ j=p \text{ or } k,\quad \hat{s}=x/L_x \text{ or } y/L_y\end{array}\right\} \tag{10.23}$$

The necessary connection between ω_n and λ_j may be obtained by introducing Eqs. 10.22 and 10.23 into the differential equation (Eq. 10.21), while λ_j may be determined from the joint solution of the relevant boundary conditions:

1) At $\hat{s}=0$ or $\hat{s}=1$ for simply supported edge: $\left\{\begin{array}{l}r_z=0 \quad\Rightarrow \varphi_j=0\\ \left.\begin{array}{l}M_x\\ M_y\end{array}\right\}=0\Rightarrow \dfrac{\partial^2\varphi_j}{\partial\hat{s}^2}=0\end{array}\right.$

2) At $\hat{s}=0$ or $\hat{s}=1$ for fixed (clamped) edge: $\left\{\begin{array}{l}r_z=0 \qquad\Rightarrow \varphi_j=0\\ \left.\begin{array}{l}\partial r_z/\partial x=0\\ \partial r_z/\partial y=0\end{array}\right\}\Rightarrow \dfrac{\partial\varphi_j}{\partial\hat{s}}=0\end{array}\right.$

3) At $\hat{s}=0$ or $\hat{s}=1$ for free edge: $\left\{\begin{array}{l}\left.\begin{array}{l}M_x=0\\ M_y=0\end{array}\right\}\Rightarrow \dfrac{\partial^2\varphi_j}{\partial\hat{s}^2}=0\\ \left.\begin{array}{l}V_{xz}=0\\ V_{yz}=0\end{array}\right\}\Rightarrow \dfrac{\partial^3\varphi_j}{\partial\hat{s}^3}=0\end{array}\right.$

where $\partial^n\varphi_j/\partial x^n=L_x^{-n}\cdot\partial^n\varphi_j/\partial\hat{s}^n$ and similarly $\partial^n\varphi_j/\partial y^n=L_y^{-n}\cdot\partial^n\varphi_j/\partial\hat{s}^n$, and

$$\begin{bmatrix}\varphi_n\\ \dfrac{1}{\lambda_j}\dfrac{\partial\varphi_n}{\partial\hat{s}}\\ \dfrac{1}{\lambda_j^2}\dfrac{\partial^2\varphi_n}{\partial\hat{s}^2}\\ \dfrac{1}{\lambda_j^3}\dfrac{\partial^3\varphi_n}{\partial\hat{s}^3}\end{bmatrix}=\begin{bmatrix}\sin(\lambda_j\hat{s}) & \sinh(\lambda_j\hat{s}) & \cos(\lambda_j\hat{s}) & \cosh(\lambda_j\hat{s})\\ \cos(\lambda_j\hat{s}) & \cosh(\lambda_j\hat{s}) & -\sin(\lambda_j\hat{s}) & \sinh(\lambda_j\hat{s})\\ -\sin(\lambda_j\hat{s}) & \sinh(\lambda_j\hat{s}) & -\cos(\lambda_j\hat{s}) & \cosh(\lambda_j\hat{s})\\ -\cos(\lambda_j\hat{s}) & \cosh(\lambda_j\hat{s}) & \sin(\lambda_j\hat{s}) & \sinh(\lambda_j\hat{s})\end{bmatrix}\begin{bmatrix}c_1\\ c_2\\ c_3\\ c_4\end{bmatrix} \tag{10.24}$$

Four boundary conditions will always render a four by four coefficient matrix which multiplied by $\begin{bmatrix} c_1 & c_2 & c_3 & c_4 \end{bmatrix}^T$ is equal to zero, and thus, to obtain a non-trivial solution the determinant of the coefficient matrix must be equal to zero.

Example 10.1 Exact Solution to the Eigenvalue Problem

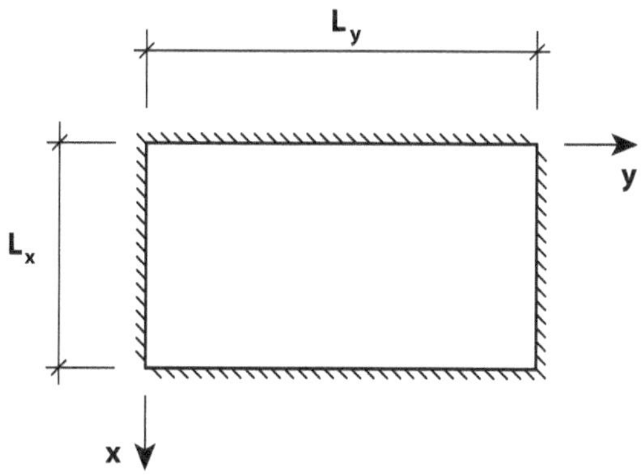

Fig. 10.8 Simply supported rectangular plate

The case of a plate with all four edges simply supported (i.e. the edges are restrained from any motion in the z *direction but free to rotate about the* x *and* y *axis) is illustrated in Fig. 10.8 above. In this case the solution is particularly simple:*

$$\varphi_n(x, y) = \sin\left(p\pi \frac{x}{L_x}\right) \cdot \sin\left(k\pi \frac{y}{L_y}\right) \quad \textit{with any combination of} \begin{cases} p = 1, 2, 3, \dots \\ k = 1, 2, 3, \dots \end{cases}$$

Introducing this into the differential equation (10.21) will then render

$$\omega_n = \left\{ \frac{\pi^4 D}{m} \left[\left(\frac{p}{L_x}\right)^2 + \left(\frac{k}{L_y}\right)^2 \right]^2 + \frac{\pi^2 N_x}{m} \left(\frac{p}{L_x}\right)^2 + \frac{\pi^2 N_y}{m} \left(\frac{k}{L_y}\right)^2 \right\}^{1/2}$$

a) *Pure bending of square plate: if* $L_x = L_y = L$ *and* $N_x = N_y = 0$ *then*

$\omega_n = \left(p^2 + k^2\right)(\pi/L)^2 \sqrt{D/m}$, *rendering*

1st eigenmode:	$p = 1$ and $k = 1$	$\Rightarrow$	$\omega_1 = 2(\pi/L)^2 \sqrt{D/m}$
2nd eigenmode:	$p = 1$ and $k = 2$	$\Rightarrow$	$\omega_2 = 5(\pi/L)^2 \sqrt{D/m}$
3rd eigenmode:	$p = 2$ and $k = 2$	$\Rightarrow$	$\omega_3 = 8(\pi/L)^2 \sqrt{D/m}$
4th eigenmode:	$p = 1$ and $k = 3$	$\Rightarrow$	$\omega_4 = 10(\pi/L)^2 \sqrt{D/m}$

and so on.

b) *Pure bending of rectangular plate: if* $L_x \neq L_y$ *but* $N_x = N_y = 0$ *then*

$$\omega_n = \left[\left(p\frac{L_y}{L_x}\right)^2 + k^2\right]\left(\frac{\pi}{L_y}\right)^2 \sqrt{\frac{D}{m}}$$

E.g., if $L_y / L_x = 2$

1st eigenmode:	$p = 1$ and $k = 1$	$\Rightarrow$	$\omega_1 = 5(\pi/L)^2 \sqrt{D/m}$
2nd eigenmode:	$p = 1$ and $k = 2$	$\Rightarrow$	$\omega_2 = 8(\pi/L)^2 \sqrt{D/m}$
3rd eigenmode:	$p = 1$ and $k = 3$	$\Rightarrow$	$\omega_3 = 13(\pi/L)^2 \sqrt{D/m}$
4th eigenmode:	$p = 2$ and $k = 1$	$\Rightarrow$	$\omega_4 = 17(\pi/L)^2 \sqrt{D/m}$

and so on.

c) *The square membrane: if* $L_x = L_y = L$, $N_x = N_y = N$ *and* $D \approx 0$ *then*

$\omega_n = \sqrt{p^2 + k^2}\,\dfrac{\pi}{L}\sqrt{\dfrac{N}{m}}$, *rendering*

1st eigenmode:	$p = 1$ and $k = 1$	$\Rightarrow$	$\omega_1 = \pi/L\sqrt{2N/m}$
2nd eigenmode:	$p = 1$ and $k = 2$	$\Rightarrow$	$\omega_2 = \pi/L\sqrt{5N/m}$
3rd eigenmode:	$p = 2$ and $k = 2$	$\Rightarrow$	$\omega_3 = \pi/L\sqrt{8N/m}$
4th eigenmode:	$p = 1$ and $k = 3$	$\Rightarrow$	$\omega_4 = \pi/L\sqrt{10N/m}$

and so on.

Galerkin's Method

By far the most effective solution to the eigenvalue problem is obtained by using the Galerkin method (of weighted residuals, see Chapter 1.7) as follows. Let us adopt an approximation in the form of a series solution

$$r_z(x,y,t)=\sum_{p=1}^{N_p} a_p\cdot\psi_p(x,y)\cdot e^{i\omega t}=\mathbf{\Psi}\cdot\mathbf{a}\cdot e^{i\omega t} \tag{10.25}$$

where
$$\left.\begin{aligned}\mathbf{\Psi}(x,y)&=\begin{bmatrix}\psi_1 & \cdots & \psi_p & \cdots & \psi_N\end{bmatrix}\\ \mathbf{a}&=\begin{bmatrix}a_1 & \cdots & a_p & \cdots & a_{N_p}\end{bmatrix}^T\end{aligned}\right\} \tag{10.26}$$

and where a_p are unknown coefficients while ψ_p are chosen known functions, whose only requirement is that they satisfy the boundary conditions of the system. Introducing this into the differential eigenvalue equation (Eq. 10.21)

$$\left\{D\left(\frac{\partial^4\mathbf{\Psi}}{\partial x^4}+2\frac{\partial^4\mathbf{\Psi}}{\partial x^2\partial y^2}+\frac{\partial^4\mathbf{\Psi}}{\partial y^4}\right)-N_x\frac{\partial^2\mathbf{\Psi}}{\partial x^2}-N_y\frac{\partial^2\mathbf{\Psi}}{\partial y^2}-\omega^2 m\mathbf{\Psi}\right\}\mathbf{a}=0 \tag{10.27}$$

pre-multiply by $\mathbf{\Psi}^T$, and integrate over the total plate area $A=L_x\cdot L_y$ will then render an eigenvalue problem

$$\left(\mathbf{K}-\omega^2\mathbf{M}\right)\mathbf{a}=\mathbf{0} \tag{10.28}$$

where
$$\left\{\begin{aligned}\mathbf{M}&=\int_A m\mathbf{\Psi}^T\mathbf{\Psi}dA=\begin{bmatrix}\ddots & & \cdot^{\cdot^{\cdot}}\\ & M_{pk} & \\ \cdot^{\cdot^{\cdot}} & & \ddots\end{bmatrix}\\ M_{pk}&=\int_A m\cdot\psi_p(x,y)\cdot\psi_k(x,y)dA\end{aligned}\right. \tag{10.29}$$

and

$$\mathbf{K} = \int_A \boldsymbol{\psi}^T \left\{ D\left(\frac{\partial^4 \boldsymbol{\psi}}{\partial x^4} + 2\frac{\partial^4 \boldsymbol{\psi}}{\partial x^2 \partial y^2} + \frac{\partial^4 \boldsymbol{\psi}}{\partial y^4} \right) - N_x \frac{\partial^2 \boldsymbol{\psi}}{\partial x^2} - N_y \frac{\partial^2 \boldsymbol{\psi}}{\partial y^2} \right\} dA = \begin{bmatrix} \ddots & & \cdot^{\cdot^{\cdot}} \\ & K_{pk} & \\ \cdot^{\cdot^{\cdot}} & & \ddots \end{bmatrix}$$

$$K_{pk} = \int_A \psi_p \left\{ D\left(\frac{\partial^4 \psi_k}{\partial x^4} + 2\frac{\partial^4 \psi_k}{\partial x^2 \partial y^2} + \frac{\partial^4 \psi_k}{\partial y^4} \right) - N_x \frac{\partial^2 \psi_k}{\partial x^2} - N_y \frac{\partial^2 \psi_k}{\partial y^2} \right\} dA \quad (10.30)$$

This will then render approximate eigenvalues ω_n and coefficients $\mathbf{a}_n$, from which the corresponding approximate eigenmodes are given by $\varphi_n(x, y) = \boldsymbol{\Psi}\mathbf{a}_n$. The accuracy of the solution will entirely depend on the ability of the chosen shape functions $\psi_p(x, y)$ to portray the real mode shapes of the system. If m and D are constants (independent of x and y) then it may be convenient to separate the variables in the shape functions

$$\psi_p(x, y) = \psi_{px}(x) \cdot \psi_{py}(y) \quad (10.31)$$

in which case

$$M_{pk} = m \int_0^{L_x} \psi_{px}(x) \cdot \psi_{kx}(x) dx \cdot \int_0^{L_y} \psi_{py}(y) \cdot \psi_{ky}(y) dy \quad (10.32)$$

$$\begin{aligned} K_{pk} = D\Bigg(& \int_0^{L_x} \psi_{px} \frac{d^4 \psi_{kx}}{dx^4} dx \int_0^{L_y} \psi_{py} \psi_{ky} dy + \\ & 2\int_0^{L_x} \psi_{px} \frac{d^2 \psi_{kx}}{dx^2} dx \int_0^{L_y} \psi_{py} \frac{d^2 \psi_{ky}}{dy^2} dy + \int_0^{L_x} \psi_{px} \psi_{kx} dx \int_0^{L_y} \psi_{py} \frac{d^4 \psi_{ky}}{dx^4} dy \Bigg) \\ & - N_x \int_0^{L_x} \psi_{px} \frac{d^2 \psi_{kx}}{dx^2} dx \int_0^{L_y} \psi_{py} \psi_{ky} dy - N_y \int_0^{L_x} \psi_{px} \psi_{kx} dx \int_0^{L_y} \psi_{py} \frac{d^2 \psi_{ky}}{dy^2} dy \end{aligned} \quad (10.33)$$

where L_x and L_y are the plate lengths in the x and y directions. To facilitate a numeric approach, this may be transformed into a normalised vector-matrix version by defining

$$\left.\begin{aligned}\hat{\mathbf{x}} = \mathbf{x}/L_x \;,\; \mathbf{x} = \begin{bmatrix} x_1 & \cdots & x_n & \cdots & x_{N_{L_x}} \end{bmatrix}^T \\ \hat{\mathbf{y}} = \mathbf{y}/L_y \;,\; \mathbf{y} = \begin{bmatrix} y_1 & \cdots & y_n & \cdots & y_{N_{L_y}} \end{bmatrix}^T \end{aligned}\right\} \tag{10.34}$$

and

$$\boldsymbol{\Psi} = \begin{bmatrix} \boldsymbol{\Psi}_{1x}\boldsymbol{\Psi}_{1y}^T & \cdots & \boldsymbol{\Psi}_{px}\boldsymbol{\Psi}_{py}^T & \cdots & \boldsymbol{\Psi}_{N_p x}\boldsymbol{\Psi}_{N_p y}^T \end{bmatrix} \tag{10.35}$$

where

$$\left.\begin{aligned}\boldsymbol{\Psi}_{px} = \begin{bmatrix} \psi_{px}(\hat{x}_1) & \cdots & \psi_{px}(\hat{x}_n) & \cdots & \psi_{px}\left(\hat{x}_{N_{L_x}}\right) \end{bmatrix}^T \\ \boldsymbol{\Psi}_{py} = \begin{bmatrix} \psi_{py}(\hat{y}_1) & \cdots & \psi_{py}(\hat{y}_n) & \cdots & \psi_{py}\left(\hat{y}_{N_{L_y}}\right) \end{bmatrix}^T \end{aligned}\right\} \tag{10.36}$$

Acknowledging that $\dfrac{\partial^j}{\partial x^j} = \dfrac{1}{L_x^j} \cdot \dfrac{\partial^j}{\partial \hat{x}^j}$ and $\dfrac{\partial^j}{\partial y^j} = \dfrac{1}{L_y^j} \cdot \dfrac{\partial^j}{\partial \hat{y}^j}$, $j = 1, 2$ and defining

$$\left.\begin{aligned} &\hat{\mathbf{M}} = \mathbf{M}/(m \cdot \Delta A) \\ &\hat{\mathbf{K}} = \mathbf{K} \cdot L_x^4/(D \cdot \Delta A) \\ &\Delta A = \Delta x \cdot \Delta y = \left(L_x/N_{L_x}\right) \cdot \left(L_y/N_{L_y}\right) \\ &\hat{\omega} = \omega\sqrt{mL_x^4/D} \end{aligned}\right\} \tag{10.37}$$

then the following normalised eigenvalue problem is obtained

$$\left(\hat{\mathbf{K}} - \hat{\omega}^2 \hat{\mathbf{M}}\right)\mathbf{a} = \mathbf{0} \tag{10.38}$$

where the content of $\hat{\mathbf{M}}$ and $\hat{\mathbf{K}}$ are given by

$$\hat{M}_{pk} = \left(\boldsymbol{\Psi}_{px}^T \boldsymbol{\Psi}_{kx}\right) \cdot \left(\boldsymbol{\Psi}_{py}^T \boldsymbol{\Psi}_{ky}\right) \tag{10.39}$$

and

Table 10.1 Some relevant shape functions

$$\psi_p(\hat{s}) = c_1 \sin\left(\hat{\lambda}_p \pi \hat{s}\right) + c_2 \sinh\left(\hat{\lambda}_p \pi \hat{s}\right) + c_3 \cos\left(\hat{\lambda}_p \pi \hat{s}\right) + c_4 \cosh\left(\hat{\lambda}_p \pi \hat{s}\right)$$

$$\hat{s} = \hat{x} \text{ or } \hat{y}, \quad p = 1, 2, 3, \ldots$$

$$\alpha_p = \frac{\sin\left(\hat{\lambda}_p \pi\right) - \sinh\left(\hat{\lambda}_p \pi\right)}{\cos\left(\hat{\lambda}_p \pi\right) - \cosh\left(\hat{\lambda}_p \pi\right)} \qquad \beta_p = \frac{\cos\left(\hat{\lambda}_p \pi\right) - \cosh\left(\hat{\lambda}_p \pi\right)}{\sin\left(\hat{\lambda}_p \pi\right) - \sinh\left(\hat{\lambda}_p \pi\right)}$$

$$\gamma_p = \frac{\sin\left(\hat{\lambda}_p \pi\right) + \sinh\left(\hat{\lambda}_p \pi\right)}{\cos\left(\hat{\lambda}_p \pi\right) + \cosh\left(\hat{\lambda}_p \pi\right)} \qquad \kappa_p = \frac{\cos\left(\hat{\lambda}_p \pi\right) - \cosh\left(\hat{\lambda}_p \pi\right)}{\sin\left(\hat{\lambda}_p \pi\right) + \sinh\left(\hat{\lambda}_p \pi\right)}$$

Case	Coefficients, c_j				Wave length, λ_p	
	c_1	c_2	c_3	c_4	$\hat{\lambda}_1$	$\hat{\lambda}_p$, $p \geq 2$
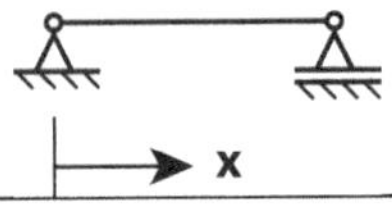	1	0	0	0	1	p
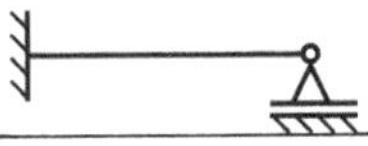	1	−1	$-\alpha_p$	α_p	$\approx \frac{5}{4}$	$\frac{4p+1}{4}$
	1	−1	κ_p	$-\kappa_p$	$\approx \frac{3}{2}$	$\frac{2p+1}{2}$
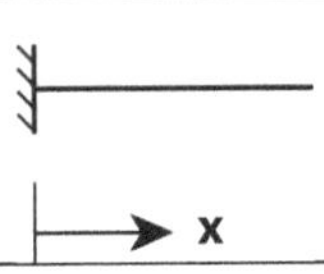	1	−1	$-\gamma_p$	γ_p	$\frac{3}{5}$	$\frac{2p-1}{2}$
	1	−1	κ_p	$-\kappa_p$	$\approx \frac{3}{4}$	$\frac{4p-1}{4}$
	1	−1	$-\alpha_p$	α_p	$\frac{1}{2}$	$\frac{2p-1}{2}$
	1	1	$-\gamma_p$	γ_p	$\approx \frac{5}{4}$	$\frac{4p+1}{4}$
	1	1	$-\alpha_p$	$-\alpha_p$	$\approx \frac{3}{2}$	$\frac{2p+1}{2}$

$$\hat{K}_{pk} = \left(\mathbf{\Psi}_{px}^T \frac{d^4 \mathbf{\Psi}_{kx}}{d\hat{x}^4} \right) \cdot \left(\mathbf{\Psi}_{py}^T \mathbf{\Psi}_{ky} \right) + 2 \left(\frac{L_x}{L_y} \right)^2 \left(\mathbf{\Psi}_{px}^T \frac{d^2 \mathbf{\Psi}_{kx}}{d\hat{x}^2} \right) \cdot \left(\mathbf{\Psi}_{py}^T \frac{d^2 \mathbf{\Psi}_{ky}}{d\hat{y}^2} \right)$$
$$+ \left(\frac{L_x}{L_y} \right)^4 \left(\mathbf{\Psi}_{px}^T \mathbf{\Psi}_{kx} \right) \cdot \left(\mathbf{\Psi}_{py}^T \frac{d^4 \mathbf{\Psi}_{ky}}{d\hat{y}^4} \right) - \frac{N_x L_x^2}{D} \left(\mathbf{\Psi}_{px}^T \frac{d^2 \mathbf{\Psi}_{kx}}{d\hat{x}^2} \right) \cdot \left(\mathbf{\Psi}_{py}^T \mathbf{\Psi}_{ky} \right) \quad (10.40)$$
$$- \frac{N_y L_y^2}{D} \left(\frac{L_x}{L_y} \right)^4 \left(\mathbf{\Psi}_{px}^T \mathbf{\Psi}_{kx} \right) \cdot \left(\mathbf{\Psi}_{py}^T \frac{d^2 \mathbf{\Psi}_{ky}}{d\hat{y}^2} \right)$$

In general, sufficient accuracy may be obtained by choosing shape functions that comply with the corresponding eigenvalue solution of simple beams whose boundary conditions are identical to that of the relevant plate system. Some useful cases are listed in Table 10.1.

Example 10.2 Galerkin Solution to the Eigenvalue Problem

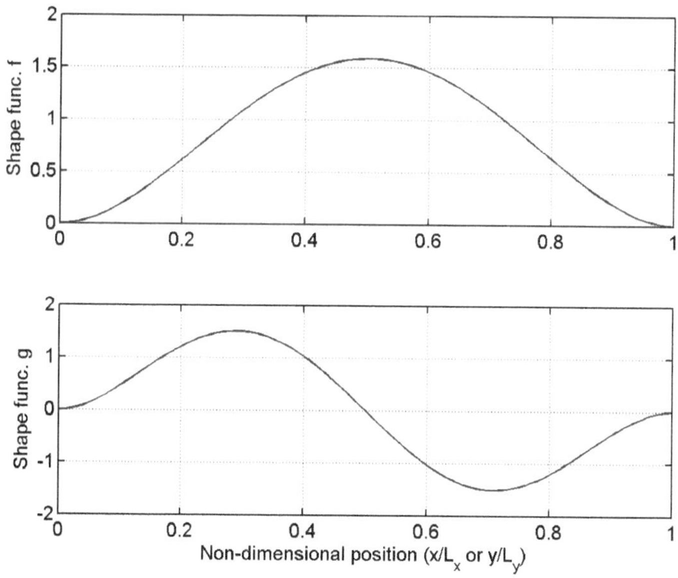

Fig. 10.9 Shape functions f and g

To illustrate the efficiency of the Galerkin method let us consider a square plate with all its edges clamped and no axial forces (i.e. $N_x = N_y = 0$*). The functions (chosen from Table 10.1 above)*

$$\left.\begin{aligned} f(\hat{s}) &= \sin(\lambda_1 \hat{s}) - \sinh(\lambda_1 \hat{s}) + \beta_1\left[\cos(\lambda_1 \hat{s}) - \cosh(\lambda_1 \hat{s})\right] \\ g(\hat{s}) &= \sin(\lambda_2 \hat{s}) - \sinh(\lambda_2 \hat{s}) + \beta_2\left[\cos(\lambda_2 \hat{s}) - \cosh(\lambda_2 \hat{s})\right] \end{aligned}\right\} \quad \hat{s} = \frac{x}{L_x} \text{ or } \frac{y}{L_y}$$

$$\beta_j = \frac{\cos(\lambda_j) - \cosh(\lambda_j)}{\sin(\lambda_j) + \sinh(\lambda_j)} \qquad j = 1 \text{ or } 2 \qquad \textit{and with} \qquad \begin{cases} \lambda_1 = 3\pi/2 \\ \lambda_2 = 5\pi/2 \end{cases}$$

are deemed relevant for the description of the first three eigenmodes

$$\psi_1(x,y) = f(\hat{x}) \cdot f(\hat{y}),\ \psi_2(x,y) = g(\hat{x}) \cdot f(\hat{y}) \textit{ and } \psi_3(x,y) = g(\hat{x}) \cdot g(\hat{y})$$

of the system. They are illustrated in Fig. 10.9.

Thus, introducing $\boldsymbol{\Psi} = \left[\left(\boldsymbol{\Psi}_{1x}\boldsymbol{\Psi}_{1y}^T\right) \ \left(\boldsymbol{\Psi}_{2x}\boldsymbol{\Psi}_{2y}^T\right) \ \left(\boldsymbol{\Psi}_{3x}\boldsymbol{\Psi}_{3y}^T\right)\right]$ *where*

$$\boldsymbol{\Psi}_{1x} = \begin{bmatrix} f(\hat{x}_1) \\ \vdots \\ f(\hat{x}_i) \\ \vdots \\ f\left(\hat{x}_{N_{L_x}}\right) \end{bmatrix} \qquad \boldsymbol{\Psi}_{1y} = \begin{bmatrix} f(\hat{y}_1) \\ \vdots \\ f(\hat{y}_j) \\ \vdots \\ f\left(\hat{y}_{N_{L_y}}\right) \end{bmatrix} \qquad \boldsymbol{\Psi}_{2x} = \begin{bmatrix} g(\hat{x}_1) \\ \vdots \\ g(\hat{x}_1) \\ \vdots \\ g\left(\hat{x}_{N_{L_x}}\right) \end{bmatrix}$$

$$\boldsymbol{\Psi}_{2y} = \begin{bmatrix} f(\hat{y}_1) \\ \vdots \\ f(\hat{y}_j) \\ \vdots \\ f\left(\hat{y}_{N_{L_y}}\right) \end{bmatrix} \qquad \boldsymbol{\Psi}_{3x} = \begin{bmatrix} g(\hat{x}_1) \\ \vdots \\ g(\hat{x}_i) \\ \vdots \\ g\left(\hat{x}_{N_{L_x}}\right) \end{bmatrix} \qquad \boldsymbol{\Psi}_{3y} = \begin{bmatrix} g(\hat{y}_1) \\ \vdots \\ g(\hat{y}_j) \\ \vdots \\ g\left(\hat{y}_{N_{L_y}}\right) \end{bmatrix}$$

into the normalised eigenvalue problem $\left(\hat{\mathbf{K}} - \hat{\omega}^2 \hat{\mathbf{M}}\right)\mathbf{a} = \mathbf{0}$ *(se Eq. 10.38)*

where the content of $\hat{\mathbf{M}}$ *and* $\hat{\mathbf{K}}$ *are given by (see Eqs. 10.39 and 10.40)*

$$\hat{M}_{pk} = \left(\boldsymbol{\Psi}_{px}^T \boldsymbol{\Psi}_{kx}\right) \cdot \left(\boldsymbol{\Psi}_{py}^T \boldsymbol{\Psi}_{ky}\right)$$

$$\hat{K}_{pk} = \left(\boldsymbol{\Psi}_{px}^T \frac{d^4 \boldsymbol{\Psi}_{kx}}{dx^4} \right) \cdot \left(\boldsymbol{\Psi}_{py}^T \boldsymbol{\Psi}_{ky} \right) + 2 \left(\frac{L_x}{L_y} \right)^2 \left(\boldsymbol{\Psi}_{px}^T \frac{d^2 \boldsymbol{\Psi}_{kx}}{dx^2} \right) \cdot \left(\boldsymbol{\Psi}_{py}^T \frac{d^2 \boldsymbol{\Psi}_{ky}}{dy^2} \right)$$

$$+ \left(\frac{L_x}{L_y} \right)^4 \left(\boldsymbol{\Psi}_{px}^T \boldsymbol{\Psi}_{kx} \right) \cdot \left(\boldsymbol{\Psi}_{py}^T \frac{d^4 \boldsymbol{\Psi}_{ky}}{dy^4} \right) - \frac{N_x L_x^2}{D} \left(\boldsymbol{\Psi}_{px}^T \frac{d^2 \boldsymbol{\Psi}_{kx}}{dx^2} \right) \cdot \left(\boldsymbol{\Psi}_{py}^T \boldsymbol{\Psi}_{ky} \right)$$

$$- \frac{N_y L_y^2}{D} \left(\frac{L_x}{L_y} \right)^4 \left(\boldsymbol{\Psi}_{px}^T \boldsymbol{\Psi}_{kx} \right) \cdot \left(\boldsymbol{\Psi}_{py}^T \frac{d^2 \boldsymbol{\Psi}_{ky}}{dy^2} \right)$$

$\left. \begin{matrix} p \\ k \end{matrix} \right\} = 1, 2, 3$ *will then render the normalised mass and stiffness matrices*

$$\hat{\mathbf{M}} = \begin{bmatrix} 10074 & -64 & 0 \\ -64 & 10036 & -64 \\ 0 & -64 & 9998 \end{bmatrix} \quad \text{and} \quad \hat{\mathbf{K}} = 10^4 \begin{bmatrix} 1294 & -33 & 0 \\ -12 & 5443 & -70 \\ 0 & -49 & 11851 \end{bmatrix}$$

from which the following eigenvalues are obtained $\begin{bmatrix} \hat{\omega}_1 \\ \hat{\omega}_2 \\ \hat{\omega}_3 \end{bmatrix} = \begin{bmatrix} 35.84 \\ 73.64 \\ 108.87 \end{bmatrix} \begin{matrix} (35.98) \\ (73.41) \\ (108.3) \end{matrix}$

More exact values are given in brackets. As can be seen, in spite of some minor unwanted off-diagonal terms in $\hat{\mathbf{M}}$ *and* $\hat{\mathbf{K}}$*, which comes from small inaccuracies in the chosen shape functions (i.e. they are not perfectly orthogonal), the solution is remarkable accurate. The corresponding eigenvectors are given by*

$$\mathbf{a}_1 = \begin{bmatrix} 1 \\ -0.0009 \\ 0 \end{bmatrix} \qquad \mathbf{a}_2 = \begin{bmatrix} 0.0004 \\ 1 \\ 0.0002 \end{bmatrix} \qquad \mathbf{a}_1 = \begin{bmatrix} 0 \\ 0.0009 \\ 1 \end{bmatrix}$$

(again, showing only minor inaccuracies) The mode shapes are illustrated in Figs. 10.10, 10.11 and 10.12.

The effect of axial forces $N_x = N_y = N$ *on the eigenvalue solution is illustrated in Fig. 10.13. The introduction of a positive external force (stretching) is to increase the plate stiffness, while a negative axial force (compression) will reduce its stiffness. As can be seen, at* $N_x L_x^2 / D \approx -52.6$ *then* ω_1 *is zero, implying that the total stiffness of the system is zero due to elastic plate buckling (a more exact value is* -49.3*).*

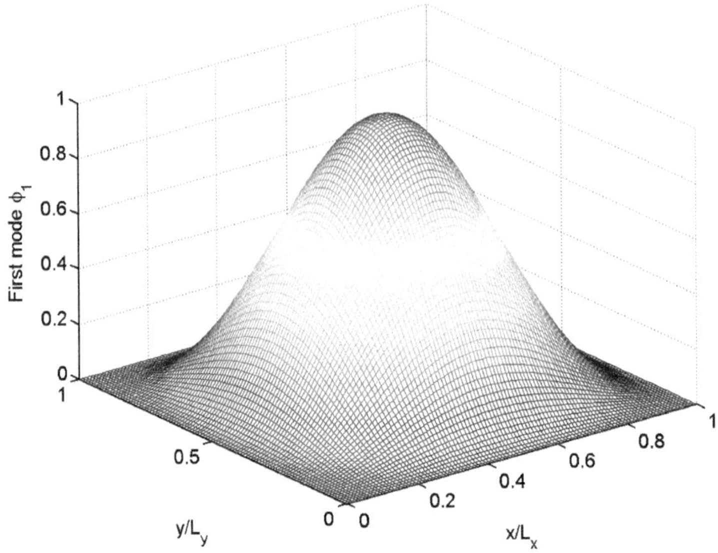

Fig. 10.10 Estimated first mode shape, $\omega_1 = 35.84\sqrt{\dfrac{D}{mL^4}}$ rad/s ($N_x = N_y = 0$)

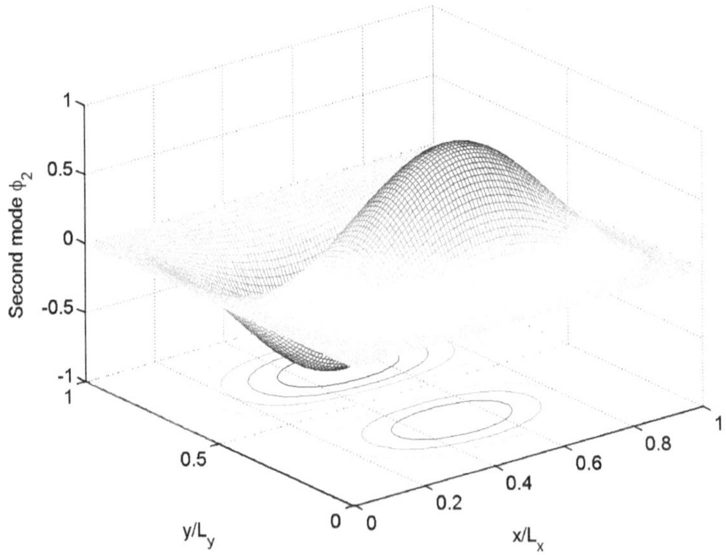

Fig. 10.11 Estimated second mode, $\omega_2 = 73.64\sqrt{\dfrac{D}{mL^4}}$ rad/s ($N_x = N_y = 0$)

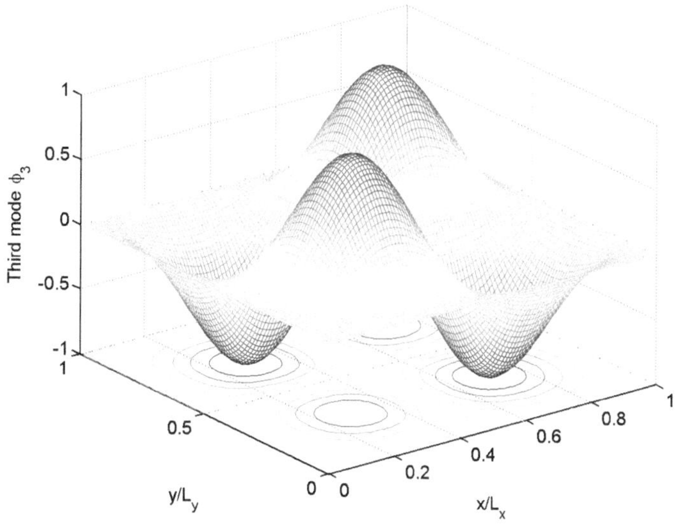

Fig. 10.12 Estimated third mode, $\omega_3 = 108.87\sqrt{\dfrac{D}{mL^4}}$ rad/s ($N_x = N_y = 0$)

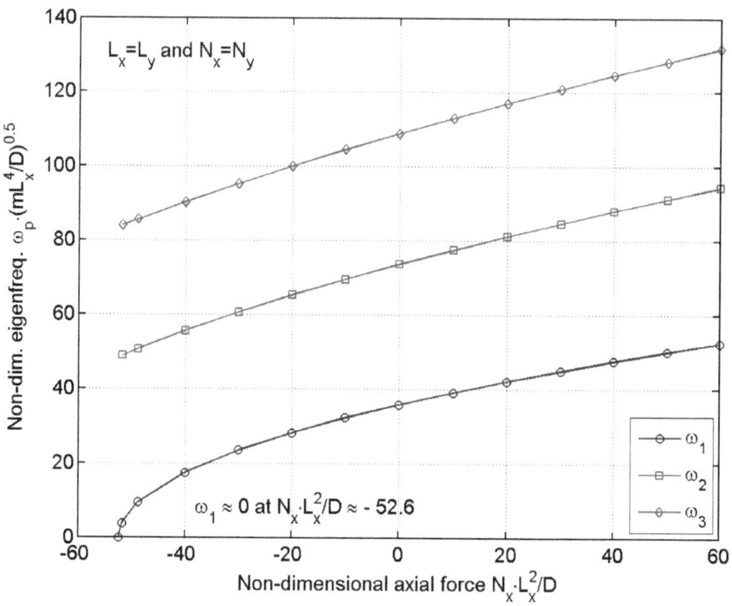

Fig. 10.13 The effects of evenly distributed biaxial external loads, N_x and N_y

10.4 Dynamic Response Calculations

Dynamic response calculations may conveniently be performed by transforming the differential equation (Eq. 10.19)

$$m\ddot{r}_z + c\dot{r}_z + D\left(\frac{\partial^4 r_z}{\partial x^4} + 2\frac{\partial^4 r_z}{\partial x^2 \partial y^2} + \frac{\partial^4 r_z}{\partial y^4}\right) - N_x \frac{\partial^2 r_z}{\partial x^2} - N_y \frac{\partial^2 r_z}{\partial y^2} = q_z \tag{10.41}$$

into a modal format by introducing the assumption that the solution may be written as a series of known mode shapes $\varphi_n(x, y)$ multiplied by unknown time domain variables $\eta_n(t)$, i.e.

$$r_z(x, y, t) = \sum_{n=1}^{N_{\text{mod}}} \varphi_n(x, y) \cdot \eta_n(t) = \mathbf{\Phi}(x, y) \cdot \boldsymbol{\eta}(t) \tag{10.42}$$

where

$$\left.\begin{aligned} \mathbf{\Phi}(x, y) &= \begin{bmatrix} \varphi_1 & \cdots & \varphi_n & \cdots & \varphi_{N_{\text{mod}}} \end{bmatrix} \\ \boldsymbol{\eta}(t) &= \begin{bmatrix} \eta_1 & \cdots & \eta_n & \cdots & \eta_{N_{\text{mod}}} \end{bmatrix}^T \end{aligned}\right\} \tag{10.43}$$

rendering

$$m\mathbf{\Phi}\ddot{\boldsymbol{\eta}} + c\mathbf{\Phi}\dot{\boldsymbol{\eta}} + \left[D\left(\frac{\partial^4 \mathbf{\Phi}_z}{\partial x^4} + 2\frac{\partial^4 \mathbf{\Phi}_z}{\partial x^2 \partial y^2} + \frac{\partial^4 \mathbf{\Phi}_z}{\partial y^4}\right) - N_x \frac{\partial^2 \mathbf{\Phi}_z}{\partial x^2} - N_y \frac{\partial^2 \mathbf{\Phi}_z}{\partial y^2}\right]\boldsymbol{\eta} = q_z \tag{10.44}$$

Pre-multiplication by $\mathbf{\Phi}^T$ and integration over the entire x, y domain will then turn Eq. 10.44 into the following modal equilibrium condition

$$\tilde{\mathbf{M}}\ddot{\boldsymbol{\eta}}(t) + \tilde{\mathbf{C}}\dot{\boldsymbol{\eta}}(t) + \tilde{\mathbf{K}}\boldsymbol{\eta}(t) = \tilde{\mathbf{R}}(t) \tag{10.45}$$

where, recalling the general property of mode shape orthogonality, the system modal quantities are given by

$$\left.\begin{aligned}
\tilde{\mathbf{M}} &= \int_A m\boldsymbol{\Phi}^T\boldsymbol{\Phi}dA = diag\left[\tilde{M}_n\right] \text{ where } \tilde{M}_n = \int_A m\varphi_n^2 dA \\
\tilde{\mathbf{C}} &= \int_A c\boldsymbol{\Phi}^T\boldsymbol{\Phi}dA = diag\left[\tilde{C}_n\right] \text{ where } \tilde{C}_n = 2\tilde{M}_n\omega_n\zeta_n \\
\tilde{\mathbf{K}} &= \int_A \boldsymbol{\Phi}^T\left[D\left(\frac{\partial^4\boldsymbol{\Phi}_z}{\partial x^4}+2\frac{\partial^4\boldsymbol{\Phi}_z}{\partial x^2\partial y^2}+\frac{\partial^4\boldsymbol{\Phi}_z}{\partial y^4}\right)-N_x\frac{\partial^2\boldsymbol{\Phi}_z}{\partial x^2}-N_y\frac{\partial^2\boldsymbol{\Phi}_z}{\partial y^2}\right]dA \\
&= diag\left[\tilde{K}_n\right] \text{ where } \tilde{K}_n = \omega_n^2\tilde{M}_n
\end{aligned}\right\} \tag{10.46}$$

and
$$\tilde{\mathbf{R}}(t) = \int_A \boldsymbol{\Phi}^T(x,y)\cdot q_z(x,y,t)dA \tag{10.47}$$

The solution to Eq. 10.45 may be pursued in time domain as well as in frequency domain. In time domain one quite simply follows one of the time step iteration methods presented in Chapter 6.3, whichever seems most suitable. In frequency domain the general method has been presented in Chapter 6.5. A frequency domain approach in modal degrees of freedom is presented below.

First, the Fourier transform is taken of the modal variable $\boldsymbol{\eta}(t)$

$$\boldsymbol{\eta}(t) = \sum_\omega \mathbf{a}_\eta(\omega)\cdot e^{i\omega t} \quad \text{where } \mathbf{a}_\eta(\omega) = \begin{bmatrix} a_1 & \cdots & a_n & \cdots & a_{N_{\text{mod}}} \end{bmatrix}^T \tag{10.48}$$

i.e. (see Eq. 10.42)

$$r_z(x,y,t) = \boldsymbol{\Phi}(x,y)\cdot\sum_\omega a_n(\omega)\cdot e^{i\omega t} = \boldsymbol{\Phi}(x,y)\cdot\mathbf{a}_\eta(\omega)\cdot e^{i\omega t} \tag{10.49}$$

and of the load

$$\tilde{\mathbf{R}}(t) = \sum_\omega \mathbf{a}_{\tilde{R}}(\omega)\cdot e^{i\omega t} \quad \text{where } \mathbf{a}_{\tilde{R}} = \int_A \boldsymbol{\Phi}^T(x,y)\cdot a_{q_z}(x,y,\omega)dA \tag{10.50}$$

and where a_{q_z} is the Fourier amplitude of the distributed load $q_z(x,y,t)$. It is a frequency domain requirement that Eq. 10.45 is satisfied at every ω setting, i.e.

$$\left(-\omega^2\tilde{\mathbf{M}} + i\omega\tilde{\mathbf{C}} + \tilde{\mathbf{K}}\right)\mathbf{a}_\eta = \mathbf{a}_{\tilde{R}} \tag{10.51}$$

Pre-multiplication with $\tilde{\mathbf{K}}^{-1}$

$$\left(\mathbf{I}-\omega^2\tilde{\mathbf{K}}^{-1}\tilde{\mathbf{M}}+i\omega\tilde{\mathbf{K}}^{-1}\tilde{\mathbf{C}}\right)\mathbf{a}_\eta=\tilde{\mathbf{K}}^{-1}\mathbf{a}_{\tilde{R}} \tag{10.52}$$

and, recalling that (see Eq. 10.46)

$$\left.\begin{aligned}\tilde{\mathbf{M}}&=diag\left[\tilde{M}_n\right]\\ \tilde{\mathbf{C}}&=diag\left[\tilde{C}_n\right]\\ \tilde{\mathbf{K}}&=diag\left[\tilde{K}_n\right]\end{aligned}\right\}\text{ where }\begin{cases}\tilde{M}_n=\int\limits_A m\varphi_n^2 dA\\ \tilde{C}_n=2\tilde{M}_n\omega_n\zeta_n\\ \tilde{K}_n=\omega_n^2\tilde{M}_n\end{cases} \tag{10.53}$$

it is seen that $$\mathbf{a}_\eta(\omega)=\hat{\mathbf{H}}_\eta(\omega)\cdot\tilde{\mathbf{K}}^{-1}\cdot\mathbf{a}_{\tilde{R}}(\omega) \tag{10.54}$$

where $$\hat{\mathbf{H}}_\eta=\left(\mathbf{I}-\omega^2\tilde{\mathbf{K}}^{-1}\tilde{\mathbf{M}}+i\omega\tilde{\mathbf{K}}^{-1}\tilde{\mathbf{C}}\right)^{-1}=diag\left[\hat{H}_{\eta_n}\right] \tag{10.55}$$

and where $$\hat{H}_{\eta_n}=\left[1-\left(\omega/\omega_n\right)^2+2i\zeta_n\,\omega/\omega_n\right]^{-1} \tag{10.56}$$

is the non-dimensional frequency response function associated with mode φ_n. Thus, the problem has been transformed into the problem of determining $\mathbf{a}_\eta$, from which $r_z(x,y,t)$ may be obtained from Eq. 10.49. As always, the advantage with a modal approach in frequency domain is that a complex problem may be reduced to one with a manageable number of degrees of freedom (N_{mod}) in which case it is possible to focus on what are the most important characteristics of the system. So far, the development is only suited to solve the case of stationary and deterministic loads, see Example 10.3 below.

Example 10.3: Deterministic Dynamic Load

Let us first consider the case of an arbitrary rectangular plate with mode shapes

$$\mathbf{\Phi}(x,y)=\left[\varphi_1\ \cdots\ \varphi_n\ \cdots\ \varphi_{N_{\text{mod}}}\right]$$

and with a single stationary and harmonic concentrated deterministic load

$$Q(t)=\text{Re}\left(a_p\cdot e^{i\omega_p t}\right)$$

at a single frequency ω_p *and at position* x_p, y_p. *Its modal frequency response matrix at* ω_p *and its modal stiffness matrix are then*

$$\hat{\mathbf{H}}_\eta(\omega_p) = diag\left[\hat{H}_{\eta_n}(\omega_p)\right] \text{ where } \hat{H}_{\eta_n}(\omega_p) = \left[1 - (\omega_p/\omega_n)^2 + 2i\zeta_n \omega_p/\omega_n\right]^{-1}$$

$$\tilde{\mathbf{K}} = diag\left[\tilde{K}_n\right] \quad \text{where} \quad \tilde{K}_n = \omega_n^2 \tilde{M}_n \quad \text{and} \quad \tilde{M}_n = \int_A m\varphi_n^2 dA$$

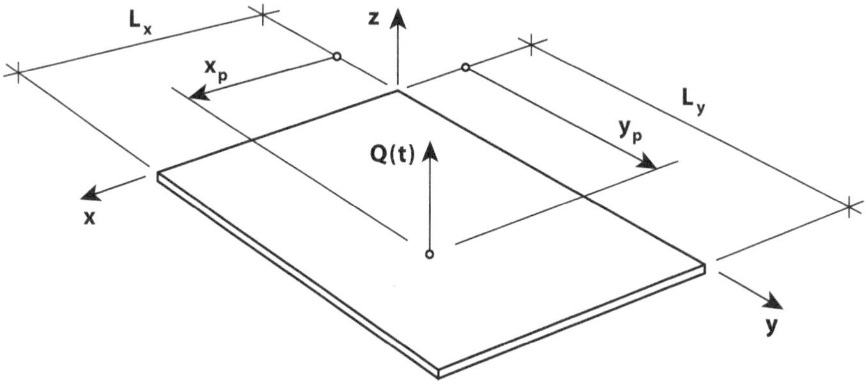

Fig. 10.14 Rectangular plate subject to force component $Q(x_p, y_p, t)$

The Fourier amplitude vector of the modal load (see Eq. 10.50) is given by $\mathbf{a}_{\tilde{R}} = \mathbf{\Phi}^T(x_p, y_p) \cdot a_p$, *and thus the Fourier amplitude vector of the modal degrees of freedom is given by*

$$\mathbf{a}_\eta(\omega_p) = \hat{\mathbf{H}}_\eta \tilde{\mathbf{K}}^{-1} \mathbf{a}_{\tilde{R}} = \left[\cdots \quad \cdots \quad a_p \frac{\hat{H}_{\eta_n}(\omega_p)}{\omega_n^2 \tilde{M}_n} \varphi_n(x_p, y_p) \quad \cdots \quad \cdots\right]$$

from which the Fourier amplitude of the stationary dynamic displacement response at an arbitrary position x_r, y_r *may be obtained by a modal superposition*

$$a_r(x_r, y_r, \omega_p) = \mathbf{\Phi}(x_r, y_r) \cdot \mathbf{a}_\eta(\omega_p) = \sum_{n=1}^{N_{\text{mod}}} a_p \cdot \frac{\hat{H}_{\eta_n}(\omega_p)}{\omega_n^2 \tilde{M}_n} \cdot \varphi_n(x_r, y_r) \cdot \varphi_n(x_p, y_p)$$

The stationary dynamic displacement response at an arbitrary position x_r, y_r *is then given by*

$$r_z\left(x_r, y_r, t\right) = \mathrm{Re}\left\{a_p \cdot \left[\sum_{n=1}^{N_{\mathrm{mod}}} \frac{\hat{H}_{\eta_n}\left(\omega_p\right)}{\omega_n^2 \tilde{M}_n} \cdot \varphi_n\left(x_r, y_r\right) \cdot \varphi_n\left(x_p, y_p\right)\right] \cdot e^{i\omega_p t}\right\}$$

$$= \sum_{n=1}^{N_{\mathrm{mod}}} \frac{\left|\hat{H}_{\eta_n}\left(\omega_p\right)\right|}{\omega_n^2} \cdot \frac{\varphi_n\left(x_r, y_r\right) \cdot \varphi_n\left(x_p, y_p\right)}{\int_A m\varphi_n^2(x, y)\, dA} \cdot a_p \cos\left(\omega_p t - \beta_n\right)$$

where $$\tan\beta_n = \frac{2\zeta_n \omega_p / \omega_n}{1 - \left(\omega_p / \omega_n\right)^2}$$

In the case of a distributed stationary and stochastic load $q_z(x, y, t)$, whose cross spectral density is given by

$$S_{qq}(\Delta s, \omega) = S_q(\omega) \cdot \hat{C}o_q(\Delta s, \omega) \tag{10.57}$$

where Δs is the distance between two arbitrary points $\left(x_1, y_1\right)$ and $\left(x_2, y_2\right)$, i.e.

$$\Delta s = \left(\Delta x^2 + \Delta y^2\right)^{1/2} \qquad \begin{cases} \Delta x = x_1 - x_2 \\ \Delta y = y_1 - y_2 \end{cases} \tag{10.58}$$

and where $S_q(\omega)$ is its single point spectral density and $\hat{C}o_q(\Delta s, \omega)$ is the corresponding normalised co-spectrum of the $q_z(x, y, t)$ process (i.e. the real value of its normalised cross spectral density, see Appendix A.4). Thus, the N_{mod} by N_{mod} spectral density matrix of the modal degrees of freedom (see Eq. 10.54) is given by

$$\begin{aligned}\mathbf{S}_\eta(\omega) &= \lim_{T\to\infty}\frac{1}{\pi T}\left[\mathbf{a}_\eta^*(\omega)\cdot\mathbf{a}_\eta^T(\omega)\right]\\ &= \lim_{T\to\infty}\frac{1}{\pi T}\left\{\left[\hat{\mathbf{H}}_\eta(\omega)\tilde{\mathbf{K}}^{-1}\mathbf{a}_{\tilde{R}}(\omega)\right]^*\cdot\left[\hat{\mathbf{H}}_\eta(\omega)\tilde{\mathbf{K}}^{-1}\mathbf{a}_{\tilde{R}}(\omega)\right]^T\right\}\\ &= \hat{\mathbf{H}}_\eta^*(\omega)\tilde{\mathbf{K}}^{-1}\lim_{T\to\infty}\frac{1}{\pi T}\left[\mathbf{a}_{\tilde{R}}^*(\omega)\cdot\mathbf{a}_{\tilde{R}}^T(\omega)\right]\left(\tilde{\mathbf{K}}^{-1}\right)^T\hat{\mathbf{H}}_\eta^T(\omega)\\ &= \hat{\mathbf{H}}_\eta^*(\omega)\tilde{\mathbf{K}}^{-1}\mathbf{S}_{\tilde{R}}(\omega)\left(\tilde{\mathbf{K}}^{-1}\right)^T\hat{\mathbf{H}}_\eta^T(\omega)\end{aligned} \tag{10.59}$$

where

$$\begin{aligned}\mathbf{S}_{\tilde{R}}(\omega) &= \lim_{T\to\infty}\frac{1}{\pi T}\left[\mathbf{a}_{\tilde{R}}^*(\omega)\cdot\mathbf{a}_{\tilde{R}}^T(\omega)\right]\\ &= \lim_{T\to\infty}\frac{1}{\pi T}\left\{\left[\int_A\mathbf{\Phi}^T(x,y)\cdot a_q(x,y,\omega)\,dA\right]^*\left[\int_A\mathbf{\Phi}^T(x,y)\cdot a_q(x,y,\omega)\,dA\right]^T\right\}\\ &= \int_A\int_A\mathbf{\Phi}^T(x_1,y_1)\cdot\mathbf{\Phi}(x_2,y_2)\lim_{T\to\infty}\frac{1}{\pi T}\left[a_q^*(x,y,\omega)\cdot a_q(x,y,\omega)\right]dA_1dA_2\end{aligned} \tag{10.60}$$

Thus, since

$$S_{qq}(\Delta s,\omega) = S_q(\omega)\cdot\hat{C}o(\Delta s,\omega) = \lim_{T\to\infty}\frac{1}{\pi T}\left[a_q^*(x,y,\omega)\cdot a_q(x,y,\omega)\right] \tag{10.61}$$

it is seen that

$$\mathbf{S}_{\tilde{R}}(\omega) = S_q(\omega)\cdot\mathbf{J}_q(\omega) \tag{10.62}$$

where

$$\mathbf{J}_q(\omega) = \int_A\int_A\mathbf{\Phi}^T(x_1,y_1)\cdot\mathbf{\Phi}(x_2,y_2)\cdot\hat{C}o_q(\Delta s,\omega)\,dA_1dA_2 \tag{10.63}$$

Then, the spectral density matrix of the modal degrees of freedom is given by

$$\mathbf{S}_\eta(\omega) = \left[\hat{\mathbf{H}}_\eta^*(\omega)\cdot\tilde{\mathbf{K}}^{-1}\cdot\mathbf{J}_q(\omega)\cdot\left(\tilde{\mathbf{K}}^{-1}\right)^T\cdot\hat{\mathbf{H}}_\eta^T(\omega)\right]\cdot S_q(\omega) \tag{10.64}$$

Since (see Eq. 10.49) $r_z(x,y,t) = \mathbf{\Phi}(x,y) \cdot \mathbf{a}_\eta(\omega) \cdot e^{i\omega t}$ then

$$a_{r_z}(x,y,\omega) = \mathbf{\Phi}(x,y) \cdot \mathbf{a}_\eta(\omega) \tag{10.65}$$

and thus, the spectral density of the dynamic response at an arbitrary position (x_r, y_r) is given by

$$\begin{aligned} S_{r_z}(x,y,\omega) &= \lim_{T\to\infty} \frac{1}{\pi T}\left[a^*_{r_z}(x_r,y_r,\omega) \cdot a_{r_z}(x_r,y_r,\omega)\right] \\ &= \lim_{T\to\infty} \frac{1}{\pi T}\left\{\left[\mathbf{\Phi}(x_r,y_r) \cdot \mathbf{a}_\eta(\omega)\right]^* \cdot \left[\mathbf{\Phi}(x_r,y_r) \cdot \mathbf{a}_\eta(\omega)\right]^T\right\} \\ &= \mathbf{\Phi}(x_r,y_r) \cdot \lim_{T\to\infty} \frac{1}{\pi T}\left[\mathbf{a}^*_\eta(\omega) \cdot \mathbf{a}^T_\eta(\omega)\right] \cdot \mathbf{\Phi}^T(x_r,y_r) \\ &= \mathbf{\Phi}(x_r,y_r) \cdot \mathbf{S}_\eta(\omega) \cdot \mathbf{\Phi}^T(x_r,y_r) \end{aligned} \tag{10.66}$$

I.e. (see Eq. 10.64)

$$S_{r_z}(x_r,y_r,\omega) = \left[\mathbf{\Phi}(x_r,y_r)\hat{\mathbf{H}}^*_\eta(\omega)\tilde{\mathbf{K}}^{-1}\mathbf{J}_q(\omega)\left(\tilde{\mathbf{K}}^{-1}\right)^T \hat{\mathbf{H}}^T_\eta(\omega)\mathbf{\Phi}^T(x_r,y_r)\right] \cdot S_q(\omega) \tag{10.67}$$

Example 10.4: Stochastic Dynamic Load

Let us consider the simple case of a square window pane ($L_x = L_y = L$) with clamped (fixed) edges subject to fluctuating horizontal wind force $q(x,y,t)$ and located at elevation 40 m above ground where the mean wind velocity is $V = 40\ m/s$ and the intensity of the stationary along-wind turbulence component $u(x,y,t)$ is $I_u = \sigma_u / V = 0.2$, where σ_u is its standard deviation. For simplicity it is assumed that V and I_u are approximately constants over the area of the plate, and that the co-spectrum of the wind turbulence is close to unity, i.e. that the fluctuating velocity $u(x,y,t)$ is close to perfectly correlated over the entire area of the pane. Thus

$$S_{uu}(\Delta s,\omega) = S_u(\omega)$$

where $S_u(\omega)$ is the single point spectral density of turbulence, defined by

$$S_u(\omega) = \sigma_u^2 \cdot \hat{S}_u(\omega) \quad where \quad \hat{S}_u(\omega) = \frac{A_u\,{}^xL_u/V}{\left(1 + 1.5\dfrac{\omega}{2\pi} A_u\,{}^xL_u/V\right)^{5/3}}$$

where $A_u = 1.08$ *and* ${}^xL_u = 160\ m$ *is the integral length scale of the turbulence.*

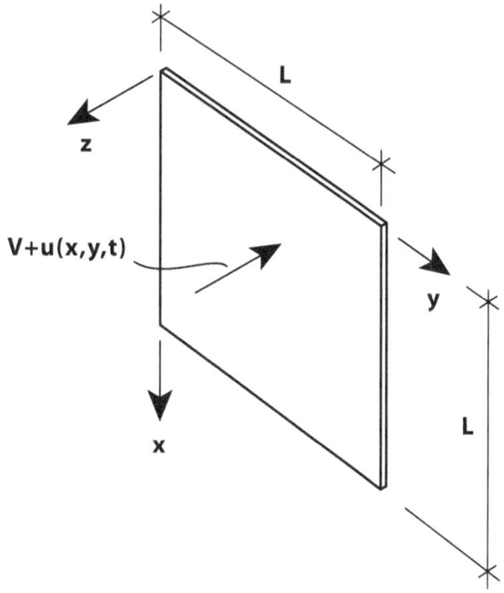

Fig. 10.15 Square window pane with clamped edges subject to fluctuating wind force

The necessary dimensions and mechanical properties of the pane are given in Table 10.2 below:

Table 10.2 Dimensions and mechanical properties

Length: $L_x = L_y = L\ (m)$	*Thickness:* $h\ (m)$	*Density:* $\rho_g\ (kg/m^3)$	*Elastic modulus* $E\ (N/m^2)$	*Poisson's ratio:* ν
2.4	$6 \cdot 10^{-3}$	$80 \cdot 10^{9}$	$2.6 \cdot 10^{3}$	0.3

As mentioned above, the window pane is clamped at its edges, and thus, the eigenfrequencies and corresponding eigenmodes may be obtained from Example 10.2, i.e.

$$\begin{bmatrix}\omega_1\\\omega_2\\\omega_3\end{bmatrix}=\begin{bmatrix}36\\73\\108\end{bmatrix}\left(\frac{D}{mL^4}\right)^{1/2}\quad and\quad \mathbf{\Phi}(x,y)=\begin{bmatrix}\varphi_1\\\varphi_2\\\varphi_3\end{bmatrix}^T=\begin{bmatrix}\varphi_{x1}\cdot\varphi_{y1}\\\varphi_{x2}\cdot\varphi_{y2}\\\varphi_{x3}\cdot\varphi_{y3}\end{bmatrix}^T\approx\begin{bmatrix}f(x)\cdot f(y)\\g(x)\cdot f(y)\\g(x)\cdot g(y)\end{bmatrix}^T$$

where $m=\rho_g h=18.2\ kg/m^2$ *and (see Example 10.2)*

$$\left.\begin{aligned}f(\hat{s})&=\sin(\lambda_1\hat{s})-\sinh(\lambda_1\hat{s})+\beta_1\left[\cos(\lambda_1\hat{s})-\cosh(\lambda_1\hat{s})\right]\\g(\hat{s})&=\sin(\lambda_2\hat{s})-\sinh(\lambda_2\hat{s})+\beta_2\left[\cos(\lambda_2\hat{s})-\cosh(\lambda_2\hat{s})\right]\end{aligned}\right\}\quad \hat{s}=\frac{x}{L_x}\text{ or }\frac{y}{L_y}$$

$$\beta_j=\frac{\cos(\lambda_j)-\cosh(\lambda_j)}{\sin(\lambda_j)+\sinh(\lambda_j)}\qquad j=1\text{ or }2\qquad and\ with\qquad \begin{cases}\lambda_1=3\pi/2\\\lambda_2=5\pi/2\end{cases}$$

The damping ratios are assumed at: $$\begin{bmatrix}\zeta_1\\\zeta_2\\\zeta_3\end{bmatrix}=10^{-3}\begin{bmatrix}1\\2\\3\end{bmatrix}$$

Before proceeding, it is necessary to develop an expression for the dynamic wind load. Adopting Bernoulli's equation for the velocity pressure at an instantaneous interpretation of the relative wind velocity, then

$$q_{tot}(x,y,t)=c_q\frac{1}{2}\rho V_{rel}^2$$

where $\rho=1.25\ kg/m^3$ *is the density of air,* c_q *is a pressure coefficient dependent on the position and geometry of the building on which the window pane is sitting (here we assume* $c_q=1.4$*) and* $V_{rel}=V+u(x,y,t)-\dot{r}_z(x,y,t)$*. Thus, it is seen that the total wind pressure*

$$q_{tot}(x,y,t)=c_q\frac{1}{2}\rho\left(V+u-\dot{r}_z\right)^2\approx\frac{c_q}{2}\rho V+c_q\rho Vu-c_q\rho V\dot{r}_z$$

may be split into a mean time invariant (static) part $c_q\rho V^2/2$*, a turbulence induced dynamic part* $c_q\rho Vu(x,y,t)$ *and a motion induced aerodynamic part* $c_q\rho V\dot{r}_z(x,y,t)$*. Thus, the fluctuating (dynamic) pressure is given by (see Eq. 10.42)*

$$q_z(x,y,t)=c_q\rho Vu(x,y,t)-c_q\rho V\dot{r}_z(x,y,t)=c_q\rho Vu(x,y,t)-c_q\rho V\mathbf{\Phi}(x,y)\dot{\boldsymbol{\eta}}(t)$$

Its Fourier transform

$$a_{q_z}(x,y,\omega) = c_q \rho V a_u(x,y,\omega) - c_q \rho V \mathbf{\Phi}(x,y) i\omega \mathbf{a}_\eta(\omega)$$

will then render (see Eq. 10.50)

$$\begin{aligned}\mathbf{a}_{\tilde{R}}(\omega) &= \int_A \mathbf{\Phi}^T(x,y) \cdot a_{q_z}(x,y,\omega) \\ &= \int_A \mathbf{\Phi}^T(x,y) c_q \rho V a_u(x,y,\omega)\, dA - \int_A \mathbf{\Phi}^T(x,y) c_q \rho V \mathbf{\Phi}(x,y)\, dA \cdot i\omega \mathbf{a}_\eta(\omega) \\ &= \int_A \mathbf{\Phi}^T(x,y) c_q \rho V a_u(x,y,\omega)\, dA - \tilde{\mathbf{C}}_{ae} \cdot i\omega \mathbf{a}_\eta(\omega)\end{aligned}$$

where (due to mode shape orthogonality)

$$\tilde{\mathbf{C}}_{ae} = \int_A \mathbf{\Phi}^T(x,y) c_q \rho V \mathbf{\Phi}(x,y)\, dA = diag\left[\tilde{C}_{ae_n}\right]$$

and $$\tilde{C}_{ae_n} = c_q \rho V \int_A \varphi_n^2(x,y)\, dA = c_q \rho V \int_0^{L_x} \varphi_{x_n}^2(x)\, dx \int_0^{L_y} \varphi_{y_n}^2(y)\, dy$$

Introducing this into Eq. 10.51

$$\left[-\omega^2 \tilde{\mathbf{M}} + i\omega\left(\tilde{\mathbf{C}} + \tilde{\mathbf{C}}_{ae}\right) + \tilde{\mathbf{K}}\right] \mathbf{a}_\eta(\omega) = c_q \rho V \int_A \mathbf{\Phi}^T(x,y) \cdot a_u(x,y,\omega)\, dA$$

and pre-multiplying by $\tilde{\mathbf{K}}^{-1}$

$$\left[\mathbf{I} - \omega^2 \tilde{\mathbf{K}}^{-1} \tilde{\mathbf{M}} + i\omega\left(\tilde{\mathbf{K}}^{-1} \tilde{\mathbf{C}} + \tilde{\mathbf{K}}^{-1} \tilde{\mathbf{C}}_{ae}\right)\right] \mathbf{a}_\eta(\omega) = c_q \rho V \tilde{\mathbf{K}}^{-1} \int_A \mathbf{\Phi}^T(x,y) a_u(x,y,\omega)\, dA$$

then (recalling that $\tilde{\mathbf{M}}$, $\tilde{\mathbf{K}}$, $\tilde{\mathbf{C}}$ *and* $\tilde{\mathbf{C}}_{ae}$ *are all diagonal) the following is obtained*

$$\mathbf{a}_\eta(\omega) = c_q \rho V \hat{\mathbf{H}}_\eta(\omega) \tilde{\mathbf{K}}^{-1} \int_A \mathbf{\Phi}^T(x,y) a_u(x,y,\omega)\, dA$$

where

$$\hat{\mathbf{H}}_\eta(\omega) = diag\left[\hat{H}_{\eta_n}\right] \text{ where } \hat{H}_{\eta_n}(\omega) = \left[1 - (\omega/\omega_n)^2 + 2i\left(\zeta_n + \zeta_{ae_n}\right)\omega/\omega_n\right]^{-1}$$

and $$\zeta_{ae_n} = \frac{\omega_n \tilde{C}_{ae_n}}{2\tilde{K}_n} = \frac{c_q \rho V \int_A \varphi_n^2\, dA}{2\omega_n \int_A m \varphi_n^2\, dA}$$

which, because m *is constant, in this case simplifies into* $\zeta_{ae_n} = \dfrac{c_q \rho V}{2\omega_n m}$.

Thus

$$\begin{aligned}
\mathbf{S}_\eta(\omega) &= \lim_{T\to\infty} \frac{1}{\pi T} \mathbf{a}_\eta^*(\omega)\mathbf{a}_\eta^T(\omega) \\
&= \lim_{T\to\infty} \frac{1}{\pi T} \left[c_q \rho V \hat{\mathbf{H}}_\eta \tilde{\mathbf{K}}^{-1} \int_A \mathbf{\Phi}^T a_u dA \right]^* \cdot \left[c_q \rho V \hat{\mathbf{H}}_\eta \tilde{\mathbf{K}}^{-1} \int_A \mathbf{\Phi}^T a_u dA \right]^T \\
&= \hat{\mathbf{H}}_\eta^* \tilde{\mathbf{K}}^{-1} \left[\left(c_q \rho V\right)^2 \lim_{T\to\infty} \frac{1}{\pi T} \left(\int_A \mathbf{\Phi}^T a_u dA \right)^* \cdot \left(\int_A \mathbf{\Phi}^T a_u dA \right)^T \right] \left(\tilde{\mathbf{K}}^{-1}\right)^T \hat{\mathbf{H}}_\eta^T \\
&= \hat{\mathbf{H}}_\eta^* \tilde{\mathbf{K}}^{-1} \mathbf{S}_{\tilde{R}} \left(\tilde{\mathbf{K}}^{-1}\right)^T \hat{\mathbf{H}}_\eta^T
\end{aligned}$$

where

$$\begin{aligned}
\mathbf{S}_{\tilde{R}}(\omega) &= \left(c_q \rho V\right)^2 \lim_{T\to\infty} \frac{1}{\pi T} \left(\int_A \mathbf{\Phi}^T a_u dA \right)^* \cdot \left(\int_A \mathbf{\Phi}^T a_u dA \right)^T \\
&= \left(c_q \rho V\right)^2 \int_A \int_A \mathbf{\Phi}^T(x_1, y_1) \mathbf{\Phi}(x_2, y_2) \lim_{T\to\infty} \frac{1}{\pi T} a_u^*(x_1, y_1, \omega) a_u(x_2, y_2, \omega) dA_1 dA_2 \\
&= \left(c_q \rho V\right)^2 \int_A \int_A \mathbf{\Phi}^T(x_1, y_1) \mathbf{\Phi}(x_2, y_2) S_{uu}(\Delta s, \omega) dA_1 dA_2
\end{aligned}$$

where Δs *is the distance between two arbitrary points* (x_1, y_1) *and* (x_2, y_2), *i.e.*

$$\Delta s = \left(\Delta x^2 + \Delta y^2\right)^{1/2} \qquad \begin{cases} \Delta x = x_1 - x_2 \\ \Delta y = y_1 - y_2 \end{cases}$$

and

$$S_{uu}(\Delta s, \omega) = \lim_{T\to\infty} \frac{1}{\pi T} a_u^*(x_1, y_1, \omega) a_u(x_2, y_2, \omega) = S_u(\omega) \cdot \hat{C}o_u(\Delta s, \omega) \approx S_u(\omega)$$

Thus, $\mathbf{S}_{\tilde{R}}(\omega) = \left(c_q \rho V\right)^2 \hat{\mathbf{J}}_q S_u(\omega)$ *where*

$$\hat{\mathbf{J}}_q = \int_A \int_A \mathbf{\Phi}^T(x_1, y_1) \mathbf{\Phi}(x_2, y_2) dA_1 dA_2 = \int_A \mathbf{\Phi}^T(x_1, y_1) dA_1 \int_A \mathbf{\Phi}(x_2, y_2) dA_2$$

whose matrix element number nm *is given by*

$$\hat{J}_{q_{nm}} = \int_0^{L_x}\int_0^{L_y} \varphi_{x_n}(x_1)\varphi_{y_n}(y_1)dx_1dy_1 \cdot \int_0^{L_x}\int_0^{L_y} \varphi_{x_m}(x_2)\varphi_{y_m}(y_2)dx_2dy_2$$

Thus,

$$\mathbf{S}_\eta(\omega) = \hat{\mathbf{H}}_\eta^* \tilde{\mathbf{K}}^{-1} \mathbf{S}_{\tilde{R}} \left(\tilde{\mathbf{K}}^{-1}\right)^T \hat{\mathbf{H}}_\eta^T = \left(c_q \rho V\right)^2 \hat{\mathbf{H}}_\eta^*(\omega) \tilde{\mathbf{K}}^{-1} \hat{\mathbf{J}}_q \left(\tilde{\mathbf{K}}^{-1}\right)^T \hat{\mathbf{H}}_\eta^T(\omega) \cdot S_u(\omega)$$

The spectral density to the response at mid-span $\left(x_r = L/2, y_r = L/2\right)$ *is then given by*

$$S_{r_z}\left(x_r = \frac{L}{2}, y_r = \frac{L}{2}, \omega\right) = \mathbf{\Phi}\left(x_r = \frac{L}{2}, y_r = \frac{L}{2}\right) \cdot \mathbf{S}_\eta(\omega) \cdot \mathbf{\Phi}^T\left(x_r = \frac{L}{2}, y_r = \frac{L}{2}\right)$$

It is seen from the expression of $\hat{\mathbf{J}}_q$ *that all its entries beyond the first are zero, because*

$$\left.\begin{matrix} \varphi_{x_n}(\hat{x}) \\ \varphi_{y_n}(\hat{y}) \end{matrix}\right\} \; n = 2 \text{ and } 3 \textit{ are all asymmetric, rendering } \left.\begin{matrix} \int_0^1 \varphi_{x_n}(\hat{x})d\hat{x} \\ \int_0^1 \varphi_{y_n}(\hat{y})d\hat{y} \end{matrix}\right\} = 0$$

I.e., mode shapes two and three will not contribute to the response (coming from the assumption of perfect turbulence correlation over the area of the plate). Thus

$$S_{r_z}(\omega) =$$

$$\left[c_q \rho V^2 I_u \cdot \varphi_{x_1}\left(\frac{L}{2}\right) \cdot \varphi_{y_1}\left(\frac{L}{2}\right) \cdot \frac{\left|\hat{H}_{\eta_1}(\omega)\right|}{\omega_n^2 m} \cdot \frac{\int_0^1 \varphi_{x_1}(\hat{x})d\hat{x}}{\int_0^1 \varphi_{x_1}^2(\hat{x})d\hat{x}} \cdot \frac{\int_0^1 \varphi_{y_1}(\hat{y})d\hat{y}}{\int_0^1 \varphi_{y_1}^2(\hat{y})d\hat{y}} \right]^2 \cdot \hat{S}_u(\omega)$$

where $\hat{S}_u(\omega) = S_u(\omega)/\sigma_u^2$ *and* $I_u = \sigma_u/V$, *and* σ_u *is the standard deviation to the turbulence component. The normalised spectral density of the dynamic response at mid-span is sown in Fig. 10.16 and a time domain simulation of the response is shown in Fig. 10.17. As can be seen, the aerodynamic damping is of significant importance in reducing the dynamic response.*

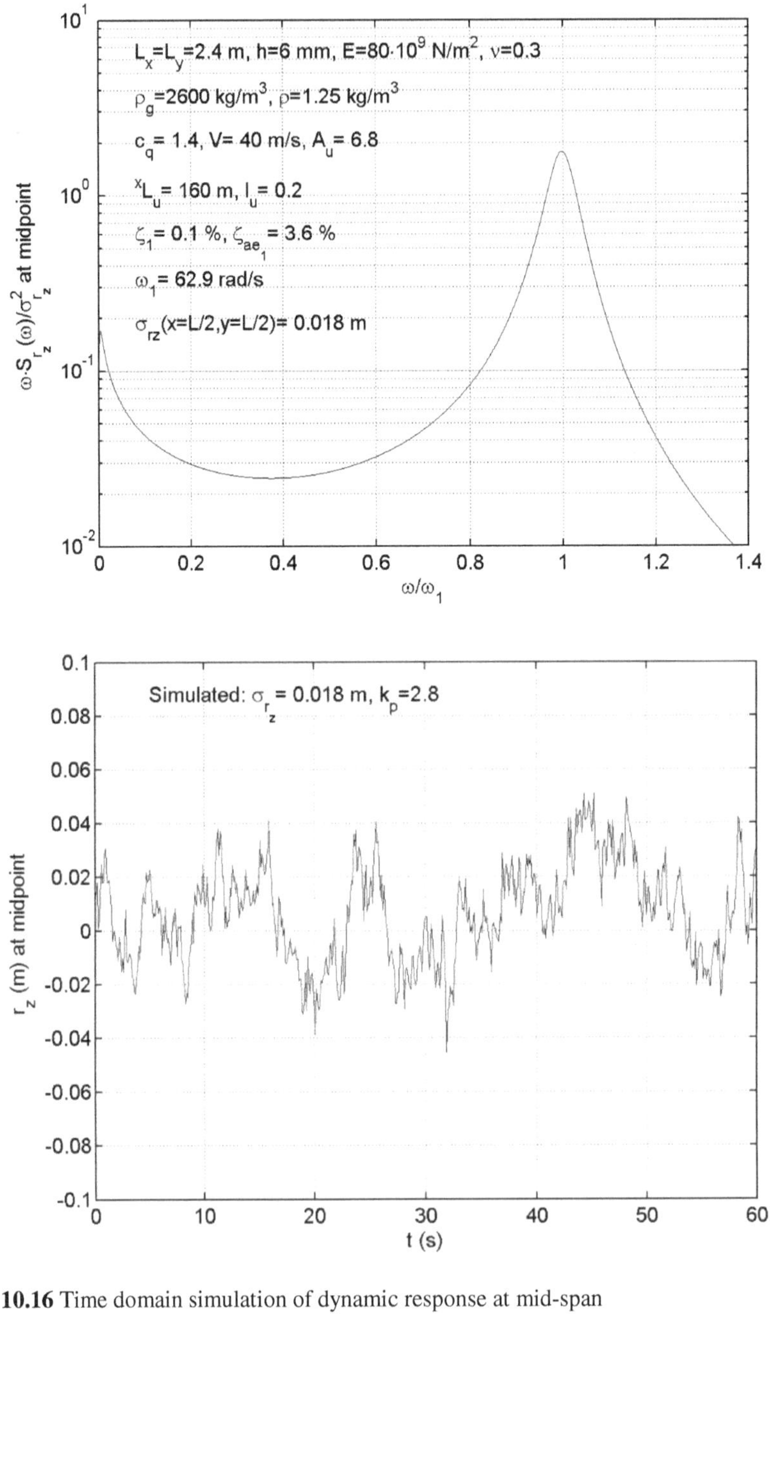

Fig. 10.16 Time domain simulation of dynamic response at mid-span

Chapter 11
Moving Loads on Beams

11.1 Concentrated Single Force

An investigation of the dynamic load effects of moving loads on beams may be required in cases of heavy vehicles or a train passing a flexible bridge, though the problem may not necessarily be that of the bridge, it may also involve uncomfortable vertical oscillations of the vehicle. The problem may also occur on moving hoisting forces on heavy cranes. Let us first consider the simple case of a concentrated single load $F(x_F, t)$, whose magnitude as well as position $x_F(t)$ are time dependent. The problem is illustrated in Fig. 11.1.

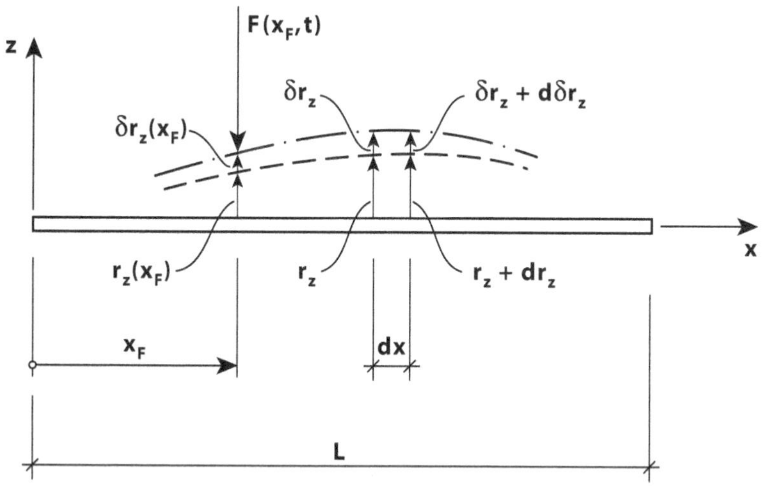

Fig. 11.1 Single concentrated force moving on a beam

The solution may most conveniently be obtained by adopting the principle of d'Alambert together with the principle of virtual work. This has comprehensively been developed in Chapter 1.6. As illustrated in Fig. 11.1 it will in this case involve the energy balance due to an arbitrary (virtual) displacement $\delta r_z(x)$ to

E.N. Strømmen, *Structural Dynamics*, Springer Series in Solid and Structural Mechanics 2, DOI: 10.1007/978-3-319-01802-7_11, © Springer International Publishing Switzerland 2014

the entire system at an instantaneous time and position $r_z(x,t)$. Thus, considering only pure bending and assuming only insignificant time invariant load effects, the following applies (see Eq. 1.119)

$$-\delta r_z(x_F)\cdot F - \int_L \delta r_z(m_z\ddot{r}_r + c_z\dot{r}_z)dx = \int_L \delta r_z'' EI_y r_z'' dx \tag{11.1}$$

where m_z, c_z and EI_y are cross sectional mass, damping coefficient and bending stiffness. Let us also assume that a single mode approach will suffice, and for simplicity that this is single component vertical mode shape $\phi_z(x)$, with corresponding eigenfrequency ω_z and damping ratio ζ_z. Then

$$r_z(x,t) = \phi_z(x)\cdot\eta(t) \quad \text{and} \quad \delta r_z(x) = \phi_z(x)\cdot\delta\eta \tag{11.2}$$

and thus, Eq. 11.1 becomes

$$\delta\eta\int_L m_z\phi_z^2 dx\ddot{\eta}_z(t) + \delta\eta\int_L c_z\phi_z^2 dx\dot{\eta}_z(t) + \delta\eta\int_L EI_y\phi_z''^2 dx\eta_z(t) = -\delta\eta\phi_z(x_F)F(x_F,t) \tag{11.3}$$

from which it is seen that $\delta\eta$ is obsolete, and thus, the following is obtained:

$$\tilde{M}_z\ddot{\eta}_z(t) + \tilde{C}_z\dot{\eta}_z(t) + \tilde{K}_z\eta_z(t) = \tilde{R}_z(t) \tag{11.4}$$

where $\tilde{M}_z = \int_L m_z\phi_z^2 dx$, $\tilde{C}_z = \int_L c_z\phi_z^2 dx = 2\tilde{M}_z\omega_z\zeta_z$, $\tilde{K}_z = \int_L EI_y\phi_z''^2 dx = \omega_z^2\tilde{M}_z$

and $\tilde{R}_z(t) = -\phi_z(x_F)F(x_F,t)$. This is identical to that which was developed in Chapter 5.3. In general, a numerical solution may be pursued, see Chapter 6.3. However, under the conditions that the mode shape is a simple harmonic sinus function and that the magnitude and velocity of the moving force are constants, then a closed form solution is presented in Example 11.1 below. Under the same conditions, except that the moving load is a harmonic cosine functions, then a closed form solution may also be obtained, as shown in Example 11.2.

Elaboration 11.1: The Corresponding Multi-mode Solution

A general multi-mode solution has been developed in Chapter 5.3. Thus, defining

$$\mathbf{F}(x_F,t) = \begin{bmatrix}0 & -F_z(x_F,t) & 0\end{bmatrix}^T$$

and

$$\mathbf{r}(x,t)=\begin{bmatrix} r_y(x,t) \\ r_z(x,t) \\ r_\theta(x,t) \end{bmatrix}=\sum_{n=1}^{N_{\text{mod}}}\begin{bmatrix} \phi_y(x) \\ \phi_z(x) \\ \phi_\theta(x) \end{bmatrix}_n \eta_n(t)=\sum_{n=1}^{N_{\text{mod}}}\boldsymbol{\varphi}_n\cdot\eta_n(t)=\boldsymbol{\Phi}(x)\cdot\boldsymbol{\eta}(t)$$

with mode shapes

$$\left.\begin{aligned} \boldsymbol{\Phi}(x)&=\begin{bmatrix}\boldsymbol{\varphi}_1 & \boldsymbol{\varphi}_2 & \cdots & \boldsymbol{\varphi}_n & \cdots & \boldsymbol{\varphi}_{N_{\text{mod}}}\end{bmatrix} \\ \boldsymbol{\eta}(t)&=\begin{bmatrix}\eta_1 & \eta_2 & \cdots & \eta_n & \cdots & \eta_{N_{\text{mod}}}\end{bmatrix}^T \end{aligned}\right\} \quad \text{where} \quad \boldsymbol{\varphi}_n(x)=\begin{bmatrix}\phi_y \\ \phi_z \\ \phi_\theta\end{bmatrix}$$

and with corresponding eigenfrequencies ω_n and damping ratios ζ_n, then

$$\tilde{\mathbf{M}}\cdot\ddot{\boldsymbol{\eta}}(t)+\tilde{\mathbf{C}}\cdot\dot{\boldsymbol{\eta}}(t)+\tilde{\mathbf{K}}\cdot\boldsymbol{\eta}(t)=\tilde{\mathbf{R}}(t)$$

where $\tilde{\mathbf{M}}=\int_L \boldsymbol{\Phi}^T\mathbf{m}_g\boldsymbol{\Phi}dx$, $\tilde{\mathbf{C}}=diag\left[2\tilde{M}\omega_n\zeta_n\right]$, $\tilde{\mathbf{K}}=diag\left[\omega_n^2\tilde{M}_n\right]$, $\mathbf{m}_g$ is defined in Eq. 5.26, and where $\tilde{\mathbf{R}}(t)=\left\{\boldsymbol{\Phi}(x_F)\right\}^T\cdot\mathbf{F}(x_F,t)$.

Example 11.1

Let us assume that a single mode single component approach will suffice, and that the mode shape is a harmonic sinus function with corresponding eigenfrequency ω_n and damping ratio ζ_n. Let us also assume that the magnitude and velocity of the moving force are constants. I.e.:

$$\phi_z=\sin(n\pi x/L) \quad \text{and} \quad \begin{cases} F(x_F,t)=F_0 \\ x_F=\dot{x}_F\cdot t \end{cases} \quad \text{where} \quad \dot{x}_F \text{ is the velocity of the}$$

force.

Then (see Eq. 11.4): $\quad \tilde{M}_z\ddot{\eta}_z(t)+\tilde{C}_z\dot{\eta}_z(t)+\tilde{K}_z\eta_z(t)=-F_0\sin(\omega_F t)$

where $\omega_F=n\pi\dot{x}_F/L$ and L is the length of the span of the beam. The total period of passage is $T=L/\dot{x}_F$. The solution to this problem has been shown in Chapter 2.3 (only with the difference that the load is positive). Thus, the response at x_r is given by

$$r(x_r,t)=-\phi_z(x_r)\frac{F_0}{\tilde{K}_z}\left|\hat{H}_z(\omega_F)\right|\left[\hat{\eta}_{tr}(t)+\hat{\eta}_{st}(t)\right] \qquad t\le T=L/\dot{x}_F$$

where

$$\begin{cases} \hat{\eta}_{tr}(t) = \dfrac{\sin\beta_p}{\cos\beta_h} e^{-\omega_n \zeta_n t} \cos(\omega_d t - \beta_h) \\ \hat{\eta}_{st}(t) = \sin(\omega_F t - \beta_p) \end{cases} \quad \text{and} \quad \begin{cases} \tan\beta_p = \dfrac{2\zeta_n \omega_F/\omega_n}{1 - \left(\dfrac{\omega_F}{\omega_n}\right)^2} \\ \tan\beta_h = \dfrac{-1 + 2\zeta_n^2 + \left(\dfrac{\omega_F}{\omega_n}\right)^2}{2\zeta_n\sqrt{1-\zeta_n^2}} \end{cases}$$

and where $\omega_d = \omega_n\sqrt{1-\zeta_n^2}$. *At* $t > T = L/\dot{x}_F$ *the response at* x_r *is simply a decaying motion in accordance with Eq. 2.28, i.e.:*

$$r(x_r,t) = -a_n(x_r)\cdot e^{-\omega_n\zeta_n(t-T)}\cdot\cos\left[\omega_d(t-T) - \beta_n\right]$$
$$t > T = L/\dot{x}_F$$

where
$$\begin{cases} a_n = \sqrt{\left[r(x_r,T)\right]^2 + \left[\dot{r}(x_r,T)/\omega_d + r(x_r,T)\cdot\zeta_n\Big/\sqrt{1-\zeta_n^2}\right]^2} \\ \tan\beta_n = \dot{r}(x_r,T)\Big/\left[r(x_r,T)\cdot\omega_d\right] + \zeta_n\Big/\sqrt{1-\zeta_n^2} \end{cases}$$

Similarly, in a multi-mode approach where

$$\phi_z = \sum_{n=1}^{N_{mod}} \phi_{z_n} = \sum_{n=1}^{N_{mod}} \sin(n\pi x/L) \quad \text{and} \quad \begin{cases} F(x_F,t) = F_0 \\ x_F = \dot{x}_F\cdot t \end{cases} \quad \text{where } \dot{x}_F$$

is the constant velocity, then

$$r(x_r,t) = -\sum_{n=1}^{N_{mod}} \phi_{z_n}(x)\frac{F_0}{\tilde{K}_{z_n}}\left|\hat{H}_{z_n}(\omega_F)\right|\left[\hat{\eta}_{tr}(t) + \hat{\eta}_{st}(t)\right] \quad at \quad t \le T = L/\dot{x}_F$$

and $$r(x_r,t) = -\sum_{n=1}^{N_{mod}} a_n e^{-\omega_n\zeta_n(t-T)}\cos\left[\omega_d(t-T) - \beta_n\right] \quad at\ t > T = L/\dot{x}_F$$

A typical response curve due to a concentrated single force $F(x_F) = Mg$ *moving along the span of a beam at a constant velocity is illustrated in the upper diagram in Fig. 11.2 (input data is given in the diagram).*

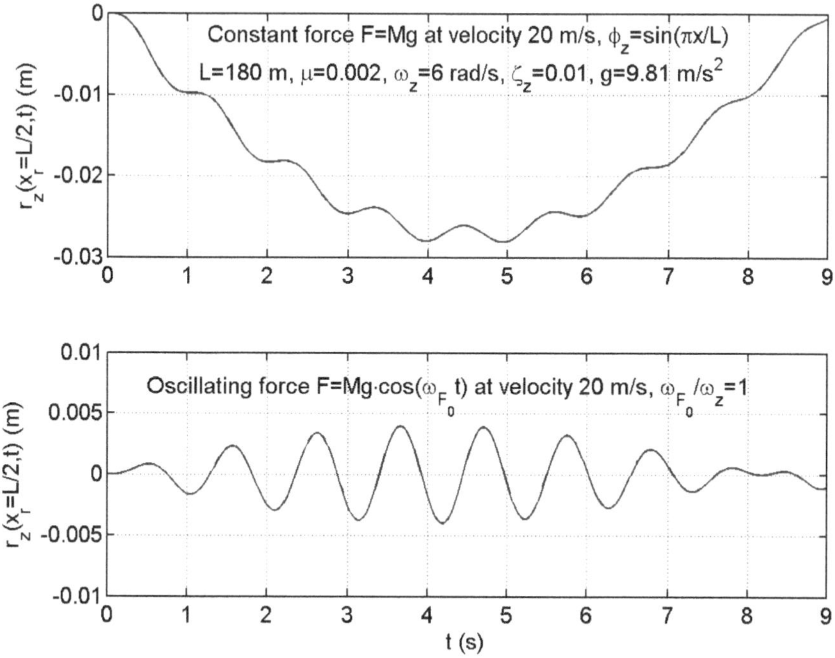

Fig. 11.2 Dynamic response at midspan of simply supported beam with modal mass $\tilde{M}$, due to a single force moving at constant velocity across its span. Upper diagram: $F(x_F) = Mg$. Lower diagram: $F(x_F, t) = Mg \cdot \cos(\omega_{F_0} t)$. $\mu = M/\tilde{M} = 0.002$

Example 11.2

Let us still consider the case of a single mode single component approach, where $\phi_z = \sin(n\pi x/L)$, with corresponding eigenfrequency ω_n and damping ratio ζ_n. Let us also assume that the load is moving with a constant velocity $\dot{x}_F$. However, in this case we assume that the magnitude of the moving force is fluctuating at a constant frequency ω_{F_0} in the shape of a cosine function. I.e.:

$$F(x_F, t) = F_0 \cos(\omega_{F_0} t) \qquad \textit{where} \qquad x_F = \dot{x}_F \cdot t$$

Then $\tilde{R}_z(t) = -\phi_z(x_F) F(t) = -F_0 \cdot \cos(\omega_{F_0} t) \cdot \sin(\omega_{F_n} t)$ where $\omega_{F_n} = \dfrac{n\pi \dot{x}_F}{L}$

Since $\cos\alpha \cdot \sin\beta = \frac{1}{2}\left[\sin(\alpha+\beta) + \sin(\alpha-\beta)\right]$ *then we may rewrite* $\tilde{R}_z(t)$ *into*

$$\tilde{R}_z(t) = -\frac{F_0}{2}\left[\sin\left(2\bar{\omega}_F t\right) + \sin\left(\Delta\omega_F t\right)\right] \quad where \quad \begin{cases} \bar{\omega}_F = \frac{1}{2}\left(\omega_{F_n} + \omega_{F_0}\right) \\ \Delta\omega_F = \omega_{F_n} - \omega_{F_0} \end{cases}$$

then (Eq. 11.4): $\tilde{M}_z \ddot{\eta}_z(t) + \tilde{C}_z \dot{\eta}_z(t) + \tilde{K}_z \eta_z(t) = -\frac{F_0}{2}\left[\sin\left(2\bar{\omega}_F t\right) + \sin\left(\Delta\omega_F t\right)\right]$

Since the principle of superposition applies the dynamic response at spanwise position x_r *is given by (see Example 11.1 above):*

$$r\left(x_r, t\right) = -\frac{\phi_z\left(x_r\right)}{2} \cdot \left[\hat{\eta}_z\left(t, \omega = 2\bar{\omega}_F\right) + \hat{\eta}_z\left(t, \omega = \Delta\omega_F\right)\right] \qquad t \le T = L/\dot{x}_F$$

where

$$\hat{\eta}_z(t,\omega) = \frac{F_0\left|\hat{H}_z(\omega)\right|}{\tilde{K}_z}\left[\hat{\eta}_{tr}(t,\omega) + \hat{\eta}_{st}(t,\omega)\right] \; and \; \hat{H}_z = \left[1 - \left(\frac{\omega}{\omega_n}\right)^2 + 2i\zeta_n \frac{\omega}{\omega_n}\right]^{-1}$$

and where

$$\begin{cases} \hat{\eta}_{tr}(t,\omega) = \frac{\sin\beta_p(\omega)}{\cos\beta_h(\omega)} e^{-\omega_n \zeta_n t} \cos\left[\omega_d t - \beta_h(\omega)\right] \\ \hat{\eta}_{st}(t,\omega) = \sin\left[\omega t - \beta_p(\omega)\right] \end{cases}$$

$$\tan\beta_p(\omega) = \frac{2\zeta_n \, \omega/\omega_n}{1 - \left(\omega/\omega_n\right)^2} \quad and \quad \tan\beta_h(\omega) = \frac{-1 + 2\zeta_n^2 + \left(\omega/\omega_n\right)^2}{2\zeta_n\sqrt{1 - \zeta_n^2}}$$

and where $\omega_d = \omega_n\sqrt{1 - \zeta_n^2}$ *. Again, At* $t > T = L/\dot{x}_F$ *the response at* x_r *is simply a decaying motion in accordance with Eq. 2.28, i.e.:*

$$r\left(x_r, t\right) = -a_n\left(x_r\right) \cdot e^{-\omega_n \zeta_n (t-T)} \cdot \cos\left[\omega_d\left(t - T\right) - \beta_n\right] \qquad t > T = L/\dot{x}_F$$

where

$$\begin{cases} a_n = \sqrt{\left[r\left(x_r, T\right)\right]^2 + \left[\dot{r}\left(x_r, T\right)/\omega_d + r\left(x_r, T\right) \cdot \zeta_n \Big/ \sqrt{1 - \zeta_n^2}\right]^2} \\ \tan\beta_n = \dot{r}\left(x_r, T\right)/\left[r\left(x_r, T\right) \cdot \omega_d\right] + \zeta_n \Big/ \sqrt{1 - \zeta_n^2} \end{cases}$$

A typical response curve due to an oscillating single force $F\left(x_F, t\right) = Mg \cdot \cos\left(\omega_{F_0} t\right)$ *moving along the span of a beam at a constant velocity is illustrated in the lower diagram in Fig. 11.2 (input data is given in the diagram).*

11.2 Rolling Single Wheel Vehicle

The case of a rolling single wheel passing along the span of a beam at arbitrary velocity is shown in Fig. 11.3. The wheel is connected to a mass M by an elastic spring with stiffness K and a damper with damping coefficient C. This is a far more realistic case than that which was presented in Chapter 11.1 above. (It may readily be expanded to include a mass carried by several wheels, or a series of consecutive masses.) The situation at an arbitrary time t (in accordance with the principle of d'Alambert) is illustrated in Fig. 11.4. The instantaneous position and velocity of the wheel is x_M and $\dot{x}_M$. It is taken for granted that the beam as well as the mass is restricted to move in the vertical z direction alone. Thus, the

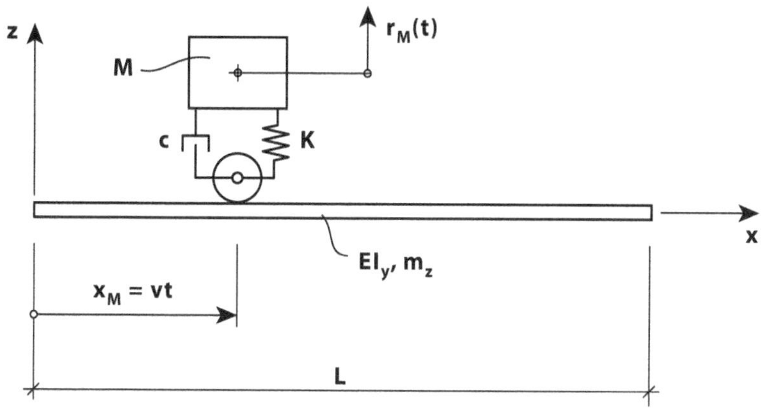

Fig. 11.3 Rolling single wheel vehicle moving on a beam

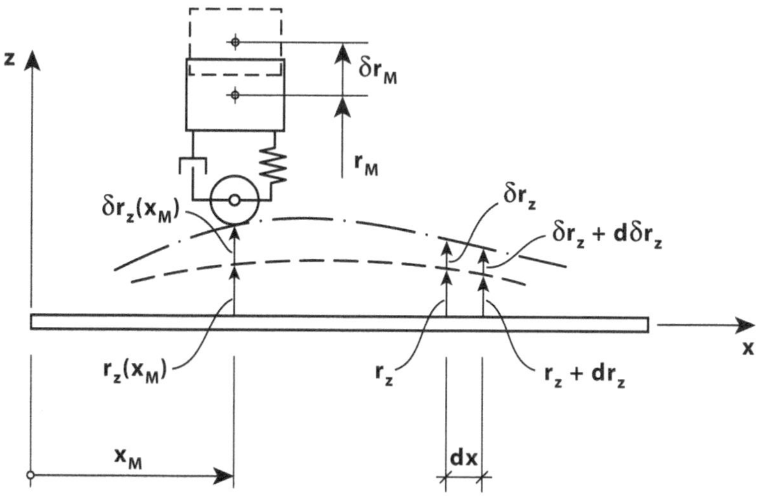

Fig. 11.4 Relevant real and virtual displacements

displacements of the beam and the mass are $r_z(x,t)$ is $r_M(t)$. Denoting $\Delta r(t)$ the difference between $r_M(t)$ and $r_z(x_M,t)$, then

$$r_M(t) = \Delta r(t) + r_z(x_M,t) \tag{11.5}$$

Since the mass is in motion both horizontally and vertically at the same time it is necessary to perform a linear expansion at $t+\Delta t$:

$$\begin{aligned} r_M(t+\Delta t) &= \Delta r(t+\Delta t) + r_z(x_M + \Delta x_M, t+\Delta t) \\ &= \Delta r(t) + \Delta\dot{r}(t)\cdot\Delta t + r_z(x_M) + r_z'(x_M,t)\cdot\Delta x_M + \dot{r}_z(x_M,t)\cdot\Delta t \end{aligned} \tag{11.6}$$

and thus

$$\begin{aligned} \dot{r}_{M_{tot}}(t) &= \frac{r_M(t+\Delta t) - r_M(t)}{\Delta t} = \Delta\dot{r}(t) + \dot{r}_z(x_M,t) + r_z'(x_M,t)\cdot\frac{\Delta x_M}{\Delta t} \\ &= \Delta\dot{r}(t) + \dot{r}_z(x_M,t) + \dot{x}_M(t)\cdot r_z'(x_M,t) \end{aligned} \tag{11.7}$$

Similarly

$$\begin{aligned} \dot{r}_{M_{tot}}(t+\Delta t) &= \Delta\dot{r}(t) + \Delta\ddot{r}(t)\cdot\Delta t + \dot{r}_z(x_M,t) + \ddot{r}_z(x_M,t)\cdot\Delta t \\ &+ \left[\dot{x}_M(t) + \ddot{x}_M(t)\cdot\Delta t\right]\cdot\left[r_z'(x_M,t) + r_z''(x_M,t)\cdot\Delta x_M\right] \end{aligned} \tag{11.8}$$

Thus

$$\ddot{r}_{M_{tot}}(t) = \frac{\dot{r}_M(t+\Delta t) - \dot{r}_M(t)}{\Delta t} = \ddot{r}_M(t) + \ddot{x}_M(t)\, r_z'(x_M,t) + \dot{x}_M^2\, r_z''(x_M,t) \tag{11.9}$$

The contact force between the wheel and the surface of the beam (see Fig. 11.5) is then given by $F_{M_2} - F_{M_1}$, where

$$F_{M_1}(t) = C\Delta\dot{r}(t) + K\Delta r(t) = C\dot{r}_M(t) + Kr_M(t) - \left[C\dot{r}_z(x_M,t) + Kr_z(x_M,t)\right] \tag{11.10}$$

$$\begin{aligned} F_{M_2} &= F_{M_1}(t) + M\ddot{r}_{M_{tot}}(t) + Mg \\ &= C\dot{r}_M(t) + Kr_M(t) - \left[C\dot{r}_z(x_M,t) + Kr_z(x_M,t)\right] \\ &+ M\ddot{r}_M(t) + M\left[\ddot{x}_M r_z'(x_M,t) + \dot{x}_M^2 r_z''(x_M,t)\right] + Mg \end{aligned} \tag{11.11}$$

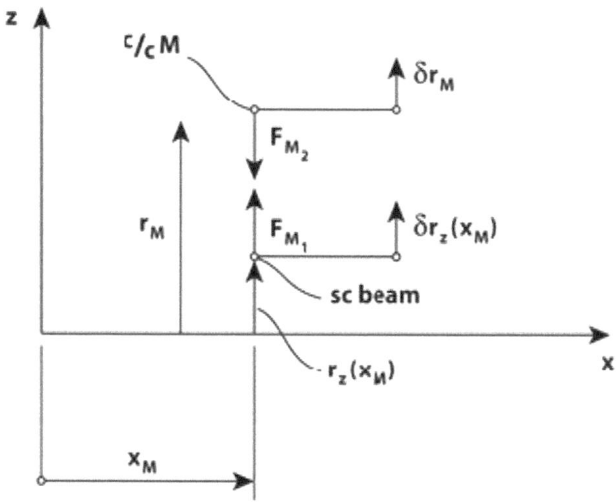

Fig. 11.5 Vehicle forces and virtual displacements

By applying a virtual displacement $\delta r_z(x)$ to the instantaneous motion of the beam and simultaneously a virtual displacement δr_M to the instantaneous motion of the mass M (see Figs. 11.4 and 11.5), and demanding work balance (see Chapter 1.6, Eq. 1.119), then

$$\begin{aligned} & F_{M_1}\delta r_z(x_M) - F_{M_2}\delta r_M \\ & -\int_L \left[m_z \ddot{r}_z(x,t) + c_z \dot{r}_z(x,t) \right] \delta r_z(x)\,dx = \int_L EI_y r_z''(x,t)\,\delta r_z''(x)\,dx \end{aligned} \tag{11.12}$$

where m_z, c_z and EI_y are mass, damping coefficient and cross sectional stiffness associated with motion in the z direction (i.e. with bending about the y axis). Introducing Eqs. 11.10 and 11.11, then

$$\begin{aligned} & \int_L m_z \ddot{r}_z(x,t)\,\delta r_z(x)\,dx + \int_L c_z \dot{r}_z(x,t)\,\delta r_z(x)\,dx + \int_L EI_y r_z''(x,t)\,\delta r_z''(x)\,dx \\ & -\left[C\dot{r}_M(t) + K r_M(t) - C\dot{r}_z(x_M,t) - K r_z(x_M,t) \right] \delta r_z(x_M) \\ & +\left[C\dot{r}_M(t) + K r_M(t) - C\dot{r}_z(x_M,t) - K r_z(x_M,t) \right] \delta r_M \\ & +M\left[\ddot{r}_M(t) + \ddot{x}_M r_z'(x_M,t) + \dot{x}_M^2 r_z''(x_M,t) + g \right] \delta r_M = 0 \end{aligned} \tag{11.13}$$

Let us for simplicity assume that a single mode $\phi_z(x)$ approach will suffice, then it is convenient to choose

$$\left.\begin{array}{lll} r_z(x,t)=\phi_z(x)\cdot\eta_z(t) & \text{and} & r_M(t)=1\cdot\eta_M(t) \\ \delta r_z(x,t)=\phi_z(x)\cdot\delta\eta_z & \text{and} & \delta r_M=1\cdot\delta\eta_M \end{array}\right\} \tag{11.14}$$

and then Eq. 11.13 may be written

$$\begin{bmatrix}\delta\eta_z\\ \delta\eta_M\end{bmatrix}^T \left\{ \begin{bmatrix}\tilde{M}_z & 0\\ 0 & M\end{bmatrix}\begin{bmatrix}\ddot{\eta}_z(t)\\ \ddot{\eta}_M(t)\end{bmatrix} + \begin{bmatrix}\tilde{C}_z+C\phi_z^2(x_M) & -C\phi_z(x_M)\\ -C\phi_z(x_M) & C\end{bmatrix}\begin{bmatrix}\dot{\eta}_z(t)\\ \dot{\eta}_M(t)\end{bmatrix} \right.$$
$$\begin{bmatrix}\tilde{K}_z+K\phi_z^2(x_M) & -K\phi_z(x_M)\\ -K\phi_z(x_M)+M\left[\ddot{x}_M\phi_z'(x_M)+\dot{x}_M^2\phi_z''(x_M)\right] & K\end{bmatrix}\begin{bmatrix}\eta_z(t)\\ \eta_M(t)\end{bmatrix} \tag{11.15}$$
$$\left. +\begin{bmatrix}0\\ Mg\end{bmatrix}\right\}=\mathbf{0}$$

where $\tilde{M}_z=\int_L m_z\phi_z^2(x)dx$, $\tilde{C}_z=2\tilde{M}_z\omega_z\zeta_z$ and $\tilde{K}_z=\omega_z^2\tilde{M}_z$ are modal mass, damping and stiffness, and where ω_z is the eigenfrequency and ζ_z is the modal damping ratio. It is readily seen that the pre-multiplication by the virtual displacement vector $\delta\boldsymbol{\eta}=\begin{bmatrix}\delta\eta_z & \delta\eta_M\end{bmatrix}^T$ may be omitted, and thus

$$\tilde{\mathbf{M}}\ddot{\boldsymbol{\eta}}(t)+\left[\tilde{\mathbf{C}}+\tilde{\mathbf{C}}_M(x_M)\right]\dot{\boldsymbol{\eta}}(t)+\left[\tilde{\mathbf{K}}+\tilde{\mathbf{K}}_M(x_M)\right]\boldsymbol{\eta}(t)=\tilde{\mathbf{R}} \tag{11.16}$$

where

$$\boldsymbol{\eta}(t)=\begin{bmatrix}\eta_z\\ \eta_M\end{bmatrix} \quad \tilde{\mathbf{M}}=\begin{bmatrix}\tilde{M}_z & 0\\ 0 & M\end{bmatrix} \quad \tilde{\mathbf{C}}=\begin{bmatrix}\tilde{C}_z & 0\\ 0 & C\end{bmatrix} \quad \tilde{\mathbf{K}}=\begin{bmatrix}\tilde{K}_z & 0\\ 0 & K\end{bmatrix} \tag{11.17}$$

and

$$\tilde{\mathbf{C}}_M=C\begin{bmatrix}\phi_z^2(x_M) & -\phi_z(x_M)\\ -\phi_z(x_M) & 0\end{bmatrix} \quad \tilde{\mathbf{R}}=\begin{bmatrix}0\\ -Mg\end{bmatrix} \tag{11.18}$$

$$\tilde{\mathbf{K}}_M=K\begin{bmatrix}\phi_z^2(x_M) & -\phi_z(x_M)\\ -\phi_z(x_M)+\dfrac{1}{\omega_M^2}\left[\ddot{x}_M\phi_z'(x_M)+\dot{x}_M^2\phi_z''(x_M)\right] & 0\end{bmatrix} \tag{11.19}$$

By pre-multiplication with $\tilde{\mathbf{M}}^{-1}$ this may alternatively (and most often more conveniently) be written on a non-dimensional format

$$\mathbf{I}\ddot{\boldsymbol{\eta}}(t)+2\omega_z\zeta_z\left[\hat{\boldsymbol{\zeta}}+\hat{\boldsymbol{\zeta}}_M(x_M)\right]\dot{\boldsymbol{\eta}}(t)+\omega_z^2\left[\hat{\boldsymbol{\omega}}+\hat{\boldsymbol{\omega}}_M(x_M)\right]\boldsymbol{\eta}(t)=\mathbf{p} \tag{11.20}$$

where $\mathbf{I}$ is the two by two identity matrix and

$$\hat{\boldsymbol{\zeta}}=\begin{bmatrix}1 & 0\\ 0 & \hat{\omega}_M\hat{\zeta}_M\end{bmatrix} \qquad \hat{\boldsymbol{\omega}}=\begin{bmatrix}1 & 0\\ 0 & \hat{\omega}_M^2\end{bmatrix} \qquad \mathbf{p}=\begin{bmatrix}0\\ -g\end{bmatrix} \tag{11.21}$$

$$\hat{\boldsymbol{\zeta}}_M=\hat{\omega}_M\hat{\zeta}_M\begin{bmatrix}\mu\phi_z^2(x_M) & -\mu\phi_z(x_M)\\ -\phi_z(x_M) & 0\end{bmatrix} \tag{11.22}$$

$$\hat{\boldsymbol{\omega}}_M=\omega_M^2\begin{bmatrix}\mu\phi_z^2(x_M) & -\mu\phi_z(x_M)\\ -\phi_z(x_M)+\dfrac{1}{\omega_M^2}\left[\ddot{x}_M\phi_z'(x_M)+\dot{x}_M^2\phi_z''(x_M)\right] & 0\end{bmatrix} \tag{11.23}$$

and where $\mu=M/\tilde{M}_z$, $\hat{\omega}_M=\omega_M/\omega_z$, $\hat{\zeta}_M=\zeta_M/\zeta_z$, $\omega_z^2=\tilde{K}_z/\tilde{M}_z$, $\omega_M^2=K/M$, $\zeta_z=\tilde{C}_z/\left(2\tilde{M}_z\omega_z\right)$ and $\zeta_M=C/\left(2M\omega_M\right)$. A closed form solution to Eq. 11.20 can in general not be obtained, i.e. it will be necessary to resort to a numeric solution strategy, e.g. as presented in Chapter 6.3.

Example 11.3

Let us still consider the case of a single mode single component approach, where $\phi_z=\sin\left(n\pi x/L\right)$, *with corresponding eigenfrequency* ω_n *and damping ratio* ζ_n. *Let us assume that the mass has a velocity of* V_0 *at the beginning of the passage, i.e.* $\dot{x}_M(t=0)=V_0$, *and that during the passage it is subject to a constant acceleration* a_M, *i.e. that* $\ddot{x}_M(t)=a_M$. *Then*

$$\dot{x}_M(t)=V_0+\int_0^t a_M\,dt=V_0+a_Mt \quad \text{and} \quad x_M(t)=\int_0^t \dot{x}_M(t)\,dt=V_0t+\frac{a_M}{2}t^2$$

while the passing period T *is obtained from*

$$L=V_0T+\frac{a_M}{2}T^2 \qquad \text{rendering} \qquad T=\frac{V_0}{a_M}\left(\sqrt{1+\frac{2a_ML}{V_0^2}}-1\right)$$

Let us for simplicity adopt the second central difference method (page 255), choosing the iteration sequence $k=\begin{bmatrix}1 & 2 & \cdots & k & \cdots & K\end{bmatrix}$ *and corresponding time series* $t=\begin{bmatrix}t_1 & \cdots & t_k & \cdots & t_K\end{bmatrix}$, *where* $t_1=0$ *and* $t_K=T$, *such that* $\Delta t=T/(K-1)$, *and thus:*

$$\mathbf{I}\ddot{\boldsymbol{\eta}}_{k+1} + 2\omega_z\zeta_z\left[\hat{\boldsymbol{\zeta}} + \hat{\boldsymbol{\zeta}}_{M_k}\right]\dot{\boldsymbol{\eta}}_k + \omega_z^2\left[\hat{\boldsymbol{\omega}} + \hat{\boldsymbol{\omega}}_{M_k}\right]\boldsymbol{\eta}_k = \boldsymbol{\rho}$$

$$\hat{\boldsymbol{\zeta}}_{M_k} = \hat{\omega}_M \hat{\zeta}_M \begin{bmatrix} \mu\phi_z^2\left(x_{M_k}\right) & -\mu\phi_z\left(x_{M_k}\right) \\ -\phi_z\left(x_{M_k}\right) & 0 \end{bmatrix}$$

$$\hat{\boldsymbol{\omega}}_{M_k} = \omega_M^2 \begin{bmatrix} \mu\phi_z^2\left(x_{M_k}\right) & -\mu\phi_z\left(x_{M_k}\right) \\ -\phi_z\left(x_{M_k}\right) + \dfrac{1}{\omega_M^2}\left[a_M\phi_z'\left(x_{M_k}\right) + \dot{x}_{M_k}^2\phi_z''\left(x_{M_k}\right)\right] & 0 \end{bmatrix}$$

$$\dot{x}_{M_k} = V_0 + a_M t_k \qquad \textit{and} \qquad x_{M_k} = V_0 t_k + \frac{a_M}{2} t_k^2$$

Thus

$$\boldsymbol{\eta}_{k+1} = \left[\mathbf{I} + \Delta t\omega_z\zeta_z\left(\hat{\boldsymbol{\zeta}} + \hat{\boldsymbol{\zeta}}_{M_k}\right)\right]^{-1} \cdot \left\{\Delta t^2\boldsymbol{\rho} + 2\left[\mathbf{I} - \frac{\Delta t^2}{2}\omega_z^2\left(\hat{\boldsymbol{\omega}} + \hat{\boldsymbol{\omega}}_{M_k}\right)\right]\boldsymbol{\eta}_k - \left[\mathbf{I} - \Delta t\omega_z\zeta_z\left(\hat{\boldsymbol{\zeta}} + \hat{\boldsymbol{\zeta}}_{M_k}\right)\right]\boldsymbol{\eta}_{k-1}\right\}$$

and $$\boldsymbol{\eta}_1 = \left\{\mathbf{I} + \Delta t\omega_z\zeta_z\left[\hat{\boldsymbol{\zeta}} + \hat{\boldsymbol{\zeta}}_M\left(x_M = 0\right)\right]\right\}^{-1}\Delta t^2\boldsymbol{\rho}$$

The dynamic response of the mass M *and of the beam at an arbitrary position* x_r *is then given by*

$$\mathbf{r}\left(x_r, t_k\right) = \begin{bmatrix} r_z\left(x_r, t_t\right) \\ r_M\left(t_k\right) \end{bmatrix} = \boldsymbol{\Phi}\left(x_r\right)\cdot\boldsymbol{\eta}\left(t_k\right) \quad \textit{where} \quad \boldsymbol{\Phi}\left(x_r\right) = \begin{bmatrix} \phi_z\left(x_r\right) & 0 \\ 0 & 1 \end{bmatrix}$$

Typical resonant response curves of such an iteration procedure are illustrated in Fig. 11.6 (input data given in the diagrams).

Elaboration 11.2: The General Multi-mode Solution

In a general multi-mode format Eq. 11.13 still applies. Thus, introducing

$$r_z\left(x,t\right) = \sum_{n=1}^{N_{\text{mod}}} \phi_{z_n}\left(x\right)\cdot\eta_{z_n}\left(t\right) = \boldsymbol{\Phi}\left(x\right)\cdot\boldsymbol{\eta}\left(t\right) \qquad \text{and} \qquad r_M\left(t\right) = 1\cdot\eta_M\left(t\right)$$

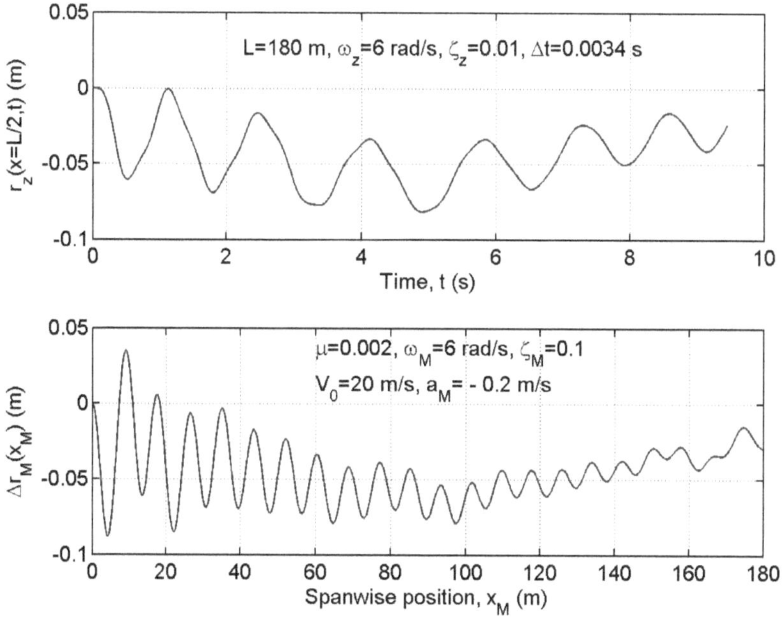

Fig. 11.6 Response curves of beam and vehicle mass during passage of a simply supported bridge system. Upper curve: the beam displacement at midspan. Lower curve: The vehicle mass motion during passage.

where $\begin{cases} \mathbf{\Phi}(x) = \begin{bmatrix} \phi_{z_1} & \cdots & \phi_{z_n} & \cdots & \phi_{z_{N_{\text{mod}}}} \end{bmatrix} \\ \mathbf{\eta}(t) = \begin{bmatrix} \eta_1 & \cdots & \eta_n & \cdots & \eta_{N_{\text{mod}}} \end{bmatrix}^T \end{cases}$ then Eq. 11.13 becomes:

$$\begin{aligned}
&\int_L \delta r_z(x) m_z \mathbf{\Phi}(x) \ddot{\mathbf{\eta}}(t) dx + \int_L \delta r_z(x) c_z \mathbf{\Phi}(x) \dot{\mathbf{\eta}}(t) dx + \int_L \delta r_z''(x) EI_y \mathbf{\Phi}''(x) \mathbf{\eta}(t) dx \\
&-\delta r_z(x_M)\left[C\dot{\eta}_M(t) + K\eta_M(t) - C\mathbf{\Phi}(x_M)\dot{\mathbf{\eta}}(t) - K\mathbf{\Phi}(x_M)\mathbf{\eta}(t)\right] \\
&+\delta r_M\left[C\dot{\eta}_M(t) + K\eta_M(t) - C\mathbf{\Phi}(x_M)\dot{\mathbf{\eta}}(t) - K\mathbf{\Phi}(x_M)\mathbf{\eta}(t)\right] \\
&+\delta r_M M\left[\ddot{\eta}_M(t) + \ddot{x}_M \mathbf{\Phi}'(x_M)\mathbf{\eta}(t) + \dot{x}_M^2 \mathbf{\Phi}''(x_M)\mathbf{\eta}(t) + g\right] = 0
\end{aligned}$$

where $\begin{cases} \mathbf{\Phi}'(x) = \begin{bmatrix} \phi'_{z_1} & \cdots & \phi'_{z_n} & \cdots & \phi'_{z_{N_{\text{mod}}}} \end{bmatrix} \\ \mathbf{\Phi}''(x) = \begin{bmatrix} \phi''_{z_1} & \cdots & \phi''_{z_n} & \cdots & \phi''_{z_{N_{\text{mod}}}} \end{bmatrix} \end{cases}$

Then we apply the virtual displacements $\begin{cases} \delta r_z(x) = \phi_{z_m} \cdot \delta\eta_m \\ \delta r_M = 1 \cdot \delta\eta_M \end{cases}$ $m = 1, 2, \ldots, N_{\text{mod}}$,

one after the other, rendering $N_{\text{mod}} + 1$ equations

$$\begin{aligned}
&\delta\eta_m \int_L m_z \phi_{z_m}^2 dx \cdot \ddot{\eta}_m(t) + \delta\eta_M M \ddot{\eta}_M(t) + \delta\eta_m \int_L c_z \phi_{z_m}^2 dx \cdot \dot{\eta}_m(t) \\
&-\delta\eta_m \phi_{z_m}(x_M) C \dot{\eta}_M + \delta\eta_m \phi_{z_m}(x_M) C \mathbf{\Phi}(x_M) \dot{\mathbf{\eta}} + \delta\eta_M C \dot{\eta}_M \\
&-\delta\eta_M C \mathbf{\Phi}(x_M) \dot{\mathbf{\eta}} + \delta\eta_m \int_L EI_y \phi_{z_m}''^2 dx \cdot \dot{\eta}_m(t) - \delta\eta_m \phi_{z_m}(x_M) K \eta_M \\
&+\delta\eta_m \phi_{z_m}(x_M) K \mathbf{\Phi}(x_M) \mathbf{\eta} + \delta\eta_M K \eta_M - \delta\eta_M K \mathbf{\Phi}(x_M) \mathbf{\eta} \\
&+\delta\eta_M M \ddot{x}_M \mathbf{\Phi}'(x_M) \mathbf{\eta} + \delta\eta_M M \dot{x}_M^2 \mathbf{\Phi}''(x_M) \mathbf{\eta} + \delta\eta_M M g = 0
\end{aligned}$$

which, by defining the following 1 by $N_{\text{mod}} + 1$ vectors

$$\begin{aligned}
\mathbf{\Psi}(x) &= [\mathbf{\Phi} \quad -1] = \begin{bmatrix} \phi_{z_1} & \cdots & \phi_{z_n} & \cdots & \phi_{z_{N_{\text{mod}}}} & -1 \end{bmatrix} \\
\mathbf{\beta} &= [0 \quad \cdots \quad 0 \quad \cdots \quad 0 \quad -1] \\
\mathbf{\Psi}'(x) &= [\mathbf{\Phi}' \quad 0] = \begin{bmatrix} \phi'_{z_1} & \cdots & \phi'_{z_n} & \cdots & \phi'_{z_{N_{\text{mod}}}} & 0 \end{bmatrix} \\
\mathbf{\Psi}''(x) &= [\mathbf{\Phi}'' \quad 0] = \begin{bmatrix} \phi''_{z_1} & \cdots & \phi''_{z_n} & \cdots & \phi''_{z_{N_{\text{mod}}}} & 0 \end{bmatrix}
\end{aligned}$$

and $N_{\text{mod}} + 1$ by $N_{\text{mod}} + 1$ matrices

$$\begin{aligned}
\tilde{\mathbf{M}} &= diag\begin{bmatrix} \tilde{M}_1 & \cdots & \tilde{M}_n & \cdots & \tilde{M}_{N_{\text{mod}}} & 0 \end{bmatrix} \\
\tilde{\mathbf{C}} &= diag\begin{bmatrix} \tilde{C}_1 & \cdots & \tilde{C}_n & \cdots & \tilde{C}_{N_{\text{mod}}} & 0 \end{bmatrix} \\
\tilde{\mathbf{K}} &= diag\begin{bmatrix} \tilde{K}_1 & \cdots & \tilde{K}_n & \cdots & \tilde{K}_{N_{\text{mod}}} & 0 \end{bmatrix} \quad \text{and} \qquad \tilde{\mathbf{R}} = Mg\mathbf{\beta}^T
\end{aligned}$$

then the following dynamic equilibrium condition is obtained:

$$\begin{aligned}
&\tilde{\mathbf{M}}\ddot{\mathbf{\eta}} + \left[\tilde{\mathbf{C}} + C\mathbf{\Psi}^T(x_M)\mathbf{\Psi}(x_M) \right] \dot{\mathbf{\eta}} \\
&+\left\{ \tilde{\mathbf{K}} + K\mathbf{\Psi}^T(x_M)\mathbf{\Psi}(x_M) - M\mathbf{\beta}^T \left[\ddot{x}_M \mathbf{\Psi}'(x_M) + \dot{x}_M^2 \mathbf{\Psi}''(x_M) \right] \right\} \mathbf{\eta} = \tilde{\mathbf{R}}
\end{aligned}$$

Obviously, a numeric approach will in general be required.

Appendix A
Basic Theory of Stochastic Processes

A.1 Introduction

A physical process is called a stochastic process if its numerical outcome at any time or position in space is random and can only be predicted with a certain probability of occurrence. Similarly, a data set of observations of a stochastic process can only be regarded as one particular set of realisations of the process, none of which can with certainty be repeated even if the conditions are seemingly the same. In fact, the observed numerical outcome of all physical processes is more or less random. The outcome of a process is only deterministic in so far as it represents a mathematical description whose input parameters has all been predetermined and remains unchanged.

The physical characteristics of a stochastic process are described by its statistical properties. If it is the cause of another process, this will also be a stochastic process. I.e. if a physical event may mathematically be described by certain laws of nature, a stochastic input will provide a stochastic output. Thus, statistics constitute a mathematical description that provides the necessary parameters for numerical predictions of the random variables that are the cause and effects of physical events. The instantaneous wind velocity pressure at a particular time and position in space is such a stochastic process. The instantaneous ground motion during the incidence of an earthquake is also random in its behaviour. This implies that an attempt to predict its value at a certain position and time can only be performed in a statistical sense. An observed set of records cannot precisely be repeated, but it will follow a certain pattern that may only be mathematically represented by statistics. Thus, the key in understanding the effects of a stochastic load is to acknowledge its random distribution in time and space, i.e. any load effect calculations, static or dynamic, will require statistical averaging in both time and space.

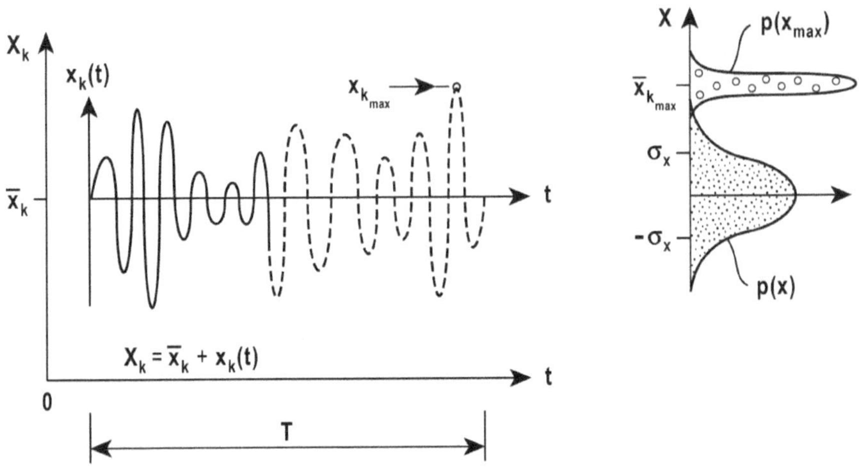

Fig. A.1 Short term stationary random process

It is necessary to distinguish between short and long term statistics, where the short term random outcome are time domain representatives for the conditions within a certain physical situation, e.g. in wind engineering a time window of $T = 10$ minutes, while the long term conditions are ensemble representatives extracted from a large set of individual short term conditions. For a meaningful use in structural engineering it is a requirement that the short term statistics are stationary and homogeneous. Thus, it represents a certain time–space–window that is short and small enough to render sufficiently constant statistical properties. The space window is usually no problem, as the weather or earthquake conditions surrounding most civil engineering structures may usually be considered homogeneous enough. Such a typical stochastic process is illustrated in Fig. A.1. It may for instance be a short term representation of the fluctuating along wind velocity, or the fluctuating structural displacement response at a certain point along its span. As can be seen, it is taken for granted that the process may be split into a constant mean and a stationary fluctuating part. There are two levels of randomness in this process. Firstly, it is random with respect to the instantaneous value within the short term period between 0 and T. I.e., regarding it as a set of successive individual events rather than a continuous function, the process observations are stored by two vectors, one containing time coordinates and another containing the instantaneous recorded values of the process. The stochastic properties of the process may then be revealed by performing statistical investigations to the sample vector of recorded values. For the fluctuating part, it is a general assumption herein that the sample vector of a stochastic process will render a Gaussian probability distribution

$$p(x) = \frac{1}{\sqrt{2\pi}\sigma_x} \exp\left[-\frac{1}{2} \left(\frac{x - \overline{x}}{\sigma_x} \right)^2 \right] \tag{A.1}$$

as illustrated to the right in Fig. A.1. This type of investigation is in the following labelled time domain statistics. The second level of randomness pertains to the simple fact that the sample set of observations shown in Fig. A.1 is only one particular realisation of the process. I.e. there are an infinite number of other possible representatives. Each of these may look similar and have nearly the same statistical properties, but they are random in the sense that they are never precisely equal to the one singled out in Fig. A.1. From each of a particular set of different realisations we may for instance only be interested in the mean value and the maximum value. Collecting a large number of different realisations will render a sample set of these values, and thus, statistics may also be performed on the mean value and the maximum value of the process. This is in the following labelled ensemble statistics.

In structural engineering $X_k = \overline{x}_k + x_k(t)$ may as mentioned above also be a representative of the displacement response at a certain position in the system. The time invariant part $\overline{x}_k$ is then the commonly known mean static response. The fluctuating part $x_k(t)$ represents the dynamic part of the response.

A.2 Time Domain and Ensemble Statistics

For a continuous random variable *X*, its probability density function $p(x)$ is defined by

$$\Pr[x \le X \le x + dx] = P(x + dx) - P(x) = \frac{dP(x)}{dx} dx = p(x) dx \tag{A.2}$$

where $P(x)$ is the cumulative probability function, from which it follows that

$$\Pr[X \le x] = P(x) = \int_{-\infty}^{x} p(x) dx \tag{A.3}$$

and that $\lim_{x \to \infty} P(x) = 1$. Similarly, for two random variables *X* and *Y* the joint probability density function is defined by

$$p(x, y) = \frac{d^2 P(x, y)}{dx\, dy} \tag{A.4}$$

where $P(x, y) = \Pr[X \le x, Y \le y]$. The mean value and variance of X are given by

$$\bar{x} = E[X] = \int_{-\infty}^{+\infty} x p(x) dx \text{ and } \sigma_x^2 = E\left[(X - \bar{x})^2\right] = \int_{-\infty}^{+\infty} (x - \bar{x})^2 p(x) dx \tag{A.5}$$

Equivalent definitions apply to a discrete random variable X. It is in the following assumed that each realisation X_k of X has the same probability of occurrence, and thus, the mean value and variance of X may be estimated from a large data set of N individual realisations:

$$\bar{x} = \lim_{N \to \infty} \frac{1}{N} \sum_{k=1}^{N} X_k \qquad \text{and} \qquad \sigma_x^2 = \lim_{N \to \infty} \frac{1}{N} \sum_{k=1}^{N} (X_k - \bar{x})^2 \tag{A.6}$$

The square root of the variance, σ_x, is called the standard deviation. Recalling that $E[X] = \bar{x}$, the expression for the variance may be further developed into

$$\sigma_x^2 = E\left[(X - \bar{x})^2\right] = E\left[X^2 - 2\bar{x}X + \bar{x}^2\right] = E\left[X^2\right] - \bar{x}^2 \tag{A.7}$$

As mentioned in above there are two types of statistics dealt with in structural engineering: *time domain statistics* and *ensemble statistics*. It will in the following be assumed that any time domain statistics are based on a continuous or discrete time variable X, which theoretically may attain values between $-\infty$ and $+\infty$ and are applicable over a limited time range between 0 and T, within which the process is stationary and homogeneous (i.e. have constant statistical properties) such that

$$X = \bar{x} + x(t) \tag{A.8}$$

Its mean value and variance are then given by

$$\bar{x} = \lim_{T \to \infty} \frac{1}{T} \int_0^T X dt \text{ and } \sigma_x^2 = \lim_{T \to \infty} \frac{1}{T} \int_0^T [x(t)]^2 dt \tag{A.9}$$

Illustrating ensemble statistics, a situation where N different recordings of a stochastic process within a time window between 0 and T are shown in Fig. A.2. These may for instance represent N simultaneous realisations of the along wind velocity in space, i.e. they represent the wind velocity variation taken simultaneously and at a certain distance (horizontal or vertical) between each of them. Extracting the recorded values at a given time from each of these realisations will render a set of data $X_k(t), k = 1, ..., N$. On this data set ensemble statistics may be performed. This is the type of statistics that provides a stochastic description of the wind field distribution in space.

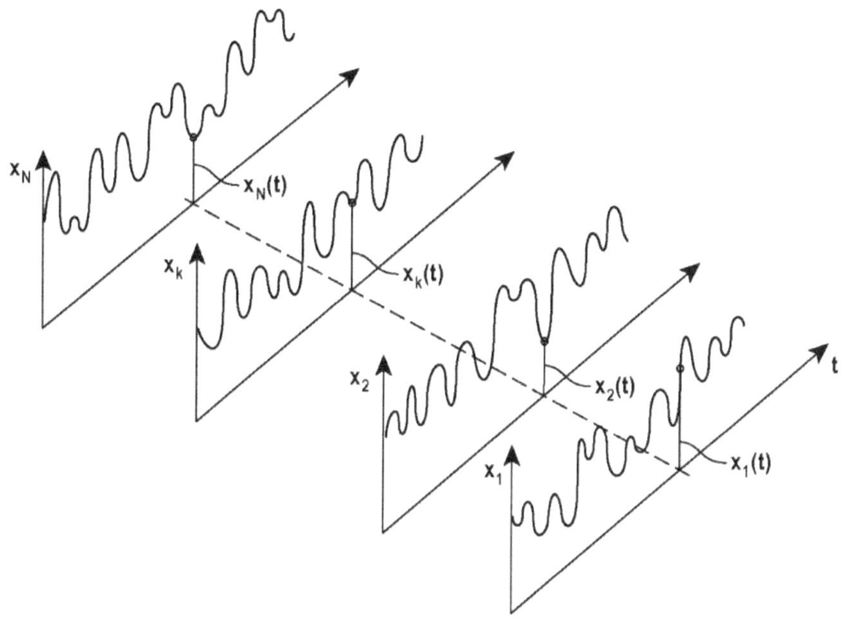

Fig. A.2 Ensemble statistics of simultaneous events

Apart from fitting the data from a random variable to a suitable parent probability distribution and estimating its mean value and variance, it is the properties of correlation and covariance that are of particular interest. These are both providing information about possible relationships in the time domain or ensemble data that have been extracted from the process. [Correlation estimates are taken on the full value of the process variable, i.e. on $X(t) = \bar{x} + x(t)$, while covariance is estimated from zero mean variables $x_i(t)$.]

Given two realisations $X_1(t) = \bar{x}_1 + x_1(t)$ and $X_2(t) = \bar{x}_2 + x_2(t)$, either two of the same process at different time or location, or of two entirely different processes. Then the correlation and covariance between these two process variables are defined by

$$R_{x_1x_2} = E\left[X_1(t) \cdot X_2(t)\right] = \lim_{T\to\infty} \frac{1}{T} \int_0^T X_1(t) \cdot X_2(t)\,dt \tag{A.10}$$

$$Cov_{x_1x_2} = E\left[x_1(t) \cdot x_2(t)\right] = \lim_{T\to\infty} \frac{1}{T} \int_0^T x_1(t) \cdot x_2(t)\,dt \tag{A.11}$$

Similarly, given two data sets of N individual and equally probable realisations that have been extracted from two random variables, X_1 and X_2, then the ensemble correlation and covariance are defined by:

$$R_{x_1x_2} = E\left[X_1 \cdot X_2\right] = \lim_{N\to\infty} \frac{1}{N} \sum_{k=1}^{N} X_{1_k} \cdot X_{2_k} \tag{A.12}$$

$$\begin{aligned} Cov_{x_1x_2} &= E\left[(X_1 - \overline{x}_1) \cdot (X_2 - \overline{x}_2)\right] \\ &= \lim_{N\to\infty} \frac{1}{N} \sum_{k=1}^{N} \left(X_{1_k} - \overline{x}_1\right) \cdot \left(X_{2_k} - \overline{x}_2\right) \end{aligned} \tag{A.13}$$

However, correlation and covariance estimates may also be taken on the process variable itself. Thus, defining an arbitrary time lag τ, the time domain auto correlation and auto covariance functions are defined by

$$R_x(\tau) = E\left[X(t) \cdot X(t+\tau)\right] = \lim_{T\to\infty} \frac{1}{T} \int_0^T X(t) \cdot X(t+\tau)\,dt \tag{A.14}$$

$$Cov_x(\tau) = E\left[x(t) \cdot x(t+\tau)\right] = \lim_{T\to\infty} \frac{1}{T} \int_0^T x(t) \cdot x(t+\tau)\,dt \tag{A.15}$$

These are defined as functions because τ is perceived as a continuous variable. As long as τ is considerably smaller than T

$$E\left[X(t)\right] = E\left[X(t+\tau)\right] = \overline{x} \tag{A.16}$$

and thus, the relationship between R_x and Cov_x is the following

$$Cov_x(\tau) = E\left[\left\{X(t) - \overline{x}\right\} \cdot \left\{X(t+\tau) - \overline{x}\right\}\right] = R_x(\tau) - \overline{x}^2 \tag{A.17}$$

There is no reason why τ may not attain negative as well as positive values, and since

$$E\left[x(t)\cdot x(t-\tau)\right]=E\left[x(t-\tau)\cdot x(t)\right]=E\left[x(t-\tau)\cdot x(t-\tau+\tau)\right] \quad \text{(A.18)}$$

then

$$Cov_x(\tau)=Cov_x(-\tau) \quad \text{(A.19)}$$

Thus, Cov_x is symmetric with respect to its variation with τ. As illustrated in Fig. A.3 the auto covariance function is the mean value of the time series multiplied by itself at a time shift equal to τ. Theoretically τ may vary between 0 and T, but the practical significance of $Cov_x(\tau)$ seizes to exist long before τ is in the vicinity of T. The reason is that while it in theoretical developments is convenient to consider $x(t)$ as a continuous function, it will in practical calculations only occur as a discrete and finite vector of random values x_k, usually taken at regular intervals Δt. If T is large and Δt is small, then the number of elements in this vector is $N \approx T / \Delta t$, in which case the continuous integral in Eq. A.15 may be replaced by its discrete counterpart

$$Cov_x(\tau=j\cdot\Delta t)=E\left[x(t)\cdot x(t+\tau)\right]=\frac{1}{N-j}\sum_{k=1}^{N-j} x_{k+j}\cdot x_k \quad \text{(A.20)}$$

from which it is seen that j must be considerably smaller than N for a meaningful outcome of the auto covariance estimate. The same is true for the auto correlation function in Eq. A.14. Similar to the definitions above, cross correlation and cross covariance functions may be defined between observations that have been obtained from two short term realisations $X_1(t)=\bar{x}_1+x_1(t)$ and $X_2(t)=\bar{x}_2+x_2(t)$ of the same process or alternatively from realisations of two different processes:

$$R_{X_1X_2}(\tau)=E\left[X_1(t)\cdot X_2(t+\tau)\right]=\lim_{T\to\infty}\frac{1}{T}\int_0^T X_1(t)\cdot X_2(t+\tau)\,dt \quad \text{(A.21)}$$

$$Cov_{x_1x_2}(\tau)=E\left[x_1(t)\cdot x_2(t+\tau)\right]=\lim_{T\to\infty}\frac{1}{T}\int_0^T x_1(t)\cdot x_2(t+\tau)\,dt \quad \text{(A.22)}$$

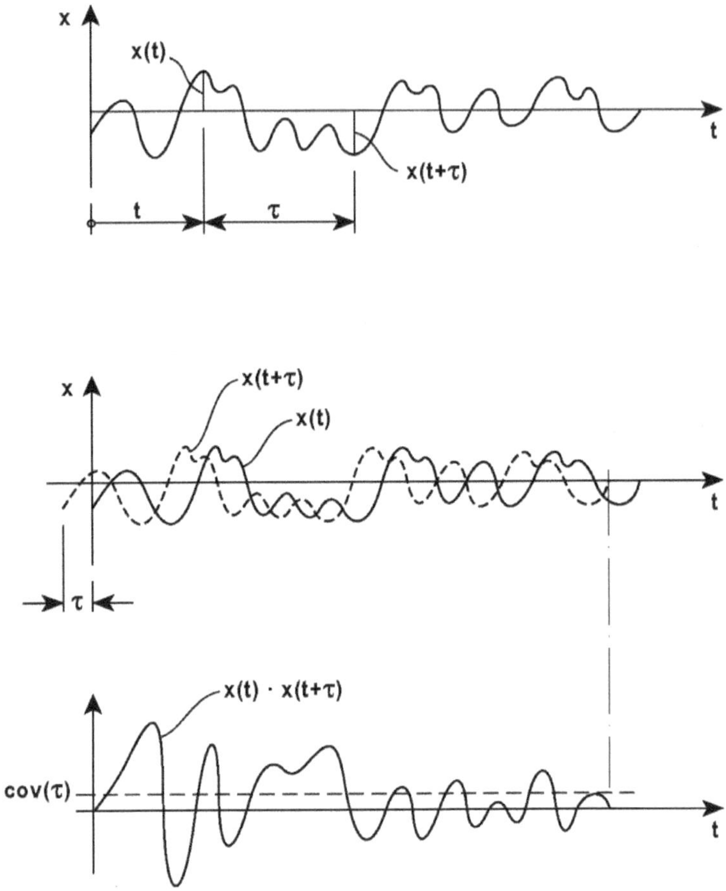

Fig. A.3 The auto covariance function

A normalised version of the cross covariance between the fluctuating parts of the realisations is defined by the cross covariance coefficient

$$\rho_{x_1 x_2}(\tau) = Cov_{x_1 x_2}(\tau) / \left(\sigma_{x_1}\sigma_{x_2}\right) \tag{A.23}$$

where σ_{x_1} and σ_{x_2} are the standard deviations of the two zero mean time variables. If such cross covariance estimates are taken from a set of simultaneous realisations of a process distributed in space, e.g. as illustrated in Fig. A.4 where the N realisations of the process are assumed to be taken at arbitrary positions y in the horizontal direction, then a cross covariance function between realisations at distance Δy may be defined:

$$Cov_{xx}(\Delta y, \tau) = E\left[x(y,t)x(y+\Delta y, t+\tau)\right] = \lim_{T\to\infty}\frac{1}{T}\int_0^T x(y,t)x(y+\Delta y, t+\tau)\,dt \quad (A.24)$$

Obviously, $Cov_{xx}(\Delta y = 0, \tau = 0) = \sigma_x^2$. In wind engineering such covariance estimates will in general be a decaying function with increasing τ or spatial separation Δs, $s = x, y$ or z, as illustrated in Fig. A.5.

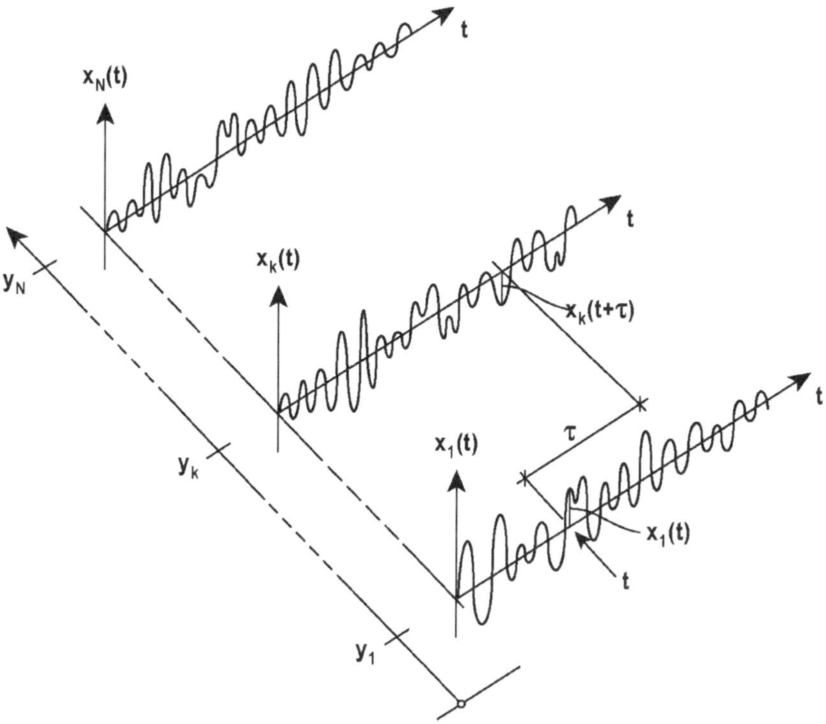

Fig. A.4 Cross covariance of time series at positions $y_k \; (k = 1, 2, \ldots, N)$

The covariance function may attain negative values at large values of Δs or τ. As previously indicated, the statistical properties defined above may also be applied to functions that are obtained from realisations of two different processes. Then, by simple arithmetic, the variance of the sum of two zero mean variables, $x_1(t)$ and $x_2(t)$, is given by

$$Var(x_1 + x_2) = E\left[(x_1 + x_2)(x_1 + x_2)\right] = Var(x_1) + Var(x_2) + 2Cov(x_1 \cdot x_2) \quad (A.25)$$

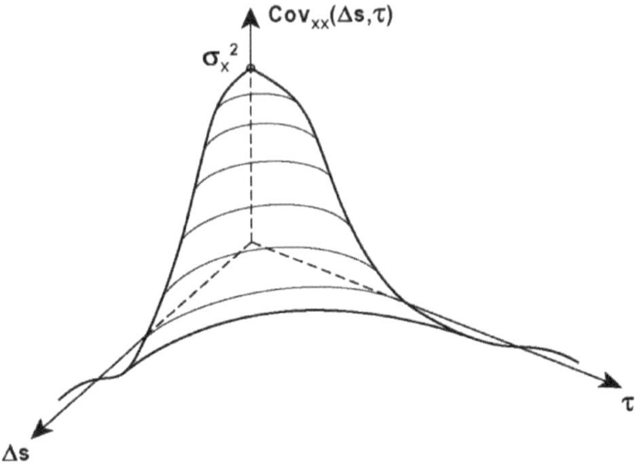

Fig. A.5 Typical spatial separation and time lag covariance function

Similarly, the variance of the sum of *N* different variables, $x_i(t)$, is given by

$$Var\left(\sum_{i=1}^{N} x_i\right) = E\left[\left(x_1 + x_2 + \ldots + x_i + \ldots + x_N\right)\cdot\left(x_1 + x_2 + \ldots + x_j + \ldots + x_N\right)\right]$$

$$\Rightarrow Var\left(\sum_{i=1}^{N} x_i\right) = \sum_{i=1}^{N}\sum_{j=1}^{N} Cov\left(x_i \cdot x_j\right) = \sum_{i=1}^{N}\sum_{j=1}^{N} \rho\left(x_i \cdot x_j\right)\cdot \sigma_i \sigma_j \tag{A.26}$$

If $x_i(t)$ are independent (i.e. uncorrelated) then the variance of the sum of the processes is the sum of the variances of the individual processes, i.e.

$$\text{if } Cov\left(x_i \cdot x_j\right) = \begin{cases} \sigma_{x_i}^2 \text{ when } i = j \\ 0 \text{ when } i \neq j \end{cases} \quad \text{then} \quad Var\left(\sum_{i=1}^{N} x_i\right) = \sum_{i=1}^{N} \sigma_{x_i}^2 \tag{A.27}$$

A.3 Threshold Crossing, Peaks and Extreme Values

A time series realisation $x(t)$ of a Gaussian stationary and homogeneous process (for simplicity with zero mean value), taken over a period T is illustrated in Fig. A.6. First we seek to develop an estimate of the average frequency $f_x(a)$ between the events that $x(t)$ is crossing the threshold a in its upward direction. Let a single upward crossing take place in a time interval Δt that is small enough to justify the approximation

$$x(t + \Delta t) \cong x(t) + \dot{x}(t)\cdot \Delta t \tag{A.28}$$

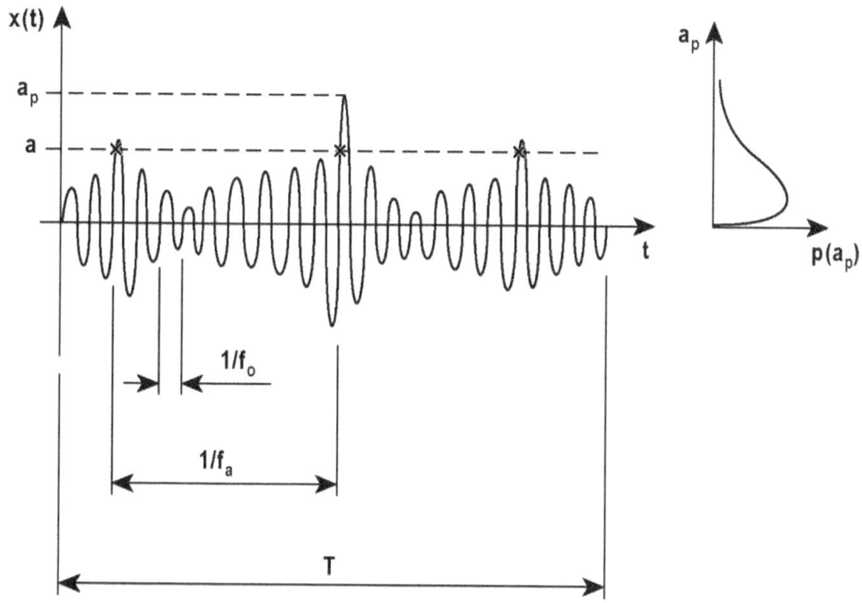

Fig. A.6 Threshold crossing and peaks

The probability of an up crossing event during Δt is then given by

$$P\left[x(t) \le a \text{ and } x(t) + \dot{x}(t) \cdot \Delta t > a\right] = f_x(a) \cdot \Delta t \tag{A.29}$$

from which it follows that

$$f_x(a) = \lim_{\Delta t \to 0} \frac{1}{\Delta t} \int_0^{\infty} \left[\int_{a - \dot{x} \cdot \Delta t}^{a} p_{x\dot{x}}(x, \dot{x})\, dx \right] d\dot{x} \tag{A.30}$$

where $p_{x\dot{x}}(x, \dot{x})$ is the probability density function for the joint events $x(t)$ and $\dot{x}(t)$. As $\Delta t \to 0$ the following approximation applies

$$\int_{a - \dot{x} \cdot \Delta t}^{a} p_{x\dot{x}}(x, \dot{x})\, dx \cong \dot{x} \cdot \Delta t \cdot p_{x\dot{x}}(a, \dot{x}) \tag{A.31}$$

For the type of processes covered herein it is a reasonable assumption that the joint events of $x(t)$ and $\dot{x}(t)$ are statistically independent, and thus, $p_{x\dot{x}}(x, \dot{x}) = p_x(x) \cdot p_{\dot{x}}(\dot{x})$. The average up crossing event that $x(t) = a$ is then given by

$$f_x(a) = \int_0^{\infty} \dot{x} \cdot p_{x\dot{x}}(a,\dot{x})\, d\dot{x} = p_x(a) \cdot \int_0^{\infty} \dot{x} \cdot p_{\dot{x}}(\dot{x})\, d\dot{x} \tag{A.32}$$

For each threshold up-crossing there is a corresponding down-crossing event, i.e. $f_x(+a) = f_x(-a)$, although there may be several consecutive positive or negative peaks in the process. Assuming that both x and $\dot{x}$ are Gaussian, then

$$f_x(a) = \frac{1}{\sqrt{2\pi}\sigma_x} \exp\left[-\frac{1}{2}\left(\frac{a}{\sigma_x}\right)^2\right] \int_0^{\infty} \dot{x} \cdot \frac{1}{\sqrt{2\pi}\sigma_{\dot{x}}} \exp\left[-\frac{1}{2}\left(\frac{\dot{x}}{\sigma_{\dot{x}}}\right)^2\right] d\dot{x}$$

$$\Rightarrow f_x(a) = \frac{1}{2\pi} \cdot \frac{\sigma_{\dot{x}}}{\sigma_x} \cdot \exp\left[-\frac{1}{2}\left(\frac{a}{\sigma_x}\right)^2\right] = f_x(0) \cdot \exp\left[-\frac{1}{2}\left(\frac{a}{\sigma_x}\right)^2\right] \tag{A.33}$$

where:

$$f_x(0) = \frac{1}{2\pi} \cdot \frac{\sigma_{\dot{x}}}{\sigma_x} \tag{A.34}$$

is the average zero up–crossing frequency of the process (see Eq. A.96). If $x(t)$ is also narrow banded, such that a zero up crossing and a peak x_p (larger than zero) are simultaneous events (as shown for the process in Fig. A.6), then the expected number of peaks $x_p > a_p$ is $f_x(a_p) \cdot T$, while the total number of peaks is $f_x(0) \cdot T$. Thus

$$\Pr\left[x_p \le a_p\right] = P(a_p) = 1 - \frac{f_x(a_p)}{f_x(0)} \tag{A.35}$$

from which it follows that the probability density distribution of a_p is given by

$$p(a_p) = \frac{d}{da_p} P(a_p) = \frac{d}{da_p}\left[1 - \frac{f_x(a_p)}{f_x(0)}\right] = -\frac{1}{f_x(0)} \cdot \frac{df_x(a_p)}{da_p}$$

$$\Rightarrow p(a_p) = \frac{a_p}{\sigma_x^2} \exp\left[-\frac{1}{2}\left(\frac{a_p}{\sigma_x}\right)^2\right] \tag{A.36}$$

This is a Rayleigh distribution. The distribution is illustrated on the right hand side of Fig. A.6.

Fig. A.7.a shows a collection of N time series, each a short term realisation of the fluctuating part $x(t)$ of a stochastic variable $X(t)=\bar{x}+x(t)$. It is assumed that they are all stationary and ergodic, and for the validity of the development below it is a necessary requirement that they are fairly broad banded. From this ensemble of realisations it may be of particular interest to develop the statistical properties of extreme values, as illustrated in Fig. A.7.b. Referring to Eq. A.33 and Fig. A.6, the extreme peak value $a_p = x_{\max}$ within each short term realisation occur when

$$\left[f_x\left(a_p\right)\right]^{-1} \rightarrow T \tag{A.37}$$

Let therefore

$$\kappa = f_x\left(x_{\max}\right) \cdot T \tag{A.38}$$

be an ensemble variable signifying the event that $x(0 \leq t \leq T)$ exceeds a given value $x_{\max}$.

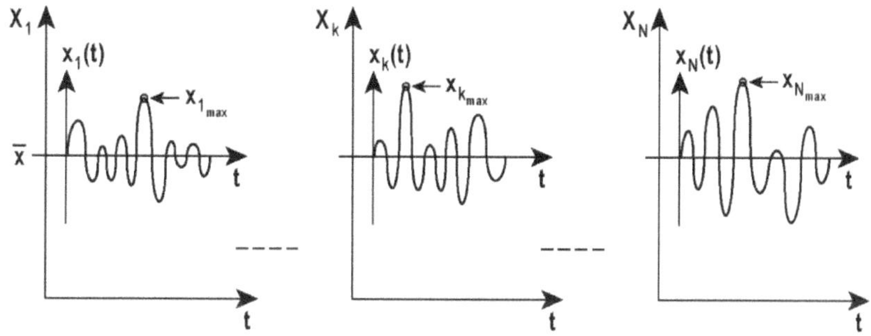

a) *N short term independent realisations*

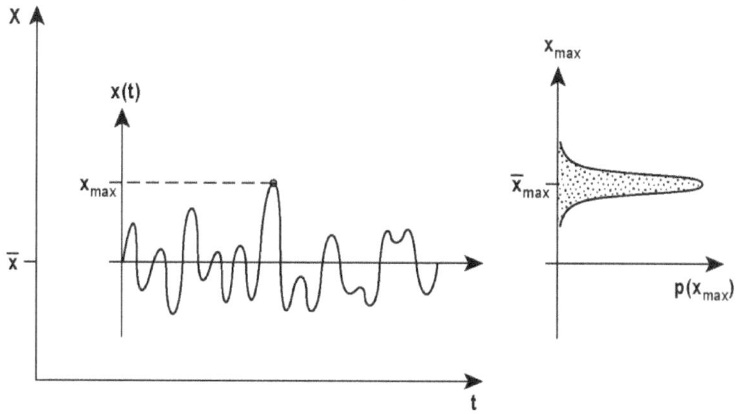

b) *The distribution of extremes*

Fig. A.7 Distribution of extreme values

The probability that κ occurs only once within each realisation is an event that coincides with the occurrence of x_{max}, i.e. they are simultaneous events. They are rare events at the tail of the peak distribution given in Eq. A.36, and for the statistics of such events it is a reasonable assumption that they will also comply to an exponential distribution, i.e. that

$$P_{\kappa}(1,T) = P_{x_{max}}\left(x_{max}\middle|T\right) = \exp(-\kappa) \tag{A.39}$$

Introducing Eqs. A.33 (with $a = x_{max}$) into A.38 and solving for x_{max}, then the following is obtained

$$x_{max} = \sigma_x\left\{2\ln\left[f_x(0)T\right] - 2\ln\kappa\right\}^{\frac{1}{2}} \approx \sigma_x\sqrt{2\ln\left[f_x(0)T\right]}\left\{1 - \frac{\ln\kappa}{2\ln\left[f_x(0)T\right]}\right\} \tag{A.40}$$

where the approximation $(1-x)^n \approx 1 - n\cdot x$ has been applied, assuming that $\ln\left[f_x(0)\cdot T\right]$ is large as compared to $\ln\kappa$. Thus, observing that $x_{max} = 0$ corresponds to $\kappa = \infty$, while $x_{max} = \infty$ corresponds to $\kappa = 0$, then the mean value of x_{max} may be estimated from

$$\begin{aligned}\bar{x}_{max} &= \int_0^\infty x_{max}\cdot\left(\frac{dP_{x_{max}}}{dx_{max}}\right)dx_{max} = \int_0^\infty x_{max}\cdot\left(\frac{dP_{x_{max}}}{d\kappa}\right)\cdot\left(\frac{d\kappa}{dx_{max}}\right)dx_{max}\\ &= \int_\infty^0 x_{max}\cdot\left[-\exp(-\kappa)\right]d\kappa = \int_0^\infty x_{max}\cdot\exp(-\kappa)\,d\kappa\end{aligned}$$

$$\Rightarrow \bar{x}_{max} = \sigma_x\cdot\sqrt{2\cdot\ln\left[f_x(0)\cdot T\right]}\cdot\left[\int_0^\infty \exp(-\kappa)\,d\kappa - \frac{\int_0^\infty \ln\kappa\cdot\exp(-\kappa)\,d\kappa}{2\cdot\ln\left[f_x(0)\cdot T\right]}\right] \tag{A.41}$$

Thus, the mean value of x_{max} is given by

$$\bar{x}_{max} = \sigma_x\cdot\left\{\sqrt{2\cdot\ln\left[f_x(0)\cdot T\right]} + \frac{\gamma}{\sqrt{2\cdot\ln\left[f_x(0)\cdot T\right]}}\right\} \tag{A.42}$$

where $\gamma = -\int_0^\infty \ln\kappa\cdot\exp(-\kappa)\,d\kappa \approx 0.5772$ is the Euler constant. Similarly, it may be shown that the variance of x_{max} is given by

$$\sigma_{x_{\max}}^2 = \frac{\pi^2}{12 \cdot \ln\left[f_x(0) \cdot T\right]} \cdot \sigma_x^2 \tag{A.43}$$

Given a stochastic variable $X(t) = \overline{x} + x(t)$, the expected value of its largest peak during a realisation with length T may then be estimated from

$$X_{\max} = \overline{x} + k_p \cdot \sigma_x \tag{A.44}$$

where the peak factor k_p is given by

$$k_p = \sqrt{2 \cdot \ln\left[f_x(0) \cdot T\right]} + \frac{\gamma}{\sqrt{2 \cdot \ln\left[f_x(0) \cdot T\right]}} \tag{A.45}$$

For fairly broad banded processes this peak factor will render values between 2 and 5. Plots of k_p and $\sigma_{x_{\max}} / \sigma_x$ are shown in Fig. A.8.

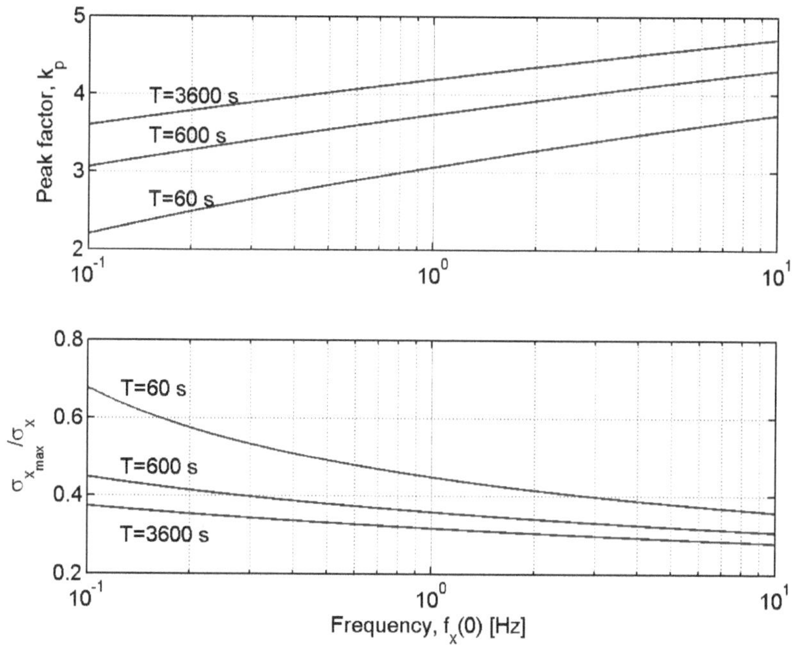

Fig. A.8 Plots of k_p and $\sigma_{x_{\max}} / \sigma_x$

It should be acknowledged that when $x(t)$ becomes ultra-narrow banded thn $k_p \rightarrow \sqrt{2}$, because for a single harmonic component

the variance $$x(t) = c_x \cdot \cos(\omega_x t),\ 0 < t < T \tag{A.46}$$

$$\sigma_x^2 = \lim_{T\to\infty} \frac{1}{T} \int_0^T \left[c_x \cos(\omega_x t) \right]^2 dt = \lim_{n\to\infty} \frac{1}{n \cdot T_x} n \int_0^{T_x} \left[c_x \cos\left(\frac{2\pi}{T_x} t \right) \right]^2 dt = \frac{c_x^2}{2} \tag{A.47}$$

and thus, for such a process $x_{\max} = c_x = \sigma_x \cdot \sqrt{2}$. Therefore, Eq. A.45 is only applicable for fairly broad banded processes.

A.4 Auto and Cross Spectral Density

The auto spectral density contains the frequency domain properties of the process, i.e. it is the frequency domain counterpart to the concept of variance. The various steps in the development of an auto spectral density function are illustrated in Fig. A.9. Given a zero mean time variable $x(t)$ with length T and performing a Fourier transformation of $x(t)$ implies that it may be approximated by a sum of harmonic components $X_k(\omega_k, t)$, i.e.

$$x(t) = \lim_{N\to\infty} \sum_{k=1}^{N} X_k(\omega_k, t) \qquad \text{where} \qquad \begin{cases} \omega_k = k \cdot \Delta\omega \\ \Delta\omega = 2\pi / T \end{cases} \tag{A.48}$$

The harmonic components in Eq. A.48 are given by

$$X_k(\omega_k, t) = c_k \cdot \cos(\omega_k t + \varphi_k) \tag{A.49}$$

where the amplitudes $c_k = \sqrt{a_k^2 + b_k^2}$ and phase angles $\varphi_k = \arctan(b_k / a_k)$, and where the constants a_k and b_k are given by

$$\begin{bmatrix} a_k \\ b_k \end{bmatrix} = \frac{2}{T} \int_0^T x(t) \begin{bmatrix} \cos \omega_k t \\ \sin \omega_k t \end{bmatrix} dt \tag{A.50}$$

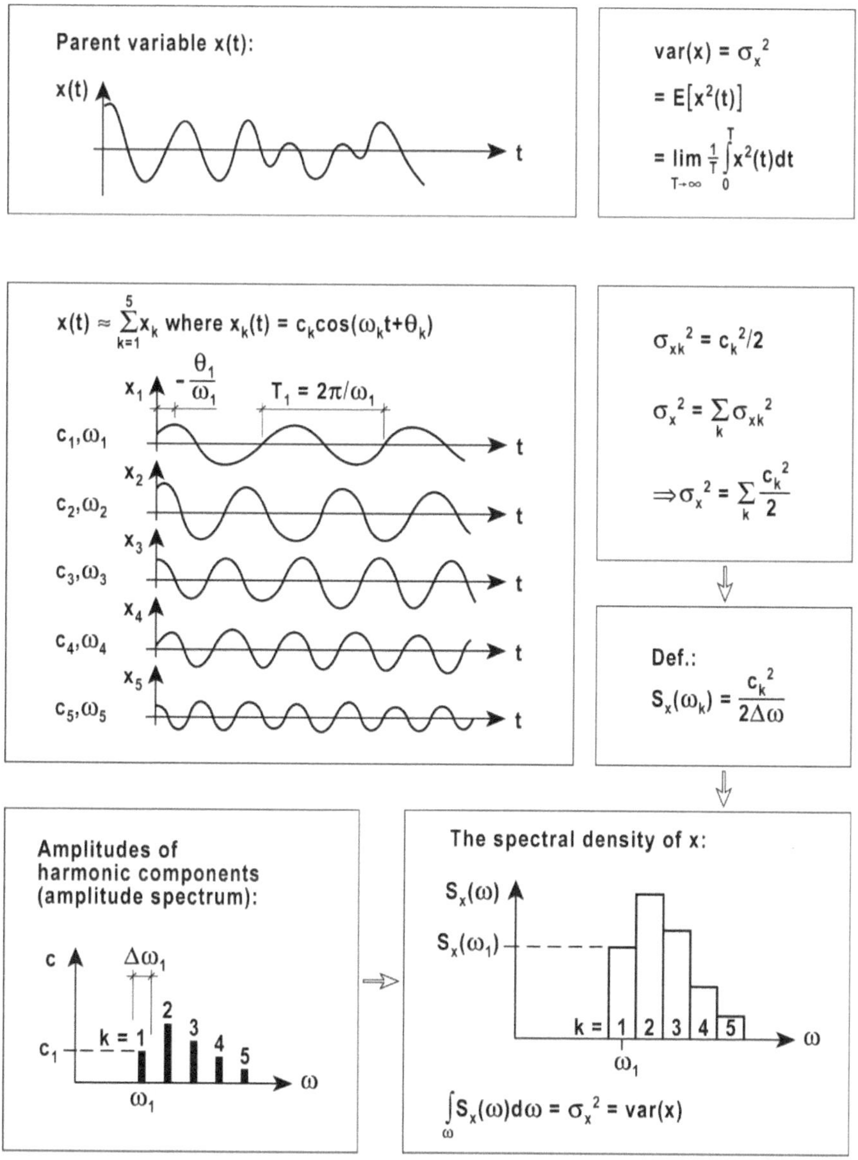

Fig. A.9 The definition of auto spectral density from a Fourier decomposition

As shown in Fig. A.9 the auto-spectral density of $x(t)$ is intended to represent its variance density distribution in the frequency domain. Hence, the definition of the single-sided auto-spectral density S_x associated with the frequency ω_k is

$$S_x(\omega_k) = \frac{E\left[X_k^2\right]}{\Delta\omega} = \frac{\sigma_{X_k}^2}{\Delta\omega} \tag{A.51}$$

which, when T becomes large, is given by

$$S_x(\omega_k) = \lim_{T\to\infty} \frac{1}{\Delta\omega}\cdot\frac{1}{T}\int_0^T \left[c_k \cos(\omega_k t + \varphi_k)\right]^2 dt \tag{A.52}$$

Introducing the period of the harmonic component, $T_k = 2\pi/\omega_k$, and replacing T with $n \cdot T_k$, and let $n \to \infty$, then the following is obtained

$$S_x(\omega_k) = \lim_{n\to\infty} \frac{1}{\Delta\omega}\cdot\frac{1}{n\cdot T_k}\cdot n\cdot\int_0^{T_k}\left[c_k \cos\left(\frac{2\pi}{T_k}\cdot t + \varphi_k\right)\right]^2 dt = \frac{c_k^2}{2\Delta\omega} \tag{A.53}$$

In Fig. A.9, the arrival at $S_x(\omega_k)$ is shown via the amplitude spectrum (or the Fourier amplitude diagram) to ease the understanding of the concept of spectral density representations. It is seen from this illustration that it is not possible to retrieve the parent time domain variable from the spectral density function alone, because it does not contain the necessary phase information (unless a corresponding phase spectrum is also established). The spectrum contains information about the variance distribution in frequency domain, i.e.

$$\sigma_x^2 = \lim_{N\to\infty}\sum_{k=1}^{N}\sigma_{X_k}^2 = \lim_{N\to\infty}\sum_{k=1}^{N}\frac{c_k^2}{2} = \lim_{N\to\infty}\sum_{k=1}^{N} S_x(\omega_k)\cdot\Delta\omega \tag{A.54}$$

In a continuous format, i.e. in the limit of both N and T approaching infinity, the single-sided auto-spectral density is defined by

$$S_x(\omega) = \lim_{T\to\infty}\lim_{N\to\infty}\frac{E\left[X^2(\omega,t)\right]}{\Delta\omega} \tag{A.55}$$

where $X(\omega,t)$ is the Fourier component of $x(t)$ at ω. In the limit $\Delta\omega \to d\omega$, and thus, the variance of the process may be calculated from

$$\sigma_x^2 = \int_0^\infty S_x(\omega)\, d\omega \tag{A.56}$$

The development above may more conveniently be expressed in a complex format. Adopting a frequency axis spanning the entire range of both positive and (fictitious) negative values, introducing the Euler formula

$$\begin{bmatrix} e^{i\omega t} \\ e^{-i\omega t} \end{bmatrix} = \begin{bmatrix} 1 & i \\ 1 & -i \end{bmatrix} \begin{bmatrix} \cos\omega t \\ \sin\omega t \end{bmatrix} \tag{A.57}$$

and defining the complex Fourier amplitude

$$d_k = \frac{1}{2}(a_k - i \cdot b_k) \tag{A.58}$$

then: $$x(t) = \sum_{-\infty}^{\infty} X_k(\omega_k, t) = \sum_{-\infty}^{\infty} d_k(\omega_k) \cdot e^{i \cdot \omega_k t} \tag{A.59}$$

Taking the variance of the complex Fourier components in Eq. A.59 and dividing by $\Delta\omega$,

$$\frac{E\left[X_k^* \cdot X_k\right]}{\Delta\omega} = \frac{1}{T}\int_0^T \frac{\left(d_k^* e^{-i\omega_k t}\right)\left(d_k e^{i\omega_k t}\right)}{\Delta\omega} dt = \frac{d_k^* d_k}{\Delta\omega} \tag{A.60}$$

which may be further developed into

$$\Rightarrow \frac{E\left[X_k^* \cdot X_k\right]}{\Delta\omega} = \frac{1}{4}\frac{(a_k + i \cdot b_k)\cdot(a_k - i \cdot b_k)}{\Delta\omega} = \frac{c_k^2}{4\Delta\omega} \tag{A.61}$$

It is seen (see Eq. A.53) that this is half the auto spectral value associated with ω_k. Thus, a symmetric double-sided auto spectrum associated with $-\omega_k$ as well as $+\omega_k$ may be defined with a value that is half the corresponding value of the single sided auto-spectrum. Extending the frequency axis from minus infinity to plus infinity and using the complex Fourier components X_k given in Eq. A.59 above, this double sided auto spectrum is then defined by

$$S_x(\pm\omega_k) = \frac{E\left[X_k^* \cdot X_k\right]}{\Delta\omega} = \frac{d_k^* d_k}{\Delta\omega} = \frac{c_k^2}{4\Delta\omega} \tag{A.62}$$

which, in the limit of T and $N \rightarrow \infty$, becomes the continuous function $S_x(\pm\omega)$, from which the variance of the process may be obtained by integration over the entire positive as well as the fictitious negative frequency range

$$\sigma_x^2 = \int_{-\infty}^{+\infty} S_x(\pm\omega)\, d\omega \tag{A.63}$$

Thus, the connection between double- and single-sided spectra is simply that $S_x(\omega) = 2 \cdot S_x(\pm\omega)$. Assuming that the process is stationary and of infinite length, such that the position of the time axis for integration purposes is arbitrary, then it is in the literature of mathematics usually considered convenient to introduce a non-normalized amplitude (which may be encountered in connection with the theory of generalised Fourier series and identified as a Fourier constant)

$$a_k(\omega_k) = \int_0^T x(t) \cdot e^{-i \cdot \omega_k t}\, dt = T \cdot d_k \tag{A.64}$$

in which case the double-sided auto-spectral density associated with $\pm\omega_k$ is defined by

$$S_x(\pm\omega_k) = \frac{d_k^* \cdot d_k}{\Delta\omega} = \frac{\left(a_k^* / T\right) \cdot \left(a_k / T\right)}{2\pi / T} = \frac{1}{2\pi T} \cdot a_k^* a_k \tag{A.65}$$

In the limit of T and $N \to \infty$ this may be written on the following continuous form

$$S_x(\pm\omega) = \lim_{T \to \infty} \lim_{N \to \infty} \frac{1}{2\pi T} \cdot a^*(\omega) \cdot a(\omega) \tag{A.66}$$

and accordingly, the single sided version is given by

$$S_x(\omega) = \lim_{T \to \infty} \frac{1}{\pi T} \cdot a^*(\omega) \cdot a(\omega) \tag{A.67}$$

where it is taken for granted that N is sufficiently large. The auto-spectral density $S_x(\omega)$ as defined by the use of circular frequency ω may be replaced by a corresponding definition $S_x(f)$ using frequency f (with unit $Hz = s^{-1}$). Since $S_x(\omega) \cdot \Delta\omega$ and $S_x(f) \cdot \Delta f$ both represent the variance of the process at ω and f, they must give the same contribution to the total variance of the process, and thus

$$S_x(f)\cdot\Delta f = S_x(\omega)\cdot\Delta\omega = S_x(\omega)\cdot(2\pi\cdot\Delta f)$$

$$\Rightarrow S_x(f) = 2\pi\cdot S_x(\omega) = \lim_{T\to\infty}\frac{2}{T}\cdot a^*(f)\cdot a(f) \tag{A.68}$$

The cross spectral density contains not only frequency domain properties but also the coherence properties between two different processes, i.e. it is the frequency domain counterpart to the concept of covariance. Given two stationary time variable functions $x(t)$ and $y(t)$, both with length T and zero mean value (i.e. $E[x(t)] = E[y(t)] = 0$), and performing a Fourier transformation (adopting a double-sided complex format) implies that $x(t)$ and $y(t)$ may be represented by sums of harmonic components $X_k(\omega_k,t)$ and $Y_k(\omega_k,t)$, i.e.

$$\begin{bmatrix} x(t) \\ y(t) \end{bmatrix} = \lim_{N\to\infty}\sum_{-N}^{N}\begin{bmatrix} X_k(\omega_k,t) \\ Y_k(\omega_k,t) \end{bmatrix} \tag{A.69}$$

where:

$$\begin{bmatrix} X_k(\omega_k,t) \\ Y_k(\omega_k,t) \end{bmatrix} = \frac{1}{T}\begin{bmatrix} a_x(\omega_k) \\ a_y(\omega_k) \end{bmatrix}\cdot e^{i\cdot\omega_k t} \text{ and } \begin{bmatrix} a_x(\omega_k) \\ a_y(\omega_k) \end{bmatrix} = \lim_{T\to\infty}\int_{-T/2}^{T/2}\begin{bmatrix} x(t) \\ y(t) \end{bmatrix}\cdot e^{-i\cdot\omega_k t}dt$$

and where $\omega_k = k\cdot\Delta\omega$ and $\Delta\omega = 2\pi/T$. The definition of the double-sided cross-spectral density S_{xy} associated with the frequency ω_k is then

$$S_{xy}(\pm\omega_k) = \frac{E\left[X_k^*\cdot Y_k\right]}{\Delta\omega} = \lim_{T\to\infty}\frac{1}{2\pi T}a_{X_k}^*\cdot a_{Y_k} \tag{A.70}$$

Since the Fourier components are orthogonal

$$E\left[X_i(\omega_i,t)\cdot Y_j(\omega_j,t)\right] = \begin{cases} S_{xy}(\omega_k,t)\cdot\Delta\omega \text{ when } i=j=k \\ 0 \text{ when } i\neq j \end{cases} \tag{A.71}$$

it follows from Eqs. A.69 and A.70 that an estimate of the covariance between $x(t)$ and $y(t)$ are given by

$$Cov_{xy} = E\left[x(t)\cdot y(t)\right] = \lim_{N\to\infty}E\left[\left(\sum_{-N}^{N}X_i\right)\cdot\left(\sum_{-N}^{N}Y_j\right)\right] = \lim_{N\to\infty}\sum_{-N}^{N}\left(E[X_k\cdot Y_k]\right)$$

$$\Rightarrow Cov_{xy} = \lim_{N\to\infty}\sum_{-N}^{N}S_{xy}(\pm\omega_k)\cdot\Delta\omega \tag{A.72}$$

In a continuous format, i.e. in the limit of both N and T approaching infinity, the double-sided cross-spectral density is defined by

$$S_{xy}(\pm\omega)=\frac{E\left[X^{*}(\omega,t)\cdot Y(\omega,t)\right]}{\Delta\omega}=\lim_{T\to\infty}\frac{1}{2\pi T}a_x^{*}(\omega)\cdot a_y(\omega) \tag{A.73}$$

The single sided version is then simply

$$S_{xy}(\omega)=2\cdot S_{xy}(\pm\omega)=\lim_{T\to\infty}\frac{1}{\pi T}a_x^{*}(\omega)\cdot a_y(\omega) \tag{A.74}$$

The corresponding single-sided version using frequency f (Hz), is defined by

$$S_{xy}(f)=2\pi\cdot S_{xy}(\omega)=\lim_{T\to\infty}\frac{2}{T}\cdot a_x^{*}(f)\cdot a_y(f) \tag{A.75}$$

Thus, the covariance between the two processes may be calculated from

$$Cov_{xy}=\int_{-\infty}^{+\infty}S_{xy}(\pm\omega)d\omega=\int_{0}^{\infty}S_{xy}(\omega)d\omega=\int_{0}^{\infty}S_{xy}(f)df \tag{A.76}$$

The cross-spectrum will in general be a complex quantity. With respect to the frequency argument, its real part is an even function labelled the co–spectral density $Co_{xy}(\omega)$, while its imaginary part is an odd function labelled the quad–spectrum $Qu_{xy}(\omega)$, i.e.

$$S_{xy}(\omega)=Co_{xy}(\omega)-i\cdot Qu_{xy}(\omega) \tag{A.78}$$

Alternatively, $S_{xy}(\omega)$ may be expressed by its modulus and phase, i.e.

$$S_{xy}(\omega)=\left|S_{xy}(\omega)\right|\cdot e^{i\cdot\varphi_{xy}(\omega)} \tag{A.79}$$

where the phase spectrum $\varphi_{xy}(\omega)=\arctan\left[Qu_{xy}(\omega)/Co_{xy}(\omega)\right]$. Auto-spectra $S_x(\omega)$ may also be calculated from the auto covariance function $Cov_x(\tau)$, see Eq. A.15. Assuming that $x(t)$ is a stationary and zero mean stochastic variable, then the following applies:

$$S_x(\omega)=\frac{E\left[X_k^* X_k\right]}{\Delta\omega}=\lim_{T\to\infty}\frac{E\left[\left(\frac{1}{T}\int_0^T x(t)e^{i\omega t}dt\right)\cdot\left(\frac{1}{T}\int_0^T x(t)e^{-i\omega t}dt\right)\right]}{2\pi/T}$$

$$=\lim_{T\to\infty}\frac{1}{2\pi T}\int_0^T\int_0^T E\left[x(t_1)\cdot x(t_2)\right]\cdot e^{-i\omega(t_2-t_1)}dt_1dt_2$$

$$\Rightarrow S_x(\omega)=\lim_{T\to\infty}\frac{1}{2\pi T}\int_0^T\int_0^T Cov_x(t_2-t_1)\cdot e^{-i\omega(t_2-t_1)}dt_1dt_2 \tag{A.80}$$

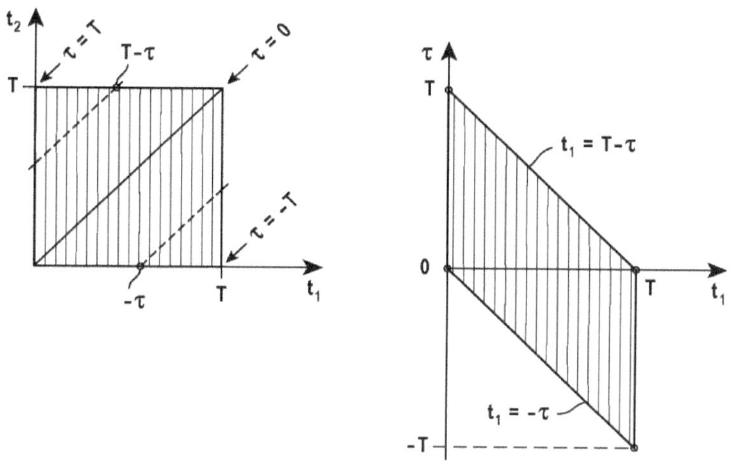

Fig. A.10 Substitution of variables and corresponding integration limits

Replacing t_2 with $t_1+\tau$, and changing the integration limits accordingly, implies (as illustrated in Fig. A.10) that

$$\int_0^T\int_0^T dt_1dt_2=\int_{-T}^{0}\int_{-\tau}^{T}dt_1d\tau+\int_0^T\int_0^{T-\tau}dt_1d\tau \tag{A.81}$$

and thus

$$S_x(\omega)=\lim_{T\to\infty}\frac{1}{2\pi T}\left[\int_{-T}^{0}\int_{-\tau}^{T}Cov_x(\tau)\cdot e^{-i\omega\tau}dt_1d\tau+\int_0^T\int_0^{T-\tau}Cov_x(\tau)\cdot e^{-i\omega\tau}dt_1d\tau\right]$$

$$=\lim_{T\to\infty}\frac{1}{2\pi}\left[\int_{-T}^{0}\left(1+\frac{\tau}{T}\right)\cdot Cov_x(\tau)\cdot e^{-i\omega\tau}d\tau+\int_0^T\left(1-\frac{\tau}{T}\right)Cov_x(\tau)\cdot e^{-i\omega\tau}d\tau\right]$$

$$\Rightarrow S_x(\omega)=\lim_{T\to\infty}\frac{1}{2\pi}\int_{-T}^{T}\left(1-\frac{|\tau|}{T}\right)Cov_x(\tau)\cdot e^{-i\omega\tau}d\tau \tag{A.82}$$

Provided the integral under the auto covariance function is finite, it is then seen that in the limit of $T \rightarrow \infty$, the following is obtained

$$S_x(\omega) = \frac{1}{2\pi} \int_{-\infty}^{+\infty} Cov_x(\tau) \cdot e^{-i\omega\tau} d\tau \tag{A.83}$$

This shows that the auto spectral density is the Fourier transform of the auto covariance function. Vice versa, it follows that the auto covariance function is the Fourier constant to the spectral density, i.e.

$$Cov_x(\tau) = \int_{-\infty}^{+\infty} S_x(\omega) \cdot e^{i\omega\tau} d\omega \tag{A.84}$$

Similarly, the cross covariance function together with the cross spectral density will also constitute a pair of Fourier transforms:

$$S_{xy}(\omega) = \frac{1}{2\pi} \int_{-\infty}^{+\infty} Cov_{xy}(\tau) \cdot e^{-i\omega\tau} d\tau \text{ and } Cov_{xy}(\tau) = \int_{-\infty}^{+\infty} S_{xy}(\omega) \cdot e^{i\omega\tau} d\omega \tag{A.85}$$

The coherence function is defined by

$$Coh_{xy}(\omega) = \frac{\left|S_{xy}(\omega)\right|^2}{S_x(\omega) \cdot S_y(\omega)} \tag{A.86}$$

If $x(t)$ and $y(t)$ are realisations of the same process, then $S_x(\omega) = S_y(\omega)$ and the cross-spectrum $S_{xy}(\omega) = S_{xx}(\omega)$ is given by

$$S_{xx}(\omega) = S_x(\omega) \cdot \sqrt{Coh_{xx}(\omega)} \cdot e^{i\varphi_{xx}(\omega)} \tag{A.87}$$

$\sqrt{Coh_{xx}(\omega)}$ is the root–coherence function and φ_{xx} is the phase spectrum (see Eq. A.79). In the practical use of cross-spectra all imaginary parts will cancel out, and thus it is only the co-spectrum that is of interest. Therefore, a normalised co-spectrum is defined by

$$\hat{C}o_{xy}(\omega) = \frac{\mathrm{Re}\left[S_{xy}(\omega)\right]}{\sqrt{S_x(\omega) \cdot S_y(\omega)}} \tag{A.88}$$

Again, if $x(t)$ and $y(t)$ are realisations of the same stationary and ergodic process, then $S_x(\omega) = S_y(\omega)$ and the real part of the cross-spectrum is given by

$$\operatorname{Re}\left[S_{xx}(\omega)\right] = S_x(\omega) \cdot \hat{C}o_{xx}(\omega) \tag{A.89}$$

It may in some cases be of interest to calculate the spectral density of the time derivatives [e.g. $\dot{x}(t)$ and $\ddot{x}(t)$] of processes. In structural engineering this is particularly relevant if $x(t)$ is a response displacement of such a character that it is necessary to evaluate as to whether or not it is acceptable with respect to human perception, in which case the design criteria most often will contain acceleration requirements. Since (see Eq. A.59)

$$\dot{x}(t) = \sum_{-\infty}^{\infty} \dot{X}_k(\omega_k, t) = \sum_{-\infty}^{\infty} i\omega_k \cdot d_k(\omega_k) \cdot e^{i \cdot \omega_k t} \tag{A.90}$$

$$\ddot{x}(t) = \sum_{-\infty}^{\infty} \ddot{X}_k(\omega_k, t) = \sum_{-\infty}^{\infty} (i\omega_k)^2 \cdot d_k(\omega_k) \cdot e^{i \cdot \omega_k t} \tag{A.91}$$

and the double sided spectral density in general is given by the complex Fourier amplitude multiplied by its conjugated counterpart (see Eq. A.62), then

$$\left.\begin{aligned} S_{\dot{x}}(\pm\omega_k) &= \frac{\left[i\omega_k d_k(\omega_k)\right]^* \left[i\omega_k d_k(\omega_k)\right]}{\Delta\omega} = \omega_k^2 \frac{d_k^* d_k}{\Delta\omega} = \omega_k^2 S_x(\pm\omega_k) \\ S_{\ddot{x}}(\pm\omega_k) &= \frac{\left[(i\omega_k)^2 d_k(\omega_k)\right]^* \left[(i\omega_k)^2 d_k(\omega_k)\right]}{\Delta\omega} = \omega_k^4 \frac{d_k^* d_k}{\Delta\omega} = \omega_k^4 S_x(\pm\omega_k) \end{aligned}\right\} \tag{A.92}$$

Similarly, cross spectral densities between a fluctuating displacement and its corresponding velocity and acceleration are given by

$$S_{x\dot{x}}(\pm\omega_k) = \frac{\left[d_k(\omega_k)\right]^* \left[i\omega_k d_k(\omega_k)\right]}{\Delta\omega} = i\omega_k \frac{d_k^* d_k}{\Delta\omega} = i\omega_k S_x(\pm\omega_k) \tag{A.93}$$

$$S_{x\ddot{x}}(\pm\omega_k) = \frac{\left[d_k(\omega_k)\right]^* \left[(i\omega_k)^2 d_k(\omega_k)\right]}{\Delta\omega} = -\omega_k^2 \frac{d_k^* d_k}{\Delta\omega} = -\omega_k^2 S_x(\pm\omega_k) \tag{A.94}$$

$$S_{\dot{x}\ddot{x}}(\pm\omega_k) = \frac{\left[i\omega_k d_k(\omega_k)\right]^* \left[(i\omega_k)^2 d_k(\omega_k)\right]}{\Delta\omega} = i\omega_k^3 \frac{d_k^* d_k}{\Delta\omega} = i\omega_k^3 S_x(\pm\omega_k) \quad \text{(A.95)}$$

In the limit of T and $N \to \infty$ this may be written on the following continuous form

$$\begin{bmatrix} S_x(\pm\omega) & S_{x\dot{x}}(\pm\omega) & S_{x\ddot{x}}(\pm\omega) \\ & S_{\dot{x}}(\pm\omega) & S_{\dot{x}\ddot{x}}(\pm\omega) \\ sym. & & S_{\ddot{x}}(\pm\omega) \end{bmatrix} = \begin{bmatrix} 1 & i\omega & -\omega^2 \\ & \omega^2 & i\omega^3 \\ sym. & & \omega^4 \end{bmatrix} \cdot S_x(\pm\omega) \quad \text{(A.96)}$$

Because $S_x(\pm\omega)$ is symmetric it is seen that for a stationary stochastic process

$$\begin{bmatrix} Cov_{x\dot{x}} \\ Cov_{\dot{x}\ddot{x}} \end{bmatrix} = \int_{-\infty}^{\infty} \begin{bmatrix} i\omega \\ i\omega^3 \end{bmatrix} \cdot S_x(\pm\omega)\, d\omega = \begin{bmatrix} 0 \\ 0 \end{bmatrix} \quad \text{(A.97)}$$

Thus, the spectral density of time derivatives of processes may be obtained directly from the spectral density of the process itself. The single sided spectrum is twice the double sided, then Eqs. A.93 – A.95 will also hold if $S_x(\pm\omega)$, $S_{\dot{x}}(\pm\omega)$ and $S_{\ddot{x}}(\pm\omega)$ are replaced by $S_x(\omega)$, $S_{\dot{x}}(\omega)$ and $S_{\ddot{x}}(\omega)$.

From $S_x(\omega)$ and $S_{\dot{x}}(\omega)$ a general expression of the average zero crossing frequency $f_x(0)$ of the process $x(t)$ may be found. Referring to Eq. A.34, A.56 and introducing $S_{\dot{x}}(\omega) = \omega^2 S_x(\omega)$, then the following applies:

$$f_x(0) = \frac{1}{2\pi} \cdot \frac{\sigma_{\dot{x}}}{\sigma_x} = \frac{1}{2\pi} \cdot \left[\frac{\int_0^{\infty} \omega^2 S_x(\omega)\, d\omega}{\int_0^{\infty} S_x(\omega)\, d\omega} \right]^{1/2} = \frac{1}{2\pi} \cdot \sqrt{\frac{\mu_2}{\mu_0}} \quad \text{(A.98)}$$

where for convenience the so–called n^{th} spectral moment $\mu_n = \int_0^{\infty} \omega^n \cdot S_x(\omega)\, d\omega$ has been introduced.

Appendix B
Time Domain Simulations

B.1 Introduction

It is in the following taken for granted that the stochastic space and time domain simulation of a process $x(t)$ implies the extraction of single point or simultaneous multiple point time series from known frequency domain cross spectral density information about the process. The process may contain coherent or non-coherent properties in space and time. Thus, a multiple point representation is associated with the spatial occurrence of the process. For a non-coherent process there is no statistical connection between the simulated time series that occur at various positions in space, and thus, the simulation may be treated as a representation of independent single point time series. This type of simulation is shown in chapter B.2. For a coherent process there is a prescribed statistical connection between each of the spatial representatives within a set of M simulated time series. E.g., if the simulated time series represent the space and time distribution of a wind field, there will be a certain statistical connection between the instantaneous values $x_m(t), m=1,2,....,M$ that matches the spatial properties of the wind field. Such a simulation is shown in chapter B.3. The simulation procedure presented below is taken from Shinozuka [23] and Deodatis [24]. Simulating time series from spectra is particularly useful for two reasons. First, there are some response calculations that render results which are more or less narrow banded (or contain beating effects), and thus, they may require separate time domain simulations to establish an appropriate peak factor for the calculation of maximum response. Secondly, if the relevant cross spectra of the wind field or the earthquake properties in frequency domain and space are known, there is always the possibility of a time domain simulation such that a time domain step-wise load effect integration may be performed, see Chapter 6.3. Thus, a time domain simulation may provide an alternative approach to that of a frequency domain solution shown in Chapters 7 and 8. The main advantage is that such an approach may contain many of the non-linear effects that will have to be simplified or discarded in the linear theory that is required for a frequency domain solution. The disadvantage is that sufficient information about all the relevant load effects may not be readily available.

B.2 Simulation of Single Point Time Series

The mathematical development from a single time series to its auto-spectral density is presented in chapter A.4. In principle, the process is illustrated on Fig. A.9. A time domain simulation is obtained by the reverse process. Let $S_x(\omega)$ be the single-sided auto spectral density of an arbitrary stochastic variable x, with zero mean value. A time domain representative, $x(t)$, can then be obtained by subdividing S_x into N segments along the frequency axis, each centred at ω_k ($k=1,2,\ldots,N$) and covering a frequency segment $\Delta\omega_k$, as shown in Fig. B.1. On a discrete form $S_x(\omega_k)$ is the variance of each harmonic component per frequency segment, as defined in Eq. A.53 (see also Fig. A.9), i.e.

$$S_x(\omega_k)=c_k^2/(2\Delta\omega_k) \tag{B.1}$$

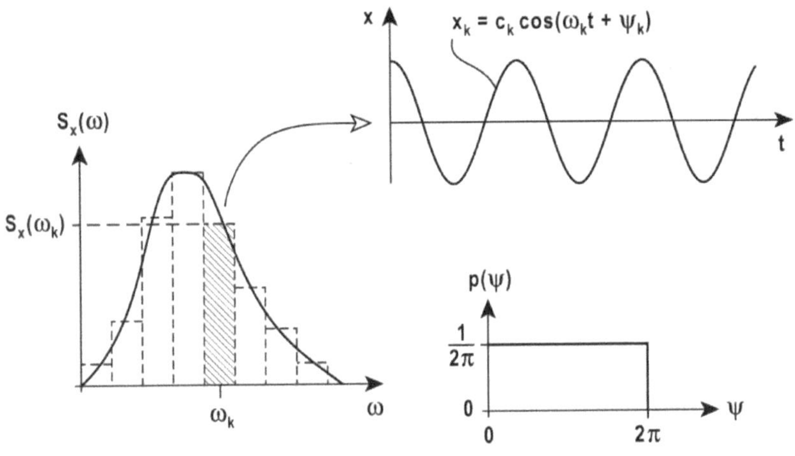

Fig. B.1 Spectral decomposition

A time series representative of x is then obtained by

$$x(t)=\sum_{k=1}^{N}c_k\cos(\omega_k t+\psi_k) \tag{B.2}$$

where
$$c_k = \left[2 \cdot S_x(\omega_k) \cdot \Delta\omega_k\right]^{1/2} \tag{B.3}$$

and where ψ_k are arbitrary phase angles between zero and 2π, one for each harmonic component. Alternatively, Eq. B.2 may be written in an exponential format (often encountered in the literature)

$$x(t) = \text{Re}\left\{\sum_{k=1}^{N} c_k \cdot \exp\left[i(\omega_k t + \psi_k)\right]\right\} \tag{B.4}$$

The variance of $x(t)$ is $\sum_{k=1}^{N} \frac{c_k^2}{2}$, which in the limit of $\Delta\omega \to 0$ and $N \to \infty$,

$$\sigma_x^2 = \sum_{k=1}^{N} \frac{c_k^2}{2} = \int_0^{\infty} S_x(\omega)\, d\omega \tag{B.5}$$

I.e., if the discretisation is sufficiently fine, then the variance of the simulated representative, $x(t)$, is equal to or close enough to the variance of the parent variable. Any number of such representatives may be simulated simply by changing the choice of phase angles. Obviously, the accuracy of such a simulation depends on the discretisation fineness, but there is also the unfavourable possibility of aliasing. Let ω_c be the upper cut-off frequency, beyond which there is none or only negligible spectral information about the process. Assuming constant frequency segments

$$\Delta\omega = \omega_c / N \tag{B.6}$$

then each simulated time series will be periodic with period

$$T = 2\pi / \Delta\omega \tag{B.7}$$

Thus, time series without aliasing will be obtained if they are generated with a time step

$$\Delta t \le 2\pi / (2\omega_c) \tag{B.8}$$

Example B.1

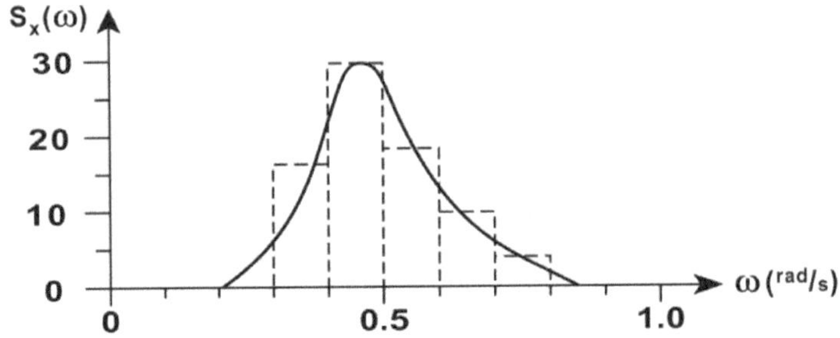

k	$S_x(\omega_k)$	ω_k	$\Delta\omega$	$c_k = \sqrt{2S_x\Delta\omega}$
1	16	0.35	0.1	1.8
2	30	0.45	0.1	2.4
3	18	0.55	0.1	1.9
4	10	0.65	0.1	1.4
5	4	0.75	0.1	0.9

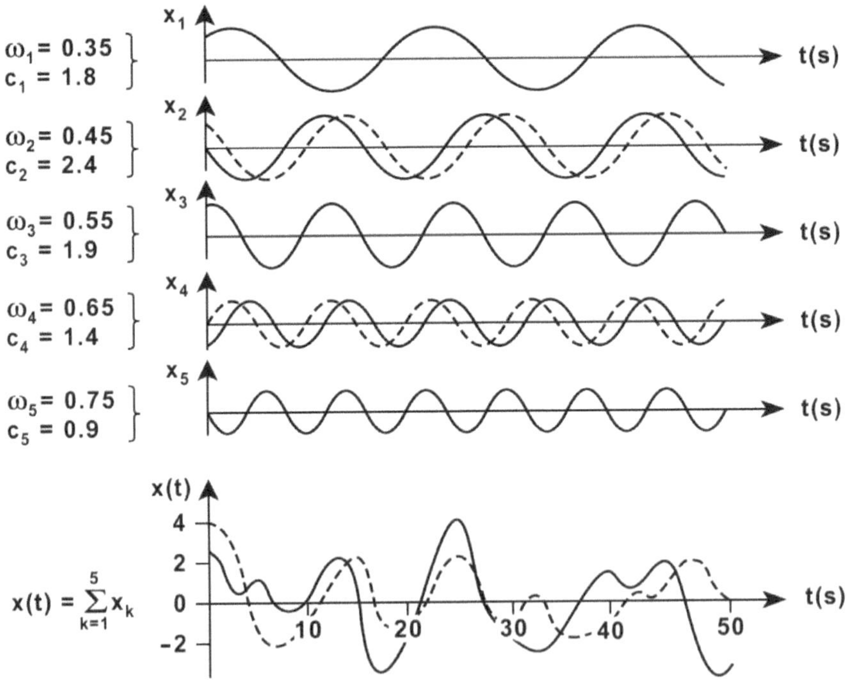

Fig. B.2 Simulation of single point time series

The top diagram in Fig. B.2 shows the single point single sided spectrum of a process x of which we wish to portray two representatives in time domain. As shown, the frequency span of the spectrum is first divided into five equal frequency segments, and the corresponding values ω_k and $S_x(\omega_k)$, $k=1,2,\ldots,5$, are read off. Thus the process is represented by five harmonic components whose amplitudes $c_k = \sqrt{2 \cdot S_x(\omega_k) \cdot \Delta\omega}$ are given in the far right hand side column in the table of Fig. B.2. Thus

$$x(t) = \sum_{k=1}^{5} \sqrt{2S_x(\omega_k)\Delta\omega} \cdot \cos(\omega_k t + \psi_k)$$

What then remains is to choose five arbitrary value of ψ_k. In Fig. B.2 the five cosine components are first shown by fully drawn lines, representing a certain choice of ψ_k values. The sum of these components as shown in the lower diagram in Fig. B.2 is an arbitrary representation of the process $x(t)$. If the second and the fourth of these components are moved an arbitrary time shift, then together with the remaining unchanged components they sum up to become another arbitrary representation of the process shown by the broken line in Fig. B.2. As can be seen, the two simulated representatives look quite different in time domain, although they come from the same spectral density. What is important is that they both have zero mean and the same variance, i.e. they have identical statistical properties up to and including the variance.

B.3 Simulation of Spatially Non-coherent Time Series

While the procedure presented above may be used to simulate single point time series representatives of $x(t)$, it is not applicable if we wish to simulate multiple point time series whose properties are expected to be distributed according to certain coherence properties. Let us assume that we wish to simulate the turbulence components

$$x\left(y_f, z_f, t\right) \qquad x = u, v \text{ or } w \tag{B.9}$$

of a stationary and homogeneous wind field at a chosen number of points M in a plane perpendicular to the main flow direction. For simplicity it is in the following assumed that cross spectra between u, v and w components are negligible, i.e. that

$$S_{xy}(\omega,\Delta s) \approx 0 \qquad \left.\begin{matrix} x \\ y \end{matrix}\right\} = u, v, w \tag{B.10}$$

where Δs is the spatial separation in the $y_f - z_f$ plane. We will then only need information about the cross spectra of the turbulence components themselves, $S_{xx}(\omega,\Delta s)$. Let $Cov_{x_m x_n}(\tau)$ be the covariance and $S_{x_m x_n}(\omega)$ the corresponding cross spectral density between two arbitrary points m and n. As shown in appendix A.4 these quantities constitute a Fourier transform pair. An M by M cross spectral density matrix

$$\mathbf{S}_{xx}(\omega) = \begin{bmatrix} S_{x_1 x_1} & \cdots & S_{x_1 x_n} & \cdots & S_{x_1 x_M} \\ \vdots & \ddots & \vdots & .\cdot\dot{} & \vdots \\ S_{x_m x_1} & \cdots & S_{x_m x_n} & \cdots & S_{x_m x_M} \\ \vdots & .\cdot\dot{} & \vdots & \ddots & \vdots \\ S_{x_M x_1} & \cdots & S_{x_M x_n} & \cdots & S_{x_M x_M} \end{bmatrix} \tag{B.11}$$

will then contain all the space and frequency domain information that is necessary for a time domain simulation of M time series with the correct statistical properties for a proper but arbitrary representation of the process. It follows from the assumptions of stationary and spatially homogeneous properties that

$$Cov_{x_m x_n} = Cov_{x_n x_m} \tag{B.12}$$

and thus (see Eq, A.85)

$$S_{x_m x_n} = S^*_{x_n x_m} \tag{B.13}$$

This implies that $\mathbf{S}_{xx}(\omega)$ is Hermitian and non–negative definite. A Cholesky decomposition of $\mathbf{S}_{xx}$ will then render a lower triangular matrix

$$\mathbf{G}_{xx}(\omega) = \begin{bmatrix} G_{x_1 x_1} & 0 & 0 & 0 & \cdots & 0 & \cdots & 0 \\ G_{x_2 x_1} & G_{x_2 x_2} & 0 & 0 & \cdots & 0 & \cdots & 0 \\ \vdots & \vdots & & \vdots & & \vdots & & \vdots \\ G_{x_m x_1} & G_{x_m x_2} & \cdots & G_{x_m x_n} & \cdots & G_{x_m x_m} & 0 & 0 \\ \vdots & \vdots & & \vdots & & \vdots & & \vdots \\ G_{x_M x_1} & G_{x_M x_2} & \cdots & G_{x_M x_n} & \cdots & G_{x_M x_m} & \cdots & G_{x_M x_M} \end{bmatrix} \tag{B.14}$$

whose properties are such that

$$\mathbf{S}_{xx}(\omega) = \mathbf{G}_{xx} \cdot \mathbf{G}_{xx}^{*T} \tag{B.15}$$

Assuming a frequency segmentation of N equidistant points, the simulated simultaneous time series at $m = 1,2,....,M$ are then given by

$$x_m(t) = \sum_{n=1}^{m} \sum_{j=1}^{N} \left| G_{mn}(\omega_j) \right| \cdot \sqrt{2\Delta\omega} \cdot \cos\left(\omega_j \cdot t + \psi_{nj}\right) \tag{B.16}$$

where j is the frequency segment number and ψ_{nj} is an arbitrary phase angle between zero and 2π. In most cases of a homogeneous wind field (see Eq. A.87)

$$S_{xx}(\omega,\Delta s) = S_x(\omega) \cdot \hat{S}_{xx}(\omega,\Delta s) \tag{B.17}$$

where S_x is the single-point spectral density of the process, $\Delta s = \left| s_{x_m} - s_{x_n} \right|$ is the spatial separation between points x_m and x_n, and where

$$\hat{S}_{xx}(\omega,\Delta s) = \sqrt{Coh_{xx}(\omega,\Delta s)} \cdot \exp\left[i\varphi_{xx}(\omega) \right] \tag{B.18}$$

Thus, defining a Cholesky decomposition $\hat{\mathbf{S}}_{xx}(\omega) = \hat{\mathbf{G}}_{xx} \cdot \hat{\mathbf{G}}_{xx}^{*T}$, then the time series at $m = 1,2,....,M$ are given by

$$x_m(t) = \sum_{n=1}^{m} \sum_{j=1}^{N} \left| \hat{G}_{mn}(\omega_j) \right| \cdot \sqrt{2S_x(\omega_j) \cdot \Delta\omega} \cdot \cos\left(\omega_j \cdot t + \psi_{nj}\right) \tag{B.19}$$

where $\hat{G}_{mn}$ is the content of $\hat{\mathbf{G}}_{xx}$ (i.e. the reduced versions of G_{mn} in Eq. B.14)

$$\hat{\mathbf{G}}_{xx}(\omega) = \begin{bmatrix} \hat{G}_{11} & 0 & 0 & 0 & \cdots & 0 & \cdots & 0 \\ \hat{G}_{21} & \hat{G}_{22} & 0 & 0 & \cdots & 0 & \cdots & 0 \\ \vdots & \vdots & & \vdots & & \vdots & & \vdots \\ \hat{G}_{m1} & \hat{G}_{m2} & \cdots & \hat{G}_{mn} & \cdots & \hat{G}_{mm} & 0 & 0 \\ \vdots & \vdots & & \vdots & & \vdots & & \vdots \\ \hat{G}_{M1} & \hat{G}_{M2} & \cdots & \hat{G}_{Mn} & \cdots & \hat{G}_{Mm} & \cdots & \hat{G}_{MM} \end{bmatrix} \tag{B.20}$$

and where a Cholesky decomposition (see Appendix B.4 below) will render

$$\hat{G}_{11}(\omega_j) = \left[\hat{S}_{xx}(\omega_j, 0)\right]^{1/2} \tag{B.21}$$

$$\hat{G}_{mm}(\omega_j) = \left[\hat{S}_{xx}(\omega_j, 0) - \sum_{k=1}^{m-1} \hat{G}_{mk}^2(\omega_j)\right]^{1/2} \tag{B.22}$$

$$\hat{G}_{mn}(\omega_j) = \frac{\hat{S}_{xx}(\omega_j, \Delta s) - \sum_{k=1}^{n-1} \hat{G}_{mk}(\omega_j) \cdot \hat{G}_{nk}(\omega_j)}{\hat{G}_{nn}(\omega_j)} \tag{B.23}$$

Example B.2

A process x *is statistically distributed in time and space. Its cross-spectrum* $S_{xx}(\omega, \Delta s)$ *is defined by the product between the single point spectrum* $S_x(\omega)$ *shown in Fig. B.3 and its root-coherence function* $\sqrt{Coh_{xx}(\omega, \Delta s)}$ *shown in Fig. B.4. I.e.,*

$$S_{xx}(\omega, \Delta s) = S_x(\omega) \cdot \sqrt{Coh_{xx}(\omega, \Delta s)}$$

The phase spectrum $\exp\left[i\varphi_{xx}(\omega)\right]$ *is assumed equal to unity for all relevant values of* ω *and* Δs. *Let us set out to simulate the process at three points in space, each a distance 10 m apart. Thus,*

$$\mathbf{\Delta s} = \begin{bmatrix} \Delta s_1 & \Delta s_2 & \Delta s_3 \end{bmatrix}^T = \begin{bmatrix} 0 & 10 & 20 \end{bmatrix}^T$$

Let us for simplicity settle with the three point frequency segmentation shown in Fig. B.3. I.e.

$$\boldsymbol{\omega} = \begin{bmatrix} \omega_1 & \omega_2 & \omega_3 \end{bmatrix}^T = \begin{bmatrix} 0.3 & 0.7 & 1.1 \end{bmatrix}^T \quad \textit{and} \qquad \Delta\omega = 0.4$$

(It should be noted that this frequency segmentation is only justified by the wish of obtaining mathematical expressions with reasonable length, such that a complete solution may be presented. For any practical purposes such a coarse segmentation will most often render unduly inaccurate results.)

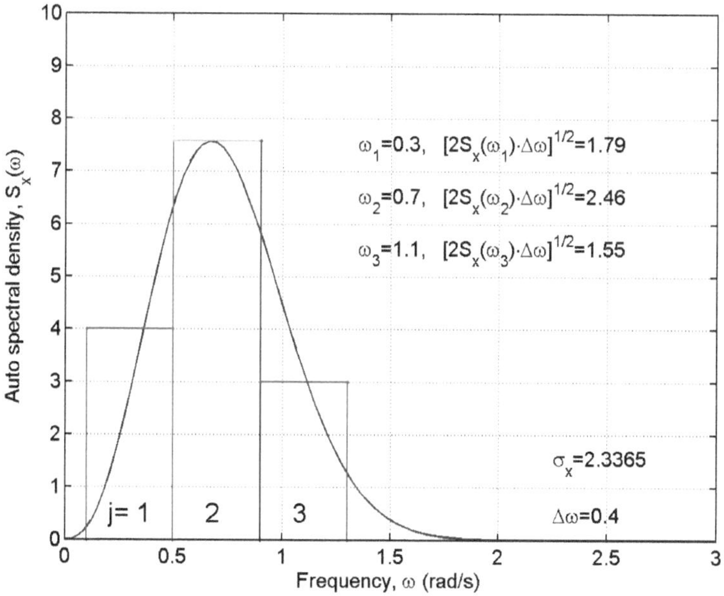

Fig. B.3 Single point spectrum

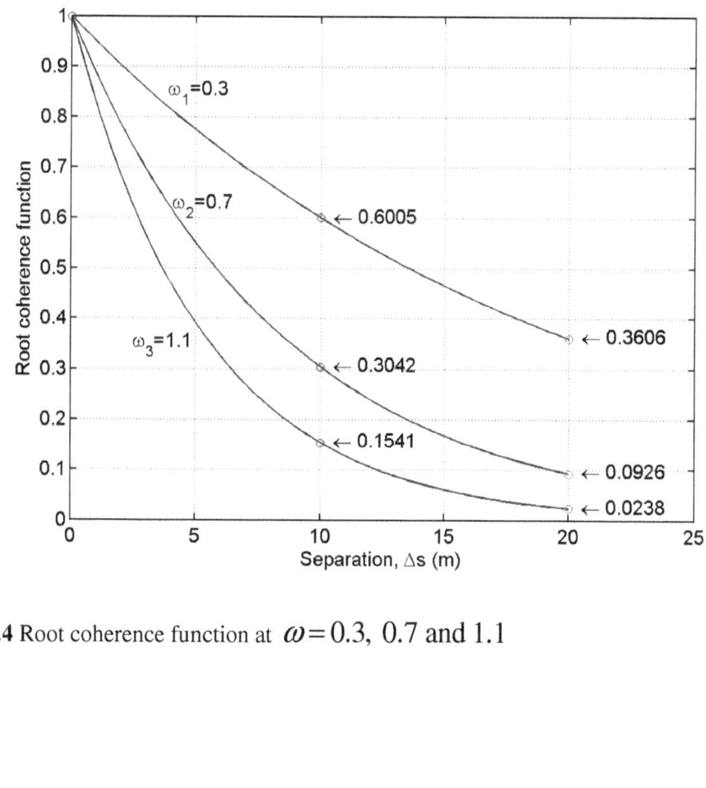

Fig. B.4 Root coherence function at $\omega = 0.3,\ 0.7$ and 1.1

The single point spectrum at these frequency settings are then (see Fig. B.3)

$$\mathbf{S}_x = \left[S_x(\omega_1) \quad S_x(\omega_2) \quad S_x(\omega_3) \right]^T = [4.0 \quad 7.6 \quad 3.0]^T$$

while the corresponding values of the root coherence function are given by (see Fig. B.4)

$\sqrt{Coh_{xx}(\omega, \Delta s)}$:		Δs		
		0	*10*	*20*
ω	*0.3*	*1.0*	*0.6005*	*0.3606*
	0.7	*1.0*	*0.3042*	*0.0926*
	1.1	*1.0*	*0,1541*	*0.0238*

Thus,
$$\hat{\mathbf{S}}_{xx}\left(\omega_j = 0.3, \Delta s_{mn}\right) = \begin{bmatrix} 1 & & sym. \\ 0.6005 & 1 & \\ 0.3606 & 0.6005 & 1 \end{bmatrix}$$

$$\hat{\mathbf{S}}_{xx}\left(\omega_j = 0.7, \Delta s_{mn}\right) = \begin{bmatrix} 1 & & sym. \\ 0.3042 & 1 & \\ 0.0926 & 0.3042 & 1 \end{bmatrix}$$

$$\hat{\mathbf{S}}_{xx}\left(\omega_j = 1.1, \Delta s_{mn}\right) = \begin{bmatrix} 1 & & sym. \\ 0.1541 & 1 & \\ 0.0238 & 0.1541 & 1 \end{bmatrix}$$

$$\hat{\mathbf{G}}_{xx}\left(\omega_j\right) = \begin{bmatrix} \hat{G}_{11} & 0 & 0 \\ \hat{G}_{21} & \hat{G}_{22} & 0 \\ \hat{G}_{31} & \hat{G}_{32} & \hat{G}_{33} \end{bmatrix}$$ *is defined such that* $\hat{\mathbf{S}}_{xx}\left(\omega_j, \Delta s_n\right) = \hat{\mathbf{G}}_{xx} \cdot \hat{\mathbf{G}}_{xx}^T$

Its content is given by (see Eqs. A.21 – A.23)

$$\hat{G}_{11}\left(\omega_j\right) = \left[\hat{S}_{xx}\left(\omega_j, \Delta s_{11} = 0\right) \right]^{1/2}$$

$$\hat{G}_{21}\left(\omega_j\right) = \hat{S}_{xx}\left(\omega_j, \Delta s_{21} = 10\right) / \hat{G}_{11}\left(\omega_j\right),$$

$$\hat{G}_{22}\left(\omega_j\right) = \left[\hat{S}_{xx}\left(\omega_j, \Delta s_{22} = 0\right) - \hat{G}_{21}^2\left(\omega_j\right) \right]^{1/2}$$

$$\hat{G}_{31}\left(\omega_j\right) = \hat{S}_{xx}\left(\omega_j, \Delta s_{31} = 20\right) / \hat{G}_{11}\left(\omega_j\right)$$

$$\hat{G}_{32}(\omega_j) = \left[\hat{S}_{xx}(\omega_j, \Delta s_{32} = 10) - \hat{G}_{31}(\omega_j) \cdot \hat{G}_{21}(\omega_j) \right] / \hat{G}_{22}(\omega_j)$$

$$\hat{G}_{33}(\omega_j) = \left[\hat{S}_{xx}(\omega_j, \Delta s_{33} = 0) - \hat{G}_{31}^2(\omega_j) - \hat{G}_{32}^2(\omega_j) \right]^{1/2}$$

Thus,

$$\omega_1 = 0.3 \Rightarrow \left\{ \begin{matrix} \hat{G}_{11} = 1 & \hat{G}_{22} = \sqrt{1 - 0.6005^2} = 0.7996 \\ \hat{G}_{21} = 0.6005 & \hat{G}_{32} = (0.6005 - 0.3606 \cdot 0.6005)/0.7996 = 0.4802 \\ \hat{G}_{31} = 0.3606 & \hat{G}_{33} = \sqrt{1 - 0.3606^2 - 0.4802^2} = 0.7996 \end{matrix} \right\}$$

$$\Rightarrow \hat{\mathbf{G}}_{xx}(\omega_1 = 0.3) = \begin{bmatrix} 1 & 0 & 0 \\ 0.6005 & 0.7996 & 0 \\ 0.3606 & 0.4802 & 0.7996 \end{bmatrix}$$

$$\omega_2 = 0.7 \Rightarrow \left\{ \begin{matrix} \hat{G}_{11} = 1 & \hat{G}_{22} = \sqrt{1 - 0.3042^2} = 0.9526 \\ \hat{G}_{21} = 0.3042 & \hat{G}_{32} = (0.3042 - 0.0926 \cdot 0.3042)/0.9526 = 0.2898 \\ \hat{G}_{31} = 0.0926 & \hat{G}_{33} = \sqrt{1 - 0.0926^2 - 0.2898^2} = 0.9526 \end{matrix} \right\}$$

$$\Rightarrow \hat{\mathbf{G}}_{xx}(\omega_2 = 0.7) = \begin{bmatrix} 1 & 0 & 0 \\ 0.3042 & 0.9526 & 0 \\ 0.0926 & 0.2898 & 0.9526 \end{bmatrix}$$

$$\omega_3 = 1.1 \Rightarrow \left\{ \begin{matrix} \hat{G}_{11} = 1 & \hat{G}_{22} = \sqrt{1 - 0.1541^2} = 0.9881 \\ \hat{G}_{21} = 0.1541 & \hat{G}_{32} = (0.1541 - 0.0238 \cdot 0.1541)/0.9881 = 0.1522 \\ \hat{G}_{31} = 0.0238 & \hat{G}_{33} = \sqrt{1 - 0.0238^2 - 0.1522^2} = 0.9881 \end{matrix} \right\}$$

$$\Rightarrow \hat{\mathbf{G}}_{xx}(\omega_1 = 1.1) = \begin{bmatrix} 1 & 0 & 0 \\ 0.1541 & 0.9881 & 0 \\ 0.0238 & 0.1522 & 0.9881 \end{bmatrix}$$

Denoting

$$\begin{bmatrix} a_1 \\ a_2 \\ a_3 \end{bmatrix} = \begin{bmatrix} \sqrt{2S_x(\omega_1 = 0.5) \cdot \Delta\omega} \\ \sqrt{2S_x(\omega_2 = 0.7) \cdot \Delta\omega} \\ \sqrt{2S_x(\omega_3 = 1.1) \cdot \Delta\omega} \end{bmatrix} = \begin{bmatrix} \sqrt{2 \cdot 4 \cdot 0.4} \\ \sqrt{2 \cdot 7.6 \cdot 0.4} \\ \sqrt{2 \cdot 3 \cdot 0.4} \end{bmatrix} \approx \begin{bmatrix} 1.79 \\ 2.46 \\ 1.55 \end{bmatrix}$$

then the three time series are given by (see Eq. B.19)

$$
\begin{aligned}
x_1(t) &= \sum_{n=1}^{1}\sum_{j=1}^{3}\left|\hat{G}_{1n}(\omega_j)\right|\cdot\sqrt{2S_x(\omega_j)\Delta\omega}\cdot\cos\left(\omega_j t+\psi_{nj}\right)\\
&= \left|\hat{G}_{11}(\omega_1)\right|a_1\cos(\omega_1 t+\psi_{11})+\left|\hat{G}_{11}(\omega_2)\right|a_2\cos(\omega_2 t+\psi_{12})\\
&\quad+\left|\hat{G}_{11}(\omega_3)\right|a_3\cos(\omega_3 t+\psi_{13})\\
&= 1.79\cdot\cos(0.3t+\psi_{11})+2.46\cdot\cos(0.7t+\psi_{12})+1.55\cdot\cos(1.1t+\psi_{13})
\end{aligned}
$$

$$
\begin{aligned}
x_2(t) &= \sum_{n=1}^{2}\sum_{j=1}^{3}\left|\hat{G}_{2n}(\omega_j)\right|\cdot\sqrt{2S_x(\omega_j)\Delta\omega}\cdot\cos\left(\omega_j t+\psi_{nj}\right)\\
&= \left|\hat{G}_{21}(\omega_1)\right|a_1\cos(\omega_1 t+\psi_{11})+\left|\hat{G}_{21}(\omega_2)\right|a_2\cos(\omega_2 t+\psi_{12})\\
&\quad+\left|\hat{G}_{21}(\omega_3)\right|\cdot a_3\cdot\cos(\omega_3 t+\psi_{13})+\left|\hat{G}_{22}(\omega_1)\right|a_1\cos(\omega_1 t+\psi_{21})\\
&\quad+\left|\hat{G}_{22}(\omega_2)\right|a_2\cos(\omega_2 t+\psi_{22})+\left|\hat{G}_{22}(\omega_3)\right|\cdot a_3\cdot\cos(\omega_3 t+\psi_{23})\\
&= 1.075\cdot\cos(0.3t+\psi_{11})+0.748\cdot\cos(0.7t+\psi_{12})+0.239\cdot\cos(1.1t+\psi_{13})\\
&\quad+1.431\cdot\cos(0.3t+\psi_{21})+2.343\cdot\cos(0.7t+\psi_{22})+1.532\cdot\cos(1.1t+\psi_{23})
\end{aligned}
$$

$$
\begin{aligned}
x_3(t) &= \sum_{n=1}^{3}\sum_{j=1}^{3}\left|\hat{G}_{3n}(\omega_j)\right|\cdot\sqrt{2S_x(\omega_j)\Delta\omega}\cdot\cos\left(\omega_j t+\psi_{nj}\right)\\
&= \left|\hat{G}_{31}(\omega_1)\right|a_1\cos(\omega_1 t+\psi_{11})+\left|\hat{G}_{31}(\omega_2)\right|a_2\cos(\omega_2 t+\psi_{12})\\
&\quad+\left|\hat{G}_{31}(\omega_3)\right|\cdot a_3\cdot\cos(\omega_3 t+\psi_{13})+\left|\hat{G}_{32}(\omega_1)\right|a_1\cos(\omega_1 t+\psi_{21})\\
&\quad+\left|\hat{G}_{32}(\omega_2)\right|a_2\cos(\omega_2 t+\psi_{22})+\left|\hat{G}_{32}(\omega_3)\right|\cdot a_3\cdot\cos(\omega_3 t+\psi_{23})\\
&\quad+\left|\hat{G}_{33}(\omega_1)\right|a_1\cos(\omega_1 t+\psi_{31})+\left|\hat{G}_{33}(\omega_2)\right|a_2\cos(\omega_2 t+\psi_{32})\\
&\quad+\left|\hat{G}_{33}(\omega_3)\right|\cdot a_3\cdot\cos(\omega_3 t+\psi_{33})\\
&= 0.646\cdot\cos(0.3t+\psi_{11})+0.228\cdot\cos(0.7t+\psi_{12})+0.039\cdot\cos(1.1t+\psi_{13})\\
&\quad+0.86\cdot\cos(0.3t+\psi_{21})+0.713\cdot\cos(0.7t+\psi_{22})+0.236\cdot\cos(1.1t+\psi_{23})\\
&\quad+1.431\cdot\cos(0.3t+\psi_{31})+2.343\cdot\cos(0.7t+\psi_{32})+1.532\cdot\cos(1.1t+\psi_{33})
\end{aligned}
$$

What then remains is to ascribe arbitrary values (between 0 and 2π) to the phase angles, ψ_{nj}. The following is chosen:

$$\boldsymbol{\psi} = 2\pi \cdot \begin{bmatrix} 0.7 & 0.6 & 0.3 \\ 0.1 & 0.4 & 0.2 \\ 0.1 & 0.7 & 0.8 \end{bmatrix}$$

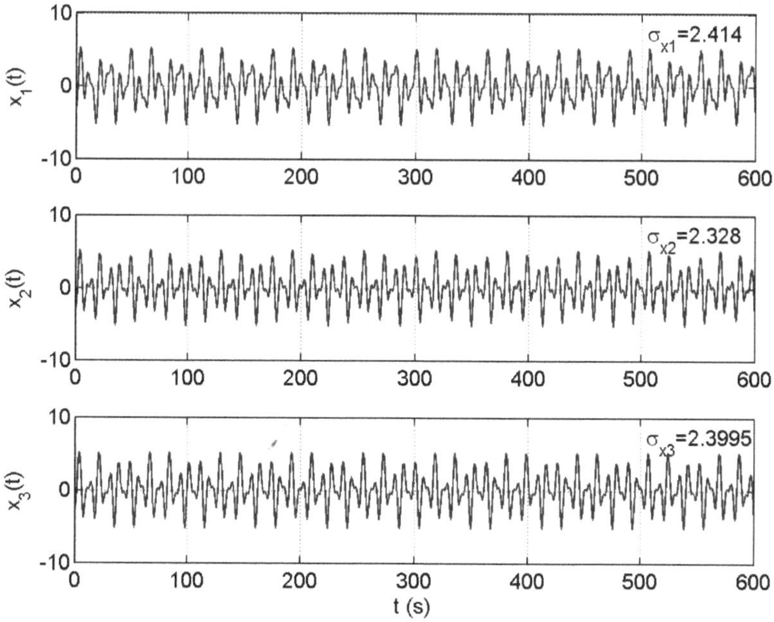

Fig. B.5 Simulated time series

The simulated time series are shown in Fig. B.5 ($T = 600$ s and $\Delta t = 0.06$ s). The standard deviation of the process as calculated from the parent spectrum is $\sigma_x = 2.3365$. The standard deviations of the three simulated time series are 2.414, 2.328 and 2.3995. The discrepancy (less than about 3 %) is caused by the unduly coarse frequency segmentation.

B.4 The Cholesky Decomposition

Given a positive definite and symmetric matrix $\mathbf{X}$, the Cholesky decomposition of $\mathbf{X}$ is defined by a lower triangular matrix $\mathbf{Y}$ of the same size that satisfies the following:

$$\mathbf{X} = \mathbf{Y}\mathbf{Y}^T \tag{B.24}$$

Expanding this equation

$$\begin{bmatrix} x_{11} & \cdots & x_{1i} & \cdots & x_{1N} \\ \vdots & & \vdots & & \vdots \\ x_{i1} & \cdots & x_{ii} & \cdots & x_{iN} \\ \vdots & & \vdots & & \vdots \\ x_{N1} & \cdots & x_{Ni} & \cdots & x_{NN} \end{bmatrix} = \begin{bmatrix} y_{11} & 0 & 0 \\ & & \\ y_{i1} & y_{ii} & 0 \\ & & \\ y_{N1} & y_{Ni} & y_{NN} \end{bmatrix} \begin{bmatrix} y_{11} & y_{1i} & y_{1N} \\ & & \\ 0 & y_{ii} & y_{iN} \\ & & \\ 0 & 0 & y_{NN} \end{bmatrix} \tag{B.25}$$

and developing the matrix multiplication column by column, it is seen that the first column renders

$$\left\{ \begin{array}{c} x_{11} = y_{11} \cdot y_{11} \\ x_{21} = y_{21} \cdot y_{11} \\ \vdots \\ x_{N1} = y_{N1} \cdot y_{11} \end{array} \right\} \Rightarrow \left\{ \begin{array}{c} y_{11} = \sqrt{x_{11}} \\ y_{21} = x_{21} / y_{11} \\ \vdots \\ y_{N1} = x_{N1} / y_{11} \end{array} \right. \tag{B.26}$$

while the second column renders

$$\left\{ \begin{array}{c} x_{22} = y_{21} \cdot y_{21} + y_{22} \cdot y_{22} \\ x_{32} = y_{31} \cdot y_{21} + y_{32} \cdot y_{22} \\ \vdots \\ x_{N2} = y_{N1} \cdot y_{21} + y_{N2} \cdot y_{22} \end{array} \right\} \Rightarrow \left\{ \begin{array}{c} y_{22} = \sqrt{x_{22} - y_{21} y_{21}} \\ y_{32} = \left(x_{32} - y_{31} y_{21} \right) / y_{22} \\ \vdots \\ y_{N2} = \left(x_{N2} - y_{N1} y_{21} \right) / y_{22} \end{array} \right. \tag{B.27}$$

and so on. This can be summarized as follows:

$$\begin{gathered} y_{11} = \left(x_{11} \right)^{1/2} \\ y_{ii} = \left(x_{ii} - \sum_{k=1}^{i-1} y_{ik}^2 \right)^{1/2} \quad \text{for } i = 2, \ldots\ldots, N-1 \\ y_{ij} = \left(x_{ij} - \sum_{k=1}^{j-1} y_{ik} y_{kj} \right) / y_{jj} \quad \text{for all } i > j \end{gathered} \tag{B.28}$$

Appendix C
Element Properties

C.1 Twelve Degree of Freedom Beam Element

The element mass matrix is given by: $\mathbf{m} = \begin{bmatrix} \mathbf{m}_{11} & \mathbf{m}_{12} \\ \mathbf{m}_{21} & \mathbf{m}_{22} \end{bmatrix}$ where $\mathbf{m}_{21} = \mathbf{m}_{12}^T$

$$\mathbf{m}_{11} = \frac{L}{420}\begin{bmatrix} 140m_x & 0 & 0 & 0 & 0 & 0 \\ & 156m_y & 0 & -147m_y e_z & 0 & 22m_y L \\ & & 156m_z & 147m_z e_y & -22m_z L & 0 \\ & & & 140m_\theta & -21m_z L e_y & -21m_y L e_z \\ & sym. & & & 4m_z L^2 & 0 \\ & & & & & 4m_y L^2 \end{bmatrix}$$

$$\mathbf{m}_{12} = \frac{L}{420}\begin{bmatrix} 70m_x & 0 & 0 & 0 & 0 & 0 \\ 0 & 54m_y & 0 & -63m_y e_z & 0 & -13m_y L \\ 0 & 0 & 54m_z & 63m_z e_y & 13m_z L & 0 \\ 0 & -63m_y e_z & 63m_z e_y & 70m_\theta & 14m_z L e_y & -14m_y L e_z \\ 0 & 0 & -13m_z L & -14m_z L e_y & -3m_z L^2 & 0 \\ 0 & 13m_y L & 0 & -14m_y L e_z & 0 & -3m_y L^2 \end{bmatrix}$$

$$\mathbf{m}_{22} = \frac{L}{420}\begin{bmatrix} 140m_x & 0 & 0 & 0 & 0 & 0 \\ & 156m_y & 0 & -147m_y e_z & 0 & -22m_y L \\ & & 156m_z & 147m_z e_y & 22m_z L & 0 \\ & & & 140m_\theta & 21m_z L e_y & 21m_y L e_z \\ & sym. & & & 4m_z L^2 & 0 \\ & & & & & 4m_y L^2 \end{bmatrix}$$

The element damping matrix is given by: $\mathbf{c} = \begin{bmatrix} \mathbf{c}_{11} & \mathbf{c}_{12} \\ \mathbf{c}_{21} & \mathbf{c}_{22} \end{bmatrix}$ where $\mathbf{c}_{21} = \mathbf{c}_{12}^T$

$$\mathbf{c}_{11} = \frac{L}{420} \begin{bmatrix} 140c_x & 0 & 0 & 0 & 0 & 0 \\ & 156c_y & 0 & 0 & 0 & 22c_yL \\ & & 156c_z & 0 & -22c_zL & 0 \\ & & & 140c_\theta & 0 & 0 \\ & sym. & & & 4c_zL^2 & 0 \\ & & & & & 4c_yL^2 \end{bmatrix}$$

$$\mathbf{c}_{12} = \frac{L}{420} \begin{bmatrix} 70c_x & 0 & 0 & 0 & 0 & 0 \\ 0 & 54c_y & 0 & 0 & 0 & -13c_yL \\ 0 & 0 & 54c_z & 0 & 13c_zL & 0 \\ 0 & 0 & 0 & 70c_\theta & 0 & 0 \\ 0 & 0 & -13c_zL & 0 & -3c_zL^2 & 0 \\ 0 & 13c_yL & 0 & 0 & 0 & -3c_yL^2 \end{bmatrix}$$

$$\mathbf{c}_{22} = \frac{L}{420} \begin{bmatrix} 140c_x & 0 & 0 & 0 & 0 & 0 \\ & 156c_y & 0 & 0 & 0 & -22c_yL \\ & & 156c_z & 0 & 22c_zL & 0 \\ & & & 140c_\theta & 0 & 0 \\ & sym. & & & 4c_zL^2 & 0 \\ & & & & & 4c_yL^2 \end{bmatrix}$$

The element stiffness matrix associated with purely material properties:

$$\mathbf{k} = \begin{bmatrix} \mathbf{k}_{11} & \mathbf{k}_{12} \\ \mathbf{k}_{21} & \mathbf{k}_{22} \end{bmatrix} \qquad \text{where} \qquad \mathbf{k}_{21} = \mathbf{k}_{12}^T \qquad \text{and}$$

$$\mathbf{k}_{11} = \begin{bmatrix} EA/L & 0 & 0 & 0 & 0 & 0 \\ & 12EI_z/L^3 & 0 & 0 & 0 & 6EI_z/L^2 \\ & & 12EI_y/L^3 & 0 & -6EI_y/L^2 & 0 \\ & & & GI_t/L & 0 & 0 \\ & sym. & & 0 & 4EI_y/L & 0 \\ & & & & & 4EI_z/L \end{bmatrix}$$

$$\mathbf{k}_{12} = \begin{bmatrix} -EA/L & 0 & 0 & 0 & 0 & 0 \\ 0 & -12EI_z/L^3 & 0 & 0 & 0 & 6EI_z/L^2 \\ 0 & 0 & -12EI_y/L^3 & 0 & -6EI_y/L^2 & 0 \\ 0 & 0 & 0 & -GI_t/L & 0 & 0 \\ 0 & 0 & 6EI_y/L^2{}_3 & 0 & 2EI_y/L & 0 \\ 0 & -6EI_z/L^2 & 0 & 0 & 0 & 2EI_z/L \end{bmatrix}$$

$$\mathbf{k}_{22} = \begin{bmatrix} EA/L & 0 & 0 & 0 & 0 & 0 \\ & 12EI_z/L^3 & 0 & 0 & 0 & -6EI_z/L^2 \\ & & 12EI_y/L^3 & 0 & 6EI_y/L^2 & 0 \\ & & & GI_t/L & 0 & 0 \\ & sym. & & & 4EI_y/L & 0 \\ & & & & & 4EI_z/L \end{bmatrix}$$

If the element length is short and the variation of time invariants $\bar{N}_n$, $\bar{M}_{y_n}$ and $\bar{M}_{z_n}$ are approximately constants $\bar{N}_n$, $\bar{\bar{M}}_{y_n}$ and $\bar{\bar{M}}_{z_n}$ along the span, then:

$$\mathbf{k}_G = \begin{bmatrix} \mathbf{k}_{G_{11}} & \mathbf{k}_{G_{12}} \\ \mathbf{k}_{G_{21}} & \mathbf{k}_{G_{22}} \end{bmatrix} \qquad \text{where} \qquad \mathbf{k}_{G_{21}} = \mathbf{k}_{G_{12}}^T \qquad \text{and}$$

$$\mathbf{k}_{G_{11}} = \begin{bmatrix} 0 & 0 & 0 & 0 & 0 & 0 \\ & 36\bar{N}_n/30L & 0 & -\bar{\bar{M}}_y & 0 & \bar{N}_n/10 \\ & & 36\bar{N}_n/30L & -\bar{\bar{M}}_z & -\bar{N}_n/10 & 0 \\ & & & \bar{N}_n e_0^2 & 0 & 0 \\ & sym. & & 0 & 2\bar{N}_n L/15 & 0 \\ & & & & & 2\bar{N}_n L/15 \end{bmatrix}$$

$$\mathbf{k}_{G_{12}} = \begin{bmatrix} 0 & 0 & 0 & 0 & 0 & 0 \\ 0 & -36\bar{N}_n/30L & 0 & \bar{\bar{M}}_y & 0 & \bar{N}_n/10 \\ 0 & 0 & -36\bar{N}_n/30L & \bar{\bar{M}}_z & -\bar{N}_n/10 & 0 \\ 0 & \bar{\bar{M}}_y & \bar{\bar{M}}_z & -\bar{N}_n e_0^2 & 0 & 0 \\ 0 & 0 & \bar{N}_n/10 & 0 & \bar{N}_n L/30 & 0 \\ 0 & -\bar{N}_n/10 & 0 & 0 & 0 & -\bar{N}_n L/30 \end{bmatrix}$$

$$\mathbf{k}_{G_{22}} = \begin{bmatrix} 0 & 0 & 0 & 0 & 0 & 0 \\ & 36\bar{N}_n/30L & 0 & -\bar{\bar{M}}_y & 0 & -\bar{N}_n/10 \\ & & 36\bar{N}_n/30L & -\bar{\bar{M}}_z & \bar{N}_n/10 & 0 \\ & & & \bar{N}_n e_0^2 & 0 & 0 \\ & sym. & & 0 & 2\bar{N}_n L/15 & 0 \\ & & & & & 2\bar{N}_n L/15 \end{bmatrix}$$

C.2 Six Degree of Freedom Beam Element

$$\mathbf{m} = \frac{L}{420}\begin{bmatrix} 140m_x & 0 & 0 & 70m_x & 0 & 0 \\ & 156m_z & -22m_zL & 0 & 54m_z & 13m_zL \\ & & 4m_zL^2 & 0 & -13m_zL & -3m_zL^2 \\ & & & 140m_x & 0 & 0 \\ & sym. & & & 156m_z & 22m_zL \\ & & & & & 4m_zL^2 \end{bmatrix}$$

$$\mathbf{c} = \frac{L}{420}\begin{bmatrix} 140c_x & 0 & 0 & 70c_x & 0 & 0 \\ & 156c_z & -22c_zL & 0 & 54c_z & 13c_zL \\ & & 4c_zL^2 & 0 & -13c_zL & -3c_zL^2 \\ & & & 140c_x & 0 & 0 \\ & sym. & & & 156c_z & 22c_zL \\ & & & & & 4c_zL^2 \end{bmatrix}$$

$$\mathbf{k} = \begin{bmatrix} EA/L & 0 & 0 & -EA/L & 0 & 0 \\ & 12EI_z/L^3 & -6EI_z/L^2 & 0 & -12EI_z/L^3 & -6EI_z/L^2 \\ & & 4EI_z/L & 0 & 6EI_z/L^2 & 2EI_z/L \\ & & & EA/L & 0 & 0 \\ & sym. & & & 12EI_z/L^3 & 6EI_z/L^2 \\ & & & & & 4EI_z/L \end{bmatrix}$$

$$\mathbf{k}_G = \frac{\bar{N}_n}{30L}\cdot\begin{bmatrix} 0 & 0 & 0 & 0 & 0 & 0 \\ & 36 & -3L & 0 & -36 & -3L \\ & & 4L^2 & 0 & 3L & L^2 \\ & & & 0 & 0 & 0 \\ & sym. & & & 36 & 3L \\ & & & & & 4L^2 \end{bmatrix}$$

References

[1] Newton, I.: Philosophiae Naturalis Principa Mathmatica. Universtity of Cambridge, UK (first published on July 5, 1687)

[2] Cohen, A., Whitman, I.B.: The Principia, A new translation. University of California press, Berkeley (1999)

[3] Lagrange, J.-L.: (1736-1813) Mecanique Analytique, Courcier, Paris 1811. Reissued by Cambridge University Press (2009) ISBN 9781108001748

[4] Navier, C.-L.: Professor at École Nationale des Ponts et Chaussées, Paris, Navier's hypothesis in theory of elasticity (first published in 1821)

[5] Timoshenko, S., Goodier, J.N.: Theory of elasticity, 3rd edn. McGraw-Hill, New York (1970)

[6] Charlton, T.M.: Energy Principles in Theory of Structures. Oxford University Press (1973)

[7] Courant, R., Hilbert, D.: Methods of Mathematical Physics, vol. I. Interscience Publishers, New York (1953)

[8] Lanczos, C.: The variational principles of mechanics, 4th edn. Dover Publications Inc., New York (1970)

[9] Ern, A., Guermond, J.L.: Theory and practice of finite elements. Springer (2004)

[10] Irvine, H.M., Caughey, T.K.: The linear theory of free vibrations of suspended cable. Proc. of the Royal Society of London, Series A, Mathematical and Physical Sciences 341(1626), 299–315 (1974)

[11] Sigbjørnsson, R., Hjorth-Hansen, E.: Along wind response of suspension bridges with special reference to stiffening by horizontal cables. Engineering Structures 3, 27–37 (1981)

[12] Bell, K.: Eigensolvers for structural problems. Delft Univ. Press (1998)

[13] Timoshenko, S., Young, D.H., Weaver, J.R.: Vibration problems in engineering, 4th edn. John Wiley & Sons (1974)

[14] Kreyzig, E.: Advanced engineering mathematics, 9th edn. John Wiley & Sons Inc. (2005)

[15] Strømmen, E.: Theory of bridge aerodynamics, 2nd edn. Springer (2010)

[16] Clough, R.W., Penzien, J.: Dynamics of structures, 2nd edn. McGraw-Hill (1993)

[17] Simiu, E., Scanlan, R.H.: Wind effects on structures, 3rd edn. John Wiley & Sons (1996)

[18] Dyrbye, C., Hansen, S.O.: Wind loads on structures. John Wiley & Sons Inc. (1999)

[19] Meirovitch, L.: Elements of vibration analysis, 2nd edn. McGraw-Hill (1993)

[20] Vickery, B.J., Basu, R.I.: Across–wind vibrations of structures of circular cross section. Part 1, Development of a mathematical model for two–dimensional conditions. Journal of Wind Engineering and Industrial Aerodynamics 12(1), 49–73 (1983)
[21] Vickery, B.J., Basu, R.I.: Across–wind vibrations of structures of circular cross section. Part 2, Development of a mathematical model for full–scale application. Journal of Wind Engineering and Industrial Aerodynamics 12(1), 79–97 (1983)
[22] Selberg, A.: Oscillation and aerodynamic stability of suspension bridges. Acta Polytechnica Scandinavica, Civil Engineering and Building Construction Series No. 13, Oslo (1961)
[23] Shinozuka, M.: Monte Carlo solution of structural dynamics. Computers and Structures 2, 855–874 (1972)
[24] Deodatis, G.: Simulation of ergodic multivariate stochastic processes. Journal of Engineering Mechanics, ASCE 122(8), 778–787 (1996)
[25] Hughes, T.J.R.: The finite element method. Prentice-Hall, Inc. (1987)
[26] Den Hartog, J.P.: Mechanical vibrations. McGraw–Hill Book Company Inc., New York (1947)
[27] Cook, R.D., Malkus, D.S., Plesha, M.E., Witt, R.J.: Concepts and applications of finite element analysis, 4th edn. John Wiley & Sons Inc. (2002)
[28] Newmark, N.M.: A method of computation for structural dynamics. Journal of Engineering Mechanics Division, ASCE 85, EM3 (1959)
[29] Hilber, H.M., Hughes, T.J.R., Taylor, R.L.: Improved numerical dissipation for time integration algorithms in structural dynamics. Earthquake Engineering and Structural Dynamics 5 (1977)
[30] Lazan, B.: Damping of materials and members in structural mechanics. Pergamon Press Inc., UK (1968)
[31] Lagomarsino, S.: Forecast models for damping and vibration periods of buildings. Journal of Wind Engineering and Industrial Aerodynamics 48, 221–239 (1993)
[32] Çelebi, M.: Comparison of damping in buildings under low amplitude and strong motion. Journal of Wind Engineering and Industrial Aerodynamics 59, 309–323 (1996)
[33] Jeary, A.P.: Damping in structures. Journal of Wind Engineering and Industrial Aerodynamics 72, 345–355 (1997)
[34] Satake, N., Suda, K.-I., Arakawa, T., Sasaki, A., Tamura, Y.: Damping Evaluation Using Full-Scale Data of Buildings in Japan. Journal of Structural Engineering, ASCE (April 2003)
[35] Wilson, E.L., Penzien, J.: Evaluation of Orthogonal Damping Matrices. International Journal of Numerical Methods in Engineering 4, 5–10 (1972)
[36] Rayleigh, J.W.S.B.: The Theory of Sound, 1st edn., vol. I and II. Macmillan, London (1887, 1888); reissued by Cambridge University Press in 2011
[37] Caughey, T.K.: Classical Normal Modes in Damped Linear Systems. International Journal of Numerical Methods in Engineering 27, 269–271 (1960)
[38] Newland, D.E.: Random vibrations and spectral analysis. Longman, London (1975)
[39] Luft, R.W.: Optimal tuned mass dampers for buildings. Journal of the Structural Division, Proc. of the American Society of Civil Engineers 105(ST12), 2766–2772 (1979)
[40] Wilkins, D.R. (ed.): William Rowan Hamilton (1805-1865): Mathematical Papers, School of Mathematics, Trinity College, Dublin, Ireland (2000); First published by Hamilton in his Philosophical Transaction of the Royal Society "On a General Method in Dynamics", Parts I and II (1834–1835)

[41] Jacobi, C.G.J.: Über ein leichtes Verfahren, die in der Theorie der Säkularstörungen vorkommenden Gleichungen numerisch aufzulösen. Crelle's Journal 30, 51–94 (1846)
[42] Davenport, A.G.: The response of slender line-like structures to a gusty wind. Proceedings of the Institution of Civil Engineers 23, 389–408 (1962)
[43] Davenport, A.G.: The prediction of the response of structures to gusty wind. In: Proceedings of the International Research Seminar on Safety of Structures under Dynamic Loading, Norwegian University of Science and Technology, Tapir, pp. 257–284 (1978)
[44] Scanlan, R.H., Tomko, A.: Airfoil and bridge deck flutter derivatives. Journal of Engineering Mechanics Division, ASCE 97(EM6) (December 1971); Proc. Paper 8609, pp. 1717–1737
[45] Love, A.E.H.: A treatise on the mathematical theory of elasticity. Cambridge University Press (First published in 1892)

Further Useful Reading

[46] Bendat, J.S., Piersol, A.C.: Random Data. Analysis and measurement procedures. John Wiley & Sons (1986)
[47] Timoshenko, S., Young, D.H.: Advanced dynamics, 1st edn. McGraw-Hill Inc. (1948)
[48] Backmann, H., et al.: Vibration problems in structures. Birkhäuser Verlag (1995)
[49] Donéa, J. (ed.): Advanced structural dynamics. Applied Science Pubisher Ltd., London (1980)
[50] Soong, T.T.: Active structural control, Theory & Practice. Longmanm Scientifc & Technical (1990)
[51] Bottega, W.J.: Engineering vibrations. CRC Taylor & Francis (2006)
[52] Worden, K., Tomlinson, G.R.: Nonlinearity in structural dynamics (University of Sheffield, UK). IoP Institute of Physics Publishing Ltd., London (2001)
[53] Bleich, F., Mccullogh, C.B., Rosecrans, R., Vincent, G.S.: The mathematical theory of suspension bridges, Department of Commerce, Bureau of Public Roads, United States Government Printing Office, Washington (1950)
[54] McNamara, R.J.: Tuned mass dampers for buildings. Journal of the Structural Division, Procedings of the American Society of Civil Engineers 103(ST9), 1785–1797 (1977)
[55] Akin, J.E.: Finite Elements for Analysis and Design. Academic Press Inc., SanDiego (1994)
[56] Gere, J.M., Timoshenko, S.P.: Mechanics of materials, 4th edn. Stanley Thornes Publishers Ltd., UK (1999)
[57] Chen, W.F., Atsuta, T.: Theory of beam-columns. Space behaviour and design, vol. 2. McGraw-Hill Inc. (1977)

Subject Index

FSC
www.fsc.org
MIX
Papier | Fördert
gute Waldnutzung
FSC® C083411

Kolibri 360 GmbH
Ferdinand-Jühlke-Straße 7
99095 Erfurt, Deutschland
produktsicherheit@kolibri360.de